Gas, Water and Solid Waste Treatment Technology

Gas, Water and Solid Waste Treatment Technology

Editors

Zhien Zhang
Avelino Núñez-Delgado
Wenxiang Zhang

MDPI • Basel • Beijing • Wuhan • Barcelona • Belgrade • Manchester • Tokyo • Cluj • Tianjin

Editors
Zhien Zhang
The Ohio State University
USA

Avelino Núñez-Delgado
University of Santiago de
Compostela
Spain

Wenxiang Zhang
University of Macau
Macau

Editorial Office
MDPI
St. Alban-Anlage 66
4052 Basel, Switzerland

This is a reprint of articles from the Special Issue published online in the open access journal *Processes* (ISSN 2227-9717) (available at: https://www.mdpi.com/journal/processes/special_issues/waste_treatment).

For citation purposes, cite each article independently as indicated on the article page online and as indicated below:

LastName, A.A.; LastName, B.B.; LastName, C.C. Article Title. *Journal Name* **Year**, *Volume Number*, Page Range.

ISBN 978-3-0365-0762-0 (Hbk)
ISBN 978-3-0365-0763-7 (PDF)

Cover image courtesy of NUÑEZ DELGADO AVELINO.

Contents

About the Editors

Zhien Zhang is a Research Fellow in William G. Lowrie Department of Chemical and Biomolecular Engineering at Ohio State University. His research interests include the advanced processes and materials for gas capture, gas separation and Carbon Capture, Utilization and Storage (CCUS). Dr. Zhang has published more than 90 peer-reviewed journal papers and 17 editorials in high-impact journals such as *Renewable & Sustainable Energy Reviews* and *Applied Energy*. He is Editorial Supervisor in *Journal of Natural Gas Science and Engineering* and Editor in some international journals, e.g., *Applied Energy*, *Processes*, *Environmental Chemistry Letters*, etc.

Avelino Núñez-Delgado (PhD). Avelino Núñez-Delgado is a Professor at the Department of Soil Science and Agricultural Chemistry at the University of Santiago de Compostela. His main research interests include: recycling, studying red mud as a sorbent material, sewage sludge, wood ash, and other types of waste and by-products, as well as different mixtures, retention/inactivation of pathogens and chemical contaminants, and potential for agronomic recycling of various types of waste and by-products. Lately, his main research has been focused on the retention of pollutants and the use of by-products to reach this end, with this facilitating the recycling of these materials. He has published 200+ scientific articles, authored several book chapters and books, and his research group has already registered eight patents in these topics.

Wenxiang Zhang is Research Fellow in University of Macau and Professor in South China Agricultural University. His research interests include water and wastewater treatment, advanced membrane materials and membrane processes and harmless disposal and recycling utilization of agricultural wastes. Dr. Zhang has published more than 50 peer-reviewed journal papers.

Preface to "Gas, Water and Solid Waste Treatment Technology"

Environmental pollution, including gas pollution, water pollution, and solid waste pollution, are the main social and environmental problems. A variety of treatment technologies, such as physical, chemical, and biological methods, have been used for the treatment of these pollutants. This book includes 34 review and research works focused on the various techniques, processes, and theories for gas, water and solid waste treatments.

Zhien Zhang, Avelino Núñez-Delgado, Wenxiang Zhang
Editors

 *processes* 

Article

Adsorption of Tetracycline and Sulfadiazine onto Three Different Bioadsorbents in Binary Competitive Systems

Raquel Cela-Dablanca [1], Manuel Conde-Cid [2], Gustavo Ferreira-Coelho [1], Manuel Arias-Estévez [2], David Fernández-Calviño [2], Avelino Núñez-Delgado [1,*], María J. Fernández-Sanjurjo [1] and Esperanza Álvarez-Rodríguez [1]

[1] Department of Soil Science and Agricultural Chemistry, Engineering Polytechnic School, Universidade de Santiago de Compostela, 27002 Lugo, Spain; raquel.cela@rai.usc.es (R.C.-D.); gf_coelho@yahoo.com.br (G.F.-C.); mf.sanjurjo@usc.es (M.J.F.-S.); esperanza.alvarez@usc.es (E.Á.-R.)

[2] Department of Plant Biology and Soil Science, Faculty of Sciences, Campus Ourense, Universidade de Vigo, 32004 Ourense, Spain; manconde@uvigo.es (M.C.-C.); mastevez@uvigo.es (M.A.-E.); davidfc@uvigo.es (D.F.-C.)

* Correspondence: avelino.nunez@usc.es; Tel.: +34-982-823-140

Abstract: Different antibiotics contained in manure, slurry, wastewater or sewage sludge are spread into the environment. The harmful effects of these antibiotics could be minimized by means of immobilization onto bioadsorbent materials. This work investigates the competitive adsorption/desorption of tetracycline (TC) and sulfadiazine (SDZ) onto pine bark, oak ash and mussel shell. The study was carried out using batch-type experiments in binary systems (with both antibiotics present simultaneously), adding 5 equal concentrations of the antibiotics (between 1 and 50 μmol L^{-1}). The adsorption percentages were higher for TC (close to 100% onto pine bark and oak ash, and between 40 and 85% onto mussel shell) than for SDZ (75–100% onto pine bark, and generally less than 10% on oak ash and mussel shell). Pine bark performed as the best adsorbent since TC adsorption remained close to 100% throughout the entire concentration range tested, while it was between 75 and 100% for SDZ. Desorption was always higher for SDZ than for TC. The results of this study could be useful to design practices to protected environmental compartments receiving discharges that simultaneously contain the two antibiotics here evaluated, and therefore could be relevant in terms of protection of the environment and public health.

Keywords: antibiotics; competitive sorption; retention/release; sorbents

Citation: Cela-Dablanca, R.; Conde-Cid, M.; Ferreira-Coelho, G.; Arias-Estévez, M.; Fernández-Calviño, D.; Núñez-Delgado, A.; Fernández-Sanjurjo, M.J.; Álvarez-Rodríguez, E. Adsorption of Tetracycline and Sulfadiazine onto Three Different Bioadsorbents in Binary Competitive Systems. *Processes* **2021**, *9*, 28. https://dx.doi.org/10.3390/pr9010028

Received: 30 November 2020
Accepted: 22 December 2020
Published: 24 December 2020

Publisher's Note: MDPI stays neutral with regard to jurisdictional claims in published maps and institutional affiliations.

1. Introduction

Antibiotics are widely used for the treatment of infectious diseases, both in humans and animals, but in addition, these drugs are administered in intensive livestock production systems to prevent infections, and in some countries, they are also used as promoters of animal growth [1–5]. Cycon et al. [2], analyzing the situation in 75 countries, indicated that between the years 2000 and 2015, the consumption of antibiotics increased by 65%, and predictions indicate that, on a global scale, in 2030, the consumption of antibiotics will be 200% higher than that in 2015.

The fact that antibiotics are discharged into the environment has led to their presence being detected in soils, waters and crops [4–7]. The reason is that antibiotics are not completely metabolized by humans and animals, and a high percentage of the administered drug is released as the parent compound through feces and urine, discharging into domestic wastewater and into the pits where slurries/manures are deposited [2]. The use of sewage sludge and animal slurries as fertilizers, as well as the irrigation of farmland with wastewater, has traditionally generated risks of chemical and microbiological contamination [8,9] and can lead to the introduction of antibiotics into the soil, and subsequently could be transferred to surface and groundwater, as well as to crops [5,10]. These pollutants can also reach waterbodies from aquaculture facilities and through wastewater effluents because many conventional treatments do not effectively remove these compounds [1]. All this implies serious environmental problems, one of the most worrying being the proliferation of bacteria resistant to antibiotics and the spread of genes resistant to antibiotics, which reduce the effectiveness of these drugs to effectively fight against pathogenic bacteria causing diseases in humans and animals [5,10]. To prevent environmental contamination due to these emerging pollutants, different procedures have been used, trying to get their degradation or removal, including techniques such as filtration, coagulation, flocculation, advanced oxidation processes, membrane processes, and combined methods [11]. Many of these methods are effective but expensive, and some of them generate products that lead to secondary contamination [11]. Therefore, it is necessary to further investigate low-cost strategies that are effective in minimizing environmental pollution due to antibiotics.

In this line, bioadsorbent materials have been used in recent years, investigating their ability to retain these compounds. In previous studies, our team studied three abundant and low-cost byproducts (pine bark, oak ash, and mussel shell) regarding their ability to adsorb tetracyclines (TCs) and sulfonamides (SAs), working in individual systems (with a single antibiotic added in each experiment). The results indicated that pine bark could be used to effectively immobilize these pollutants due to its high adsorption and low desorption capacities [12].

Despite the fact that antibiotics of these two groups (tetracyclines and sulfonamides) are widely used in veterinary medicine [6,13] and can be present simultaneously in discharges that reach soils and water bodies, there is a lack of studies on the characteristics of their eventual competition for adsorption sites in different soils and bioadsorbents [13].

In view of this, the objective of this work is to shed light on the eventual competition for adsorption sites of the bioadsorbents pine bark, oak ash and mussel shell, taking place between tetracycline (TC) and sulfadiazine (SDZ). To do this, batch-type adsorption/desorption experiments were carried out in binary systems (with both antibiotics present simultaneously). The results of the research could be useful to design systems or management strategies/alternatives for the treatment of pollution affecting environmental compartments where both antibiotics are spread simultaneously, which can also be considered relevant at the public health level.

2. Materials and Methods

2.1. Bioadsorbent Materials

Three different low-cost and abundant byproducts were used in this study. Specifically, two of them were from the forestry industry (pine bark and oak ash), and one from the food industry (mussel shell). A complete characterization of these materials is detailed in Quintáns-Fondo et al. [14], and the Supplementary Material (Tables S1 and S2) includes methodological details regarding the analyses carried out.

2.2. Chemicals

Both antibiotics were supplied by Sigma-Aldrich (Barcelona, Spain), TC with a purity of 95%, and SDZ with a purity of 99.7%. The reagents used were of analytical grade, with a high degree of purity, and were supplied by Panreac (Barcelona, Spain), except for acetonitrile (HPLC grade), which was supplied by Fisher Scientific (Madrid, Spain). All solutions were prepared using Milli-Q water (Millipore, Madrid, Spain).

2.3. Adsorption/Desorption Experiments

2.3.1. Adsorption

Batch-type binary experiments (with TC and SDZ present simultaneously) were carried out, putting in contact, for each of the bioadsorbents, 1 g of sorbent material with 40 mL of a solution containing 0.005 M $CaCl_2$ (added as background electrolyte to keep constant the ionic strength) together with both antibiotics, containing the same concentration for both, specifically 1, 3, 5, 25 and 50 μmol L^{-1} for each of them. The suspensions were shaken for 24 h (which was enough time to reach equilibrium, as demonstrated in previous kinetic studies), in the dark and at room temperature (25 ± 2 °C) on a rotary shaker, at 50 rpm, and then centrifuged (2665× g, 15 min) and filtered (using 0.45 μm nylon syringe filters, Fisher Scientific, Madrid, Spain). All the adsorption tests were done in natural (not adjusted) pH. HPLC-UV equipment was used to quantify the concentration of the antibiotics in the equilibrium solution (see details below), while pH was measured by means of a combined glass electrode (Crison, Barcelona, Spain). The amount of antibiotic adsorbed was calculated as the difference between that initially added and that present in the solution at equilibrium.

2.3.2. Desorption

To quantify desorption of the previously adsorbed antibiotics, the solids remaining from the adsorption experiments were first weighed to calculate the volume of the occluded solution. Next, 40 mL of 0.005 M $CaCl_2$ were added, and the suspensions were shaken, centrifuged, filtered and analyzed in the same way as described for adsorption. In parallel, blanks without byproducts (containing just antibiotic) were performed to quantify the possible loss of antibiotic due to degradation and/or adsorption to the tubes or filters, obtaining in all cases that the loss of antibiotic was very low (<3%). In addition, the biodegradation of both antibiotics has been ruled out in previous research [15,16]. All experiments were carried out in triplicate.

2.4. Quantification of the Antibiotics

Tetracycline (TC) was determined as indicated in Fernández-Calviño et al. [17,18] using HPLC equipment (Dionex Corporation, Sunnyvale, CA, USA) with a P680 quaternary pump, an ASI-100 autosampler, a TCC-100 thermostated column compartment and a UVD170U detector. Chromatographic separations were carried out by means of a Luna C18 column (5 μm particle size; 4.6 mm internal diameter; 150 mm long) and a guard column (5 μm particle size; 2 mm i.d.; 4 mm long) packed with the same material, both from Phenomenex (Madrid, Spain). The flow rate was 1.5 mL min^{-1}, for an injection volume of 50 μL, with a mobile phase constituted by acetonitrile (phase A) and 0.02 mol L^{-1} oxalic acid/0.01 mol L^{-1} triethylamine (phase B). A linear gradient elution program was run within 10.5 min, from 5% to 32% of phase A (and the rest to 100% of phase B). After 2 min, the initial conditions were reestablished and then held for 2.5 min. The total time for analysis was 15 min, using 8.0 min as the retention time. The wavelength for detection of TC was 360 nm.

Regarding the quantification of sulfadiazine (SDZ), it was carried out after passing the suspensions through 0.45 μm nylon filters (Panreac, Spain), using the same HPLC equipment as above, as well as the same flow rate and volume of injection. In this case, the mobile phase A was also acetonitrile, and phase B was 0.01 M phosphoric acid. The linear gradient was also run in 10.5 min, from 5% to 32% of phase A. Furthermore, coincident, the total time for analysis was 15 min, but the retention time was 5.2 min, and the wavelength for detection was 270 nm.

2.5. Data Analysis and Statistical Treatment

To describe the experimental adsorption data, the fitting to the following models were tested: linear Equation (1), Freundlich Equation (2) and Langmuir Equation (3):

$$Q_a = K_d Ceq, \tag{1}$$

$$Q_a = K_F C_{eq}^n, \tag{2}$$

$$Q_a = \frac{K_L C_{eq} q_m}{1 + K_L C_{eq}}, \tag{3}$$

In these equations, Q_a (expressed in μmol kg^{-1}) is the amount of antibiotic adsorbed once the equilibrium is reached; C_{eq} (μmol L^{-1}) is the concentration of antibiotic remaining in solution in the situation of equilibrium; K_d (L kg^{-1}) is the distribution coefficient; K_F (L^n μmol^{1-n} kg^{-1}) is the Freundlich's affinity coefficient; n (dimensionless) is the Freundlich's linearity index; K_L (L μmol^{-1}) is a Langmuir parameter related to the adsorption energy; q_m (μmol kg^{-1}) is the Langmuir's maximum adsorption capacity.

In addition, considering the possibility of eventual competition for adsorption sites, the linear and Freundlich models could be adapted to be used in binary competitive systems [19]. In this regard, the first step is to use Equations (4) and (5) to focus on the total amount of adsorbed of both antibiotics (TC and SDZ).

$$(Q_{aTC} + Q_{aSDZ}) = K_d\,(C_{eqTC} + C_{eqSDZ}), \tag{4}$$

$$(Q_{aTC} + Q_{aSDZ}) = K_F\,(C_{eqTC} + C_{eqSDZ})^n, \tag{5}$$

where Q_a is the individual amount adsorbed of each of both antibiotics; C_{eq} is the concentration of each of both antibiotics in the equilibrium solution; K_d is the distribution coefficient, and K_F and n are the Freundlich parameters indicated above.

As the second step, the model developed by Murali and Aylmore [20] was also used to deepen the study of TC and SDZ competitive adsorption in the binary system Equations (6) and (7).

$$Q_{a1} = (K_{F1}C_{eq1}{}^{n1+1})/(C_{eq1} + a_{12}C_{eq2}), \tag{6}$$

$$Q_{a2} = (K_{F2}C_{eq2}{}^{n2+1})/(C_{eq2} + a_{21}C_{eq1}), \tag{7}$$

In these equations, Q_a is the amount adsorbed of each of both antibiotics; C_{eq} is the concentration in the equilibrium solution for TC (1) and SDZ (2); K_F and n are the Freundlich parameters obtained from the individual (not binary) experiments, and a_{12} and a_{21} are parameters related to the competition between TC and SDZ for adsorption sites. Taking into account that a_{12} and a_{21} are placed in the denominator of the quotients, higher a_{12} and a_{21} values will give lower Q_a results, indicating lower adsorption for the antibiotics.

Desorption was presented as percentages, which were calculated in relation to the amounts previously adsorbed after determination of values for desorbed TC and SDZ expressed as μmol kg^{-1}.

The fitting of experimental data to the adsorption models was performed by using R statistical software, version 3.1.3 and the nlstools package for R. In addition, any further statistical treatments were carried out by means of the SPSS 21.0 software.

3. Results and Discussion

3.1. Adsorption of TC and SDZ onto the Bioadsorbents in Binary Systems

Figure 1 shows TC and SDZ adsorption when both are added simultaneously and at the same concentration (with added concentration values between 1 and 50 μmol L^{-1} for each antibiotic) to each of the three bioadsorbents (oak ash, pine bark and mussel shell).

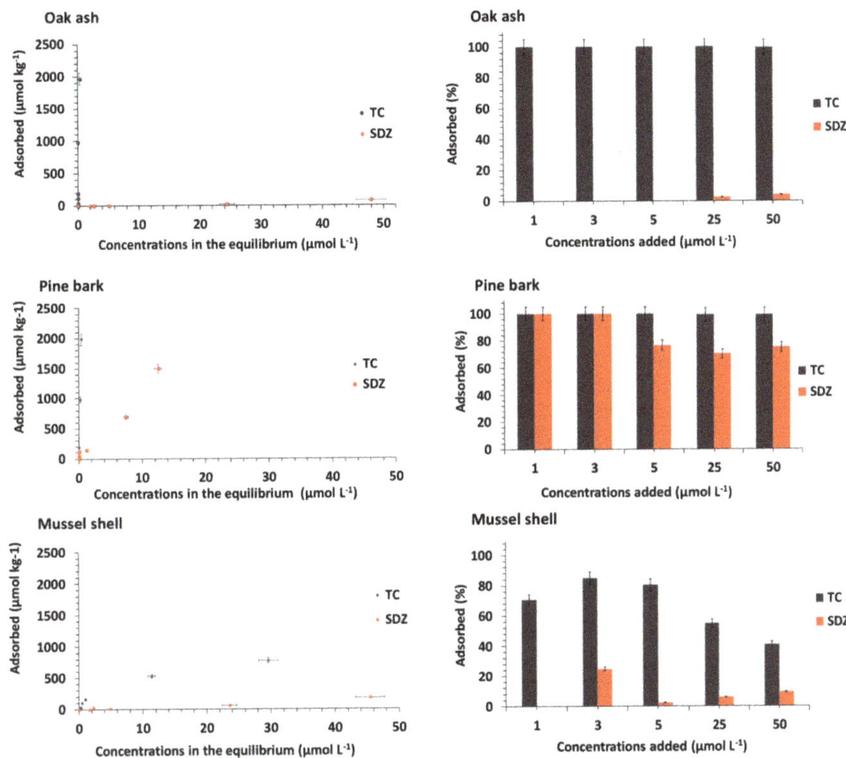

Figure 1. Tetracycline (TC) and sulfadiazine (SDZ) adsorption onto the three bioadsorbents studied (oak ash, pine bark and mussel shell) in binary competitive systems. Adsorption curves on the left, and percentage adsorption on the right. Average values ($n = 3$) with error bars indicating that coefficients of variation were <5%.

As shown in Figure 1, TC adsorption reached really high scores on the three bioadsorbents in these binary systems. Oak ash and pine bark have similar adsorption values (maxima of 1966 and 1983 μmol kg^{-1}, respectively), while mussel shell adsorbed much less (maximum of 777 μmol kg^{-1} of TC). Regarding SDZ adsorption, the most effective bioadsorbent was pine bark, reaching a maximum value of 1499 μmol kg^{-1}, while the other two bioadsorbents adsorbed less than 175 μmol kg^{-1} when the maximum concentration of antibiotics (50 μmol L^{-1}) was added.

When the results are expressed in percentage values, it is shown that pine bark adsorbed 100% of both antibiotics when they were added at the two lowest doses (1 and 3 μmol L^{-1}), while this sorbent was able of adsorbing 100% of added TC and about 75% of added SDZ when the highest doses of antibiotics (25 and 50 μmol L^{-1}) were applied. Regarding oak ash, this sorbent was able of retaining 100% of the TC added, but the retention percentages were poor for SDZ (always lower than 5%), causing that it could not be considered as an effective bioadsorbent for simultaneous retention of the antibiotics evaluated in the competitive binary systems. Finally, mussel shell adsorbed about 80% of the added TC when the doses of antibiotic were low, but adsorption decreased to 40% when the highest doses of TC were added, while SDZ adsorption onto mussel shell was always less than 20%.

In a previous study, Conde-Cid et al. [21] worked with binary systems that included different tetracyclines (not sulfonamides), finding that pine bark was able to adsorb most of the antibiotics added, with low desorption in most cases. Oak ash showed high adsorption for each antibiotic individually, but retention clearly decreased in the binary systems, and finally, the mussel shell showed poor adsorption in most cases. In addition, in another research, Conde-Cid et al. [22] studied the adsorption of sulfonamides (including SDZ) onto agricultural soils after being amended with the same byproducts here used, but in individual (not competitive) experiments. These authors found that the pine bark amendment caused an increase in the adsorption (and also decreased desorption) for the three sulfonamide antibiotics they studied. However, mussel shell and oak ash did not increase adsorption. These authors postulated that the positive effect of pine bark could be due to its high organic carbon content and to its acidic pH value. In fact, as previously found by Conde-Cid et al. [21], pine bark has the highest organic matter content (48.7%) among all three bioadsorbents, while oak ash and mussel shell showed values clearly lower (13.23 and 11.43%, respectively). These authors also indicate that oak ash showed an alkaline value for pH in water (11.31), as mussel shell (9.39), while it was acidic for pine bark (3.99), with differences between pH in KCl and pH in water indicating that oak ash had more positive charges, while negative charges dominate in pine bark and mussel shell. Conde-Cid et al. [21] also reported that oak ash showed an exchangeable Ca value (95 $cmol_c$ kg^{-1}) that was higher than those found for mussel shell (24.75 $cmol_c$ kg^{-1}) and pine bark (5.38 $cmol_c$ kg^{-1}), and BET surface area was higher for oak ash, followed by mussel shell and by pine bark.

In the current work, when analyzing the results and making a comparison focused on the antibiotics, it can be observed that TC is generally more adsorbed than SDZ. Similarly, other authors [13,23,24] found higher adsorption of tetracyclines in relation to sulfonamides, this fact being responsible for the higher persistence of TCs in the environment, and specifically in solid media.

It is noteworthy that both antibiotics (TC and SDZ) are amphoteric and can be present as cations, zwitterions or as anions, depending on the pH of the medium [6,12,21,22]. However, the higher affinity for TC shown by the three bioadsorbents here assessed could be related to the adsorption mechanisms of each antibiotic. Thus, while TC can be adsorbed by various mechanisms, such as electrostatic interactions, complexation, cationic bridges, H bonds [25], SDZ has electrostatic attractions as the predominant and almost exclusive mechanism for adsorption [26], which reduces the possibility of interacting with the functional groups of sorbents.

For example, at neutral pH, SDZ is mainly in its neutral form (SDZ^0), and the adsorption mechanism would imply the intervention of weak hydrophobic forces [13,27], while TC molecules can be present as different chemical species, which is of great relevance since they can be adsorbed by different mechanisms [13,17,28], also affected by their Log K_{OW} [29] and pK_a values [30].

Focusing on the differences regarding adsorption onto the three bioadsorbents, the most remarkable is the higher adsorption of SDZ taking place in pine bark, which could be explained based on the pH and organic matter content of this sorbent, as previously indicated by Conde-Cid et al. [22].

In fact, pH influences the speciation of both antibiotic molecules and the charge on the surface of adsorbent materials, which can undergo protonation/deprotonation reactions depending on the pH value [31], thus affecting the interaction between sorbate and sorbent. In this regard, Białk-Bielińska et al. [32] indicate that as the pH decreases, the adsorption capacity of sulfonamides increases, with maximum adsorption taking place at pH 4–4.25, with these antibiotics being in the cationic form [33].

In view of this, it must be taken into account that pine bark has a pH of 3.99, and the cationic species of SDZ will be adsorbed on the deprotonated carboxylic groups of the organic matter of this bioadsorbent through electrostatic interactions [26]. On the contrary, at the alkaline pH values characterizing oak ash and mussel shell, anionic species will predominate for SDZ, which will be repelled by the negative charges of the organic matter of these two bioadsorbents, causing that the SDZ/adsorbent bond would have to occur through a cationic bridge, which is a rare mechanism for sulfonamides [26].

However, TC is fully adsorbed both when the sorbent is acidic (pine bark) or strongly alkaline (oak ash). As previously noted, TC has several mechanisms for binding to sorbents. Thus, at an acidic pH, such as that found in pine bark (3.99), there are cationic species of TC that will interact electrostatically with the carboxylic groups of the organic matter of the bark, which begins to deprotonate from pH 3 [34,35]. At a strongly alkaline pH, like that of oak ash (11.31), TC will be in anionic form, and the organic matter (13.23% carbon) and non-crystalline components (8323.0 and 4233.0 mg kg^{-1} Al and Fe extracted with ammonium oxalate) present in the ash will also be negatively charged, causing that adsorption will occur through a cationic bridge, in which cations such as Ca^{2+} can intervene [13,25,35], facilitated by its abundance in this sorbent (95 cmol$_C$ kg^{-1} of exchangeable Ca, clearly higher than 24.75 and 5.38 cmol$_C$ kg^{-1} in mussel shell and pine bark, respectively). The lower adsorption of TC in mussel shell can be related to the lower content of organic matter (11.43% carbon) and non-crystalline components (178.33 and 171.0 mg kg^{-1} Al and Fe extracted with ammonium oxalate) characterizing this sorbent.

Based on the adsorption data obtained under the conditions of this study, it can be stated that pine bark could be used as an effective bioadsorbent for TC and SDZ in a binary system (with both antibiotics present), while oak ash could be used to effectively remove TC (but not SDZ) from solution, and finally, mussel shell would not be recommended as a bioadsorbent for these two antibiotics, due to its limited efficacy.

3.2. Fitting of TC and SDZ Adsorption Data to Different Models

Tables 1–3 show results on the fitting of TC and SDZ adsorption onto the three bioadsorbents, in binary systems, to three different models.

Table 1. Parameters of the linear model for tetracycline (TC) and sulfonamide (SDZ) adsorption onto the three biosorbents studied.

Bioadsorbent	Antibiotic	K_d (L kg^{-1})	*Error*	R^2
Oak ash	TC	-	-	-
	SDZ	1.51	0.16	0.939
	TC + SDZ	42.06	0.46	0.999
Pine bark	TC	4805.01	275.58	0.979
	SDZ	113.09	7.17	0.975
	TC + SDZ	256.45	12.42	0.982
Mussel shell	TC	29.04	4.18	0.858
	SDZ	3.50	0.37	0.934
	TC + SDZ	13.49	0.96	0.955

Table 2. Parameters of the Langmuir model for tetracycline (TC) adsorption onto mussel shells. All other values for both antibiotics and the three biosorbents studied had too high error values for fitting.

| Bioadsorbent | Antibiotic | Langmuir Parameter | | | | |
		Q_m (µmol kg^{-1})	*Error*	K_L (L µmol^{-1})	*Error*	R^2
Mussel shell	TC	934.35	97.63	0.14	0.05	0.986

Q_m: maximum adsorption capacity; K_L: constant related to the strength of interaction adsorbent/adsorbate; R^2: coefficient of determination.

Table 3. Parameters of the Freundlich model for tetracycline (TC) and sulfonamide (SDZ) adsorption onto the three biosorbents studied.

Sorbent	Antibiotic	Freundlich Parameter				
		K_F (L^n µmol^{1-n} kg^{-1})	*Error*	*n*	*Error*	R^2
Oak ash	TC	-	-	-	-	-
	SDZ	0.07	0.01	1.82	0.05	1
	TC + SDZ *	35.16	3.12	1.05	0.02	1
Pine bark	TC	6036.53	1411.70	1.22	0.22	0.985
	SDZ	46.07	23.98	1.37	0.21	0.988
	TC + SDZ *	123.29	54.40	1.30	0.6	0.991
Mussel shell	TC	148.88	18.95	0.50	0.04	0.992
	SDZ	-	-	1.67	0.37	0.997
	TC + SDZ *	50.18	9.16	0.70	0.04	0.997

K_F: parameter related to the adsorption capacity; *n*: parameter related to the heterogeneity of the sorbent; R^2: coefficient of determination.
* Obtained from Equation (5); -: error values too high for fitting.

Table 1 shows that the adjustments to the linear model are generally good, with high R^2 values. However, in the case of TC adsorption onto oak ash, the fitting was not possible to the linear or any other model due to the high adsorption capacity of the sorbent for TC (100% in most cases), causing that no TC was detected in the equilibrium solution, making not possible to apply adsorption models. The best fit to the linear model corresponded to pine bark for both antibiotics. In addition, it is noteworthy that K_d values were much higher for TC (between 29.04 and 4805.01 L kg^{-1}) than for SDZ (between 1.51 and 113.09 L kg^{-1}), which indicates that the bioadsorbents studied have a higher affinity for TC [13]. Furthermore, relevant, pine bark showed higher affinity for both antibiotics, compared to oak ash and mussel shell (Table 1), which is in agreement with the experimental results (Figure 1).

Just for comparison, we could take into account that Conde-Cid et al. [36] obtained K_d values between 0.40 and 9.43 L kg^{-1} for SDZ in soils of Galicia, which are of the same order as those obtained for other soils by Sukul et al. [37] (between 0.1 and 24.3 L kg^{-1}), and by Leal et al. [38] (from 0.8 to 14.3 L kg^{-1}), while Hu et al. [39] found lower values (1.54 and 3.41 L kg^{-1}) in other agricultural soils. Compared to these results, the K_d values obtained in the present study for SDZ in oak ash and mussel shell (1.51 and 3.50 L kg^{-1}) are of the order of those reported in the soils above, while K_d was clearly higher in pine bark (113.09 L kg^{-1}), which indicates a high affinity of this material for SDZ, and its potential feasibility to be used to retain this antibiotic in soils, slowing its passage into water bodies and the food chain.

Regarding TC, K_d values obtained in previous studies for Galician soils ranged between 53 and 30,237 L kg^{-1} [40], while Bao et al. [41] reported values between 838 and 15,278 L kg^{-1} for other soils. In the present study, K_d values for TC in mussel shell were lower than those obtained in the lower range of the soils reported above, while K_d was high in pine bark (4805 L kg^{-1}), suggesting that this biosorbent could be used to increase the retention of TC in soils with low adsorption capacity.

Table 1 also shows that when the linear model is applied to the set of the two antibiotics in the binary system (TC + SDZ), R^2 improves slightly compared to when considering the antibiotics separately, and it is also observed that K_d values for TC + SDZ are lower than those obtained for TC alone but higher than those for SDZ alone (Table 1).

Regarding the Langmuir model, Table 2 shows that it is not satisfactory to explain the adsorption of TC and SDZ onto oak ash and pine bark in binary systems, just fitting for TC adsorption onto mussel shell. In all other cases, error values were too high for fitting.

As regards the Freundlich model, Table 3 shows the model parameters for TC and SDZ separately and for the set of the two antibiotics in the binary system. In fact, as previously commented, the Freundlich model can be adapted to be used in binary competitive systems, as shown in Equation (5). The fits to this model are generally good, with R^2 values between 0.985 and 1 (Table 3).

The values of the Freundlich K_F parameter (related to the adsorption capacity of a certain adsorbent, [42]) were higher for TC than for SDZ (Table 3). These results also indicate a higher affinity of the bioadsorbents used in this study (specifically of pine bark) for TC than for SDZ, and also a higher adsorption capacity of pine bark for both antibiotics, compared to oak ash and mussel shell, which coincides with the experimental results (Figure 1).

Using the Freundlich model adapted for the binary systems (Equation (5)), the resulting K_F values for the set of the two antibiotics (TC + SDZ) were lower than those obtained for TC and higher than those obtained for SDZ (Table 3), so that the sequence for the three bioadsorbents, considering both the antibiotics individually and in binary systems, was: TC> TC + SDZ> SDZ.

Regarding possible comparisons between the K_F values of the present study and those from previous research, it is hard to perform it properly due to the differences affecting the experimental conditions, including the initial concentrations added. Therefore, at the level of data comparison among different publications and adsorbent materials, it is preferable to use the partition coefficient (K_d) discussed above for the linear model, which is less sensitive to variations in the initial concentration than the adsorption capacity [43–45].

As regards the n parameter, it is related to the reactivity and heterogeneity of the active sites of the adsorbents. Specifically, when $n = 1$ the adsorption is linear, when $n > 1$ the adsorption process is mainly chemical, and when $n < 1$ physical adsorption is dominant, with heterogeneous high-energy sites present, strong interactions taking place between adsorbent and sorbate, and with high-energy sites being the first to be occupied [46–48]. In the present study, n values were generally > 1 (denoting dominance of chemisorption), with the exception of TC and the sum of TC + SDZ in mussel shell (Table 3). Other authors also obtained values of $n > 1$ in chlortetracycline + SDZ binary systems, which suggest a strong interaction between these two antibiotics and the high-energy sites of the bioadsorbents [13]. On the contrary, in studies dealing with soils in individual (non-binary) systems, values of $n < 1$ were obtained for both TC [40] and SDZ [36], which suggests that the simultaneous presence of both antibiotics in binary systems could cause modifications in the mechanisms of interaction of these compounds with adsorbent surfaces.

Table 4 shows the results of the adsorption adjustments of TC and SDZ to the model of Murali and Aylmore [20], with R^2 values ranging between 0.725 and 0.991. This model is used to study the competition between different sorbates for adsorption sites, which is expressed by the competitiveness index a. The lower the value of a, the lower is the competition between sorbates for the adsorption sites of the sorbents. Comparing the index a_{12} (index of competition of SDZ to eventually occupy adsorption sites before TC) and a_{21} (index of competition of TC to eventually occupy adsorption sites before SDZ), it is observed that the former are always lower than the latter (Table 4); therefore TC would be more likely to displace SDZ and occupy first the adsorption sites for which they compete.

Table 4. Fitting of adsorption results to the Murali and Aylmore equation, using TC and SDZ solutions in relation 1:1, with concentrations of the antibiotic adsorbed expressed in $\mu mol\ kg^{-1}$ and concentrations of antibiotic present in the equilibrium solution in $\mu mol\ L^{-1}$.

Bioadsorbent	a_{12}	R^2	a_{21}	R^2
Oak ash	−0.956	0.907	0	-
Pine bark	−0.555	0.991	−0.013	0.975
Mussel shell	−0.501	0.725	−0.045	0.98

a: parameter related to competition between antibiotics; a_{12}: index of competition of SDZ with TC; a_{21}: index of competition of TC with SDZ.

3.3. Desorption of TC and SDZ from the Three Bioadsorbents in Binary Systems

Figure 2 represents the percentage of desorption for SDZ and TC, calculated taking into account the amounts previously adsorbed in the binary systems in each of the three bioadsorbents studied (oak ash, pine bark and mussel shell).

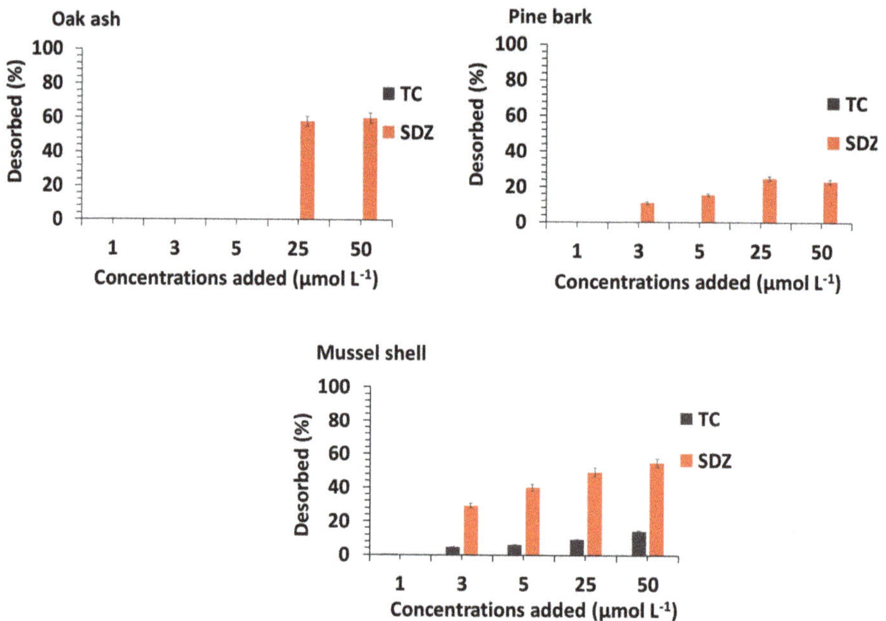

Figure 2. Desorption of TC and SDZ from the three different bioadsorbents studied (oak ash, pine bark and mussel shell) in binary systems. Average values ($n = 3$), with error bars indicating that coefficients of variation were <5%.

Figure 2 shows that TC was irreversibly adsorbed (no desorption taking place) onto pine bark and oak ash and intensely adsorbed onto mussel shell (desorption <10%). Regarding desorption of SDZ, it was always higher than that of TC, and pine bark was the material desorbing the least, with percentages always lower than 20%. In oak ash, the adsorption of SDZ was irreversible up to a concentration of antibiotic added of 5 $\mu mol\ L^{-1}$, but for the two highest concentrations added, the desorption values exceeded 50% (Figure 2). Finally, mussel shell desorbed up to 60% of SDZ for the three highest concentrations added. A similar behavior, with higher desorption of different sulfonamides in relation to tetracyclines, was found in previous studies using individual (non-binary) systems in soils [12,49].

Taking into consideration adsorption and desorption data commented above, it is clear that, in those cases where TC and SDZ are present simultaneously in a binary system, pine bark is promising as bioadsorbent. However, it must be taken into account that from concentrations of antibiotic added reaching values of 3 μmol L^{-1}, SDZ adsorption drops to 75% and its desorption increases to 20%. In any case, pine bark is the adsorbent material performing best among the three evaluated in the binary systems since oak ash and mussel shell show very low adsorption capacity for SDZ, also desorbing practically half of the previously retained amount.

In the present work, the good performance of pine bark to retain the antibiotics studied in binary systems can be due to its high organic matter content, with an abundance of adsorption sites, with a high affinity for these compounds, as previously shown for tetracyclines [21]. To be noted that pine bark has a markedly acidic pH (<4) at which these antibiotics have a positive or neutral charge, favoring binding to the negatively charged carboxylic groups contained in the organic matter of the bark, that dissociate at pH between 3 and 6 [21].

4. Conclusions

The results of this study indicate that the three bioadsorbents evaluated (oak ash, pine bark and mussel shell) have a higher affinity for tetracycline (TC) than for sulfadiazine (SDZ). When TC and SDZ were incorporated together in a binary system, pine bark was the most suitable sorbent for the retention of both antibiotics, but the efficacy against SDZ decreased when the antibiotics were added in high concentrations due to lower adsorption and higher desorption values. Oak ash and mussel shells were not efficient enough to be recommended for adsorption of TC and SDZ in binary systems, as they have limited affinity for SDZ. In the binary systems, adsorption data corresponding to each individual antibiotic and to both present simultaneously showed the highest K_d and K_F values (after fitting to the linear and Freundlich models) for pine bark, which indicates that this material has highly reactive adsorption sites for TC and SDZ, that hardly could be saturated. It was also shown that competition for adsorption sites between TC and SDZ was favorable to TC, with the highest competition taking place in mussel shells. Specifically, adsorption percentages for TC were close to 100% onto pine bark and oak ash and between 40 and 85% onto mussel shell, thus being higher than for SDZ (75–100% onto pine bark, and generally less than 10% on oak ash and mussel shell). TC adsorption onto pine bark remained close to 100% throughout the entire concentration range tested, while it was between 75 and 100% for SDZ. In addition, desorption was always higher for SDZ than for TC. The results of this study could be useful when evaluating management strategies related to situations with risks of simultaneous contamination by TC and SDZ, which may have environmental and public health relevance. Future studies could consider other tetracycline and sulfonamide antibiotics in competition, as well as other concentrations than that used here, and even more than two antibiotics present simultaneously. In addition, studying additional low-cost adsorbents, as well as modified sorbents, would give a broader view regarding alternatives to retain/remove antibiotics from environmental compartments.

Supplementary Materials: The following are available online at https://www.mdpi.com/2227-9717/9/1/28/s1, Table S1. Main characteristics of the three bioadsorbents studied. Average values (*n* = 3), with coefficients of variation always <5%. Table S2. Data corresponding to the BET surface area results for the three bioadsorbent materials studied. Mean values (*n* = 3) with coefficients of variation always <5%.

Author Contributions: Conceptualization, E.Á.-R., M.A.-E., M.J.F.-S. and A.N.-D.; methodology, E.Á.-R., M.A.-E., D.F.-C. and M.C.-C.; software, G.F.-C., R.C.-D. and M.C.-C.; validation, E.Á.-R., M.A.-E., M.J.F.-S. and A.N.-D.; formal analysis, G.F.-C., R.C.-D., D.F.-C. and M.C.-C.; investigation, R.C.-D., D.F.-C. and M.C.-C.; resources, E.Á.-R., M.A.-E., M.J.F.-S., D.F.-C. and A.N.-D.; data curation, E.Á.-R., M.A.-E., M.J.F.-S., D.F.-C. and A.N.-D.; writing—original draft preparation, G.F.-C., R.C.-D.,

M.J.F.-S. and E.Á.-R.; writing—review and editing, A.N.-D.; visualization, E.Á.-R., M.A.-E., M.J.F.-S. and A.N.-D.; supervision, E.Á.-R., M.A.-E., M.J.F.-S., D.F.-C. and A.N.-D.; project administration, E.Á.-R., M.A.-E.; funding acquisition, E.Á.-R., M.A.-E. All authors have read and agreed to the published version of the manuscript.

Funding: This research was funded by the Spanish Ministry of Science, Innovation and Universities, grant numbers RTI2018-099574-B-C21 and RTI2018-099574-B-C22. It also received funds from the European Regional Development Fund (ERDF) (FEDER in Spain), being a complement to the previous grants, without additional grant number. M. Conde-Cid holds a pre-doctoral contract (FPU15/0280, Spanish Government). The research of Dr. Gustavo F. Coelho was also supported by the Improving Coordination of Senior Staff (CAPES), Post-Doctoral Program Abroad (PDE) Process number {88881.172297/2018-01} of the Brazilian Government.

Informed Consent Statement: Not applicable.

Conflicts of Interest: The authors declare no conflict of interest. The funders had no role in the design of the study; in the collection, analyses, or interpretation of data; in the writing of the manuscript, or in the decision to publish the results.

References

1. Albero, B.; Tadeo, J.L.; Escario, M.; Miguel, E.; Pérez, R.A. Persistence, and availability of veterinary anbitiotics in soil and soil-manure systems. *Sci. Total Environ.* **2018**, *643*, 1562–1570. [CrossRef] [PubMed]
2. Cycon, M.; Mrozik, A.; Piotrowska-Seget, Z. Anbitiotics in the Soil Environment—Degradation and Their Impact on Microbial Activity and Diversity. *Front. Microbiol.* **2019**, *10*, 338. [CrossRef] [PubMed]
3. Hanna, N.; Sun, P.; Sun, Q.; Li, X.W.; Yang, X.W.; Ji, X.; Zou, H.Y.; Ottoson, J.; Nilsson, L.E.; Berglund, B.; et al. Presence of antibiotic residues in various environmental compartments of Shandong province in eastern China: Its potential for resistance development and ecological and human risk. *Environ. Int.* **2018**, *114*, 131–142. [CrossRef] [PubMed]
4. Tasho, R.P.; Cho, J.Y. Veterinary antibiotics in animal waste, its distribution in soil and uptake by plants: A review. *Sci. Total Environ.* **2016**, *563*, 366–376. [CrossRef]
5. Zhang, G.; Zhao, Z.; Zhu, Y. Changes in abiotic dissipation rates and bound fractions of antibiotics in biochar-amended soil. *J. Clean. Prod.* **2020**, *256*, 120312. [CrossRef]
6. Conde-Cid, M.; Álvarez-Esmorís, C.; Paradelo-Núñez, R.; Novoa-Muñoz, J.C.; Arias-Estevez, M.; Álvarez-Rodríguez, E.; Fernández-Sanjurjo, M.J.; Núñez-Delgado, A. Occurrence of tetracyclines and sulfonamides in manures, agricultural soils and crops from different areas in Galicia (NW Spain). *J. Clean. Prod.* **2018**, *197*, 491–500. [CrossRef]
7. Yang, Y.Y.; Song, W.J.; Lin, H.; Wang, W.B.; Du, L.N.; Xing, W. Antibiotics and antibiotic resistance genes in global lakes: A review and meta-analysis. *Environ. Int.* **2018**, *116*, 60–73. [CrossRef]
8. Núñez-Delgado, A.; López-Periago, E.; Díaz-Fierros Viqueira, F. Chloride, sodium, potassium and faecal bacteria levels in surface runoff and subsurface percolates from grassland plots amended with cattle slurry. *Bioresour. Technol.* **2002**, *82*, 261–271. [CrossRef]
9. Pousada-Ferradás, Y.; Seoane-Labandeira, S.; Mora-Gutierrez, A.; Núñez-Delgado, A. Risk of water pollution due to ash–sludge mixtures: Column trials. *Int. J. Environ. Sci. Technol.* **2012**, *9*, 21–29. [CrossRef]
10. Ben, Y.J.; Fu, C.X.; Hu, M.; Liu, L.; Wong, M.H.; Zheng, C.M. Human health risk assessment of antibiotic resistance associated with antibiotic residues in the environment: A review. *Environ. Res.* **2019**, *169*, 483–493. [CrossRef]
11. Ding, H.J.; Wu, Y.X.; Zou, B.C.; Lou, Q.; Zhang, W.H.; Zhong, J.Y.; Lu, L.; Dai, G.F. Simultaneous removal and degradation characteristics of sulfonamide, tetracycline, and quinolone antibiotics by laccase-mediated oxidation coupled with soil adsorption. *J. Hazard. Mater.* **2016**, *307*, 350–358. [CrossRef] [PubMed]
12. Conde-Cid, M.; Ferreira-Coelho, G.; Arias-Estévez, M.; Fernández-Calvinho, D.; Núñez-Delgado, A.; Álvarez-Rodríguez, E.; Fernández- Sanjurjo, M.J. Adsorption/desorption of three tetracycline antibiotics on different soils in binary competitive systems. *J. Environ. Manag.* **2020**, *262*, 110337. [CrossRef] [PubMed]
13. Jiang, Y.; Zhang, Q.; Deng, X.; Nan, Z.; Liang, X.; Wen, H.; Huang, K.; Wu, Y. Single and competitive sorption of sulfadiazine and chlortetracycline on loess soil from Northwest China. *Environ. Pollut.* **2020**, *263*, 114650. [CrossRef]
14. Quintáns-Fondo, A.; Ferreira-Coelho, G.; Paradelo-Núñez, R.; Nóvoa-Muñoz, J.C.; Arias-Estévez, M.; Fernández-Sanjurjo, M.J.; Álvarez-Rodríguez, E.; Núñez-Delgado, A. As(V)/Cr(VI) pollution control in soils, hemp waste, and other by-products: Competitive sorption trials. *Environ. Sci. Pollut. Res.* **2016**, *23*, 18182–19192. [CrossRef]
15. Conde-Cid, M.; Fernández-Calviño, D.; Nóvoa-Muñoz, J.C.; Arias-Estévez, M.; Díaz-Raviña, M.; Fernández-Sanjurjo, M.J.; Núñez-Delgado, A.; Álvarez-Rodríguez, E. Biotic and abiotic dissipation of tetracyclines using simulated sunlight and in the dark. *Sci. Total Environ.* **2018**, *635*, 1520–1529. [CrossRef] [PubMed]
16. Conde-Cid, M.; Fernández-Calviño, D.; Nóvoa-Muñoz, J.C.; Arias-Estévez, M.; Díaz-Raviña, M.; Núñez-Delgado, A.; Fernández-Sanjurjo, M.J.; Álvarez-Rodríguez, E. Degradation of sulfadiazine, sulfachloropyridazine and sulfamethazine in aqueous media. *J. Environ. Manag.* **2018**, *228*, 239–248. [CrossRef] [PubMed]

17. Fernández-Calviño, D.; Bermúdez-Couso, A.; Arias-Estévez, M.; Nóvoa-Muñoz, J.C.; Fernández- Sanjurjo, M.J.; Álvarez-Rodríguez, E.; Núñez-Delgado, A. Kinetics of tetracycline, oxytetracycline and chlortetracycline adsorption and desorption on two acid soils. *Environ. Sci. Pollut. Res.* **2015**, *22*, 425–433. [CrossRef]
18. Fernández-Calviño, D.; Bermúdez-Couso, A.; Arias-Estévez, M.; Nóvoa-Muñoz, J.C.; Fernández-Sanjurjo, M.J.; Álvarez-Rodríguez, E.; Núñez-Delgado, A. Competitive adsorption/desorption of tetracycline, oxytetracycline and chlortetracycline on two acid soils: Stirred flow chamber experiments. *Chemosphere* **2015**, *134*, 361–366. [CrossRef]
19. Arias, M.; Pérez-Novo, C.; López, E.; Soto, B. Competitive adsorption and desorption of copper and zinc in acid soils. *Geoderma* **2006**, *133*, 151–159. [CrossRef]
20. Murali, V.; Aylmore, A.G. Competitive adsorption during solute transport in soils: 1. Mathematical models. *Soil Sci.* **1983**, *135*, 143–150. [CrossRef]
21. Conde-Cid, M.; Fernández-Sanjurjo, M.J.; Ferreira-Coelho, G.; Fernández-Calviño, D.; Arias-Estevez, M.; Núñez-Delgado, A.; Álvarez-Rodríguez, E. Competitive adsorption and desorption of three tetracycline antibiotics on bio-sorbent materials in binary systems. *Environ. Res.* **2020**, *190*, 110003. [CrossRef] [PubMed]
22. Conde-Cid, M.; Fernández-Calviño, D.; Núñez-Delgado, A.; Fernández-Sanjurjo, M.J.; Arias-Estévez, M.; Álvarez-Rodríguez, E. Influence of mussel shell, oak ash and pine bark on the adsorption and desorption of sulfonamides in agricultural soils. *J. Environ. Manag.* **2020**, *261*, 110221. [CrossRef] [PubMed]
23. Sarmah, A.K.; Meyer, M.T.; Boxall, A.B. A global perspective on the use, sales, exposure pathways, occurrence, fate and effects of veterinary antibiotics (VAs) in the environment. *Chemosphere* **2006**, *65*, 725–759. [CrossRef] [PubMed]
24. Carvalho, I.T.; Santos, L. Antibiotics in the aquatic environments: A review of the European scenario. *Environ. Int.* **2016**, *94*, 736–757. [CrossRef] [PubMed]
25. Wang, S.; Wang, H. Adsorption behavior of antibiotic in soil environment: A critical review. *Front. Environ. Sci. Eng.* **2015**, *9*, 565–574. [CrossRef]
26. Wegst-Uhrich, S.R.; Navarro, D.A.G.; Zimmerman, L.; Aga, D.S. Assessing antibiotic sorption in soil: A literature review and new case studies on sulfonamides and macrolides. *Chem. Cent. J.* **2014**, *8*, 5. [CrossRef] [PubMed]
27. Rath, S.; Fostier, A.H.; Pereira, L.A.; Dioniso, A.C.; Ferreira, F.D.O.; Doretto, K.M. Sorption behaviors of antimicrobial and antiparasitic veterinary drugs on subtropical soils. *Chemosphere* **2019**, *214*, 111–122. [CrossRef]
28. Teixidó, M.; Granados, M.; Prat, M.D.; Beltrán, J.L. Sorption of tetracyclines onto natural soil: Date analysis and prediction. *Environ. Sci. Pollut. Res.* **2012**, *19*, 3087–3095. [CrossRef]
29. Chen, M.; Yi, Q.; Hong, J.; Zhang, L.; Lin, K.; Yuan, D. Simultaneous determination of 32 antibiotics and 12 pesticides in sediment using ultrasonic-assisted extraction and high performance liquid chromatography-tandem mass spectrometry. *Anal. Methods* **2015**, *7*, 1896–1905. [CrossRef]
30. Babić, S.; Horvat, A.J.M.; Mutavdžić-Pavlović, D.; Kaštelan-Macan, M. Determination of pKa values of active pharmaceutical ingredients. *Trends Anal. Chem.* **2007**, *26*, 1043–1061. [CrossRef]
31. Figueroa-Diva, R.A.; Vasudevan, D.; MacKay, A.A. Trends in soil sorption coefficients within common antimicrobial families. *Chemosphere* **2010**, *79*, 786–793. [CrossRef] [PubMed]
32. Białk-Bielińska, A.; Stolte, S.; Matzke, M.; Fabiańska, A.; Maszkowska, J.; Kołodziejska, M.; Liberek, B.; Stepnowski, P.; Kumirska, J. Hydrolysis of sulphonamides in aqueous solutions. *J. Hazard. Mater.* **2012**, *221*, 264–274. [CrossRef] [PubMed]
33. Ahmed, M.B.; Zhou, J.L.; Ngo, H.H.; Guo, W.S.; Johir, M.A.H.; Sornalingam, K. Single and competitive sorption properties and mechanism of functionalized biochar for removing sulfonamide antibiotics from water. *Chem. Eng. J.* **2017**, *311*, 348–358. [CrossRef]
34. Edwards, M.; Benjamin, M.M.; Ryan, J.N. Role of organic acidity in sorption of natural organic matter (NOM) to oxide surfaces. *Colloids Surf. Physicochem. Eng. Asp.* **1996**, *107*, 297–307. [CrossRef]
35. Conde-Cid, M.; Ferreira-Coelho, G.; Arias-Estévez, M.; Álvarez-Esmorís, C.; Nóvoa-Muñoz, J.C.; Núñez-Delgado, A.; Fernández-Sanjurjo, M.J.; Álvarez-Rodríguez, E. Competitive adsorption/desorption of tetracycline, oxytetracycline and chlortetracycline on pine bark, oak ash and mussel shell. *J. Environ. Manag.* **2019**, *250*, 109509. [CrossRef] [PubMed]
36. Conde-Cid, M.; Ferreira-Coelho, G.; Fernández-Calviño, D.; Núñez-Delgado, A.; Fernández-Sanjurjo, M.J.; Arias-Estévez, M.; Álvarez-Rodríguez, E. Single and simultaneous adsorption of three sulfonamides in agricultural soils: Effects of pH and organic matter content. *Sci. Total Environ.* **2020**, *744*, 140872. [CrossRef]
37. Sukul, P.; Lamshöft, M.; Zühlke, S.; Spiteller, M. Sorption and desorption of sulfadiazine in soil and soil-manure systems. *Chemosphere* **2008**, *73*, 1344. [CrossRef]
38. Leal, R.M.P.; Alleoni, L.R.F.; Tornisielo, V.L.; Regitano, J.B. Sorption of fluoroquinolones and sulfonamides in 13 Brazilian soils. *Chemosphere* **2013**, *92*, 979. [CrossRef]
39. Hu, S.; Zhang, Y.; Shen, G.; Zhang, H.; Yuan, Z.; Zhang, W. Adsorption/desorption behavior cand mechanisms of sulfadiazine and sulfamethoxazole in agricultural soil systems. *Soil Tillage Res.* **2019**, *186*, 233. [CrossRef]
40. Conde-Cid, M.; Fernández-Calviño, D.; Nóvoa-Muñoza, J.C.; Núñez-Delgado, A.; Fernández-Sanjurjo, M.J. Arias-Estévez, M.; Álvarez-Rodríguez, E. Experimental data and model prediction of tetracycline adsorption and desorption in agricultural soils. *Environ. Res.* **2019**, *177*, 108607. [CrossRef]
41. Bao, Y.J.; Ding, H.S.; Bao, Y.Y. Effects of Temperature on the Adsorption and Desorption of Tetracycline in Soils. *Adv. Mater. Res.* **2013**, *726–731*, 344–347. [CrossRef]

42. Bhaumik, R.; Mondal, N.K.; Das, B.; Roy, P.K.C. Eggshell powder as an adsorbent for removal of fluoride from aqueous solution: Equilibrium, kinetic and thermodynamic studies. *J. Chem.* **2012**, *9*, 1457–1480. [CrossRef]
43. Na, C.J.; Yoo, M.J.; Tsang, D.C.W.; Kim, H.W.; Kim, K.H. High-performance materials for effective sorptive removal of formaldehyde in air. *J. Hazard. Mater.* **2019**, *366*, 452–465. [CrossRef] [PubMed]
44. Szulejko, J.E.; Kim, K.H.; Parise, J. Seeking the most powerful and practical realworld sorbents for gaseous benzene as a representative volatile organic compound based on performance metrics. *Sep. Purif. Technol.* **2019**, *212*, 980–985. [CrossRef]
45. Vikrant, K.; Kim, K.H. Nanomaterials for the adsorptive treatment of Hg (II) ions from water. *Chem. Eng. J.* **2019**, *358*, 264–282. [CrossRef]
46. Khezami, L.; Capart, R. Removal of chromium (VI) from aqueous solution by activated carbons: Kinetic and equilibrium studies. *J. Hazard. Mater.* **2005**, *123*, 223–231. [CrossRef]
47. Foo, K.Y.; Hameed, B.H. Insights into the modeling of adsorption isotherm systems. *Chem. Eng. J.* **2010**, *156*, 2–10. [CrossRef]
48. Behnajady, M.A.; Bimeghdar, S. Synthesis of mesoporous NiO nanoparticles and their application in the adsorption of Cr(VI). *Chem. Eng. J.* **2014**, *239*, 105–113. [CrossRef]
49. Conde-Cid, M.; Fernandez-Calviño, D.; Fernandez-Sanjúrjo, M.J.; Núñez-Delgado, A.; Álvarez-Rodríguez, E.; Arias-Estevez, M. Adsorption/desorption and transport of sulfadiazine, sulfachloropyridazine, and sulfamethazine, in acid agricultural soils. *Chemosphere* **2019**, *234*, 978. [CrossRef]

Article

Novel Study for Energy Recovery from the Cooling–Solidification Stage of Synthetic Slag Manufacturing: Estimation of the Potential Energy Recovery

Francisco M. Baena-Moreno *, Mónica Rodríguez-Galán, Benito Navarrete and Luis F. Vilches

Chemical and Environmental Engineering Department, Technical School of Engineering, University of Seville, C/Camino de los Descubrimientos s/n, 41092 Sevilla, Spain; mrgmonica@us.es (M.R.-G.); bnavarrete@us.es (B.N.); luisvilches@us.es (L.F.V.)
* Correspondence: fbaena2@us.es

Received: 6 November 2020; Accepted: 25 November 2020; Published: 2 December 2020

Abstract: Herein, a novel method for energy recovery from molten synthetic slags is analyzed. In this work, the potential energy that could be recovered from the production of synthetic slag is estimated by means of an integrated experimental–theoretical study. The energy to be recovered comes from the cooling–solidification stage of the synthetic slag manufacturing. Traditionally, the solidification stage has been carried out through quick cooling with water, which does not allow the energy recovery. In this paper, a novel cooling method based on metal spheres is presented, which allows the energy recovery from the molten slags. Two points present novelty in this work: (1) the method for measuring the metal spheres temperature (2) and the estimation of the energy that could be recovered from these systems in slag manufacturing. The results forecasted that the temperature achieved by the metal spheres was in the range of 295–410 °C in the center and 302–482 °C on the surface. Furthermore, we estimated that 325–550 kJ/kg of molten material could be recovered, of which 15% of the energy consumption is in the synthetic slag manufacturing process. Overall, the results obtained confirmed the potential of our proposal for energy recovery from the cooling–solidification stage of synthetic slag manufacturing.

Keywords: sustainable synthetic slag production; energy recovery; metal spheres; fixed bed regenerator; waste and energy nexus

---

1. Introduction

1.1. Background

The future challenges related to the known circular economy policy need to intensify the research of more environmental processes and less energy-intensive industrial processes [1,2]. In this sense, one of the key points is the development of new product manufacturing through the recovery of waste/by-products in applications with high added value and easy technology transfer to the industrial sector [3,4]. The use of waste to form new vitreous materials offer a potential possibility for waste valorization. Even though the idea of valorizing waste as constituents of cements is documented in the literature through patents and research works [5–10], novelties can be studied in different ways such as (1) exploring new materials and seeking similar properties to those presented by blast furnace slags; (2) the obtaining of synthetic slags exclusively from waste mixtures that could be managed in non-hazardous waste landfills; and (3) the energy-efficient production of synthetic slag manufacturing.

The first and second options pointed out above seek to provide value-added products from waste based on compliance with regulatory requirements. Nevertheless, the third one is considered associated to the viability of the manufacturing process of these new materials from an energy perspective. Therefore, the energy-efficiency study of the production process is considered one of the keys to obtain its technical–economic viability, together with the minimization of waste transport costs. In a previous work of this research group, cement substitutive with properties similar to blast furnace slags (synthetic slags) was obtained. The typical composition of blast furnace slag can be seen in Table 1. In this process, a mix of waste was proposed to obtain the synthetic slag material [11]. As announced in previous work, a deep energy recovery study is needed to estimate the viability of this novel production method.

Table 1. Typical composition ranges of blast furnace slags. Own elaboration based on [12–14].

Composition Range	SiO_2	CaO	Al_2O_3	MgO	Fe_2O_3	SO_3	Na_2O	K_2O
	27–40	30–50	5–15	1–15	0.2–2.5	1–2.5	0.1–3	0.1–3

In accordance with EN 15167-1 [15], granulated blast furnace slag is a vitrified material produced by rapid cooling of a molten slag of suitable composition. This kind of slag is obtained by melting iron ore in a blast furnace, containing at least two-thirds of crystalline slag mass and having hydraulic properties when activated properly. The parameters that influence the hydraulic behavior of the final slag are the vitreous phase content, the chemical composition, the fineness, the additives, and the methods and/or substances for the activation stage, presenting what is called latent or potential hydraulic capacity [16]. The chemical composition of the blast furnace slags is one of the properties that marks its hydraulic potential, since the greater the basicity of the slags, the better the hydraulic behavior. However, the main characteristic related to hydraulic behavior is the proportion of the vitreous phase of the slag, which must exceed 70% to guarantee the hydraulic behavior [17].

The efficient production of the manufacture of synthetic slags needs a study of the energy recovery in its manufacturing process. These studies have been conducted in depth by numerous authors [18–22], focusing especially on the energy that can be recovered in the cooling stage of the molten material. Slag cooling systems can be classified according to the cooling fluid or material used for their impact on the energy efficiency of the process (energy recovery in the form of steam and/or warm gases). Thus, basically, all cooling systems consist of a heat exchange between the molten material and the selected fluid (water and/or air) or solid material (metallic material). In any case, the technologies applicable to the development of the cooling systems must perform a quick cooling, as well as being capable of harnessing the potential energy contained in molten material during the cooling process. Below, some methods for energy recovery from the molten material are explained, which are mainly based on wet systems (water as cooling fluid) or dry systems (where the cooling fluid used is air/gases).

Cooling with water achieves the objectives regarding the vitreous properties of the material. However, the disadvantages of water-based systems are related to their low energy efficiency and contamination with particles and other pollutants, generating a highly corrosive vapor due to the presence of acid gases that increase treatment, equipment maintenance, and energy recovery costs. In the dry cooling system, a transmission occurs of heat from the molten material to a stream of air, which is in contact with recovery boilers. In many of the air-cooling techniques, a prior atomization of the molten solid could be performed. Subsequently, a contacting stage with a stream of air in a fluidized or non-fluidized bed is carried out, which absorbs the heat, achieving solidification of the melt. Other granulation options arise when the molten material is fragmented by rotating devices [23]. Another possibility of cooling could be by contact of the molten material with solid materials such as spherical bodies. This option requires that the solid material that not react chemically with the molten slag (i.e., metal spheres [24] or cooling on rotating metal drums [25–27]). With these example systems, the energy recovery from the molten slag is possible. In addition, combining the energy efficiency of these energy recovery systems with the improvement of slag properties, the overall energy recovery efficiency of cooling processes can be increased [28].

This paper studies the potential energy recovery of the cooling process of synthetic slags production by means of a mass cooling system with metal spheres. Conceptually, the mass cooling system by means of thermal regeneration with metal spheres achieves rapid cooling of the molten material as a consequence of the transfer of energy from the molten material to the metal spheres. Subsequently, there is the possibility of a recovery of the energy contained in the spheres by means of an air current. In addition, the characteristics associated with the properties of the solid elements and the solidified material, generally of low thermal resistance, allow heat to dissipate rapidly. Once the metal spheres and the material have been cooled by the air flow in the regenerator and a posterior separation stage of the solidified material from the balls, the metal spheres would come back into contact with the synthetic slags in a continuous cycle. The conceptual scheme of the energy recovery system is presented in Figure 1. In this system, the slag is separated from the metal spheres after cooling through screening by the size of the spheres and solidified slags. This separation is possible as the slag are not adhered to the surface of the metal spheres.

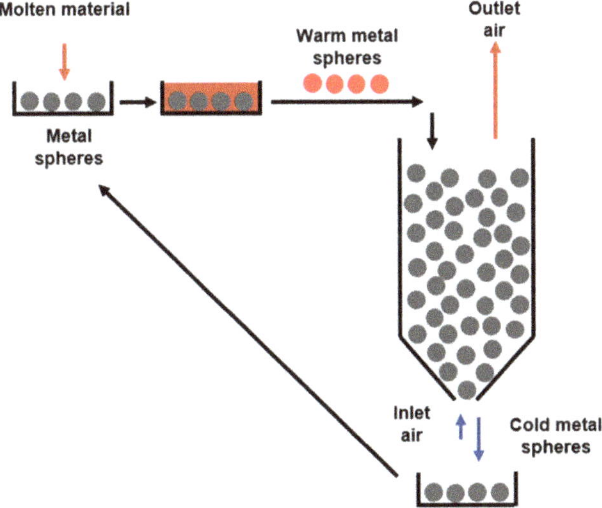

Figure 1. Conceptual scheme of the energy recovery process proposed in this work.

1.2. Goal and Scope

Based on the novel concept of developing synthetic slags from waste, as explained in our previous work [11], the main objective of this study is to evaluate the properties of synthetic slags and the energy recovery capacity of the aforementioned cooling system described in Figure 1. Likewise, the energy recovered in the cooling air stream was estimated in relation to the energy possessed by the molten material in order to analyze the possible energy savings that would result in the manufacture of slags. Two main points present novelty in this work. The first one is the method for measuring the temperatures that the metal spheres achieve. To the knowledge of the authors, this is the first time that both the methodology and the real temperature data are published. The second novelty of this work is that knowing the real temperature achieved in the metal spheres, it was possible to estimate the energy that could be recovered by these systems in slag manufacturing.

The estimation of energy recovery from the molten material consisted in several steps, which are indicated below:

- Justification of the temperature reached by the metal spheres at various molten mass (m_m)/metal spheres (m_s) mass ratio.

- Corroboration of the industrial feasibility of the proposed system in Figure 1 by means of fixed bed (or regenerator) height calculation. For this purpose, the calculation of the convective heat transfer coefficient by solid–air convection is needed. The definition of this step is crucial for checking that the process proposed is of industrial interest.
- Estimation of the potential energy recovered per kg of molten material.

To meet this end, this work is organized as follows. First, experiments for measuring the maximum temperature that could be reached by the metal spheres were performed. Inasmuch than the study aims to be useful for industrial purposes, the experiments were performed at different m_m/m_s ratios to reproduce real industrial scenarios. Subsequently, the molten material that was poured over the metal spheres was analyzed by means of X-ray diffraction (XRD) to verify the formation of the characteristic vitreous phase. Afterwards, before estimating the potential energy recovery, the technical feasibility of the proposed cooling process was characterized. To this end, the fixed bed height was calculated following the methodology explained in Section 2.2.4. For its estimation, the previous determination of the convective heat transfer coefficient (h) was necessary. The methodology employed for h estimation can be seen in Appendix A. Two experiments were performed to estimate the metal spheres—air flow temperature profiles in a real fixed bed. The experiments allowed the estimation of h following the assumptions explained in Section 2.2.4. Once the industrial fixed bed height was obtained and analyzed as feasible, the energy recovery per kg of molten material was estimated. To estimate the approximate percentage of the energy recovery in the overall slag manufacturing process, a comparison of the vitreous–mineral phases between our molten material and traditional clinker was done. This comparison allowed ensuring that no significant differences are found and hence that the total energy consumption of our molten material can be approximate to the one of clinker production.

2. Materials and Methods

In order to study the energy recovery efficiency of the cooling and solidification system for the molten synthetic slags, a three-stage laboratory experiment was proposed:

(1) Melt the waste mixture in an oven in appropriate proportions to achieve a composition similar to blast furnace slags. This stage was addressed in depth in our previous work [11].
(2) Obtain the temperature reached by the metal spheres when the molten material is poured with different melting/metal mass ratios. Determination of the characteristics of the vitreous phase obtained in order to ensure the vitreous properties of the material solidified.
(3) Evaluate the energy that can be extracted from a fixed bed of metal spheres at the temperature reached in the previous phase by exchanging energy with an air current.

Below, the materials and methods employed to fulfill this scheme are explained in depth.

2.1. Materials

The mix of waste for manufacturing synthetic slags was defined in a previous work of our group [11]. Three different wastes were selected as raw materials, which were construction and demolition waste, the solid waste stream generated in an aluminum recovery plant, and mussel shell waste from the aquiculture industry. For more information, please see [11].

2.2. Experimental Setup

2.2.1. Melting Furnace

The mix of waste was melted by employing the melting furnace (4 kW power) schemed in Figure 2, which was the same one employed in our previous work [11]. In brief, the melting furnace consists of six electrical resistances located inside a hexagonal fusion chamber. The discharge of the molten material is carried out from the bottom of the furnace.

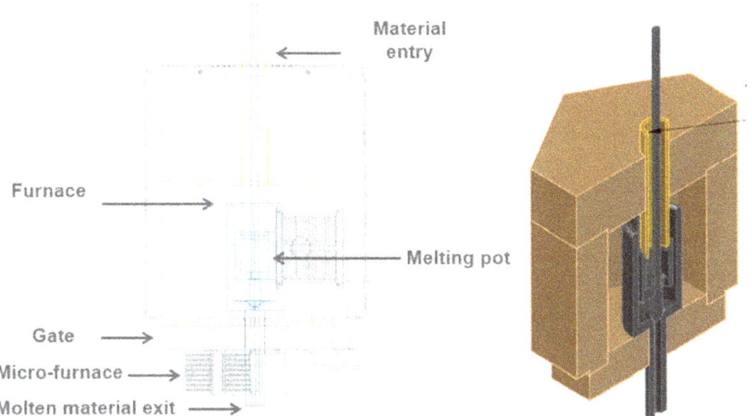

Figure 2. Melting furnace scheme.

2.2.2. Metal Spheres: Cooling of the Molten Material

The cooling stage to energy recovery from the molten material was based on metal spheres. The first stage was the study of the temperature evolution inside a single metal sphere. To this end, the system shown in Figure 3 was designed. In this system, a single metal sphere welded to a plate was built. The sphere was designed with two perforations in the lower part in which two thermocouples were inserted that allow measuring the temperature data in the center and on the surface of the sphere. All thermocouples were connected to the corresponding data acquisition system. For sake of safety, the sphere was assembled in a metal container and insulated by a refractory material, as can be seen in Figure 3. The system was placed under the melting furnace, and the molten material was poured directly from the furnace onto the sphere. In order to reproduce an industrial environment close to real cases, two sizes of spheres of 40 and 50 mm diameter were used. Figure 3 shows the characteristic dimensions of the 50 mm diameter metal sphere and the ducts in which the thermocouples were inserted. Furthermore, photos of the plates aforementioned with the spheres coupled can be seen. A detailed explanation of properties and composition of the spheres can be found in reference [29].

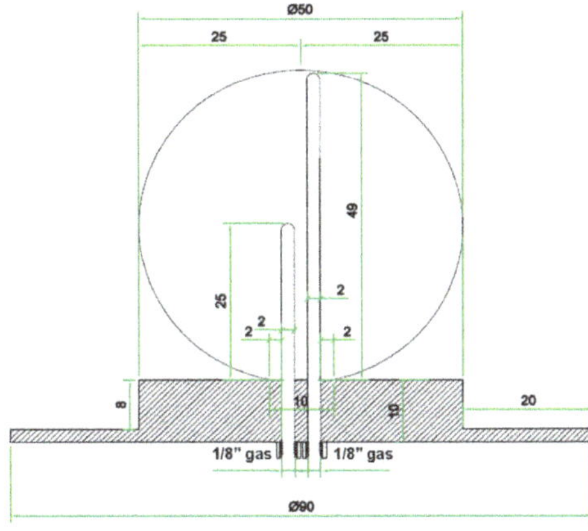

Figure 3. *Cont.*

Figure 3. Scheme of the device for cooling molten material with metal spheres and views of the containment vessel.

Figure 4 shows a real photo in which the coupling between the melting furnace outlet and the cooling device with the metal sphere can be observed.

Figure 4. Coupling of the melting furnace outlet and the cooling device.

2.2.3. Metal Spheres Packing Device for Energy Recovery

To estimate the potential energy recovery from the metal spheres with air, a fixed bed composed by two baskets filled with metal spheres was designed. The baskets are made of stainless steel and have a mesh that allows the gas to pass through the bed. The needed thermocouples (for temperature measurements) were connected to the data acquisition system. Figure 5 shows an image of the fixed bed and metal sphere baskets.

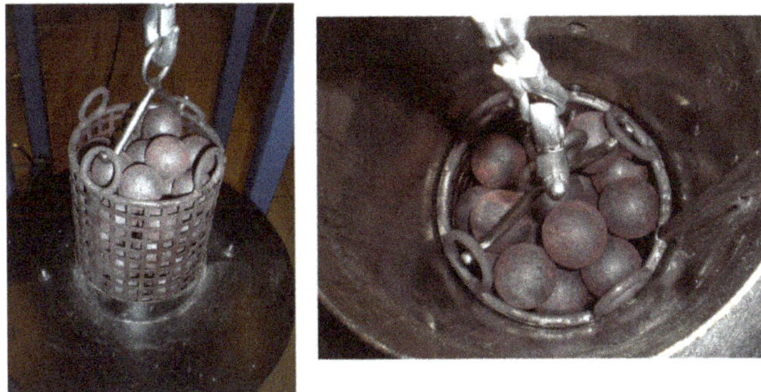

Figure 5. Metal spheres packing device.

The metal sphere packing, after being heated in an oven to the desired temperature, was deposited in a thermally insulated vessel that can be seen in Figure 6.

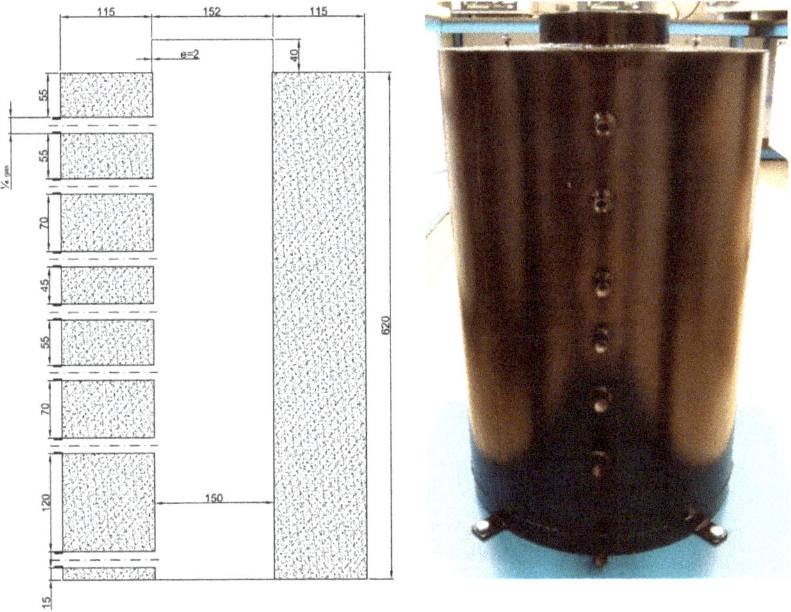

Figure 6. Thermally insulated vessel (dimensions in mm).

Moreover, the designed device incorporates an air preheater by means of electrical resistance, which allowed performing tests at different initial air temperatures. All the elements described in this section are integrated in the final experimental setup shown in Figure 7.

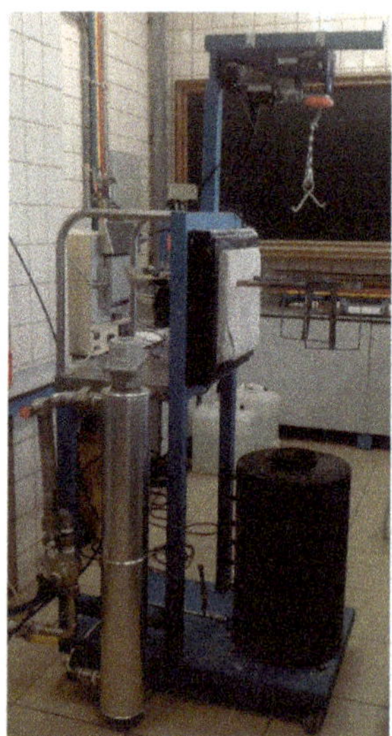

Figure 7. Complete metal spheres packing device for energy recovery.

2.2.4. Fixed Bed Height Estimation

The conceptual idea for the fixed bed of metal spheres herein applied consists of the following steps. First, the molten material is poured on the metal spheres. After that, the spheres together with the vitrified material are located in a fixed bed in which the spheres move by gravity slowly (following a plug flux model). Thus, the spheres go from the upper part of the fixed bed to the bottom, cooling with countercurrent air from 400 °C to room temperature. As will be explained in Section 3.1, a temperature of 400 °C can be reached in the spheres, for example with an m_m/m_s ratio of 0.4 and with a sphere diameter of 50 mm.

The estimation of the potential energy recovery from the proposed process needs the previous calculations of some parameters. Therefore, the purpose of this section was to estimate the residence time of the spheres to produce a temperature decrease from 400 °C to room temperature, as well as the necessary air flow and the temperature that would be reached in the air. The proposed method to obtain these results is a semi-empirical model variation of the temperature of the spheres in the bed. Please see Appendix A for more information. This allowed estimating h′, which is a variation of the original h under the assumptions imposed, which are explained in Appendix A.

Once h′ was known, it was possible to estimate the dimensions of the equipment. To this end, a steady-state balance was performed in the fixed bed regenerator shown in Figure 8. Equations (1) and (2) collect the components of these balance equations. The residence time necessary for achieving a decrease in the metal spheres temperature from 400 to 40 °C was calculated also by means of these balance equations. This calculation allowed obtaining the temperature profile along the fixed bed regenerator (Figure 8).

$$m_s \cdot \left(\frac{kg}{h}\right) \cdot c_s\left(\frac{kJ}{kg \cdot K}\right) \cdot (T_s(x - \Delta x) - T_s(x))(K) = m_a\left(\frac{kg}{h}\right) \cdot c_a\left(\frac{kJ}{kg \cdot K}\right) \cdot (T_a(x - \Delta x) - T_a(x))(K) \qquad (1)$$

$$h'(x)\left(\frac{kJ}{h \cdot m^2 \cdot K}\right) \cdot A(\Delta x)(m^2) \cdot \left(\frac{T_s(x-\Delta x)+T_s(x)}{2} - \frac{T_a(x-\Delta x)+T_a(x)}{2}\right)(K) = m_a\left(\frac{kg}{h}\right) \cdot c_a\left(\frac{kJ}{kg \cdot K}\right) \cdot (T_a(x - \Delta x) - T_a(x))(K) \qquad (2)$$

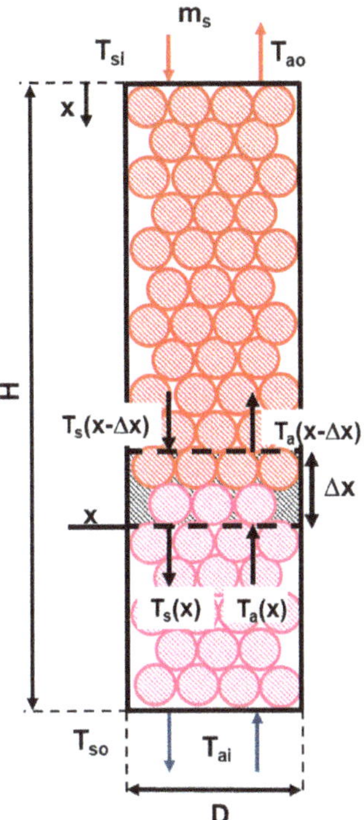

Figure 8. Fixed bed regenerator scheme.

The equations were solved following a standalone path. Imposing that Ts(x), Ts(x − "Δ" x), and Ta (x − "Δ" x) are known, it is possible to obtain Ta(x) from Equation (1), as well as "Δ" x from Equation (2).

2.3. Physicochemical Characterization

The physicochemical characterization was carried out by means XRD analysis, model AXIOS from Panalytical. The 2θ angle was increased by 0.05°, with a 450 time per step over a range of 10–90°. Then, diffraction patterns were recorded at 40 mA and 45 kV in the cases in which it was necessary, using Cu Kα radiation ($\lambda = 0.154$ nm).

2.4. Experimental Plan

To fulfill the scope of the work previously defined, the following experimental plan was designed. The whole experimental tests were divided into two main groups of experiments, corresponding to the same previous division between the justification of the temperature that metal spheres can achieve and the measurement of the air temperature in the fixed bed packing for estimating the potential energy recovery. Table 2 shows the tests performed in both stages. The objective of the first stage was to cover all the possibilities regarding metal spheres diameters and molten quantity. However, the design of the second stage was aimed to acquire data for energy recovery estimation. All the experiments included in Table 2 were conducted twice, ensuring that the results are reproducible with an overall experimental error of ±2%.

Table 2. Experimental plan.

First Experimental Stage			
Test	Sphere mass (g)	Melt mass (g)	m_m/m_s
1	260	157.5	0.61
2	260	122.5	0.47
3	510	157.5	0.31
4	510	122.5	0.24
5	510	192.5	0.38
Second Experimental Stage			
Test	Metal spheres diameter (mm)	Air inlet temperature (°C)	Experimental time (s)
6	50	20–25	3000
7	50	150	2500

After performing the tests included in the second experimental stage, the estimation explained in Section 2.2.4 was carried out to fulfill the aforementioned purpose.

3. Results

The results obtained in the two stages of the experimentation are shown and discussed in this section. Furthermore, chemical and vitreous characterization of the material obtained with this cooling system are given and compared with the ones obtained in our previous work [11].

3.1. Energy Transfer from the Molten Material to Metal Spheres

The maximum temperature that the spheres of 40 and 50 mm (industrial application diameters) were determined based on the melt, which was poured over them. To this end, different m_m/m_s ratio were tested. This ratio was adjusted between 0.24 and 1 based on characteristic values found for slag cooling [30]. In all the experiments, the temperature of the molten material poured over the spheres was 1500 °C. Table 3 shows the characteristic parameters of each of the tests performed as well as the maximum temperatures reached in the center and on the surface of the spheres according to the direct reading of the two thermocouples placed in the spheres (see Figure 3). Additionally, Figure 9 show the evolution of the sphere's temperature versus time of each test performed.

Table 3. Main characteristics of the different tests performed and temperatures reached in the metal spheres.

Test	Sphere Mass (g)	Melt Mass (g)	m_m/m_s	Temperature in the Center of the Sphere (°C)	Temperature in the Surface of the Sphere (°C)
1	260	157.5	0.61	410	482
2	260	122.5	0.47	350	405
3	510	157.5	0.31	340	383
4	510	122.5	0.24	295	302
5	510	192.5	0.38	390	438

In Figure 9, it can be seen that the time needed to reach the maximum temperature is 2–3 min. From an industrial point of view, this could be a characteristic time interval between the time of melting fall on the spheres and the time in which the spheres, together with the molten material, are deposited in the spheres packing for energy recovery with air. Figure 10 plotted the maximum temperature reached at the surface and center of the spheres as a function of its diameter and the ratio m_m/m_s.

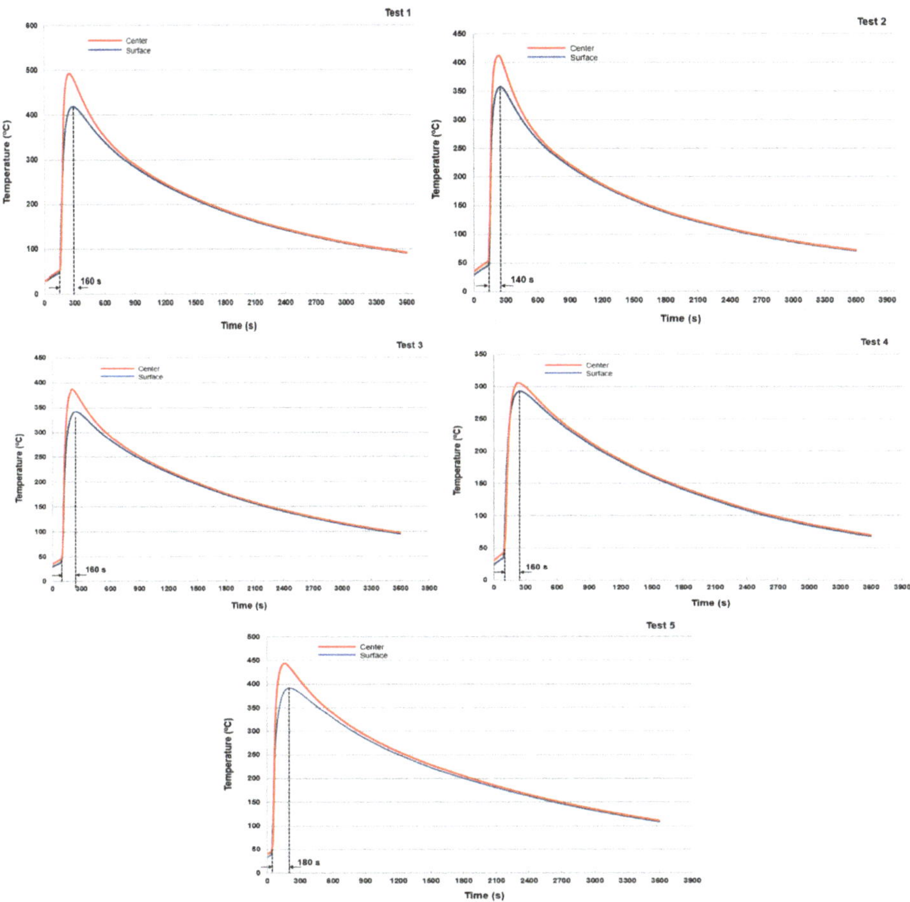

Figure 9. Temperature evolution.

From the data presented above, some considerations can be discussed. In the ranges of tested variables, the maximum temperature reached by the spheres is less than 500 °C. Thus, in comparison with the molten material temperatures (1500 °C), the maximum temperature means several energy losses. To obtain a certain temperature (i.e., 400 °C), spheres of smaller diameter need less melt mass (higher ratio m_m/m_s, but lower m_m). Therefore, the energy transfer from the molten material seems more efficient to smaller sphere diameters. This may be caused by the whole covering of the melting material in these cases. Assuming that the material on the metal surface is in thermal equilibrium, it can be deduced that in all the cases tested, the material reaches 400 °C at a cooling rate exceeding 300 °C/min.

Another interesting fact needs some explanation. For the same sphere diameter and equal molten/metal heat transfer coefficient, as the ratio m_m/m_s increases, the gradient between the maximum surface and center temperature increases. This fact shows that, possibly, the thickness of the melt mass on the surface of the sphere influences the rate of energy transfer from the molten material to the metal sphere. This result needs further optimization, which opens new horizons for future works aimed to increase the energy transfer velocity and minimize the energy losses to the environment.

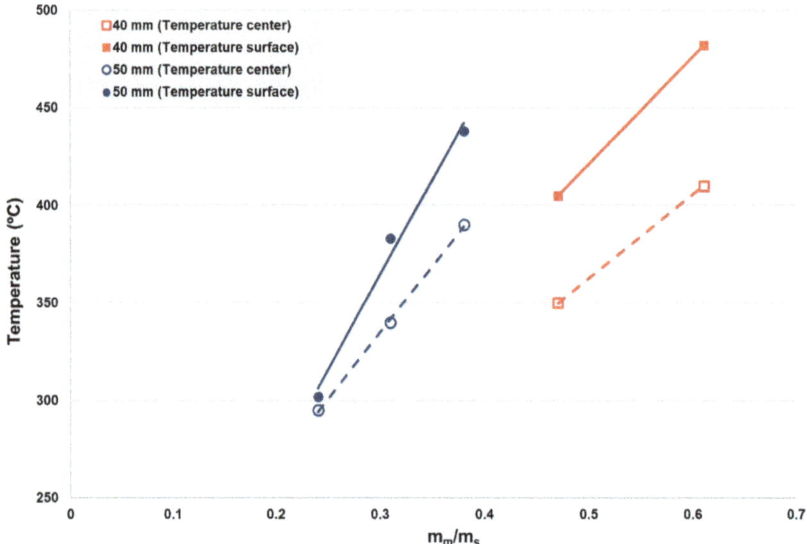

Figure 10. Maximum temperature reached at the surface and center of the spheres as a function of its diameter and the ratio m_m/m_s.

3.2. Properties of the Cooled Material

To check the existence of a vitreous phase of the cooled material obtained at m_m/m_s for each ratio tested, an XRD analysis was performed after each test. Figure 11 shows that no differences are found between the properties of the cooled material obtained in each test. The cooled materials exceed 85% of the vitreous phase, confirming that all the m_m/m_s ratios tested, spheres of 40 and 50 mm produce a material with suitable vitreous properties.

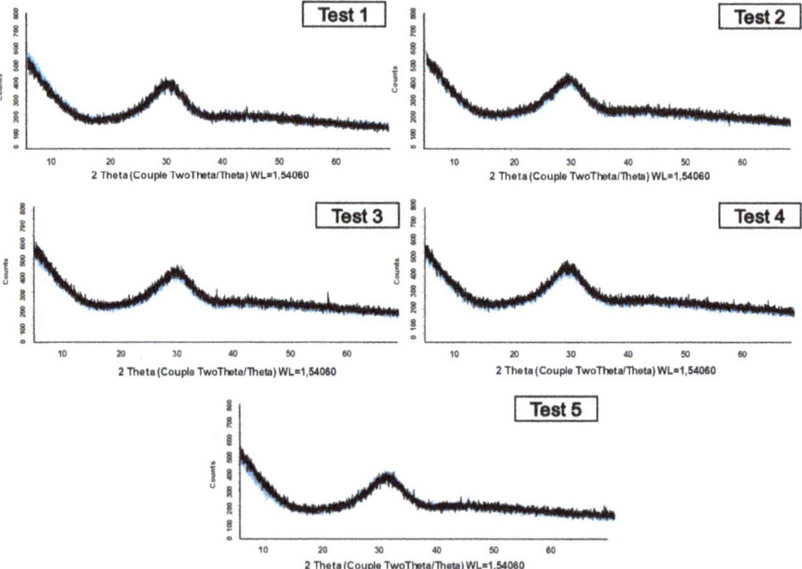

Figure 11. Diffractograms of tests 1 to 5.

The chemical composition of the waste mix before melting and after melting–cooling is presented in Table 4. Chemical composition was determined using an X-ray fluorescence spectrometer (XRF, Model AXIOS, Panalytical). Thus, it is possible to verify how the material obtained is within the typical ranges of composition of the blast furnace slags (see Table 1). Some losses happen due to the calcination of some components, mainly calcium carbonate, which is converted into calcium oxide and carbon dioxide. This loss in weight causes the variation in the composition after fusion. The sum of the oxides of silicon, calcium, and magnesium is 82.12%, that is, more than two-thirds of the total mass. Moreover, the ratio $(CaO + MgO)/SiO_2$ is 1.16. These two specifications, together with a vitreous phase result, meet the requirements of blast furnace slags to be used as additions to cement, according to EN 15167-1.

Table 4. Chemical composition of waste before–after melting and solidification (%) (C.L: calcination losses).

	SiO_2	Al_2O_3	Fe_2O_3	MnO	MgO	CaO	Na_2O	K_2O	TiO_2	P_2O_5	SO_3	C.L.
Before melting	25.2	8.4	1.39	0.04	1.12	30.93	1.03	0.66	0.2	0.07	0.28	27.81
After melting	38.05	11.42	1.77	0.07	1.44	42.63	1.36	0.79	0.32	0.1	0.68	0.21
Deviations	±0.68	±1.13	±0.34	±0.00	±0.06	±0.49	±0.07	±0.04	±0.01	±0.01	±0.21	±0.31

3.3. Estimation of the Energy Recovery Using a Fixed Bed of Metal Spheres with Air

In the case of 50 mm diameter spheres, 23 spheres (≈ 0.18 m^2) were deposited in the bed of dimensions shown in Figure 6. The tests were performed at constant inlet air flow equal to 26 m^3/h, which was at room temperature. The inlet air temperature was later varied to 150 °C using the preheater in order to analyze various temperature ranges. The experimental results obtained are shown in Figures 12 and 13, in which the rapid heat dissipation obtained can be verified. Thus, for an inlet air temperature of 20 °C, outlet temperatures greater than 300 °C were achieved, with air residence times in the fixed bed less than one second (Figure 12).

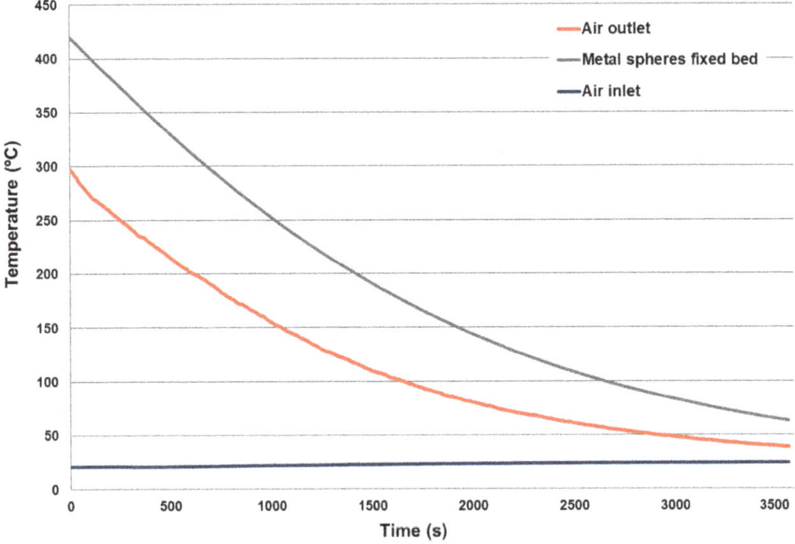

Figure 12. Evolution of temperatures in the system for entering air at room temperature.

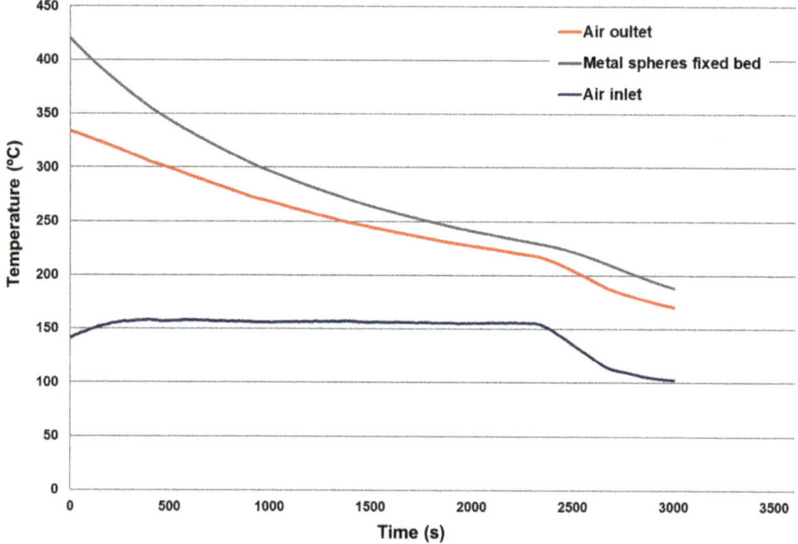

Figure 13. Evolution of temperatures in the system for inlet air at 150 °C.

An estimation of h′ was carried out from the data obtained in Figures 12 and 13, splitting the results in equal time interval of 500 s. For spheres temperatures higher than 225 °C, the results obtained in Figure 13 were used; whereas for lower temperatures, the data from Figure 12 were selected. Thus, the following expression was obtained for h′ (Equation (3)), where h′ units are W/m^2·K and T_s in °C.

$$h'(T_s) = 17.545 \cdot e^{0.0066 \cdot T_s} \left(R^2 = 0.9522\right) \tag{3}$$

Once h′ was calculated, it is possible to obtain a real dimension's estimation of the energy regenerator equipment in steady state. To this end, a steady-state balance was performed, giving as a result the following results under the assumptions indicated in Section 2.2.4: 30.26 kg/h air flow; 68.8 kg/h metal spheres flow. The transfer area metal spheres–air was calculated as A(Δx) = 0.943 · Δx. Figure 14 presents the fixed bed regenerator scheme solved.

To obtain the residence time needed for the metal spheres temperature to decrease from 400 to 40 °C and, therefore, to obtain the temperature profiles (both spheres and air) along the fixed bed regenerator, Equations (1) and (2) were employed. The profiles resulted can be seen plotted in Figure 15, needing 2.9 m of fixed bed height and 2.5 h to achieve the transfer imposed.

With the experimental data obtained and to corroborate the hypotheses made, Equation (1) as readjusted, obtaining the following expression:

$$T_s(x) = T_a(x) + 19.162 \cdot e^{-0.005 \cdot h'(x)} \left(R^2 = 0.9476\right). \tag{4}$$

With these results, in the section of the equipment studied, the molten material that could be cooled (imposing m_m/m_s = 0.4) is 27.52 kg/h, recovering 11,215 kJ/h. Thus means that the maximum energy recovered by air in the equipment per kilogram of molten material would be 408 kJ/kg at 395 °C. The losses that this kind of equipment may have are about 20%. Therefore, it can be deduced that with the dimensions of the equipment obtained, 325 kJ/kg of molten material can be recovered at a temperature higher than 350 °C. In any case, the energy to be recovered would be greater, as long as in this estimation, it has not been considered that the vitrified solid that comes with the metal spheres represent an additional energy recovery potential of about 225 kJ/kg of molten material (employing a Cp value of 0.63 kJ/kg·K [31]). Therefore, it is reasonable that with this equipment, an energy greater than 550 kJ/kg of molten material can be recovered with air at a temperature higher than 350 °C using m_m/m_s ratios of 0.4. The results obtained in this section are quite conservative, because it is more than likely that a more efficient energy recovery could be obtained by optimizing, among other variables, the mass air flow per unit area.

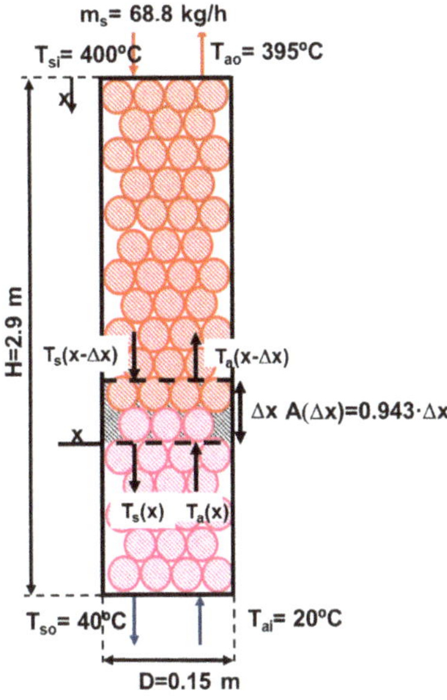

Figure 14. Fixed bed regenerator scheme solved.

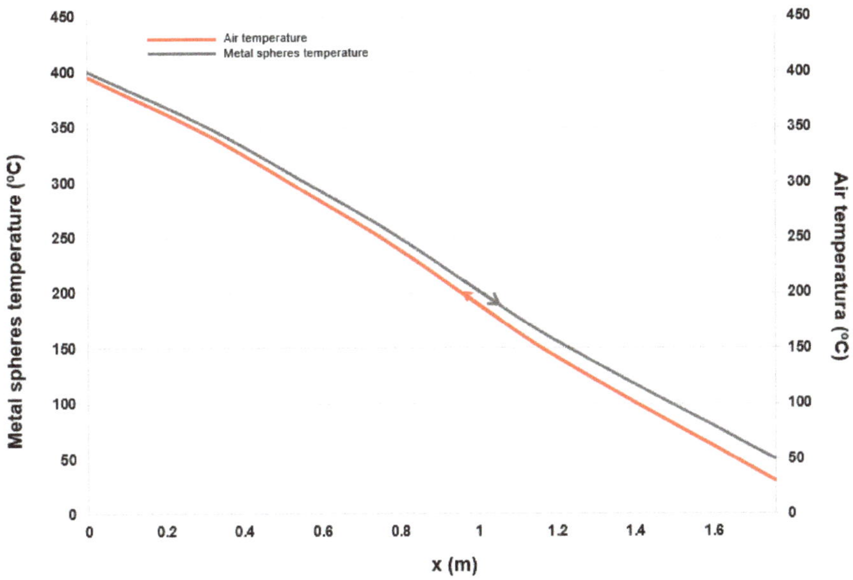

Figure 15. Temperature profiles of metal spheres and air in the fixed bed.

3.4. Potential Use of the Energy Recovered in the Regenerator in the Manufacture of Synthetic Slags

A possible vitreous material production scheme could correspond to a typical clinker manufacturing process [31], in which, after the rotary kiln, an oven is included where the complete fusion of the material occurs. The process would consist of a preheater, in which part of the calcination reactions would occur; a rotary kiln, in which the decarbonation of the material would be finished and that would take the material to 1300 °C; and finally, a melting or reverberating furnace, where the material that arrives from the rotary kiln would be molten and heat up to 1500 °C to guarantee its fluidity. After melting, the material would be cooled quickly in the fixed bed of metal spheres to achieve the desired vitreous properties.

To confirm the proposed production process, an XRD analysis of the mixed material was performed in the proportions indicated in Section 2.1 and heated to temperatures prior to the melting point (1250–1280 °C). As shown in Figure 16 in which the XRD results are presented, at those temperatures, crystalline phases appear similar to those that occur in the clinkerization process (C2S—Dicalcium Silicate, C3S—Tricalcium Aluminate, C3A—Tricalcium Silicate, CAF—Tetracalcium Aluminoferrite).

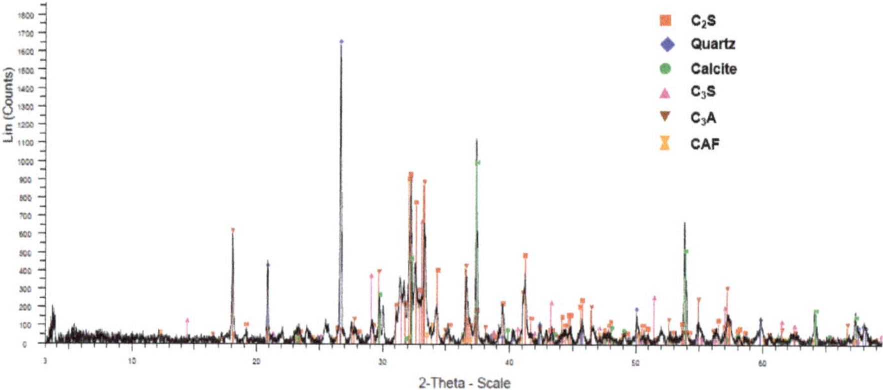

Figure 16. Diffractogram with crystalline phases of the waste mixture (1250–1280 °C).

Therefore, it can be concluded that synthetic slags have characteristics and mineral phases that are typically found in the raw materials used in the manufacture of the clinker. Furthermore, it can be affirmed that the energy consumption for its manufacture will be of the same order as that consumed in the manufacture of clinker. In this context, the energy recovered in the fixed bed regenerator presented in this work could be used as air at about 350–400 °C. For example, savings in fuel consumption could occur if the regenerator air stream was used as the combustion air inlet stream. Broadly, if 10% higher than the energy consumption of clinker production (about 3800 kJ/kg clinker [32]) is taken as a reference for the energy consumption of synthetic slags, the regenerator's air flow could result in energy savings in the manufacture of synthetic slags that could be around 15% of the energy needed for its production.

4. Conclusions

In this work, a novel method for energy recovery from the cooling–solidification stage of a synthetic slag manufacturing process was satisfactory studied. The study was evaluated by means of an experimental study that was complemented by theoretical calculations to close the overall performance.

The experimental stage was divided into two differentiated group of tests. The first group aimed to obtain the temperature value of the metal spheres when the molten material was poured over them. The second experimental was carried out to evaluate the metal spheres–air flow temperature profiles in a real fixed bed. During the experimental tests, the scope was to analyze the temperatures obtained in the center and on the surface of the metal spheres. Thanks to this experimental study, it was possible to measure a temperature in the range of 295–410 °C in the center of the sphere and 302–482 °C over the surface. Thus, it allowed setting up a temperature value for the second experimental stage, which consequently allowed the theoretical calculation of the energy recovery per kg of molten slag. Moreover, the molten material that was poured over the metal spheres was analyzed by means of XRD, and the formation of the characteristic vitreous phase was verified.

The main purpose of the theoretical approach was to achieve a fixed bed regenerator height, which proved that the process is viable industrially. To meet this end, h was previously estimated with the help of the values obtained in the second experimental stage. The fixed bed height calculated proved to be industrially achievable. Once the potential industrial fixed bed height was obtained and analyzed as feasible, the energy recovery per kg of molten material was estimated. We estimated that between 325 and 550 kJ/kg of molten material could be recovered, which is 15% of the energy consumption in synthetic slag manufacturing process.

Overall, the results obtained confirmed the potential of our proposal for energy recovery from the cooling–solidification stage of synthetic slag manufacturing. Our work herein studied proved to be of great interest for the industrial scale. The energy recovery for high-energy consumptions industries is one of the key points to achieve a more sustainable industrial model.

Author Contributions: Conceptualization, L.F.V., B.N. and M.R.-G.; methodology, M.R.-G., F.M.B.-M.; software, M.R.-G., F.M.B.-M.; validation, M.R.-G., F.M.B.-M.; formal analysis, F.M.B.-M.; investigation, L.F.V., B.N., M.R.-G., F.M.B.-M.; resources, L.F.V., B.N.; data curation, M.R.-G.; writing—original draft preparation, L.F.V., F.M.B.-M.; writing—review and editing, L.F.V., M.R.-G., F.M.B.-M.; visualization, M.R.-G., F.M.B.-M., X.X.; supervision, L.F.V., F.M.B.-M.; project administration, L.F.V., B.N.; funding acquisition, L.F.V., B.N. All authors have read and agreed to the published version of the manuscript.

Funding: This work was supported by University of Seville through V PPIT-US. Financial support for this work was also provided by MAVIT project (FEDER-INNTERCONECTA).

Conflicts of Interest: The authors declare no conflict of interest.

Abbreviations

m_m	molten mass
m_s	spheres mass
Tsi	spheres temperature inlet
Tso	spheres temperature outlet
Tai	air temperature inlet
Tao	air temperature outlet
Ts	spheres temperature
Ta	air temperature
H	fixed bed regenerator height
D	fixed bed regenerator diameter
h	convective heat transfer coefficient
h′	convective heat transfer coefficient variation
Cs	spheres calorific value
Ca	air calorific value
A("Δ" x)	transfer area metal spheres–air

Appendix A

Existing models for thermal regenerators require a very rigorous adjustment of various parameters such as speed, Reynolds number, Prandtl number, or the characteristic diameter of the spheres. In addition, when considering the balance equations that allow determining the behavior of the equipment, it is necessary to solve a nonlinear equation system that requires approximate numerical techniques for its resolution. However, semi-empirical bed models are usually resolved under the consideration that the sphere internal resistance is negligible to heat conduction [33]. This hypothesis allows easily estimating the cooling temperature along the fixed bed with an exponentially function as follows (Equation (A1)):

$$T(t) \ = \ T_\infty + (T_i - T_\infty) \cdot e^{\frac{-h}{\rho_s \cdot c_s \cdot D_p} \cdot t} \tag{A1}$$

where T(t) is the sphere temperature at time t (°C); T is the ambient temperature that surrounds the sphere (°C); T_i is the sphere temperature in t = 0 (°C); ρ_s is the sphere density (kg/m^3); c_s is the specific heat of the sphere (J/kg·K); h is the convective heat transfer coefficient by solid–air convection (W/m^2·K); and D_p is the sphere diameter (m).

At a constant descent rate of the spheres along the fixed bed and in a steady state, Equation (1) can be expressed as follows (Equation (A2)):

$$T(x) = A(x) + B(x) \cdot e^{C(x)}. \tag{A2}$$

Thus, for a coordinate x in the bed, T(x) is the temperature of the spheres; A(x) is the temperature of the air surrounding the spheres inside the bed; B(x) is a constant function of A(x); and C(x) has the expression the following expression (Equation (A3)):

$$C(x) = \frac{h'(x)}{\rho_s \cdot c_s \cdot D_p} \tag{A3}$$

where h′(x) is the convective heat transfer coefficient at coordinate x. Since ρ_s, c_s, and D_p are constant, it can be affirmed that C(x) only depends on h′(x). This allows finding expressions that explain the variation of h′(x) knowing the temperatures of the sphere and the surrounding air. Furthermore, h′(x) depends on both air properties–flow conditions and the geometry of the metal spheres of the bed.

Strictly, the h coefficient should be calculated under a complete experimental design in a steady state of the temperatures of the spheres and varying the following parameters: Reynolds number (Re), Prandtl number (Pr), and sphere diameter to bed diameter ratio. Nevertheless, due to the estimative purposes of this work, the calculation of h′(x) was done with a fixed bed of metal spheres similar to the one shown in Figure A1. In this system, the following parameters were measured: the temperature of the spheres (T_s) and the air temperatures at the inlet and outlet of the bed (T_{ai} and T_{ao}, respectively), as well as the mass air flow.

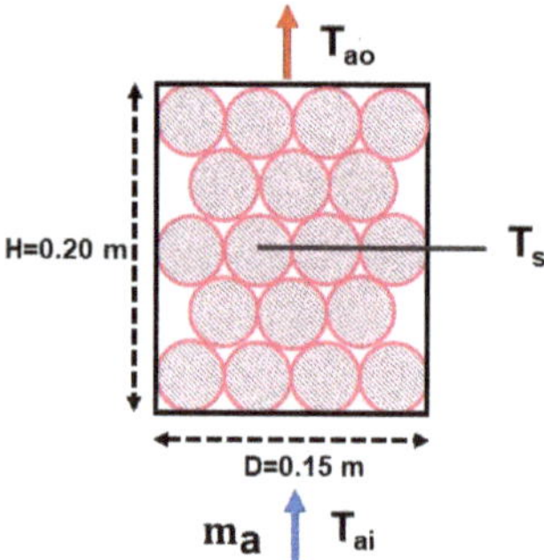

Figure A1. Scheme of the experimental system used to estimate h′(x).

With this system, the energy transfer balance for a time of experimentation t approximates the expression indicated by Equation (A4):

$$h' \cdot A \cdot \left(\overline{T}_S - \overline{T}_a\right) = m_a \cdot c_g \cdot \left(\overline{T}_{ai} - \overline{T}_{ao}\right) \tag{A4}$$

where h′ is the convective heat transfer coefficient (W/m^2·K); A is the transfer area (m^2); $\overline{T}_s$ is the average spheres temperature in the experimentation time t (°C); $\overline{T}_a$ is the average temperature between the inlet air and outlet air in the experimentation time t (°C); $\overline{T}_{ao}$ is the average outlet air temperature in the experimentation time t (°C); $\overline{T}_{ai}$ is the average inlet air temperature in the experimentation time t (°C); m_a is the inlet air mass flow (kg/s); and C_g is the air-specific heat (J/kg·K).

References

1. Royo, P.; Acevedo, L.; Ferreira, V.J.; García-Armingol, T.; López-Sabirón, A.M.; Ferreira, G. High-temperature PCM-based thermal energy storage for industrial furnaces installed in energy-intensive industries. *Energy* **2019**, *173*, 1030–1040. [CrossRef]
2. Baena-Moreno, F.M.; Rodríguez-Galán, M.; Vega, F.; Reina, T.R.; Vilches, L.F.; Navarrete, B. Regeneration of sodium hydroxide from a biogas upgrading unit through the synthesis of precipitated calcium carbonate: An experimental influence study of reaction parameters. *Processes* **2018**, *6*, 205. [CrossRef]
3. Peceño, B.; Leiva, C.; Alonso-Fariñas, B.; Gallego-Schmid, A. Is recycling always the best option? Environmental assessment of recycling of seashell as aggregates in noise barriers. *Processes* **2020**, *8*, 776. [CrossRef]

4. González-Arias, J.; Sánchez, M.E.; Martínez, E.J.; Covalski, C.; Alonso-Simón, A.; González, R.; Cara-Jiménez, J. Hydrothermal carbonization of olive tree pruning as a sustainableway for improving biomass energy potential: Effect of reaction parameters on fuel properties. *Processes* **2020**, *8*, 1201. [CrossRef]

5. Philip, P.T.; Hannaford, A.L.; Konick, E. Methods of Making Cementitious Compositions from Waste Products. U.S. Patent 4756761A, 16 June 1986.

6. Iglesias, J.; Tuya, A.; Peña, F. Procedure for Obtaining Calcium Aluminate From Waste Obtained Following Treatment of Saline Dross From the Production of Secondary Aluminium. U.S. Patent 20110293494A1, 18 July 2011.

7. Akiyama, K. Process for the Manufacture of Aluminous Cement From Aluminum Smelting Residue. U.S. Patent 4071373, 6 July 1976.

8. Leiva, C.; Arenas, C.; Alonso-Fariñas, B.; Vilches, L.F.; Peceño, B.; Rodriguez-Galán, M.; Baena, F. Characteristics of fired bricks with co-combustion fly ashes. *J. Build. Eng.* **2016**, *5*. [CrossRef]

9. Karellas, S.; Leontaritis, A.D.; Panousis, G.; Bellos, E.; Kakaras, E. Energetic and exergetic analysis of waste heat recovery systems in the cement industry. *Energy* **2013**, *58*, 147–156. [CrossRef]

10. Hara, T.; Shima, H.; Yoshida, Y.; Matsuhashi, R. Model analysis of an inter-industrial and inter-regional waste recycling system in Japan. *Energy* **2007**, *32*, 609–618. [CrossRef]

11. Rodríguez-Galán, M.; Alonso-Fariñas, B.; Baena-Moreno, F.M.; Leiva, C.; Navarrete, B.; Vilches, L.F. Synthetic slag production method based on a solid waste mix vitrification for the manufacturing of slag-cement. *Materials* **2019**, *12*, 208. [CrossRef]

12. Motz, H.; Geiseler, J. Products of steel slags an opportunity to save natural resources. *Waste Manag.* **2001**. [CrossRef]

13. Zhu, J.; Zhong, Q.; Chen, G.; Li, D. Effect of particlesize of blast furnace slag on properties of Portland cement. *Procedia Eng.* **2012**, *27*, 231–236. [CrossRef]

14. Puertas, F. Escorias de alto horno: Composición y comportamiento hidráulico. *Mater. Construcción* **1993**, *43*, 37–48. [CrossRef]

15. BSI. *Granulated Slag Ground from High Blast Furnace for Use in Concrete, Mortar and Paste*; British Standards Institution: London, UK, 2006.

16. Wang, K.S.; Lin, K.L.; Tzeng, B.Y. Latent hydraulic reactivity of blended cement incorporating slag made from municipal solid waste incinerator fly ash. *J. Air Waste Manag. Assoc.* **2003**. [CrossRef] [PubMed]

17. Reino García, H. Supercem®. Experiencia de Holcim (España) con cementos con escorias de alto horno altamente adicionados. *Patol. Cim. Estruct. Hormig.* **2013**.

18. Li, J.; Mou, Q.; Zeng, Q.; Yu, Y. Experimental study on precipitation behavior of spinels in stainless steel-making slag under heating treatment. *Processes* **2019**, *7*, 487. [CrossRef]

19. Zhang, H.; Wang, H.; Zhu, X.; Qiu, Y.J.; Li, K.; Chen, R.; Liao, Q. A review of waste heat recovery technologies towards molten slag in steel industry. *Appl. Energy* **2013**, *112*, 956–966. [CrossRef]

20. Xiong, B.; Chen, L.; Meng, F.; Sun, F. Modeling and performance analysis of a two-stage thermoelectric energy harvesting system from blast furnace slag water waste heat. *Energy* **2014**, *77*, 562–569. [CrossRef]

21. Wang, Q.; Wang, Q.; Tian, Q.; Guo, X. Simulation study and industrial application of enhanced arsenic removal by regulating the proportion of concentrates in the SKS copper smelting process. *Processes* **2020**, *8*, 385. [CrossRef]

22. Ishaq, H.; Dincer, I.; Naterer, G.F. Exergy and cost analyses of waste heat recovery from furnace cement slag for clean hydrogen production. *Energy* **2019**, *172*, 1243–1253. [CrossRef]

23. Barati, M.; Esfahani, S.; Utigard, T.A. Energy recovery from high temperature slags. *Energy* **2011**, *36*, 5440–5449. [CrossRef]

24. Company, N. Metodo Para Enfriar Material Abrasivo Aluminoso Fundido. ES Patent ES399446, 3 February 1972.

25. Hulek, A.; Ritzberguer, F. Method of Continuously Producing Vitreous Blast Furnace Slag. U.S. Patent 6250109B1, 10 July 1998.

26. Jirou, K.; Yasuto, T.; Ohkoshi, K. Apparatus for Manufacturing Vitreous Slag. U.S. Patent 4330264A, 18 May 1982.

27. Nakatani, G.; Kanai, K.; Itoh, H.; Takasaki, Y.; Ohkoshi, K.; Yanagida, Y. Apparatus for Manufacturing Rapidly Cooled Solidified Slag. U.S. Patent 4420304A, 13 December 1983.

28. Unit, A.G. Low Energy Slag and Cement Production. U.S. Patent 9233485B1, 12 January 2016.

29. Gresesqui-Lobaina, E.; Rodríguez-González, I.; Fernández-Columbié, T. Characterization of steel 70XL used in the manufacture of balls for the clinker's milling. *Min. Geol.* **2017**. Available online: https://www.redalyc.org/jatsRepo/2235/223553249009/html/index.html (accessed on 24 November 2020).
30. Shakurov, A.G.; Shkol'nik, Y.S.; Parshin, V.M.; Chertov, A.D.; Zhuravlev, V.V. Cooling and solidification of slag melt in spherical packing. *Steel Transl.* **2012**. [CrossRef]
31. Mills, K.C. The estimation of slag properties. In Proceedings of the Southern African Pyrometallurgy International Conference, Johannesburg, South Africa, 6–9 March 2011.
32. Castillo Neyra, P. *Manual Práctico de Combustión y Clinkerización.* 1990. Available online: https://dl-manual.com/download/manual-practico-de-combustion-y-clinkerizacion-6vj3xerk20oe?hash=168fd0ddbedc13b8ad71e8ec972e7a70 (accessed on 30 November 2020).
33. Incropera, F.P.; deWitt, D.P. *Fundamentos de Transferencia de Calor*; Pearson Prentice Hall: Upper Saddle River, NJ, USA, 1999; ISBN 970-17-0170-4.

Publisher's Note: MDPI stays neutral with regard to jurisdictional claims in published maps and institutional affiliations.

 *processes*

Article

Equilibrium, Kinetic and Thermodynamic Studies for Sorption of Phosphate from Aqueous Solutions Using ZnO Nanoparticles

Tra Huong Do [1,*], Van Tu Nguyen [2], Quoc Dung Nguyen [1], Manh Nhuong Chu [1], Thi Cam Quyen Ngo [3,4] and Lam Van Tan [3,4,*]

[1]   Thai Nguyen University of Education, Thai Nguyen University, 20 Luong Ngoc Quyen, Thai Nguyen City 250000, Vietnam; dungnq@tnue.edu.vn (Q.D.N.); chumanhnhuong@dhsptn.edu.vn (M.N.C.)

[2]   Institute of Chemistry and Materials, Institute of Military Science and Technology, 17 Hoang Sam, Cau Giay, Hanoi 100000, Vietnam; nguyenvantu882008@yahoo.com

[3]   NTT Hi-Tech Institute, Nguyen Tat Thanh University, Ho Chi Minh City 700000, Vietnam; ntcquyen@ntt.edu.vn

[4]   Center of Excellence for Green Energy and Environmental Nanomaterials, Nguyen Tat Thanh University, Ho Chi Minh City 700000, Vietnam

*   Correspondence: huongdt.chem@tnue.edu.vn (T.H.D.); lvtan@ntt.edu.vn (L.V.T.)

Received: 28 September 2020; Accepted: 27 October 2020; Published: 2 November 2020

Abstract: In this study, ZnO nanoparticles were fabricated by using the hydrothermal method for adsorption of phosphate from wastewater. The obtained ZnO nanorods were characterized by powder X-ray diffraction spectroscopy (XRD), scanning electron microscopy (SEM), specific surface area (BET) and energy dispersive X-ray spectroscopy (EDS). The ZnO materials were applied for adsorption of phosphate from water using batch experiments. The effects of pH (4–10), adsorption time (30–240 min), the amount of adsorbent (0.1–0.7 g/L) and initial concentration of phosphate (147.637–466.209 mg/L) on the adsorption efficiency were investigated. The optimum condition was found at pH = 5 and at an adsorption time of 150 min. The adsorption was fitted well with the Langmuir isotherm and the maximum adsorption capacity was calculated to be 769.23 mg/g. These results show that ZnO nanomaterial would highly promising for adsorbing phosphate from water. The adsorption of phosphate on ZnO nanomaterials follows the isothermal adsorption model of Langmuir, Tempkin and Freundlich with single-layer adsorption. There is weak interaction between the adsorbent and the adsorbate. Phosphate adsorption of the ZnO nanomaterials follows Lagergren's apparent second-order kinetic model and was spontaneous and exothermic.

Keywords: nanoparticles; ZnO; equilibrium; kinetic; thermodynamic; phosphate; aqueous solution

---

1. Introduction

The wide direct bandgap and large exciton binding energy of zinc oxide (ZnO) have conferred the material unique electrical and photovoltaic properties and enabled its applications in the fields of fluorescence [1], photocatalytic catalysis [2], electric fire, gas sensors [3,4], electrochemical sensors [5] and solar cells [6]. ZnO can be synthesized by using various methods, including sputtering [7], sol-gel [8], co-precipitation [9], gel burning [10] and hydrothermal [11]. Depending on the synthesis routine, the morphologies of ZnO could be varied, ranging from nanorods [2], nanospheres [4], nanopores [1], nanoparticles [9] and many kinds of structures of nanorods, such as nanobelts, nanocombs and nanoforests [5]. In most synthesis routines, ZnO plays the role as an *n*-type semiconductor with electrical amphoteric properties [12]. In addition, selection of an appropriate synthesis method is also important in ensuring operational ease and in avoiding the technical difficulties of the fabricating process. At the nano scale, ZnO finds a wide range of applications in environmental remediation, bactericidal agents and in electronic devices, of which the use as adsorbent materials for removal of cation and anion pollutants features prominently due to its largely improved surface area compared to its bulk counterparts [13].

Treatment of phosphate represents an important aspect of wastewater treatment. Excessive phosphate causes overgrowth of photosynthetic cyanobacteria and algae and in turn accelerated eutrophication in water ecosystems [14]. Phosphorus could enter water bodies through effluents from households, industrial or agricultural activities, and exists in the form of $H_2PO_4^-$, HPO_4^{-2} and PO_4^{-3} organic phosphate and polyphosphate [14–17]. In water, hydrolysis could occur, transforming polyphosphates and organic phosphate into other phosphate forms [18]. Therefore, industrial and agricultural discharge management, especially in terms of treatment of ammonium and phosphate, is pivotal in ensuring environmental stability and preventing water-derived health risks [19,20]. Because treatment of environmental pollution requires a large amount of adsorbents, scalability of a treatment method largely depends on its ability to easily fabricate bulk materials with a high adsorption efficiency. The hydrothermal method has been shown to be an appropriate method to synthesize ZnO due to a number of advantages, including inexpensive chemicals and ease of mass production. However, the maximum adsorption efficiency of fabricated ZnO requires optimization of all factors of adsorption, such as pH, time, the volume of the adsorbent, the initial concentration as well as careful investigation of the adsorption isothermal model, adsorption kinetic and adsorption thermal dynamic, which are relatively scarce in the literature [14–17].

In this report, we present the results of studies on isothermal adsorption, kinetics, thermodynamics of phosphate adsorption (PO_4^{3-}) removal in aqueous environments by using ZnO nanomaterials fabricated by the hydrothermal method.

2. Materials and Methods

2.1. Preparation of Materials

Zinc oxide nanomaterials were prepared by the hydrothermal method from a mixture of 25 mL $Zn(NO_3)_2$ 0.1 M + NaOH 0.1 M + 20 mL C_2H_5OH (ratio C_2H_5OH:H2O = 1:1), pH = 11. Briefly, 0.6525 g of Zn $(NO_3)_2$.4H$_2$O was weighed accurately and diluted to 25 mL of Zn $(NO_3)_2$ 0.1 M to produce Solution A. An amount of 0.1 g of NaOH was dissolved with water in a 25 mL volumetric flask, forming a NaOH solution with a 0.1 M concentration (Solution B). Solution A was added to Solution B. Afterwards, 20 mL of C_2H_5OH (ratio C_2H_5OH: H_2O = 1: 1) was added to the reaction mixture above. The pH of the solution was maintained at 11. The reaction mixture was stirred on the stirrer for 20 min, and then the whole reaction mixture was heated in an autoclave at 180 °C for 24 h. The samples were taken for centrifugation, followed by washing to a pH = 7 to obtain a white, viscous gel. After obtaining the ZnO nanoparticles, the samples were heated in an air atmosphere at 350 °C for 10 h to remove the organic impurities.

2.2. Characterization

The crystalline structure and morphology of the sample were examined by using a powder X-ray diffraction spectrometer (XRD, X'Pert Pro Panalytical, Almelo, The Netherlands) equipped with Cu-K$_\alpha$ radiation (λ = 1.5418 Å) and an emission scanning electron microscope (SEM, S4800, JEOL, Tokyo, Japan), respectively. To determine the specific surface area, the Brunauer, Emmett and Teller (BET) method was used, using Tri Start 3000 equipment, Micromeritics. The distribution size of the sample was determined by the laser scattering method on HORIBA Laser Scattering Particle Size Distribution. The concentration of PO_4^{3-} before and after adsorption were determined by the UV-Vis method (Hitachi UH5300) at the University of Medicine and Pharmacy, Thai Nguyen University.

2.3. Assessment of the Prepared ZnO Nanopowder for Phosphorus Decontamination

A series of batch experiments was carried out to assess the adsorptivity of the as-synthesized ZnO nanoparticles against phosphorus. The stock solution containing potassium phosphate at a concentration of 1000 mg PO_4 per liter was first prepared to simulate wastewater. The synthesized ZnO nanoparticle was introduced into 25 mL of the diluted synthetic waste solution at a predetermined ratio and phosphorus concentration, and then mixed for a certain time period. The reaction took place in 50 mL polypropylene plastic vials under end-over-end shaking at 40 rpm. Samples were taken after specific intervals from the tube for the measurement of phosphorus concentrations using the ascorbic acid-molybdate blue method [21]. The method is based on the principle that ascorbic acid could reduce phosphomolybdic acid, which is formed by the orthophosphate and molybdate interaction, to form a complex with a blue color that is measurable by using a spectrophotometer at 885 nm. Adsorbed phosphorus would be quantified based on the initial (C_0) and final measured concentration (C). The contact time, initial phosphorus concentration, material dosage and solution pH were varied to find out its effect on the adsorption capacity of the prepared nano-ZnO.

2.4. Adsorption Capacity of the PO_4^{3-}

The adsorption experiments were investigated under a varying pH, time, weight of ZnO and initial (PO_4^{3-}) concentration. The fixed conditions included the room temperature and a shaking rate of 200 revolutions per minute (rpm).

The effect of the initial pH was measured by adding 0.025 g of ZnO to 50.0 mL of the PO_4^{3-} concentration of 150 mg/L, with the different pH values ranging from 1 to 10, followed by shaking for 150 min. The pH solution was adjusted with either by NaOH or HNO_3 solutions (0.1 M). The effect of the adsorption time was examined by adding 0.025 g ZnO to 50.0 mL PO_4^{3-} solution with an initial concentration of 150 mg/L, for a period that ranged from 30 to 240 min, at pH = 5. The effect of the ZnO mass was achieved by gradually increasing the mass of ZnO from 0.005 to 0.035 g into a 50.0 mL PO_4^{3-} solution with the initial concentration of 162.1978 mg/L, in 120 min, at pH = 5. The effect of the initial concentration of PO_4^{3-} was examined by adding 0.025 g ZnO to 50.0 mL PO_4^{3-} solution with a concentration varying from 150 to 450 mg/L, in 150 min, at pH = 5. After the above processes, the samples were centrifuged at a rate of 4000 rpm for 10 min. The concentration of PO_4^{3-} before and after adsorption were determined by the UV-Vis method (Hitachi UH5300) at the University of Medicine and Pharmacy, Thai Nguyen University.

The adsorption capacity and adsorption efficiency were calculated according to the formula

$$q = \frac{(C_0 - C_{cb})V}{M} \tag{1}$$

$$H\% = \frac{(C_0 - C_{cb})}{C_0} \times 100\% \tag{2}$$

where V is the volume of the solution (L), M is the mass of the adsorbent (g), C_0 is the initial solution concentration (mg/L), Ccb is the solution concentration when the adsorption is at equilibrium (mg/L), q is the adsorption capacity at the time of equilibrium (mg/g), and H is the adsorption efficiency (%).

2.5. Equilibrium and Kinetic Modeling of the Phosphorus Decontamination Process onto ZnO

Equilibrium and kinetic modeling are essential in shedding light onto adsorption capacity. A number of different models were included: the pseudo first-order equation and second-order equation, as well as the Langmuir, Freundlich, Temkin, and Dubinin–Radushkevich equilibrium isotherm adsorption models.

3. Results and Discussion

3.1. Factors Affecting the Morphology and Surface Structure of ZnO Nanoparticles

3.1.1. SEM Image Analysis

The SEM images of the as-synthesized ZnO are shown in Figure 1. Visually, the ZnO nanoparticle structures are interwoven with some nanorods. This may be due to the developing reaction of the ZnO nanoparticles to the nanorod structures. Figure 1 shows that the ZnO materials are of high porosity and had particle sizes widely distributed due to the integration of the ZnO nanoparticles and ZnO nanorods. This distribution can be confirmed by the laser diffraction method.

Figure 1. SEM image of the ZnO nanomaterials.

3.1.2. Determination of Surface Area by the BET Method

The ZnO samples were measured by N_2 isotherm adsorption at $-196\ ^\circ C$. The adsorption isotherm and N_2 adsorption are shown in the Figure 2. The BET graph of the ZnO samples, obtained from the software, is also shown in Figure 3. The surface areas of the ZnO samples are 17.05 m^2/g.

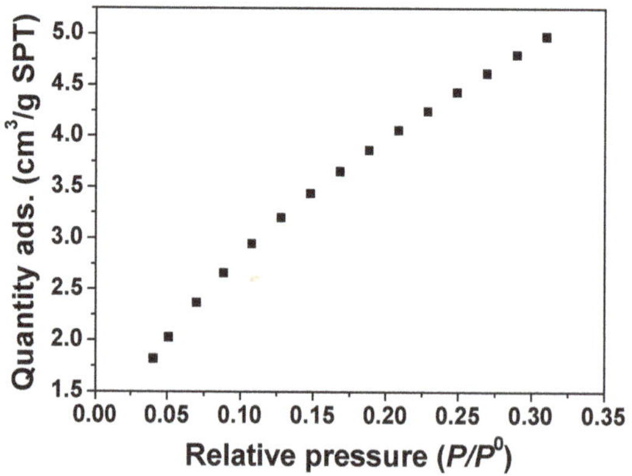

Figure 2. Adsorption isotherm–N_2 adsorption of nano ZnO, in the as-synthesis condition at a temperature of 180 $^\circ C$.

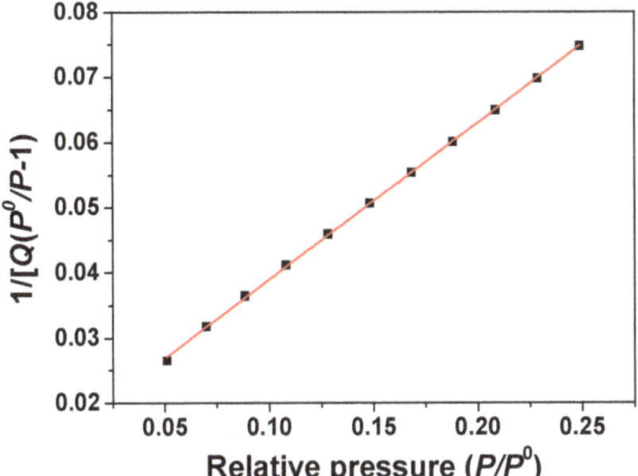

Figure 3. BET graph of the as-synthesized ZnO nanomaterials obtained at a synthesis temperature of 180 °C.

3.2. Structure and Morphology

3.2.1. XRD Analysis

The XRD patterns of the nano ZnO is shown in Figure 4, showing characteristic peaks at (100); (002); (101); (102); (110); (103); (112) and (201), which correspond to ZnO diffractions of its wurtzite structure.

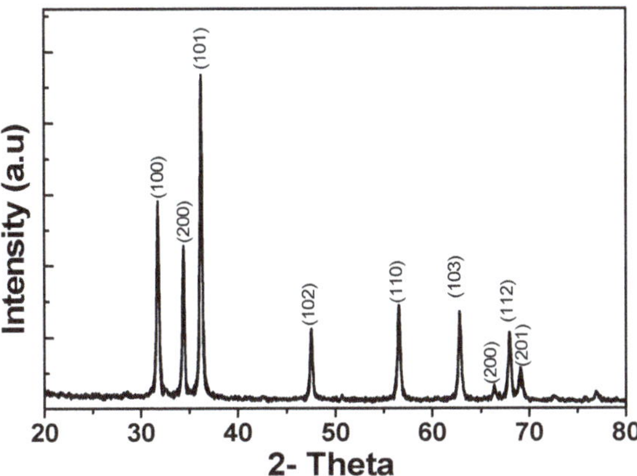

Figure 4. Results of the XRD patterns of ZnO.

3.2.2. EDS Analysis

The chemical composition of the prepared ZnO samples was analyzed by the EDS method and the results are shown in Figure 5 and Table 1. From the EDS results, it is indicated that the surface composition of the ZnO is very pure, constituting about 99.99% of the Zn and O, and that only a small quantity of carbon impurities is present in the composition, possibly due to the incomplete absorption of CO_2 from the air or incomplete decomposition of the organic matter.

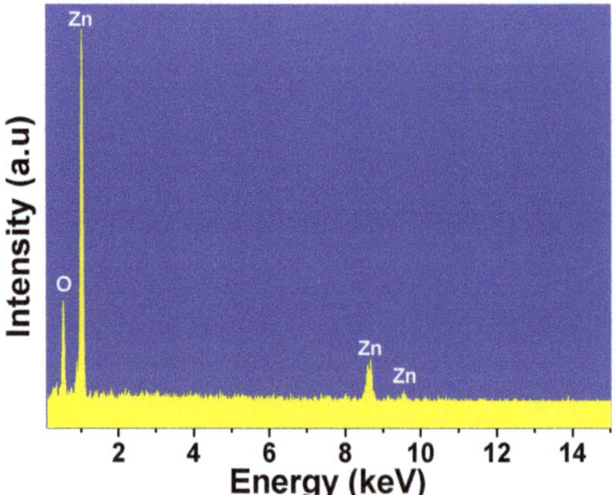

Figure 5. EDS pattern of the ZnO nanorods.

Table 1. Elemental analysis of the synthesized ZnO nanorods.

Element	Theory	Result
Zn	80.34	80.32
O	19.66	19.67
Total	100.00	99.99

3.2.3. Particle-Size Distribution by Laser Diffraction

The particle-size distribution in Figure 6, obtained by a laser diffraction analyzer, indicated that the majority of the ZnO particles, obtained at a synthesis temperature of 180 °C, were sized around 0.814 µm. The range in which the ZnO particle size was measured was also low, suggesting that particles synthesized under this temperature condition was relatively stable in size. The first distribution peak is related to the ZnO nanoparticles. The second distribution peak is related to the nanorod.

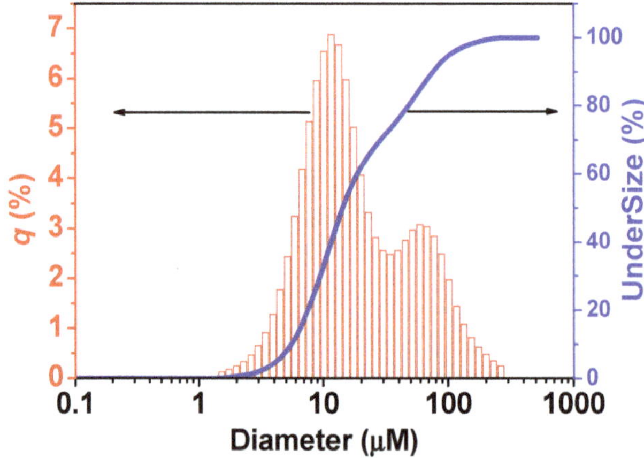

Figure 6. The synthesized ZnO nanoparticle-size distribution, as per the laser diffraction analyzer.

3.3. Determination of the Isoelectric Point of ZnO

To determine the isoelectric point of ZnO, ten 0.1 M NaCl solutions were prepared with an initial pH (pH_i) ranging from 1 to 10. In each flask, 0.025 g of prepared ZnO material was added, followed by

50 mL of the NaCl solution prepared above. The mixture was allowed to stand for 48 h before being filtered and re-measured for pH. The results from Figure 7 suggests that the isoelectric point of ZnO is approximately 7.1. As a result, when the pH < pH_{pzc}, the surface of the ZnO material is positively charged; otherwise, the surface of the ZnO material is negatively charged.

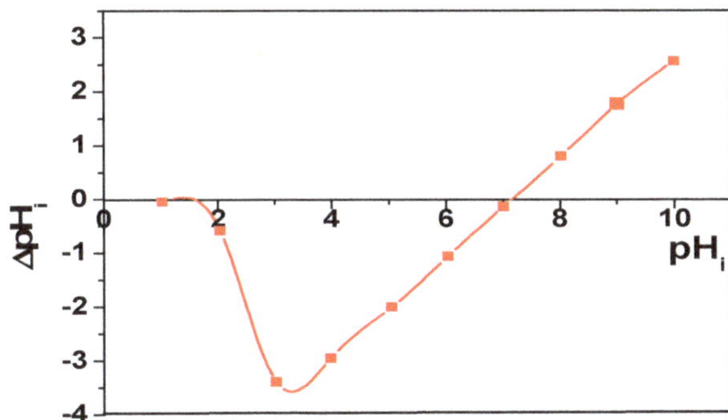

Figure 7. The graph for determination of the isoelectric point of ZnO.

3.4. Adsorption of Phosphate by ZnO Nanoparticles

3.4.1. Effect of pH

Controlling the pH of the media is critical for the efficiency of the adsorption process as it plays a key role in mediating the adsorbate behaviors and adsorption capacity of the adsorbent. As illustrated in Figure 8, the adsorption efficiency was improved gradually and then eventually peaked when increasing the pH from 1 to 5, with the maximum phosphorus adsorption efficiency reaching approximately 90%. Afterwards, the efficiency dropped rapidly and at a pH of 10; the efficiency was reduced to 44.1%. Therefore, the pH value at which the phosphorus removal was maximal was recorded within the range of 5–6. According to the thermodynamic calculations, three forms of phosphate, including $H_2PO_4^-$, HPO_4^{-2} and PO_4^{3-}, with a pK of 2.15, 7.20 and 12.33, respectively, existed in the aqueous solution and the ratio of the three forms is largely determined by pH of the solution. On the other hand, the point of zero charge (pH_{pzc}), which could influence the surface adsorption capacity and active centers, is lower than the pH when the pH was in the range of 2 to 12. As a result, the electrostatic repulsion and the competition between the OH^- ions in occupying the active sites are also the driving force for the adsorption against phosphate. Conversely, in the case of pH < pH_{pzc}, a positive charge of the sorbent surface might induce enhanced adsorption of the anions [19]. The reduced phosphorus adsorption at very high pH values could be explained by the competition of excessive OH^- and phosphate anions, such as $H_2PO_4^-$, HPO_4^{-2} and PO_4^{3-}, onto the ZnO adsorption sites [22]. Another explanation for the pH-dependent adsorption activity of ZnO is that protonation/deprotonation of the nano-ZnO surface is highly sensitive to the pH of the solution due to the amphoteric properties of the metal oxide surfaces. As a result, in the media that have a pH lower than the isoelectrical point of the ZnO, the adsorption of the H^+ ions is promoted [23], thus leading to a positively charged ZnO surface and in turn facilitating the uptake of phosphate anions [24]. On the contrary, increasing the solution's pH further causes accumulation of the negative charges on both the nano-ZnO and the adsorbate, enhancing the electrical repulsive force between the two phases. Based on those observations, we proposed the adsorption mechanism of the phosphorus onto the ZnO particles as follows [25]:

$$= \text{Zn-OH} + HPO_4^{-2} \longleftrightarrow = \text{Zn-} PO_4^{-2} + H_2O \tag{3}$$

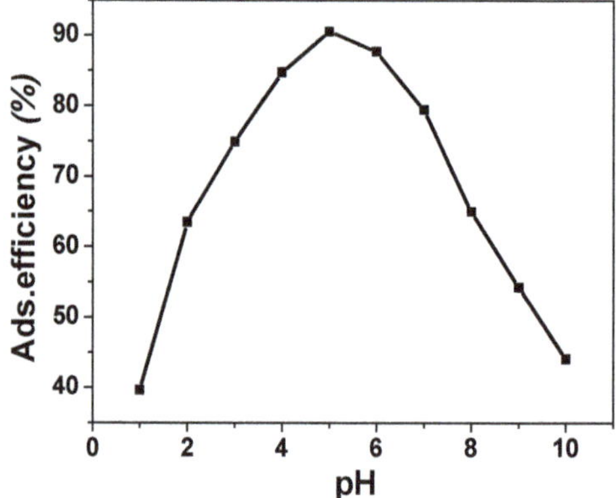

Figure 8. Effect of pH on the PO_4^{3-} adsorption efficiency.

It was shown that a higher solution pH is associated with degraded P adsorption. This suggests that the adsorption of phosphate anions could be hindered by a repulsive force caused by strong hydroxyl ions in the solution. Therefore, a pH = 5 was chosen for the next experiment.

3.4.2. Effect of Contact Time

Adsorption efficiencies with respect to contact duration are shown as in Figure 9. Generally, prolonging the stirring was positively correlated with the adsorption efficiency. In particular, a rapid adsorption improvement of PO_4^{3-} was observed in the first 150 min. Afterwards, the adsorption efficiency increased marginally, insignificantly during the later period.

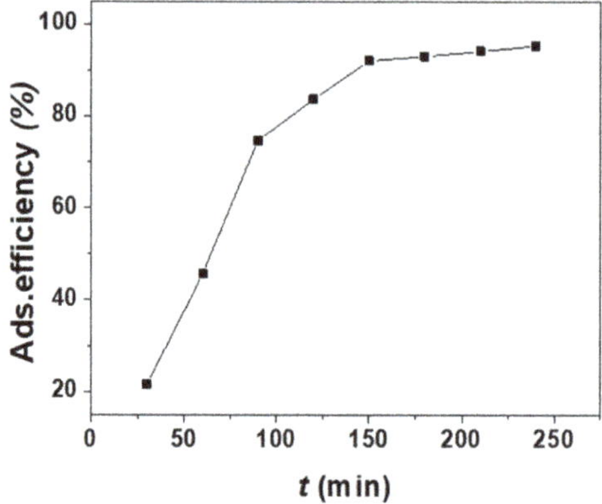

Figure 9. Effect of time on the PO_4^{3-} adsorption efficiency.

The accelerated PO_4^{3-} uptake in the beginning period is possibly due to the abundance of adsorption holes situated on the ZnO material surface. As those holes were occupied during the process, stagnant efficiency was observed in the later stage and then the adsorption reached the equilibrium state where no adsorption improvement took place with prolonged time. Therefore, the time needed for adsorption equilibrium was selected as 150 min for subsequent experiments.

3.4.3. Effect of ZnO Dosage

The dosage of the adsorbent is responsible for both adsorption efficiency and capacity. Figure 10 presents the phosphorus removal with respect to varying ZnO dosages. Other parameters included a contact time of 150 min, initial solute concentration of 50 ppm and solution pH of 5. Increasing the dosage from 0.1 to 0.7 g/L caused significant improvement in phosphorus adsorption, from 24.7% to 96.3%, possibly due to enhanced availability of adsorption active sites that facilitates P ion binding [26]. On the other hand, further increasing the ZnO dose higher than 0.5 g/L induced no clear P removal enhancement. This is possibly explained by saturated adsorption sites on the material surface caused by abundant active sites and the accumulation of material particles as effective surface area is lowered. Therefore, the optimal absorbent was selected as 0.5 g/L.

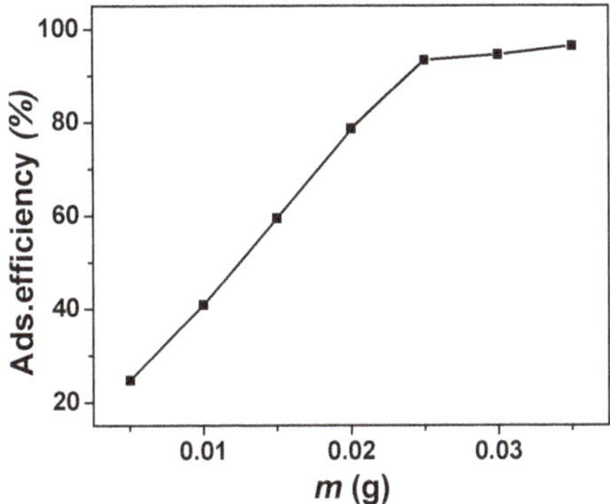

Figure 10. Effect of the adsorbent dose on the PO_4^{3-} adsorption efficiency.

3.4.4. Effect of Temperature

Figure 11 shows the effect of temperature on the adsorption efficiency of PO_4^{3-}. It can be observed that the removal of PO_4^{3-} gradually increased when elevating the temperature from 298 K to 323 K. This can be explained by the reduction in adsorption sites at higher temperatures. With the temperature increase from 298 to 323 K, the adsorption efficiency declined slowly from 93 to 89%. However, the amount of PO_4^{3-} adsorbed per unit volume of adsorbent will decrease as the temperature increases, suggesting that the adsorption of PO_4^{3-} by the ZnO nanomaterials is an exothermic process.

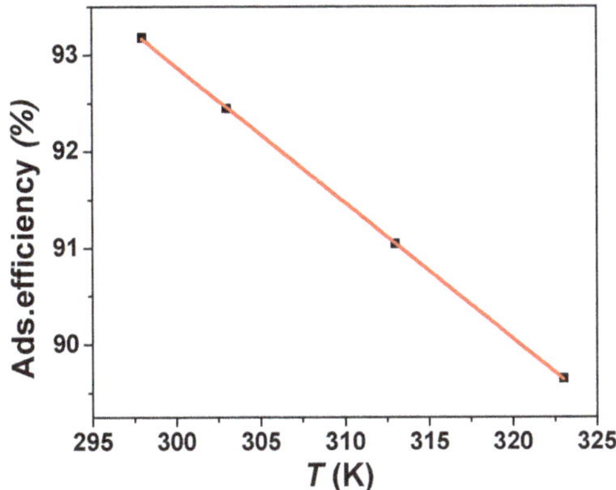

Figure 11. The effect of temperature on the efficiency of the PO_4^{3-} adsorption.

3.5. Adsorption Isotherm

3.5.1. Langmuir Isothermal Adsorption Model

Adsorption isothermal analysis plays a very important role for the purpose of designing experiments and manufacturing adsorbents. Experimental data are analyzed with Langmuir isothermal models because they are classical and simple for describing the equilibrium between the adsorbed ions on adsorbents and ions in solution at a constant temperature.

The Langmuir equation can be written as follows:

$$\frac{C_e}{q} = \frac{1}{q_{max}K} + \frac{C_e}{q_{max}} \tag{4}$$

where q is specific adsorbance, which is the number of mg of adsorbent per 1 g of adsorbent at equilibrium (mg/g); q_{max}: the maximum adsorption capacity (mg/g); C_e: the concentration of the adsorbent in solution at equilibrium (mg/L); and K: Langmuir's constant.

When substituting a and b with

$$a = \frac{1}{q_{max}} \tag{5}$$

$$b = \frac{1}{K.q_{max}} \tag{6}$$

the above equation becomes y = ax + b. From empirical calculations, it is possible to calculate the K constant and the maximum adsorption capacity (q_{max}). The graph of the Langmuir model is presented as in Figure 12. Inferring from the results, the maximum adsorption capacity q_{max} and the constant K was 769.23 mg/g and 0.12 L/mg, respectively. Although the q_{max} values might not represent an indication of adsorption capacity in the long term, it could serve as a measure for a comparative study of adsorbents. As indicated in Table 2, the prepared ZnO material showed an adsorption capacity of 769.23 mg P/g, which far exceeded that of previously reported adsorbents, implying that ZnO may be a promising candidate adsorbent for treatment of phosphate from aqueous environments.

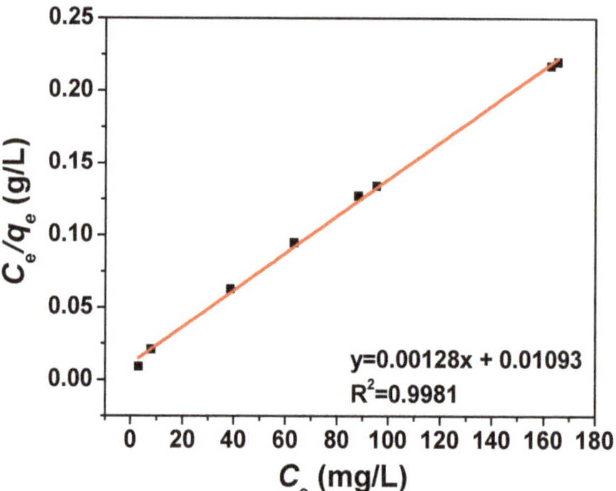

Figure 12. C_e/q and C_e correlations.

Table 2. Maximum adsorption capacities of phosphorus onto various adsorbents.

No.	Adsorbent	q_{max} (mg/g)	Ref.
1	ZnO	168.4	[27]
2	Zn-Al LDO (573 K, Curea $\frac{1}{4}$ 0.4 M)	232.0	[28]
3	Zn-Al LDH (Curea $\frac{1}{4}$ 0.4 M)	76.1	[28]
4	ZnO Nanorods	89.0	[25]
5	SnO_2	21.5	[29]
6	WO_3	19.0	[29]
7	Fe(III)–Cu(II) binary oxides	35.2	[30]
8	Silver nanoparticle-loaded activated carbon	4.5	[31]
9	Magnetite-enriched particles (MEP)	6.4	[32]
10	ZnO	769.23	Our work

The estimated results indicated that the Langmuir isotherm adsorption model described the adsorption process of PO_4^{3-} onto ZnO material, demonstrated by the high coefficient of determination (0.9983). Accordingly, the adsorption process is suggested to adhere to monolayer and chemical adsorption. Notably, those findings are larger than the previously published results [19–29].

In order to further justify the single layer adsorption of the PO_4^{3-} of ZnO, we evaluated the suitability through the R_L equilibrium parameter, which is calculated as

$$R_L = \frac{1}{1 + K_L C_O} \tag{7}$$

where C_o: initial concentration of the substance (mg/g); and K_L: the Langmuir constant (L/mg).

The R_L parameter calculated from Figure 13 ranged from 0.123 to 0.256, which is lower than 1. Thus, it can be confirmed that the Langmuir model is suitable for the absorption of PO_4^{3-} by ZnO nanomaterials.

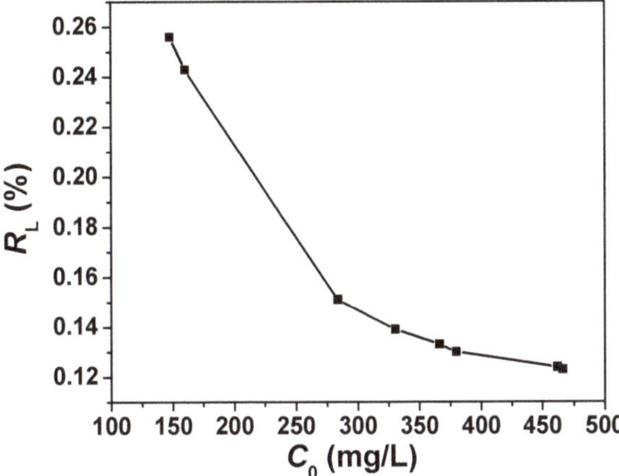

Figure 13. The dependence of R_L on C_o in the Langmuir model.

3.5.2. Freundlich Isothermal Adsorption Model

The Freundlich isothermal adsorption model is an empirical equation based on adsorption on heterogeneous surfaces of materials. The linear equation is usually expressed as

$$\log q_e = \log k + \frac{1}{n} \log C_e \tag{8}$$

where C_{cb} is the concentration at the time of equilibrium (mg/L); and q_e: the adsorption volume (mg/g). The constant n is the exponent in the Freundlich equation, which characterizes the energy heterogeneity of the adsorbed surface; and k: the Freundlich constant, to show the relative adsorption capacity of the adsorbent materials.

The calculation results of the reactive red adsorption process of PO_4^{3-} on ZnO by the Freundlich adsorption isothermal model are displayed in Figure 14.

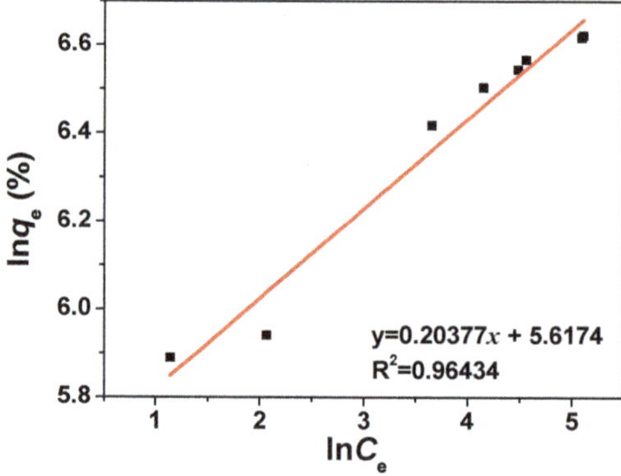

Figure 14. Dependence of logq on logC$_e$ on PO_4^{3-}.

Figure 14 expresses the absorption of PO_4^{3-} on ZnO materials according to the Freundlich isothermal adsorption model. The calculated K_F was 275.145 (mg/g)(L/mg)$^{1/n}$ and the constant of n was 4.907. The correlation coefficient, R^2, was 0.9697. The R^2 value of the Freundlich isotherm is lower than that of the Langmuir model. The value n gives an indication of the favorability of the adsorption process. When the value of n is less than 1, the adsorption is said to be unfavorable, but if it is greater than 1 it is favorable [33]. The value of n obtained in this study is greater than 1, which indicates that the adsorption process is favorable.

3.5.3. Dubinin–Radushkevich Isothermal Adsorption Model

The Dubinin–Radushkevich (D–R) isotherm is more inclusive than the Langmuir isotherm for it does not assume a homogenous surface or constant sorption potential [34]. It is used for the estimation of the characteristic porosity of the biomass as well as the mean free energy of the adsorption. The linear form of the D–R model is presented as follows:

$$\ln q = \ln q_{max} - \beta.\varepsilon^2 \tag{9}$$

where:

q: adsorption volume (mg/g);
q_{max}: maximum adsorption volume (mg/g);
β: constant of the adsorption energy (mol^2/J^2);

ε: Polanyi, described as follows:

$$\varepsilon = RT\ln\left(1 + \frac{1}{Ccb}\right) \tag{10}$$

where:

T: solution temperature (K);
R: gas constant (8.314×10^{-3} kJ/mol.K).

The value of the average adsorption energy, E (kJ/mol), can be calculated from D–R according to parameter β as follows:

$$E = \frac{1}{\sqrt{-2\beta}} \tag{11}$$

The value of the average adsorption energy indicates the nature of the adsorption process. When the E value is less than 8 kJ/mol, the adsorption process is physical adsorption and when the E value ranges from 8 to 16 kJ/mol, the adsorption is chemical.

The Dubinin–Radushkevich isothermal graph shown in Figure 15 reveals that phosphate adsorption does not follow the Dubinin–Radushkevich model, evidenced by the low R^2.

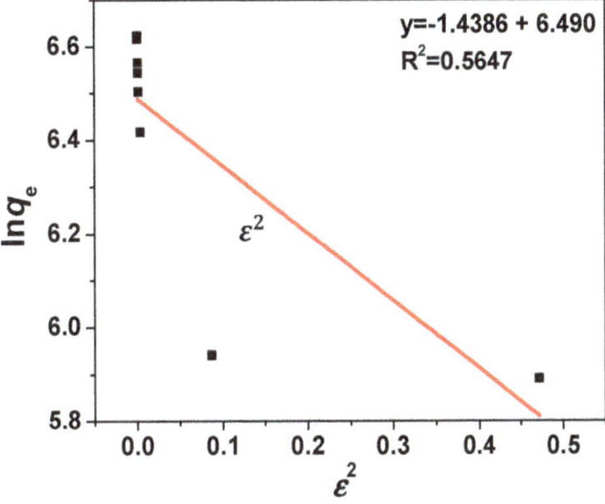

Figure 15. Dependence of lnq on ε^2 on PO_4^{3-} adsorption.

3.5.4. Tempkin Isothermal Adsorption Model

The isothermal model of Tempkin assumes that the adsorption heat of all the molecules on the surface of the material decreases linearly with the density of the coverage due to the interaction between the adsorbent and the substrate, and the adsorption is characterized by the uniform distribution of linked energy sources, up to a maximum number of associated energies. Temkin isotherms are represented by the following equation:

$$q = \frac{RT}{b_T}\ln(K_T.C_{cb}) \tag{12}$$

The equation can be expressed in linear form as

$$q = B\ln K_T + B\ln C_{cb} \tag{13}$$

where

$$B = \frac{RT}{b_T} \tag{14}$$

T: is the absolute temperature (K);

R: gas constant (valued by 8.314×10^{-3} (kJ/mol.K);

b_T: Tempkin constant (kJ/mol).

Determination of Temkin model parameters through plotting was described in Figure 16 and parameters of the models were summarized in Table 3.

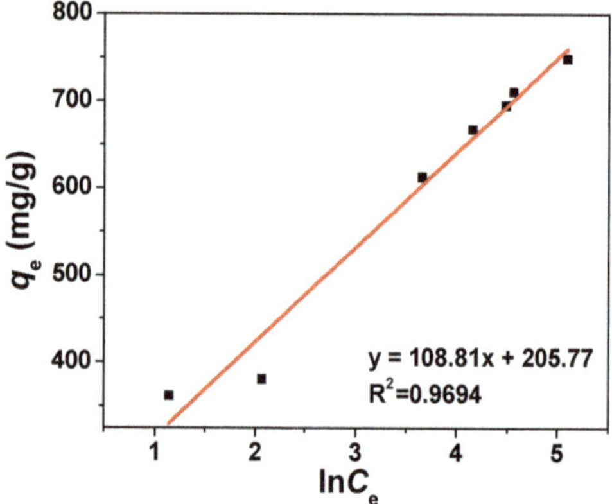

Figure 16. Dependence of q on ln C_{cb} on the adsorption of PO_4^{3-}.

Table 3. Isothermal parameters: Langmuir, Freundlich, Tempkin and Dubinin–Radushkevich.

Isothermal Line Model	Parameters	
Langmuir	K_L (L/mg)	0.12
	q_{max} (mg/g)	769.23
	R^2	0.99
Freundlich	K_F (mg/g)(mg/L)$^{1/n}$	275.145
	N	4.907
	R^2	0.97
Tempkin	K_T	6.887
	b_T (kJ/mol)	0.023
	R^2	0.9694
Dubinin–Radushkevich	q_{max} (mg/g)	658.392
	β (mol^2/J^2)	−1.439
	R^2	0.6276
	E (kJ/mol)	0.5894

From Table 3, the value of the Tempkin constant b_T of 0.023 indicates the presence of a weak interaction between the adsorbents and supports the physical adsorption of phosphate by ZnO nanomaterials. The determination coefficient R^2 of the Langmuir, Freundlich and Tempkin models are nearly equivalent (0.99, 0.97 and 0.97) whereas the Dubinin–Radushkevich model is 0.628. Thus, the phosphate adsorption of the ZnO nanomaterials do not seem to follow the Dubinin–Radushkevich model, but the isothermal adsorption models of Langmuir, Freundlich and Tempkin. The adsorption of phosphate onto the ZnO nanomaterials is a single layer under the condition of a heterogeneous surface and there is a weak interaction between the adsorbent and adsorbate.

3.6. Study on the Adsorption Kinetics of PO_4^{3-} on ZnO Nanomaterials

The adsorption process of a solute could be conventionally described by the Lagergren rate equation, which could be presented in pseudo-first-order form as follows:

$$\frac{dq_t}{dt} = k_1 \left(q_e - q_t\right) \tag{15}$$

By taking the integral on both sides of the equation, with the boundary t from 0 to t and q from 0 to q_t, the following equation is obtained:

$$g(q_e - q_t) = \lg q_e - \frac{k_1}{2.303}t \tag{16}$$

where q is the amount of adsorbed PO_4^{3-} (mg/g). The subscript t and q denote the state at equilibrium and time t (min), respectively. k_1 is the rate constant. Plotting $\ln(q_e - q_t)$ against t gives the graph whose slope could be interpreted as the rate constant of the pseudo-first-order model.

The pseudo-second-order equation based on the adsorption equilibrium capacity can be expressed as

$$\frac{dq_t}{dt} = k_2(q_e - q_t)^2 \tag{17}$$

By taking a similar approach, one could derive the following integrated linear form of the pseudo-second-order as follows:

$$\frac{t}{q_t} = \frac{1}{k_2 \cdot q_e^2} + \frac{1}{q_e}t \tag{18}$$

By linearly plotting time against $\log(q_e - q_t)$, the pseudo-first-order rate constant and sorption capacity could be calculated (Figure 17). On the other hand, the pseudo-second-order model could be plotted by considering t/q_t as a function of t (Figure 18). The model parameters are summarized as in Table 4. It was showed that the coefficient of determination of the pseudo-second-order is higher than that of the first-order model. In addition, the second-order model showed high agreement between the experimental ($q_{e(exp)}$) and calculated ($q_{e(cal)}$) capacities, suggesting that the second-order model would be more appropriate to describe the sorption process.

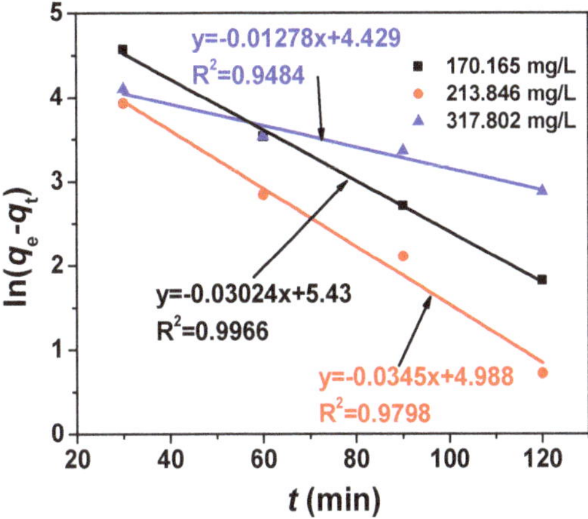

Figure 17. Pseudo-first-order non-linear method for adsorption of PO_4^{3-} onto ZnO.

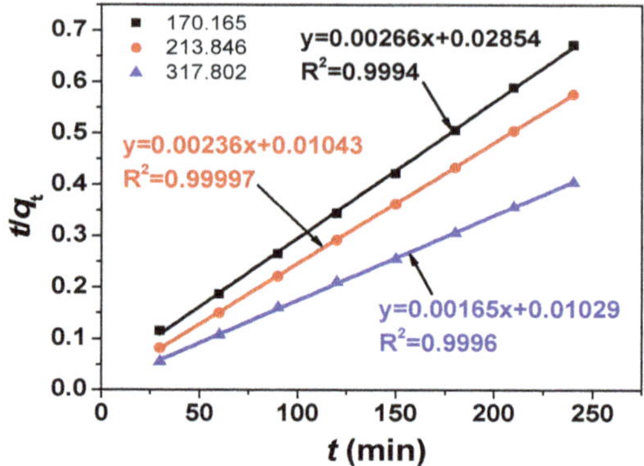

Figure 18. Pseudo-second-order non-linear method for adsorption of PO_4^{3-} onto ZnO.

Table 4. Coefficients of the pseudo-first and second-order adsorption kinetic models and intraparticle diffusion model.

Concentration (mg/L)	$q_{e.exp}$ (mg/g)	$q_{e.cal}$ (mg/g)	k_1 (min^{-1})	R^2
		Pseudo-first-order model		
51.765	29.984	20.289	0.0589	0.872
71.350	39.436	30.033	0.0564	0.927
103.546	53.768	42.897	0.0640	0.960
		Pseudo-second-order model		
Concentration (mg/L)	$q_{e.exp}$ (mg/g)	$q_{e.cal}$ (mg/g)	k_2 (g/mg.min)	R^2
51.765	30.073	30.959	0.015	0.999
71.350	39.436	40.650	0.010	0.999
103.546	53.590	54.645	0.010	0.999

Given that the adsorption process obeys the second-order apparent kinetics model of Lagergren, the activation energy of the absorption can be determined by the formula

$$k_2 = k_0 \exp(- E_a/RT) \tag{19}$$

where k_2 is adsorption rate constant (g/mg.min); k_0 is the initial rate constant; E_a is the activation energy (kJ/mol); R is the gas constant (R = 8.314 J/mol.K); and T is the absolute constant (K). In the above equation, k_2 can be replaced by $h = k_2 q_e^2$, reflecting the initial adsorption rate constant when q_t/t reaches zero, thus:

$k_2 = h.\exp(-E_a/RT)$

As a result,

$$E_a = RT (\ln h - \ln k_2). \tag{20}$$

As previously reported [35], physisorption is due to weak van der Waal forces between the adsorbent and the adsorbate. The magnitude of the energy of physisorption is in the region of less than 20 kJ/mol. The same study also further stated that a covalent bond exists between the adsorbate and the adsorbent in chemisorption in which the substrate (adsorbent) is limited to monolayer coverage. The value of E, the mean free energy of the biosorption obtained in this study, as reflected in Table 5, was higher than 20 kJ/mol, and the Langmuir isotherm (monolayer coverage) best fitted the surface coverage in this study.

Table 5. Activation energy of PO_4^{3-} adsorption on ZnO nanomaterials.

Concentration (mg/L)	h	k_2 (g/mg.min)	E_a (kJ/mol)
170.165	26.150	2.074×10^{-4}	23.394
213.846	34.519	2.198×10^{-4}	23.938
317.802	52.322	2.238×10^{-4}	24.923

3.7. Thermal Dynamic of Adbsorption of PO_4^{3-} onto ZnO Nanomaterials

The free energy (ΔG^0), enthalpy (ΔH^0) and entropy variation (ΔS^0) of the PO_4^{3-} adsorption process was calculated using the following equations:

$$K_D = \frac{q_e}{C_{cb}} \tag{21}$$

$$\Delta G^0 = -RT\ln K_D \tag{22}$$

$$\ln K_D = -\frac{\Delta G^0}{RT} = -\frac{\Delta H^0}{RT} + \frac{\Delta S^0}{R} \tag{23}$$

where:

K_D: equilibrium constant;
q_e: adsorption volume at equilibrium (mg/g);
C_{cb}: concentration of the substrate at equilibrium (mg/L);
R: gas constant;
T: temperature (K).

According to the calculated parameters in Table 6, the nature of the PO_4^{3-} adsorption process onto the ZnO material was spontaneous, evidenced by negativity of the change of the free energy (ΔG^0). The increase in ΔG^0 could be attributed to the elevating temperature under which the adsorption of phosphorus took place. On the other hand, the negative value of ΔH^0 suggests the reaction was endothermic. The negative value of ΔS^0 also indicated the stability of the solid–solution interface during the adsorption process [36].

Table 6. Values of the thermodynamic parameters (kJ/mol) for the adsorption of PO_4^{3-} onto ZnO.

T (K)	ΔG^0 (kJ/mol)	ΔH^0 (kJ/mol)	ΔS^0 (kJ/mol.K)
298	−8.73		
303	−8.63	−14.55	−0.02
313	−8.44		
323	−8.24		

4. Conclusions

A high adsorption efficiency of ZnO nanorods for phosphate ions was found by using the hydrothermal method. The ZnO showed a wurtzite structure with a specific surface area of 17.05 m^2/g and its purity reached 99%. The as-synthesized ZnO particles were used in the adsorption of phosphate in aqueous media. The ZnO materials were applied for adsorption of phosphate from water using batch experiments. The effects of pH (4–10), adsorption time (30–240 min), the amount of adsorbent (0.1–0.7 g/L) and the initial concentration of phosphate (147.637–466.209 mg/L) on the adsorption efficiency were investigated. The optimum condition was found at pH = 5 and at an adsorption time of 150 min. The adsorption fitted well with the Langmuir isotherm and the maximum adsorption capacity was calculated to be 769.23 mg/g. These results show that ZnO nanomaterial is highly promising for adsorbing phosphate from water. The adsorption of phosphate onto ZnO nanomaterials follows the isothermal adsorption model of Langmuir, Tempkin and Freundlich with single-layer adsorption. There is a weak interaction between the adsorbent and adsorbate.

Author Contributions: Investigation, D.T.H., N.V.T., N.Q.D., C.M.N., N.T.C.Q. and L.V.T.; writing—original draft, D.T.H; writing—review and editing, L.V.T. All authors have read and agreed to the published version of the manuscript.

Funding: This study received no external funding.

Conflicts of Interest: The authors declare no conflict of interest.

References

1. Lee, S.-C.; Park, H.-H.; Kim, S.-H.; Koh, S.-H.; Han, S.-H.; Yoon, M.-Y. Ultrasensitive Fluorescence Detection of Alzheimer's Disease Based on Polyvalent Directed Peptide Polymer Coupled to a Nanoporous ZnO Nanoplatform. *Anal. Chem.* **2019**, *91*, 5573–5581. [CrossRef] [PubMed]
2. Zhou, T.; Hu, M.; He, J.; Xie, R.; An, C.; Li, C.; Luo, J. Enhanced catalytic performance of zinc oxide nanorods with crystal plane control. *CrystEngComm* **2019**, *21*, 5526–5532. [CrossRef]
3. Chen, W.; Han, J.; Yi, J. Nanostructured SnO_2- and ZnO-Based Gas Sensors for Early Warning of Electrical Fires. *Meet. Abstr.* **2020**, *MA2020-01*, 2176. [CrossRef]
4. Chao, J.; Chen, Y.; Xing, S.; Zhang, D.; Shen, W. Facile fabrication of ZnO/C nanoporous fibers and ZnO hollow spheres for high performance gas sensor. *Sens. Actuators B Chem.* **2019**, *298*, 126927. [CrossRef]
5. ZnO-based nanostructured electrodes for electrochemical sensors and biosensors in biomedical applications. *Biosens. Bioelectron.* **2019**, *141*, 111417. [CrossRef]
6. Kim, J.Y.; Vincent, P.; Jang, J.; Jang, M.S.; Choi, M.; Bae, J.-H.; Lee, C.; Kim, H. Versatile use of ZnO interlayer in hybrid solar cells for self-powered near infra-red photo-detecting application. *J. Alloy. Compd.* **2020**, *813*, 152202. [CrossRef]
7. Minami, T.; Nanto, H.; Takata, S. Highly Conductive and Transparent Aluminum Doped Zinc Oxide Thin Films Prepared by RF Magnetron Sputtering. *Jpn. J. Appl. Phys.* **1984**, *23*, L280–L282. [CrossRef]
8. Demirci, S.; Dikici, T.; Tünçay, M.M.; Kaya, N. A study of heating rate effect on the photocatalytic performances of ZnO powders prepared by sol-gel route: Their kinetic and thermodynamic studies. *Appl. Surf. Sci.* **2020**, *507*, 145083. [CrossRef]
9. Thambidurai, S.; Gowthaman, P.; Venkatachalam, M.; Suresh, S. Natural sunlight assisted photocatalytic degradation of methylene blue by spherical zinc oxide nanoparticles prepared by facile chemical co-precipitation method. *Optik* **2020**, *207*, 163865. [CrossRef]
10. Irshad, K.; Khan, M.T.; Murtaza, A. Synthesis and characterization of transition-metals-doped ZnO nanoparticles by sol-gel auto-combustion method. *Phys. B: Condens. Matter* **2018**, *543*, 1–6. [CrossRef]
11. Thakur, S.; Mandal, S.K. Effect of dilution in a hydrothermal process and post-synthetic annealing on the tailoring of hierarchical ZnO nanostructures. *CrystEngComm* **2020**, *22*, 3059–3069. [CrossRef]
12. Shokry Hassan, H.; Kashyout, A.B.; Morsi, I.; Nasser, A.A.A.; Raafat, A. Fabrication and characterization of gas sensor micro-arrays. *Sens. Bio-Sens. Res.* **2014**, *1*, 34–40. [CrossRef]
13. F Elkady, M.; Shokry Hassan, H. Equilibrium and dynamic profiles of azo dye sorption onto innovative nano-zinc oxide biocomposite. *Curr. Nanosci.* **2015**, *11*, 805–814. [CrossRef]
14. Rizzi, V.; D'Agostino, F.; Gubitosa, J.; Fini, P.; Petrella, A.; Agostiano, A.; Semeraro, P.; Cosma, P. An Alternative Use of Olive Pomace as a Wide-Ranging Bioremediation Strategy to Adsorb and Recover Disperse Orange and Disperse Red Industrial Dyes from Wastewater. *Separations* **2017**, *4*, 29. [CrossRef]
15. Li, N.; Tian, Y.; Zhao, J.; Zhan, W.; Du, J.; Kong, L.; Zhang, J.; Zuo, W. Ultrafast selective capture of phosphorus from sewage by 3D Fe_3O_4@ZnO via weak magnetic field enhanced adsorption. *Chem. Eng. J.* **2018**, *341*, 289–297. [CrossRef]
16. Nakarmi, A.; Bourdo, S.E.; Ruhl, L.; Kanel, S.; Nadagouda, M.; Kumar Alla, P.; Pavel, I.; Viswanathan, T. Benign zinc oxide betaine-modified biochar nanocomposites for phosphate removal from aqueous solutions. *J. Environ. Manag.* **2020**, *272*, 111048. [CrossRef]
17. Liu, Z.; Lu, Y.; Li, X.; Chen, H.; Hu, F. Adsorption of phosphate from wastewater by a ZnO-ZnAl hydrotalcite. *Int. J. Environ. Anal. Chem.* **2019**, *99*, 1415–1433. [CrossRef]

18. Laliberté, G.; Lessard, P.; de la Noüe, J.; Sylvestre, S. Effect of phosphorus addition on nutrient removal from wastewater with the cyanobacterium Phormidium bohneri. *Bioresour. Technol.* **1997**, *59*, 227–233. [CrossRef]

19. Li, M.; Liu, J.; Xu, Y.; Qian, G. Phosphate adsorption on metal oxides and metal hydroxides: A comparative review. *Environ. Rev.* **2016**, *24*, 319–332. [CrossRef]

20. Lalley, J.; Han, C.; Li, X.; Dionysiou, D.D.; Nadagouda, M.N. Phosphate adsorption using modified iron oxide-based sorbents in lake water: Kinetics, equilibrium, and column tests. *Chem. Eng. J.* **2016**, *284*, 1386–1396. [CrossRef]

21. Almeelbi, T.; Bezbaruah, A. Aqueous phosphate removal using nanoscale zero-valent iron. In *Proceedings of the Nanotechnology for Sustainable Development*; Diallo, M.S., Fromer, N.A., Jhon, M.S., Eds.; Springer International Publishing: Cham, Switzerland, 2014; pp. 197–210.

22. Tu, Y.-J.; You, C.-F.; Chang, C.-K.; Chen, M.-H. Application of magnetic nano-particles for phosphorus removal/recovery in aqueous solution. *J. Taiwan Inst. Chem. Eng.* **2015**, *46*, 148–154. [CrossRef]

23. Akyol, A.; Yatmaz, H.C.; Bayramoglu, M. Photocatalytic decolorization of Remazol Red RR in aqueous ZnO suspensions. *Appl. Catal. B Environ.* **2004**, *54*, 19–24. [CrossRef]

24. Mohd Omar, F.; Abdul Aziz, H.; Stoll, S. Aggregation and disaggregation of ZnO nanoparticles: Influence of pH and adsorption of Suwannee River humic acid. *Sci. Total Environ.* **2014**, *468–469*, 195–201. [CrossRef]

25. Elkady, M.F.; Shokry Hassan, H.; Salama, E. Sorption Profile of Phosphorus Ions onto ZnO Nanorods Synthesized via Sonic Technique. *J. Eng.* **2016**, *2016*, 1–9. [CrossRef]

26. Elkady, M.F.; El-Sayed, E.M.; Farag, H.A.; Zaatout, A.A. Assessment of Novel Synthetized Nanozirconium Tungstovanadate as Cation Exchanger for Lead Ion Decontamination. *J. Nanomater.* **2014**, *2014*, 1–11. [CrossRef]

27. Luo, Z.; Zhu, S.; Liu, Z.; Liu, J.; Huo, M.; Yang, W. Study of phosphate removal from aqueous solution by zinc oxide. *J. Water Health* **2015**, *13*, 704–713. [CrossRef]

28. Zhou, J.; Yang, S.; Yu, J.; Shu, Z. Novel hollow microspheres of hierarchical zinc–aluminum layered double hydroxides and their enhanced adsorption capacity for phosphate in water. *J. Hazard. Mater.* **2011**, *192*, 1114–1121. [CrossRef]

29. Mahdavi, S.; Hassani, A.; Merrikhpour, H. Aqueous phosphorous adsorption onto SnO_2 and WO_3 nanoparticles in batch mode: Kinetic, isotherm and thermodynamic study. *J. Exp. Nanosci.* **2020**, *15*, 242–265. [CrossRef]

30. Li, G.; Gao, S.; Zhang, G.; Zhang, X. Enhanced adsorption of phosphate from aqueous solution by nanostructured iron(III)–copper(II) binary oxides. *Chem. Eng. J.* **2014**, *235*, 124–131. [CrossRef]

31. Trinh, V.T.; Nguyen, T.M.P.; Van, H.T.; Hoang, L.P.; Nguyen, T.V.; Ha, L.T.; Vu, X.H.; Pham, T.T.; Nguyen, T.N.; Quang, N.V.; et al. Phosphate Adsorption by Silver Nanoparticles-Loaded Activated Carbon derived from Tea Residue. *Sci. Rep.* **2020**, *10*, 1–13. [CrossRef]

32. Shahid, M.K.; Kim, Y.; Choi, Y.-G. Adsorption of phosphate on magnetite-enriched particles (MEP) separated from the mill scale. *Front. Environ. Sci. Eng.* **2019**, *13*, 71. [CrossRef]

33. Xiong, L.; Yang, Y.; Mai, J.; Sun, W.; Zhang, C.; Wei, D.; Chen, Q.; Ni, J. Adsorption behavior of methylene blue onto titanate nanotubes. *Chem. Eng. J.* **2010**, *156*, 313–320. [CrossRef]

34. Plaza Cazón, J.; Bernardelli, C.; Viera, M.; Donati, E.; Guibal, E. Zinc and cadmium biosorption by untreated and calcium-treated Macrocystis pyrifera in a batch system. *Bioresour. Technol.* **2012**, *116*, 195–203. [CrossRef] [PubMed]

35. Ajaelu, C.J.; Nwosu, V.; Ibironke, L.; Adeleye, A. Adsorptive removal of cationic dye from aqueous solution using chemically modified African Border Tree (*Newbouldia laevis*) bark. *J. Appl. Sci. Environ. Manag.* **2017**, *21*, 1323–1329. [CrossRef]

36. Zhu, N.; Yan, T.; Qiao, J.; Cao, H. Adsorption of arsenic, phosphorus and chromium by bismuth impregnated biochar: Adsorption mechanism and depleted adsorbent utilization. *Chemosphere* **2016**, *164*, 32–40. [CrossRef]

Publisher's Note: MDPI stays neutral with regard to jurisdictional claims in published maps and institutional affiliations.

Article

Influences of Ash-Existing Environments and Coal Structures on CO_2 Gasification Characteristics of Tri-High Coal

Lang Liu [1],*, Qingrui Jiao [2], Jian Yang [2], Bowen Kong [2], Shan Ren [2] and Qingcai Liu [2]

[1] Chemical Engineering Institute, Guizhou Institute of Technology, Guiyang 550003, China
[2] College of Materials Science & Engineering, Chongqing University, Chongqing 400044, China; 201809021080@cqu.edu.cn (Q.J.); skyinjune@cqu.edu.cn (J.Y.); BoWen.Kong@cisdi.com.cn (B.K.); shan.ren@cqu.edu.cn (S.R.); liuqc@cqu.edu.cn (Q.L.)
* Correspondence: l.liu@git.edu.cn

Received: 26 August 2020; Accepted: 24 October 2020; Published: 28 October 2020

Abstract: Two kinds of tri-high coals were selected to determine the influences of ash-existing environments and coal structures on CO_2 gasification characteristics. The TGA results showed that the gasification of ash-free coal (AFC) chars was more efficient than that of corresponding raw coal (RC) chars. To uncover the reasons, the structures of RCs and AFCs, and their char samples prepared at elevated temperatures were investigated with SEM, BET, XRD, Raman and FTIR. The BET, SEM and XRD results showed that the Ash/mineral matter is associated with coal, carbon forms the main structural framework and mineral matters are found embedded in the coal structure in the low-rank tri-high coal. The Raman and FTIR results show that the ash can hinder volatile matters from exposing to the coal particles. Those results indicate that the surface of AFC chars has more free active carbon sites than raw coal chars, which are favorable for mass transfer between C and CO_2, thereby improving reactivity of the AFC chars. However, the gasification reactivity was dominated by pore structure at elevated gasification temperatures, even though the microcrystalline structure, functional group structure, and increase in the disorder carbon were improved by acid pickling.

Keywords: ash-free coal; CO_2 gasification; coal structure; tri-high coal

---

1. Introduction

Excessive carbon dioxide emissions have caused serious environmental problems, especially for global warming [1]. Synthetic gas (H_2 + CO) via gasification is the most important intermediate product in the highly efficient technologies [2,3]. Steam and carbon dioxide (CO_2) are two regular gasification agents and can usually control the overall conversation process [4]. Although it has been widely used as a gasification agent, CO_2 gasification is important as it is the slowest among gasification reactions and considered as the rate determining step, as well as the key to making a balance between air or oxygen and steam to generate optimum heat for driving endothermic gasification reactions [5,6]. The process of coal CO_2 gasification can be divided into pyrolysis and char gasification reaction [7], and the char gasification formed in situ from pyrolysis process is the rate-determining step [8]. Therefore, the kinetics of char gasification are vital in the design and operation of the gasifier. The gasification of tri-high coal chars often proceeds under the effect of the mineral content in particular, by significantly affecting the gasification rate, which will complicate the kinetics [9,10]. There are several combined chemical and physical steps involved in the conversion rate of coal char [11–14], such as the external mass transfer, inter-particle diffusion and surface reactions. All these steps are associated with changes in the pore structure and the chemical composition of the coal char. However, in some conditions, with the effects of high-ash contents on the chemical composition and porous structure during coal gasification, the gasification characteristics will vary significantly. In southwest of China, tri-high coal, as the most representative type of coal, is characterized by high-ash content, high-sulfur content and high-ash fusion, which will limit its use in gasification process. However, it is uncertain whether the existing environment of ash is really a factor determining the rate of gasification [15]. In our previous work [3], the derivations and variations of char structures throughout the char gasification process were studied, indicating that both of the porous structure and carbon crystallites can affect the char CO_2 gasification kinetics [16].

The most reasonable way to quantify the influence of mineral content on the char gasification kinetics and its relationship with the initial coal structure is to remove the mineral content from the coal (named ash-free coal), and then investigate the gasification characteristics and structures of the ash-free coal [17–19]. There are two methods of char preparation to investigate the influence of mineral content on the char gasification: (1) removal of mineral content by pickling of char derived from pyrolysis, and (2) pyrolysis of coal, of which mineral content has been removed by pickling or other methods [20–22]. However, coal pyrolysis is the initial stage of coal gasification, and closely related to coal composition and structure, which will significantly affect the char gasification characteristics. The mineral content has a great influence on the formation of coal char, such as the cracking of organic matter in coal, volatilization of low-molecular-weight pyrolysis products, polycondensation of cracking residues, decomposition and combination of volatile products during emission, and further decomposition and repolycondensation of the polycondensation products during the pyrolysis process. The composition and structure of coal are directly related to the coal gasification kinetics [9,10,23]. Thus, the method (2) was chosen to produce ash-free coals (AFCs) in this research.

In this paper, two tri-high coals were selected to remove mineral contents by pickling method. The CO_2 gasification characteristics of raw coals and ash-free coals were investigated by using a thermo-gravimetric analyzer. Meanwhile, the structure of the raw coals (RCs), ash-free coals (AFCs) and their chars were characterized by SEM, BET, XRD, Raman and FTIR spectroscopy. The specific method can be obtained in our previous work [3].

2. Materials and Methods

2.1. Materials

Two tri-high coal samples were collected for the experiments from Guizhou Province in southwest of China. The AFC samples were produced by acid pickling. The nitric acid solution was added to the RC at a solution to coal ratio of 20:1 by weight, and the slurry was stirred for 24 h to ensure the coal wetting. After filtration, the filtered coal was mixed with hydrofluoric acid, and then stirred for 24 h and again filtered. Mixing with deionized water, stirring and filtering were repeated until the PH = 7 to ensure complete removal of coal ash. All the agents used above were analytical reagents and the concentration was not diluted. For comparison, the RCs with similar particle size to the AFCs (75–106 μm) were used as well. The char was prepared at 950 °C under a nitrogen atmosphere in a horizontal tubular furnace. During the char preparation process, approximately 20 g coal was placed in a corundum crucible, and then heated at 20 °C/min to the designed temperature under a nitrogen atmosphere. Finally, the sample was held at 950 °C for 30 min. The proximate and ultimate analyses of the AFCs and the corresponding RCs are summarized in Table 1.

On a dry basis, RC-I and AFC-I samples had 9.42 wt.% and 13.67 wt.% of volatile, respectively, while RC-II and AFC-II samples had 13.79 wt.% and 52.34 wt.% of volatile, respectively. The ash contents of RC-I, AFC-I, RC-II and AFC-II were 21.45 wt.%, 0.19 wt.%, 25.26 wt.% and 0.16 wt.%, respectively. The increased Fixed carbon of AFC-I was consistent with the result of Rubiera et al. [24], while that of AFC-II decreased. This probably due to the nitric acid showing poor effect on the mineral content, hydrofluoric acid dominated the acid pickling process. In addition, nitric acid played a role of oxidizing agent led to the increase in O content after pickling.

Table 1. Proximate and ultimate analysis of raw coals and ash-free coals.

Sample	Proximate Analysis (wt.%, db)			Ultimate Analysis (wt.%, daf)				
	Fixed Carbon	Volatile	Ash	C	H	N	O [a]	S
RC-I	69.13	9.42	21.45	89.2	2.25	0.52	5.56	2.47
AFC-I	86.14	13.67	0.19	88	1.76	1.01	8.39	0.84
RC-II	60.95	13.79	25.26	86.95	3.73	1.89	5.12	2.31
AFC-II	47.5	52.34	0.16	71.01	2.65	5.33	20.28	0.73

db, dry basis; and daf, dry and ash-free; [a], oxygen content by difference.

2.2. CO_2 Gasification

Coal gasification consists of both the coal pyrolysis and char gasification. Meanwhile, char plays a role in the rate-determining step, because of its much lower speed than that of coal pyrolysis, throughout the coal gasification process. STA449F3 thermo-gravimetric analyzer (TGA) was used for the char gasification. To evaluate the gasification efficiency, Equation (1) was proposed to calculate the carbon conversion.

$$x = \frac{m_0 - m_t}{m_0 - m_{ash}}, \tag{1}$$

where m_0 and m_t are the initial char mass and the instantaneous char mass at reaction time t, respectively, and m_{ash} is the mass of the ash.

2.3. Sample Characterization

The BET was employed to obtain the specific surface areas and pores volumes. The crystallization phase and components of the samples were detected by X-ray diffraction (XRD). Thermo Scientific DXR Raman spectrometer and FTIR spectrometer were applied to determine the band positions, intensities, widths and areas. The specific method was referred to our previous work [3].

3. Results and Discussion

3.1. Experimental Results

3.1.1. Gasification Characteristics

To investigate the influence of ash on the CO_2 gasification kinetics, TGA experiments were carried out at different temperatures (950 °C–1200 °C). The results showed that the temperature effects on the char gasification were straightforward, and the elevation of gasification temperature generally resulted in increased carbon conversion efficiency. Figure 1 also shows that the acid pickling was conducive to the CO_2 gasification. It was observed that the total conversion time of coal was shortened after ash removal. For example, the total conversion time of RC-I and AFC-I at 1150 °C was nearly 50 min and 30 min, respectively, while that of RC-II and AFC-II at 1150 °C was more than 80 min and nearly 5 min, respectively. It was illustrated that the existence of ash did have negative impacts on the gasification of coal char.

It is known that the particle size and the pore structure of the RC chars and AFC chars were not fixed. Therefore, with the purpose to quantitatively evaluate the reactivity of RC chars and AFC chars, the reactivity index $R_{0.5}$ ($R_{0.5} = 0.5/t_{50}$) [25] was used in this study, t_{50} is the time at which the carbon conversion reached 50%. The reactivity index for RC chars and AFC chars were shown in Figure 2. For all chars, it is obvious that the reactivity index $R_{0.5}$ increased with the gasification temperature. It suggests that the increasing gasification temperature is favor to the gasification reactivity. Furthermore, the reactivity index of RC-I char, AFC-I char, RC-II char, AFC-II char at temperatures of 950 °C–1200 °C were 0.0081–0.18182 min^{-1}, 0.01111–0.11905 min^{-1}, 0.00676–0.08264 min^{-1}, 0.06849–0.43478 min^{-1}, respectively. It can be seen that the $R_{0.5}$ of ARC-II was obviously higher than that of RC-II while the

AFC-I showed a contrary trend. In addition, the $R_{0.5}$ of AFC-I was similar to that of RC-I, and then the $R_{0.5}$ of RC-I was higher than that the former at elevated temperature.

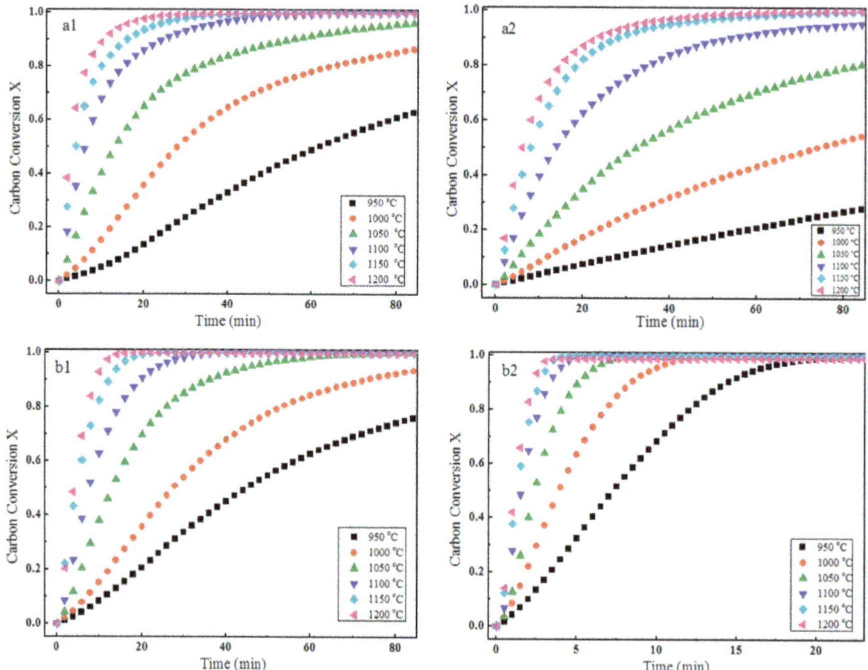

Figure 1. The profiles of carbon conversion vs. gasification time, in response to variation of the temperatures (**a1**: RC-I, **b1**: RC-II; **a2**: AFC-I; **b2**: AFC-II).

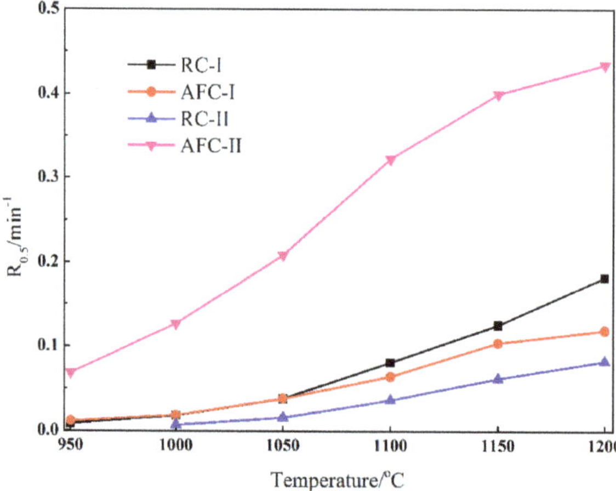

Figure 2. The reactivity index for raw coals (RCs), ash-free coals (AFCs).

The ash content has a great influence on the composition and structure of coal during the pyrolysis process, which is directly relative to the char gasification kinetics. Thus, the structure of the coal and char samples prepared at elevated temperatures and their influences on the gasification characteristics were studied in the next section.

3.1.2. Pore Structure

Measurements of BET specific surface area and total pore volume for the RCs, AFCs and their coal chars are presented in Table 2. Figure 3 shows the surface topography of each sample.

Table 2. BET surface areas and total pore volume of the selected samples.

Sample	RC-I	RC-I Char	AFC-I	AFC-I Char	RC-II	RC-II Char	AFC-II	AFC-II Char
BET surface area (m^2/g)	20.37	31.21	9.87	18.97	6.68	5.89	18.34	12.67
Total pore volume (ml/g)	0.043	0.083	0.023	0.042	0.030	0.021	0.050	0.043

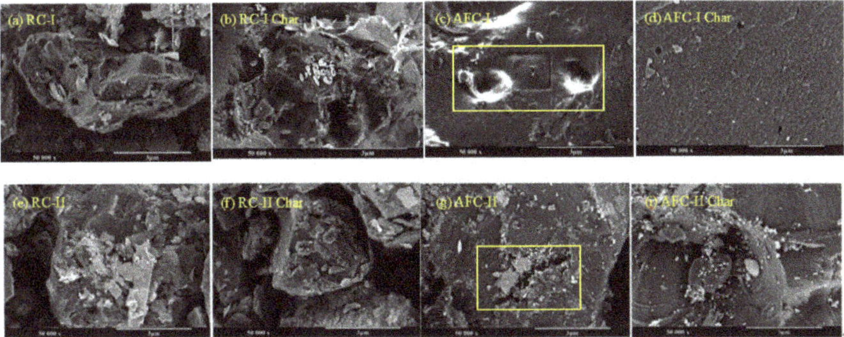

Figure 3. SEM of RCs, AFCs and their Chars (**a–d**: coal I, **e–i**: coal II).

Both the surface area and total pore volume of AFC-I were smaller than those of RC-I, which may result from the collapse of pore structure without ash embedded in the coal. Figure 3c reveals that the channels caused by pickling deeply penetrated into the sample surface, which means the ash is embedded in the coal and plays the role of skeleton of pore structure. In the research of Ni et al. [26], nitric acid can increase the surface area of coal, but nitric acid shows poor effect on the RC-I. Hydrofluoric acid removed the ash content and led to the collapse of RC-I pore structure. Therefore, the surface area and total pore volume of RC-I char were larger than those of AFC-I char, which is because the pore structure under sample surface may have been reserved when not to remove the ash. However, the surface area and total pore volume of RC-II and AFC-II were opposite compared with RC-I and AFC-I. Figure 3g illustrates that RC-II has a more stable structure, and ash removal contributes to the pore development, which is absolutely different with RC-I. The ash dissolution produced a large amount of pores during acid pickling [26], which resulted in an increase in surface area for AFC-II char. The larger surface area promoted the mass transfer of CO_2; therefore, the carbon conversion of AFC-II was significantly higher than that of RC-II. However, the conversion of AFC-I was lower than that of RC-I because of the surface area was decreased.

3.1.3. XRD Patterns Analysis

The XRD patterns and crystalline structure of RCs, AFCs and their chars are shown in Figure 4 and Table 3, respectively. It was noted that no ash content was detected in the AFCs and their chars, indicating the ash content was absolutely removed. The presence of a clear (002) band at ~26° and (100) band in the neighborhood of the graphite at ~43° suggested the existence of some graphite-like structures (crystalline carbon) in RCs and AFCs, as shown in Figure 4, which indicated that the crystallites in the samples had intermediate structures between graphite and the amorphous state. The presence of the clear asymmetric (002) band around 26° suggested the existence of another band (γ) on its left-hand side, which was attached to the periphery of carbon crystallites. It was observed that the γ peaks of RCs decreased after pickling, meaning that pickling contributes to the crack of aliphatic side chains.

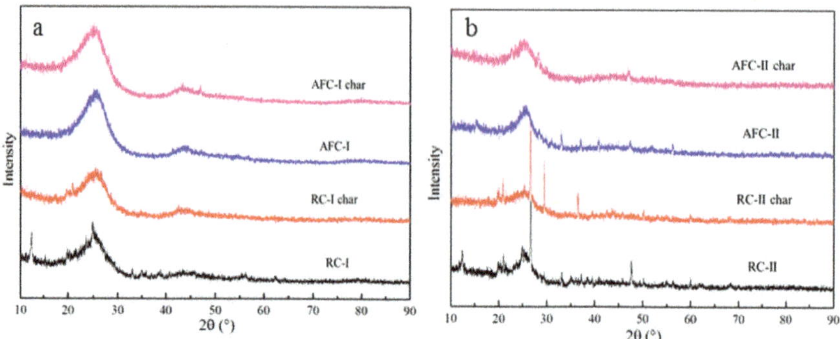

Figure 4. XRD patterns of RCs, AFCs and their Chars ((**a**): I, (**b**): II).

Table 3. Crystalline structure fitting results of RCs, AFCs and their Chars.

Sample	d_{002} (Å)	L_c (Å)
RC-I	3.52	3.066
RC-I Char	3.50	3.229
AFC-I	3.48	3.370
AFC-I Char	3.53	3.107
RC-II	3.49	4.335
RC-II Char	3.53	3.770
AFC-II	3.55	3.516
AFC-II Char	3.49	3.159

Compared with RC-I, the graphite layer spacing of AFC-I was decreased, while the crystal thickness rose. It indicated the presence of minerals between graphite layers. When acid pickling was executed to remove the ash, condensation of graphite layers occurred, which is consistent with the BET and SEM results. Compared with RC-II, the graphite layer spacing of AFC-II rose, while the vertical dimension (L_c) width decreased. AFC-II and its char have more disordered microcrystalline structures than RC-II and its chars. This may be because the fracture effect of the graphite layer produced much smaller layers during removing the ash in the layers by combining analysis with the BET and SEM results. Meanwhile, decomposition products of aromatic lamellar lubricated the fractured graphite layers, which resulted in the graphite layer spacing of AFC-II rising. Meanwhile, the microcrystalline structures of RC-II, AFC-II and their char were more ordered than those of RC-I, AFC-I and their chars.

3.1.4. Raman Spectra Analysis

The typical first-order region Raman spectra profile between 800 and 2000 cm^{-1} of the selected sample are shown in Figure 5. Figure 5 exhibits two characteristic peaks at ~1330 cm^{-1} (D band) and ~1590 cm^{-1} (G band) in Raman spectra [27]. The Raman spectra were subjected to peak fitting using a curve fitting software, Peakfit4.2, to resolve the spectra into three Lorentzian bands (designated as the G, D_1 and D_4 bands, respectively) and one Gaussian band (for the D_3 band), as shown in Figure 5. The D_1 band at ~1350 cm^{-1} is the broadening of the G peak resulting from the introduced disorder carbon, the D_3 band at ~1500 cm^{-1} refers to the amorphous sp^2-bonded forms of carbon, D_4 at ~1250 cm^{-1} is considered to be caused by the amorphous mixed sp^2-sp^3 bonded forms of carbon, and G band refers to graphitic band [28–30].

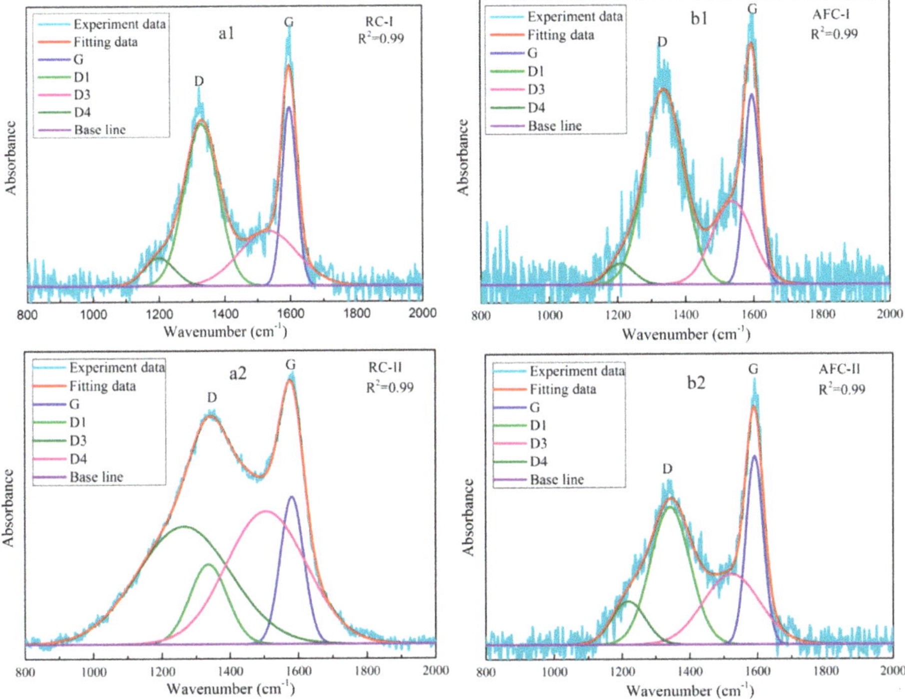

Figure 5. Typical first-order region Raman spectra and the bands of the selected samples (**a1**: RC-I; **b1**: AFC-I; **a2**: AFC-I; **b2**: AFC-II).

The band area ratios of the D_1, D_3 and D_4 to the G (denoted as I_{D1}/I_G, I_{D3}/I_G and I_{D4}/I_G) and the G band relative to the integrated area under the spectra (denoted as I_G/I_{All}) of each sample are shown in Table 4. Compared with the RCs, the I_{D1}/I_G of the AFC chars increased, indicating that the acid pickling removed the ash is not conducive to the orderly development of RCs [31]. The ratios of I_{D3}/I_G and I_{D4}/I_G of AFC chars were less than those of the RC chars, which was due to the hydrolysis of small aromatic structures to aromatic C=C and some aliphatic groups to C–O in phenols, alcohols, ethers and esters bands, decreasing the relative contents of sp^2 and sp^2-sp^3 bonding carbon atoms in AFC chars. The gasification reactivity of RC-I and RC-II was improved by the acid pickling due to more disordered carbon forming and being exposed to the surface.

Table 4. I_{D1}/I_G, I_{D3}/I_G, I_{D4}/I_G and I_G/I_{All} of each sample.

Sample	I_{D1}/I_G	I_{D3}/I_G	I_{D4}/I_G	I_G/I_{All}
RC-I	1.72	0.91	0.49	24.24
RC-I char	2.12	1.22	0.33	21.43
AFC-I	2.03	0.79	0.26	24.5
AFC-I char	2.54	1.11	0.19	20.62
RC-II	1.43	1.64	0.95	19.75
RC-II char	0.88	3.09	3.23	12.19
AFC-II	1.98	1.2	0.35	22.03
AFC-II char	1.66	1.21	0.44	23.23

3.1.5. FTIR Spectra Analysis

FTIR was carried out to understand the carbon functional groups of the selected samples, and the FTIR spectra are shown in Figure 6. The spectra showed six principal bands at 3900–3200 cm^{-1}, 3200–3000 cm^{-1}, 2960–2850 cm^{-1}, 1630–1540 cm^{-1}, 1390 cm^{-1}, 1250–1000 cm^{-1} and 900–700 cm^{-1}, respectively. The bands at 3900–3200 cm^{-1} were assigned to –OH stretching and organic compounds having oxygen functional groups found in coal including phenols, alcohols and carboxylic acid, the bands between 3150 and 3000 cm^{-1} were assigned to C–H bonds in aromatics, the bands at 2960–2850 cm^{-1} were assigned to aliphatic C–H stretching, the bands at 1630–1540 cm^{-1} were assigned to aromatic C=C stretching, the bands at ~1400 cm^{-1} to aliphatic –CH$_3$ bending, the bands between 1250 and 1000 cm^{-1} were assigned to C–O in phenols, alcohols, ethers and esters, and the bands between 900 and 700 cm^{-1} were assigned to aromatic out-of-plane C–H bending [29,32–34].

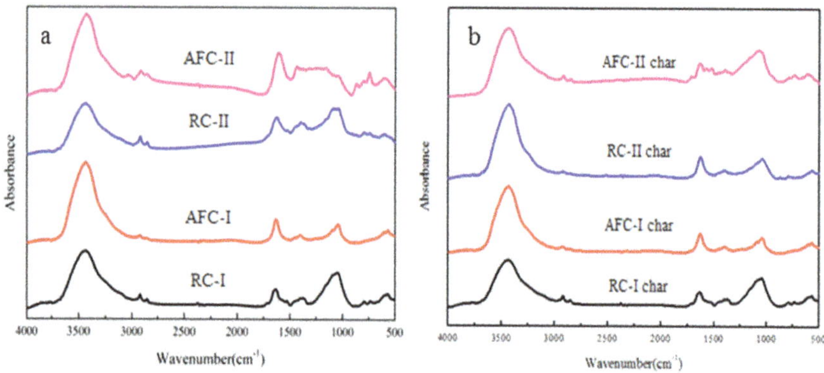

Figure 6. FTIR spectra of the selected samples ((**a**): raw coals, (**b**): chars).

The fitted FTIR spectra of the samples at the selected regions (4000–2600 cm^{-1} and 1800–650 cm^{-1}) are shown in Figure 7, and the band ratios of the samples at each region are shown in Table 5. For the range of 4000–2600 cm^{-1}, there was a short and narrow absorption above 3600 cm^{-1} in all of the samples, suggesting that free hydroxyl groups exist in the samples. It was also observed that there was a broad and strong absorption peak at ~3450 cm^{-1} in the samples, which is assigned to hydrogen-bonded hydroxyl group vibrations (poly –OH1). The absorption at ~3210 cm^{-1} assigned to wagging vibrations of hydroxyl group (poly –OH2) was also present. Owing to the loss of a part of hydroxyl groups (poly –OH1 and –OH2) dissolved in acid solution, the hydroxyl group ratios between two AFCs were lower than those of the corresponding RCs, and the relative content of aromatic C–H was increased. However, the band ratios of aliphatic C–H between 2960 and 2850 cm^{-1} of the two AFCs decreased, which was due to the reactions between acid solution and ash in coals during the acid pickling process, which releases a lot of heat, resulting in the cracking of aliphatic C–H to low-molecular-weight groups, such as C–O in phenols, alcohols, ethers and esters.

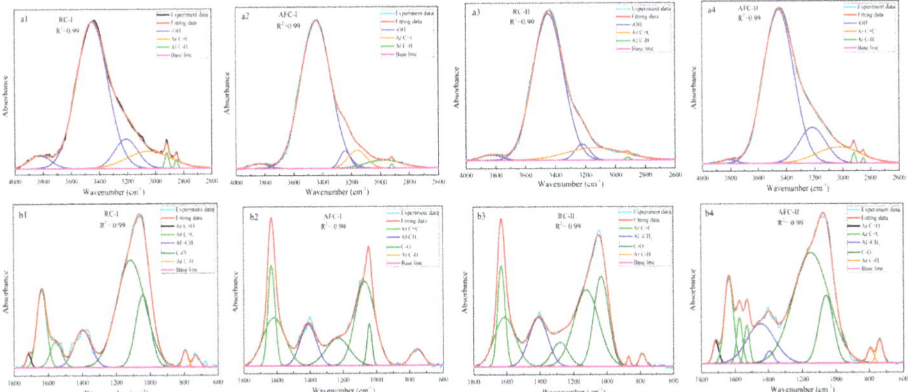

Figure 7. Infrared spectra of selected samples with the corresponding curve fitted bands in the ranges 4000–2600 cm^{-1} (a_1–a_4) and 1800–650 cm^{-1} (b_1–b_4); Ar, aromatic and Al, aliphatic.

Table 5. The band ratios of the samples at each region (%, dry basis).

Range, cm^{-1}	Band Position, cm^{-1}	Sample							
		RC-I	RC-I Char	AFC-I	AFC-I Char	RC-II	RC-II Char	AFC-II	AFC-II Char
4000–2600	poly –OH (~3400–3200)	82.67	87.42	80.67	86.31	96.02	87.04	81.32	88.12
	aromatic C–H (3150–3000)	16.08	11.58	18.57	13.37	-	12.72	17.28	11.13
	aliphatic C-H (2960–2850)	1.25	1.28	0.76	0.32	3.98	0.24	1.44	0.75
1800–600	aromatic carboxyl C=O (1710)	2.46	1.06	1.13	-	-	-	-	1.7
	aromatic C=C (1630–1440)	16.95	17.25	28.19	34.53	21.62	27.94	23.99	16.05
	aliphatic –CH$_3$ (~1400)	10.56	10.38	6.25	13.09	20.56	14.03	12.22	1.27
	C-O (1250–1050)	66.83	68.45	61.29	48.41	51.73	56.38	63.79	78.16
	aromatic C–H (700–900)	3.2	3.86	3.13	3.97	6.09	1.65	5.21	2.92

For the range 1800–600 cm^{-1}, more aromatic C=C is exposed to the surface of particles via acid pickling, which results in a higher aromatic C=C ratio of AFC-I char than that of RC-I char. More aliphatic groups are also exposed to the particle surface via acid pickling, but some of aliphatic groups crack to small molecular structure under heat produced from the reactions between acid solution and ash, which results in a lower band ratio of aliphatic groups of AFC-I than that of RC-I. However, a higher band ratio of aliphatic –CH$_3$ of AFC-I char than that of RC-I char was obtained after devolatilization, and the band ratio of low-molecular-weight group C-O was greatly reduced.

More aromatic and aliphatic groups are also exposed to the particle surface of particles via acid pickling, which results in a higher aromatic ratio of AFC-II than that of RC-II. Meanwhile, acid pickling produced more structure defects, leading to much easier cracking of aromatic C=C and aliphatic –CH$_3$ to low-molecular-weight group C–O. Thus, the band ratio of aromatic C=C and aliphatic –CH$_3$ of AFC-II char was lower than that of RC-II char, and the band ratio of low-molecular-weight group C–O of AFC-II was higher than that of RC-II.

In brief, the lower rank the coal is, the more volatile matters will be hindered by the ash while exposed to the coal particle surface.

4. Conclusions

Two kinds of tri-high coals were studied to determine the influence of ash-existing environments on CO_2 gasification characteristics. The results illustrated that the ash embedded in high rank tri-high coal. The ash usually hinders volatile matter exposed to the surface of coal particles. Acid pickling could improve the microcrystalline structure and functional group structure, and increase the disorder carbon, but the gasification reactivity was also dominated by pore structure at elevated gasification temperatures. The lower the rank the tri-high coal is, the more obstruction effects the ash has. In other words, removing the ash of the low rank tri-high coal can help to promote CO_2 gasification efficiency.

Author Contributions: L.L.: Conception, editing, obtain research funding; Q.J.: Experiment, data analysis, editing; B.K.: Experiment, data analysis; J.Y., S.R., Q.L.: Conception, writing review. All authors have read and agreed to the published version of the manuscript.

Funding: This study was supported by the Science and Technology Planning Project of Guizhou province (Qiankehejichu [2018]1066), Young Teachers Training Project of Guizhou Institute of Technology ([2017]5789-05), Guizhou Science and Technology Support Project (Contract no: Qiankehe Zhicheng [2019]2872) and the Research Start-up Funding Project for High-level Talents of Guizhou Institute of Technology.

Conflicts of Interest: The authors declare no conflict of interest.

References

1. Ren, S.; Aldahri, T.; Liu, W.; Liang, B. CO_2 mineral sequestration by using blast furnace slag: From batch to continuous experiments. *Energy* **2021**, *214*, 118975. [CrossRef]
2. Pérez-Fortes, M.; Bojarski, A.D.; Velo, E.; Nougués, J.M.; Puigjaner, L. Conceptual model and evaluation of generated power and emissions in an IGCC plant. *Energy* **2009**, *34*, 1721–1732. [CrossRef]
3. Liu, L.; Cao, Y.; Liu, Q. Kinetics studies and structure characteristics of coal char under pressurized CO_2 gasification conditions. *Fuel* **2015**, *146*, 103–110. [CrossRef]
4. Wang, M.; Zhang, J.; Zhang, S.; Wu, J.; Yue, G. Experimental studies on gasification of the Shenmu coal char with CO 2 at elevated pressures. *Korean J. Chem. Eng.* **2008**, *25*, 1322–1325. [CrossRef]
5. Huo, W.; Zhou, Z.; Chen, X.; Dai, Z.; Yu, G. Study on CO_2 gasification reactivity and physical characteristics of biomass, petroleum coke and coal chars. *Bioresour. Technol.* **2014**, *159*, 143–149. [CrossRef] [PubMed]
6. Nowicki, L.; Siuta, D.; Markowski, M.J.E. Carbon Dioxide Gasification Kinetics of Char from Rapeseed Oil Press Cake. *Energies* **2020**, *13*, 2318. [CrossRef]
7. Kang, T.J.; Park, H.; Namkung, H.; Xu, L.H.; Fan, S.; Kim, H.T. Comparison of catalytic pyrolysis and gasification of Indonesian low rank coals using lab-scale bubble fluidized-bed reactor. *Korean J. Chem. Eng.* **2017**, *34*, 1238–1249. [CrossRef]
8. Sawettaporn, S.; Bunyakiat, K.; Kitiyanan, B.J. CO_2 gasification of Thai coal chars: Kinetics and reactivity studies. *Korean J. Chem. Eng.* **2009**, *26*, 1009–1015. [CrossRef]
9. Franklin, H.D.; Peters, W.A.; Howard, J.B. Mineral matter effects on the rapid pyrolysis and hydropyrolysis of a bituminous coal: 2. Effects of yields of C3-C8 hydrocarbons. *Fuel* **1981**, *61*, 1213–1217. [CrossRef]
10. Ehrburger, P.; Addoun, F.; Donnet, J.B. Effect of mineral matter of coals on the microporosity of charcoals. *Fuel* **1988**, *67*, 1228–1231. [CrossRef]
11. Lin, L.; Strand, M. Investigation of the intrinsic CO_2 gasification kinetics of biomass char at medium to high temperatures. *Appl. Energy* **2013**, *109*, 220–228. [CrossRef]
12. Ion, I.V.; Popescu, F.; Rolea, G.G. A biomass pyrolysis model for CFD application. *J. Therm. Anal. Calorim.* **2013**, *111*, 1811–1815. [CrossRef]
13. Tomaszewicz, M.; Łabojko, G.; Tomaszewicz, G.; Kotyczka-Morańska, M. The kinetics of CO_2 gasification of coal chars. *J. Therm. Anal. Calorim.* **2013**, *113*, 1327–1335. [CrossRef]
14. Fan, D.; Zhu, Z.; Na, Y.; Lu, Q. Thermogravimetric analysis of gasification reactivity of coal chars with steam and CO2 at moderate temperatures. *J. Therm. Anal. Calorim.* **2013**, *113*, 599–607. [CrossRef]
15. Feng, B.; Bhatia, S.K. Variation of the pore structure of coal chars during gasification. *Carbon* **2003**, *41*, 507–523. [CrossRef]

16. Yoon, S.P.; Deng, L.; Namkung, H.; Fan, S.; Kang, T.J.; Kim, H.T. Coal structure change by ionic liquid pretreatment for enhancement of fixed-bed gasification with steam and CO_2. *Korean J. Chem. Eng.* **2017**, *35*, 445–455. [CrossRef]

17. Kopyscinski, J.; Habibi, R.; Mims, C.A.; Hill, J.M. K_2CO_3-Catalyzed CO_2 gasification of ash-free coal: Kinetic study. *Energy Fuels* **2013**, *27*, 4875–4883. [CrossRef]

18. Wu, X.; Jie, W. K_2CO_3-catalyzed steam gasification of ash-free coal char in a pressurized and vertically blown reactor. Influence of pressure on gasification rate and gas composition. *Fuel Process. Technol.* **2017**, *159*, 9–18. [CrossRef]

19. Sharma, A.; Saito, I.; Takanohashi, T. Catalytic Steam Gasification Reactivity of HyperCoals Produced from Different Rank of Coals at 600–775 °C. *Energy Fuels* **2008**, *22*, 3561–3565. [CrossRef]

20. Kong, Y.; Kim, J.; Chun, D.; Lee, S.; Rhim, Y.; Lim, J.; Choi, H.; Kim, S.; Yoo, J. Comparative studies on steam gasification of ash-free coals and their original raw coals. *Int. J. Hydrog. Energy* **2014**, *39*, 9212–9220. [CrossRef]

21. Kim, J.; Choi, H.; Lim, J.; Rhim, Y.; Chun, D.; Kim, S.; Lee, S.; Yoo, J. Hydrogen production via steam gasification of ash free coals. *Int. J. Hydrog. Energy* **2013**, *38*, 6014–6020. [CrossRef]

22. Wu, X.; Jia, T.; Jie, W. A new active site/intermediate kinetic model for K_2CO_3-catalyzed steam gasification of ash-free coal char. *Fuel* **2016**, *165*, 59–67. [CrossRef]

23. Ellis, N.; Masnadi, M.S.; Roberts, D.G.; Kochanek, M.A.; Ilyushechkin, A.Y. Mineral matter interactions during co-pyrolysis of coal and biomass and their impact on intrinsic char co-gasification reactivity. *Chem. Eng. J.* **2015**, *279*, 402–408. [CrossRef]

24. Rubiera, F.; Arenillas, A.; Arias, B.; Pis, J.J.; Suarez-Ruiz, I.; Steel, K.M.; Patrick, J.W.J. Combustion behaviour of ultra clean coal obtained by chemical demineralisation. *Fuel* **2003**, *82*, 2145–2151. [CrossRef]

25. Takarada, T.; Tamai, Y.; Tomita, A.J. Reactivities of 34 coals under steam gasification. *Fuel* **1985**, *64*, 1438–1442. [CrossRef]

26. Ni, G.; Li, S.; Rahman, S.; Xun, M.; Wang, H.; Xu, Y.; Xie, H. Effect of nitric acid on the pore structure and fractal characteristics of coal based on the low-temperature nitrogen adsorption method. *Powder Technol.* **2020**, *367*, 506–516. [CrossRef]

27. Bai, Y.; Wang, Y.; Zhu, S.; Fan, L.I.; Xie, K. Structural features and gasification reactivity of coal chars formed in Ar and CO_2 atmospheres at elevated pressures. *Energy* **2014**, *74*, 464–470. [CrossRef]

28. Liu, X.; Ying, Z.; Liu, Z.; Ding, H.; Huang, X.; Zheng, C.J. Study on the evolution of the char structure during hydrogasification process using Raman spectroscopy. *Fuel* **2015**, *157*, 97–106. [CrossRef]

29. Cole-Clarke, P.A.; Vassallo, A.M. Infrared emission spectroscopy of coal. *Fuel* **1992**, *71*, 469–470. [CrossRef]

30. Chabalala, V.P.; Wagner, N.; Potgieter-Vermaak, S.J. Investigation into the evolution of char structure using Raman spectroscopy in conjunction with coal petrography; Part 1. *Fuel Process. Technol.* **2011**, *92*, 750–756. [CrossRef]

31. Yu, J.; Guo, Q.; Ding, L.; Gong, Y.; Yu, G. Studying effects of solid structure evolution on gasification reactivity of coal chars by in-situ Raman spectroscopy. *Fuel* **2020**, *270*, 117603. [CrossRef]

32. Sonibare, O.O.; Haeger, T.; Foley, S.F. Structural characterization of Nigerian coals by X-ray diffraction, Raman and FTIR spectroscopy. *Energy* **2010**, *35*, 5347–5353. [CrossRef]

33. Sharma, R.K.; Wooten, J.B.; Baliga, V.L.; Lin, X.; Geoffrey Chan, W.; Hajaligol, M.R. Characterization of chars from pyrolysis of lignin. *Fuel* **2004**, *83*, 1469–1482. [CrossRef]

34. Zhang, Y.; Kang, X.; Tan, J.; Frost, R.L. Influence of Calcination and Acidification on Structural Characterization of Anyang Anthracites. *Energy Fuels* **2013**, *27*, 7191–7197. [CrossRef]

Publisher's Note: MDPI stays neutral with regard to jurisdictional claims in published maps and institutional affiliations.

Article

Exploring E-Waste Resources Recovery in Household Solid Waste Recycling

Muhammad Mobin Siddiqi [1], Muhammad Nihal Naseer [1,*], Yasmin Abdul Wahab [2,*], Nor Aliya Hamizi [2], Irfan Anjum Badruddin [3], Mohd Abul Hasan [4], Zaira Zaman Chowdhury [2], Omid Akbarzadeh [2], Mohd Rafie Johan [2] and Sarfaraz Kamangar [3]

[1]   Department of Applied Sciences, National University of Sciences and Technology (NUST), Islamabad 44000, Pakistan; mobinsiddiqi@hotmail.com
[2]   Nanotechnology & Catalysis Research Centre, Deputy Vice Chancellor (Research & Innovation) Office, University of Malaya, Kuala Lumpur 50603, Malaysia; aliyahamizi@um.edu.my (N.A.H.); dr.zaira.chowdhury@um.edu.my (Z.Z.C.); omid@um.edu.my (O.A.); mrafiej@um.edu.my (M.R.J.)
[3]   Department of Mechanical Engineering, College of Engineering, King Khalid University, P.O. Box 394, Abha 61421, Saudi Arabia; magami.irfan@gmail.com (I.A.B.); sarfaraz.kamangar@gmail.com (S.K.)
[4]   Department of Civil Engineering, College of Engineering, King Khalid University, P.O. Box 394, Abha 61421, Saudi Arabia; mohad@kku.edu.sa
*   Correspondence: nihal.me@pnec.nust.edu.pk (M.N.N.); yasminaw@um.edu.my (Y.A.W.); Tel.: +92-305-429-8988 (M.N.N.)

Received: 30 July 2020; Accepted: 21 August 2020; Published: 27 August 2020

Abstract: The ecosystem of earth, the habitation of 7.53 billion people and more than 8.7 million species, is being imbalanced by anthropogenic activities. The ever-increasing human population and race of industrialization is an exacerbated threat to the ecosystem. At present, the global average waste generation per person is articulated as 494 kg/year, an enormous amount of household waste (HSW) that ultimately hits 3.71×10^{12} kg of waste in one year. The ultimate destination of HSW is a burning issue because open dumping and burning as the main waste treatment and final disposal systems create catastrophic environmental limitations. This paper strives to contribute to this issue of HSW management that matters to everyone's business, specifically to developing nations. The HSW management system of the world's 12th largest city and 24th most polluted city, Karachi, was studied with the aim of generating possible economic gains by recycling HSWs. In this regard, the authors surveyed dumping sites for sample collection. The sample was segregated physically to determine the content type (organic, metals, and many others). Afterward, chemical analysis on AAS (Atomic Absorption Spectrophotometry) of debris and soil from a landfill site was performed. HSW is classified and quantified into major classes of household materials. The concentrations of e-waste [Cu], industrial development indicator [Fe], and the main component of lead-acid storage batteries [Pb] are quantified as 199.5, 428.5, and 108.5 ppm, respectively. The annual generation of the aforementioned metals as waste recovery is articulated as 1.2×10^6, 2.6×10^6 and 6.5×10^5 kg, respectively. Significantly, this study concluded that a results-based metal recovery worth 6.1 million USD is discarded every year in HSW management practices.

Keywords: household solid waste; metal recovery value; socio-economic benefits; waste composition of Karachi-Pakistan; waste management; waste recycling

---

1. Introduction

Solid waste management is one of the most critical issues being faced by urban areas of the world [1]. The intensity of this issue is meager in developed countries because authenticated data of MSW (Municipal Solid Waste) is available and being collected and evaluated on a daily basis [2]. Contrarily, the developing countries, characterized by uncontrolled population growth with gravitation to industrialization accompanied by no substantial heed to environmental suitability, lacks the appropriate and authenticated data of MSW, making its management more critical and worse. This missing data is incredibly crucial to effectively and efficiently allot a sustainable destination to MSW. This data also defines the economic status of a nation [3].

Unfortunately, just like other developing countries, Pakistan is not an exception where adequate data is a nonentity. This gap of knowledge has been charging Pakistan by not only exacerbating the standard of urban life but also by refraining industries from its incineration for power acquisition purposes. The imperativeness of MSW converged the attention of researchers and substantial improvement has been observed in the last decade. Various urban regions of Pakistan such as Lahore [4–6], Gujranwala [7], Hyderabad [8], Faisalabad [9], and Karachi [1,10,11] has been evaluated on various bases and remarkable data has been acquired. This study contributes to filling the knowledge gap by providing experimentally calculated authenticated data of Karachi by focusing on improved collection systems, supported and organized recycling to develop state-of-the-art WMS. The main focus of previous studies of Karachi was on the concentration of combustible matter in MSW, its potential to generate energy and environmental impacts [11–13]. This paper reports detailed information about e-waste present in MSW of Karachi for the first time and also provides a significant evaluation of its annual economical worth.

1.1. Conceptual Basis

1.1.1. Global Household Solid Waste Generation

The destination of MSW is one of the most trending topics of discussion in this century. Its significance can be accessed from the ever-increasing attention of researchers towards it, as depicted in Figure 1. This paper is also a constituent of the same congregation.

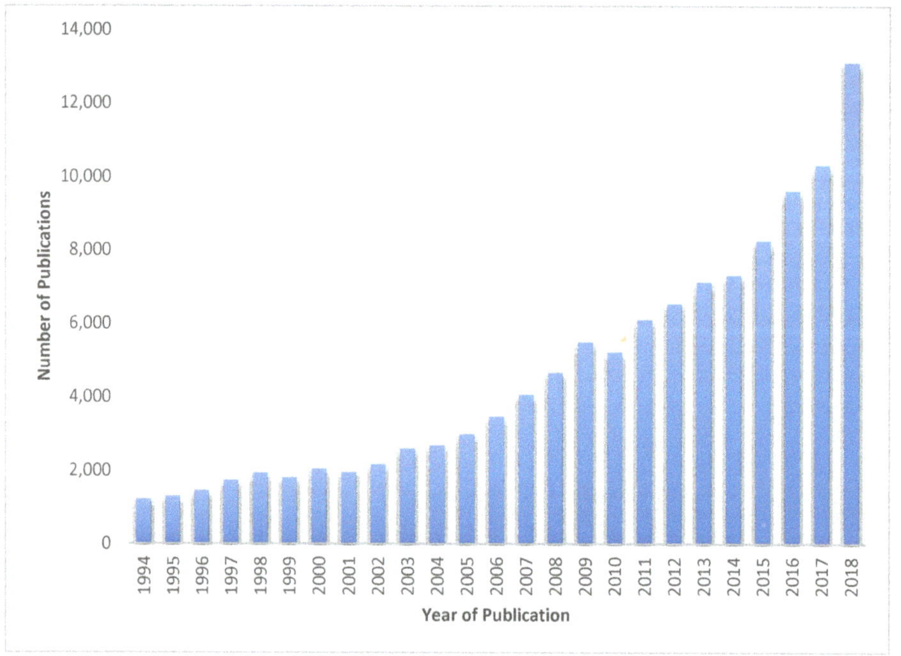

Figure 1. Ever-increasing attention of researchers towards environmental sustainability and destination of waste demonstrated by the number related publications per year (Web of Science).

The consequences of the rapid growth of the human population have remained a point of concern for environmentalists and economists. According to the UN, the human population across the globe is about 7.71 billion (Figure 2) and is increasing with an average growth rate of 1.1% per year. Since 1950, the population has almost tripled (Figure 2) and is on the rise, mainly due to better health facilities and low mortality rates. As people living on Earth are increasing in number, so is the household solid waste generated by them. The average generation of household solid waste is directly related to the population [14]. Every person does not generate the same amount of household solid waste, rather it depends upon lots of factors including the economic status and lifestyle of the individual. Per capita household solid waste generation values have been determined by researchers for selected regions; however, country-wise collection of data for per capita household solid waste generation is an enormous task and is beyond the scope of individual researchers or research consortiums. For country-wise data of household solid waste generation, available sources are mostly government organizations or international independent bodies. Existing data of household solid waste generation of different regions [15,16] have been analyzed to calculate the global average household solid waste generation per person per year. At present, the global average waste generation is 494 kg per person per year. As the population of the globe is 7 billion, thus 3.5×10^{11} kg of waste is estimated to be generated annually only from households.

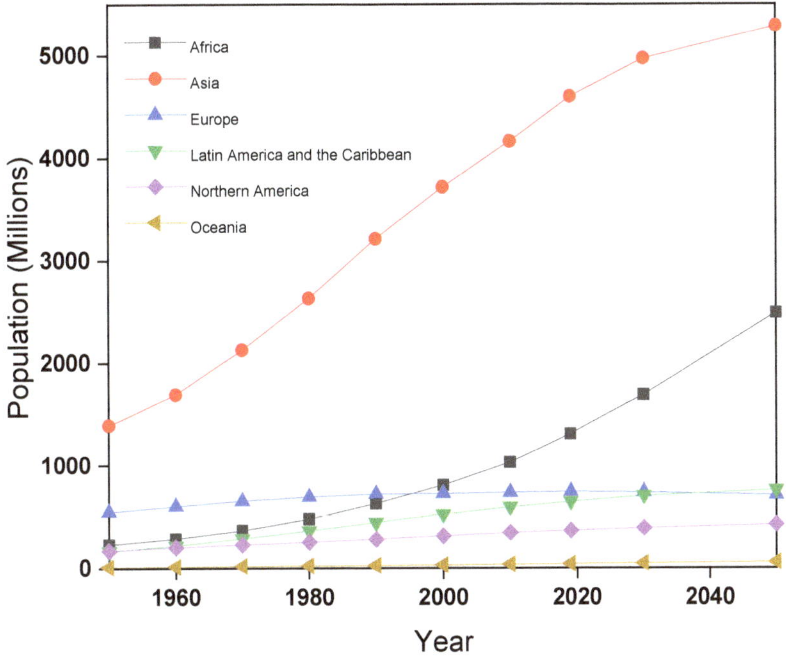

Figure 2. World population, both sexes combined (Million) [17].

1.1.2. Relationship between Socio-Economic Conditions and HSW Generation

The generation of household solid waste per capita has a multi-variant dependency. Among the most influential variables, affecting the quantity and composition of household solid waste, are the socio-economic conditions of the inhabitants of the area under consideration. Data mentioned in Figure 3 shows the relationship between GDP (Gross Domestic Product) per capita and household solid waste generation of countries of interest. It shows that Brunei Darussalam, France, and Germany have high GDP values, and their per capita household solid waste generation is also high. Economically more stable and high-income societies have high availability of resources to utilize for maintaining better living standards [18]. As a result, the amount of household waste generated is also higher in quantity and comprises more synthetic and metallic articles compared to low-income areas where the composition of household solid waste comprises more kitchen and organic waste. In addition to socio-economic conditions, education, and awareness about environmental concerns also have a significant effect on household solid waste generation [19,20].

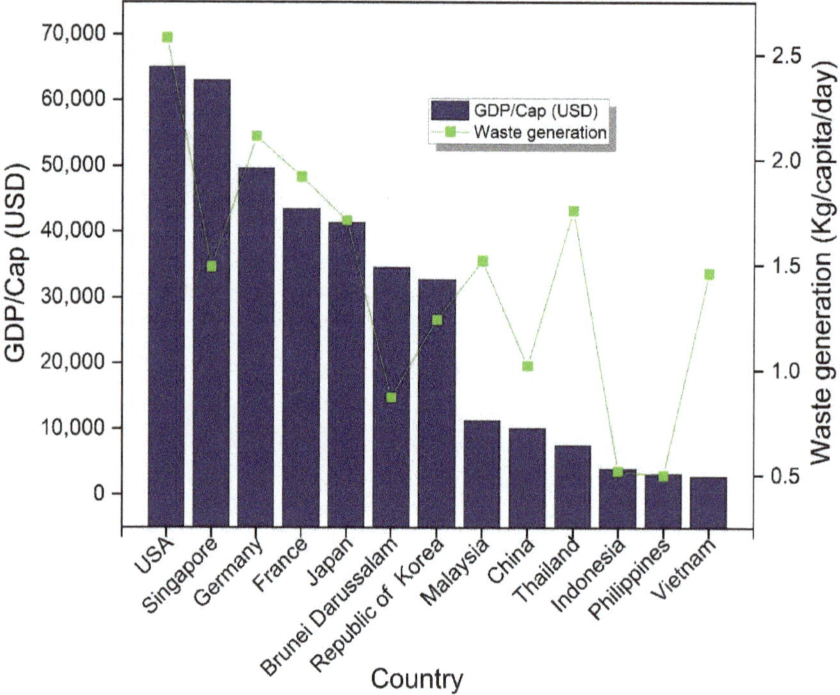

Figure 3. Relationship between GDP per capita and HSW (Household Solid Waste) generation [16,21].

1.1.3. HSW Generation by Developed and Developing Countries

Technologically advanced countries have better resources and therefore accentuate on recycling of waste into other reusable items. The percentage of recycled materials out of total HSW generated in the USA is 45%, and 44% in the UK, whereas the percentage of recycling HSW in developing countries is very low due to the unavailability of resources and limited realization of environmentally-friendly waste management. The population in developing regions of the world is increasing faster as compared to the population of developed regions, as depicted in Figure 2. Population growth leads to increment in utilities' demand in order to meet basic needs. Ultimately, developing countries will increase the rate of industrialization, paying no heed to HSW management, mainly due to low budget. Hence, the total amount of household solid waste generation is expected to rise with population growth and economic development. Statistics show that in 2016, 2.01 billion tons of MSW were generated globally and projections evaluated this quantity to be 3.40 billion tons by 2050. Developing countries are forecasted to triple their waste by 2020 [22].

Thus, a time will come when developing regions will join developed regions in GDP, and their per capita household solid waste generation is also expected to increase. According to the above facts, household solid waste generation is expected to increase in the future. Due to the unavailability of complete data, determination of percentage of HSW recycling with accuracy is not possible, however close estimates based on published data lead to recycling percentages of Malaysia = 5%, Pakistan = 5% and Thailand = 11%. The point of concern is comparatively increased in the population of developing regions as compared to the developed regions.

1.1.4. A New Class of Waste, the E-waste

Electronic waste is a relatively new environmental problem. Its disposal procedures and recycling techniques are inviting researchers and investors towards new scientific and business opportunities. The unprecedented ingress of electronic and IoT components in daily life are giving rise to an emerging problem, which demands early solution. E-waste contains 60% valuable materials like iron, copper, aluminum, gold and other metals, whereas it comprises 2.70% pollutants and hazardous materials [23], especially Pb, Sb, Hg, Cd, Ni, polybrominated diphenyl ethers (PBDEs), and polychlorinated biphenyls (PCBs). A study carried out in 2008 concluded that the quantity of rare and noble metals in e-waste is more than in typical metal mines [24]. The rapid development in software pushes manufacturers to develop new technologies and tempt consumers to upgrade hardware at a continuously increasing rate. In 1994, the average life of a personal computer was 4–5 years and it was estimated that approximately 20 million PCs (about 7 million tons) became obsolete in that year. By 2004, average PC life decreased to 2 years and in that year approximately 100 million PCs (about 35 million tons) became obsolete. As an estimate, 100 million PCs contain approximately 574,400 tons of plastics, 143,600 tons of lead, 273 tons of cadmium and 57 tons of mercury [25]. Since 2005, the use of handheld electronic devices like mobile phones and tablets have gained popularity, and the average life before getting obsolete for mobile phones has become less than two years. Nevertheless, the quantity of WEEE (Waste from Electrical and Electronic Equipment) generated constitutes one of the fastest-growing waste fractions, accounting for 8% of all municipal waste [26]. Exposure of the general public to these e-wastes is a routine matter, but the severity of this exposure, either direct or indirect, is high. Untreated e-waste dumped at landfill sites has all the means to come into indirect contact through soil, water, air, animal and plant sources. People can come into contact with e-waste materials, and associated pollutants, through contact with contaminated soil, dust, air, water, and through food sources, including meat [27–39]. Multi Criteria Analysis (MCA) of e-waste management and Applications of Life Cycle Assessment (ALCA) have recommended recycling as the best course of action for e-waste disposal [40–42]. The task of e-waste disposal and recycling is enormous and requires the intervention of governments in the form of legislation, direct investment and subsidiaries to small recycling enterprises. Apart from the government's responsibility for protecting the environment from pollution by e-waste recycling, manufacturers should also be made accountable for reducing environmental hazards from their products. The principle of Extended Producer Responsibility (EPR), which obliges producers to cover the cost of collection, recycling, and disposal, should be effectively and sufficiently implemented [42–48].

1.1.5. General Concept about HSW Recycling

The Earth has a limited supply of resources available for humans to explore and utilize. Our future generations depend on these resources. For future survival of the human race, a reduction in the consumption of new resources and recycling of available resources is imminent. On the other hand, poor solid waste management will keep on adding to the pollutants, and leave Earth as an inhabitable planet. In both cases, recycling is unavoidable for the survival of the human race, until some other habitable planet is discovered at a reachable distance in the universe. Notwithstanding the above arguments, today's profit-oriented system is more focused on the gains of today than the fortunes of tomorrow. In order to determine whether the recycling of household solid waste was economically worth performing, Karachi city was selected as subject location. The fieldwork research was carried out and the results depicted that it would be significantly expensive.

2. Materials and Methods

2.1. Demography of Karachi

To evaluate the possible economic benefits obtainable from recycling, the HSW management system of the world's 12th largest city Karachi was selected for the study. Karachi is a 3527 km^2 metropolitan area and its population is estimated at over 14.91 million as of 2017. The average population density is about 6300 people/km^2. It is situated 129 km due west of the present Indus River delta. The boundary of the city lies between the geographical co-ordinates 24°45′ N to 25°38′ N and 66°40′ E to 67°34′ E. For the city, six officially designated landfill sites are available, whereas a number of unofficially used sites for landfill are in use.

2.2. Survey and Site Selection

Among the six officially designated landfill facilities, the largest in terms of area and load dumped per day is the Deh Jam Chakro landfill facility. Household solid waste from 14 out of 18 towns of Karachi is dumped at this site. The site is located at 25°01′ N and 67°01′ E, surrounded by the thickly populated areas of New Karachi, Orangi Town and Surjani town. For this study, the authors performed the survey in the first week of every month from April 2017 to March 2018. Household solid waste was collected from 54 different sites of Karachi. These sites were carefully selected to give equal representation of all types of socio-income groups.

Sample Collection for Chemical Analysis

Samples were collected from a zone of 0.3048 m (12 inches) from the surface of the soil ash mixture residue, leftover after the burning of household solid waste. To ensure that the sample was a true representative sample, a systematic distribution approach was adopted for sample collection. For this purpose, 3 replicates of samples were collected from each corner of the 20.9032 sq meter grid of square geometry. In this way the total area of 209 sq meter area of the Jam Chakro landfill facility was covered during sample collection. This resulted in 30 samples collected from 10 sites at the landfill facility. Three replicates of samples were collected from each of the 10 sites. Collected samples were sealed in airtight containers and transported to the laboratory for analysis. All the equipment used in the process of sample collection was non-metallic in nature and mostly made up of plastic.

2.3. Quantification and Characterization Technique

As per the recommendations from the industrial peers of the case study area, industrial level extraction of the metals needs to focus on the physical segregation of metals. Hence the first step of experiment was the physical segregation of HSW; afterwards, chemical analysis was also performed. For segregation purposes, the collected household solid waste was mixed to obtain a representative sample of the whole city. Constituents of the sample were then separated as per the defined categories and weighed, as discussed in the results. Segregation of HSW was followed by grinding of segregated content, digestion in acidic medium, dilution and, finally, analysis with AAS.

In order to determine the metal content of HSW at the Jam Chakro landfill facility, open air burning of HSW was carried out. During this process, most of the combustible substances were burnt and non-combustible materials including metal were left behind in the ash–soil mixture. Heavy metals were selected to analyze in the ash–soil mixture collected from the Jam Chakro landfill facility. The purpose was to estimate the amount of metals that can be extracted from the site and recycled. Due to limited resources, the study was restricted to detailed estimations of Cu, Fe and Pb only and generation of these metals as waste in kg/yr was determined. The following sections provide deep insight into the experimentation section.

2.3.1. Sample Pretreatment

All the collected samples were mixed in laboratory with a plastic trowel. A testing sample, weighing 0.02 kg, was withdrawn. To pass a selected sample through the mesh of a 0.841 mm grid opening, it was finely grounded. A total of 5.0 g of the sieved sample was transferred into an Erlenmeyer flask.

2.3.2. Blank Preparation

In order to analyze the sample on Atomic Absorption Spectrometry (ASS), separate blanks were made ready form soil, dust, plant and water analysis. The preparation methodology for the blanks was quite similar to that of sample pretreatment, but the respective samples were not added.

2.3.3. Chemical Analysis of Samples

All chemicals used during the analysis were AnalR grade. An aqua regia solution was prepared for digestion of sample by mixing equal volumes of 0.05 N HCl and 0.025 N H_2SO_4. A total of 20 mL of Aqua regia was added to the Erlenmeyer flask, the mixture was stirred for 20 min using a magnetic stirrer and filtered through a Whatman® No 42-filter paper into a 50 mL volumetric flask and made up to the mark with aqua regia solution. The analytical reagent blanks containing only acids (Aqua Regia) were prepared for baseline calibration and for AAS cleaning between the sample runs. The concentration of heavy Cu was determined by Flame Atomic Absorption Spectrometer (Perkins Elmer model 2380) [23].

Atomic absorption spectroscopy was used to analyze the samples of debris and soil from the landfill site. The set of parameters used during analysis are tabulated in Table 1.

Table 1. AAS (Atomic Absorption Spectroscopy) parameters used for analysis.

Element	Cu	Fe	Pb
Wavelength (nm)	324.8	248.3	283.3
Slit (nm)	0.7	0.2	0.7
Mode	AA	AA	AA
Flame	Air-Ac	Air-Ac	Air-Ac
Burner Head	10 cm	10 cm	10 cm
Lamp	HCL (Hollow Cathode Lamp)	HCL	HCL
Spiked conc. (mg/L)	2.5	2.5	2.5
Read Time (seconds)	3	3	3
Replicates	3	3	3
Air Flow (L/min)	17	17	17
Acetylene flow (L/min)	1.5	1.5	1.5

3. Results

Contents of the sample of HSW generated in Karachi have been separated and classified (Section 2.3 refers) into major classes of substances present in it. Figure 4 depicts the percentage-wise distribution of different substances present in this sample. It is observed that the percentage of organic material is highest followed by metal content. Out of 11% metal content, further separation of different metals is performed and 51% of the total metal waste is determined to be iron, which has considerable recycling worth. Table 2 shows the weight of per day generated waste of selected materials in the household solid waste of Karachi.

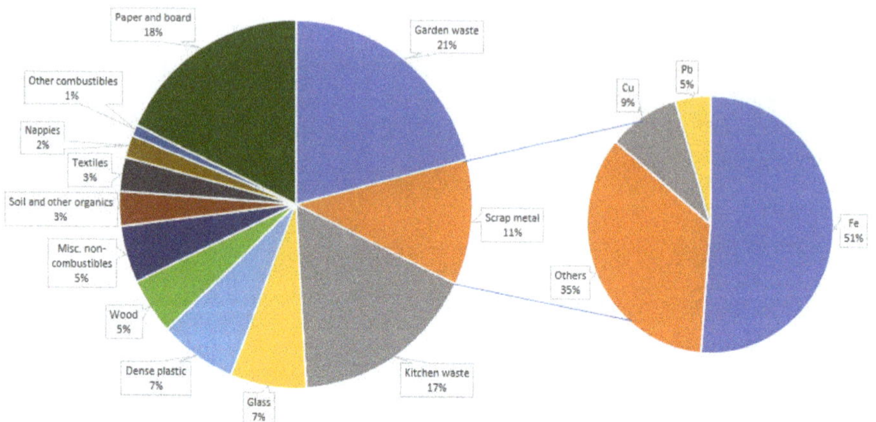

Figure 4. Classification of HSW generated in Karachi.

Table 2. Classification of HSW generated in Karachi.

Class of Material	Composition in HSW kg/Day	Metals	Composition in HSW kg/Day
Garden waste	18,480		
Paper and board	15,840		
Kitchen waste	14,960		
Metal	9680		
		Sub Classification of Metals	Composition in HSW kg/day
		Cu	906
		Fe	4958
		Pb	447
		Others	3369
Glass	6160		
Dense plastic	6160		
Textiles, Cotton, Nappies	4400		
Wood	4420		
Misc. non-combustibles	4390		
Soil and other organics	2640		
Other combustibles	880		

The estimated weight of household solid waste generated in Karachi city is 88,000 kg per day and the amount of copper in this household solid waste is 906 kg/day, the amount of Iron is 4958 kg/day, the amount of lead is 447 kg/day, and other metallic contents are 3369 kg/day.

The major constituents of E-waste i.e., [Cu] were determined to be 199.5 ppm, industrial development indicator i.e., [Fe] were determined to be 428.5 ppm, and the main components of lead-acid storage batteries i.e., [Pb] were determined to be 108.5 ppm, as detailed in Table 3.

Table 3. Concentration of metals in the soil of the Jam-Chakro landfill facility.

Metal	Number of Samples Run (n)	Sample Average Concentration ppm	Standard Deviation within Batch	Standard Error of the Mean	Confidence Interval	95% Confidence Interval Min	95% Confidence Interval Max
Cu	10	199.5	3.140534	0.993124	2.246446433	196.4257	202.5743
Fe	10	428.5	2.64928	0.837776	1.895049162	425.2	431.8
Pb	10	108.5	2.736506	0.865359	1.957442291	105.946	111.054

The amount of metal present in total household solid waste is 11%. The amount of recyclable metals estimated in Table 2 is a considerable amount, such as about 5000 kg/day of iron, about 900 kg/day of lead, 450 kg/day of copper and 3400 kg/day of other metals. These amounts have been converted into their worth per annum and results are very much promising to encourage the industry to take up the recycling of metals from household solid waste as a profitable business (Table 4).

Table 4. Annual worth estimation of Cu, Fe and Pb.

Metal	Per Day Metal Weight kg	Annual Metal Weight kg	Worth of Metal USD per kg	Annual Worth of Metal USD
Cu	906	330,849	6.38	2,109,802.20
Fe	4957	1,809,540	1.21	2,189,700.70
Pb	447	163,257	1.95	318,152.25
Total annual worth of 65.2% HSW metals generated in Karachi				4,617,655.15

The soil of Deh Jam Chakro landfill facility was analyzed, to determine concatenation of heavy metals (Cu, Fe, Pb) and the concentrations were found to be higher than unpolluted soil values, as detailed in Table 4. These metals will not be confined to the limits of soil, but will transfer to underground aquifers through leaching and will find a way to the food chain.

4. Discussion

A major problem being faced by the major cities of the world is municipal solid waste management. There exists a direct relationship between human population size and household solid waste generation. In the case of Karachi, limited resources and inadequate budget allocation make HSW management a strenuous issue. Moreover, ignorance about environmental problems and a lack of civic sense among the masses make HSW management a challenging task. Here, the household waste is collected by the garbage collectors, working on a public–private sharing basis. Due to limited resources, the Karachi Metropolitan Corporation is incapable of collecting household waste from all the houses. In this situation, vacant plots, underneath bridges and similar areas are illegally used as garbage dumping points from where garbage is collected and transported to landfill sites. Due to resource limitations, only 70% of household solid waste is transported to landfill sites, where it is combusted in the open air. Point source segregation of the household solid waste is a very convenient and inexpensive segregation technique, which is further helpful in recycling and the proper management of waste. In the city of the case study, materials worth direct sale value or recycle value are collected by scavengers, however a proper system of segregation of household solid waste does not exist.

The classification of household solid waste according to its composition, revealed that a major component of the waste is organic in nature with a considerable amount of kitchen and garden waste. These materials, if segregated at source, will be saved from hazardous chemical contamination thereby caused after mixing with other substances of chemically or physiologically harmful properties. Thus, they may be re-used at cattle farms (as cattle feed) without any major process involved. Paper and board can be recycled to regenerate paper products and packaging material. Plastics can be recycled to make new plastic products at very low cost.

The scope of the study was restricted to a detailed estimation of only Cu, Fe and Pb in household solid waste. Subsequently, the generation of Cu, Fe and Pb as waste in kg/yr is determined to be 3.3×10^5, 1.8×10^6 and 1.6×10^5 respectively. Detailed results of the analysis are tabulated in Table 2. In August 2020, the market price of copper in the international market is 6.38 USD/kg (United States dollar per kilogram), iron is 1.21 USD/kg and lead is 1.95 USD/kg. Thus, it is estimated that the annual worth of three metals in the household solid waste of Karachi is USD 4.6 M. It is pertinent to mention that these three metals comprise only 65.2% of total metal present in household solid waste generated in Karachi; therefore, segregation and recycling of the remaining 34.8% of metals present in the HSW will increase the worth of recyclable material present in the household solid waste.

During the study of the transportation cycle of HSW in Karachi, observations show that the task of HSW segregation is being performed to some extent by scavengers. These scavengers collect HSW of resale value and can be classified as informal agents of recycling. The results of the study also conclude that the formal and planned establishment of industries for recycling of reusable material present in household solid waste will be profitable in economic terms and will make development sustainable in environmental terms.

Apart from the capital worth of recyclable materials in household solid waste, another very important aspect is sustainable development. As the resources are limited, the development must be futuristically sustainable. This is only possible if the available energy and material resources are recycled and reused, in various applications such as CMOS (Complementary Metal Oxide Semiconductor) technologies [23–25]. These recycled materials will be a crucial base for electronic and thermo-photovoltaic applications [26,27]. If recycling is not given due importance, the available resources will deplete, thus pushing environmental pollution to increase over the years. The rise in population accompanied by increasing waste generation is a threat to the sustainability of life on Earth.

5. Conclusions

Metal resources for extraction, throughout the globe, are limited, but in the form of waste, they are persistent in nature for considerably long periods of time and their recycling is economically worth processing. In this regard, the recycling of household solid waste, in the mentioned case study area, is proven to be profitable when used along with point source segregation system. In this study, the amount of metals determined to be present in waste when translated in terms of worth of recovered metal will be more than USD 4.7 Million per annum. Based on the HSW segregation and classification results, it is estimated that the total annual worth of all the metals discarded in HSW will be around USD 27 Million per year. In developing countries, budget constraints faced by governments restrain investment in environmental protection and recycling projects. This study indicates the presence of a significant opportunity for profitable investment while protecting the environment from global pollution of HSW.

Household solid waste is not given much importance when sources of pollution are considered, but, as proven here, despite being less in volume, HSW is a constant source of pollution and, on an annual scale, the amount of waste produced is massive. A misconception among the masses is that HSW does not cause much harm to the environment due to the domestic nature of its origin. The factual figure is otherwise, as HSW contains several toxic components and elements like lead, which are non-degradable and have the ability to leach into underground water streams, making their way to the human food chain, through animal or plant sources.

Electronic-waste generation is on the rise, and the positive side is that most components in e-waste are recyclable. E-waste is hazardous for animal and plant life; it is sourced from depleting natural resources and can have a disastrous impact if not recycled. There is evidence that e-waste-associated contaminants may be present in some agricultural or manufactured products for export.

In general, the household waste generated is expected to increase with population, and so will the need for efficient recycling, as astronomers and astrophysicists have not yet been able to find any habitable planet in outer space.

Author Contributions: Conceptualization, M.M.S.; Data curation, M.M.S., M.N.N., N.A.H. and M.A.H.; Formal analysis, M.M.S. and M.N.N.; Funding acquisition, Y.A.W.; Investigation, M.N.N. and N.A.H.; Methodology, M.M.S. and M.N.N.; Project administration, M.M.S., M.A.H., O.A. and M.R.J.; Resources, M.M.S., I.A.B., M.A.H., O.A., M.R.J. and S.K.; Software, Y.A.W., I.A.B., Z.Z.C. and M.R.J.; Validation, M.M.S., M.N.N., N.A.H., Z.Z.C. and O.A.; Visualization, M.N.N., Y.A.W., I.A.B., Z.Z.C. and S.K.; Writing—original draft, M.N.N.; Writing—review & editing, M.M.S. and Y.A.W. All authors have read and agreed to the published version of the manuscript.

Funding: This research received no external funding.

Acknowledgments: The authors extend their appreciation to the Deanship of Scientific Research at King Khalid University for funding this work through the research groups program under grant number (R.G.P 2/85/41), and also the University of Malaya for funding this work under grant number RU001-2018 and ST030-2019.

Conflicts of Interest: The authors declare no conflict of interest.

References

1. Abbasi, H.N.; Lu, X.; Zhao, G. An overview of Karachi solid waste disposal sites and environs. *J. Sci. Res. Rep.* **2015**, *6*, 294–303. [CrossRef]
2. Karak, T.; Bhagat, R.; Bhattacharyya, P. Municipal solid waste generation, composition, and management: The world scenario. *Crit. Rev. Environ. Sci. Technol.* **2012**, *42*, 1509–1630. [CrossRef]
3. Shekdar, A.V. Sustainable solid waste management: An integrated approach for Asian countries. *Waste Manag.* **2009**, *29*, 1438–1448. [CrossRef] [PubMed]
4. Batool, S.A.; Ch, M.N. Municipal solid waste management in Lahore city district, Pakistan. *Waste Manag.* **2009**, *29*, 1971–1981. [CrossRef] [PubMed]
5. Azam, M.; Jahromy, S.S.; Raza, W.; Raza, N.; Lee, S.S.; Kim, K.-H.; Winter, F. Status, characterization, and potential utilization of municipal solid waste as renewable energy source: Lahore case study in Pakistan. *Environ. Int.* **2020**, *134*, 105291. [CrossRef] [PubMed]
6. Batool, S.A.; Chuadhry, M.N. The impact of municipal solid waste treatment methods on greenhouse gas emissions in Lahore, Pakistan. *Waste Manag.* **2009**, *29*, 63–69. [CrossRef]
7. Mahmood, K.; Ul-Haq, Z.; Faizi, F.; Tariq, S.; Naeem, M.A.; Rana, A.D. Monitoring open dumping of municipal waste in Gujranwala, Pakistan using a combination of satellite based bio-thermal indicators and GIS analysis. *Ecol. Indic.* **2019**, *107*, 105613. [CrossRef]
8. Korai, M.S.; Mahar, R.B.; Uqaili, M.A. Optimization of waste to energy routes through biochemical and thermochemical treatment options of municipal solid waste in Hyderabad, Pakistan. *Energy Convers. Manag.* **2016**, *124*, 333–343. [CrossRef]
9. Usman, M.; Yasin, H.; Nasir, D.; Mehmood, W. A case study of groundwater contamination due to open dumping of municipal solid waste in Faisalabad, Pakistan. *Earth Sci. Pak.* **2017**, *1*, 15–16. [CrossRef]
10. Ahmed, N.; Zurbrugg, C. *Urban Organic Waste Management in Karachi, Pakistan*; Loughborough University: Loughborough, UK, 2002.
11. Siddiqi, M.M.; Naseer, M.N.; Abdul Wahab, Y.; Hamizi, N.A.; Badruddin, I.A.; Chowdhury, Z.Z.; Akbarzadeh, O.; Johan, M.R.; Khan, T.; Kamangar, S. Evaluation of Municipal Solid Wastes Based Energy Potential in Urban Pakistan. *Processes* **2019**, *7*, 848. [CrossRef]
12. Korai, M.S.; Mahar, R.B.; Uqaili, M.A. The feasibility of municipal solid waste for energy generation and its existing management practices in Pakistan. *Renew Sustain. Energy Rev.* **2017**, *72*, 338–353. [CrossRef]
13. Shahid, M.; Nergis, Y.; Siddiqui, S.A.; Choudhry, A.F. Environmental impact of municipal solid waste in Karachi city. *World Appl. Sci. J.* **2014**, *29*, 1516–1526.
14. Mosler, H.J.; Drescher, S.; Zurbrügg, C.; Rodríguez, T.C.; Miranda, O.G. Formulating waste management strategies based on waste management practices of households in Santiago de Cuba, Cuba. *Habitat Int.* **2006**, *30*, 849–862. [CrossRef]
15. United Nations. *Environmental Indicators: Waste*; United Nations: New York, NY, USA, 2011.
16. Hoornweg, D.; Bhada-Tata, P. *What a Waste: A Global Review of Solid Waste Management*. 2012. Available online: https://openknowledge.worldbank.org/handle/10986/17388?source=post_page (accessed on 21 August 2020).
17. United Nations. *World Population Prospects 2019*; United Nations: New York, NY, USA, 2019.
18. Afroz, R.; Hanaki, K.; Tudin, R. Factors affecting waste generation: A study in a waste management program in Dhaka City, Bangladesh. *Environ. Monit. Assess.* **2011**, *179*, 509–519. [CrossRef] [PubMed]
19. De Feo, G.; De Gisi, S. Public opinion and awareness towards MSW and separate collection programmes: A sociological procedure for selecting areas and citizens with a low level of knowledge. *Waste Manag.* **2010**, *30*, 958–976. [CrossRef] [PubMed]
20. Zhang, H.; Wen, Z.-G. Residents' Household Solid Waste (HSW) Source Separation Activity: A Case Study of Suzhou, China. *Sustainability* **2014**, *6*, 6446–6466. [CrossRef]
21. Times, S. *Projected GDP per Capita Ranking (2015–2020)*. Available online: http://statisticstimes.com/economy/projected-world-gdp-capita-ranking.php (accessed on 21 August 2020).
22. Kaza, S.; Yao, L.; Bhada-Tata, P.; Van Woerden, F. *What a Waste 2.0: A Global Snapshot of Solid Waste Management to 2050*; The World Bank: Washington, DC, USA, 2018.
23. Widmer, R.; Oswald-Krapf, H.; Sinha-Khetriwal, D.; Schnellmann, M.; Böni, H. Global perspectives on e-waste. *Environ. Impact Assess. Rev.* **2005**, *25*, 436–458. [CrossRef]

24. Hagelueken, C.; Meskers, C. Mining our computers-opportunities and challenges to recover scarce and valuable metals from endof-life electronic devices. In *Electronics Goes Green 2008+*; Electronics Goes Green: Stuttgart, Germany, 2008.

25. Puckett, J.; Smith, T. *Exporting Harm: The High-Tech Trashing of Asia The Basel Action Network*; Seattle7 Silicon Valley Toxics Coalition: San Jose, CA, USA, 2002.

26. Clifton, R.; Simmons, J. *The Economist Brands and Branding*; Profile Books: London, UK, 2003; p. 56.

27. Robinson, B.H. E-waste: An assessment of global production and environmental impacts. *Sci. Total Environ.* **2009**, *408*, 183–191. [CrossRef]

28. Agency for Toxic Substances and Disease Registry (ATSDR). *Toxicological Profile for Polycyclic Aromatic Hydrocarbons (PAHs)*; US Department of Health and Human Services, Health Service: Atlanta, GA, USA, 1995.

29. Agency for Toxic Substances and Disease Registry (ATSDR). *Toxicological Profile for Chlorinated Dibenzo-P-Dioxins (CDDs)*; US Department of Health and Human Services, Health Service: Atlanta, GA, USA, 1998.

30. Agency for Toxic Substances and Disease Registry (ATSDR). *Toxicological Profile for Mercury*; US Department of Health and Human Services, Health Service: Atlanta, GA, USA, 1999.

31. Agency for Toxic Substances and Disease Registry (ATSDR). *Toxicological Profile for Polychlorinated Biphenyls (PCBs)*; US Department of Health and Human Services, Health Service: Atlanta, GA, USA, 2000.

32. Agency for Toxic Substances and Disease Registry (ATSDR). *Toxicological Profile for Polybrominated Biphenyls and Polybrominated Diphenyl Ethers*; US Department of Health and Human Services, Public Health Service: Atlanta, GA, USA, 2004.

33. Agency for Toxic Substances and Disease Registry (ATSDR). *Toxicological Profile for Lead*; US Department of Health and Human Services, Health Service: Atlanta, GA, USA, 2007.

34. Agency for Toxic Substances and Disease Registry (ATSDR). *Toxicological Profile for Cadmium*; US Department of Health and Human Services, Health Service: Atlanta, GA, USA, 2012.

35. Agency for Toxic Substances and Disease Registry (ATSDR). *Toxicological Profile for Beryllium*; US Department of Health and Human Services, Health Service: Atlanta, GA, USA, 2002.

36. Agency for Toxic Substances and Disease Registry (ATSDR). *Toxicological Profile for Zinc*; US Department of Health and Human Services, Health Service: Atlanta, GA, USA, 2005.

37. Agency for Toxic Substances and Disease Registry (ATSDR). *Toxicological Profile for Nickel*; US Department of Health and Human Services, Health Service: Atlanta, GA, USA, 2005.

38. Agency for Toxic Substances and Disease Registry (ATSDR). *Toxicological Profile for Barium*; US Department of Health and Human Services, Health Service: Atlanta, GA, USA, 2007.

39. Agency for Toxic Substances and Disease Registry (ATSDR). *Toxicoloigcal Profile for Chromium*; US Department of Health and Human Services, Health Service: Atlanta, GA, USA, 2012.

40. Scharnhorst, W.; Althaus, H.-J.; Classen, M.; Jolliet, O.; Hilty, L.M. The end of life treatment of second generation mobile phone networks: Strategies to reduce the environmental impact. *Environ. Impact. Assess. Rev.* **2005**, *25*, 540–566. [CrossRef]

41. Rousis, K.; Moustakas, K.; Malamis, S.; Papadopoulos, A.; Loizidou, M. Multi-criteria analysis for the determination of the best WEEE management scenario in Cyprus. *Waste Manag.* **2008**, *28*, 1941–1954. [CrossRef]

42. Schnoor, J.L. Extended producer responsibility for e-waste. *Environ. Sci. Technol.* **2012**, *46*, 7927. [CrossRef] [PubMed]

43. Mobin, S.M.; Azmat, R. Impact of Improper Household Solid Waste Management on Environment: A Case Study of Karachi City, Pakistan. *Asian J. Chem.* **2015**, *27*, 4523. [CrossRef]

44. Wahab, Y.A.; Fadzil, A.; Soin, N.; Fatmadiana, S.; Chowdhury, Z.Z.; Hamizi, N.A.; Pivehzhani, O.A.; Sabapathy, T.; Al-Douri, Y. Uniformity improvement by integrated electrochemical-plating process for CMOS logic technologies. *J. Manuf. Process.* **2019**, *38*, 422–431. [CrossRef]

45. Sagadevan, S.; Venilla, S.; Marlinda, A.R.; Johan, M.R.; Wahab, Y.A.; Zakaria, R.; Umar, A.; Hegazy, H.H.; Algarni, H.; Ahmad, N. Effect of Synthesis Temperature on the Morphologies, Optical and Electrical Properties of MgO Nanostructures. *J. NanoSci. Nanotechnol.* **2020**, *20*, 2488–2494. [CrossRef] [PubMed]

46. Wahab, Y.A.; Fatmadiana, S.; Naseer, M.N.; Johan, M.R.; Hamizi, N.A.; Sagadevan, S.; Akbarzadeh, O.; Chowdhury, Z.Z.; Sabapathy, T.; Al Douri, Y. Metal oxides powder technology in dielectric materials. In *Metal Oxide Powder Technologies*; Elsevier: Amsterdam, The Netherlands, 2020; pp. 385–399.

47. Chowdhury, Z.Z.; Krishnan, B.; Sagadevan, S.; Rafique, R.F.; Hamizi, N.A.B.; Abdul Wahab, Y.; Khan, A.A.; Johan, R.B.; Al-Douri, Y.; Kazi, S.N. Effect of temperature on the physical, electro-chemical and adsorption properties of carbon micro-spheres using hydrothermal carbonization process. *Nanomaterials* **2018**, *8*, 597. [CrossRef] [PubMed]
48. Wahab, Y.A.; Soin, N.; Naseer, M.N.; Hussin, H.; Osman, R.A.M.; Johan, M.R.; Hamizi, N.A.; Pivehzhani, O.A.; Chowdhury, Z.Z.; Sagadevan, S. Junction engineering in two-stepped recessed SiGe MOSFETs for high performance application. In Proceedings of the AIP Conference Proceedings, Putrajaya, Malaysia, 22 August 2019; p. 020033.

processes  

Communication

Kinetic Analysis of Algae Gasification by Distributed Activation Energy Model

Guozhao Ji [1,2], Abdul Raheem [3], Xin Wang [1], Weng Fu [4], Boyu Qu [1], Yuan Gao [1,2], Aimin Li [1,2], Ming Zhao [3], Weiguo Dong [5,*] and Zhien Zhang [6,*]

[1]   School of Environmental Science and Technology, Dalian University of Technology, Dalian 116024, China; guozhaoji@dlut.edu.cn (G.J.); wangxin87@dlut.edu.cn (X.W.); qby_@mail.dlut.edu.cn (B.Q.); gaoyuan1988@dlut.edu.cn (Y.G.); leeam@dlut.edu.cn (A.L.)

[2]   Key Laboratory of Industrial Ecology and Environmental Engineering, Ministry of Education, Dalian 116024, China

[3]   School of Environment, Tsinghua University, Beijing 100084, China; channa.raheem@gmail.com (A.R.); ming.zhao@tsinghua.edu.cn (M.Z.)

[4]   School of Chemical Engineering, The University of Queensland, St Lucia 4072, QLD, Australia; w.fu1@uq.edu.au

[5]   School of Management, China University of Mining and Technology (Beijing), Beijing 100083, China

[6]   William G. Lowrie Department of Chemical and Biomolecular Engineering, The Ohio State University, Columbus, OH 43210, USA

*   Correspondence: dwg_gasification@foxmail.com (W.D.); zhang.4528@osu.edu (Z.Z.)

Received: 30 June 2020; Accepted: 21 July 2020; Published: 2 August 2020

Abstract: Conversion of algal biomass into energy products via gasification has attracted increasing research interests. A basic understanding of the gasification kinetics of algal biomass is of fundamental importance. Distributed activation energy model (DAEM), which provides the information of energy barrier distribution during the gasification process, is a promising tool to study the kinetic process of algae gasification. In this study, DAEM model was used to investigate *Chlorella vulgaris* and *Spirulina* gasification. The activation energy of *Chlorella vulgaris* gasification was in the range from 370 to 650 kJ mol^{-1}. The range of activation energy for *Spirulina* gasification was a bit wider, spanning from 330 to 670 kJ mol^{-1}. The distribution of activation energy for both *Chlorella vulgaris* and *Spirulina* showed that 500 kJ mol^{-1} had the most components, and these components were gasified at around 300 °C. The DAEM algorithm was validated by the conversion and conversion rate from experimental measurement, demonstrating that DAEM is accurate to describe the kinetics of algal biomass gasification.

Keywords: algal biomass; gasification; kinetics; activation energy distribution

1. Introduction

The dependency of fossil fuels in recent centuries has significantly contributed to the carbon emission and the consequent global warming. Climate change, together with the declining reservation of fossil fuels, is urging the environment and energy research community to seek alternative energy sources that are renewable and have lower environmental impact. Among possible options, biomass, which could be thermally converted into various bio-fuels, for instance oil and syngas via pyrolysis and gasification, have attracted great research attention, not only because of the renewability but also due to the carbon neutrality throughout its life cycle.

Biomass could be generally divided into three generations. The first generation of biomass includes those from terrestrial plants such as rice, potato, corn, bean, wheat, maize, oil palm, sugarcane and food wastes [1]. However, extensive conversion of first-generation biomass to bio-fuel might endanger our food stock security [2]. Energy production at the expense of food supply should never be encouraged. The majority of second-generation bio-resources are non-food waste and lingo-cellulosic materials such as grass, husk, wood, municipal solid waste and sewage sludge [3]. These feedstocks do not sacrifice the food supplies. In addition, the bio-fuel yield from this type of biomass is generally higher than that from the first generation of biomass [1]. In spite of these advantages, they still have some barriers to reaching commercial-scale conversion, such as the transportation, collection networks and the cost-effective pre-treatment. The third generation is algal biomass, which could avoid the deficiencies mentioned above for the first- and second-generation biomass. Cultivating the third-generation biomass does not require cultivatable land. They do not compete with traditional food crops either [4]. Compared to normal plants, microalgae show exceptionally rapid growth rates and exceptional photosynthetic efficiency [5]. Microalgae can be grown in ponds or photo-bioreactors with nutrients [6] or wastewater supply [7–9], which is another advantage over the first- and second-generation biomass that heavily rely on agricultural resources such as fertile soil and fertilizer.

In view of the advantages of algal biomass over other types of biomass, converting algal biomass into value-added energy products has attracted increasing research attention. A great number of studies has successfully converted algal biomass to syngas, hydrocarbon oils and biochar via some common thermochemical techniques such as liquefaction [10], torrefaction [11], pyrolysis [12] and gasification [13]. Among them, producing hydrogen-rich syngas by gasification is of great interest in the energy and environmental community. Díaz-Rey et al. gasified *Scenedesmus almeriensis* algae with a Ni-based catalyst below 700 °C and produced hydrogen-rich syngas, which has a calorific value of about 25 MJ Nm^{-3} [14]. Onwudili et al. conducted hydrothermal gasification of *Chlorella vulgaris*, *Spirulina platensis* and *Saccharina latissima* at 500 °C and 36 MPa, and the hydrogen yield was more than 10 mol kg^{-1}-algae [15]. Duman et al. employed steam gasification of algae and obtained almost 45 mol hydrogen from per gram *Fucus serratus* [16].

Gasification of algal biomass is an endothermic process that requires considerable thermal energy input. Biomass generally consists of large molecules including cellulose, hemicellulose and lignin. The bonds in those large molecules need to be ruptured before converting to oil or gases, and there is always some energy barrier to activate the bond breakage. The energy barrier called activation energy is a good indicator of energy requirement for a process to occur. Because of the complex nature of real biomass, the value of activation energy generally spreads in a large range instead of being a fixed number [17]. Therefore, studying the activation energy distribution is essential in many aspects for algal biomass gasification, such as evaluating the conversion efficiency and assessing the effectiveness of the catalyst [18]. This study will apply distributed activation energy model (DAEM) to inspect the distribution of activation energy for *Chlorella vulgaris* and *Spirulina* gasification.

2. Method

2.1. Experimental Test

Chlorella vulgaris and *Spirulina* powders, purchased from Xi'an Snooker Biotech Co., Ltd., China, were selected as algal biomass samples in this study. The characteristics of the *Chlorella vulgaris* and *Spirulina* including bio-chemical composition, proximate analysis and ultimate analysis could be found in our previous study [19]. The gasification rate was measured in the TA-SDT-Q600 thermo-gravimetric analysis (TGA) instrument. In each test, around 2–3 mg sample in a fine powder state was loaded in the TGA crucible. The temperature was raised from 30 °C to 600 °C with different heating rates of 10, 20 and 30 °C min^{-1}. The conversion of biomass gasification was defined as

$$\alpha = \frac{m - m_0}{m_\infty - m_0} \tag{1}$$

where m is the mass of sample during gasification, m_0 is the original mass before the gasification and m_∞ is the final mass after gasification. The gasifying agent was composed oxygen and argon with a volumetric ratio of 20%:80%, and the flow rate of gas gasifying agent was maintained at 500 mL min^{-1}. More experimental details about the kinetic test were described in our previous publication [20].

2.2. Model Development

The early development of distributed activation energy model was to study the activation energy for coal pyrolysis [21,22]. However, the method to obtain the activation energy distribution is universally applicable to thermal treatment of other complex organics. The basic assumption for algae gasification is that the algae is composed of a series of constituents that have different activation energies.

If the total volatile of algae is denoted as V^*, the constituent of which activation energy equals E_i is V_i^*.

$$V_i^* = V^* f(E_i) \Delta E \tag{2}$$

where $f(E_i)$ is the activation energy distribution. According to Arrhenius law and the hypothesis of first order law, the decomposition rate of this constituent is

$$\frac{\mathrm{d}(V_i/V_i^*)}{\mathrm{d}t} = k_i \left(\frac{V_i^* - V_i}{V_i^*} \right) = A_i \exp\left(-\frac{E_i}{RT}\right)\left(\frac{V_i^* - V_i}{V_i^*} \right) \tag{3}$$

where V_i is the decomposed volatile in V_i^*, T is temperature and R is gas constant. Upon integration of Equation (3), we get

$$V_i = V_i^* \left\{ 1 - \exp\left[-\int_0^t A_i \exp\left(-\frac{E_i}{RT}\right) \mathrm{d}t \right] \right\} \tag{4}$$

Summation of all the constituents together leads to

$$V = V^* - \sum_{i=1}^{n} \exp\left[-\int_0^t A_i \exp\left(-\frac{E_i}{RT}\right) \mathrm{d}t \right] V^* f(E_i) \Delta E \tag{5}$$

Since the total conversion $\alpha = V/V^*$, Equation (4) is rearranged to

$$1-\alpha = \sum_{i=1}^{n} \exp\left[-\int_0^t A_i \exp\left(-\frac{E_i}{RT}\right) \mathrm{d}t \right] f(E_i) \Delta E \tag{6}$$

If the temperature is raised at a constant rate β, Equation (6) is then rewritten as

$$1-\alpha = \sum_{i=1}^{n} \exp\left[-\frac{A_i}{\beta} \int_{T_0}^{T} \exp\left(-\frac{E_i}{RT}\right) \mathrm{d}T \right] f(E_i) \Delta E \tag{7}$$

For simplicity, a function $\phi(E,T)$, which means the unconverted fraction of group with activation energy E at temperature T, is defined as

$$\phi(E, T) = \exp\left[-\int_0^T \frac{A}{\beta} \exp\left(-\frac{E}{RT}\right) \mathrm{d}T \right] \tag{8}$$

By converting the continuous temperature data to discretization form, Equation (7) is then transformed to matrix form

$$
\begin{bmatrix} 1\text{-}\alpha_1 \\ 1\text{-}\alpha_2 \\ \cdot \\ 1\text{-}\alpha_j \\ \cdot \\ 1\text{-}\alpha_m \end{bmatrix} = \begin{bmatrix} \phi(E_1,T_1) & \phi(E_2,T_1) & \cdots & \phi(E_n,T_1) \\ \phi(E_1,T_2) & \phi(E_2,T_2) & & \cdot \\ \cdot & & \cdot & \cdot \\ \cdot & & & \cdot \\ \cdot & & \cdot & \cdot \\ \phi(E_1,T_m) & \cdot & \cdots & \phi(E_n,T_m) \end{bmatrix} \begin{bmatrix} f(E_1)\Delta E \\ f(E_2)\Delta E \\ \cdot \\ \cdot \\ \cdot \\ f(E_n)\Delta E \end{bmatrix} \tag{9}
$$

If the activation energy and temperature are divided into the same number of divisions ($m = n$), the reverse matrix of $[\phi(E,T)]$ could be mathematically obtained as $[\phi(E,T)]^{-1}$. Theoretically, by multiplying $[\phi(E,T)]^{-1}$ on both sides of Equation (9), the distribution of activation energy could be expressed as Equation (10)

$$
[\phi(E,T)]^{-1}[1\text{-}\alpha] = [f(E)\Delta E] \tag{10}
$$

However, a number of studies pointed out that $\phi(E,T)$ is almost a step function to E, which means progressing from 0 to 1 sharply, so $[\phi(E,T)]$ is nearly a singular matrix. As a consequence, $[\phi(E,T)]^{-1}$ could not be calculated, thus obtaining distribution of activation energy by Equation (10) is mathematically impossible. However, if the distribution of activation energy is already known, calculating the algae conversion by the matrix method is very convenient.

The actual activation energy distribution is continuous rather than discrete, and the integration form of Equation (7) is

$$
1\text{-}\alpha = \int_0^\infty \phi(E,T)f(E)dE \tag{11}
$$

Differentiating Equation (11) with respect to E, together with the boundary conditions $f(0) = f(\infty) = 0$, $\phi(0,T) = 0$ and $\phi(\infty,T) = 1$, we get

$$
\frac{d\alpha}{dE} = -\frac{d}{dE}\int_0^\infty \phi(E,T)f(E)dE = -\int_0^0 \phi(E,T)d[f(E)] - \int_1^0 f(E)d[\phi(E,T)] = f(E) \tag{12}
$$

Thus, the distribution of activation energy could be obtained from Equation (12).

After obtaining the distribution of activation energy, it is used to calculate the conversion by Equation (9), and the conversion rate could be obtained subsequently. To validate the model, the conversion rates computed from DAEM model and measured from the experiment are compared. The root-mean-square error is defined as

$$
RMSE = \sqrt{\frac{\sum_{i=1}^{n}\left(\left(\frac{d\alpha}{dt}\right)_{DAEM} - \left(\frac{d\alpha}{dt}\right)_{EXP}\right)^2}{n}} \tag{13}
$$

3. Results and Analysis

3.1. Thermo-Gravimetric Analysis

The profiles of thermogravimetry (TG) and differential thermogravimetry (DTG) data during the gasification are shown in Figure 1. The TG of *Chlorella vulgaris* was almost stable when the temperature was above 500 °C, down to the minimum of ~10 wt.%. The TG of *Spirulina* after 500 °C also hardly presented any mass variation and stabilized around 14 wt.%. The final values of TG were in good agreement with the ash and fixed carbon values in proximate analysis [20]. The DTG profiles showed three major mass loss processes at around 300, 420 and 450 °C for *Chlorella vulgaris*. For *Spirulina* gasification, there were two major peaks at ~300 and 450 °C and one minor peak at ~420 °C. From the perspective of distributed activation energy, the phenomenon that volatile components are decomposed or gasified at different temperatures was due to the different activation energies. The volatiles gasified at higher temperature generally have higher energy barriers to be activated than those gasified at lower temperature [23].

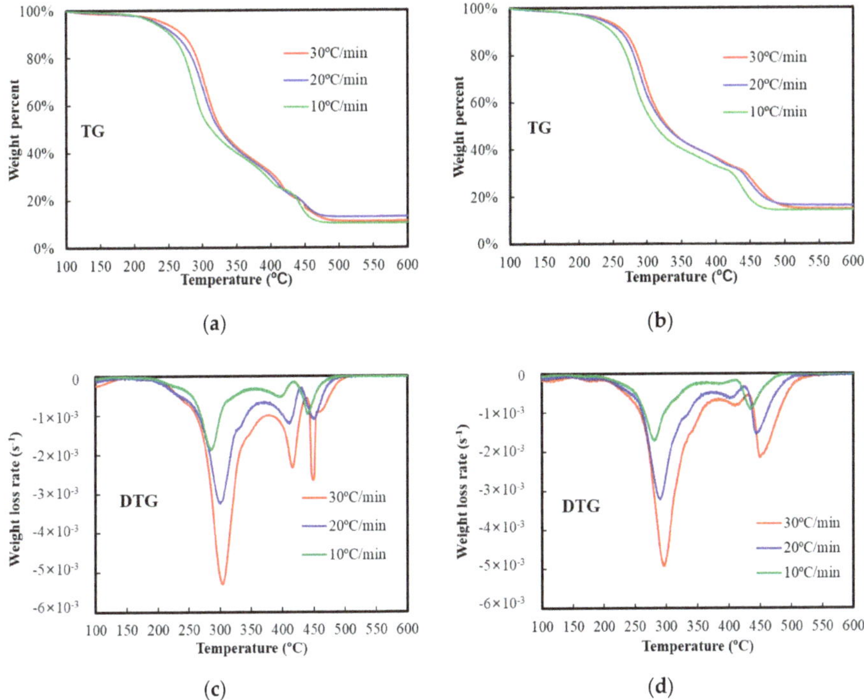

Figure 1. Thermogravimetry (TG) and differential thermogravimetry (DTG) profiles of *Chlorella vulgaris* and *Spirulina* gasification. (**a**) TG of *Chlorella vulgaris*; (**b**) TG of *Spirulina*; (**c**) DTG of *Chlorella vulgaris*; (**d**) DTG of *Spirulina*.

3.2. The Distribution of Activation Energy

With the thermogravimetry data and the DAEM model developed above, the distributions of activation energy for *Chlorella vulgaris* and *Spirulina* gasification could be obtained. As displayed in Figure 2, the activation energy of *Chlorella vulgaris* gasification covered the range from 370 to 650 kJ mol^{-1}. The activation energy range for *Spirulina* gasification was a bit wider, spanning from 330 to 670 kJ mol^{-1}. Both *Chlorella vulgaris* and *Spirulina* had a significant number of constituents with activation energy around 500 kJ mol^{-1}. Owing to the lower activation energy, this part of algae was relatively easier to be gasified, and the gasification of this part corresponded to the first major peak in DTG. The range of activation energy presented in Figure 2 covered the main activation energy values for decomposition of *Chlorella vulgaris* (208–546 kJ mol^{-1}) [24] and *Chlorella variabilis* (133–876 kJ mol^{-1}) [25] but was slightly higher than the activation energy values reported in thermal treatment of other algae such as *Cladophora glomerata* (177 kJ mol^{-1}) [26], *Chlorococcum humicola* (140–240 kJ mol^{-1}) [27] and *Kappaphycus alvarezii* (61–312 kJ mol^{-1}) [28]. Since the feedstocks are not the same, the difference was likely due to the variations of compound types and contents. In addition, the different algorithms in DAEM model and iso-conversional model-free method also lead to the difference in activation energy.

In the iso-conversional method assuming a reaction model is not necessary, but in DAEM the reaction model is under the order-law frame [18].

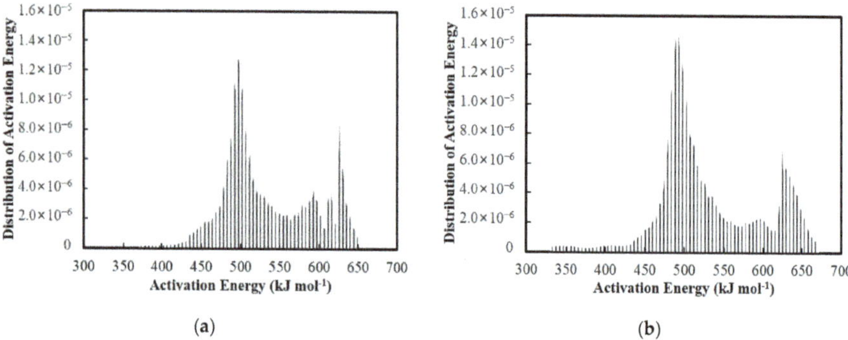

Figure 2. The distributed activation energy of (**a**) *Chlorella vulgaris* and (**b**) *Spirulina* gasification.

The upper bound of *Spirulina* gasification was a bit higher than *Chlorella vulgaris* gasification, indicating that *Spirulina* had some constituents that were difficult to decompose. As observed in the TG and DTG profiles (Figure 1), the gasification of *Chlorella vulgaris* completed before 500 °C, but there was still ongoing gasification above 500 °C for *Spirulina*. This was highly likely due to the constituents with activation energy higher than 650 kJ mol^{-1} in Figure 2b. According to our previous study, *Spirulina* has 60.23% protein, and *Chlorella vulgaris* has 51.51% protein [19]. The higher protein content in *Spirulina* might be owing to its higher activation energy.

3.3. DAEM Model Validation

To evaluate the validity of the derived distribution of activation energy, the conversion calculated from DAEM model should be compared with the experimental results. Once we have the distribution of activation energy, the conversion of the gasification process can be computed from the activation energy distribution according to Equation (9) and then compared with the conversion measured by TGA. The comparison between experimental conversion and modeling data is displayed in Figure 3. The conversions match each other well with marginal deviations only for the case of 10 °C min^{-1}, indicating the reliability of activation energy distribution derived above.

Another pathway to validate the activation energy distribution is to examine the conversion rate. The conversion rate is actually the first-order derivative of conversion with respect to time. After performing the differentiation of α to t, the conversion rates are further compared in Figure 4, and the DAEM conversion rates also closely fit the experimental conversion rates. Figure 4 showed some insignificant difference. The root-mean-square errors are only 3.04×10^{-4} (10 °C min^{-1}), 2.44×10^{-4} (20 °C min^{-1}) and 4.17×10^{-4} (30 °C min^{-1}) for *Chlorella vulgaris*. For *Spirulina*, the values are 3.17×10^{-4} (10 °C min^{-1}), 2.04×10^{-4} (20 °C min^{-1}) and 3.09×10^{-4} (30 °C min^{-1}). The insignificant difference between experimental and DAEM model values might arise from the neglect of thermal-lag effect due to the mismatch of temperature values. The peaks of experimental conversion rates did not appear at the same temperature, and it seemed that the peaks were delayed at higher heating rates. This phenomenon is generally because the heat transfer inside the biomass particle is not sufficiently quick. In DAEM model, the thermal-lag effect was not considered, thus almost all the conversion rate peaks arrived at the same temperature. As a consequence, the conversion rate from DAEM model cannot perfectly fit all the conversion rates from the three heating rates.

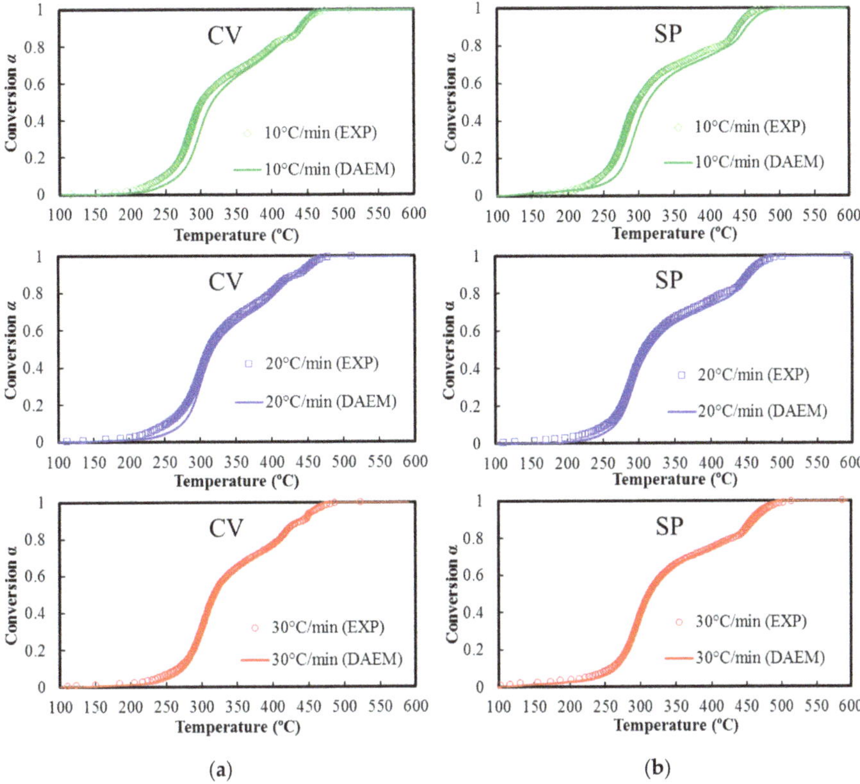

Figure 3. Comparison of conversion value between experimental measurement and DAEM model. (**a**) *Chlorella vulgaris*; (**b**) *Spirulina*. EXP: experimental data; DAEM: model data from DAEM model.

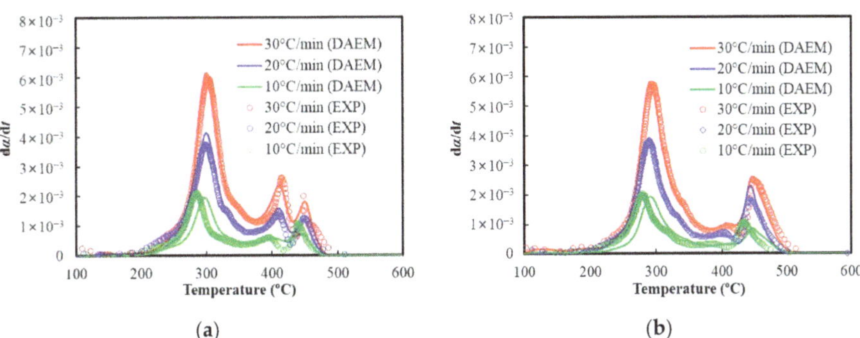

Figure 4. Comparison of conversion rate between experimental measurement and DAEM model. (**a**) *Chlorella vulgaris*; (**b**) *Spirulina*.

4. Conclusions

Distributed activation energy model (DAEM) was applied to study the gasification of *Chlorella vulgaris* and *Spirulina*. Both *Chlorella vulgaris* and *Spirulina* showed complex behaviors during gasification, as the DTG profiles presented multiple major peaks at different temperature ranges. By using the DAEM model, it was found that both *Chlorella vulgaris* and *Spirulina* had a significant number of components with activation energy close to 500 kJ mol^{-1}, and this part of components was gasified around 300 °C. *Spirulina* had more components with high activation energy, which explained the ongoing gasification above 500 °C. The conversion and conversion rate from DAEM model could accurately reproduce the experimental conversion and conversion rate, demonstrating the validity of applying DAEM to algal biomass gasification.

Author Contributions: Conceptualization, G.J. and Z.Z.; Methodology, G.J., A.R. and W.D.; Validation, G.J., B.Q. and M.Z.; Formal Analysis, X.W. and Y.G.; Investigation, G.J. and Z.Z.; Resources, Z.Z.; Writing-Original Draft Preparation, G.J. and Z.Z.; Writing-Review & Editing, W.F. and W.D.; Visualization, G.J.; Supervision, A.L.; Funding Acquisition, G.J. and Z.Z. All authors have read and agreed to the published version of the manuscript.

Funding: This research was funded by [Open Foundation of Key Laboratory of Industrial Ecology and Environmental Engineering, Ministry of Education] grant number [KLIEEE-19-01] and [the Fundamental Research Funds for the Central Universities] grant number [DUT18RC(3)036].

Conflicts of Interest: The authors declare no conflict of interest.

Nomenclature

A	Pre-exponential factor
E	Activation energy
$f(E)$	Distribution of activation energy
m	Mass of algal biomass
m_0	Initial mass of algal biomass
m_∞	Final mass of algal biomass
R	Gas constant
t	Time
T	Temperature
T_0	Initial temperature
V	Quantity of gasified biomass
V^*	Total convertible biomass

Greek letters

α	Biomass conversion
β	Heating rate
ϕ	The unconverted fraction of biomass

References

1. Vassilev, S.V.; Vassileva, C.G. Composition, properties and challenges of algae biomass for biofuel application: An overview. *Fuel* **2016**, *181*, 1–33. [CrossRef]

2. Guo, M.; Song, W.; Buhain, J. Bioenergy and biofuels: History, status, and perspective. *Renew. Sustain. Energy Rev.* **2015**, *42*, 712–725. [CrossRef]

3. Rodionova, M.V.; Poudyal, R.S.; Tiwari, I.; Voloshin, R.A.; Zharmukhamedov, S.K.; Nam, H.G.; Zayadan, B.K.; Bruce, B.D.; Hou, H.J.M.; Allakhverdiev, S.I. Biofuel production: Challenges and opportunities. *Int. J. Hydrogen Energy* **2017**, *42*, 8450–8461. [CrossRef]

4. Laurens, L.M.L.; Markham, J.; Templeton, D.W.; Christensen, E.D.; Van Wychen, S.; Vadelius, E.W.; Chen-Glasser, M.; Dong, T.; Davis, R.; Pienkos, P.T. Development of algae biorefinery concepts for biofuels and bioproducts; a perspective on process-compatible products and their impact on cost-reduction. *Energy Environ. Sci.* **2017**, *10*, 1716–1738. [CrossRef]

5. Zhang, Q.; Hong, Y. Comparison of growth and lipid accumulation properties of two oleaginous microalgae under different nutrient conditions. *Front. Environ. Sci. Eng.* **2014**, *8*, 703–709. [CrossRef]

6. Chang, H.; Quan, X.; Zhong, N.; Zhang, Z.; Lu, C.; Li, G.; Cheng, Z.; Yang, L. High-efficiency nutrients reclamation from landfill leachate by microalgae Chlorella vulgaris in membrane photobioreactor for bio-lipid production. *Bioresour. Technol.* **2018**, *266*, 374–381. [CrossRef]

7. Wang, X.; Lin, L.; Lu, H.; Liu, Z.; Duan, N.; Dong, T.; Xiao, H.; Li, B.; Xu, P. Microalgae cultivation and culture medium recycling by a two-stage cultivation system. *Front. Environ. Sci. Eng.* **2018**, *12*, 14. [CrossRef]

8. Gopalakrishnan, K.; Roostaei, J.; Zhang, Y. Mixed culture of Chlorella sp. and wastewater wild algae for enhanced biomass and lipid accumulation in artificial wastewater medium. *Front. Environ. Sci. Eng.* **2018**, *12*, 14. [CrossRef]

9. Zhan, J.; Zhang, Q.; Qin, M.; Hong, Y. Selection and characterization of eight freshwater green algae strains for synchronous water purification and lipid production. *Front. Environ. Sci. Eng.* **2016**, *10*, 548–558. [CrossRef]

10. Zhou, D.; Zhang, L.; Zhang, S.; Fu, H.; Chen, J. Hydrothermal Liquefaction of Macroalgae Enteromorpha prolifera to Bio-oil. *Energy Fuels* **2010**, *24*, 4054–4061. [CrossRef]

11. Yu, K.L.; Lau, B.F.; Show, P.L.; Ong, H.C.; Ling, T.C.; Chen, W.-H.; Ng, E.P.; Chang, J.-S. Recent developments on algal biochar production and characterization. *Bioresour. Technol.* **2017**, *246*, 2–11. [CrossRef] [PubMed]

12. Chaiwong, K.; Kiatsiriroat, T.; Vorayos, N.; Thararax, C. Study of bio-oil and bio-char production from algae by slow pyrolysis. *Biomass Bioenergy* **2013**, *56*, 600–606. [CrossRef]

13. Watanabe, H.; Li, D.; Nakagawa, Y.; Tomishige, K.; Watanabe, M.M. Catalytic gasification of oil-extracted residue biomass of Botryococcus braunii. *Bioresour. Technol.* **2015**, *191*, 452–459. [CrossRef] [PubMed]

14. Díaz-Rey, M.R.; Cortés-Reyes, M.; Herrera, C.; Larrubia, M.A.; Amadeo, N.; Laborde, M.; Alemany, L.J. Hydrogen-rich gas production from algae-biomass by low temperature catalytic gasification. *Catal. Today* **2015**, *257*, 177–184. [CrossRef]

15. Onwudili, J.A.; Lea-Langton, A.R.; Ross, A.B.; Williams, P.T. Catalytic hydrothermal gasification of algae for hydrogen production: Composition of reaction products and potential for nutrient recycling. *Bioresour. Technol.* **2013**, *127*, 72–80. [CrossRef]

16. Duman, G.; Uddin, M.A.; Yanik, J. Hydrogen production from algal biomass via steam gasification. *Bioresour. Technol.* **2014**, *166*, 24–30. [CrossRef]

17. Yang, H.; Ji, G.; Clough, P.T.; Xu, X.; Zhao, M. Kinetics of catalytic biomass pyrolysis using Ni-based functional materials. *Fuel Process. Technol.* **2019**, *195*, 106145. [CrossRef]

18. Cai, J.; Wu, W.; Liu, R. An overview of distributed activation energy model and its application in the pyrolysis of lignocellulosic biomass. *Renew. Sustain. Energy Rev.* **2014**, *36*, 236–246. [CrossRef]

19. Raheem, A.; Liu, H.; Ji, G.; Zhao, M. Gasification of lipid-extracted microalgae biomass promoted by waste eggshell as CaO catalyst. *Algal Res.* **2019**, *42*, 101601. [CrossRef]

20. Zhao, M.; Raheem, A.; Memon, Z.M.; Vuppaladadiyam, A.K.; Ji, G. Iso-conversional kinetics of low-lipid micro-algae gasification by air. *J. Clean. Prod.* **2019**, *207*, 618–629. [CrossRef]

21. Miura, K. A new and simple method to estimate f (E) and k0 (E) in the distributed activation energy model from three sets of experimental data. *Energy Fuels* **1995**, *9*, 302–307. [CrossRef]

22. Miura, K.; Maki, T. A Simple Method for Estimating f(E) and k0(E) in the Distributed Activation Energy Model. *Energy Fuels* **1998**, *12*, 864–869. [CrossRef]

23. Vyazovkin, S. *Isoconversional Kinetics of Thermally Stimulated Processes*; Springer: Berlin/Heidelberg, Germany, 2014.

24. Chen, C.; Ma, X.; He, Y. Co-pyrolysis characteristics of microalgae Chlorella vulgaris and coal through TGA. *Bioresour. Technol.* **2012**, *117*, 264–273. [CrossRef] [PubMed]

25. Maurya, R.; Ghosh, T.; Saravaia, H.; Paliwal, C.; Ghosh, A.; Mishra, S. Non-isothermal pyrolysis of de-oiled microalgal biomass: Kinetics and evolved gas analysis. *Bioresour. Technol.* **2016**, *221*, 251–261. [CrossRef]

26. Plis, A.; Lasek, J.; Skawińska, A.; Zuwała, J. Thermochemical and kinetic analysis of the pyrolysis process in Cladophora glomerata algae. *J. Anal. Appl. Pyrolysis* **2015**, *115*, 166–174. [CrossRef]

27. Kirtania, K.; Bhattacharya, S. Application of the distributed activation energy model to the kinetic study of pyrolysis of the fresh water algae Chlorococcum humicola. *Bioresour. Technol.* **2012**, *107*, 476–481. [CrossRef]

28. Das, P.; Mondal, D.; Maiti, S. Thermochemical conversion pathways of Kappaphycus alvarezii granules through study of kinetic models. *Bioresour. Technol.* **2017**, *234*, 233–242. [CrossRef]

Article

Is Recycling Always the Best Option? Environmental Assessment of Recycling of Seashell as Aggregates in Noise Barriers

Begoña Peceño [1], Carlos Leiva [2], Bernabé Alonso-Fariñas [2,*] and Alejandro Gallego-Schmid [3,*]

[1] Facultad de Ciencias del Mar, Escuela de Prevención de Riesgos y Medioambiente, Universidad Católica del Norte, Larrondo 1281, Coquimbo 1780000, Chile; begopc@ucn.cl

[2] Departamento de Ingeniería Química y Ambiental, Escuela Técnica Superior de Ingeniería, Universidad de Sevilla, Camino de los Descubrimientos s/n, 41092 Seville, Spain; cleiva@us.es

[3] Tyndall Centre for Climate Change, School of Mechanical, Aerospace and Civil Engineering, The University of Manchester, Pariser Building, Sackville Street, Manchester M13 9PL, UK

* Correspondence: bernabeaf@us.es (B.A.-F.); alejandro.gallegoschmid@manchester.ac.uk (A.G.-S.)

Received: 5 June 2020; Accepted: 28 June 2020; Published: 2 July 2020

Abstract: Waste recycling is an essential part of waste management. The concrete industry allows the use of large quantities of waste as a substitute for a conventional raw material without sacrificing the technical properties of the product. From a circular economy point of view, this is an excellent opportunity for waste recycling. Nevertheless, in some cases, the recycling process can be undesirable because it does not involve a net saving in resource consumption or other environmental impacts when compared to the conventional production process. In this study, the environmental performance of conventional absorption porous barriers, composed of 86 wt % of natural aggregates and 14 wt % cement, was compared with barriers composed of 80 wt % seashell waste and 20 wt % cement through an attributional cradle-to-grave life cycle assessment. The results show that, for the 11 environmental impact categories considered, the substitution of the natural aggregates with seashell waste involves higher environmental impacts, between 32% and 267%. These results are justified by the high contribution to these impacts of the seashell waste pre-treatment and the higher cement consumption. Therefore, the recycling of seashells in noise barrier manufacturing is not justified from an environmental standpoint with the current conditions. In this sense, it could be concluded that life cycle assessments should be carried out simultaneously with the technical development of the recycling process to ensure a sustainable solution.

Keywords: life cycle assessment; circular economy; environmental sustainability; mollusk shell; porous concrete; construction

---

1. Introduction

The construction industry is one of the most important sectors for economic development in the European Union [1]. The sector demands high amounts of natural resources and energy, consuming around 40% of the global energy demand [2,3]. In light of this fact, many potential solutions have been studied, highlighting among them the use of waste instead of natural resources. The concrete industry, for instance, allows the use of large quantities of waste as a substitute for a conventional raw material, without sacrificing the technical properties of the product [4,5]. From a circular economy point of view, the benefit of this proposal is two-fold: The decrease of natural resource consumption [6,7], and the reduction of air, soil, and groundwater pollution associated with landfilling [8], which can negatively affect human health, environment, and biota [9,10]. However, the recycling process could be undesirable from an environmental perspective if there is no real saving in energy, raw materials, and water consumption, or in pollutant emissions [11].

Natural aggregates, one of the primary ingredients in concrete, can be obtained from: (1) Naturally occurring unconsolidated sand and gravel; and (2) quarries, by crushing bedrock to obtain crushed stone and sand [12]. The first alternative can cause irreparable damage to the river ecosystems, such as channel degradation and the incision of the entire fluvial system [13]. For this reason, in countries, such as Spain, the extraction of aggregates from gravel pits is restricted by environmental laws to protect the biodiversity and biological resources of rivers [14]. Currently, natural aggregates are mainly obtained from quarries [15]. In 2015, 2660 million tons of aggregates were produced in Europe from quarries, and 47% were crushed gravel, gravel, and sand [16].

On the one hand, an increasing amount of different types of waste is being recycled as fine and coarse aggregates in concrete manufacturing as a substitution for natural aggregates. Some examples are ashes, bottom ashes, and slags [17]; construction and demolition wastes [18]; and farming wastes, such as corn, wheat, bamboo, coconut shells, and olive stone [19,20]. On the other hand, the aquaculture industry produces between 6,000,000 and 8,000,000 tons of waste worldwide annually and only 25% is recycled [21], and the rest is usually dumped in coastal waters or landfills [22]. Part of the seashell waste generated by the aquaculture industry is recycled as lime substitute, wastewater decontaminant, soil conditioner, fertilizer constituent, feed additive, and liming agent [23]. However, these recycling options are not able to consume a significant amount of seashell waste produced annually [24].

Several authors have proposed the recycling of seashell waste as an aggregate in mortars, substituting natural aggregates totally replacing limestone aggregates from quarry [25] and employing it alone [26,27] or with other waste, such as fly ash [28], for silica sand substitution. Due to the large amount of organic matter and chloride present in the waste [29], and to comply with the requirements of construction material standards [30], seashell waste pre-treatment, consisting of washing and calcination, is required, regardless of the specific application: Insulation construction materials [31], both conventional [32] and marine concretes [33]; cement mortars for masonry and plastering [34]; and cement-based bricks [35]. This pre-treatment involves the use of significant amounts of resources and energy. However, the specific quantification of the potential environmental benefits of the recycling of seashell waste for natural aggregate substitution is still lacking in the literature. Only two environmental impact assessments were reported in the literature, with opposite conclusions in favor of and against seashell waste recycling as a source of $CaCO_3$ [36,37].

The World Health Organization (WHO) recommends reducing noise levels produced by road traffic below 53 decibels (dB) as road traffic noise above this level is associated with adverse health effects [38]. The high level of noise pollution led the European Union to develop a specific directive to manage environmental noise [39]. The growing number of these types of policies has led to an increase in the production of noise barriers [40]. Related to this, the design of noise barriers with porous concrete (PC) as a mitigation measure has started to be incorporated into new road construction projects [40]. The present study is part of a research project where different alternatives for seashell recycling in construction materials are technically and environmentally assessed. Noise barriers were chosen within the scope of this research project because local industries could consume the full amount of waste annually produced. In a recently published work, concrete-based noise barriers with recycled seashell waste were obtained with similar technical properties to barriers made with natural aggregates [41].

This study aimed to analyze the environmental impacts of the use of seashell as an aggregate in porous concrete for noise barriers through life cycle assessment (LCA) for the first time, and to identify the potential advantages of this proposed recycling process. The use of LCA makes it possible to determine whether this recycling alternative is a good option for seashell waste management in terms of environmental sustainability. The study focused on the mussel-canning industries from Galicia (NW Spain), which produce 267,000 tons of mussels annually [42] and around 150,000 tons of seashell waste [43].

2. Materials and Methods

This LCA was conducted according to the guidelines in ISO 14040/44 [44,45]. The data and main assumptions considered are detailed in the following sections.

2.1. Goal and Scope

The main goal of this study was to assess the life cycle environmental sustainability of a new noise barrier manufactured from porous concrete produced with seashell waste (PCSW), and compare it with a conventional noise barrier made of concrete produced with natural aggregates from quarry (PCNA).

The scope of the study was from 'cradle to grave' and the main stages for PCSW and PCNA systems are the following (Figure 1):

- Extraction and treatment of raw materials:
    - Cement.
    - Natural aggregates (sand, gravel and crushed gravel) for PCNA.
    - Seashell waste for PCSW, which needs a pre-treatment consisting of storing, washing, calcination, and milling to obtain fine and coarse aggregates.
- Production of the noise barrier:
    - Mixing and kneading: Cement and aggregates or seashell waste are mixed with water.
    - Molding: Concrete is molded (see Figure S1 in Supplementary Materials (SI)).
    - Cured: Hardening process.
    - Dismantling: Removal from the mold where the mixture is supported.
- Use. This stage is considered to have a negligible impact in the comparison according to different suppliers of acoustic barriers [46,47].
- End-of-life. Landfilling of the waste from the extraction and treatment of raw materials, the production process, and the end-of-life of the noise barrier.
- Transport. It includes the transport of the seashell waste to the pre-treatment facility, of the raw materials to the noise barrier factory, of the porous concrete panel to the location of use, and of the waste to the landfill.

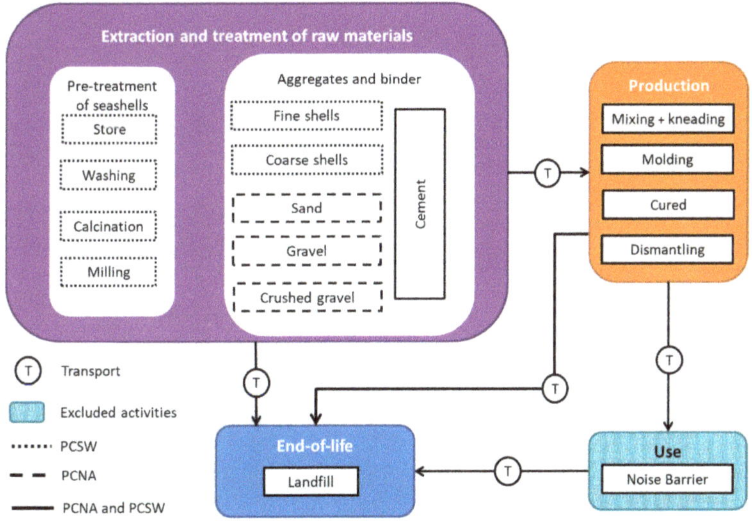

Figure 1. System boundaries for the life cycle of noise barriers using porous concrete with natural aggregates (PCNA) and aggregates from seashell waste (PCSW).

The defined functional unit (FU) is the production of 1 m^2 of noise barrier. This FU was selected to compare the environmental impact of both products for the same area of sound absorption and mechanical properties. As can be seen in our previous work [41], both types of panels (PCNA and PCSW) can be classified as the A2 category of the acoustic absorption assessment index, according to EN 1793-1 [48], for the same thickness. In addition, the mechanical and durability tests showed similar results for both materials [41].

2.2. Life Cycle Inventory

Table 1 shows the inventory data for both types of noise barriers, with natural aggregates (PCNA) and seashell waste (PCSW). Inventory data for the seashell waste pre-treatment were adapted from Iribarren et al. [37]. Consumptions for PCSW production were obtained from Peceño et al. [41]. The raw material and energy consumption of PCNA were determined from the manufacturers' data. The background data were sourced from the Ecoinvent v3.1 database [49].

Table 1. Life cycle inventory data for 1 m^2 of noise barrier using porous concrete from natural aggregates (PCNA) and seashell waste (PCSW).

Parameters	PCNA	PCSW
Extraction and treatment of raw materials		
Pre-treatment of seashells [1]		
Seashell waste (kg)	-	301.8
Propane (kg)	-	6.3
Diesel (g)	-	140.0
Aluminum sulphate (g)	-	7.2
Chlorine dioxide (g)	-	1.8
Water (kg)		280.0
Electricity, low voltage (kWh)	-	29.4
Aggregates and cement		
Gravel (kg)	48.0	-
Sand (kg)	96.0	-
Crushed gravel (kg)	69.1	
Fine shells (kg)	-	90.4
Coarse shells (kg)	-	90.4
Cement (kg)	35.2	45.2
Production		
Water (kg)	14.6	27.1
Electricity (low voltage) (kWh)	0.5	0.4
End-of-life		
Landfill (including waste from raw material treatment, production and end-of-life) (kg)	262	240
Transport		
Seashell waste to pre-treatment facility (tkm)	-	3.02
Raw materials to the factory (tkm)	11.2	18.5
From factory to use place (tkm)	52.4	48.6
From use place to landfill (tkm)	1.0	0.9

[1] Only the main consumptions for the pre-treatment of seashell are included here. A detailed inventory table for this process, adapted from [37], is given in Supplementary Materials (Table S1).

2.2.1. Raw Materials

As mentioned above, previously used as aggregates, seashell waste from the canning industry must be pre-treated to satisfy the requirements of the EN 1744-1 standard [30] regarding the maximum quantity of organic matter and chloride [29]. The first step was the washing of the seashell waste. Wash water was treated subsequently with chlorine dioxide in a ratio of 6 mg per kg seashell waste. To remove suspended solids from the wash water, aluminum sulphate was added as a coagulant at a ratio of 24 mg per kg seashell waste. Afterwards, to remove the organic matter left, the seashell waste was calcined at 550 °C for 15 min. The calcination temperature was limited to 550 °C to avoid the decomposition of $CaCO_3$ (90 wt % of the seashell) to CaO and CO_2. Finally, the washed and calcined seashell waste were milled to obtain two different kinds of aggregates, fine and coarse, with particle sizes lower and higher than 2 mm, respectively [37].

2.2.2. Production

Aggregates, cement and water were mixed and kneaded mechanically to produce the concrete mass. Afterwards, the concrete is poured into a mold of the required dimensions (see Figure S1 in Supplementary Materials). The electricity consumption reported for PCNA for these processes was also considered for PCWS but proportionally adapted to the consumed mass of each component.

For each panel, two masses were prepared: The porous concrete mass, made with fine aggregates, and the structural concrete mass, produced with coarse aggregates. Firstly, the porous concrete mass was poured into the mold. Then, a steel frame was placed, and the structural concrete was poured. The steel reinforcement was not included in the inventory since it was equal for both the PCNA and PCSW. The next step was the curing, which is the hardening process of the panels and where water evaporation is produced. The curing process was considered finished after 28 days, when the mold was dismantled.

For PCNA, while gravel and sand were used for a layer of structural concrete, crushed gravel was used for a layer of porous concrete. The structural concrete mass was composed of 57 wt % sand, 29 wt % gravel, and 14 wt % cement. The amount of water added was 46 wt % with respect to the amount of cement. The porous concrete mass was composed of 86%wt of crushed gravel and 14 wt % cement. Then, 30 wt % water was added with respect to the amount of cement. The mass of the noise panel was 262 kg/m² per functional unit.

In the case of PCSW, a mass ratio of 80:20 seashell waste: cement was added for both layers of concrete, porous (with coarse shells) and structural (with fine shells). The amount of water added was 60%wt with respect to the amount of cement, to ensure the workability of the mass [41]. The total mass of the noise panel was 240 kg/m² per functional unit.

2.2.3. End-of-Life

After their use, both PCNA and PCSW panels are considered as construction and demolition waste (CDW), according to section 17 of the European Waste Catalogue [50]. Following the most recent data for CDW treatment in Spain, 100% of the waste was assumed to be sent to the landfill [51].

2.2.4. Transport

For both types of alternatives, the production of the concrete was assumed to be in the concrete plant of General de Hormigones in Meis (Galicia, NW Spain). This company commercializes different concrete-made products with natural aggregates, including porous concrete. Both the closest gravel extraction quarry and cement factory of General de Hormigones were chosen for the study: Gravel extraction in Salcedo (22 km from Meis) and cement industry in Ouras (186 km from Meis). According to Barros et al. [52], seashell waste is pre-treated in Caliza Marina Company, located in Boiro (39 km from Meis), while mussel-canning industries are located in Rianxo [52] (10 km from Boiro). Distances were measured using the Google MapsTM distance measurement tool [53]. Once it was produced, the noise barrier of porous concrete was assumed to travel a distance of 200 km to the final installation on a road. Regarding the end-of-life, generated waste was considered to be transported a distance of 15 km to the landfill [54]. All transport was done by road, and a 16-32-tonne Euro 6 truck was considered, according to the classification included in Regulation (EC) No 715 (2007) [55].

2.3. Sensitivity Analysis

To reduce the necessary heat consumption for seashell waste calcination during its pre-treatment, other authors have proposed lower temperatures than 550 °C, reported by Barros et al. [56]. On the one hand, Neves and Manos [57] and Sahari and Aniza Mijan [58] calcined seashell waste at 250 °C. On the other hand, Martinez-Garcia et al. [59] calcined seashell waste at 135 °C. Due to the expected significant influence that heat production for calcination could have on the results of the environmental assessment, the present work also studied the environmental impact of PCSW for the two other calcination temperatures aside from 550 °C through a sensitivity analysis: 135 °C (PCSW-135) and 250 °C (PCSW-250). Life cycle inventories of PCSW-135 and PCSW-250 are provided in Table S1 in Supplementary Materials.

2.4. Environmental Impact Assessment

The LCA software SimaPro version 8.0.4 9 (PRé Consultants, B.V.: Amersfoort, The Netherlands, 2016) [60] was used for life cycle modelling, and the CML-IA (version 3.03) mid-point impact assessment method [61] was applied to calculate the environmental impacts. The following impacts were considered: Abiotic depletion potential of elements (ADPe), abiotic depletion potential of fossil resources (ADPf), acidification potential (AP), eutrophication potential (EP), global warming potential (GWP), human toxicity potential (HTP), marine aquatic ecotoxicity potential (MAETP), freshwater aquatic ecotoxicity potential (FAETP), ozone depletion potential (ODP), photochemical oxidant creation potential (POCP), and terrestrial ecotoxicity potential (TETP).

3. Results and Discussion

3.1. Comparison of PCNA and PCSW

As illustrated in Figure 2, PCNA has the lowest values in all impact categories considered in this study, despite the recycling of the seashell waste in the PCSW material. Overall, the impacts associated with PCSW are between 32% (TETP) and 267% (ODP) higher in comparison to PCNA.

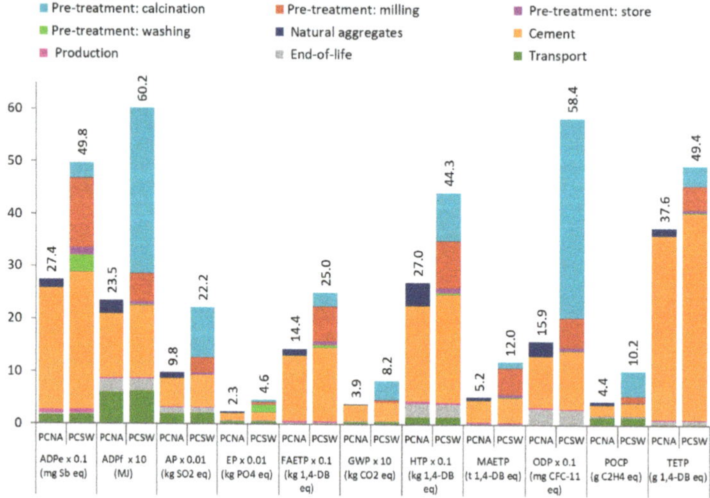

Figure 2. Life cycle environmental impacts of PCNA and PCSW noise barriers. Some impacts should be multiplied by the factor shown to obtain the original values. ADPe: abiotic depletion potential of elements, ADPf: abiotic depletion potential of fossil resources, AP: acidification potential, EP: eutrophication potential, FAETP: freshwater aquatic ecotoxicity potential, GWP: global warming potential, HTP: human toxicity potential, MAETP: marine aquatic ecotoxicity potential, ODP: ozone layer depletion potential, POCP: photochemical oxidants creation potential, TETP: terrestrial ecotoxicity potential, PCNA: porous concrete with natural aggregates and PCSW: porous concrete with seashell waste calcined at 550 °C.

The seashell waste pre-treatment is the main contributor to seven impact categories in the case of PCSW barriers: ODP (76%), ADPf (63%), POCP (59%), AP (58%), MAETP (57%), EP (51%), and GWP (50%). The ADPf impact is mainly associated with the consumption of fossil fuels during calcination. In the case of the ODP, fossil fuel transportation produces emissions of bromotrifluoride and halon 1301 that contribute to this impact. For AP, the impact is mostly caused by the SO_x, NO_x, and NH_3, emissions generated from the burning of propane to produce the heat used for calcination The SO_2, CO, C_4H_{10}, and CH_4 emissions also released in the combustion of propane generate a significant contribution to the POCP impact. The same cause, the propane combustion, contributed significantly to the GWP impact due to the emissions of CO_2, CH_4, and N_2O. In the case of EP, the impact is mainly associated with the aquatic emissions of PO_4^{3-}, NO, COD, and NO_2 released in the washing process of seashell waste. Finally, for MAETP, the impact is mainly caused by the HF, Be, Ni, and Ba emissions during the mining of the coal burned for the generation of electricity necessary for the milling. The pre-treatment of the shells is the second most important contributor to the other four impact categories: ADPe (42%), FAETP (41%), HTP (44%), and TETP (18%). Most of these impacts are associated with the electric consumption in the milling process and, more specifically, from the extraction of coal used for electricity generation. The consumption of electricity in the milling of fine shells used in the structural layer of concrete is three times higher than in the milling of the coarse shells for the porous concrete layer.

In the case of PCNA, the extraction of natural aggregates only contributes significantly to ODP (18%) due to the emissions associated with the consumption of diesel used in heavy machinery in the quarry. The impacts associated with natural aggregates are much lower than those associated with recycled seashells. The pre-treatment of the seashell waste generates between 4 (HTP) and 22 (GWP) times higher impacts compared to the natural aggregates' production.

Cement production is the most significant contributor in all categories (47–93%) for PCNA and, for PCSW, in those four impacts in which the pre-treatment does not entail most of the burdens (HTP (46%), ADPe (52%), FAEP (56%), and TETP (80%)). Cement contributes significantly to the impacts of TETP (>80%), FAETP (>56%), ADPe (>52%), and HTP (>46%) for both types of barriers. The impacts per FU due to the cement consumption are higher in the case of PCSW because more cement is needed to produce the noise barrier (35.20 kg/FU in the case of PCNA versus 45.20 kg/FU in the case of PCSW as illustrated in Table 1). The higher consumption of cement in PCSW is necessary to accomplish the required mechanical properties due to the shape and porosity of shell wastes [41,62]. The TETP and HTP impacts are associated with the emissions of heavy metals, such as chromium, zinc, tin, and lead, and dioxins in the production of clinker. The impacts of FAETP and ADPe are mainly caused by the mining and extraction of fossil fuels used in the generation of heat and electricity consumed in the cement production process.

The contribution of the production stage to the impacts is relatively low (<5% for PCNA and <2% for PCSW). In this stage, there is a reduction in all the impacts of PCSW compared to PCNA except for the ADPe. The higher value for ADPe is caused by the higher demand for water consumption in PCSW production (27.10 kg/FU for PCSW compared to 14.60 kg/FU, see Table 1). The lower impacts in the rest of the categories are caused by a slightly lower consumption of electricity in the case of PCSW (0.43 kWh/FU for PCSW compared to 0.47 kWh/FU for PCNA). There is a mass difference per 1 m^2 (240 kg/FU for PCSW versus 262 kg/FU for PCNA), and therefore, the electrical consumption for molding and kneading is lower for PCSW panels.

The end-of-life stage does not contribute significantly to the impacts (<19% for PCNA and <5% for PCSW). The impacts of landfilling are mainly generated by the consumption of diesel in the use of heavy machinery needed in the disposal. The difference in weight between both barriers (262 kg/FU for PCNA versus 240 kg/FU for PCSW) makes the diesel consumption lower in the case of PCSW.

The transport only contributes significantly for POCP (33% for PCNA and 15% for PCSW) and ADPf (25% for PCNA and 10% for PCSW) due to the emissions of volatile organic compounds and NO_x

from fuel combustion. Although the global distances considered for PCSW are slightly lower than for PCRG (450 km vs. 467 km), the higher amount of consumed cement increases the impacts for PCSW.

To summaries, the results show that the manufacture of the PCSW noise barrier has no environmental advantages over the PCNA noise barrier due to the high environmental burden of the pre-treatment of the seashells, the increase in the consumption of cement, and the low environmental load of the natural aggregates. Therefore, to apply wastes instead of natural aggregates for the production of porous concrete, it is necessary to look for sources of waste that require more sustainable pre-treatment and do not involve an increase in cement consumption.

As far as we are aware, there are no other comprehensive LCA studies of the use of recycled seashells for noise barriers, so a direct comparison with the literature is not possible. The most similar studies are from Iribarren et al. [37] and Alvarenga et al. [36], which analyzed the life cycle of seashell waste management, considering the recycling of seashells as $CaCO_3$ versus landfilling. The comparison of the results should only be considered as illustrative due to the difference in the raw materials (e.g., different types of seashells and final products), processes, and LCA parameters (e.g., system boundaries, FU, or impact categories) considered. The LCA results reported by Iribarren et al. [37] showed that the recycling of seashells as $CaCO_3$ presents disadvantages compared with landfilling. Still, the authors noted that, despite these results from the LCA analysis, social concerns could be an argument against landfilling [37]. However, in the case of Alvarenga et al. [36], the recycling of seashell waste as $CaCO_3$ generated lower environmental impacts than landfilling in all the impacts (between 1% and 85%). These results are justified because only energy and water consumption required for the pre-treatment of shells were considered, thus reducing the environmental burdens. In addition, the recycled seashell is credited with the avoided environmental impacts of the production of $CaCO_3$ [36].

3.2. Sensitivity Analysis

As shown in Figure 3, the reduction in the calcination temperature involves a reduction in the environmental impacts for all categories under study. The percentage of reduction versus the base case (PCSW), with a calcination temperature of 550 °C, depends on the impact category and ranges from 3% (PCSW-250)-5% (PSW-135) for ADPe to 36% (PCSW-250)-52% (PSW-135) for ODP. These differences are justified by the different percentages of contribution of the calcination process to the environmental burdens of each impact category.

The lowest reductions in the impacts (<10% for PCSW-135 and PCSW-250) are for TETP, FAETP, ADPe, and MAETP. As shown in Section 3.1, these impacts are mainly associated with the production of cement and the electricity consumption of the milling process but not with the calcination process. Despite this, and if only the pre-treatments are compared, there is a reduction in the four impact categories mentioned by 8–23% for PCSW-250 and by 11–33% for PCSW-135, compared to PCSW.

Moderate reductions are observed for EP and HTP when the heating temperature is reduced. In the case of the EP, the impact is reduced between 18% (PCSW-250) and 26% (PCSW-135) compared with PCSW. The reduction is between 11% (PCSW-250) and 16% (PCSW-135) for HTP. The minimization of propane consumption decreases aerial NO_x emissions and, therefore, the EP impact. This reduction of consumption also decreases the PO_4^{3-}, NO_3^-, Se, and Ni emissions in the extracting, processing and the use of propane and, therefore, causes a reduction in the HTP impact.

Finally, the ADPf, AP, POCP, GWP, and ODP impacts have the most significant decreases (>22% for PCSW-250 and >32% for PCSW-135). These reductions are justified by the decline in the consumption of propane between 60% (PCSW-250) and 78% (PCSW-135), which is the main contributor to these impacts (see Section 3.1). If only the pre-treatment is compared, the decreases in these impacts are >38% for PCSW-250 and >56% for PCSW-135 compared to PCSW.

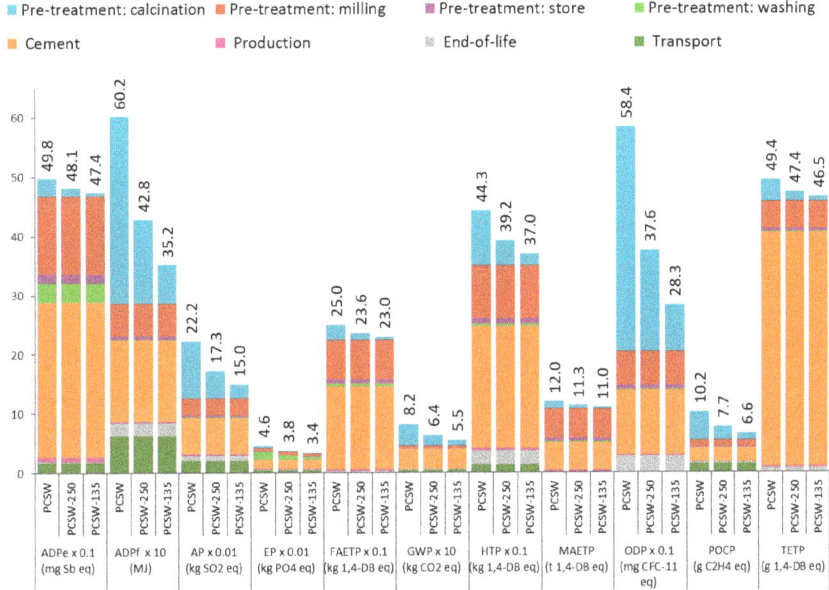

Figure 3. Environmental influence of different calcination temperatures in the pre-treatment of seashell waste. Some impacts should be multiplied by the factor shown to obtain the original values. PCSW-250: porous concrete with seashell waste heated at 250 °C, PCSW-135: porous concrete with seashell waste heated at 135 °C. For the rest of the nomenclatures, consult Figure 2.

The results for PCNA and the best heating temperature alternative, PCSW-135, are compared in Figure 4. The reduction of the calcination temperature to 135 °C is not enough to allow the recycling seashells to be a better alternative compared with the use of natural aggregates for any impact category. TETP shows the closest impact between PCSW-135 and PCNA, with a 24% difference. ODP (28.3 mg CFC-11 for PCSW-135 compared to 15.9 mg CFC-11 for PCNA) and MAETP (11.0 t 1.4 DB eq. for PCSW-135 and 5.2 t 1.4 DB eq. for PCNA) have the most significant differences with respect to PCNA, 78% and 113%, respectively. These differences are mainly caused by the higher consumption of fossil fuels for cement production and the demand for fossil fuels in the pre-treatment of the seashell waste. Due to the reduction in the calcination temperature, the contribution of propane consumption to MAETP is only 24% compared with the total contribution of the complete pre-treatment. However, the pre-treated shells still have 10 times more MAETP impact if compared to the natural aggregates (5.94 t DB eq. versus 0.56 t DB eq.), mainly due to the electric consumption of the milling.

The results show that even with the lower calcination temperature (135 °C), the use of natural aggregates is the best alternative from an environmental standpoint. These results initially go against some of the principles of the circular economy, like preserving and enhancing natural capital [63] by controlling finite stocks [64] and designing out waste [65]. Circular economy principles have been considered by new European Union policies that promote waste prevention, recycling, and eco-design to reduce the environmental impact [66]. Several tools have been developed for circular economy principle implementation [67], and they have been used by researchers in order to attain more eco-friendly designs and processes [68]. However, in some cases, as shown in the present study, the recycling of waste can imply an increase in the environmental impacts, due to higher consumption in other resources (cement) and energy (for the pre-treatment) compared with the extraction of virgin resources (natural aggregates). Therefore, case-by-case research of the environmental implications is needed to achieve sustainable solutions. In the particular case of seashell waste, it is necessary to research the possibilities of replacing other materials with higher environmental impacts or trying to reduce the environmental footprint of the recycling process.

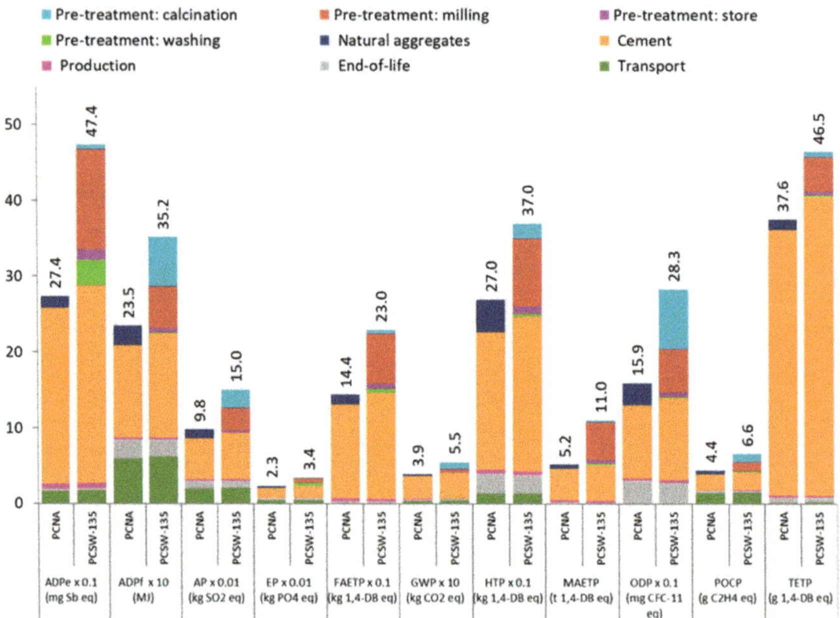

Figure 4. Life cycle environmental impacts of PCNA and PCSW-135 noise barriers. Some impacts should be multiplied by the factor shown to obtain the original values. PCNA: porous concrete with natural aggregates, PCSW-135: porous concrete with seashell waste heated at 135 °C. For the rest of the nomenclatures, consult Figure 2.

4. Conclusions

This study provides the first environmental life cycle assessment of seashell waste as an aggregate in porous concrete panels in noise barriers. The results reveal that the noise barriers manufactured with seashell waste have higher impacts (between 32% and 267%) than barriers based on porous concrete made with natural aggregates in the 11 categories considered. These increases are justified by the contribution of the seashell waste pre-treatment, including storage, washing, calcination, and milling, and the higher cement consumption, 13% compared with acoustic barriers with natural aggregates. The impacts caused by the end-of-life and production stages are low in both alternatives (<19%), and the transport has a significant influence only in the generation of photochemical oxidants in both types of barriers (>15%).

A decrease in the calcination temperature of seashell waste from 550 °C to 250 and 135 °C causes a reduction of 3–36% and 5–52% in the different impacts, respectively. However, even the alternative with a lower calcination temperature (135 °C) still has higher impacts in all the considered categories than the noise barrier made with natural aggregates. The decrease in the calcination temperature turns the consumption of cement into the primary contributor in all of the impact categories. In this sense, future research studies should analyze the technical feasibility of reducing cement consumption when natural aggregates are substituted with seashell waste. Thus, it could be concluded that life cycle assessments should be carried out simultaneously with the technical development of the product and not at the end. This approach would simultaneously make it possible to optimize the technical properties of the materials and minimize their environmental impact and, combined with a socio-economic analysis, will ensure the sustainability of the recycling process.

Supplementary Materials: The following are available online at http://www.mdpi.com/2227-9717/8/7/776/s1, Figure S1: Acoustic noise barrier dimensions, Table S1: Inventory data for the seashell waste pre-treatment per 1 m^2 of noise barrier when different calcination temperatures are applied (PCSW, 550 °C; PCSW-250, 250 °C; PCSW-135, 135 °C).

Author Contributions: Conceptualization, B.A.-F., A.G.-S. and C.L.; methodology, B.P., A.G.-S. and B.A.-F.; software, B.P., A.G.-S. and B.A.-F.; validation, B.P., A.G.-S., B.A.-F. and C.L.; formal analysis, B.P., A.G.-S. and B.A.-F.; investigation, B.P., A.G.-S., B.A.-F. and C.L.; resources, A.G.-S., B.A.-F. and C.L.; data curation, B.P., A.G.-S. and B.A.-F.; writing—original draft preparation, B.P., A.G.-S. and B.A.-F.; writing—review and editing, B.P., A.G.-S., B.A.-F. and C.L.; visualization, B.P., A.G.-S. and B.A.-F.; supervision, A.G.-S., C.L. and B.A.-F.; project administration, B.P.; funding acquisition, B.P and C.L. All authors have read and agreed to the published version of the manuscript.

Funding: This research was funded by Innovation Funds for competitiveness from Regional Government of Coquimbo (Chile), grant number BIP 40014353-0.

Acknowledgments: The authors would like to acknowledge Carmen Barros and Diego Iribarren for their contribution to the article.

Conflicts of Interest: The authors declare no conflict of interest.

References

1. Estanqueiro, B.; Dinis Silvestre, J.; De Brito, J.; Duarte Pinheiro, M. Environmental life cycle assessment of coarse natural and recycled aggregates for concrete. *Eur. J. Environ. Civ. Eng.* **2016**, *22*, 429–449. [CrossRef]
2. Ingrao, C.; Arcidiacono, C.; Bezama, A.; Ioppolo, G.; Winans, K.; Koutinas, A.; Gallego-Schmid, A. Sustainability issues of by-product and waste management systems to produce building material commodities. *Resour. Conserv. Recycl.* **2017**, *126*, 4–5. [CrossRef]
3. Ingrao, C.; Arcidiacono, C.; Bezama, A.; Ioppolo, G.; Winans, K.; Koutinas, A.; Gallego-Schmid, A. Sustainability issues of by-product and waste management systems, to produce building material commodities: A comprehensive review of findings from a virtual special issue. *Resour. Conserv. Recycl.* **2019**, *146*, 358–365. [CrossRef]
4. Alonso-Fariñas, B.; Rodríguez-Galán, M.; Arenas, C.; Arroyo Torralvo, F.; Leiva, C. Sustainable management of spent fluid catalytic cracking catalyst from a circular economy approach. *J. Waste Manag.* **2020**, *110*, 10–19. [CrossRef]
5. Rodríguez-Galán, M.; Alonso-Fariñas, B.; Baena-Moreno, F.M.; Leiva, C.; Navarrete, B.; Vilches, L.F. Synthetic slag production method based on a solid waste mix vitrification for the manufacturing of slag-cement. *Materials* **2019**, *12*, 208. [CrossRef]
6. Gallego-Schmid, A.; Chen, H.M.; Sharmina, M.; Mendoza, J.M.F. Links between circular economy and climate change mitigation in the built environment. *J. Clean. Prod.* **2020**, *260*, 121115. [CrossRef]
7. Srivastava, R.R.; Kim, M.S.; Lee, J.C.; Jha, M.K.; Kim, B.S. Resource recycling of superalloys and hydrometallurgical challenges. *J. Mater. Sci.* **2014**, *49*, 4671–4686. [CrossRef]
8. Kim, T.G.; Srivastava, R.R.; Jun, M.; Kim, M.; Lee, J.C. Hydrometallurgical recycling of surface-coated metals from automobile-discarded ABS plastic waste. *J. Waste Manag.* **2018**, *80*, 414–422. [CrossRef]
9. Sattar, R.; Ilyas, S.; Bhatti, H.N.; Ghaffar, A. Resource recovery of critically-rare metals by hydrometallurgical recycling of spent lithium ion batteries. *Sep. Purif. Technol.* **2014**, *49*, 4671–4686. [CrossRef]
10. Nekouei, R.K.; Maroufi, S.; Assefi, M.; Pahlevani, F.; Sahajwalla, V. Thermal isolation of a clean alloy from waste slag and polymeric residue of electronic waste. *Processes* **2020**, *8*, 53. [CrossRef]
11. Zaimes, G.G.; Vora, N.; Chopra, S.S.; Landis, A.E.; Khanna, V. Design of sustainable biofuel processes and supply chains: Challenges and opportunities. *Processes* **2015**, *3*, 634–663. [CrossRef]
12. Langer, W. Sustainability of aggregates in construction. In *Sustainability of Construction Materials*, 2nd ed.; Khatib, J.M., Ed.; Elselvier: London, UK, 2016; pp. 181–207.
13. Sreebha, S.; Padmalal, D. Environmental impact assessment of sand mining from the small catchment rivers in the Southwestern Coast of India: A case study. *Environ. Manag.* **2011**, *47*, 130–140. [CrossRef] [PubMed]
14. BOE. Law 7/1992, de 24 de July, about River Fishing, of Comunidad Autónoma de Galicia, BOE 1992, Pages 34618 to 34625. Available online: https://www.boe.es/eli/es-ga/l/1992/07/24/7 (accessed on 21 February 2020).
15. Hill, A.R.; Dawson, A.R.; Mundy, M. Utilisation of aggregate materials in road construction and bulk fill. *Resour. Conserv. Recycl.* **2001**, *32*, 305–320. [CrossRef]

16. Union Européenne des Producteurs de Granulat (UEPG). Stadistic. 2018. Available online: http://www.uepg.eu/statistics/estimates-of-production-data/data-2015 (accessed on 23 July 2018).

17. Arenas, C.; Luna-Galiano, Y.; Leiva, C.; Vilches, L.F.; Arroyo, F.; Villegas, R.; Fernández-Pereira, C. Development of a fly ash-based geopolymeric concrete with construction and demolition wastes as aggregates in acoustic barriers. *Constr. Build. Mater.* **2017**, *134*, 433–442. [CrossRef]

18. Xu, G.; Shen, W.; Zhang, B.; Li, Y.; Ji, X.; Ye, Y. Properties of recycled aggregate concrete prepared with scattering-filling coarse aggregate process. *Cem. Concr. Compos.* **2018**, *93*, 19–29. [CrossRef]

19. Mo, K.H.; Alengaram, U.J.; Jumaat, M.Z.; Yap, S.P.; Lee, S.C. Green concrete partially comprised of farming waste residues: A review. *J. Clean. Prod.* **2016**, *117*, 122–138. [CrossRef]

20. Chandni, T.J.; Anand, K.B. Utilization of recycled waste as filler in foam concrete. *J. Build. Eng.* **2018**, *19*, 154–160. [CrossRef]

21. Yan, N.; Chen, X. Sustainability: Don't waste seafood waste. *Nature* **2015**, *524*, 155–157. [CrossRef]

22. Lu, J.; Cong, X.; Li, Y.; Hao, Y.; Wang, C. Scalable recycling of oyster shells into high purity calcite powders by the mechanochemical and hydrothermal treatments. *J. Clean. Prod.* **2018**, *172*, 1978–1985. [CrossRef]

23. Lu, J.; Lu, Z.; Li, X.; Xu, H.; Li, X. Recycling of shell wastes into nanosized calcium carbonate powders with different phase compositions. *J. Clean. Prod.* **2015**, *92*, 223–229. [CrossRef]

24. Yao, Z.; Xia, M.; Li, H.; Chen, T.; Ye, Y.; Zheng, H. Bivalve Shell: Not an abundant useless waste but a functional and versatile biomaterial. *Crit. Rev. Environ. Sci. Technol.* **2014**, *44*, 2502–2530. [CrossRef]

25. Ballester, P.; Mármol, I.; Morales, J.; Sánchez, L. Use of limestone obtained from waste of the mussel cannery industry for the production of mortars. *Cem. Concr. Res.* **2007**, *37*, 559–564. [CrossRef]

26. Yang, E.I.; Yi, S.T.; Leem, Y.M. Effect of oyster shell substituted for fine aggregate on concrete characteristics: Part I. Fundamental properties. *Cem. Concr. Res.* **2005**, *35*, 2175–2182. [CrossRef]

27. Yang, E.I.; Kim, M.Y.; Park, H.G.; Yi, S.T. Effect of partial replacement of sand with dry oyster shell on the long-term performance of concrete. *Constr. Build. Mater.* **2010**, *24*, 758–765. [CrossRef]

28. Yoon, H.; Park, S.; Lee, K.; Park, J. Oyster shell as substitute for aggregate in mortar. *Waste Manag. Res.* **2004**, *22*, 158–170. [CrossRef] [PubMed]

29. Adewuyi, P.; Adegoke, T. Exploratory study of periwinkle shells as coarse aggregates in concrete works. *ARPN J. Eng. Appl. Sci.* **2008**, *3*, 1–5.

30. European Committee for Standardization (EN). *Tests for Chemical Properties of Aggregates—Part 1: Chemical Analysis*; EN 1744-1; EN: Brussels, Belgium, 2009.

31. Felipe-Sesé, M.; Eliche-Quesada, D.; Corpas-Iglesias, F.A. The use of solid residues derived from different industrial activities to obtain calcium silicates for use as insulating construction materials. *Ceram. Int.* **2011**, *37*, 3019–3028. [CrossRef]

32. Foti, D.; Cavallo, D. Mechanical behavior of concretes made with non-conventional organic origin calcareous aggregates. *Constr. Build. Mater.* **2018**, *179*, 100–106. [CrossRef]

33. Chen, H.Y.; Li, L.G.; Lai, Z.; Kwan, A.; Chen, P.; Ng, P.L. Effects of crushed oyster shell on strength and durability of marine concrete containing fly ash and blastfurnace slag. *J. Mater. Sci.* **2019**, *25*, 97–107.

34. Lertwattanaruk, P.; Makul, N.; Siripattarapravat, C. Utilization of ground waste seashells in cement mortars for masonry and plastering. *J. Environ. Manag.* **2012**, *111*, 133–141. [CrossRef]

35. Li, G.; Xu, X.; Chen, E.; Fan, J.; Xiong, G. Properties of cement-based bricks with oyster-shells ash. *J. Clean. Prod.* **2015**, *91*, 279–287. [CrossRef]

36. Alvarenga, R.A.F.; De Galindro, B.M.; Helpa, C.F.; Soares, S.R. The recycling of oyster shells: An environmental analysis using Life Cycle Assessment. *J. Environ. Manag.* **2012**, *106*, 102–109. [CrossRef] [PubMed]

37. Iribarren, D.; Moreira, M.T.; Feijoo, G. Implementing by-product management into the life cycle assessment of the mussel sector. *Resour. Conserv. Recycl.* **2010**, *54*, 1219–1230. [CrossRef]

38. World Health Organization (WHO). *Environmental noise guidelines for the European region. Proceedings of the Institute of Acoustics*; World Health Organization Regional Office for Europe: Copenhagen, Denmark, 2018; p. 181.

39. European Parliament and Council of the European Union (EC). Directive 2002/49/Ec of the European Parliament and of the Council of 25 June 2002 relating to the assessment and management of environmental noise. *Off. J. Eur. Communities* **2002**, *L189*, 12–25. Available online: http://eur-lex.europa.eu/legal-content/EN/TXT/PDF/?uri=CELEX:32002L0049andfrom=EN (accessed on 3 June 2020).

40. Kotzen, B.; English, C. *Book Environmental Noise Barriers—A Guide to Their Acoustic and Visual*, 2nd ed.; Taylor & Francis: New York, NY, USA, 2009; pp. 15–38.

41. Peceño, B.; Arenas, C.; Alonso-Fariñas, B.; Leiva, C. Substitution of coarse aggregates with mollusk-shell waste in acoustic-absorbing concrete. *J. Mater. Civ. Eng.* **2019**, *31*, 04019077. [CrossRef]

42. Galician Statistical Institute (IGE). Production of Marine Aquaculture in Galicia. 2017. Available online: https://www.ige.eu/web/mostrar_actividade_estatistica.jsp?idioma=esandcodigo=0301004 (accessed on 12 June 2018).

43. Eurostat. Catches—Major Fishing Areas (from 2000 Onwards). 2019. Available online: https://appsso.eurostat.ec.europa.eu/nui/show.do?dataset=fish_ca_mainandlang=en (accessed on 21 February 2020).

44. International Standards Organization (ISO). *Environmental Management—Life Cycle Assessment—Principles and Framework, ISO 14040*; ISO: Geneva, Switzerland, 2006.

45. International Standards Organization (ISO). *Environmental Management—Life Cycle Assessment—Requirements and Guidelines, ISO 14044*; ISO: Geneva, Switzerland, 2006.

46. ACH. Noise Barriers. Available online: http://www.barrieresach.com/pantallas-barreras-acusticas-ACH (accessed on 18 August 2019).

47. Pan Rodo. Noise Barriers. Available online: https://www.obralia.com/dir/minisites/catalogos/419295/catalogo.pdf (accessed on 18 June 2018).

48. European Committee for Standardization (EN). *Road Traffic Noise Reducing Devices—Test Method for Determining the Acoustic Performance—Part 1: Intrinsic Characteristics of Sound Absorption under Diffuse Sound Field Conditions; EN 1793-1*; EN: Brussels, Belgium, 2017.

49. Moreno Ruiz, E.; Valsasina, L.; Brunner, F.; Symeonidis, A.; FitzGerald, D.; Treyer, K.; Bourgault, G.; Wernet, G. *Documentation of Changes Implemented in Ecoinvent Data 3.1*; Ecoinvet: Zurich, Switzerland, 2014; p. 70.

50. European Commission (EC). Commission Decision on the European List of Waste (COM 2000/532/EC). *Off. J. Eur. Communities* **2000**, *50*, 1–31.

51. Galvez-Martos, J.L.; Styles, D.; Schoenberger, H.; Zeschamar-Lahl, B. Construction and demolition waste best management practice in Europe. *Resour. Conserv. Recycl.* **2018**, *136*, 166–178. [CrossRef]

52. Barros, M.C.; Bello, P.M.; Bao, M.; Torrado, J.J. From waste to commodity: Transforming shells into high purity calcium carbonate. *J. Clean. Prod.* **2009**, *17*, 400–407. [CrossRef]

53. Google Maps. Distances. 2020. Available online: https://www.google.com/maps/ (accessed on 2 March 2020).

54. Mercante, I.T.; Bovea, M.D.; Ibáñez-Forés, V.; Arena, A.P. Life cycle assessment of construction and demolition waste management systems: A Spanish case study. *Int. J. Life Cycle Assess.* **2012**, *17*, 232–241. [CrossRef]

55. European Parliament and of the Council (EC). Regulation (EC) No 715/2007 of the European Parliament and of the Council of 20 June 2007 on Type Approval of Motor Vehicles with Respect to Emissions from Light Passenger and Commercial Vehicles (Euro 5 and Euro 6) and on Access to Vehicle Repair and Maintenance Information. Available online: https://eur-lex.europa.eu/eli/reg/2007/715/oj (accessed on 16 April 2019).

56. Barros, C.; Bello, P.; Valiño, S.; Bao, M.; Arias, J. Odours prevention and control in the shell waste valorisation. In Proceedings of the International Symposium on EcoTopia Science (ISETS07), Nagoya, Japan, 23–25 November 2007; pp. 890–895.

57. Neves, N.M.; Mano, J.F. Structure/mechanical behavior relationships in crossed-lamellar seashells. *Mater. Sci. Eng. C* **2015**, *25*, 113–118. [CrossRef]

58. Sahari, F.; Aniza, N. Cockle shell as an alternative construction material for artificial reef. In Proceedings of the International Conference on Creativity and Innovation for Sustainable Development, Sarawak, Malaysia, 12–14 September 2011.

59. Martínez-García, C.; González-Fonteboa, B.; Martínez-Abella, F.; Carro-López, D. Performance of mussel shell as aggregate in plain concrete. *Constr. Build. Mater.* **2017**, *139*, 570–583.

60. PRé Consultants. *SimaPro 8.0.4.LCA Software and Database Manual*; PRé Consultants, B.V.: Amersfoort, The Netherlands, 2016.

61. Guinee, J.B.; Gorrèe, M.; Heijungs, R.; Huppes, G.; Kleijn, G.R.; van Oers, R.L.; Wegener, L.; Sleeswijk, A.; Suh, S.; Udo de Haes, H.A.; et al. *Life Cycle Assessment, an Operational Guide to the ISO Standards. Part 2a*; Guide Kluwer Academic Publishers: Dordrecht, The Netherlands, 2001.

62. Silva, R.V.; De Brito, J.; Dhir, R.K. Fresh-state performance of recycled aggregate concrete: A review. *Constr. Build. Mater.* **2018**, *178*, 19–31. [CrossRef]

63. Heyes, G.; Sharmina, M.; Mendoza, J.M.F.; Gallego-Schmid, A.; Azapagic, A. Developing and implementing circular economy business models in service-oriented technology companies. *J. Clean. Prod.* **2018**, *177*, 621–632. [CrossRef]

64. Mendoza, J.M.F.; Sharmina, M.; Gallego-Schmid, A.; Heyes, G.; Azapagic, A. Integrating backcasting and eco-design for the circular economy: The BECE framework. *J. Ind. Ecol.* **2017**, *21*, 526–544. [CrossRef]

65. Mendoza, J.M.F.; Gallego-Schmid, A.; Azapagic, A. Building a business case for implementation of a circular economy in higher education institutions. *J. Clean. Prod.* **2019**, *220*, 553–567. [CrossRef]

66. European Commission (EC). Towards a Circular Economy: A Zero Waste Programme for Europe. 2014. Available online: https://ec.europa.eu/environment/circular-economy/pdf/circular-economy-communication.pdf (accessed on 3 June 2020).

67. Kalmykova, Y.; Sadagopan, M.; Rosado, L. Circular economy—From review of theories and practices to development of implementation tools. *Resour. Conserv. Recycl.* **2018**, *135*, 190–201. [CrossRef]

68. Ianni, F.; Segoloni, E.; Blasi, F.; Di Maria, F. Low-molecular-weight phenols recovery by eco-friendly extraction from *Quercus* spp. wastes: An analytical and biomass-sustainability evaluation. *Processes* **2020**, *8*, 387. [CrossRef]

Article

The Development of a Waste Tyre Pyrolysis Production Plant Business Model for the Gauteng Region, South Africa

Nhlanhla Nkosi [1,*], Edison Muzenda [1,2,3], Tirivaviri A. Mamvura [2], Mohamed Belaid [1] and Bilal Patel [3]

[1] Department of Chemical Engineering Technology, University of Johannesburg, Johannesburg 2001, South Africa; muzendae@biust.ac.bw (E.M.); mbelaid@uj.ac.za (M.B.)
[2] Department of Chemical, Materials and Metallurgical Engineering, Botswana International University of Science and Technology, Private Mail Bag 16, Palapye, Botswana; mamvurat@biust.ac.bw
[3] Department of Chemical Engineering, College of Science, Engineering and Technology, University of South Africa, Christian de Wet and Pioneer Avenue, Private Bag X6, Florida, Johannesburg 1710, South Africa; patelb@unisa.ac.za
* Correspondence: 200670341@student.uj.ac.za or nkosinhlanhla1@gmail.com

Received: 11 March 2020; Accepted: 8 May 2020; Published: 30 June 2020

Abstract: Some of today's modern life challenges include addressing the increased waste generation and energy deficiencies. Waste tyres have been identified as one of the key environmental concerns due to their non-biodegradable nature and bulk storage space demand. Pyrolysis is a thermochemical process with the potential to address the growing waste tyre problem, energy deficits, and material recovery by converting waste tyres to pyrolysis oil that can be used as a fuel. This study seeks to critically evaluate the feasibility of constructing and operating a waste tyre processing facility and then subsequently marketing and selling the pyrolysis secondary end products by developing a financial business model. The model encompasses costing, procurement, installation, commissioning, and operating a batch pyrolysis plant in Gauteng, South Africa. To achieve the study objectives, an order of magnitude costing method was used for model construction. The results showed the feasibility and sustainability of operating a 3.5 tonne per day batch waste tyre pyrolysis plant in Gauteng Province, South Africa, with a 15-year life span and a projected payback period of approximately 5 years. It was concluded that for the pyrolysis plant to be successful, further treatment steps are required to improve the process economics; also, a stable and sustainable product market should exist and be regulated in South Africa.

Keywords: batch pyrolysis; business model; South Africa; waste tyres

---

1. Introduction

South Africa is burdened with 30 million waste tyres that are either landfilled or illegally dumped in open fields [1]. The generation rates are escalating at a rate of approximately 200,000 tonnes, equating to a million waste tyres annually [2]. In the 2016/2017 financial year, The Department of Environmental Affairs (DEA) reported that 31% of waste tyres have been diverted from landfills to be repurposed in reuse, recycling, and material or energy recovery industries [1]. In addition, through the Recycling and Economic Development Initiative of South Africa (REDISA), approximately 170,000 tonnes of waste tyres were recycled in 2016 [3]. In 2013, approximately 16,037 waste tyres were channeled to recycling, recovery, and reuse initiatives. Subsequently, in 2014 and 2015, the recorded figures of waste tyres allocated to recycling, energy recovery and reuse initiatives were 31,448 and 71,806, respectively [3], showing a significant increase over the years. South Africa is successfully directing waste tyres to different markets of which 25% of waste tyres are reused, 23% are designated for cutting and shredding to be used for spongy mats or playground material, 18% are converted to oil and carbon black through pyrolysis for various applications, and 16% are incinerated for energy recovery in cement or brick manufacturing kilns [3]. The remainder, which is about 18% of waste tyres, are redirected to landfills; this is still very high, and it gives more evidence to the case for repurposing more waste tyres. The characteristics of tyres, waste tyre pyrolysis, the associated primary and valuable secondary products, and the waste tyre management challenges have been comprehensively discussed by several authors in the literature. Muzenda [4] comprehensively detailed the various thermochemical processes such as pyrolysis, gasification, and liquefaction that waste tyres can undergo. The literature was utilised to formulate a detailed understanding of the different thermochemical technologies. Rodriguez [5], Islam [6], and Williams [7] discussed the waste tyre pyrolysis process in great detail as well as the different forms of products that the process can yield. Furthermore, Parthasarathy et al. [8] investigated the effect of process conditions on the product yield of waste tyre pyrolysis. Laresgoiti et al. [9] critically analysed the gases obtained in tyre pyrolysis; Cunliffe and Williams [10] as well as Islam et al. [11] assessed the composition of oils resulting from the pyrolysis of tyres, and Shah et al. [12] examined waste tyre-derived carbon black and its use as an adsorbent. The referenced literature has assisted the authors with the understanding of waste tyre pyrolysis chemistry, final product quantity, and quality as well as possible markets and applications. This paper aims to assess the economic viability of operating a waste tyre batch pyrolysis plant as well as the potential of producing high-value primary and secondary final products. This can significantly contribute towards addressing South Africa's waste tyre challenges, energy, and material recovery, thus contributing to socioeconomic development. In addition, this study taps into the current socioeconomic ills currently experienced in South Africa, such as unemployment and the lack of successful small medium and micro enterprises (SMMEs) [13]. The results of this study can also be adopted by developing countries such as Thailand, Nigeria, and Brazil, who also face the same waste tyre problem as South Africa. In 2012, Thailand reported the energy recovery of waste tyres in the form of pyrolysis to be at 30.23% [14]. Similarly, Nigeria and Brazil are assessing the socioeconomic benefits to be derived from waste tyre management. Waste tyre pyrolysis is yet to be fully explored in these countries [15,16].

1.1. Waste Tyre Pyrolysis Product Compositions, Characteristics, and Application

The waste tyre pyrolysis process yields a gaseous fraction of mainly non-condensable gases, an oily fraction mainly composed of organic substances, and a solid fraction that comprises of mainly carbon, metal, and other inert material. The composition of the primary pyrolysis products is influenced by process operating parameters such as the feed size, temperature and pressure, residence time, heating rate, and reactor configuration, as well as the presence of catalytic medium [17]. Reactor design has been reported as one of the significant factors that affects product output, gas and oil characteristics, and process parameters [18]. The most generally used designs are fixed-bed, rotary/screw kiln, stirred tank, vacuum and fluidised-bed reactor types [18]. Fixed-bed reactors are commonly utilised for slow pyrolysis in batch systems with oil yield ranging from 35% to 50%, while fluidised bed reactors are commonly employed in the fast pyrolysis process and require small particle sizes, with oil yields ranging from 65% to 70% [18]. A rotary kiln reactor is slightly inclined (1°–10°) to progress the waste material forward; the added advantages are that the processing speed of turning, the extent of filling, and particle dimensions can be optimised to improve product yield [19]. Stirred tank reactors are designed for processing whole tyres, resulting in a considerable energy saving on size reduction costs [20]. The vacuum pyrolysis reactor is designed to accommodate larger tyre particles at low pressure and minimum temperature [21].

The approximate yield of gas from waste tyre pyrolysis is about 10–30% by weight and has a heating value of around 30–40 MJ/Nm^3 [22]. Pyrolytic gas is commonly used as a source of fuel and can be adequate to provide the energy required to run a small-scale pyrolysis plant. The gas has high concentrations of methane (CH_4, 44.50 vol %) and ethane (C_2H_4, 4.4 vol %), akin to natural gas

(84.6 vol % CH_4 and 6.4 vol % C_2H_4). However, the large quantities of carbon monoxide (2.41 vol %) hinders its blending with natural gas (0 vol % CO) [23]. The tyre-derived oil (40–50%) is composed mainly of alkylated benzenes, naphthalenes, phenanthrenes, n-alkanes from C_{11} to C_{24}, and alkenes from C_8 to C_{15}, with small quantities of nitrogen (N_2), sulphur (S), and oxygenated compounds [23]. The liquid fraction contains valuable chemicals such as aromatics, d-limonene, and BTX (benzene, toluene, xylenes). BTX compounds play a critical role in the production of chemicals, dyes, plastics, and synthetic fibres; their markets are projected to upsurge at a compound annual growth rate of 5.9% from 2019 to 2027 [24]. Additionally, tyre-derived oil has the potential to be used as automotive fuel after the removal of metal and metalloid contaminants such as zinc, aluminium, iron, titanium, sodium, lead, nickel, and traces of arsenic, chromium, and cobalt [23]. Pyrolytic oil has a high calorific value of about 44 MJ/kg as compared to waste tyres, 33 MJ/kg [23]; wood, 12.4 MJ/kg; coal, 30.2 MJ/kg; and it is close to that of diesel oil, which is 45.5 MJ/kg [25]. Pyrolytic char is a valuable energy source and contributes 30–35% of the product mix with a heating value close to 30.5 MJ/kg [12]. Its heating value is higher than that of South African lignite coals (16.7 MJ/kg) and compares well with petroleum coke (34.9 MJ/kg) [26]. Thus, the char can substitute coal as a source of energy in briquettes and industrial boilers, or it can be co-fired together with coal. Carbon black has the potential to be used as an adsorbent in the removal of heavy metals from industrial and municipal wastewater and as a precursor for activated carbon generation [27]. Furthermore, carbon black can be used as a filler and pigment for making printing inks, tyres, etc. [28]. The commercial viability of steel wire derived from the pyrolytic process which constitutes 10–15 wt % of product mix depends on its cleanliness, quantity, and packaging. Steel with less than 10% of rubber contamination is considered commercially viable [29].

These overwhelming advantages have motivated the authors to undertake this study with the aim of determining the business model for waste tyre pyrolysis in South Africa.

1.2. Waste Tyre Pyrolysis in South Africa

The waste tyre pyrolysis process is not a new concept in South Africa; however, it has not been fully explored to yield any significant successes. Several attempts have been made to operate profitable plants that adhere to environmental laws. Such attempts include the Pretoria-based pyrolysis plant (Innovative Recycling Pty Ltd.), which used to process 25 tonnes of waste tyres daily. The company capitalised on the opportunity of using excess waste tyres by erecting a waste rubber and plastic conversion to fuel plant. The facility ceased operations because it failed to adhere to environmental regulations and laws. In late 2018, another waste tyre pyrolysis facility, Milvinetix in Rosslyn Pretoria was shut down. This plant operated intermittently from March 2012 and processed 10 tonnes per day of the waste tyre employing a batch operation. It produced 40 wt % pyrolysis oil, 30–35 wt % carbon black, 15 wt % steel cords, and 10 wt % uncondensed gases [30]. The oil was sold as crude for industrial applications, such as a source of fuel for kilns and furnaces [30]. In recent years, more pyrolysis businesses have been established that produce a variety of end products. The IRR Manufacturing facility processes waste tyres, waste wood, and waste polyolefin plastics in a 1000 kg/hour pyrolysis plant in Rosslyn, Pretoria, to produce pyrolytic gas, oil, and carbon black [31]. Recor-Waste to Energy Solutions is a South-Gauteng-based pyrolysis plant that converts a variety of wastes such as municipal solids, agricultural, medical, abattoir, and sawdust, converting them to energy. Additionally, waste tyres are converted to energy, oil, and char, while plastic and waste oils are processed to yield a variety of diesel grades [32]. Lastly, Trident Fuels Pty Ltd., located in Germiston, Gauteng, processes waste tyres to produce crumb rubber and carbon black [33].

The pyrolysis oil specifications and properties of the Milvinetix oil were benchmarked against local and international oil standards before application. As a result, an oil sample obtained from Milvinetix was analysed against international standards. Moreover, an additional sample of crude pyrolytic oil was obtained from Pace Oil, which is an oil refinery company that purifies crude oils obtained from different sources. Oil specifications for both Milvinetix and Pace oil are presented in Table 1. A comparison between the different oils shows that the pyrolysis characteristics conform to the American Society for Testing and Material International (ASTM) standards [34] for density and viscosity. However, both oils exhibit a low flash point requiring specific storage measures to meet insurance and fire prevention requirements. As a result, additives and blending would be required to increase the flashpoint and reduce the water content [35]. The two oils were observed as containing excess contaminants, although there was minute metallic contamination. Furthermore, both oils are out of specification regarding the octane index and micro carbon residue, thus limiting their application. The Milvinetix oil is not further treated due to the economics of the business model not being sound when purified [31,36]; thus, it is sold in its crude form.

Carbon black obtained from Milvinetix showed a high calorific value, making it possible for fuel applications, as shown in Table 2. However, it does not conform to the ASTM standards [37] for ash and volatile matter content, meaning the char cannot be considered for industrial use in its virgin form after pyrolysis. The fixed carbon content is about 60%, which is an indication that the char will require a longer combustion time. Consequently, further chemical or physical purification of the carbon black is required to improve its marketing and standardisation potential. Activation agents such as $ZnCl_2$, KOH, and H_3PO_4 are frequently used to chemically treat the carbon black; also, traditional gases such as steam/H_2O, CO_2 and air/O_2 are employed during physical activation [27].

Table 1. Pyrolysis oil specifications.

Test Description	Test Method [34]	Specification [34]	Milvinetix [38]	Pace Oil [39]
Density @20 °C, kg/L	ASTM D4052	0.800 min	0.895	0.8772
Viscosity @40 °C, cSt	ASTM D445	2.2–5.3	2.868	2.1
Flash point, °C	ASTM D93	55 min	<25	26
Water Content, ppm	ASTM 6304	500 max	673	600
Sulphur content, ppm	ASTM 6304	500 max	8100	12,400
Total Contamination number, mg/kg	ASTM 6304	24 max	31	38.6
Distillation °C: 90% Recovery, °C	ASTM D 86	362 max	378.8	360
Micro Carbon Residue	ASTM D4530	0.2 max	4.5	2
Cetane Index	ASTM D4737	51 min	32.01	34.2

Table 2. Pyrolysis carbon black specifications.

Test Description	Test Method [37]	Milvinetix [40]	ASTM Test Method (ASTM, 1996)
Calorific Value	SANS 1928:2009	31.18 MJ/kg	
Moisture Content	SANS 15325:2007	1.30%	
Ash content	ISO 1171:2010	14.50%	0.5% Max
Volatile Matter Content	ISO 582:2010	24.30%	0.3% Max
Fixed Carbon	By difference	59.90%	
Total Sulphur	ASTM 4239:2010	2.61%	

2. Materials and Methods

This work seeks to critically evaluate the profitability of constructing and operating a waste tyre processing facility and subsequently marketing the pyrolysis secondary end products. This was achieved through a desktop study involving comprehensive literature analysis, the evaluation of waste tyre treatment options from the literature as well as local companies doing pyrolysis, and in-depth studies of the pyrolysis process, as well as pyrolysis plant model construction through the costing of major equipment. Telephonic interviews to local-based companies were also used where appropriate and convenient to both parties.

2.1. Costing Data

Before commencing with the costing of equipment, a simplified process flow diagram, Figure 1, was prepared for the aim of identifying all the process steps and all the major equipment to be sized. This was deduced from the literature search for both local and international companies, plant visits, and/or telephonic interviews with local company representatives. Capital cost estimates are differentiated into three types based on their precision and purpose. The Authorisation (Budgeting)

Estimates, with 10–15% precision, is utilised when a preliminary Piping and Instrumentation Diagram (P&ID) of all major equipment is available. The Detailed (Quotation) Estimates, with 5–10% precision, is based on acquired quotations for all equipment and estimations of construction costs [41]. Lastly, The Order of Magnitude Estimate is used in the initial feasibility studies of the project and is based on production rates and major equipment requirements. It may also be used as a tool to estimate the profitability of the project and has ±30% estimation precision [1]. The corrections used in the study and the calculation data were obtained from references [41,42]. The order of magnitude costing method was utilised due to the unavailability of literature information and lack of documented data as well as the existence of a small pyrolysis process industry that is still in its infancy stages in South Africa. In addition, the currently operating pyrolysis facilities are independently owned and ideally protect their intellectual property.

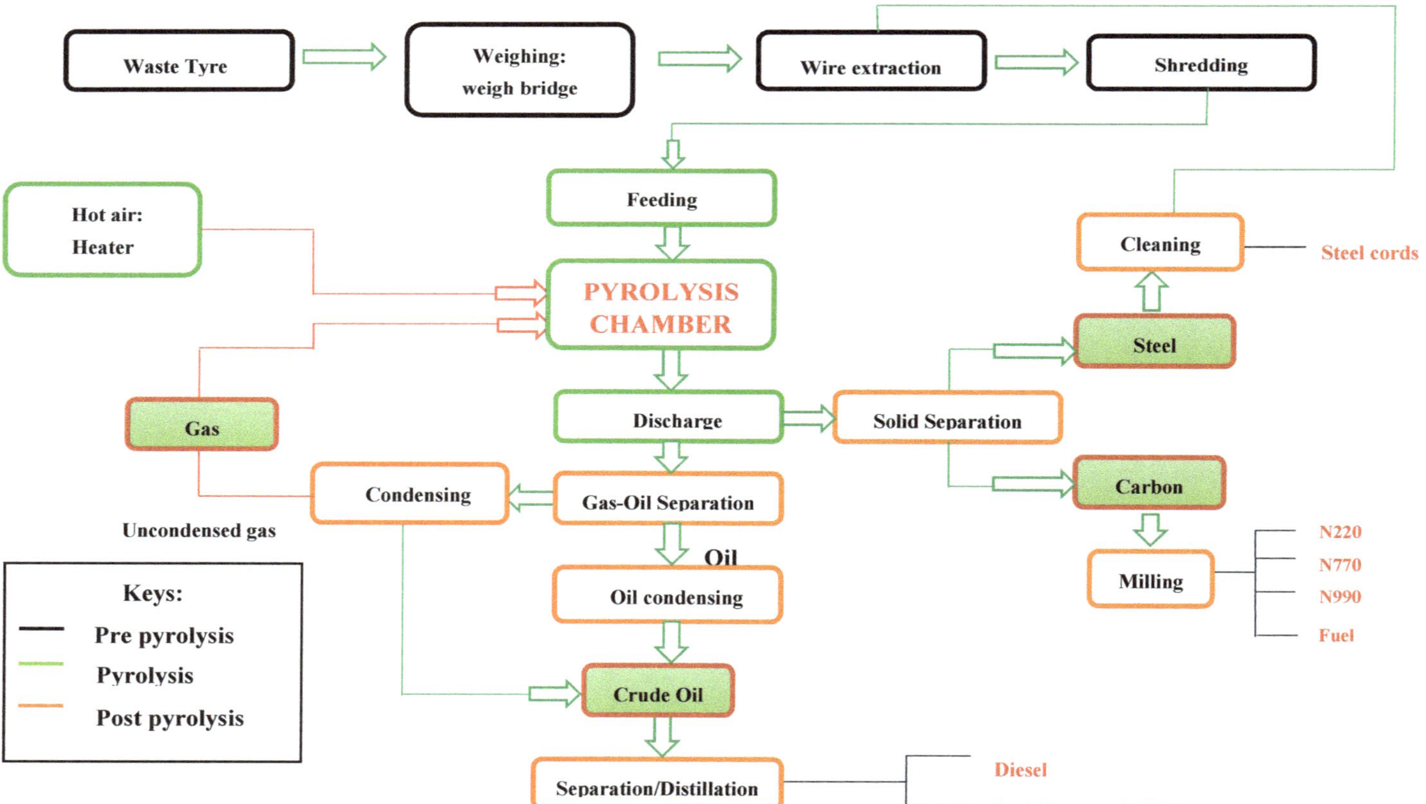

Figure 1. Process flow diagram.

Literature from Sinnott [41] and Sila [42] was used to obtain the base cost equipment data as presented by Equations (1) to (4). Equation (1) is used to obtain the base cost of major equipment using [41], while Equation (2) is utilised to obtain the costs of all equipment except the heat exchanger from [42]. Furthermore, the temperature, pressure, and material of construction correction factors when sizing equipment were derived as provided by [42], as shown in Table 3. From Equation (1), the size parameter (S) is limited to a specific equipment size range and should be factored in during equipment costing. Figure 2 shows the temperature correction factor to be considered during the costing of a heat exchanger, as provided by [41]. The cost literature contains equipment costs at specified capacities; thus, to scale the equipment cost to the required capacities, the standard design limit (Q_1) is introduced in Equation (3) as provided by [42]. To obtain the recent year cost of equipment, such as shell and tube heat exchangers, reactors, boilers etc., for a certain year, the base year index and recent year index are considered. Equation (4) is utilised to obtain a recent price using the Engineering Plant Cost Index (CEPCI). In this regard, the mid-2004 CEPCI values, used as the base year by [41], together with the annual average CEPCI values for 2019 were used to obtain the cost of plant equipment. The final equipment costs are presented in Table 4. Additionally, detailed quotations were obtained for certain process equipment utilised mainly during pre-pyrolysis.

Equipment costing equations

$$C_E = CS^n \tag{1}$$

$$C_E = F_T\ F_P\ F_M\ C_2 \tag{2}$$

$$C_1 = C_b\ (Q_1/Q_b)^n \tag{3}$$

$$C_2 = C_1(I_2/I_1) \tag{4}$$

Table 3. Temperature (F_T), pressure (F_p), and material of construction correction factor (F_M) [42].

Design Pressure, (atm)	Correction Factor
0.005	1.3
0.014	1.2
0.048	1.1
0.54 to 6.8	1.0
48	1.1
204	1.2
408	1.3

Design Temp, °C	Correction Factor
−80	1.3
0	1.0
100	1.05
600	1.1
5000	1.2
1000	1.4

Material of Construction	Correction Factor
Carbon steel (mild)	1.0
Bronze	1.05
Carbon/molybdenum Steel	1.065
Aluminium	1.075
Cast steel	1.11
Stainless steel	1.28 to 1.5
Worthite alloy	1.41
Hastelloy C alloy	1.54
Monel alloy	1.65
Titanium	2.0

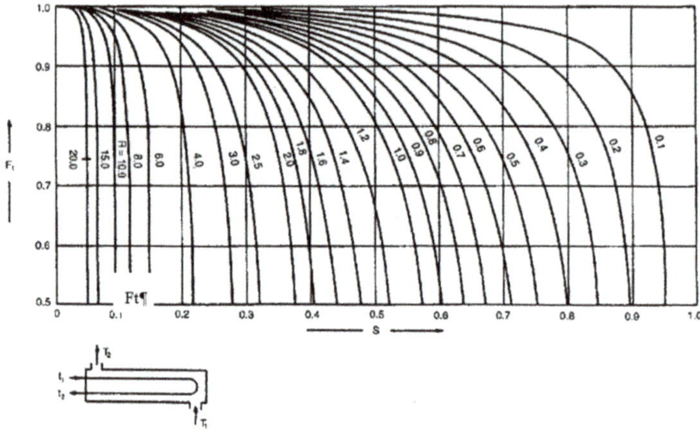

Figure 2. Temperature correction factor: two shell passes, four or multiples of four tube passes [41].

Table 4. Plant major equipment cost.

	Description	Cost	Comments	Costing Reference
		Pre-Pyrolysis		
1.1	Magnetic separator	R60,976.80	1.1 kW motor, 550 kg	[Quote]
1.2	Tyre shredding	R315,900.00	60–80 tyres per hour, 15 hp	[Quote]
1.3	weighing system	R10,649.88	SABS approved, with readout screen	[Quote]
	Total	R387,526.68		
		Pyrolysis		
2.1	Reaction chamber	R1,635,822.26	D2.8 m, L6 m, 15 KW, 380 V, struc steel	[42]
2.2	Heat exchanger	R475,422.06	CS tubes, TEMA B std, 16.88 KW	[41]
2.3	Heat exchanger	R394,254.99	CS tubes, TEMA B standards	[41]
2.4	Reboiler	R405,998.85	High carbon steel	[41]
2.5	Condenser	R596,748.87	Mild carbon steel	[41]
2.6	Storage tanks	R206,162.66	Steel fuel tank and plastic water tank	[41]
2.7	Cooling tower	R1,125,640.00	Stainless steel/galvanised steel	
2.8	Pumps	R207,777.26	Carbon steel	[42]
2.9	Instrumentation and controls	R2,058,487.70		[41]
2.10	Interconnecting pipes and valves	R230,216.33		[41]
	Total	R7,336,530.97		
		Post -Pyrolysis		
3.1	Distillation column	R335,142.20	Carbon steel/stainless steel	[41]
3.2	Column plates	R149,795.00	Plate (bubble cap)	[41]
3.3	Micro ball mill	R839,878.00	Stainless steel	[Quote]
3.4	Gas scrubber	R685,844.00	Carbon steel/stainless steel	[41]
	Total	R2,010,659.2		
	Total equipment	R9,734,716.85		

2.2. Economic Feasibility Analysis

The project's economic feasibility analysis is performed subsequent to the costing of all major equipment. This is selectively done to determine the financial performance of the project during its lifespan. Several methods are used to determine the profitability of the project and these are also compared with values provided in literature. The following project profitability parameters are used and represented by Equations (5) to (11).

Net present value (NPV)—the present time value of money used in investment planning to analyse the profitability of a project.

Rate of return (ROR)—the gain or loss of an investment compared to the cost of an initial investment, expressed as a percentage.

Return of assets (ROA)—indicates the profitability of a company relative to its total assets.

Depreciation—the rate at which fixed assets are reduced uniformly over their useful life.

$$\text{Labour costs} = \text{hourly rate} \times \text{total number of hours worked} \tag{5}$$

$$\text{Operating costs} = \text{labour costs} + \text{Fixed costs} + \text{variable cost} \tag{6}$$

$$\text{Revenue} = \text{quantity} \times \text{price} \tag{7}$$

$$\text{Cumulative cash flow} = \text{revenue} - \text{operating costs} - \text{taxes} \tag{8}$$

$$\text{Return of asset} = (\text{net income})/(\text{total assets}) \tag{9}$$

$$\text{Net present value} = (\text{initial investment})/((\sum_{i=1}^{time}(\text{Cash flow})/(1+\text{discounted rate})) \tag{10}$$

$$\text{Rate of return} = \text{cumulative cash and the end of project}/(\text{project life} \times \text{original investment}) \tag{11}$$

3. Results and Discussions

3.1. Pyrolysis Parameters and End Products

The literature analysis regarding the feasibility of constructing and operating a batch pyrolysis plant in Gauteng as previously highlighted shows that to a large extent, the reactor temperature, heating rate, and reactor design determine the yield of solid, gas, and liquid products. In this regard, the operation of a fixed-bed reactor at 550 °C, ambient pressure and constant heating rate would allow for the production of 45% oil, 5% pyrolysis gas, 35% carbon black, and 15% steel wires as final products. Upon completion of the pyrolysis process, pyrolytic gas is cooled and further condensed to form pyrolytic oil. This oil is classified as No. 6 oil, which is a thick, syrupy heavy crude oil and has an acid smell. To increase the economics of this product that would facilitate a viable waste tyre pyrolysis facility, purification steps, such as desulphurisation and/or distillation will be integrated into the process. This treatment step improves the quality of the oil, thus increasing its application potential. The crude pyrolysis oil is distilled to form light tyre-derived distillate fuel and residual fuel oil fractions. The residual oil is blended with diesel, while the distillate fraction is utilised in agricultural vehicles. Alternatively, the oil can be treated to derive some important chemicals such as BTX and D-limonene and this value addition contributes to the improvement of the process economics. The uncondensed gases are recycled back to the system for use as fuel to sustain the process, thereby reducing the energy input. This contributes to making the process thermally self-sufficient. The emissions from the reactor burners are chemically treated using a gas absorption process. The solid fractions consist of a mixture of carbon black and steel wires. A magnetic separation is used to remove ferrous metals to isolate the two components. The carbon black is further milled to obtain different grades and fractions of the char. The following fractions can be obtained: N220 (24–33 nm) used in rubber and rubber products; N770 (70–96 nm) used in paints and pigments; N990 (250–350 nm) used as activated carbon, and the residue fraction can be used for briquettes. The steel component is sold to steel manufacturers or dealers. However, in the South African context, no records of such applications have been reported.

3.2. Process Description

The pyrolysis process P&ID is shown in Figure 3, and the process description is explained in accordance with the process streams in the P&ID. Waste tyres are transported onsite and stored in a designated storage area. At the beginning of each shift, tyres are quantified using a weighing system, and the obtained figures are kept as record. Reinforcing wire is subsequently removed from the tyre, which is represented by process line 2 in the P&ID, and it is baled to be sold to recyclers as a form of revenue. Then, tyres are allowed to pass through a mechanical shredder, which will minimise their size to particle sizes of between 15 and 40 mm as recommended by [43] and denoted by process line 3. The shredded tyre chips are introduced into the reactor chamber where nitrogen is purged into the system to allow for an inert environment, as shown in process line 4. Pyrolysis is an endothermic process; thus, an electrical heater will be utilised to provide thermal energy up to 550 °C, which is required to initiate the decomposition of the tyre chips in the pyrolysis reactor. The vaporised gases are cooled using two heat exchangers in series (process line 5 and 14); thereafter, condensation of the oil is facilitated and is represented by process lines 15 and 20. The non-condensable gases, as shown in process line 18, are recycled back into the reactor to sustain the process. The oil fraction is further distilled into light distillate oils, as shown in process line 29, for application in agricultural vehicle or blended with commercial diesel containing low sulphur content. The heavy residue oils, as shown in process line 26, will be sold to fire up furnaces, boilers, and kilns. The solid char is collected at the end of each batch process and later segregated from the unextracted steel cords using a magnetic separator, as shown in process line 8. The char will be further milled into different grades, as seen in process line 11, and sold for profit. The flue gas that remains in the reactor chamber after each batch will be channeled to the absorption column for abatement, as presented by line 12 in the P&ID.

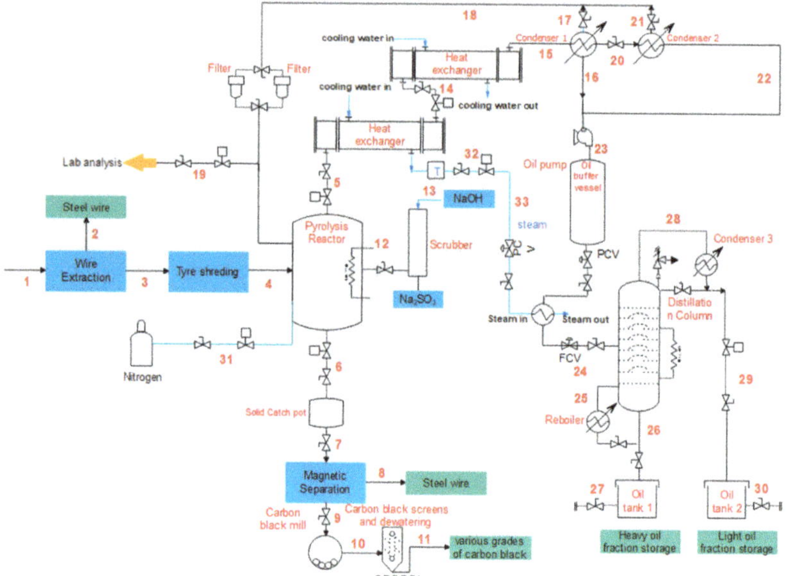

Figure 3. Pyrolysis process piping and instrumentation diagram.

The material and energy balances are presented in Table 5. The material balance gives an indication that a material feed rate of 145 kg/h produces a gas fraction at 50.86 kg/h. Consequently, approximately 54 and 20 litres per hour of distillate and heavy oils will be produced respectively during the process. In addition, steel wire and carbon black will be produced at a rate of 21.79 and 50.86 kg/h respectively. Apart from the thermal energy provided in the reactor to facilitate the reaction, the feed material and products generated during pyrolysis will also aid in providing the process energy requirements, as highlighted by the energy balance in Table 5.

Table 5. Pyrolysis mass and energy balances.

Stream	Description	Mass Flow (kg/hour)	Process Conditions (Temperature and Pressure)	Volume Flow (m³/hour)	Energy, Q_{in} (MJ/s)	Energy, Q_{out} (MJ/s)
1	Tyres	145.31	25 °C 87 kPa		1.29	
2 + 8	Steel wire	21.79	25 °C, 87 kPa			
3	Tyre chips	130.78	25 °C, 87 kPa		1.16	1.16
5	Pyrolysis gas	72.66	550 °C, 300 kPa [44]	90.83	0.5	0.5
9	Carbon black	50.86	25 °C, 87 kPa		0.4	0.4
12	Flue gas	0.727	150 °C, 200 kPa	0.91		
18	Uncondensed pyrolysis gas	7.27	105 °C, 200 kPa	9.10		0.0076
24	Pyrolysis oil	65.39	350 °C, 350 kPa	0.074		0.73
26	Residual oils	18.08	40 °C, 200 kPa	0.02		0.2
29	Distillate oils	47.31	40 °C, 200 kPa	0.054		0.53

3.3. Pyrolysis Utilities

The process requires process water for cooling equipment such as heat exchangers, cooling towers, condensers, and carbon black wet grinding as well as general use in the plant. Approximately 8500 L of water will be used monthly in the plant. The energy requirement for the plant for all the different process energy demands is shown in Table 6. The available fuel is the 5% gas produced during pyrolysis, which is subsequently recycled back into the process to help sustain the process. This is power supplied to components such as the tyre shredder, heaters, pumps, and control systems as well as large mechanical and heating equipment. Sodium hydroxide (NaOH) is used as the scrubbing reagent at a cost of R9.55/kg per bag. Nitrogen is required in the process to purge out the oxygen (O_2) and create an inert environment; the N_2 cost estimate is approximately R485.79 per 13 kg cylinder.

Table 6. Total plant power and energy requirement.

Energy Type	Amount	Unit
Heating	121.05	kW
Mechanical	151.79	kW
Cooling	10.5	kW
Electrical	81.75	kW
Available fuel	139.56	kW
Energy efficiency	0.8	
Total supply	317.53	kW
Yearly energy requirements	1,846,957.70	kWhr/year

3.4. Gas Emission Scrubbing

The flue gas emissions are confined and scrubbed through a gas absorption tower using NaOH solution. The typical flue gas composition is sulphur dioxide (SO_2), carbon dioxide (CO_2), nitrogen oxides (NOx), hydrocarbons (H-C), moisture (H_2O), CO, N_2, and O_2 [45]. Carbon dioxide is reacted with sodium hydroxide to form water and sodium carbonate, as denoted by Equation (12). In the presence of excess SO_2, sodium sulfite (Na_2SO_3) is formed as a final stable product, as shown in Equation (13). Sodium sulfite is primarily used in the pulp and paper industry, food preservatives, and pharmaceuticals [46]. The sale of Na_2SO_3 can potentially add another income stream into the business model.

$$NaOH + CO_2 \rightarrow Na_2CO_3 + H_2O \tag{12}$$

$$SO_2 + H_2O + Na_2CO_3 \rightarrow Na_2SO_3 + CO_2 \tag{13}$$

3.5. Pyrolysis Production Plant Model

The economic model is based on a 15-year pyrolysis plant life span consisting of 4 rotating shifts with 3 shifts operating daily. The processing facility operates as a batch process of 303 days per year;

the remaining days are utilised for maintenance as shown in Table 7. The available plant capacity is 1056.5 tonnes/year, with a shutdown period of 62 days/year. The initial process assumptions, input, and input costs for the project are also given in Table 7.

Four income streams will be the core revenues to be considered for the project. In this regard, a tyre gate fee of R14.10 per tyre is collected. Distilled pyrolysis oil is sold at R14.65 per litre, taking into account the annual average 2019 fuel price of R15.81 per litre with fluctuation allowance. The oil will be mainly sold to agricultural businesses for use in agricultural vehicles and machinery. Overall revenues of R2.21 million and R1.85 million per annum of pyrolysis oil and various carbon black grades (N220, N770, N990, and briquettes) respectively will be collected at the end of the first financial year. Lastly, the residual steel wires are sold to appropriate dealers at a rate of R2500 per tonnes, thus contributing approximately R0.5 million towards the first-year revenues. The 5% uncondensed gasses are recycled back into the process to sustain the plant.

As highlighted previously, the order of magnitude estimate method was used to cost all major equipment. Using project evaluation methods, represented by Equations (1) to (4), it can be concluded that the project is worth investing in with a projected payback period of approximately 5 years, as shown in Figure 4a. Consequently, the project requires a capital incentive of R 17.5 million during year 0; this total includes all major equipment, plant assessment costs (accounting for 12.5% of the capital expenditure, Capex); building and structure (accounting for 25% of Capex), engineering and construction (accounting for 30% of Capex) as well as other costs such as contingency fees (accounting for 15% of the total initial investment). The required Capex can be solely funded by a financial institution with a pay period of 5 years based on an annual interest rate of 10%.

Table 7. Pyrolysis plant process assumptions, inputs, and costs.

Variable	Unit	Value
Weight of rimless tyre	Kg	7.75
Single tyre gate fee	R/tyre	13
Operating hours	Days/year	302.95
Plant shut down time	Days/year	62.05
Operating shifts	Per day	3
Loading cycles	Per shift	3
Treated tyres	ton/day	3.5
Annual working hours	hour/year	8760
Downtime	hour/year	1489.2
Plant operating time	hour/year	7270.8
Actual plant capacity	ton/hour	1056.54
Available plant capacity	ton/hour	1272.94
Actual annual production	ton/hour	1056.5
Process input costs		
Electrical energy	R/kWh	1.18
Water	R/L	0.029
NaOH (scrubber reagent)	R/ton	R9550.00
Nitrogen	R/13 kg	R485.79

During year 1, annual operating costs that can be further categorised into fixed costs and variable costs need to be factored in before operations can commence. Fixed operating costs are inclusive of maintenance (labour and materials); operating labour; land rental, laboratory costs; supervision; rates (and any other local taxes), medical and insurance cost. An incentive of approximately R10 million is required during year 1 to accommodate all fixed and variable operating costs. A tax rate and value-added tax (VAT) rate of 28% and 15% respectively are considered for the annual revenues. An annual straight-line depreciation rate is applied over the 15-year plant lifespan. This tally's up to initial cash injection of R27.5 million required to successfully set up the business. Detailed operating costs, labour costs, revenues, and revenues after-tax projection from year 0 to the end of the project are shown in Table 8.

Figure 4a shows a general increasing trend for all the plotted variables (cumulative cash flow, operating costs, and annual revenue) from year 0 to year 15. A steady increase in the net profit is projected predominantly around year 5; this is due to the business ending its Capex loan repayment period after year 4. The cumulative cash flow can be used to indicate the projected plant life. Revenues are flowing into the business; however, the cumulative cash flow remains negative until the initial investment required to design, build, and start up the plant is paid off. Year 0 to year 4 is classified as the payback period of the project. As a result, the business only breaks even after 5 years due to the high capital investment accompanied by an annual 10% interest on loan repayment. However, due to the availability of raw materials at no cost and the sale of high-value end products, the plant is seen to be highly profitable thereafter. In the literature, [41] highlights that a payback time of 2 to 5 years is expected for projects estimates using cost data [41], thus showing that the obtained results agree with the literature. An average inflation rate of 7.3% for the end-products sale price, operating costs, and raw material prices is factored in annually.

Figure 4b provides the profitability predictions of the project; it shows stabilisation in the accumulated annual revenue and production costs towards the end of the project. This is expected, as a high increase in revenue is realised from year 5 onwards. The net present value (NPV) of money has also been used to determine whether the project is a worthwhile investment, in terms of profit yield and breakeven period. In this regard, Figure 4b shows that the project is worth investing in due to the NPV proceeding from being negative to being positive over the increasing duration of the project; this is also depicted in Table 8. In addition, the NPV curve also indicates the projected plant life, which agrees very well with the plotted plant life curve given in Figure 4a. The return of assets (ROA) curve in Figure 4b is also in agreement with the NPV and shows an increasing trend in Table 8. The project is shown to gradually depreciate over time from year 1 to year 15 at an acceptable rate. This is highly influenced by the scheduled plant maintenance, which will ensure proper plant preservation and the regular servicing of all plant equipment. According to [41], 20% to 30% for the rate of return (ROR) can be used roughly as a guide for evaluating small projects and the higher a project's return, the more attractive it is [41]. The ROR for this project is projected to be 31.9%, as shown in Table 8; this figure is in strong agreement with the literature, thus confirming that the project is a viable investment that will yield profits and longevity during its projected lifespan.

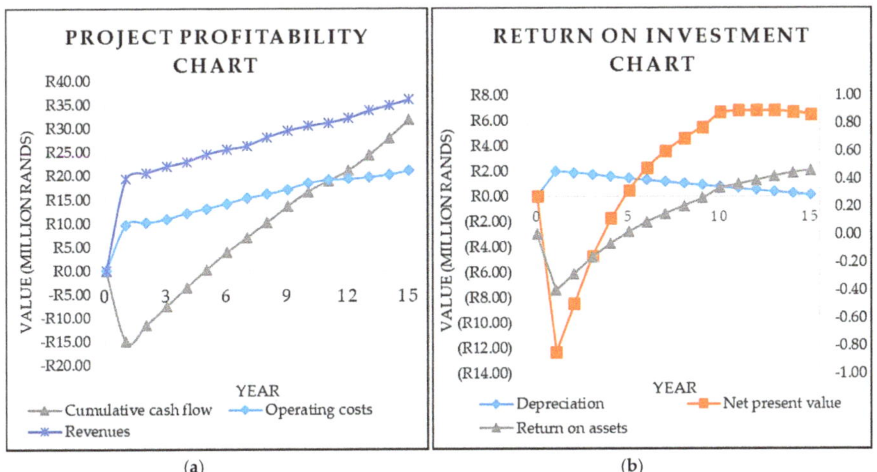

(a)

(b)

Figure 4. Project cost evaluation: (**a**) Cumulative cash flow, revenues and operating costs projections; (**b**) Net present value, return of assets, and depreciation rate projection.

Table 8. Annual project profitability projections.

Year	Annual Cash Flow ($\times 10^6$)	Revenues After Tax ($\times 10^6$)	Cumulative Cash Flow ($\times 10^6$)	Operating and Labour Costs ($\times 10^6$)	Revenues ($\times 10^6$)	NPV ($\times 10^6$)	Depreciation ($\times 10^6$)	Depreciation %	ROA	ROR %
0	−R17.49	R0.00	R0.00	R0.00	R0.00	R0.00	R0.00	0	R0.00	31.91
1	−R28.93	R13.96	−R14.96	R15.88	R19.39	−R13.60	R2.19	0.125	R2.19	
2	−R26.40	R14.90	−R11.50	R16.55	R20.70	−R9.51	R2.04	0.117	R2.04	
3	−R23.36	R15.90	−R7.46	R17.88	R22.08	−R5.61	R1.89	0.108	R1.89	
4	−R20.14	R16.61	−R3.53	R19.83	R23.06	−R2.41	R1.75	0.100	R1.75	
5	R13.93	R17.74	R0.28	R20.76	R24.64	R0.18	R1.60	0.092	R1.60	
6	R14.86	R18.45	R3.88	R22.20	R25.63	R2.19	R1.46	0.083	R1.46	
7	R15.91	R19.02	R6.99	R23.95	R26.42	R3.59	R1.31	0.075	R1.31	
8	R17.10	R20.30	R10.19	R25.51	R28.19	R4.75	R1.17	0.067	R1.17	
9	R17.88	R21.29	R13.59	R27.42	R29.57	R5.76	R1.02	0.058	R1.02	
10	R18.94	R22.00	R16.65	R29.74	R30.55	R6.42	R0.87	0.050	R0.87	
11	R20.23	R22.49	R18.91	R31.10	R31.24	R6.63	R0.73	0.042	R0.73	
12	R20.92	R23.19	R21.18	R31.41	R32.20	R6.75	R0.58	0.033	R0.58	
13	R21.12	R24.30	R24.36	R32.04	R33.75	R7.06	R0.44	0.025	R0.44	
14	R21.50	R25.13	R28.00	R33.00	R34.90	R7.37	R0.29	0.017	R0.29	
15	R22.09	R25.96	R31.87	R34.32	R36.06	R7.63	R0.15	0.008	R0.15	

4. Conclusions

Based on the pyrolysis plant business model and all relevant data collected through literature analysis, site visits as well as telephonic, and personal interviews, the following conclusions can be drawn. (i) The waste tyre plant business model shows that a 3.5 tonne per day plant yields a reasonable payback period of 5 years and a plant life of 15 years is projected. (ii) Further treatment steps are required to improve the process economics by creating valuable products; however, further optimisation studies can be performed with the aim of increasing plant productivity. (iii) For a successful business model, a stable and sustainable product market should exist and be regulated in South Africa. For further work, the authors recommend a detailed analysis of the environmental impact and policy framework on operating pyrolysis plants in South Africa.

Author Contributions: This paper is part of the work presented in the first author's Master's dissertation at the University of Johannesburg. N.N. conceptualised the research work presented under the supervision and guidance of E.M., M.B. and B.P. The methodology followed during the compilation of the final work included literature research, site visits, telephonic and personal interviews. A combined effort by N.N. and E.M. was undertaken to validate and authenticate the obtained data. Formal data analysis and interpretation was carried out for the aim of fitting the work within the defined scope of the project. N.N. and E.M. were involved in the writing of the original draft for this work; subsequently, M.B., B.P., and T.A.M. reviewed and edited the final draft submitted. All authors have read and agreed to the published version of the manuscript.

Funding: This research was funded by the University of Johannesburg, The Global Stature 4.0.

Acknowledgments: The Universities of Johannesburg and South Africa and as well as the Botswana International University of Science and Technology.

Conflicts of Interest: The authors declare no conflict of interest.

Nomenclature

C_E = Purchased equipment cost (US Dollar)
F_T = Temperature correction factor (degrees celsius)
F_P = Pressure correction factor (atmosphere)
f_M = Material of construction correction factor
n = Equipment type index
C = Cost constant
S = Size parameter
C_b = Base cost (US Dollar)
C_1 = Base year cost (US Dollar)
C_2 = Recent year cost (US Dollar)
Q_b = Equipment design value
lQ_1 = Standard design limit
I_1 = Base year index
I_2 = Recent year index

References and Notes

1. Targets for Diverting Waste Tyres from Landfill Sites, Parliament of the Republic of South Africa. Available online: http://pmg-assets.s3-website-eu-west-1.amazonaws.com/180417Waste_Tyre.pdf (accessed on 14 July 2019).
2. Burger, S. Creamer Media's Engineering News. Available online: https://www.engineeringnews.co.za/searchadvanced_en.php?is_id=81&sortOrder=DESC&st=i&searchAll=on&searchStartDate=&searchEndDate=&searchSortBy=sr_date&searchString=false+start+Redisa (accessed on 27 February 2020).
3. Hartley, F.; Caetano, T.; Daniels, R.C. *Economic Benefits of Extended Producer Responsibility Initiatives in South Africa: The Case of Waste Tyres*; Springer: Berlin, Germany, 2016.
4. Muzenda, E. A comparative review of waste tyre pyrolysis, gasification and liquefaction (PGL) processes. In Proceedings of the International Conference on Chemical Engineering Advanced Computational Technologies, Pretoria, South Africa, 24–25 November 2014.

5.   Rodriguez, E.; Laresgoiti, M.F.; Torres, A.; Chomo'n, M.J.; Caballero, B. Pyrolysis of scrap tyres. *Fuel Process. Technol.* **2001**, *72*, 9–22. [CrossRef]

6.   Islam, R.M. Innovation in pyrolysis technology for management of scrap tire: A solution of energy and environment. *Int. J. Environ. Sci. Dev.* **2010**, *1*, 89–96. [CrossRef]

7.   Williams, P.T. Pyrolysis of waste tyres: A review. *Waste Manag.* **2013**, *33*, 1714–1728. [CrossRef]

8.   Parthasarathy, P.; Choi, H.S.; Park, H.C.; Hwang, J.G.; Yoo, H.S.; Lee, B.K.; Upadhyay, M. Influence of process conditions on product yield of waste tyre pyrolysis-A review. *Korean J. Chem. Eng.* **2016**, *33*, 2268–2286. [CrossRef]

9.   Laresgoiti, M.F.; Marco, I.; Torres, A.; Caballero, B.; Cabrero, M.A.; Chomo´n, M.J. Chromatographic analysis of the gases obtained in tyre pyrolysis. *J. Anal. Appl. Pyrolysis* **2000**, *55*, 43–54. [CrossRef]

10. Cunliffe, A.M.; Williams, P.T. Composition of oils derived from the batch pyrolysis of tyres. *J. Anal. Appl. Pyrolysis* **1998**, *44*, 131–152. [CrossRef]

11. Islam, M.R.; Haniu, H.; Beg, M.R.A. Liquid fuels and chemicals from pyrolysis of motorcycle tire waste: Product yields, compositions and related properties. *Fuel* **2008**, *8*, 3112–3122.

12. Shah, J.; Jan, M.R.; Mabood, F.; Shahid, M. Conversion of waste tyres into carbon black and their utilization as adsorbent. *J. Chin. Chem. Soc.* **2006**, *53*, 1085–1089. [CrossRef]

13. Chimucheka, T. Overview and performance of the SMMEs sector in South Africa. *Mediterr. J. Soc. Sci.* **2013**, *4*, 783–795. [CrossRef]

14. Jacob, P.; Kashyap, P.; Suparat, T.; Visvanathan, C. Dealing with emerging waste streams: Used tyre assessment in Thailand using material flow analysis. *Waste Manag. Res.* **2014**, *32*, 918–926. [CrossRef]

15. Obi, C. Environmental impact of end of life tyre (ELT) or scrap tyre waste pollution and the need for sustainable waste tyre disposal and transformation mechanism in Nigeria. *Nnamdi Azikiwe Univ. J. Int. Law Jurisprud.* **2019**, *10*, 60–70.

16. Neto, G.C.; Chaves, L.E.; Pinto, L.F.; Santana, J.C.; Amorim, M.P.; Rodrigues, M.J. Economic, environmental and social benefits of adoption of pyrolysis process of tires: A feasible and ecofriendly mode to reduce the impacts of scrap tires in Brazil. *Sustainability* **2019**, *11*, 2076. [CrossRef]

17. Fortuna, F. Pilot-scale experimental pyrolysis plant: Mechanical and operational aspects. *J. Anal. Appl. Pyrolysis* **1997**, *40*, 403–417. [CrossRef]

18. Martínez, J.D. Waste tire pyrolysis–A review. *Renew. Sust. Energ. Rev.* **2013**, *23*, 179–213. [CrossRef]

19. Alsaleh, A.; Sattler, M.L. Waste Tire Pyrolysis: Influential Parameters and Product Properties. *Curr. Sustain. Renew. Energy Rep.* **2014**, *1*, 129–135. [CrossRef]

20. Bianchi, M.; Bortolabi, G.; Cavazzoni, M.; de Pascale, A.; Montanari, I.; Peretto, A.; Tosi, C.; Vecci, R. Preliminary design and numerical analysis of a scrap tires pyrolysis system. *Energy Procedia* **2014**, *45*, 111–120. [CrossRef]

21. Rowhani, A.; Thomas, J. Scrap tyre management pathways and their use as a fuel—A review. *Energies* **2016**, *9*, 888. [CrossRef]

22. Czajczyńska, D.; Krzyżyńska, R.; Jouhara, H.; Spencer, N. Use of pyrolytic gas from waste tire as a fuel: A review. *Energy* **2017**, *134*, 1121–1131.

23. Roy, C.; Chala, A.; Darmstadt, H. The vacuum pyrolysis of used tires End-uses for oil and carbon black products. *J. Anal. Appl. Pyrolysis* **1999**, *51*, 201–221. [CrossRef]

24. Report and Data. Available online: http://www.reportsanddata.com/report-detail/btx-benzene-toluene-and0xylene-matket (accessed on 28 March 2020).

25. Ahmed, R.; van de Klundert, A.; Lardinois, I. *Rubber Waste Options for Small-Scale Resource Recovery in Urban Solid Waste*; Waste: Gouda, The Netherlands, 1996.

26. Edward, L.; Danny, C.; Gordon, M. Production of active carbons from waste tyres—A review. *Carbon* **2004**, *42*, 2789–2805.

27. Hageman, N.; Spokas, K.; Schmidt, H.T.; Ralf, R.; Böhler, M.A.; Bucheli, T.D. Activated carbon, biochar and charcoal: Linkages and synergies across pyrogenic Carbon's ABCs. *Water* **2018**, *10*, 182. [CrossRef]

28. Li, S.Q.; Yao, Q.; Chi, Y.; Yan, J.H.; Cen, K.F. Pilot-scale pyrolysis of scrap tires in a continuous rotary kiln reactor. *Ind. Eng. Chem. Res.* **2004**, *43*, 5133–5145. [CrossRef]

29. California Integrated Waste Management Board. *Technology Evaluation and Economic Analysis of Waste Tire Pyrolysis, Gasification, and Liquefaction*; University of California Riverside: Riverside, CA, USA, 2006.

30. Nkosi, N.P. *Site Visit at Milvinetix, Rosslyn and Interview with Brandon*; Pretoria, South Africa, 2013.

31. IRR Manufacturing. Available online: http://irrmanufacturing.com/pyrolysis/ (accessed on 2 March 2020).

32. Recor. Available online: https://www.recor.co.za/waste-to-energy-solutions (accessed on 2 March 2020).

33. Trident Fuels. Available online: http://www.tridentfuels.co.za/ (accessed on 2 March 2020).

34. American Society for Testing and Material International. *ASTM D86-12 in Standard Test Method for Distillation of Petroleum Products at Atmospheric Pressure*; ASTM International: West Conshohocken, PA, USA, 1996.

35. Li, D.; Zhen, H.; Xingcai, L.; Wu-gao, Z.; Jian-guang, Y. Physico-chemical properties of ethanol–diesel blend fuel and its effect on performance and emissions of diesel engines. *Renew. Ener.* **2005**, *30*, 967–976. [CrossRef]

36. Nkosi, N.P. *Site Visit and Telephonic Interview with Pace Oil Personnel: Justice*; 2013.

37. American Society for Testing and Material International. *International Designation Index, Rubber Standards, in D5900–13, Standard Specification for Physical and Chemical Properties of Industry Reference Materials (IRM)*; ASTM International: West Conshohocken, PA, USA, 1996.

38. *SGS Pyrolysis Diesel Oil Analysis Report*; SGS: Randburg, Johannesburg, South Africa, 2012.

39. *Wear Check Diesel Analysis Report*; Wear Check: Isando, Johannesburg, South Africa, 2012.

40. Coal and Minerals Laboratory. *Carbon Black Sample*; Council of Science and Industrial Research: Lynwood, Pretoria, South Africa, 2012.

41. Sinnott, R.K. Costing and project evaluation. In *Chemical Engineering Design, Coulson & Richardson's Chemical Engineering*, 4th ed.; Elsevier Butterworth-Heinemann: London, UK, 2005; Volume 6, pp. 243–280.

42. Silla, H. Production and capital cost estimation. In *Chemical Process Engineering: Design and Economics*; Heinemann, H., Ed.; Marcel Dekker, Inc.: New York, NY, USA, 2003.

43. Oyedun, A.; Lam, K.L.; Fittkau, M.; Hui, C.W. Optimisation of particle size in waste tyre pyrolysis. *Fuel* **2012**, *96*, 417–424. [CrossRef]

44. Murillo, R.; Aranda, A.; Aylo´n, E.; Calle´n, M.S.; Mastral, A.N. Process for the separation of gas products from waste tire pyrolysis. *Ind. Eng. Chem. Res.* **2006**, *45*, 1734–1738.

45. Schrenk, H.H.; Berger, L.B. Composition of Diesel Engine Exhaust Gas. *Am. J. Public Health Nations Health* **1941**, *31*, 669–681. [CrossRef]

46. Weil, E.D.; Sandler, R. *Kirk-Othmer Concise Encyclopedia of Chemical Technology*, 4th ed.; John Wiley & Sons: New York, NY, USA, 1999.

Article

The Effects of pH Change through Liming on Soil N₂O Emissions

Muhammad Shaaban [1,2], **Yupeng Wu** [1], **Lei Wu** [3], **Ronggui Hu** [1,*], **Aneela Younas** [4], **Avelino Nunez-Delgado** [5,*], **Peng Xu** [1], **Zheng Sun** [6,7], **Shan Lin** [1], **Xiangyu Xu** [1] **and Yanbin Jiang** [1]

[1] College of Resources and Environment, Huazhong Agricultural University, Wuhan 430070, China; shabanbzu@hotmail.com (M.S.); wyp19851205@mail.hzau.edu.cn (Y.W.); xupengstu192@163.com (P.X.); linshan@mail.hzau.edu.cn (S.L.); xuxiangyu2004@sina.com (X.X.); jiangyanbin@mail.hzau.edu.cn (Y.J.)

[2] Department of Soil Science, Faculty of Agricultural Sciences and Technology, Bahauddin Zakariya University, Multan 60800, Punjab, Pakistan

[3] Institute of Agricultural Resources and Regional Planning, Chinese Academy of Agricultural Sciences, Beijing 100081, China; wuleies@163.com

[4] College of Plant Science and Technology, Huazhong Agricultural University, Wuhan 430070, China; aneelayounas1@yahoo.com

[5] Department of Soil Science and Agricultural Chemistry, Engineering Polytechnic School, Campus University, University Santiago de Compostela, 27002 Lugo, Spain

[6] CNRS, EPHE, Sorbonne Universités, UPMC University Paris 06, UMR 7619 METIS, 4 Place Jussieu, CEDEX 05, 75252 Paris, France; zheng.sun@sorbonne-universite.fr

[7] Geosciences Division, IFP Energies Nouvelles, 1 et 4 Avenue de Bois-Préau, CEDEX 92852 Rueil-Malmaison, France

[*] Correspondence: rghu@mail.hzau.edu.cn (R.H.); avelino.nunez@usc.es (A.N.-D.)

Received: 19 April 2020; Accepted: 15 June 2020; Published: 17 June 2020

Abstract: Nitrous oxide (N₂O) is an overwhelming greenhouse gas and agricultural soils, particularly acidic soils, are the main source of its release to the atmosphere. To ameliorate acidic soil condition, liming materials are added as an amendment. However, the impact of liming materials has not been well addressed in terms of exploring the effect of soil pH change on N₂O emissions. In the present study, a soil with pH 5.35 was amended with liming materials ($CaMg(CO_3)_2$, $CaCO_3$, $Ca(OH)_2$ and CaO) to investigate their effects on N₂O emissions. The results indicate that application of liming materials reduced the magnitudes of N₂O emissions. The maximum reduction of soil N₂O emissions took place for $Ca(OH)_2$ treatment when compared to the other liming materials, and was related to increasing soil pH. Mineral N, dissolved organic C, and microbial biomass C were also influenced by liming materials, but the trend was inconsistent to the soil pH change. The results suggest that N₂O emission mitigation is more dependent on soil pH than C and N dynamics when comparing the different liming materials. Moreover, ameliorating soil acidity is a promising option to mitigate N₂O emissions from acidic soils.

Keywords: lime; mineral nitrogen; soil pH; organic carbon; microbial biomass; N₂O

1. Introduction

Soil acidity is a master variable that hinders plant growth by limiting nutrient availability and thus impacts both the quantity and quality of crops. Soil acidification occurs very slowly naturally as soil is weathered, but this process is accelerated by intensive agriculture [1,2]. Soil acidity is expressed in terms of pH, and its extent and degree impact a wide range of soil biogeochemical properties. Soil acidity also has marked effects on soil microbial communities and their pertinent processes. Soil acidification is a natural and very slow process that takes over hundreds of years to develop. However, it may reach its greatest expression within a few years under intensive agricultural practices and in humid regions where rainfall is sufficient to leach down the nutrients [3]. Thus, although most processes developing soil acidification are natural, anthropogenic activities have a major impact on some of them. In fact, several reasons may contribute to soil acidification and excessive use of nitrogen (N) is one of them [3].

To obtain high crop production in intensive crop-growing areas, excessive application of N fertilizers has been carried out for years, but when it is excessive, it leads to soil acidification [4]. According to estimations [3], the application of nitrogen fertilizer in arable lands of China usually ranges from 200 to 500 kg N ha^{-1} yr^{-1}. Aside from the beneficial effects of high N fertilizer application, devastating impacts and environmental risks have also been observed including eutrophication, nitrous oxide (N_2O) emissions, and soil acidity [5,6]. Researchers have demonstrated that nitrate and ammonium applied to soils can generate 20 to 33 kmol hydrogen ions (H^+) ha^{-1} yr^{-1} under exhaustive growing systems [3]. This indicates that the application of N can drive soil acidification.

Soil acidity can be offset with alkaline materials that provide conjugate bases such as CO_3^{2-} and OH^- of weak acids. These anionic bases react with H^+ and form weak acids. For example:

$$CO_3^{2-} + 2H^+ \rightarrow H_2CO_3 \tag{1}$$

Generally, liming materials are applied in the forms of hydroxides or oxides containing magnesium (Mg) or calcium (Ca), which form hydroxide ions in water.

$$CaO + H_2O \rightarrow Ca(OH)_2 \rightarrow Ca^{2+} + 2OH^- \tag{2}$$

Most liming materials, whether they are carbonates, hydroxides, or oxides, react with CO_2 and H_2O to generate bicarbonates (HCO_3^-) when added to acidic soils. As a result, partial pressure of CO_2 in the soil is high enough to proceed such reactions forward, for instance:

$$CaMg(CO_3)_2 + 2H_2O + 2CO_2 \rightleftharpoons Ca^{2+} + Mg^{2+} + 4HCO_3^- \tag{3}$$

The resultant bicarbonates, Ca and Mg, counteract the acidity.

Liming acidic soils not only raises soil pH, but also alters biochemical processes and nutrient cycling. The rise in soil pH following lime application substantially triggers the N transformation processes [7], markedly controls the microbial processes of nitrification and denitrification, and thus influences N_2O production and emission. However, the subsequent effects of lime application on N_2O emissions is ambiguous and contrary hypotheses have been proposed by the scientific community. For instance, a laboratory incubation study proposed that increasing soil pH may substantially decrease emissions of N_2O from acidic agricultural soils [8]. In contrast, some scientists have reported that lime application and subsequent rise in soil pH caused increased soil N_2O emissions from arable acidic soils [9,10].

Keeping in mind the importance of liming acidic soils, we hypothesized that the application of lime materials can trigger N transformations following soil pH change and subsequently influence N_2O emissions in a way that would be interesting to further elucidate. Thus, the current study was conducted with the aim to examine and shed further light on the pH change effects of various liming materials on N_2O emissions from acidic agricultural soils.

2. Materials and Methods

2.1. Soil and Liming Materials

Soil was obtained from a rapeseed-rice cropping system, located in Xianing (a city of central China; 29°88′209″ N, 114°39′416″ E). According to Soil Survey staff [11], the soil is classified as Ultisol. Soil (0–20 cm) was sampled from the selected field after rice harvest from multiple-points. A composite soil sample was made by mixing subsamples. Plant residues (straw and roots) were separated from soil. After shifting in the laboratory, soil was dried in the open air, crumbled, and then sieved through a 2 mm sieve. The basic soil chemical and physical analysis [12] was performed prior to onset of the experiment. Soil texture was silty clay loam. The main characteristics of soil are given in Table 1. Different liming materials (dolomite $CaMg(CO_3)_2$, calcium hydroxide ($Ca(OH)_2$, calcium carbonate $CaCO_3$, and calcium oxide CaO) used in the present study were purchased from Xinjing Chemicals Co. Ltd. (Xiaogan, Hubei, China).

Table 1. Some selected physical and chemical characteristics of the tested soil.

$pH_{(H2O)}$	Total C (g/kg)	Total N (g/kg)	NO_3^--N (mg/kg)	NH_4^+-N (mg/kg)	Bulk Density (g/cm^3)	Cation Exchange Capacity (cmol$_c$/kg)
5.35	11.5	1.2	15.6	46.9	1.39	10.8

2.2. Experimental Setup for Collection and Analysis of N_2O

Initially, air dried soil without any amendment was incubated in a plastic (polyethylene terephthalate) tub with 20% gravimetric water content (60% water filled pore space) at a temperature of 25 ± 1 °C for one week (Figure 1). After one week of initial incubation, incubated wet soil (100 g on dry basis) from the tub was placed in glass jars. Liming materials were added separately to the soil. The application dose of each liming material was 1 g kg^{-1} dry soil with a particle size ≤0.3 mm. Treatments for the present study were as follows: (i) $Ca(OH)_2$, (ii) CaO, (iii) $CaCO_3$, (iv) $CaMg(CO_3)_2$, and (v) control (soil without any amendment). Each treatment had three replicates. Treated soils in jars were placed in an electric-automated chamber (S-400-HP) and incubated at 25 ± 0.5 °C in the dark for four weeks (28 days). During incubation, a plastic sheet with pin holes (about 30) was used on the top of each jar to reduce water loss, but permit gas exchange. Soil water content in each jar was sustained at 20% throughout the study by weighing jars and refilling with distilled water on a daily basis.

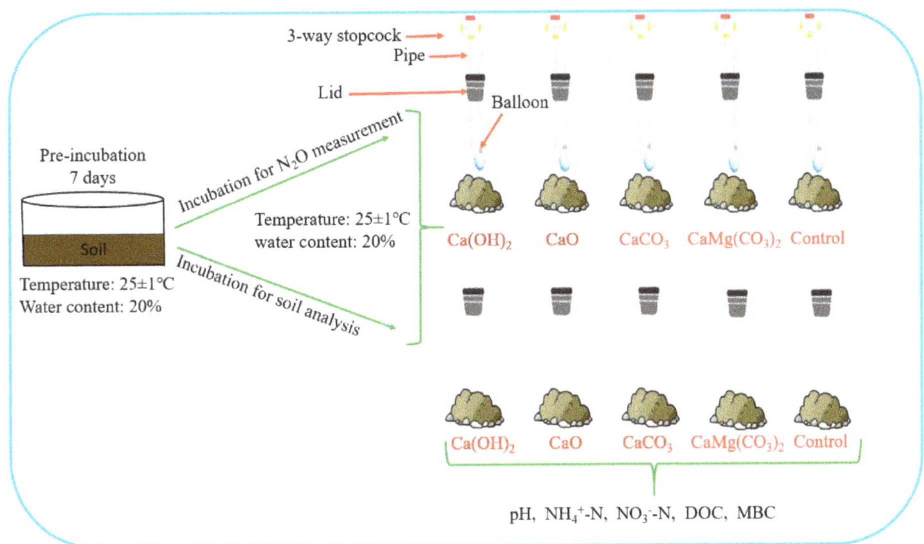

Figure 1. Schematic diagram of the experimental setup.

Gas from the headspaces of jars, equipped with air-tight lids holding a 3-mm diameter pipe, was collected at days 1, 3, 5, 7, 10, 13, 16, 20, 24, and 28 using a special air-tight syringe made for sampling purposes. On gas sampling day, the tops of jars were uncovered prior to gas sampling and soil in the jars was allowed to be exposed to ambient air for 30 to 40 min. After that, the jars were closed with air-tight lids and gas samples were taken immediately after closure to know the initial concentration of gas in the jars. Another gas sample from headspace was collected after 60 min to know the change in gas production. The gas samples were analyzed for N_2O concentration using a gas chromatograph system (7890A, Agilent technology, Santa Clara, CA, USA). The concentration of N_2O in the gas sample was calculated using the equation as given below [13].

$$F = \rho \times V/W \times \Delta c/\Delta t \times 273/(T + 273) \tag{4}$$

In Equation (1), F denotes the rate of N_2O–N emission ($\mu g\ kg^{-1}\ h^{-1}$); ρ denotes the density ($kg\ m^{-3}$) of N_2O gas; V denotes the volume (m^3) of headspace of jars; W denotes soil weight (kg); Δc denotes change in gas concentration during closure time of jars; Δt denotes the time period of closure (h) of the treatment jars; and T denotes the temperature at which the experiment was conducted (25 °C).

The cumulative emissions of N_2O ($\mu g\ kg^{-1}$) for the whole period of study were calculated based on the following formula [14].

$$\text{Cumulative N2O emission} = \sum_{i=1}^{n}(Ri \times 24 \times Di) \tag{5}$$

where Ri is the N_2O emission rate ($\mu g\ kg^{-1}\ h^{-1}$); Di are days between the sampling periods; and n is the number of samples.

2.3. Experimental Setup for Soil Analysis

A separate experiment to that for gas analysis was concurrently performed for soil analysis. Treatments, pre-incubation period, temperature, and moisture conditions for the soil analysis study were identical as that for the gas analysis setup. After pre-incubation, a weight of 200 g soil was incubated after being placed in 1000 mL beakers. Soil sub-samples from jars were taken after one day of imposing treatments and then on a weekly basis over 28 days.

For pH determination of the soil-sub samples, a soil slurry was made by performing a 1:2.5 ratio suspension of soil:distilled water [12]. The slurry was shaken in an orbital shaker for 40 min, and the pH was tested using a pH-meter (2FPHS, Wincoms Co. Ltd., Shanghai, China) after 30 min of shaking. Soil was subjected to specific extraction for the subsequent determination of the mineral contents of soil nitrogen (NO_3^-–N and NH_4^+–N) by adding 1 M KCl (5 mL for 1 g soil), shaking for 60 min, and subsequently using a flow injector system analyzer (SEAL Co. Ltd., Henstedt-Ulzburg, Germany) [15]. Chloroform fumigation specific extraction method was adopted for testing microbial biomass C [16]. Dissolved organic C content in soil was determined by extracting the soil with distilled water (1:5, soil:distilled water) and using Elementar system analysis (Vario, Elementar-CN, Hanau, Germany).

2.4. Data Analysis

Data pertinent to soil and gas parameters were analyzed using Analysis of Variance (ANOVA) one-way analysis of variance. Tukey's test was employed to identify significant differences for treatments of their mean results. The Kolmogorov–Smirnov test for the normality distribution of variables was performed before proceeding further for ANOVA [17]. All data were statistically evaluated using Windows-based software Statistical Package for the Social Sciences (SPSS) Statistics 23.

3. Results

3.1. Soil pH

Soil pH was statistically significantly ($p \leq 0.01$) different among the treatments of liming materials. Soil pH before the immediate day of adding liming materials was 5.35, and liming of soil rapidly increased pH (Figure 2). On day 1, soil pH in all treatments was substantially higher than that of the control and thereafter continued to gradually increase up until the end of the study. The highest value of soil pH corresponded to Ca(OH)$_2$ treatment on day 28. The pH values were 7.21, 6.99, 6.70, 6.43, and 5.30 for Ca(OH)$_2$, CaO, CaCO$_3$, CaMg(CO$_3$)$_2$, and the control, respectively, on day 28 of the experiment.

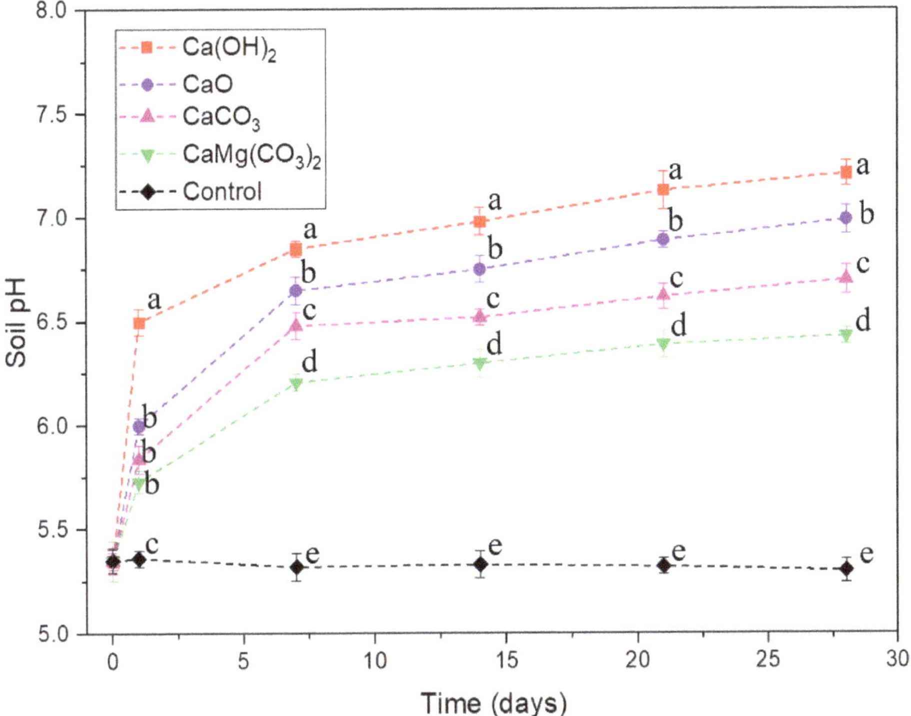

Figure 2. Dynamics of soil pH following the application of liming materials. Vertical bars denote the error bars of the mean of three replicates. Values at the same time followed by different letters indicate significant differences between different treatments ($p < 0.05$).

3.2. Soil Mineral–N (NH$_4^+$–N and NO$_3^-$–N)

Soil NH$_4^+$–N concentrations were highly and significantly ($p \leq 0.01$) influenced by the addition of liming materials. NH$_4^+$–N concentration before the immediate day of adding liming materials was 35 mg kg^{-1}, whereas the addition of liming materials caused diverse patterns of NH$_4^+$–N concentrations (Figure 3). The CaCO$_3$, CaMg(CO$_3$)$_2$ and control treatments showed continuous decline of NH$_4^+$–N concentrations throughout the study period. However, NH$_4^+$–N concentration in the Ca(OH)$_2$ and CaO treatments declined on day 1 of the onset of the study, increased on day 2, and afterward speedily decreased throughout until the end of the experiment. The NH$_4^+$–N concentrations were 9.1, 10.9, 6.8, 12.0, and 20.2 mg kg^{-1} in the Ca(OH)$_2$, CaO, CaCO$_3$, CaMg(CO$_3$)$_2$ and control treatments, respectively, on day 28 of the study.

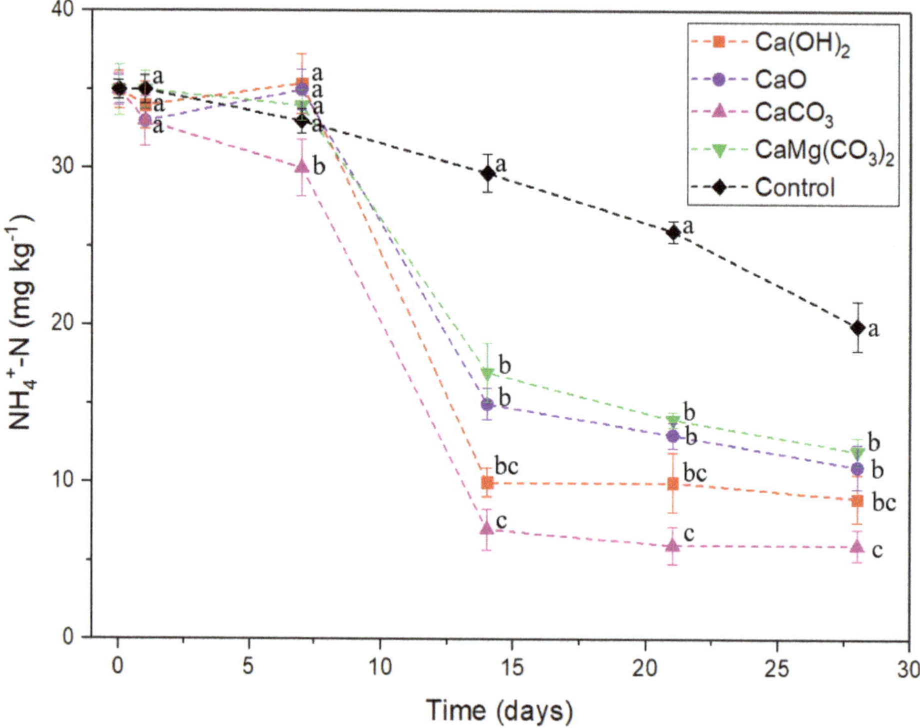

Figure 3. Variations in soil ammonium concentrations following the application of liming materials. Vertical bars denote the error bars of the mean of three replicates. Values at the same time followed by the different letters indicate significant differences between different treatments ($p < 0.05$).

The amendment of liming materials significantly ($p \leq 0.01$) augmented soil NO_3^-–N concentrations (Figure 4). The trend of increase of NO_3^-–N concentration kept continuing in all treatments until the end of the study. The maximum concentration of NO_3^-–N was observed in the $CaCO_3$ treatment on day 28 of the study. The NO_3^-–N concentrations were 60.2, 52.4, 75.1, 50.0, and 43.9 mg kg^{-1} in the $Ca(OH)_2$, CaO, $CaCO_3$, $CaMg(CO_3)_2$, and control, respectively, on day 28 of the study.

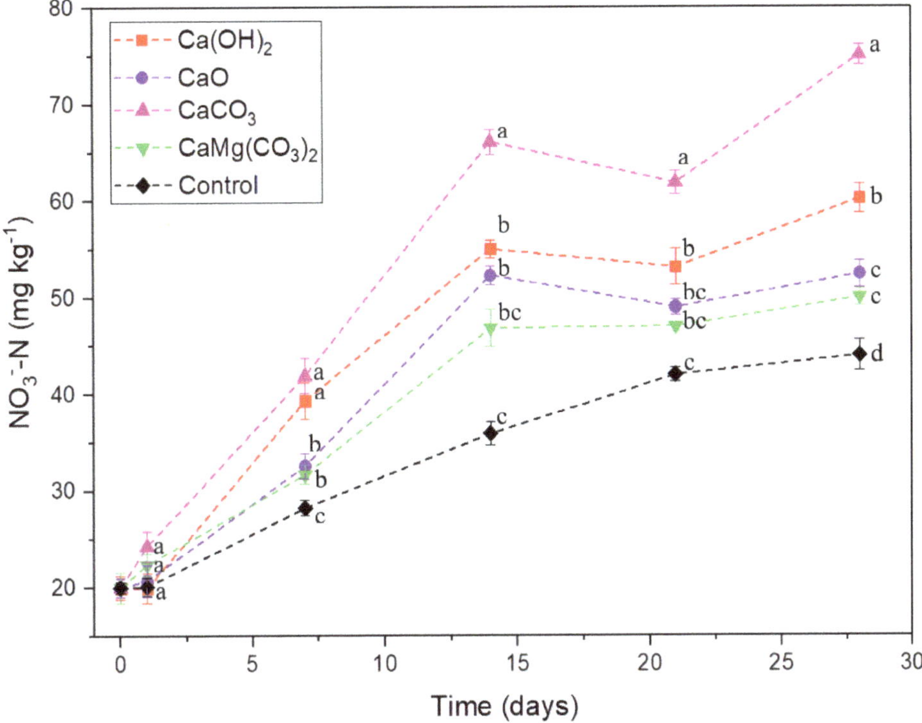

Figure 4. Variations in soil nitrate concentrations following the application of liming materials. Vertical bars denote the error bars of the means of three replicates. Values at the same time followed by different letters indicate significant differences between different treatments ($p < 0.05$).

3.3. Dissolved Organic C and Microbial Biomass C

Addition of liming materials significantly ($p \leq 0.01$) impacted the microbial biomass C (MBC) as well as dissolved organic C (DOC) in soil. Before the addition of liming materials, the DOC content was 25 mg kg^{-1} and instantly increased on day 1 in all treatments, except for the control (Figure 5). The DOC contents reached maximum values of 38.5, 33.2, 30.1, 28, 24.9 mg kg^{-1} on day 7 in the Ca(OH)$_2$, CaO, CaCO$_3$, CaMg(CO$_3$)$_2$, and control treatments, respectively, and afterward declined until the end of the study.

In the case of MBC contents, all the liming treatments showed an increment on day 1, while a divergent trend was observed afterward (Figure 6). Only Ca(OH)$_2$ treatment showed a rise in MBC content after day 1, reached the maximum at 59 mg kg^{-1} on day 14, and after that gradually declined and reached 49.1 mg kg^{-1} at the end of the experiment, whereas MBC contents decreased in all other treatments of CaO, CaCO$_3$, and CaMg(CO$_3$)$_2$ as well as the control over the entire study period.

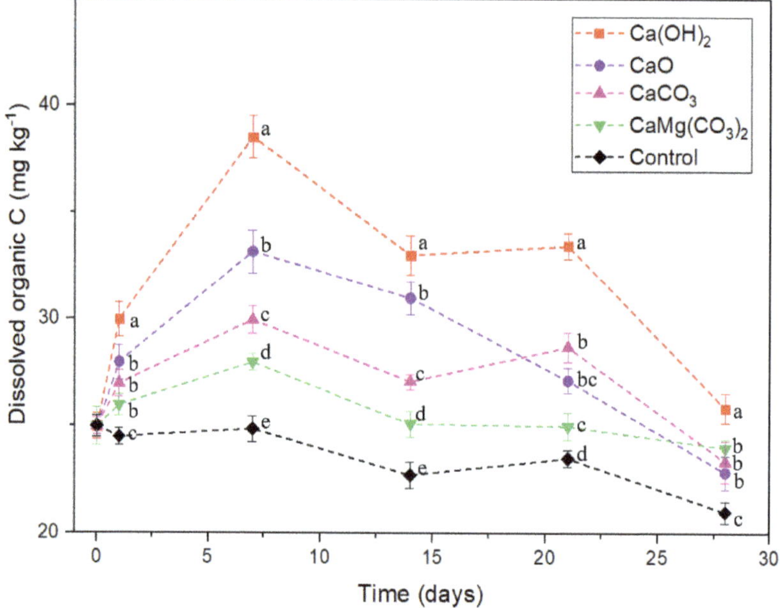

Figure 5. Fluctuations in soil dissolved organic carbon following the application of liming materials. Vertical bars denote error bars of mean of three replicates. Values at the same time followed by different letters indicate significant differences between different treatments ($p < 0.05$).

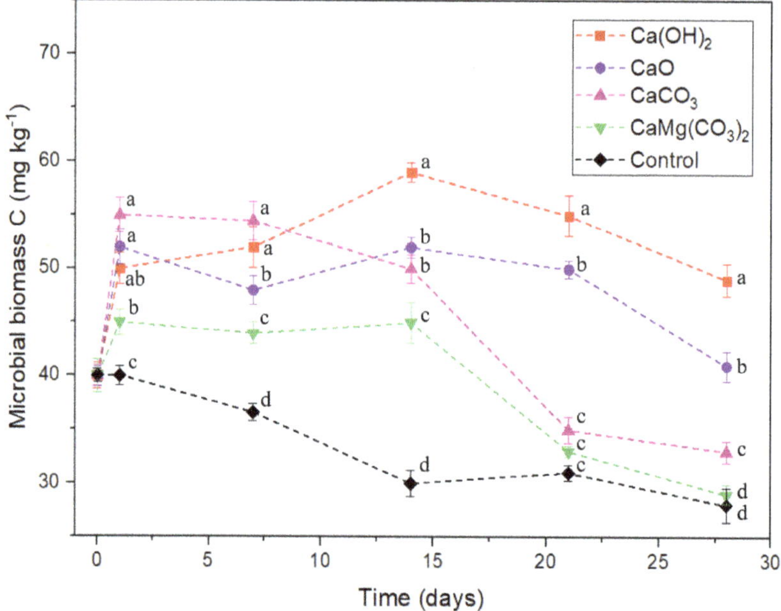

Figure 6. Fluctuations in soil microbial biomass carbon following the application of liming materials. Vertical bars denote the error bars of the means of three replicates. Values at the same time followed by different letters indicate significant differences between different treatments ($p < 0.05$).

3.4. Nitrous Oxide (N_2O) Emissions

Nitrous oxide emissions were significantly ($p \leq 0.01$) affected by the application of liming materials. The N_2O emission rate and the cumulative soil N_2O emissions (329.52 µg kg^{-1}) were highest in the control among all the treatments (Figures 7 and 8). The N_2O emissions increased on day 1 following the addition of liming materials, and then started to decline, with variant magnitudes (Figure 7). The decrease in N_2O emissions was sharper in Ca(OH)$_2$ than that of the other liming treatments, and indeed in the control. The lowest emission rate and cumulative N_2O emissions were observed in the Ca(OH)$_2$ treatment. The cumulative N_2O emissions in the Ca(OH)$_2$, CaO, CaCO$_3$, CaMg(CO$_3$)$_2$, and control treatments were 208.08, 237.60, 261.72, 287.64, and 329.52 µg kg^{-1}, respectively (Figure 8).

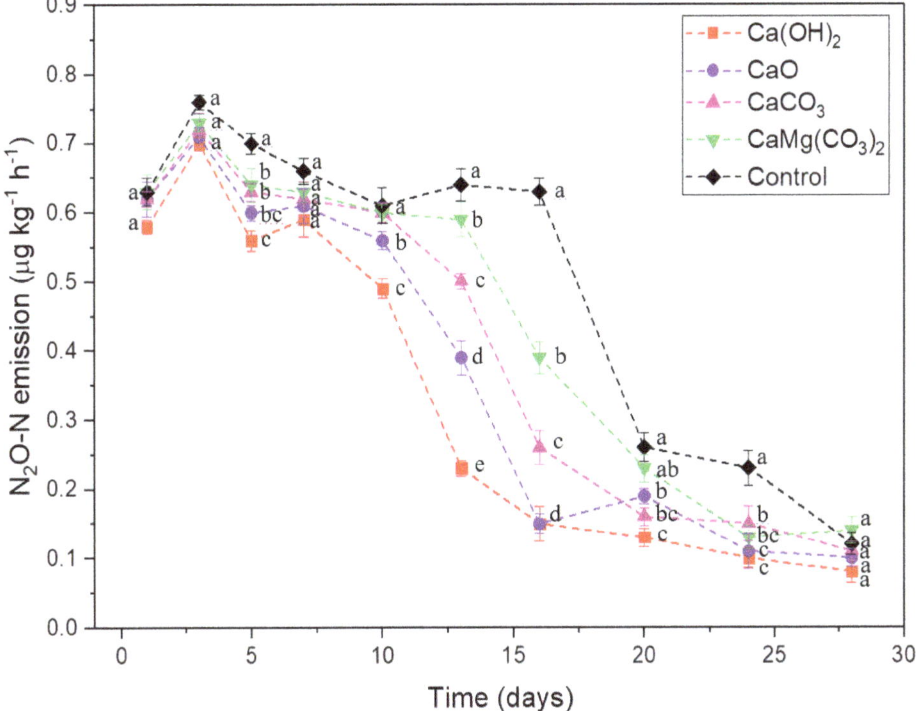

Figure 7. Soil N_2O emissions following the application of liming materials. Vertical bars denote the error bars of the means of three replicates. Values at the same time followed by different letters indicate significant differences between different treatments ($p < 0.05$).

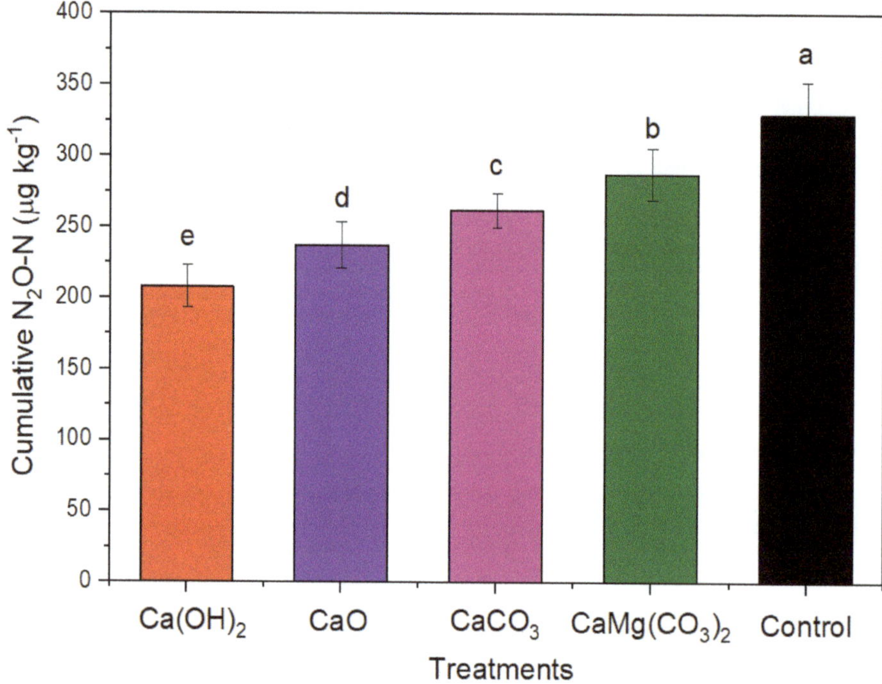

Figure 8. Cumulative soil N_2O emissions following the application of liming materials. Vertical bars denote the error bars of the means of three replicates. Different letters (from *a* to *e*) denote significant differences ($p \leq 0.05$) among the means of treatments.

4. Discussion

Acidic soils are generally considered as less efficient for high crop production. To ameliorate acidic soils, farmers usually amend them with liming materials. Application of liming materials has dual benefits of raising soil pH as well as supply essential elements, mainly Ca and Mg. In the present study, the liming materials used were $Ca(OH)_2$, CaO, $CaCO_3$, and $CaMg(CO_3)_2$. Application of all these liming materials obviously influenced N_2O emissions, but the magnitudes of the N_2O emissions dramatically altered with soil pH. In fact, high N_2O emissions were observed at low pH levels (without lime application) in the acidic soil in the present study.

High magnitudes of N_2O emissions from low pH soils can be explained by incomplete denitrification and less activity or complete inhibition of N_2O–reductase. Nitrous oxide reductase (N_2O–R) is the sole enzyme of the denitrification process, which reduces N_2O to N_2 at neutral, near neutral, or above 7 pH [18]. Therefore, higher magnitudes of N_2O emissions are expected from soils at low pH relative to higher pH values because of the incomplete denitrification process [19,20]. In the present study, N_2O emissions were perceptibly mitigated by the application of all selected liming materials. However, the highest decline in cumulative as well as emission rates of N_2O occurred in the $Ca(OH)_2$ treatment, and this was possible due to the highest pH value. Kunhikrishnan et al. [21] also indicated that the pH value could prominently control N_2O production and emissions, and Bakken et al. [22] commented that the possible mechanism involved in low magnitudes of N_2O emissions in limed soils at high pH values was pertinent to the activities of N_2O–reductase. It has been shown that the application of liming materials improved the activities of N_2O–reductase for N_2O reduction [23], and magnitudes of soil N_2O emissions are unswervingly controlled by pH [8,24]. These studies demonstrated that N_2O–reductase was functional at higher pH relative to low pH, which led to a complete denitrification process and low N_2O emissions at high pH levels.

Results of raising soil pH regarding the effect on N_2O emissions have been proposed by several researchers. Stevens and Laughlin [25] reported that raising the soil pH from 6.5 to 8 eminently reduced N_2O emissions. Qu et al. [26] reported that acidic soils produced higher magnitudes of N_2O emissions, whereas neutral pH soils showed less magnitudes of N_2O emissions. Khan et al. [27] found that the application of calcium hydroxide to soil at the dose of 5.63 g kg^{-1} soil significantly decreased N_2O emissions by increasing soil pH from 5.2 to 7.6. Additionally, an 80-day laboratory study revealed that $Ca(OH)_2$ amendment (1.1 to 5.6 g kg^{-1} soil) substantially reduced N_2O emission [28]. Moreover, some other experiments showed the following: application of $Ca(OH)_2$ at the dose of 7.3 g kg^{-1} soil mitigated cumulative emissions of N_2O from 547 g ha^{-1} to 46 g ha^{-1} in a soil with a pH of 4.71 [29]. A 2-year research showed that increasing the pH from 4 to 5.5 by $CaCO_3$ application dwindled N_2O emissions from 0.96 mg m^{-1} d^{-1} to 0.88 mg m^{-1} d^{-1} [30]. The mitigation of N_2O emissions from limed soils showed that pH plays an imperative role in regulating such N_2O release to the atmosphere [31].

In the present study, the addition of liming materials greatly impacted mineral N concentrations displaying a quick decline of $NH_4{}^+$–N with time, indicating that the nitrification process sped up, as linked to the concurrent rise of $NO_3{}^-$–N concentrations. Higher $NO_3{}^-$–N concentrations at relatively higher soil pH levels advocate that microbes consumed N_2O as an electron acceptor instead of $NO_3{}^-$–N. It can be observed from these results that complete denitrification occurred, rendering N_2O to N_2 conversion in all liming material treated soils, and thus correspondingly, low magnitudes of N_2O emissions occurred. Moreover, it is interesting to report herein that the trend and behavior of N_2O release from liming material amended soils corresponded with the changes in $NH_4{}^+$–N and $NO_3{}^-$–N concentrations, but the degree of the mitigation of N_2O emissions did not follow the same pattern. The most rapid changes of mineral N dynamics were observed in the $CaCO_3$ treatment, whereas the highest reduction of N_2O emissions occurred in the $Ca(OH)_2$ treatment. The discrepancies between the degrees of N_2O emission magnitudes following liming material application is plausible because of the potential of soil pH manipulation.

In addition to mineral N dynamics, the application of all liming materials influenced dissolved organic C, which is believed to be a readily available C substrate for microbial growth prolongation and proliferation, leading to processing nitrification and denitrification producing N_2O [32]. It is interesting to note that the changes in MBC comparing the end values with the starting values: ca. −10 mg/kg soil for the control versus are ca. +10 mg/kg for treatment $Ca(OH)_2$. Dissolved organic C acted as a substrate for microbes, conjecturing that available C favored N_2O reduction. Furthermore, high contents of MBC in the liming material added soils were detected when compared to the control, which indicated the likely high reduction of N_2O emissions.

5. Conclusions

The present research showed that the application of liming materials reduced magnitudes of N_2O emissions. The pronounced and maximum reduction of soil N_2O emissions occurred in the $Ca(OH)_2$ treatment through increasing soil pH when compared to the other liming materials tested. The results suggest that N_2O emission mitigation is more dependent on soil pH than on C and N dynamics when capering different liming materials. Moreover, ameliorating the soil acidity condition is a promising option to alleviate N_2O emissions from acidic soils. The results can be considered of environmental relevance, and further research in this regard could be interesting, especially in the current context of global warming due to a variety of greenhouse gases released to the atmosphere from different compartments and due to various anthropogenic activities.

Author Contributions: Conceptualization, M.S. and R.H.; Formal analysis, M.S., Y.W., L.W., R.H. and S.L.; Investigation, M.S.; Methodology, M.S., Y.W., L.W., P.X., Z.S., X.X. and Y.J.; Software, M.S. and A.Y.; Writing—original draft, M.S.; Writing—review & editing, M.S., A.Y., A.N.-D. and S.L. All authors have read and agreed to the published version of the manuscript.

Funding: The authors would like to thank the funding bodies of the National Science Foundation of China (417501 10485), China Post-Doctoral Science Foundation (2017 M 622478), and National Key R&D Program (2017 YFD 0800102) for financially assisting the present research.

Conflicts of Interest: The authors declare no conflicts of interest.

References

1. von Uexküll, H.R.; Mutert, E. Global extent, development and economic impact of acid soils. *Plant. Soil* **1995**, *171*, 1–15. [CrossRef]

2. Goulding, K. Soil acidification and the importance of liming agricultural soils with particular reference to the United Kingdom. *Soil Use Manag.* **2016**, *32*, 390–399. [CrossRef] [PubMed]

3. Guo, J.; Liu, X.; Zhang, Y.; Shen, J.; Han, W.; Zhang, W.; Christie, P.; Goulding, K.; Vitousek, P.; Zhang, F. Significant acidification in major Chinese croplands. *Science* **2010**, *327*, 1008–1010. [CrossRef] [PubMed]

4. 4 Shaaban, M.; Peng, Q.; Lin, S.; Wu, Y.; Zhao, J.; Hu, R. Nitrous oxide emission from two acidic soils as affected by dolomite application. *Soil Res.* **2014**, *52*, 841–848. [CrossRef]

5. Smith, V.H.; Tilman, G.D.; Nekola, J.C. Eutrophication: Impacts of excess nutrient inputs on freshwater, marine, and terrestrial ecosystems. *Environ. Pollut.* **1999**, *100*, 179–196. [CrossRef]

6. Erisman, J.W.; Galloway, J.N.; Seitzinger, S.; Bleeker, A.; Dise, N.B.; Petrescu, A.R.; Leach, A.M.; de Vries, W. Consequences of human modification of the global nitrogen cycle. *Philos. Trans. R. Soc. B Biol. Sci.* **2013**, *368*, 20130116. [CrossRef]

7. Shaaban, M.; Peng, Q.; Hu, R.; Lin, S.; Wu, Y.; Ullah, B.; Zhao, J.; Liu, S.; Li, Y. Dissolved organic carbon and nitrogen mineralization strongly affect CO_2 emissions following lime application to acidic soil. *J. Chem. Soc. Pak.* **2014**, *36*, 875–879.

8. Shaaban, M.; Hu, R.; Wu, Y.; Younas, A.; Xu, X.; Sun, Z.; Jiang, Y.; Lin, S. Mitigation of N_2O emissions from urine treated acidic soils by liming. *Environ. Pollut.* **2019**, *255*, 113237. [CrossRef]

9. Nägele, W.; Conrad, R. Influence of pH on the release of NO and N_2O from fertilized and unfertilized soil. *Biol. Fertil. Soils* **1990**, *10*, 139–144.

10. Feng, K.; Yan, F.; Hütsch, B.W.; Schubert, S. Nitrous oxide emission as affected by liming an acidic mineral soil used for arable agriculture. *Nutr. Cycl. Agroecosyst.* **2003**, *67*, 283–292. [CrossRef]

11. Soil Survey Staff. *Keys to Soil Taxonomy*, 11th ed.; USDA Natural Resources Conservation Service: Washington, DC, USA, 2010.

12. Tan, K.H. *Soil Sampling, Preparation, and Analysis*; Marcel Dekker: New York, NY, USA, 1996.

13. Hu, R.; Hatano, R.; Kusa, K.; Sawamoto, T. Soil respiration and net ecosystem production in an onion field in central Hokkaido, Japan. *Soil Sci. Plant Nutr.* **2004**, *50*, 27–33. [CrossRef]

14. Jin, T.; Shimizu, M.; Marutani, S.; Desyatkin, A.R.; Iizuka, N.; Hata, H.; Hatano, R. Effect of chemical fertilizer and manure application on N_2O emission from reed canary grassland in Hokkaido, Japan. *Soil Sci. Plant Nutr.* **2010**, *56*, 53–65. [CrossRef]

15. Kachurina, O.; Zhang, H.; Raun, W.; Krenzer, E. Simultaneous determination of soil aluminum, ammonium-and nitrate-nitrogen using 1 M potassium chloride extraction. *Commun. Soil Sci. Plant. Anal.* **2000**, *31*, 893–903. [CrossRef]

16. Vance, E.D.; Brooks, P.C.; Jenkinson, D.S. An extraction method for measuring soil microbial biomass. *Soil Biol. Biochem.* **1987**, *19*, 703–707. [CrossRef]

17. Razali, N.M.; Wah, Y.B. Power comparisons of Shapiro–Wilk, Kolmogorov–Smirnov, Lilliefors and Anderson–Darling tests. *J. Stat. Model. Anal.* **2011**, *2*, 21–33.

18. Flessa, H.; Wild, U.; Klemisch, M.; Pfadenhauer, J. Nitrous oxide and methane fluxes from organic soils under agriculture. *Eur. J. Soil Sci.* **1998**, *49*, 327–335.

19. Šimek, M.; Cooper, J. The influence of soil pH on denitrification: Progress towards the understanding of this interaction over the last 50 years. *Eur. J. Soil Sci.* **2002**, *53*, 345–354. [CrossRef]

20. Shaaban, M.; Wu, Y.; Peng, Q.-a.; Lin, S.; Mo, Y.; Wu, L.; Hu, R.; Zhou, W. Effects of dicyandiamide and dolomite application on N_2O emission from an acidic soil. *Environ. Sci. Pollut. Res.* **2016**, *23*, 6334–6342. [CrossRef] [PubMed]

21. Kunhikrishnan, A.; Thangarajan, R.; Bolan, N.S.; Xu, Y.; Mandal, S.; Gleeson, D.B.; Seshadri, B.; Zaman, M.; Barton, L.; Tang, C.; et al. Functional relationships of soil acidification, liming, and greenhouse gas flux. *Adv. Agron.* **2016**, *139*, 1–71.

22. Bakken, L.R.; Bergaust, L.; Liu, B.; Frostegard, A. Regulation of denitrification at the cellular level: A clue to the understanding of N$_2$O emissions from soils. *Philos. Trans. R. Soc. B-Biol. Sci.* **2012**, *367*, 1226–1234. [CrossRef]

23. Samad, M.S.; Bakken, L.R.; Nadeem, S.; Clough, T.J.; de Klein, C.A.M.; Richards, K.G.; Lanigan, G.J.; Morales, S.E. High-resolution denitrification kinetics in pasture soils link N$_2$O emissions to pH, and denitrification to C mineralization. *PLoS ONE* **2016**, *11*, e0151713. [CrossRef] [PubMed]

24. Senbayram, M.; Chen, R.; Mühling, K.H.; Dittert, K. Contribution of nitrification and denitrification to nitrous oxide emissions from soils after application of biogas waste and other fertilizers. *Rapid Commun. Mass Spectrom.* **2009**, *23*, 2489–2498. [CrossRef] [PubMed]

25. Stevens, R.; Laughlin, R. Measurement of nitrous oxide and di-nitrogen emissions from agricultural soils. *Nutr. Cycl. Agroecosyst.* **1998**, *52*, 131–139. [CrossRef]

26. Qu, Z.; Wang, J.; Almøy, T.; Bakken, L.R. Excessive use of nitrogen in Chinese agriculture results in high N$_2$O/(N$_2$O+N$_2$) product ratio of denitrification, primarily due to acidification of the soils. *Glob. Chang. Biol.* **2014**, *20*, 1685–1698. [CrossRef] [PubMed]

27. Khan, S.; Clough, T.; Goh, K.; Sherlock, R. Influence of soil pH on NO$_x$ and N$_2$O emissions from bovine urine applied to soil columns. *N. Z. J. Agric. Res.* **2011**, *54*, 285–301. [CrossRef]

28. Clough, T.J.; Kelliher, F.M.; Sherlock, R.R.; Ford, C.D. Lime and soil moisture effects on nitrous oxide emissions from a urine patch. *Soil Sci. Soc. Am. J.* **2004**, *68*, 1600–1609. [CrossRef]

29. Mkhabela, M.; Gordon, R.; Burton, D.; Madani, A.; Hart, W. Effect of lime, dicyandiamide and soil water content on ammonia and nitrous oxide emissions following application of liquid hog manure to a marshland soil. *Plant. Soil* **2006**, *284*, 351–361. [CrossRef]

30. Galbally, I.E.; Meyer, C.P.; Wang, Y.-P.; Smith, C.J.; Weeks, I.A. Nitrous oxide emissions from a legume-pasture and the influences of liming and urine addition. *Agric. Ecosyst. Environ.* **2016**, *230*, 353. [CrossRef]

31. Page, K.; Allen, D.; Dalal, R.; Slattery, W. Processes and magnitude of CO$_2$, CH$_4$, and N$_2$O fluxes from liming of Australian acidic soils: A review. *Soil Res.* **2010**, *47*, 747–762. [CrossRef]

32. Di, H.; Cameron, K. How does the application of different nitrification inhibitors affect nitrous oxide emissions and nitrate leaching from cow urine in grazed pastures? *Soil Use Manag.* **2012**, *28*, 54–61. [CrossRef]

Article

Characterizations of Biomasses for Subsequent Thermochemical Conversion: A Comparative Study of Pine Sawdust and Acacia Tortilis

Gratitude Charis [1],*, Gwiranai Danha [1] and Edison Muzenda [1,2]

[1] Department of Chemical, Materials and Metallurgical Engineering, Botswana International University of Science and Technology, Palapye P Bag 016, Botswana; danhag@biust.ac.bw (G.D.); muzendae@biust.ac.bw (E.M.)

[2] Department of Chemical Engineering Technology, School of Mining, Metallurgy and Chemical Engineering, University of Johannesburg, P O Box, 17011, Doornfontein 2094, Johannesburg, South Africa

* Correspondence: gratitude.charis@studentmail.biust.ac.bw; Tel.: +267-72-483-242

Received: 14 March 2020; Accepted: 22 April 2020; Published: 8 May 2020

Abstract: The bioenergy production potential from biomasses is dependent on their characteristics. This study characterized pine sawdust samples from Zimbabwe and acacia tortilis samples from Botswana using conventional and spectrometry techniques. The ultimate analysis results for pine were 45.76% carbon (C), 5.54% hydrogen (H), 0.039% nitrogen (N), 0% sulphur (S) and 48.66% oxygen (O) and, for acacia, were 41.47% C, 5.15% H, 1.23% N, 0% S and 52.15% O. Due to the low N and S in the biomasses, they promise to provide cleaner energy than fossil-based sources. Proximate analysis results, on a dry basis, for acacia were 3.90% ash, 15.59% fixed carbon and 76.51% volatiles matter and 0.83%, 20% and 79.16%, respectively, for pine. A calorific value of 17.57 MJ/kg was obtained for pine, compared with 17.27 MJ/kg for acacia, suggesting they are good thermochemical feedstocks. Acacia's bulk energy density is five times that of pine, making it excellent for compressed wood applications. Though the ash content in acacia was much higher than in pine, it fell below the fouling and slagging limit of 6%. In pyrolysis, however, high ash contents lead to reduced yields or the quality of bio-oil through catalytic reactions. Fourier transform infrared spectrometry indicated the presence of multiple functional groups, as expected for a biomass and its derivatives.

Keywords: biomass; characterization; lignocellulosic; bioenergy

1. Introduction

1.1. Background and Purpose of the Study

A sustainable biomass-to-bioenergy conversion requires adequate knowledge of both the supply capacity and quality of the biomass [1]. The utilization of a biomass for bio-energy is limited by the biomass' low bulk and energy density, resulting in increased transportation and handling costs per unit energy potential or produced [2]. Most biomasses also have high moisture and mineral contents, along with wide-ranging, nonuniform physical and chemical properties [3]. These physical and chemical properties affect biomass-to-bioenergy conversion processes and the entire value chain, as further described in Table 1. Therefore, it is crucial to assess the quality of a biomass by evaluating these physicochemical properties and their corresponding effects on the thermochemical conversion process and quality of products. Such characterization results would aid the design and development of necessary biomass pretreatment methods that improve its quality, efficiency of the conversion process and quality of products.

The properties of lignocellulosic biomasses widely vary with the diversity of the tree species. Even for the same species from different geographical origins, there can be notable differences that could be caused by the different climatic and soil conditions, as demonstrated by several studies made on pine biomass in various forms [4–8]. For instance, studies of pine biomass from Spain and Canada yielded different results for the ultimate analysis, with those produced in Spain giving a lower threshold carbon (C), hydrogen (H), nitrogen (N), Oxygen (O) and sulphur (S) composition of 48.3%, 5.2%, 0.16%, 46.2% and 0.14%, compared with those grown in Canada yielding 49.0%, 6.4%, 0.14%, 44.45% and 0.01%, respectively [7,8].

The main motivation of this study is to investigate the different types of biomass materials from different geographical locations. Thus, we chose *Acacia tortilis*, a "new" biomass yet to be explored for thermochemical processes, and made a comparison with the widely studied pine. Even so, characterizing the Zimbabwean pine will give specific facts about the feedstock in this particular geographical location, while providing a point of reference for comparison with acacia. Moreover, the pine studied in this research is in the form of sawmill residues, whose properties will also vary with different storage conditions. This further justifies the need to characterize this feedstock from specified locations.

The properties of the two biomasses are useful in predicting the most efficient logistic, pretreatment and biomass-to-bioenergy conversion methods and conditions [9]. Ultimately, the characterization of feedstocks would help explain experimental biomass-to-bioenergy conversion profiles and how product yields and quality relate to biomass properties. This knowledge will be used for subsequent pyrolysis experiments in this study. Regardless, the characterization study was generalized to include the effects of biomass properties on other thermochemical biomass-to-bioenergy conversions, such as combustion, torrefaction and gasification.

1.2. The Significance of Biomass Characterisation for Bioenergy Applications

Biomass physicochemical properties are relevant in many ways at various points of a biomass process flow, especially for upstream activities and midstream conversion stages [10]. Figure 1 shows the process flow steps affected by the physicochemical properties of a biomass.

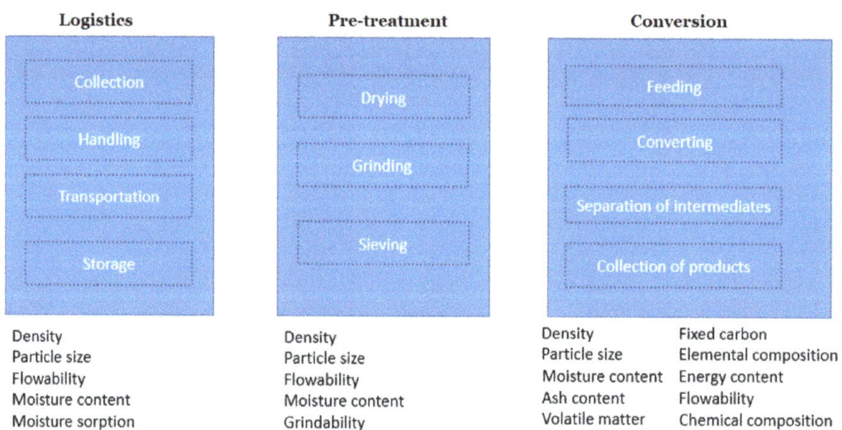

Figure 1. Relevant physicochemical properties of a biomass at different process flow steps. Reproduced with permission from Cai et al. [9], Review of physicochemical properties and analytical characterization of lignocellulosic biomass; published by Elsevier, 2017.

Álvarez-Álvarez et al. [1] investigated the thermochemical properties of four short-rotation woody coppices to determine their energy production potential, with an emphasis on characterization as a critical requirement for the efficient and sustainable exploitation of biomass resources. Tumuluru et al. [11] and Mani et al. [12] examined the grinding energy requirements and physical properties of hammer-milled cereals and energy crops. They found that physical properties like the moisture content and initial particle size affected grinding energy requirements, while bulk and tapped density affected the storage requirements, size of the handling systems and conversion efficiency. Several studies have examined the efficacy of particular pretreatment methods such as torrefaction, chemical or biological treatments in enhancing specific desirable fuel or physical properties [11–14]. Biomass characterization has been used to determine the effect of such a pretreatment on physicochemical and thermal properties. In Table 1, we present a summary discussion on the relevance of selected properties to logistic and unit processes of the biomass process flow chart as provided in the literature.

Table 1. Selected physicochemical properties and their relevance to various process flow steps.

Physicochemical Property	Relevance/Significance	Literature
Density (particle, bulk and tapped)	Useful for design of storage, handling and transportation facilities. Bulk and tapped density also help determine flowability and compressibility of biomass through the *Carr* and *Hausner* indices.	[1,9,11]
Compressibility	Useful for the design of storage, handling and feeding facilities. Higher compressibility shows that more feedstock can be stored or transported in a fixed volume.	[9,15]
Moisture content	A useful parameter in giving drying, storage, feeding and handling specifications. It affects the energy input in the conversion process. Higher moisture implies higher energy requirements.	[9,16]
Ash content	Indicates a potential risk of fouling or slagging during combustion or gasification. Higher ash/mineral content indicates a higher risk for fouling and slagging in combustion and gasification, the critical value being 6%. Ash also catalyzes the breakdown of pyrolysis vapors into water, carbon dioxide and other simpler molecules. This reduces yields and the quality of bio-oil.	[13,16,17]
Volatile matter	Useful in showing the potential conversion into a desirable gaseous or liquid product. For instance, in pyrolysis, a high VM indicates there can be higher bio-oil recoveries.	[9]
Elemental (CHNS-O) [1] composition	Determines conversion efficiency and product composition. Significant N and S contents indicate a polluting effect.	[13]
Functional group composition (as investigated by FTIR)	Useful in predicting the dominant behavior of a substance and its conversion products by analyzing the reactive groups. For instance, the dominance of aliphatic functional groups in a compound can be a good indication of the good fuel properties of liquid derivatives from a feedstock.	[13,17]
Fuel/Thermal properties	Indicates the thermochemical conversion efficiency and heating capabilities of fuel (especially in combustion or exothermic phases).	[13,16]

[1] CHNS-O—carbon, hydrogen, nitrogen, sulphur, and oxygen. FTIR: Fourier-transform infrared spectroscopy.

2. Materials and Methods

The physicochemical characterization carried out on the biomasses in this study comprised proximate, ultimate, thermogravimetric and functional group analyses. Compositional analyses were used to determine the fuel potential of the feedstock. The thermal and fuel properties, in terms of heating value, were measured using a bomb calorimeter. Physical properties, namely the particle, bulk and tapped densities, were also determined. The physicochemical, thermal and fuel properties indicate how a feedstock behaves during a thermochemical conversion and the expected energy output from the process. On the other hand, physical properties generally affect the storage and handling logistics.

2.1. Sources of Biomasses Studied

Two lignocellulosic biomass feedstocks from Zimbabwe and Botswana were considered for comparative purposes, given their potential bioenergy benefits to these two nations. Pine sawdust (*Pinus patula spp.*) is a waste product from sawmilling activities in Manicaland Province, situated on the eastern part of Zimbabwe. Sawdust and wood shavings represent the most unutilized waste fractions from the sawmill operations, accumulating at approximately 70,000 tonnes per annum [18]. Heaps of such waste fractions are scattered all over Manicaland Province, marring its aesthetic appeal and posing various ecological threats, including fire, greenhouse gas emissions and wood residue leachate that has high concentrations of dissolved organic matter [19,20]. The valorization opportunities that have been identified for the sawmill waste include combined heat and power generation from pyrolysis and the gasification/co-gasification of liquid and solid fuels and wood-engineered products [18].

Acacia tortilis is an encroacher species that is debushed from rangelands and urban circles of Botswana. Together with other encroachers, it has significantly reduced the size of quality rangeland available for both domestic and wild animals in Botswana. It costs the local and national government millions of *pulas* annually, since bush control programs have to be carried out to regenerate rangeland grass or improve the aesthetic appeal in villages, towns, cities and highways [21]. However, drought-resistant *Acacia tortilis* quickly rejuvenates after debushing, encouraged by overgrazing. Charis et al. [21] noted that the bush encroachment situation in Botswana was similar to that of Namibia, a country that is successfully valorizing its vast encroacher bush in charcoal, heat and power-generation schemes. They recommended that some of the ideas and other novel schemes, encompassing pyrolysis and gasification of the biomass, be adopted in Botswana.

2.2. Compositional Analysis and Thermal Properties

2.2.1. Thermogravimetric Analyzer

The LECO thermogravimetric analyzer (TGA) 701 (Leco TGA, St Joseph, MI, USA) was used to determine the ash composition, inherent moisture content (MC), fixed carbon (FC) and volatile matter (VM) present in the pine and acacia samples. The TGA was set to comply with the American Society for Testing and Materials (ASTM) E1131-03 and to analyze the MC, ash, VM and FC compositions as described under the ensuing subheadings [22]. The instrument was programmed to perform all these tests at various temperatures. The LECO TGA 701 automatically weighed the samples, giving real-time changes in mass. For the *Acacia tortilis*, the samples were placed in two categories, "with bark" and "debarked", to determine if there would be a significant difference in the TGA profiles and proximate composition for the two groups of acacia biomasses.

Moisture Content

Samples of 1–3 g were put into the automatic sampling containers in the TGA. The exact sample masses were then determined using the automatic balance and, subsequently, ramped the TGA at 10 °C/min to 105 °C to remove all moisture. The drying process was carried out in a dynamic atmosphere of nitrogen gas until a constant "moisture mass" was achieved; then, the TGA automatically engaged the next stage.

Volatiles Content

The TGA was set at 550 °C for volatization, such that, after removing the moisture, the temperature automatically adjusted to the set point within 10 min. Volatization occurred in an inert environment of nitrogen gas until a near-constant mass was achieved.

Ashing

After burning off the volatiles, the lids were removed from the samples, and the biomass samples were reheated to 550 °C in an oxygen-rich environment. The mass that was lost during the ashing process was the fixed carbon, which reacted with oxygen.

2.2.2. Ultimate Analysis

The ultimate analysis was conducted using the Flash 2000 CHNS elemental analyzer (ThermoFisher Scientific, Waltham, MA, USA). The biomass samples were first weighed in the sample containers with the equipment's auto-sampler system, then introduced into the combustion reactor. After combustion in an oxygen-rich environment, the gases given off were carried by a helium flow past a copper-filled layer, through gas chromatography (GC) column where the combustion gases were separated and detected by a thermal conductivity detector [23]. The oxygen content was calculated from the difference between the cumulative C, H, N and S percentage composition and 100%.

2.2.3. Fourier Transform Infrared Spectroscopy (FTIR) Analysis

The biomass samples were grounded using a hammer-chopper mill fitted with a screen size of 3 mm. The mill product was further ground and sieved to obtain samples of less than 1-mm particle size. An infrared (IR) scan was then conducted for the wavenumber of 4000 to 500 cm^{-1} for the underflow samples from the sieves using a Vertex70v FTIR spectrometer (Bruker, Ettlingen, Germany) [24].

2.2.4. Calorimetry

The high heating value (HHV) of the biomasses was determined using a bomb calorimeter, Bomb CAL2K-2 (Digital Data Systems, Randburg, South Africa), according to DIN 51,900 T3 standards for the "testing of solid and liquid fuels—determination of gross calorific value". A crucible with a sample of about 1.00-g biomass was placed in the calorimeter. The bomb was then closed and filled with oxygen pressurized to 30 bars. The sample was covered in an adiabatic jacket along with some known quantity of water. It was then ignited electrically, resulting in a rise of the water temperature, enabling automatic evaluation of the HHV of the sample. The calorimeter was calibrated using benzoic acid.

Table 2 summarizes the standard analyses conducted on the biomass and instruments that were used.

Table 2. Standard compositional and thermal analyses and the instruments used. HHV: high heating value and TGA: thermogravimetric analyzer.

Test Conducted	Standards Used/Reference	Instrument	Manufacturer
Proximate analysis	ASTM E1131-03	LECO TGA 701	LECO, St Joseph, MI, USA
Ultimate (CHNS-O)	[23]	Flash 2000 CHNS elemental analyzer	ThermoFisher Scientific, Waltham, MA, USA
Vertex70v FTIR	[24]	Vertex70v FTIR spectrometer	Bruker, Ettlingen, Germany
HHV	DIN 51,900 T3	Bomb CAL2K-2	Digital Data Systems, Randburg, South Africa

2.3. Physical Properties

The MC, in particular, was measured "as received" and also after solar drying for at least three weeks. All the compositional and calorimetric characterizations were only done for dried biomasses in the condition they would be when introduced into the thermochemical conversion system. This criterion would also make the two cases comparable; otherwise, a significant difference in moisture contents would lead to a bias in many properties of the biomasses.

2.3.1. Particle Density

Since a pycnometer was not available, we measured the particle density using relatively large biomass particles whose dimensions could be measured according to the method used by Lam et al. [15]. For pine sawdust, particles with a rectangular geometry were chosen, while cylindrical twigs were more suitable in the *Acacia tortilis* case. In both cases, the particles where further filed on the ends to correctly simulate the relevant geometry. The length, width and height ($L \times W \times H$) of the rectangular sawdust particles were measured, while only the length and diameter (L and D_p) of the *Acacia tortilis* cylindrical particles where determined. Vernier caliper, accurate to the nearest 0.1 mm, was used to measure all the dimensions. The mass of the particles was measured using a scale accurate to the nearest 0.01 g. Particle density (D_p) was determined as:

$$D_p = \frac{M_p}{V_p} \tag{1}$$

where M_p was the mass and V_p the volume of the cylindrical particle. The particle densities of three particles were obtained and used to calculate the mean as the final particle density of the biomasses.

2.3.2. Bulk and Tapped Density

The bulk density for the selected size classes was obtained through sieving. The size ranges (x) that were used are:

- Mixed (as-received basis)—Sawdust comes with a mixed size, while any first grind of *Acacia tortilis* similarly would contain a mixed size range. Since this mixed size is often handled or stored before further pretreatment or conversion, it is useful to determine the bulk and tapped density for this category. A sieve of 10 mm was used for the 1st grind of *Acacia tortilis*.
- Five millimeters $< x \leq 7.1$ mm, 3.35 mm $< x \leq 5$ mm and 1.70 mm $< x \leq 3.35$ mm—The pine sawdust and Acacia tortilis particles were subjected to a sieve classification using 9-mm, 7.1-mm, 5-mm, 3.35-mm and 1.70-mm sieves. The three classes of interest in the subsequent thermochemical conversion were used for the bulk and tapped density determinations.

Feedstock will inevitably be handled in the form it is received before being ground into a final size, depending on the optimum feed particle size stipulated for the conversion equipment and process. For instance, some studies specify a size range of between 1.5–3 mm for fluidized bed pyrolysis or gasification, while larger sizes are acceptable for fixed bed pyrolysis or combustion [25]. In the case where the conversion plant is distant from the storage and pretreatment sites, it would, therefore, be good to know whether to grind the biomass to the size required by the conversion process at the beginning or to handle it in its received state until just before the conversion process.

A digital mass balance was used to weigh a container of known volume (0.280 L), then tared, subsequently. Biomass samples from each size class were then poured into the container from a constant height and leveled off. The actual mass of the biomass on balance was recorded. The bulk density, D_b, was calculated as follows:

$$D_b = \frac{\sum M_p}{V_c} \tag{2}$$

where $\sum M_p$ is the total particle mass, and V_c is the volume of the container. Tapped density D_t was determined by tapping the poured sample 30 times, adding more biomass, then tapping 20 times again. The samples were leveled off, and the mass of the compressed biomass was measured. Tapped density D_t was then calculated as follows:

$$D_t = \frac{\sum M_{tp}}{V_c}. \tag{3}$$

where $\sum M_{tp}$ is the total mass of the tapped particles. Four sets of experiments were done to obtain a more accurate averaged result, and an error analysis was conducted.

2.3.3. Flowability and Compressibility: The Hausner Ratio and *Carr* Index

The Hausner ratio and *Carr* index are derived properties calculated from the bulk and tapped densities, indicating the flowability and compressibility of the biomass. The higher the Hausner or *Carr* indices, the more compressible the biomass [11]. The Hausner and *Carr* indices were calculated as follows, using the already defined parameters D_t and D_b:

$$Hausner = \frac{D_t}{D_b} \tag{4}$$

$$Carr = \frac{100(D_t - D_b)}{D_b} \tag{5}$$

3. Results and Analysis

This section catalogues the results obtained from the tests in Section 2. It is mainly divided into two subsections to cover "Compositional and thermal properties" and "Physical properties".

The analyses and discussion of the results were done as they are presented, and then a summary discussion is also supplied in Section 4.

3.1. Compositional and Thermal Properties

3.1.1. Calorimetry, Thermogravimetry, Ultimate and Proximate Analyses

Figure 2 shows a typical TGA profile for a pine sample. The proximate composition was determined from the plateau regions and is shown in Table 2.

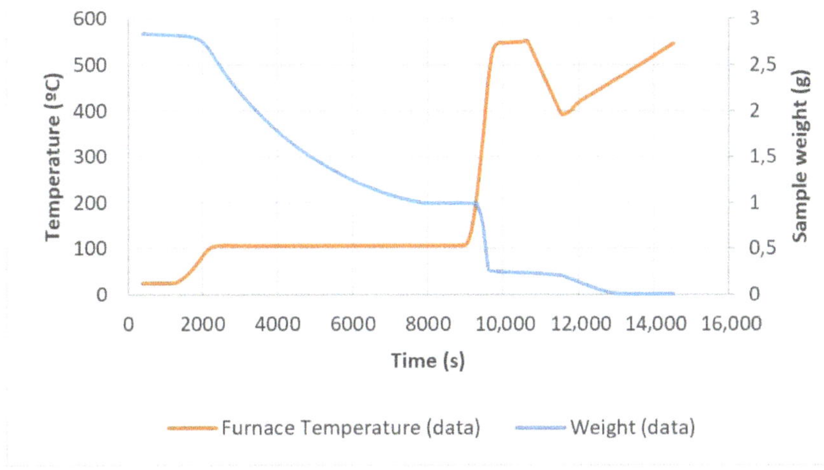

Figure 2. Thermogravimetric analyzer (TGA) weight loss curves for pine dust for a sample of 2.80 g.

Figure 2 shows the thermal degradation profile for weight loss against time and temperature rise against time for the sample. As can be seen from the profile, the initial weight loss began after 33 min from the beginning and continued for just over an hour until the 2.80-g sample reduced to 1 g. This long drying duration shows that the pine dust samples had high moisture contents. When the temperature was ramped up to 550 °C, still within the inert environment, another weight-loss profile occurred for 30 min, until the TGA automatically stopped after registering a constant mass. The lids were then taken off to allow free air circulation during the ashing process in an oxygen environment. The exposure to a cool atmosphere explains the sudden drop in temperature from 550 °C to 400 °C. After removing the caps, the temperature was ramped back to 550 °C. The ashing immediately commenced when the temperature had just reached the set point and continued until the sample reached a constant weight; then, the TGA automatically stopped.

Figure 3 shows the TGA weight-loss curves for the debarked acacia and that with bark.

The profiles for the various forms of acacia ("debarked" and "with bark") were mostly the same, with slight variations in the final proximate compositions. Figure 3 shows that there is little loss of weight on the acacia-drying profile duel to the low MC of 3.72% compared to 65.41% for the pine sawdust on an "as-received" basis. No air-drying had yet been done, and the pine dust was coming from humid storage conditions. When the two biomasses' weight-loss profiles were compared, it was noted that the drying step takes slightly less than 30 min for the acacia, while it takes 100 min for the pine. The volatization and ashing steps take 17 and 50 min, respectively. The volatization step is longer for pine dust (30 min), attesting to the higher VC, while the ashing period is almost the same.

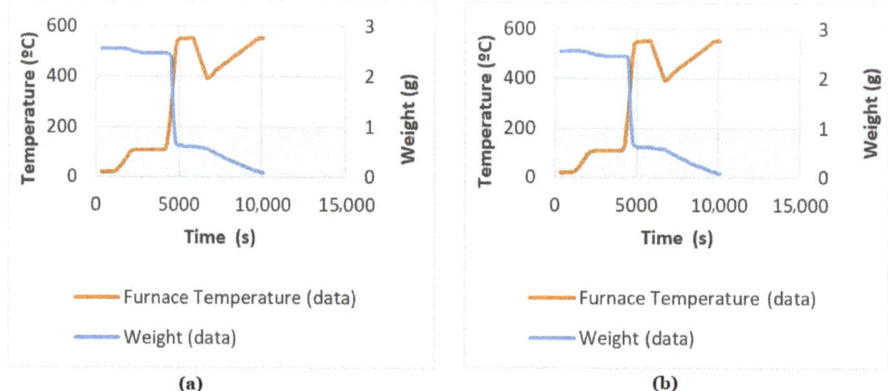

Figure 3. TGA weight-loss curves for (**a**) debarked acacia and (**b**) acacia with bark.

It was further observed that there were random differences in the proximate composition between debarked samples and those that had bark, with no particular trend. The thinness of the acacia shrub bark layer could make its contribution to the results insignificant. Other authors such as Olatunji et al. [26], however, observed higher ash contents and a lower HHV for samples with bark (from mature trees) compared to the debarked biomass. As a consequence of our findings, only the proximate compositions for the samples with bark were adopted, since it is also quite unlikely that the *Acacia tortilis* will be debarked before conversion or valorization exercises. The calorimetry, thermogravimetry proximate and ultimate analyses results are summarized in Table 3.

Table 3. Results of ultimate and proximate analyses (dry basis) for acacia, compared to the common pine. FC—fixed carbon, VM—volatile matter and MC—moisture content.

Lignocellulosic Biomass	Ultimate Analysis (%)—An Average of 2				Proximate Analysis (%) (Dry Basis)—Averages				HHV (MJ/kg)
	C	H	N	O [1]	Ash	FC	VM	MC	
ACACIA	41.47	5.15	1.23	52.15	3.90	19.59	76.51	3.72	17.267
PINE DUST	45.76	5.54	0.039	48.66	0.83	20.00	79.16	65.41 [2]/6.50 [3]	17.568

[1] Oxygen calculated by difference, considering that S = 0. [2] Moisture content as received. [3] Moisture content after solar-drying for at least 3 weeks.

Table 3 shows that the HHV for *Acacia tortilis* is not very different from that of pine dust, though the acacia has a considerably higher ash content and a slightly lower fixed carbon. The higher fixed carbon of the pine dust explains why it would have a higher HHV compared to acacia. The FC of *Acacia tortilis* corresponds with the low C content in the ultimate analysis (41%), which is also smaller in comparison to the 44% in *Acacia Holosericea* [17]. The higher VM in the pine dust also indicates it may have a more favorable conversion efficiency, with potential for a higher yield of bio-oil in pyrolysis, for instance. On the other hand, the higher ash content (3.90%) in *A. tortilis* shows there is a higher mineral content, making it more prone to fouling and slagging compared to the pine; however, the figure is below the critical 6% stipulated for gasification systems [13]. This figure is comparable to *Acacia Holosericea* with an ash content of 3.91%. It would be interesting to investigate if the mineral content in the acacias would not have a significant impact on the yield and quality of bio-oil obtained through pyrolysis. The pine residues also have more VM (79.16%) than *Acacia tortilis* (76.51%). However, the latter's VM is higher than other acacia species studied: *Acacia holosericea* (65.32%), *Acacia mangium* (65.2%) and *Acacia auriculiformis* (65.37%). Overall, it does seem that pine dust would be a better feedstock for thermochemical conversions compared to acacia, although this would further depend on the particular conversion and the targeted products.

The ultimate analysis results corroborate with the proximate study, which shows a higher FC content in the pine residues compared to acacia. There was no S detected for both biomass types, though acacia registered a significant N content (1.23%) due to its nitrogen-fixing and retaining capabilities. This N content is considerably higher compared to 0.25% in *Acacia Holosericea*, and in both acacias, S was not detected. Charis et al. [10] noted that the pine dust from Zimbabwe has relatively lower N and S contents compared to the Canadian (N: 0.14% and S: 0.01%) and Spanish (N: 0.16%–1.6% and S: 0.14%–0.45%), making it a cleaner thermochemical feedstock [5,7]. In general, the low amounts of N and S in the pine dust and acacia imply that the two biomass types have a lower environmental burden compared to fossil fuels. A further comparison of the ultimate, proximate and calorimetric analyses of the sawdust with other pine residues from the literature was made. The HHV for pine sawdust (17.568 MJ/kg) was higher than the value of 15.01 MJ/kg quoted by Braza and Crnkovic [4] but lower than that found by Olatunji et al. [26] of 20.54 MJ/kg. The range is, however, acceptable, and the differences could be due to varying MCs. The pine dust in this study has the lowest C content at 45.76% compared to 48.62% by Braza and Crnkovic [4] and 50.54% by Olatunji et al. [26], as well as the highest O content at 48.66% compared to 43.20% and 41.11% from the two other studies, respectively. Most proximate parameters were found to have values somewhere between those obtained by Braza and Crnkovic [4] and Olatunji et al. [26] except for the MC and fixed carbon, which were the highest for the pine dust in this study. The differences in compositional values for the pine sawdust show how variations can occur due to various storage and environmental factors or as a result of the different soil and climatic conditions in which the trees were nurtured.

3.1.2. FTIR Analysis

Figures 4 and 5 show the peaks obtained from the FTIR scan of the *Acacia tortilis* and pine dust samples, respectively. The functional groups most possibly represented by the peaks were determined by referring to the literature, as shown in Table 4.

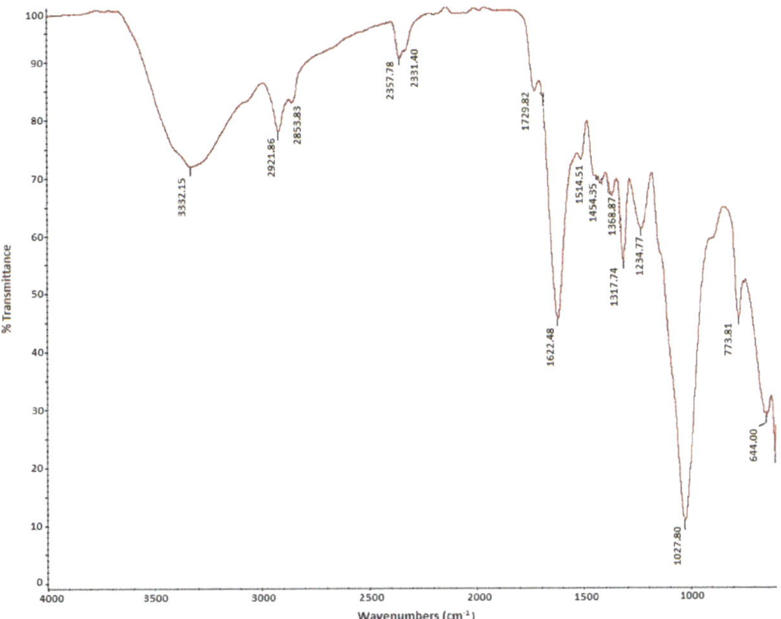

Figure 4. Fourier-transform infrared spectroscopy (FTIR) peaks for *Acacia tortilis*.

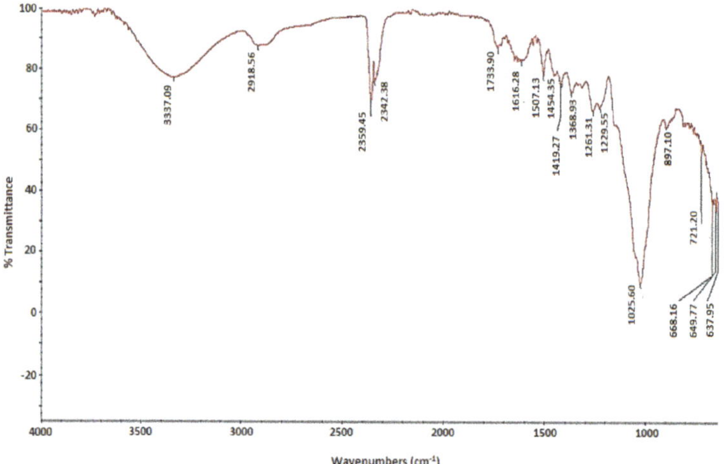

Figure 5. FTIR peaks for pine dust.

Table 4. Compounds represented by the acacia and pine FTIR peaks.

Absorption Peak-Acacia	Absorption Peak-Pine	Possible Compounds/Chains Rationale	Type of Vibration Causing the IR Absorption	Literature
3332.15	3337.09	Phenols, carboxylic acids and alcohols (broader peak)	Hydrogen-bonded OH stretch in cellulose and lignin	[17,27]
2921.86	2918.56	Alkanes or primary amine Alkanes and acids	H-C-H asymmetric and symmetric stretch	[17,27]
2853.83		Alkanes or aldehydes	H-C-H asymmetric and symmetric stretch; C-H branch off C=O	[17,27]
	2359.45	Alkanes and acids	H-C-H asymmetric and symmetric stretch; C-H branch off C=O	[28,29]
2357.78 and 2331.40	2342.38	Carbon dioxide	CO_2	[28,29]
1729.82	1733.90	Ketones and aldehydes Carboxylic acid	C=O stretch in hemicellulose	[17,27]
1622.48	1616.28	Alkenes	sp^2 C-X double bond	[28,29]
1514.51	1507.13	Secondary Amine confirmed	N-H bend	[28,29]
1454.35	1454.35	Alkanes confirmed	H-C-H bend	[28,29]
1422.69 (actual: 1425)		Lignin and wood	CH in-plane deformation	[17]
	1419.27	Alkanes	sp^3 C-H bend	[28,29]
1368.87 and 1317.74	1368.93		C–H deformation in cellulose and hemicellulose	[17]
1234.77 and 1027.80	1261.31, 1229.55 and 1025.60	Esters or ethers	Alkoxy C-O stretch and C-O stretch in cellulose, hemicellulose or lignin	[28,29]
	897.10 and 721.20	Aromatic rings and meta di-substituted compounds	sp^2 C-H-bending patterns	[28,29]
773.81, 644.00 and 604.05	668.16 to 604.56	Fingerprint regions not easy to interpret for a previously unknown compound		[17]

The FTIR image for the pine dust in Figure 5 seems to have more functional groups (16) depicted by the peaks below the 2200-cm^{-1} wavenumber, compared to 12 groups for the acacia in Figure 4.

Table 4 shows the interpretation of the peaks obtained in the FTIR for the two biomasses. The functional groups at those peaks were identified by comparing with the standards derived from the literature.

Using the FTIR analyses, we identified many functional groups in both biomass samples, attesting to the heterogeneous nature of the biomasses. The FTIR analyses indicated the presence of a wide range of compounds within the cellulose, hemicellulose and lignin structures [30]. Reza et al. [17] explicitly linked the C-H bond stretch in the region 2960–2850 cm^{-1} to the presence of lignin, hemicellulose and cellulose in the biomasses. They further associated the area between 2000 and 1650 cm^{-1} to the C-H bonding in cellulose and hemicellulose, while the region 1750–1630 cm^{-1} was mainly linked with the C=O in hemicellulose. This finding is in tandem with literary works like Ahmed et al. [27] and Naik et al. [7]. The functional groups identified in pine dust are mostly the same as those in *Acacia tortilis*, though in varying quantities. The peak at wave 2853.83-cm^{-1} frequency seemed to suggest a more substantial presence of aldehydes or alkanes in the acacia, while the one at 1422.69 cm^{-1} suggested a stronger lignin presence in this hardwood, as expected and compared to the pine softwood. Pine dust also had prominent peaks at 2359.45 cm^{-1} and 1419.27 cm^{-1}, which suggested a strong presence of alkanes as well. Further analyses would be required to tell the exact comparative alkane compositions. The other peaks at 897.10 cm^{-1} and 721.20 cm^{-1} for the pine dust suggested a significant presence of aromatic rings, meta di-substituted compounds. The analyses of pyrolysis products in the latter stages of the bioenergy project using FTIR and other scanning calorimetry methods could help to confirm these results and simulate the possible conversion pathways. The other difference in the FTIR results for the two biomasses was the presence of a broader fingerprint region (six clumped peaks) of unknown functional groups for the pine dust, in the range 604.56–668.16 cm^{-1}. Meanwhile, the *Acacia tortilis'* lower-end fingerprint region only comprised three unknown peaks at 603.05, 644 and 773.81 cm^{-1}. For both lignocellulosics, the FTIR analyses showed more functional groups than those reported by Reza et al. [17] and Ahmed et al. [27]. The acacia's FTIR results agree with the studies on other acacias published by Reza et al. [17] and Ahmed et al. [27], especially at the peaks depicted by wavenumbers 3332.15 cm^{-1}, 2921.86 cm^{-1}, 2853.83 cm^{-1}, 1729.82 cm^{-1} and 1422.69 cm^{-1} and the corresponding functional groups.

3.2. Physical Properties

The MC has already been covered under the compositional analysis. The results for the other physical properties, mainly the particle, bulk and tapped densities are presented in this section. Using these values, the Hausner and Carr indices, which indicate the compressibility and flowability of the biomasses, were calculated.

3.2.1. Particle Density

The average density from three particles, as calculated by Equation (1) from Section 2.2.1, was 478.8 kg/m^3 for pine and 867.3 kg/m^3 for the *Acacia tortilis* particles. The average density obtained for the pine dust was close to the 490 kg/m^3 stipulated by the *pinus patula* datasheet [31]. Logically, pine residues that have been under various storage conditions and exposed to different weather elements would differ in some physical properties from freshly harvested pine trees. Furthermore, the density of the *Acacia tortilis* (867.3 kg/m^3) was almost double that of the pine (478.8 kg/m^3), as expected for a hardwood compared to the pine softwood.

3.2.2. Bulk and Tapped Density

Figure 6 shows the mean values obtained from four runs on bulk density and tapped density for both biomasses and the relative standard error associated with the type of biomass. These mean values were then used to calculate the Hausner ratios and Carr's indices for various size ranges.

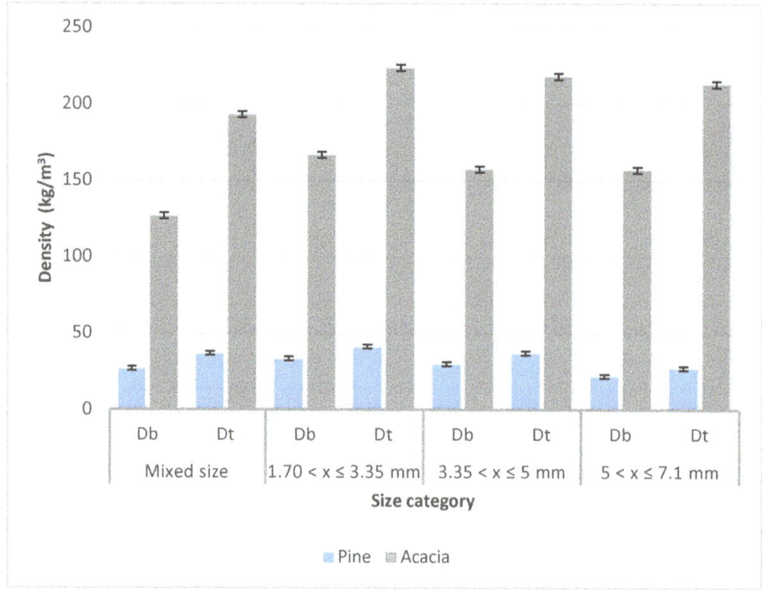

Figure 6. Bulk and tapped densities for various size classes of pine and acacia samples. Db: bulk densities and Dt: trapped densities.

The bulk and tapped densities of the acacia were at least four times larger than for pine, because they are a function of the number of particles contained in the fixed volume and the mass of each particle, according to Equation (2) in Section 2.2.2. These densities are an essential consideration when the biomass is being transported or stored for subsequent thermochemical conversion. In this case, the energy stored or transported per unit volume would be a lot larger for the *Acacia tortilis* compared to the pine dust. For example, in the size range 3.35 mm < x ≤ 5 mm, one cubic metre of pine dust stored or transported would have equivalent energy of 522.1 MJ compared to 2709.7 MJ for the same size range for acacia. The formula for calculating the bulk energy density (ED_b) of the biomass is given by Equation (6):

$$ED_b = D_b \times HHV \tag{6}$$

3.2.3. The Hausner Ratio and Carr index

Figures 7 and 8 show the Hausner ratios and Carr calculated from the mean values of the bulk and tapped densities from Figure 6. The calculations were made according to Equations 4 and 5 provided in Section 2.2.2.

For both samples, the mixed size (as-received basis) had the highest Hausner ratio and Carr's index, indicating higher compressibility. This result is probably because, during tapping, the smaller particles fill in the voids left by the bigger particles. Consequently, it may be desirable to store and transport mixed size fractions (mostly as-received) within a specific range to make optimal use of the space. The acacia biomass also has higher Hausner's ratios and Carr indices than pine, which shows it is a more compressible biomass type.

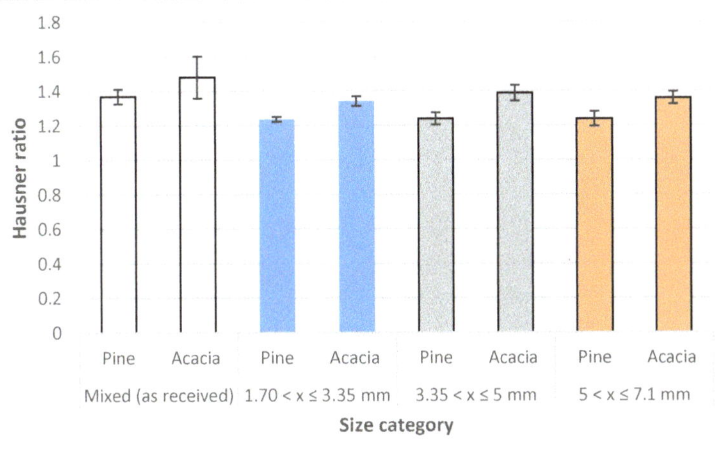

Figure 7. Hausner ratios for pine and acacia.

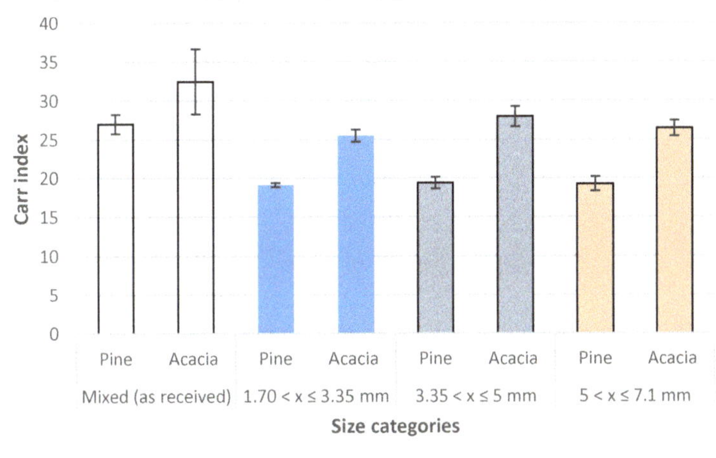

Figure 8. Carr indices for pine and acacia.

4. Discussion

The characterization of the two biomass feedstocks makes it possible to predict their behaviors at various steps of the supply chain and add any necessary pretreatments. Table 5 discusses the results obtained from the previous section with regards to their implications on supply-chain (SC) dynamics like transport, handling, storage, conversion and the overall economics, as hinted in Table 1.

Table 5. The implications of the characterization results on the supply-chain (SC) dynamics, economics and the environment.

Physicochemical Properties		Comparative Status of the Properties for Acacia and Pine Dust	Potential Effects on SC Dynamics, Economics and Environment
Elemental composition (CHNS-O)		Higher N content in *A. tortilis* compared to pine dust. Both biomasses have no S detected. Higher O/C ratio in *A. tortilis* (1.25) compared to pine dust (1.06).	Both biomasses have low N and S contents, implying a lower environmental impact compared to fossils [13]. Costs of scrubbing are circumvented. Higher O/C ratio would affect the quality of some products or by-products, such as pyrolysis oil, though marginally compared to the pine dust [32].
Proximate composition	Moisture content	The MC of pine dust (65.41%) on an "as-received" basis is way higher than for acacia (3.72%). Solar-dried pine goes down to 6.50%.	Solar-drying can be used to bring pine dust MC down, but that exposes it to contamination with dust, etc., which affects the quality of conversion products. Other methods of drying would mean extra costs. High MC lowers the HHV of a biomass [33].
	Ash content	Higher ash content in acacia (3.90%) compared to pine dust (0.83%).	The ash in acacia is below the critical fouling and slagging level (6%). It could, however, act as a catalyst in processes like pyrolysis, decreasing the yield and quality of bio-oil [34]. This would, however, need to be confirmed with the actual pyrolysis set-up.
	Volatile matter	Higher VM in pine dust (79.16%) compared to acacia (76.51%).	There could be higher yields/recoveries of liquid or gaseous products from conversions like pyrolysis and gasification [9]. It does not, however, indicate the quality of the gas or liquid product.
	Fixed carbon	A slightly lower FC in acacia (19.59%) compared to pine dust (20.00%).	The difference is not significant, though it could explain why the acacia has a slightly lower HHV compared to pine dust.
HHV		A slightly lower HHV in acacia (17.267MJ/kg) compared to pine dust (17.568 MJ/kg).	The moderately high HHV means both can be exploited well by thermochemical means, including upgrading their energy value through torrefaction [14]. The latter would bring their energy value closer to that of coal.
Densities and compressibility		The particle, bulk and tapped densities of acacia are about 2–4 times higher than for pine dust.	This fact brings an advantage for acacia in terms of storage and energy density. For the size range 3.35 mm < x ≤ 5 mm, acacia has an energy density of 2709.7 MJ/m^3 compared to 522.1 MJ/m^3 for pine dust. More energy values can be recovered for the same volume of stored, transported or fed acacia (Section 3.2.2). Figures 7 and 8 show that the acacia generally has a higher compressibility, and, coupled with the energy density factor, it has a better potential for compressed wood products [18].
Functional groups		Multiple functional groups for both biomass types.	This result proves there are many compounds in the biomasses, which is also characteristic of some of the conversion products like bio-oil [34]. Consequently, separation and purification to derive chemicals or fuels from the oil becomes a costly and challenging task.

5. Conclusions and Recommendations

Various literature sources have confirmed the importance of the physicochemical characterizations of biomasses, a primary step in exploring the feasibility of exploiting them for bioenergy. For this particular study, the characterization results concurred with general expectations emanating from the tree classifications. They also compared well with studies of similar biomasses from the literature. The comparative study showed that pine dust has better fuel properties per unit weight than *Acacia tortilis* due to its higher HHV and fixed carbon, with low ash and oxygen contents. However, since *Acacia tortilis'* thermochemical properties are not very far from those of pine, and the ash content (3.61%) is about half the critical slagging and fouling value of 6%, it can also be effectively exploited thermochemically [13]. Moreover, the energy density (energy per unit volume) of *Acacia tortilis* is about four times larger than for the pine dust due to its higher bulk density, giving it an advantage in storage and the actual thermochemical conversions. The considerably higher particle, bulk and tapped densities for *Acacia tortilis* compared to pine are expected, since the former is a hard wood, while the latter is a soft wood. Such high bulk-energy densities of *Acacia tortilis* also mean that it can be compressed into pellets or briquettes of higher energy values compared to pine dust. Since both biomasses are wastes in their local contexts, and their thermochemical properties are within the recommended ranges, they can be converted by any of the combustion, gasification and pyrolysis routes. They can also be torrefied or carbonized into charcoal to obtain solid fuels of higher heating values. Further experiments can be carried out for the various conversion routes to determine the optimal conditions and the properties of the products. For storage purposes, it seems more convenient to store the feedstocks as they are received, or after the first grind, while they have a mixed size range that is more compressible than uniform sizes.

There were also some limitations in the study, mainly in the FTIR analysis and particle density measurements. It is ideal to have a library of standards attached to the FTIR spectrometer to identify the functional groups directly; however, the equipment used did not have such a library. Therefore, the literature was used to make inferences, using findings done before by other authors. A pycnometer is recommended for densities of small particles; however, since it was not available, particle dimensions were used as the next most-appropriate method.

Although feedstock characterizations give a broad picture of the expected behavior of a material in subsequent stages, further tests still need to be done to confirm such predictions and derive specific results for the actual logistic and conversion stages.

Author Contributions: Conceptualization, data curation and writing—original draft preparation, G.C.; writing—review and editing, G.D. and E.M. and supervision, G.D. and E.M. All authors have read and agreed to the published version of the manuscript.

Funding: This research received no external funding.

Acknowledgments: We acknowledge the support of the Botswana International University of Science and Technology and the University of Johannesburg.

Conflicts of Interest: The authors declare no conflicts of interest.

Abbreviations and Symbols

Abbreviation	Explanation
ASTM	American Society for Testing and Materials
CHNS-O	Carbon, hydrogen, nitrogen and sulphur; oxygen determined by the difference
D_b	Bulk density of particles
D_t	Tapped density
ED_b	Bulk energy density
FC	Fixed carbon
FTIR	Fourier-transform infrared spectroscopy
GC	Gas chromatography
HHV	High heating value
L and D_p	Length and diameter of cylindrical acacia particles
$(L \times W \times H)$	Length, width and height dimensions for rectangular pine particles
MC	Moisture content
$\sum M_p$	Total particle mass
M_p and V_p	Mass and volume of the cylindrical particles
$\sum M_{tp}$	The total mass of tapped particles
SC	Supply chain
TGA	Thermogravimetric analyzer

References

1. Álvarez-Álvarez, P. Evaluation of tree species for biomass energy production in Northwest Spain. *Forests* **2018**, *9*, 160. [CrossRef]
2. Amundson, J.; Sukumara, S.; Seay, J.; Badurdeen, F. Decision Support Models for Integrated Design of Bioenergy Supply Chains. *Handb. Bioenergy* **2015**, 163–190. [CrossRef]
3. Eason, J.P.; Cremaschi, S. A multi-objective superstructure optimization approach to biofeedstocks-to-biofuels systems design. *Biomass Bioenergy* **2014**, *63*, 64–75. [CrossRef]
4. Braza, C.E.M.; Crnkovic, P.M. Physical—Chemical characterization of biomass samples for application in pyrolysis process. *Chem. Eng. Trans.* **2014**, *37*, 523–528. [CrossRef]
5. Feria, M.J. Energetic characterization of lignocellulosic biomass from Southwest Spain. *Int. J. Green Energy* **2011**, *8*, 631–642. [CrossRef]

6. Mishra, R.K.; Mohanty, K. Thermal and catalytic pyrolysis of pine sawdust (Pinus ponderosa) and Gulmohar seed (Delonix regia) towards production of fuel and chemicals. *Mater. Sci. Energy Technol.* **2019**, *2*, 139–149. [CrossRef]

7. Naik, S.; Goud, V.V.; Rout, P.K.; Jacobson, K.; Dalai, A.K. Characterization of Canadian biomass for alternative renewable biofuel. *Renew. Energy* **2010**, *35*, 1624–1631. [CrossRef]

8. Núñez-Regueira, L.; Proupín-Castiñeiras, J.; Rodríguez-Añón, J.A. Energy evaluation of forest residues originated from shrub species in Galicia. *Bioresour. Technol.* **2004**, *91*, 215–221. [CrossRef]

9. Cai, J. Review of physicochemical properties and analytical characterization of lignocellulosic biomass. *Renew. Sustain. Energy Rev.* **2017**, *76*, 309–322. [CrossRef]

10. Charis, G.; Danha, G.; Muzenda, E. A critical taxonomy of socio-economic studies around biomass and bio-waste to energy projects. *Detritus* **2018**, *3*, 47–57. [CrossRef]

11. Tumuluru, J.S.; Tabil, L.G.; Song, Y.; Iroba, K.L.; Meda, V. ScienceDirect Grinding energy and physical properties of chopped and hammer-milled barley, wheat, oat, and canola straws. *Biomass Bioenergy* **2013**, *60*, 58–67. [CrossRef]

12. Mani, S.; Tabil, L.G.; Sokhansanj, S. Grinding performance and physical properties of wheat and barley straws, corn stover and switchgrass. *Biomass Bioenergy* **2004**, *27*, 339–352. [CrossRef]

13. Anukam, A.I.; Mamphweli, S.N.; Reddy, P.; Okoh, O.O. Characterization and the effect of lignocellulosic biomass value addition on gasification efficiency. *Energy Explor. Exploit.* **2016**, *34*, 865–880. [CrossRef]

14. Mamvura, T.A.; Pahla, G.; Muzenda, E. Torrefaction of waste biomass for application in energy production in South Africa. *S. Afr. J. Chem. Eng.* **2018**, *25*, 1–12. [CrossRef]

15. Lam, P.S. Physical characterization of wet and dry wheat straw and switchgrass–bulk and specific density. In Proceedings of the 2007 ASABE Annual International Meeting, Minneapolis, MN, USA, 17–20 June 2007; ASABE: St. Joseph, MI, USA, 2008; Volume 300.

16. Kirsanovs, V.; Blumberga, D.; Dzikevics, M.; Kovals, A. Design of Experimental Investigations on the Effect of Equivalence Ratio, Fuel Moisture Content and Fuel Consumption on Gasification Process. *Energy Procedia* **2016**, *95*, 189–194. [CrossRef]

17. Reza, S. Acacia Holosericea: An Invasive Species for Bio-char, Bio-oil, and Biogas Production. *Bioengineering* **2019**, *6*, 33. [CrossRef]

18. Charis, G.; Danha, G.; Muzenda, E. A review of timber waste utilization: Challenges and opportunities in Zimbabwe. *Procedia Manuf.* **2019**, *35*, 419–429. [CrossRef]

19. Effah, B.; Antwi, K.; Boampong, E.; Asamoah, J.N.; Asibey, O. The management and disposal of small scale sawmills residues at the Sokoban and Ahwia wood markets in Kumasi–Ghana. *Int. J. Innov. Sci. Res.* **2015**, *19*, 15–23.

20. Arimoro, F.O.; Ikomi, R.B.; Osalor, E.C. The impact of sawmill wood wastes on the water quality and fish communities of Benin River, Niger Delta area, Nigeria. *World J. Zool.* **2006**, *1*, 94–102.

21. Charis, G.; Danha, G.; Muzenda, E. Waste valorisation opportunities for bush encroacher biomass in savannah ecosystems: A comparative case analysis of Botswana and Namibia. *Procedia Manuf.* **2019**, *35*, 974–979. [CrossRef]

22. Acar, S.; Ayanoglu, A. Determination of higher heating values (HHVs) of biomass fuels. *Uluslararası Yakıtlar Yanma Ve Yangın Dergisi* **2016**, *3*, 1–3.

23. Krotz, L.; Giazzi, G. Elemental Analysis: CHNS/O Characterization of Biomass and Bio-fuels, 2017. Thermo Fischer Scientific. Available online: https://assets.thermofisher.com/TFS-Assets/CMD/Application-Notes/an-42151-oea-chnso-biomass-biofuels-an42151-en.pdf (accessed on 20 January 2019).

24. Liu, Y.; Zeng, F.; Sun, B.; Jia, P.; Graham, I.T. Structural Characterizations of Aluminosilicates in in Two Types of Fly Ash Samples from Shanxi Province, North China. *Minerals* **2019**, *9*, 358. [CrossRef]

25. Bridgwater, A.V. Review of fast pyrolysis of biomass and product upgrading. *Biomass Bioenergy* **2011**, *38*, 68–94. [CrossRef]

26. Olatunji, O.; Akinlabi, S.; Oluseyi, A.; Peter, M.; Madushele, N. Experimental investigation of thermal properties of Lignocellulosic biomass: A review. *IOP Conf. Ser. Mater. Sci. Eng.* **2018**, *413*, 012054. [CrossRef]

27. Ahmed, A.; Bakar, M.S.A.; Azad, A.K.; Sukri, R.S.; Phusunti, N. Intermediate pyrolysis of Acacia cincinnata and Acacia holosericea species for bio-oil and biochar production. *Energy Convers. Manag.* **2018**, *176*, 393–408. [CrossRef]

28. MIT. *FTIR Spectrophotometer. Qual Instrumentation*; MIT: Cambridge, MA, USA, 2003; pp. 1–31.

29. Spectroscopy DataTables, "Infrared Tables". Available online: z:%5Cfiles%5Cclasses%5Cspectroscopy%5Ctypicalspectracharts.DOC (accessed on 20 January 2019).

30. Liu, C.; Wang, H.; Northwest, P.; Sun, J. Catalytic fast pyrolysis of lignocellulosic biomass. *Chem. Soc. Rev.* **2014**, *43*, 7594–7623. [CrossRef]

31. Dvorak, W.S.; Hodge, G.R.; Kietzka, J.E.; Malan, F.; Osorio, L.F.; Stanger, T.K. Pinus patula. In *Conservation and Testing of Tropical and Subtropical Forest Tree Species by the CAMCORE Cooperative*; CAMCORE: Raleigh, NC, USA, 2000; pp. 148–173.

32. Pratap, A.; Chouhan, S. Critical Analysis of Process Parameters for Bio-oil Production via Pyrolysis of Biomass: A Review. *Recent Pat. Eng.* **2013**, *7*, 98–114. [CrossRef]

33. DECOSA. *Support to De-Bushing Project Value Added End-Use Opportunities for Namibian Encroacher Bush*; DECOSA: Windhoek, Namibia, 2015.

34. Bridgwater, T. Challenges and Opportunities in Fast Pyrolysis of Biomass: Part I. *Johns. Matthey Technol. Rev.* **2018**, *62*, 118–130. [CrossRef]

Article

Utilization of Steel-Making Dust in Drilling Fluids Formulations

Musaab I. Magzoub [1,2], **Mohamed H. Ibrahim** [1], **Mustafa S. Nasser** [1], **Muftah H. El-Naas** [1,*] and **Mahmood Amani** [3]

[1]   Gas Processing Center, College of Engineering, Qatar University, P.O. Box 2713 Doha, Qatar;
      magzoub@ou.edu (M.I.M.); m.ibrahim@qu.edu.qa (M.H.I.); m.nasser@qu.edu.qa (M.S.N.)
[2]   Well Construction Technology Center, University of Oklahoma, Norman, OK 73069, USA
[3]   Petroleum Engineering, Texas A&M University at Qatar, P.O. Box 23874 Doha, Qatar;
      mahmood.amani@qatar.tamu.edu
*   Correspondence: muftah@qu.edu.qa

Received: 27 March 2020; Accepted: 30 April 2020; Published: 3 May 2020

Abstract: Steelmaking is an energy-intensive process that generates considerable amounts of by-products and wastes, which often pose major environmental and economic challenges to the steel-making industry. One of these by-products is steel dust that is produced during the separation of impurities in the smelting and refining of metals in steel-making furnaces. In this study, electric arc furnace (EAF) dust has been evaluated as a potential, low-cost additive to increase the viscosity and weight of drilling muds. Currently, the cost of drilling operations typically accounts for 50 to 80% of the exploration costs and about 30 to 80% of the subsequent field development costs. Utilization of steelmaking waste in drilling fluids formulations is aimed to produce new and optimized water-based drilling formulations, which is expected to reduce the amount of bentonite and other viscosifier additives used in the drilling formulations. The results showed that in a typical water-based drilling fluid of 8.6 ppg (1030.51 kg/m^3), the amount of standard drilling grade bentonite could be reduced by 30 wt.% with the addition of the proposed new additive to complete the required mud weight. The mixture proved to be stable with no phase separation.

Keywords: steelmaking; recycling; bentonite; solid waste management; sustainable materials

---

1. Introduction

The steel-making process produces large amounts of steel dust, which are reported to be as high as 2–4 tons for each ton of steel produced [1]. Consequently, waste management and processing of these by-products is becoming a major environmental issue [2]. The steel dust solid material is formed as a result of interactions between impurities such as silica and lime at various stages of steel production [3], and during the separation of flux and impurities in the smelting and refining of metals processes in steel-making furnaces [4]. Most of the produced steel dust is used for land filling [5,6] and many civil engineering applications including cement production [7–11]. It is also used as an asphalt mixture additive in the surface layer of roads or airport pavements [12–15]. Bentonite clays and steel dust share similar chemical composition. Both contain aluminum silicates and various combinations of oxides, in addition to sodium, calcium and magnesium ions [16,17]. As shown in the general formula below [18], bentonite clays consist of aluminum silicates with the presence of other ions, such as Na$^+$, Ca^{+2}, and Mg^{+2}, alongside other elements like iron oxide (FeO), manganese oxide (MnO), and magnesium oxide (MgO) [19]. Some compounds such as CaO, Fe$_2$O$_3$, SiO$_2$, Al$_2$O$_3$, TiO$_2$ and MgO also exist in the steel-making by-products with little variations in mass ratios [20,21].

Therefore, many studies propose using steel dust in applications where clays are similarly used. They are used as adsorbents to remove toxic and heavy metals from water [22–28], as catalyst [29–31], and for CO_2 sequestration [32,33]. In addition, steel dust was also reported to have been used for drill cuttings disposal [34,35]. The drilled cuttings were disposed through solidification by combining the drill cuttings with water and blast furnace slag to form highly concentrated drilled cutting wastes. The blast furnace slag was compatible with both oil and water-based drilling muds. The drill cutting wastes solidified by blast furnace slag were hard and unreachable [35].

Bentonite clays are also widely used in many industries, such as cosmetics and medical products, paints, water treatment [36–39], pharmaceuticals [40], dyes [41,42], and papermaking [43–45]. In drilling fluids formulation, bentonite mostly makes 80 wt.% of the drilling fluids. Circulation of viscous heavy fluids, such drilling fluids (drilling mud) is essential for successful drilling operation [46,47]. The favorable chemical composition and physical properties of bentonite increases mud viscosity and reduces filtration loss that occur due to differences in pressure between the column of drilling fluids and the formation pore pressure [48]. The fluid is injected through a hollow drill-string, and then flushed out of the well lifting the drilled cutting through annulus space between the drilling string and the wall of the well [49]. The mud circulation serves as hole cleaner by lifting rock cuttings, providing a reasonable hydrostatic pressure to suppress the overburden pressure of the formation and preventing formation fluids from flowing into the well while drilling [46,50]. This mud shares about 50% of the drilling cost, in addition to the overall field development cost due to mud related problems [51]. The proposed formula with the utilization of steel-making waste in drilling fluids is expected to reduce the amount of bentonite and other costly additives used in the drilling formulations.

Worldwide, the total production of steel increased recently to hundreds of million metric tons according to world steel association reports [52]. A typical integrated steel plant produces about 90 to 100 kg of steel slag per ton of steel during the refining of hot metal from the blast furnace. This generates alkaline solid residues about 10 wt.% to 15 wt.% of the produced steel, depending on the characteristics of the manufacturing process. The dust is extensively available and can be supplied as raw material from many steel industries [53]. Composition of the dust varies depending on the type of steel being manufactured, raw materials, cooling and crash methods. Various slag types are produced as by-products in metallurgical processes or as residues in incineration processes in large amounts, which can be classified into three categories according to its origins and the characteristics: ferrous slag, non-ferrous slag, and incineration slag [54]. The main types of steel dust produced at steelmaking process are Ladle Furnace (LF) slag, Bag house dust (BHD) and Cyclone silo dust. The bag house dust (BHD) is the type evaluated in this study and proposed as a drilling fluids additive [51].

Waste management focuses on managing and monitoring of waste and by-product materials, mainly collection, transportation, processing or disposal [55,56]. Steelmaking by-products, such as steel dust is being processed and utilized in many applications. However, to the best knowledge of the author, steel dust has never been evaluated as drilling fluid additives or used in drilling fluids formulations. In this study, the steelmaking by-products have been evaluated to make up to 30 wt.% of the drilling fluid base formulation.

2. Materials

Two types of steel by-products: Ladle Furnace (LF) slag and Bag house dust (BHD) were selected for the study. The samples were collected from the open-to-the atmosphere yard at a local steel factory. The samples were sieved through 200 mesh sieves (75 μm) to remove any coarse parts. Commercial bentonite was purchased from Sigma-Aldrich Company Ltd., Germany. The chemical compositions for the two types of steel dust and the commercial bentonite are shown in Figure 1, highlighting the major elements content in mass percentage.

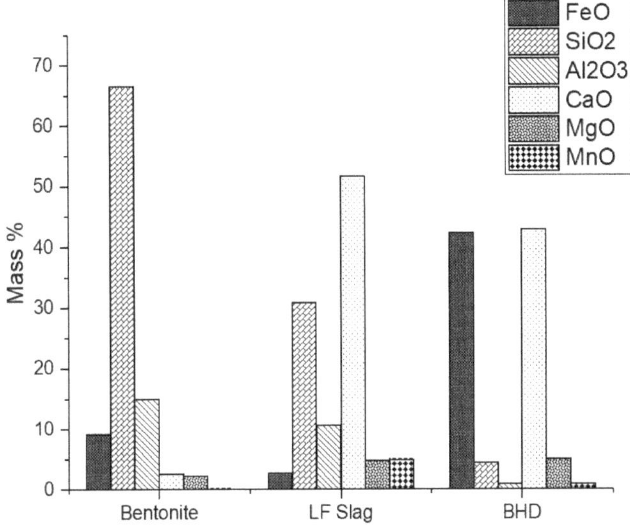

Figure 1. Chemical composition of bentonite and Ladle Furnace (LF) slag, Bag house dust (BHD) and Cyclone.

3. Methodology

The physical and chemical properties of the steel dust were investigated according to API recommendation, considering the chemical composition, dispersion stability, rheological properties and the standard drilling fluid testing.

3.1. Screening and Characterization

Dispersion stability was evaluated for screening and selecting the best steel dust type by measuring the turbidity of 150 mg/L suspension in Nephelometric Turbidity Units (NTU) at various pH values, using a Hach 2100N turbidity meter. Moreover, the stability of the suspension was assessed by measuring the electrokinetic potential (commonly known as Zeta Potential (ζ), that quantifies the magnitude of surface electric charges of the suspended particles) using Zetasizer ZEN3600 (Malvern Instruments Ltd., Worcestershire, UK). Analysis of particle size distribution was conducted using a laser diffraction particle size analyzer (Mastersizer 3000, Malvern Instruments Ltd., UK).

3.2. Sample Preparation and Drilling Fluid Testing

For viscosity and yield point measurements, a speed dial viscometer (Fann Model 35 Viscometer) was employed. The samples were prepared at room temperature by adding bentonite and steel dust in different ratios into 350 mL of distilled water while stirring in a mud mixer for 2 min to form dispersions, then the samples were stirred for another 20 min. Any powder at the wall of the container is scraped using spatula every 5 min to make sure that all powder is suspended in the mixture. Bentonite and steel dust mixture was then aged for 16 h at room temperature. Subsequently, the samples were stirred for 5 min to condition before testing. Dial readings at 600, 300, 200, 100, 6, 3 rpm were recorded when the reading was stabilized at each rotational speed. The low pressure/low temperature API filtration was used to evaluate water control efficiency of the drilling fluid under 100 psi differential pressure. The conductivity of the mud cake and filtrates were recorded, and a mud balance (Fann Model 140) was used to check the drilling fluid density. The commercial standard drilling grade bentonite was used as a reference for comparison.

3.3. Rheological Measurement

A flow sweep test at 25 °C was carried out using a controlled stress and strain instrument (Anton Paar rheometer, MCR302).Co-centric cylinder geometry was used with a cup radius of 15 mm, configured with DIN rotor of a 14 mm radius and a height of 42 mm. The shear viscosity for drilling fluids samples of 6 wt.% solid in water, at different steel dust to bentonite ratios, was determined from the readings of shear stress over a wide range of shear rates (1 e s^{-1} to 1000 s^{-1}). In addition, dynamic sweep test was conducted for the formulated samples of drilling fluid to obtain information about the structure and elastic properties of the mixture. The test was carried out at room temperature over a range of oscillation frequencies (0.1 to 100 rad/s) at constant oscillation amplitude. An equilibrium time of 5 min was given for the sample before applying any stress. For each experiment, three repeated runs were carried out and the results were reproducible with an average experimental error of less than 5%.

4. Results & Discussion

4.1. Dispersion Stability and Suspension Properties

Stability of the drilling fluids formula is an essential parameter to maintain the original properties of the injected drilling fluid and provide the main functions of the mud circulation throughout the drilling of oil and gas wells. Conducting zeta potential analyses provides evidence on the quality of suspension properties. In drilling fluids, high negative ZP values (more than −20 mV) represent high suspension stability [18,57]. Turbidity measurements also reflect the quantity of suspended particles at a given pH and concentration. Based on this, the results illustrated in Figures 2 and 3 show that bag house dust (BHD) seems to be a good candidate for consideration as drilling fluid additive. Although Ladle Furnace (LF) slag type seems to have a high silica and low iron content (see Figure 1), its suspension in water exhibited poor stability evident from turbidity and zeta potential measurements. The LF slag produced low zeta potential (≈ −15 mV) and low turbidity values over all pH ranges (Figure 2). On the other hand, the bag house dust (BHD) dispersions exhibited high negative zeta potential (≈ −35 mV) and high turbidity values, especially at higher pH values (the typical drilling fluid condition), displaying high dispersion stability (Figure 3).

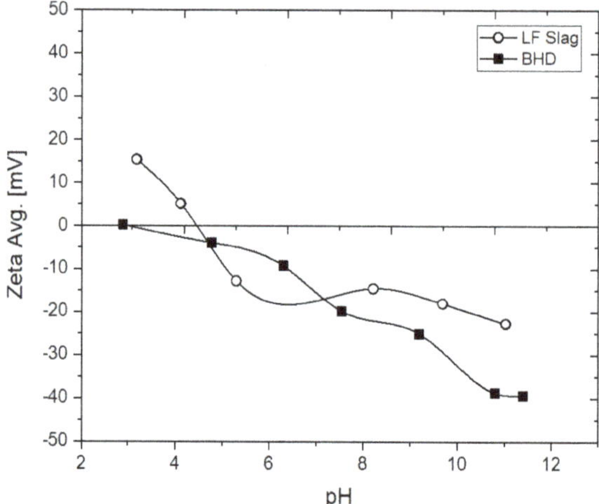

Figure 2. Zeta potential of steel dust LF and BHD in distilled water.

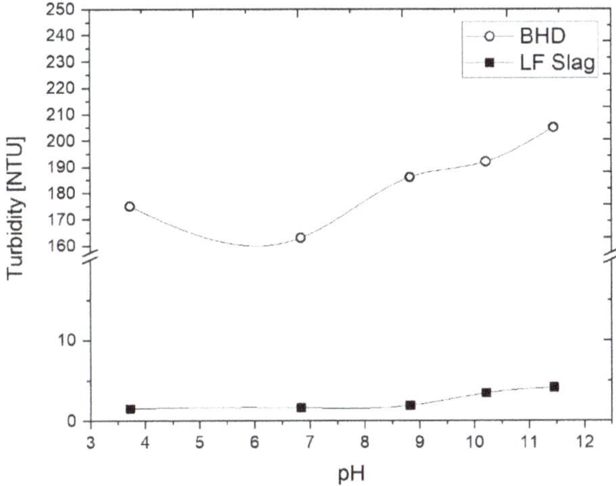

Figure 3. Turbidity of steel dust LF and BHD in distilled water.

Particle size and distribution (PSD) greatly influences dispersion stability and water filtration control. Drilling grade bentonite should have a narrow range of PSD with 96% of particles less than 75 µm [58]. In order to have a uniform PSD for the steel dust and bentonite mixture, the steel dust samples were sieved through 200 mesh (75 µm) to remove course particles. This makes all particles less than 75 µm, with 90% of particles (d.90) less than 59.8 µm and gives an average particle size (d.50) of 19.6 µm for steel dust compared to 7.9 µm of bentonite. The particle dispersity index of 1.7 for steel dust and 1.5 for bentonite was obtained. The full profiles of particle size distribution (PSD) are shown in Figures 4 and 5.

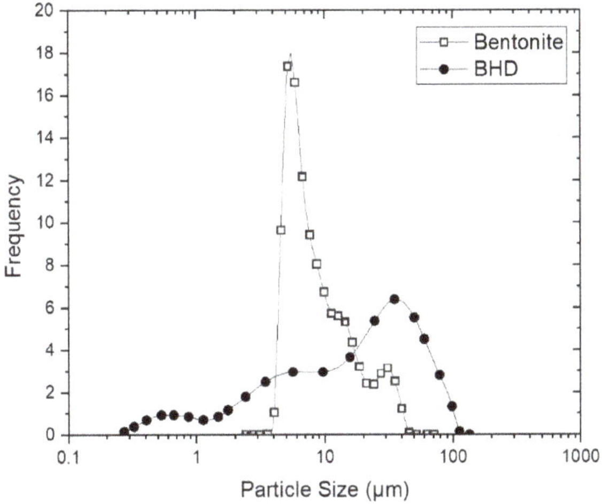

Figure 4. Particle size distribution of steel dust (BHD) compared to bentonite.

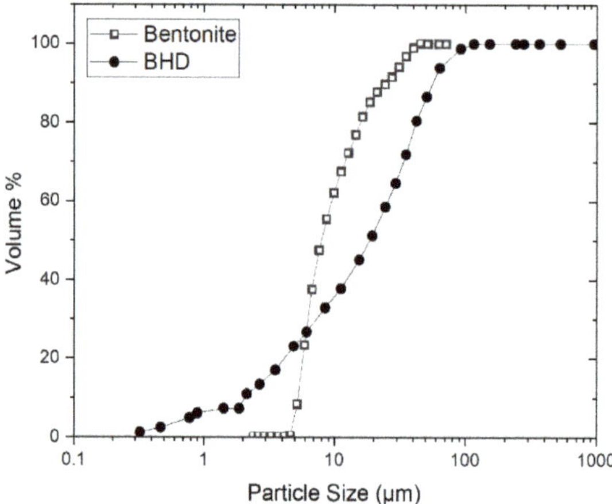

Figure 5. Mass percent of size distribution for steel dust (BHD) compared to bentonite.

4.2. Rheological Measurement

Rheological characteristics of drilling fluids are of high importance during drilling operations as the column of the drilling fluids inside the well may be subjected to different forces and stress [59,60]. High-speed mixers are used at well sites to prepare drilling fluids, where high shear rates are applied to formulate the mud prior to being pumped into the well. Inside the well, drilling fluids are also subjected to high shear resulting from rotation of the drilling string, in addition to the shear applied by circulation speeds and pressures. Viscosity of the fluids depends on the rate of shear that it endures. Results of viscosity at shear rates ranging from $1\ \text{s}^{-1}$ to $1000\ \text{s}^{-1}$ are shown in Figure 6. While keeping the total solid content in the drilling fluid fixed at 6 wt.%, the addition of steel dust (BHD) increased the viscosity of the drilling fluid up to 30 wt.% of BHD in bentonite, following which the viscosity began to decrease. This indicates that a maximum of 30 wt.% steel dust can be used in a mixture with bentonite. Figure 7 shows the changes in shear viscosity with the addition of steel dust. The effect is more obvious under high shear, where the viscosity increases from 43.6 mPa.s to 87.3 mPa.s and almost steadies up to 30 wt.% of BHD in bentonite.

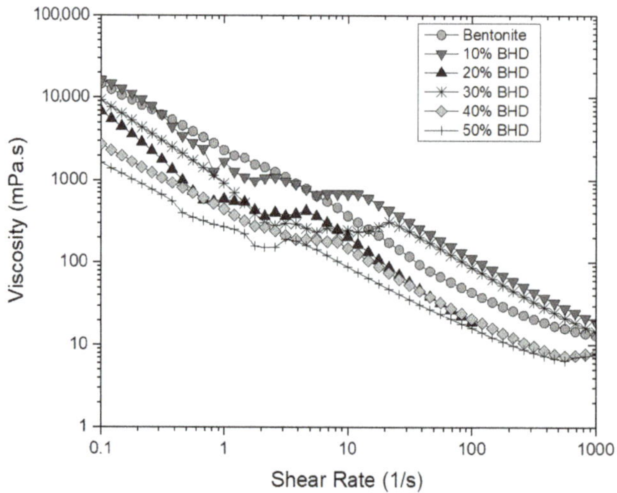

Figure 6. Flow sweep test, viscosity for different steel dust (BHD) ratios.

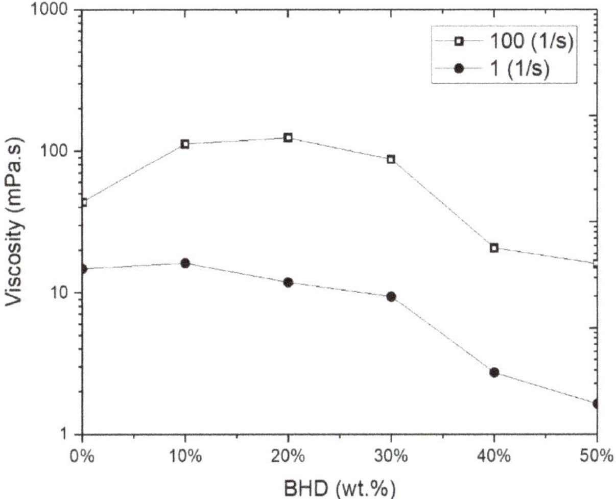

Figure 7. Effect of steel dust addition on viscosity at $1\ \text{s}^{-1}$ & $100\ \text{s}^{-1}$.

Structure, elasticity and gel strength have great influence on the calculation of circulation hydraulics and hole cleaning efficiency. Efficient rheological properties maintain whole cleaning by insuring cutting removal and prevent cutting from settling or packing the well during downtime, tripping or pulling out of the hole (POOH). The results of dynamic flow test showed that the storage and loss modulus of bentonite suspensions increased in orders of magnitude with replacement of steel dust (BHD) in ratio of 30 wt.% to bentonite. Figure 8 shows that both elastic and viscous moduli increased. Large storage modulus reflects that the addition of steel dust to drilling fluids renders the dispersion to exhibit more solid-like behavior with increased elasticity, resulting in an enhanced cutting suspension property of the mud and improved hole cleaning.

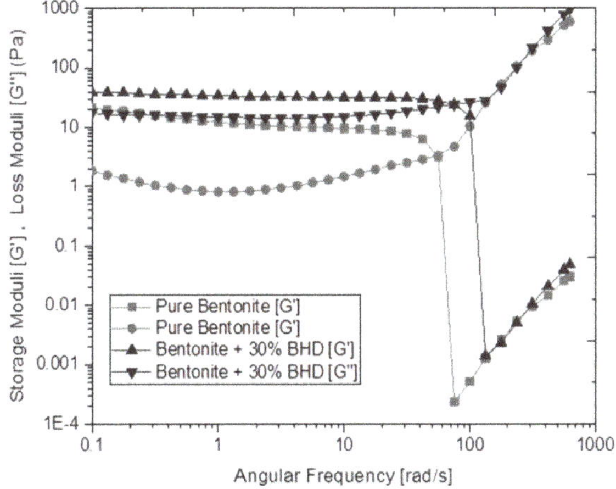

Figure 8. Dynamic flow test.

4.3. Drilling Fluid Testing

Measuring flow parameters of the drilling mud, such as apparent viscosity, plastic viscosity, and yield point is important for controlling frictional pressure drop and solids-bearing capacity. The results of 20 wt.% and 30 wt.% compared to 100% bentonite are shown in Figures 9 and 10. The viscometer readings increased with steel dust addition at 20 wt.% and then decreased at 30wt%. Subsequently, the apparent viscosity increased from 24.5 cP to 44 cP and then decreased to 38 cP at 30 wt.%. The structure of the mixture seems to be enforced by the presence of steel dust (BHD). The apparent viscosity was successfully maintained at 38 cP despite using less bentonite (70% of the initial mass), indicated by higher apparent viscosity and higher yield point. The plastic viscosity was maintained between 16 to 21 cP.

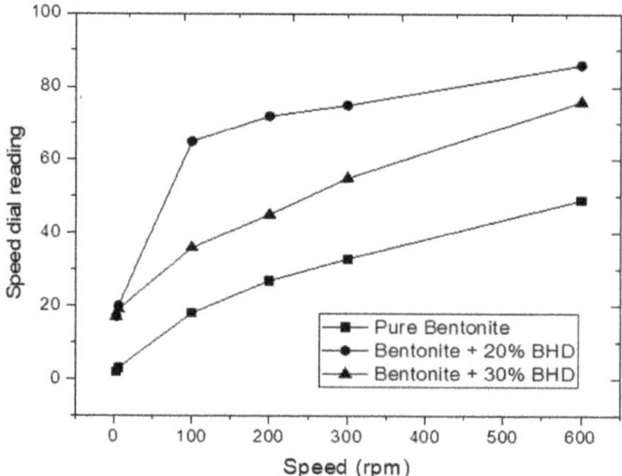

Figure 9. Viscometer readings.

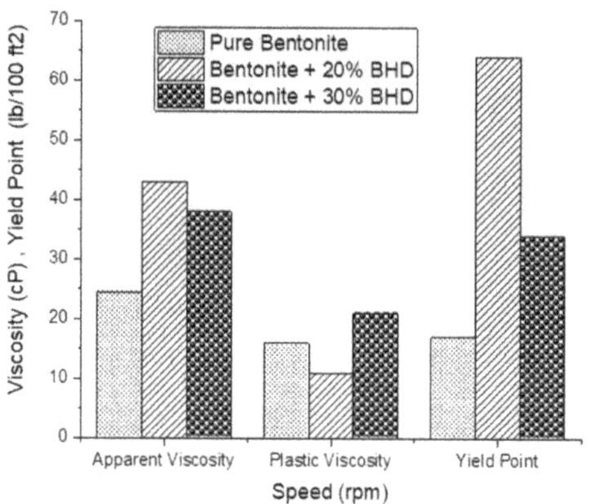

Figure 10. Apparent and plastic viscosities and yield point of bentonite and steel dust mixture (BHD).

Low pressure/low temperature API filtration test is also used to evaluate the drilling fluid quality. Filtrates volume and mud cake thickness are shown in Figure 11. The mixture of bentonite and steel dust exhibited higher filtration rates. However, the mud cake formed on the filter paper thickness did not increase and remained around 3 mm, suitable for preventing drill pipe stuck due to tight spots inside the borehole. If a high swelling formation is being drilled, the high filtrations rates may cause damage to the formation as well as pipe blockage. This may limit the amounts of steel dust to 20 wt.% to be used as drilling fluid additive to formulate the water-based drilling fluid and reduces the amount of bentonite needed to 80%. Since the filtration control is affected by the viscosity as well as the particle size, it is highly recommended to use additional additives to maintain the viscosity of the mud such as Xanthan gum, before increasing the amount of bag house dust in the formula. In addition, the steel dust can be used in some non-conventional drilling practices, where control of water filtration to formation is not required, such as underbalanced drilling (UBD). In this type of drilling, "the hydrostatic head of a drilling fluid is intentionally designed to be lower than the pressure of the formation being drilled", and the formation fluids are allowed to flow into the wellbore up to the surface [61].

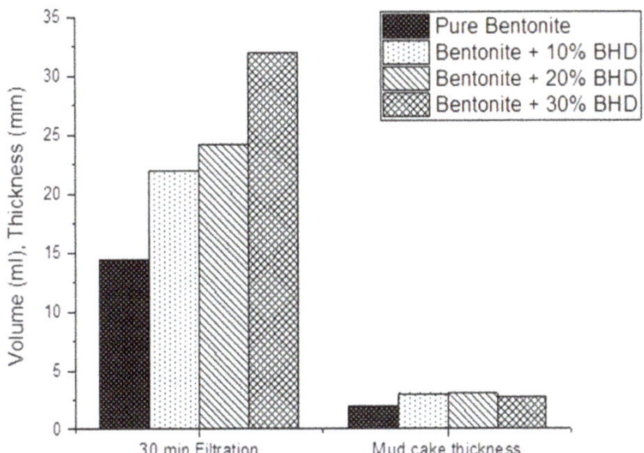

Figure 11. API filtration test for different ratios of steel dust (BHD) and bentonite.

The steel dust could also be used to formulate killing fluids in case of influx from formation fluids (Oil/Gas) into the wellbore, to prevent a blowout. Kill procedures typically involve pumping of higher density mud into the wellbore. In the case of an induced kick, steel dust of high specific gravity can provide mud density that is sufficient to kill the well [61]. Another possible utilization of steel dust is in the case of loss circulation (partial or total loss). In such cases, a very large amount of expensive fluids are lost into the formation due to highly permeable or fractured zones [62]. Steel dust is a low-cost material abundantly available that can be potentially used with lost circulation materials (LCM) to regain control of the well.

5. Conclusions

Steel dust powders were evaluated in this study and found to have a high potential for utilization as drilling fluid additives. Characterization and analysis of the steel dust showed that processing and pretreatment of the dust by cleaning, grinding, and sieving is recommended. The pretreatment is needed to homogenize the particle size distribution and enhance water filtration control in addition to improving dispersion stability. Evaluation of the proposed formula helped to conclude that the amount of standard drilling grade bentonite can be reduced by 30 wt.% to produce stable drilling fluids with sufficient density and rheological properties that compete well with standard formulations. The apparent viscosity and plastic viscosity increased from 24.5 cP to 38 cP and from 16 cP to 21 cP respectively. Moreover, the elastic properties were significantly improved as the storage modulus G′ increased from 8.93 Pa to 132 Pa, which is expected to improve cutting lifting and hole cleaning efficiency. The mud cake resulted from API filtration test was 2 to 3 mm, which is less than the maximum allowable thickness of 4 mm. This guarantees the absence of tight spots through the drilled sections and prevents drill pipes from getting mechanically stuck. However, the water filtration was higher (22 to 32 mL). Therefore, to avoid formation damage, the new additive is recommended to be used for drilling the main hole (usually more than 90% of the total depth of the well) from the surface to the depth before the hydrocarbon zone. Special cases such as underbalanced drilling can also be a good application for steel dust drilling fluids. They can also be used in case of partial or total loss circulation in fractured zones, where a massive amount of expensive drilling fluids is lost.

Author Contributions: Conceptualization, M.I.M., M.S.N. and M.H.E.-N. Methodology, M.I.M.; Formal analysis, M.I.M., M.H.I. and M.A.; Investigation, M.I.M., M.S.N., M.A. and M.H.E.-N.; Writing—original draft preparation,

M.I.M.; Writing—review and editing, M.I.M., M.H.I., M.S.N. and M.H.E.-N.; supervision, M.S.N. and M.H.E.-N. All authors have read and agreed to the published version of the manuscript.

Funding: This research received no external funding.

Conflicts of Interest: The authors declare no conflict of interest.

References

1. Das, B.; Prakash, S.; Reddy, P.; Misra, V. An overview of utilization of slag and sludge from steel industries. *Resour. Conserv. Recycl.* **2007**, *50*, 40–57. [CrossRef]
2. Yusuf, M.; Chuah, L.; Khan, M.A.; Choong, T.S.Y. Adsorption of Nickel on Electric Arc Furnace Slag: Batch and Column Studies. *Sep. Sci. Technol.* **2014**, *49*, 388–397. [CrossRef]
3. Teir, S.; Eloneva, S.; Fogelholm, C.-J.; Zevenhoven, R. Dissolution of steelmaking slags in acetic acid for precipitated calcium carbonate production. *Energy* **2007**, *32*, 528–539. [CrossRef]
4. Bölükbaşı, Ö.; Tufan, B. Steelmaking slag beneficiation by magnetic separator and impacts on sinter quality. *Sci. Sinter.* **2014**, *46*, 331–344. [CrossRef]
5. Teir, S.; Kuusik, R.; Fogelholm, C.-J.; Zevenhoven, R. Production of magnesium carbonates from serpentinite for long-term storage of CO_2. *Int. J. Miner. Process.* **2007**, *85*, 1–15. [CrossRef]
6. Teir, S.; Revitzer, H.; Eloneva, S.; Fogelholm, C.-J.; Zevenhoven, R. Dissolution of natural serpentinite in mineral and organic acids. *Int. J. Miner. Process.* **2007**, *83*, 36–46. [CrossRef]
7. Abdel-Ghani, N.T.; El-Sayed, H.A.; El-Habak, A.A. Utilization of by-pass cement kiln dust and air-cooled blast-furnace steel slag in the production of some "green" cement products. *HBRC J.* **2018**, *14*, 408–414. [CrossRef]
8. Machado, A.; Valenzuela-Diaz, F.; De Souza, C.; de Andrade Lima, L. Structural ceramics made with clay and steel dust pollutants. *Appl. Clay Sci.* **2011**, *51*, 503–506. [CrossRef]
9. Colorado, H.A.; Garcia, E.; Buchely, M. White Ordinary Portland Cement blended with superfine steel dust with high zinc oxide contents. *Constr. Build. Mater.* **2016**, *112*, 816–824. [CrossRef]
10. Rosales, J.; Cabrera, M.; Agrela, F. Effect of stainless steel slag waste as a replacement for cement in mortars. Mechanical and statistical study. *Constr. Build. Mater.* **2017**, *142*, 444–458. [CrossRef]
11. Jiang, Y.; Ling, T.-C.; Shi, C.; Pan, S.-Y. Characteristics of steel slags and their use in cement and concrete—A review. *Resour. Conserv. Recycl.* **2018**, *136*, 187–197. [CrossRef]
12. Sayadi, M.; Hesami, S. Performance evaluation of using electric arc furnace dust in asphalt binder. *J. Clean. Prod.* **2017**, *143*, 1260–1267. [CrossRef]
13. Alsheyab, M.A.; Khedaywi, T.S. Dynamic creep analysis of Electric Arc Furnace Dust (EAFD)–Modified asphalt. *Constr. Build. Mater.* **2017**, *146*, 122–127. [CrossRef]
14. Loaiza, A.; Colorado, H.A. Marshall stability and flow tests for asphalt concrete containing electric arc furnace dust waste with high ZnO contents from the steel making process. *Constr. Build. Mater.* **2018**, *166*, 769–778. [CrossRef]
15. Wang, J.; Guo, M.; Tan, Y. Study on application of cement substituting mineral fillers in asphalt mixture. *Int. J. Transp. Sci. Technol.* **2018**, *7*, 189–198. [CrossRef]
16. Murray, H.H. Structure and composition of the clay minerals and their physical and chemical properties. *Dev. Clay Sci.* **2006**, *2*, 7–31.
17. Saboori, R.; Sabbaghi, S.; Kalantariasl, A. Improvement of rheological, filtration and thermal conductivity of bentonite drilling fluid using copper oxide/polyacrylamide nanocomposite. *Powder Technol.* **2019**, *353*, 257–266. [CrossRef]
18. Magzoub, M.I.; Nasser, M.S.; Hussein, I.A.; Benamor, A.; Onaizi, S.A.; Sultan, A.; Mahmoud, M. Effects of sodium carbonate addition, heat and agitation on swelling and rheological behavior of Ca-bentonite colloidal dispersions. *Appl. Clay Sci.* **2017**, *147*, 176–183. [CrossRef]
19. Temraz, M.G.; Hassanien, I. Mineralogy and rheological properties of some Egyptian bentonite for drilling fluids. *J. Nat. Gas Sci. Eng.* **2016**, *31*, 791–799. [CrossRef]
20. Hameedi, S.M. Carbon Dioxide Capture by Steel-Making Residues in a Fluidized Bed Reactor. Master's Thesis, UAE University, Al-Ain, United Arab Emirates, 2016.
21. Sadeghalvaad, M.; Sabbaghi, S. The effect of the TiO2/polyacrylamide nanocomposite on water-based drilling fluid properties. *Powder Technol.* **2015**, *272*, 113–119. [CrossRef]

22. Ma, A.; Zheng, X.; Li, S.; Wang, Y.; Zhu, S. Zinc recovery from metallurgical slag and dust by coordination leaching in NH_3–CH_3COONH_4–H_2O system. *R. Soc. Open Sci.* **2018**, *5*, 180660. [CrossRef] [PubMed]
23. Omran, M.; Fabritius, T. Utilization of blast furnace sludge for the removal of zinc from steelmaking dusts using microwave heating. *Sep. Purif. Technol.* **2019**, *210*, 867–884. [CrossRef]
24. Wang, J.; Wang, Z.; Zhang, Z.; Zhang, G. Removal of zinc from basic oxygen steelmaking filter cake by selective leaching with butyric acid. *J. Clean. Prod.* **2019**, *209*, 1–9. [CrossRef]
25. Blahova, L.; Mucha, M.; Navratilova, Z.; Budzyń, S.; Tora, B. Industrial Wastes as Potentional Sorbents of Heavy Metals. *Inżynieria Miner.* **2018**, *19*, 61–66.
26. Pająk, M.; Dzieniszewska, A.; Kyzioł-Komosińska, J.; Chrobok, M. Use of Metallurgical Dust for Removal Chromium Ions from Aqueous Solutions. In *E3S Web of Conferences*; EDP Sciences: Les Ulis, France, 2018; p. 01029.
27. Bouabidi, Z.B.; El-Naas, M.H.; Cortes, D.; McKay, G. Steel-Making dust as a potential adsorbent for the removal of lead (II) from an aqueous solution. *Chem. Eng. J.* **2018**, *334*, 837–844. [CrossRef]
28. Sun, L.; Chen, D.; Wan, S.; Yu, Z. Performance, kinetics, and equilibrium of methylene blue adsorption on biochar derived from eucalyptus saw dust modified with citric, tartaric, and acetic acids. *Bioresour. Technol.* **2015**, *198*, 300–308. [CrossRef]
29. Lee, J.-M.; Kim, J.-H.; Chang, Y.-Y.; Chang, Y.-S. Steel dust catalysis for Fenton-like oxidation of polychlorinated dibenzo-p-dioxins. *J. Hazard. Mater.* **2009**, *163*, 222–230. [CrossRef]
30. Shan, R.; Lu, L.; Shi, Y.; Yuan, H.; Shi, J. Catalysts from renewable resources for biodiesel production. *Energy Convers. Manag.* **2018**, *178*, 277–289. [CrossRef]
31. Borisov, V.; Nedosekov, A.; Sigayeva, S.; Suprunov, G.; Vershinin, V.; Tsyrulnikov, P. Deep Oxidation of Methane on Palladic Catalysts on Suppliers ZrO_2, CeO_2, ZrO_2-CeO_2, CeO_2-CuO on Stainless Steel Prepared with the Method of Plasma Drawing. *Procedia Eng.* **2015**, *113*, 124–130. [CrossRef]
32. Dreillard, M.; Broutin, P.; Briot, P.; Huard, T.; Lettat, A. Application of the DMXTM CO2 Capture Process in Steel Industry. *Energy Procedia* **2017**, *114*, 2573–2589. [CrossRef]
33. El-Naas, M.H.; El Gamal, M.; Hameedi, S.; Mohamed, A.-M.O. CO_2 sequestration using accelerated gas-solid carbonation of pre-treated EAF steel-making bag house dust. *J. Environ. Manag.* **2015**, *156*, 218–224. [CrossRef]
34. Hale, A.H. Well Drilling Cuttings Disposal. U.S. Patent No. 5,341,882, 30 August 1994.
35. Nahm, J.J. Well Drilling Cuttings Disposal. U.S. Patent No. 5,277,519, 11 January 1994.
36. de Figueirêdo, J.; Araújo, J.; da Silva, I.A.; Cartaxo, J.M.; Neves, G.A.; Ferreira, H.C. Purified smectite clays organofilized with ionic surfactant for use in oil-based drilling fluids. *Mater. Sci. Forum* **2014**, *798*, 21–26. [CrossRef]
37. Zhuang, G.; Zhang, Z.; Guo, J.; Liao, L.; Zhao, J. A new ball milling method to produce organo-montmorillonite from anionic and nonionic surfactants. *Appl. Clay Sci.* **2015**, *104*, 18–26. [CrossRef]
38. Vipulanandan, C.; Mohammed, A. Hyperbolic rheological model with shear stress limit for acrylamide polymer modified bentonite drilling muds. *J. Pet. Sci. Eng.* **2014**, *122*, 38–47. [CrossRef]
39. Zhang, L.-Y.; Cai, S.-Y.; Mo, J.-H.; Wei, G.-T.; Li, Z.-M.; Ye, R.-C.; Xie, X.-M. Study on the Preparation of H3PW12O40–TiO2/Bentonite Composite Material. *Mater. Manuf. Process.* **2015**, *30*, 279–284. [CrossRef]
40. Cara, S.; Carcangiu, G.; Padalino, G.; Palomba, M.; Tamanini, M. The bentonites in pelotherapy: Chemical, mineralogical and technological properties of materials from Sardinia deposits (Italy). *Appl. Clay Sci.* **2000**, *16*, 117–124. [CrossRef]
41. Al-Hussaini, A.; Eldars, W. Non-conventional synthesis and antibacterial activity of poly (aniline-co-o-phenylenediamine)/bentonite nanocomposites. *Des. Monomers Polym.* **2014**, *17*, 458–465. [CrossRef]
42. Bergaya, F.; Lagaly, G. General introduction: Clays, clay minerals, and clay science. In *Developments in Clay Science*; Elsevier: Amsterdam, The Netherlands, 2013; Volume 5, pp. 1–19.
43. Li, Y.; Yue, Q.; Li, W.; Gao, B.; Li, J.; Du, J. Properties improvement of paper mill sludge-based granular activated carbon fillers for fluidized-bed bioreactor by bentonite (Na) added and acid washing. *J. Hazard. Mater.* **2011**, *197*, 33–39. [CrossRef]
44. Yoon, D.H.; Jang, J.W.; Seo, S.S.; Jeong, S.H.; Cheong, I.W. Preparation of cationic poly (acrylamide)/bentonite nanocomposites from in situ inverse emulsion copolymerization and their performance in papermaking. *Adv. Polym. Technol.* **2013**, *32*. [CrossRef]

45. Shih, P.-H.; Wu, Z.-Z.; Chiang, H.-L. Characteristics of bricks made from waste steel slag. *Waste Manag.* **2004**, *24*, 1043–1047. [CrossRef]

46. Bourgoyne, A.T., Jr.; Chenevert, M.E.; Millheim, K.K.; Young, F., Jr. Applied Drilling Engineering, second printing. *Textb. Ser. Sperichardsontexas* **1991**, *2*, 502.

47. Afolabi, R.O.; Paseda, P.; Hunjenukon, S.; Oyeniyi, E.A. Model prediction of the impact of zinc oxide nanoparticles on the fluid loss of water-based drilling mud. *Cogent Eng.* **2018**, *5*, 1514575. [CrossRef]

48. Nasser, M.; Onaizi, S.A.; Hussein, I.; Saad, M.; Al-Marri, M.; Benamor, A. Intercalation of ionic liquids into bentonite: Swelling and rheological behaviors. *Colloids Surf. A Physicochem. Eng. Asp.* **2016**, *507*, 141–151. [CrossRef]

49. González, J.; Quintero, F.; Arellano, J.; Márquez, R.; Sánchez, C.; Pernía, D. Effects of interactions between solids and surfactants on the tribological properties of water-based drilling fluids. *Colloids Surf. A Physicochem. Eng. Asp.* **2011**, *391*, 216–223. [CrossRef]

50. Song, K.; Wu, Q.; Li, M.; Ren, S.; Dong, L.; Zhang, X.; Lei, T.; Kojima, Y. Water-based bentonite drilling fluids modified by novel biopolymer for minimizing fluid loss and formation damage. *Colloids Surf. A Physicochem. Eng. Asp.* **2016**, *507*, 58–66. [CrossRef]

51. Gahan, C.S.; Cunha, M.L.; Sandström, Å. Comparative study on different steel slags as neutralising agent in bioleaching. *Hydrometallurgy* **2009**, *95*, 190–197. [CrossRef]

52. Charles, J. *Past, Present and Future of Duplex Stainless Steels*; Duplex Conference: Grado, Italy, 2007; pp. 18–20.

53. Morone, M.; Costa, G.; Polettini, A.; Pomi, R.; Baciocchi, R. Valorization of steel slag by a combined carbonation and granulation treatment. *Miner. Eng.* **2014**, *59*, 82–90. [CrossRef]

54. Shen, H.; Forssberg, E. An overview of recovery of metals from slags. *Waste Manag.* **2003**, *23*, 933–949. [CrossRef]

55. Oluwasola, E.A.; Hainin, M.R.; Aziz, M.M.A.; Yaacob, H.; Warid, M.N.M. Potentials of steel slag and copper mine tailings as construction materials. *Mater. Res. Innov.* **2014**, *18* (Suppl. 6), 250–254. [CrossRef]

56. Ennaciri, Y.; Bettach, M. Procedure to convert phosphogypsum waste into valuable products. *Mater. Manuf. Process.* **2018**, *33*, 1727–1733. [CrossRef]

57. Tunç, S.; Duman, O. The effect of different molecular weight of poly (ethylene glycol) on the electrokinetic and rheological properties of Na-bentonite suspensions. *Colloids Surf. A Physicochem. Eng. Asp.* **2008**, *317*, 93–99. [CrossRef]

58. API. *API-13A Specification for Drilling Fluid Materials*, 5th ed.; API: Washington, DC, USA, 1993.

59. Raheem, A.M.; Vipulanandan, C. Salt contamination and temperature impacts on the rheological and electrical resistivity behaviors of water based drilling mud. *Energy Sources Part A Recover. Util. Environ. Eff.* **2019**, *42*, 344–364. [CrossRef]

60. Afolabi, R.O.; Orodu, O.D.; Efeovbokhan, V.E.; Rotimi, O.J. Optimizing the rheological properties of silica nano-modified bentonite mud using overlaid contour plot and estimation of maximum or upper shear stress limit. *Cogent Eng.* **2017**, *4*, 1287248. [CrossRef]

61. Bennion, D.B.; Thomas, F.B.; Bietz, R.F.; Bennion, D.W. Underbalanced Drilling, Praises and Perils. In Proceedings of the Permian Basin Oil and Gas Recovery Conference, Midland, TX, USA, 27–29 March 1996; Society of Petroleum Engineers: Richardson, TX, USA, 1996.

62. Gockel, J. Lost Circulation Drilling Fluid. U.S. Patent No. 4,498,995, 12 February 1985.

Article

Modeling the Municipal Waste Collection Using Genetic Algorithms

Elisabete Alberdi [1,*,†], Leire Urrutia [1,†], Aitor Goti [2,†] and Aitor Oyarbide-Zubillaga [3,†]

[1] Department of Applied Mathematics, University of the Basque Country UPV/EHU, 48013 Bilbao, Bizkaia, Spain; leire.urru@gmail.com

[2] Deusto Digital Industry Chair, University of Deusto, 48007 Bilbao, Bizkaia, Spain; aitor.goti@deusto.es

[3] Department of Industrial Mechanics, Design and Organization, University of Deusto, 48007 Bilbao, Bizkaia, Spain; aitor.oyarbide@deusto.es

* Correspondence: elisabete.alberdi@ehu.eus; Tel.: +34-946-017-790

† These authors contributed equally to this work.

Received: 12 March 2020; Accepted: 21 April 2020; Published: 27 April 2020

Abstract: Calculating adequate vehicle routes for collecting municipal waste is still an unsolved issue, even though many solutions for this process can be found in the literature. A gap still exists between academics and practitioners in the field. One of the apparent reasons why this rift exists is that academic tools often are not easy to handle and maintain by actual users. In this work, the problem of municipal waste collection is modeled using a simple but efficient and especially easy to maintain solution. Real data have been used, and it has been solved using a Genetic Algorithm (GA). Computations have been done in two different ways: using a complete random initial population, and including a seed in this initial population. In order to guarantee that the solution is efficient, the performance of the genetic algorithm has been compared with another well-performing algorithm, the Variable Neighborhood Search (VNS). Three problems of different sizes have been solved and, in all cases, a significant improvement has been obtained. A total reduction of 40% of itineraries is attained with the subsequent reduction of emissions and costs.

Keywords: waste collection route planning; traveling salesman problem; genetic algorithms

---

1. Introduction

The improvement of the waste collection process can be considered aligned with the 11th Sustainable Development Goal (SDG) "Sustainable cities and communities" [1]. Nevertheless, this process is still far from being in an optimal status. As experts state [2], inadequate vehicle routes are among the problems the process should tackle. The optimization of the routes for waste collection vehicles with time window is known as the Waste Collection Vehicle Routing Problem (WCVRP).

As indicated by Caria, Todde, and Pazzona [3], the WCVRP is a specific case of the whole class of problems, known as the Vehicle Routing Problem (VRP). The oldest VRP type problem in the transport history is the Travelling Salesman Problem (TSP), solved for the first time by Lin [4], where the aim is to find the shortest route visiting each member of a collection of locations and returning to the starting point. The TSP has evolved towards solving similar problems with different and additional restrictions and objectives, including the WCVRP presented herein.

The WCVRP needs to organize routes, vehicles, and customers, while respecting the constraints that are imposed by the system. VRP allows for reaching the goals that are referred to as transport logistics, as well as the minimization of costs and carbon dioxide (CO_2) emissions [3]. On top of the overall VRP considerations, WCVRP is concerned with finding cost optimal routes for garbage trucks such that all garbage bins are emptied and the waste is driven to disposal sites while respecting customer time windows and ensuring that drivers are given the breaks that the law requires. The first waste collection problem was probably presented by Beltrami and Bodin [5]. Since then, the problem has evolved into the herein presented WCVRP. Many relevant references have studied this problem so far, offering different approaches for solving the same challenge (or similar ones). Concerning the most relevant and recent ones, Buhrkal, Larsen, and Ropke [6] propose an adaptive large neighborhood search algorithm for solving the problem. Babaee Tirkolaee, Mahdavi, and Seyyed Esfahani [7] apply an improved hybrid simulated annealing algorithm (SA) for optimizing the mathematical model developed. Hannan et al. [8] propose a particle swarm optimization algorithm to optimize a model that considers not only typical distance, total waste, collection efficiency, etc. parameters but also Threshold Waste Level (TWL) tightness factors. De Bruecker et al. [9] present an enhanced model for the WCVRP that models distinct labor cost types and collecting speeds; thus, they differentiate between cheaper regular shifts during traffic peak hours, against higher collection speeds but with expensive, non-regular shift-rates. Rodrigues Pereira Ramos, Soares de Moraisa, and Barbosa-Póvoa [10] study three operational management approaches to define dynamic optimal routes based on sensoring, Radio Frequency Identification (RFID), and considering the access to real-time information on the bins' fill-levels. Benjamin and Beasley [11] develop a model for the waste collection vehicle routing problem with time windows, driver rest period, and multiple disposal facilities as a differential aspect.

It is worth mentioning that the sometimes explicit, but always present idea of this optimization problem is the reduction of CO_2 emissions, as the current logistic methods used in this routing problems depend strongly on fossils fuels [12].

The fact of not having generalized optimized routes for the WCVRP is striking, as many solutions for this process can be found in the literature. When analyzing the potential causes of this breach between the practitioners and the academia, it could be stated that generic routing tools are considered incomplete for this variant of the traveling salesman problem [13], whereas more specific tools (such as [14,15]) have a related cost that often entities are not willing to afford. Additionally, many researchers developed their solutions to the problem, but normally they are more focused on publishing them than in offering them to entities that manage the waste collection. Thus, even though the problem has been deeply studied from a theoretical point of view, its application to real-world problems is scarce.

Within this context, the aim of this work has been to show the development and implementation of a procedure for the optimization for a local commonwealth for tackling the WCVRP of a region.

In the final solutions, Genetic Algorithms (GAs) have been applied. GAs belong to the more general classification of evolutionary optimization techniques or evolutionary programs and they are surely the most widely known type of Evolutionary Algorithms. They are based on selection, crossover, and mutation principles of Darwin's theory of evolution. In the last few years, there has been a growing effort to apply GAs to general constrained optimization problems as most of engineering optimization problems often see their solution constrained by a number of restrictions imposed on the decision variables [16].

The research method had three major steps: first, we performed a literature review of optimization algorithms applied to this case that is detailed in Section 1. The aim of this study was to analyze how this problem was modeled in the past, and to pre-select which types of algorithms could fit best to the end user requirements in terms of user friendliness, cost, and quality of results. Second, the characteristics of this specific optimization case (detailed in Section 2) were analyzed to offer the company a selection of the algorithm fitting best to its requirements. Finally, the designed solution was successfully implemented and tested with the pre-selected types of algorithms, to choose the one performing best, as it is detailed in Section 3 Methodology, Section 4 Algorithm, and Section 5 Results, remarking that the obtained results raised the interest of the surrounding communities.

2. Optimization Problems

Combinatorial optimization problems can be written mathematically as:

$$\text{minimize } f(x), \tag{1}$$

$$\text{subject to} \quad h_i(x) \geq a_i, \quad i = 1, \ldots, m, \tag{2}$$

$$f_j(x) = b_j, \quad j = 1, \ldots, s, \tag{3}$$

where $f : \mathbb{R}^n \to \mathbb{R}$ is the objective function, and $h_i : \mathbb{R}^n \to \mathbb{R}, i = 1, \ldots, m$ and $f_j : \mathbb{R}^n \to \mathbb{R}, j = 1, \ldots, s$ are the constraints.

The optimization problem has been written as a minimization problem, but after some modifications it can be written as a maximization problem:

$$\min g(x) = \max -g(x). \tag{4}$$

In the same way, the equality constraints $f_j(x) = b_j$ can be written as inequalities:

$$f_j(x) \geq b_j \text{ and } f_j(x) \leq b_j. \tag{5}$$

The simplest constrained optimization problem arises when the objective function $f(x)$ and the restrictions are linear functions. This type of problem is a linear programming problem, and it can be solved quite efficiently by the simplex algorithm. However, in the majority of the optimization problems, neither the objective function nor the restrictions are linear functions. The vast majority of these problems are NP-complete problems, which means that there is no any solving algorithm that can be executed in polynomial time in relation to the size of the problem. In complexity theory, NP-complete denotes the set of problems that are not solvable by a deterministic polynomial time algorithm. The feasible solutions' space is so large that the computation of the exact solution requires a lot of time. NP-complete problems can be solved by a restricted class of brute force search algorithms and they can be used to simulate any other problem with a similar algorithm. Genetic algorithms are also a good and efficient choice to find an approximate solution of these problems [17]. A classical NP-complete problem is the Travelling Salesman Problem (TSP), in which the shortest route for a traveling salesman starting and finishing in the same point and visiting every city once has to be found.

An efficient way to solve these types of problems is using genetic algorithms. The basic principles of genetic algorithms were established by Holland [18], and they are well described in several texts [19–23]. Owing to their simplicity, flexibility, ease of implementation, minimal requirements, fast convergence towards close-to-optimal solutions, and global perspective, GAs are successfully used in a wide variety of problems [16]. As these characteristics are essential for practitioners to have a utilizable solution, and as the performed literature review did not bring better solutions, the chosen option was to implement the simple GA (SGA) presented herein.

Traveling Salesman Problem

In the TSP problem, a collection of n cities are given. The objective is to determine the shortest route that a salesman has to follow, in which each city is visited once and then the salesman returns to the starting point of the route. This problem can be defined mathematically as follows:

Given an integer n and an $n \times n$ matrix $D = (d_{ij})$ in which each d_{ij} is the distance between two cities, the cyclic permutation π of the integers $i = 1, 2, \ldots, n$ that minimizes the distances has to be determined. As a first approximation, the feasible search space is formed by all cyclic permutations of the numbers $\{1, 2, \ldots, n\}$:

$$F = \left\{ (\pi_1 \pi_2 \ldots \pi_n) \,|\, \pi_i \in \{1, 2, \ldots, n\} \text{ and } \pi_i \neq \pi_j \forall i \neq j \right\}. \tag{6}$$

The number of elements of this space is $n!$, being n the number of cities. In this way, the length of a permutation $\pi = (\pi_1, \pi_2, \ldots, \pi_n)$ can be expressed as:

$$\sum_{i=2}^{n} d_{\pi_{i-1}, \pi_i} + d_{\pi_n, \pi_1}. \tag{7}$$

For each permutation π, there are $(n-1)$ permutations that starting in a different city are similar to the given one. That is to say, the distance of these n permutations is the same. Taking into account this consideration, the size of the feasible space is $n!/n = (n-1)!$. Moreover, if the distance matrix D is symmetric, the distance of each permutation in both directions is the same, and the size of the feasible space is reduced to $(n-1)!/2$. As it can be observed, the main difficulty of this problem is the huge number of possible tours.

3. Methodology

The objective of this work is to find an optimal itinerary for the waste collection. This methodology will be applied to the data of Sopelana, a municipality in the province of Biscay, autonomous community of the Basque Country (Spain). Sopelana is located in the region of Plencia-Munguia or Uribe, and it is part of the Commonwealth of Services of Uribe Coast. It has an extension of 8.40 km^2 and a population of 13,510 inhabitants, a figure that in summer is usually multiplied by four due to summer visitors.

It is the responsibility of the commonwealth everything related to waste management. There are various trash cans to collect waste:

- Restwaste: 317 trash cans. They are distributed in three routes. There are 186 trash cans in the first route, 72 trash cans in the second route, and 59 in the third one. The trash of the first route is collected everyday. The second route is done four days per week and the third route three days per week.
- Organic waste: 29 trash cans. This waste is collected once per week.
- Small recipients of plastic and metal: 85 trash cans. This waste is collected twice per week.
- Glass: 69 trash cans. Glass is collected once per month.
- Paper: 62 trash cans. These trash cans are divided in two groups depending on their filling frequency. Paper is collected everyday, but not all the containers are emptied everyday.
- Oil: 4 trash cans. This waste is collected twice per month.
- Reusable waste: 7 trash cans. From September to June, it is collected once per week, and in July and August twice per week.
- Batteries: 31 trash cans. They are collected twice per month.

Other types of waste such as big volume wastes are collected once per week following the same route as for the restwaste; there are eight points to collect lamps and fluorescent lightings and they are collected when the container is full; there are some locations in which CDs are collected when the containers are full. All the aforementioned data were given by the local government.

In this work, three problems of different sizes will be analyzed. In the three cases, the aim is to improve the waste collection itinerary in the sense of obtaining a shorter route than the one followed nowadays. A small problem of reusable waste with seven trash cans, a medium problem of organic waste consisting of 29 trash cans, and a big problem consisting of the first route of restwaste with a total of 186 trash cans will be considered. In the case of restwaste, there are several trash cans in the same location. Specifically, the 186 trash cans are distributed in 147 different locations; thus, these locations will be considered in the problem. The data of these problems can be seen in Table 1.

Table 1. Three problems analyzed.

Waste-Type	Number of Locations	Collecting Frequency
Reusable waste	7	September to June once per week, July and August twice per week
Organic waste	29	Once per week
Restwaste	147	Everyday

According to the data given by the enterprise in charge of restwaste collection, between 8000 kg and 13,000 kg of restwaste are collected everyday in Sopelana. A truck is enough to carry out this collection. The truck used has a load of 13 tons, and it has a compaction mechanism. Each container has 1.1 m^3, which is reduced to 0.183 m^3 after compaction. This means that the volume of 125 trash cans that are full can fit in the truck (125×0.183 m^3 = 22.875 m^3). In the event that the truck was filled during the collection of the 186 containers (147 location points), it would be emptied in the dump intended for it and, after that, it would continue with the route. However, this event is not very probable as it means that 125 trash cans out of 186 are full up (which is the 67.20%).

4. Algorithms

For the three problems, the coordinates of the locations and the distance matrices have been calculated using Google Maps (https://www.google.com/maps, data from October and November 2019). The coordinates of all locations and the distance matrices can be found in Appendix A. The smallest distance between two locations has been considered in the distance matrices. The distance matrices are not symmetric, as the way back and forth from a location to another may be different.

The three problems have been solved using a VNS and a GA. The smallest problem (reusable waste, seven locations) has also been solved using the brute force algorithm. These two algorithms were chosen among all the set of alternatives for developing the final solution for maintainability reasons. The developers of the solutions are academic institutions, so we cannot provide maintenance services. As both literature [24] and the news [25] state, there is a lack of availability of computer science technicians. Because of this fact, the final user of the solution wanted a competitive but mainly easy to maintain algorithm. These algorithms being two of the most frequently applied when developing the syllabuses of subjects related to programming, the research team decided to compare them and offer the most effective possible solution. The distances that correspond to all the permutations of $n = 7$ elements have been calculated that is 7! = 5040. The process has been implemented as it is explained in Algorithm 1.

Algorithm 1 Algorithm to determine solutions using brute force.

1: **procedure** BRUTEFORCE(distance matrix of $n = 7$ problem)
2:     Create all the permutations of $n = 7$ elements
3:     **for** i=1:7! **do**
4:         Calculate the distance of the permutation
5:     **end for**
6:     Calculate the shortest distance among all the permutations
7: **end procedure**

We have started by trying the VNS algorithm [26] for all the problems. The VNS consists of applying a local search method repeatedly in the neighborhood $\mathcal{N}_k$ of the actual solution. When a local optimal is reached, the algorithm changes the system of neighborhood with the aim of escaping from local optima and reaching a better one. For this reason, it can be said that the VNS performs well both when searching local and global optima. In Algorithm 2, the process followed is explained. We have implemented two systems of neighborhoods: the 2-opt neighborhood and a swap-based neighborhood. The neighborhood structure 2-opt consists of changing a pair of edges between cities [27]. The swap-based neighborhood that we have created swaps the first element of the permutation with all the rest.

Algorithm 2 Algorithm to determine solutions using VNS.

1: **procedure** VARIABLE NEIGHBORHOOD SEARCH(distance matrix)
2:      Choose a set of neighborhood structures $\mathcal{N}_k$ for $k = 1, 2, \ldots, k_{max}$
3:      Generate the initial solution
4:      Consider the initial solution as the best one
5:      $k \leftarrow 1$
6:      **while** $k \leq k_{max}$ **do**
7:          **while** There is an improvement **do**
8:              Choose the neighborhood system that corresponds to k
9:              Find the best solution among all the neighbors
10:             **if** The solution is improved **then**
11:                 Update the best solution and its evaluation function value
12:                 Continue to search with $k \leftarrow 1$
13:             **else**
14:                 Continue to search with $k \leftarrow k + 1$
15:             **end if**
16:         **end while**
17:     **end while**
18:     Output the shortest distance and the corresponding route of the overall process
19: **end procedure**

Additionally, for all the problems $n = 7$, $n = 129$, and $n = 147$, another two algorithms have been implemented. In Algorithm 3, an initial population of m different individuals (permutations) has been created. The evaluation function (distance of the route) of each individual has been calculated. In each generation, the process of selecting two parents randomly, the crossover operator, the posterior correction of the individual in order to be a permutation, and the mutation process have been performed $m/2$ times. If the new descendants are different from the individuals of the population and if their evaluation function (fitness function) is smaller than the worst (the largest) of the population, the worst individuals are replaced by these new descendants. The fourth algorithm implemented only differs from the third in the fact that the routes followed nowadays in the trash collection (one route in each problem) are inserted as a seed in the initial population.

Algorithm 3 Algorithm to determine solutions using SGA.

1: **procedure** SIMPLE GENETIC ALGORITHM(distance matrix, generations, population size, probabilities)
2:      Create an initial population of m different permutations randomly
3:      Compute the evaluation function of each permutation
4:      **for** i=1:generations **do**
5:          **for** j=1:m/2 **do**
6:              Select two parents randomly from the population
7:              Cross with a certain probability to produce two descendants
8:              Correct the descendants to be permutations
9:              Mutate each individual with a certain probability
10:             Compute the evaluation function of each descendant
11:             **if** evaluation function of the descendants smaller than the largest evaluation **then**
12:                 **if** the descendants are not repeated in the population **then**
13:                     Replace the descendants by the permutations with largest evaluation
14:                 **end if**
15:             **end if**
16:         **end for**
17:         Output the shortest distance and the corresponding route of each generation
18:     **end for**
19:     Output the shortest distance and the corresponding route of the overall process
20: **end procedure**

The crossover operator used is the classical one [18]. A crossover point is selected randomly from which the two strings of the parents are broken into separate parts. The new descendants are formed by recombination of these parts. For example, consider $n = 7$ and the routes:

$$(1\,2\,3\,4\,5\,6\,7)$$
$$(2\,4\,5\,3\,7\,1\,6)$$

Randomly, a crossover point in which the strings are broken into separate parts is selected. Suppose that the crossover point is chosen between the third and the fourth bit:

$$(1\,2\,3 \mid 4\,5\,6\,7)$$
$$(2\,4\,5 \mid 3\,7\,1\,6)$$

Combining the head of the first route with the tail of the second route and vice versa, the result is the following:

$$(1\,2\,3 \mid 3\,7\,1\,6)$$
$$(2\,4\,5 \mid 4\,5\,6\,7)$$

If the resulting descendants are not permutations, then a correction algorithm is applied in order to obtain two individuals that belong to the feasible space. Each resulting individual is repaired, by calculating the repeated values and their positions, and by inserting the missing values randomly on the positions where the repeated values are located.

The exchange mutation operator, also called the swap mutation operator, has been applied, in which two positions of the route are selected randomly and the cities on those positions are exchanged [28]—for example, if we consider the route:

$$(1\,2\,3\,5\,7\,4\,6),$$

and if we choose the fourth and seventh positions, the cities on those positions are interchanged:

$$(1\,2\,3\,6\,7\,4\,5)$$

5. Results

In this section, the results obtained for each of the problems are presented.

For $n = 7$, the brute force algorithm has been applied first. The shortest itinerary has 7.67 km and the longest 13.2 km. In Figure 1, the distances of all permutations ($7! = 5040$) are shown. All the permutations are represented in the horizontal axis (that is, $7! = 5040$), and the distance that corresponds to each of them is plotted in the vertical axis.

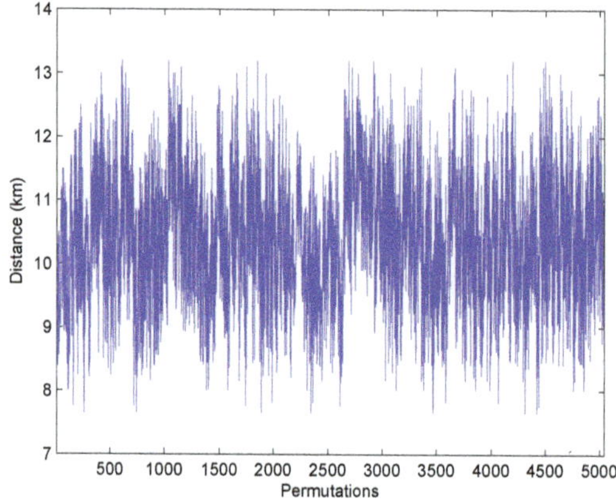

Figure 1. Brute force application, problem $n = 7$.

For the three problems, the VNS algorithm has been applied next. Ten executions have been done for each problem. The best, the worst, the mean, and the median distances of this executions are presented in Tables 2–4. In all the executions, an initial solution has been generated randomly.

Table 2. Results of 10 executions $n = 7$ problem, VNS.

$n = 7$	VNS
Best distance (km)	7.6700
Worst distance (km)	8.4000
Mean distance (km)	7.7690
Median distance (km)	7.6700

Table 3. Results of 10 executions $n = 29$ problem, VNS.

$n = 29$	VNS
Best distance (km)	19.4170
Worst distance (km)	25.5570
Mean distance (km)	22.5475
Median distance (km)	21.9450

Table 4. Results of 10 executions $n = 147$ problem, VNS.

$n = 147$	VNS
Best distance (km)	61.5470
Worst distance (km)	71.3560
Mean distance (km)	66.8944
Median distance (km)	66.9507

In addition, finally, the SGA (without and with a seed) has been performed. For $n = 7$, taking into account that the average of the shortest and the longest route is 10.4350 km, the route $(6, 7, 3, 4, 2, 5, 1)$ of 10.45 km has been chosen as seed. In the cases of $n = 29$ and $n = 147$, the actual itineraries have been chosen as seed. These data are available in Table 5. Notice that, if the VNS algorithm is started, taking the actual itinerary as initial solution, all the executions have the same performance. In this case, for $n = 7$, a route of 8.2500 km is obtained; for $n = 29$, a route of 18.7670 km; and, for $n = 147$, a route of 34.4280 km.

Table 5. Average distance route ($n = 7$) and actual routes ($n = 29$, $n = 147$).

	$n = 7$	$n = 29$	$n = 147$
Route	$(6,7,3,4,2,5,1)$	$(26,24,23,18,15,11,12,13,16,-$ $17,20,21,22,19,9,7,27,28,-$ $14,25,10,8,6,2,3,4,5,29,1)$	$(135,136,140,141,139,138,137,107,142,147,-$ $110,42,40,44,101,102,104,146,105,103,-$ $108,109,144,143,106,145,111,112,115,113,-$ $41,114,116,121,126,127,132,131,133,134,-$ $130,129,128,125,120,118,119,117,124,122,-$ $123,58,57,69,70,96,97,100,99,98,-$ $68,67,66,49,48,47,45,46,55,56,-$ $65,64,62,95,94,77,76,75,74,73,-$ $61,60,59,36,35,37,84,85,88,89,-$ $86,7,10,9,8,3,2,1,4,6,-$ $5,11,14,13,15,16,18,17,12,90,-$ $91,92,93,78,79,80,81,87,19,83,-$ $82,71,29,27,26,28,23,22,21,20,-$ $24,25,72,30,31,32,34,33,38,43,-$ $50,52,63,54,53,51,39)$
Distance (km)	10.4500	22.9170	46.2890

An execution for each problem has been performed. Parameters have been chosen according to [29]: a relatively high crossover rate (≥ 0.6), small mutation rate (range $[0.001, 0.1]$) and a moderate population size. Data and results are presented in Tables 6–8. In all the cases, the execution in which the seed is considered obtains better solutions.

Table 6. Results of the $n = 7$ problem.

$n = 7$	SGA without Seed	SGA with Seed
Generations	20	20
Population size	30	30
Crossover probability	0.8	0.8
Mutation probability	0.01	0.01
Best route	$(3,2,1,6,7,4,5)$	$(7,4,5,3,2,1,6)$
Best distance (km)	7.6700	7.6700

Table 7. Results of the $n = 29$ problem.

$n = 29$	SGA without Seed	SGA with Seed
Generations	4000	4000
Population size	200	200
Crossover probability	0.8	0.8
Mutation probability	0.01	0.01
Best route	$(18,16,28,27,14,25,13,12,11,9,-$ $7,4,3,2,5,8,6,10,17,20,-$ $21,23,24,26,22,19,1,29,15)$	$(10,12,13,17,23,24,26,22,19,21,-$ $20,18,15,11,9,7,28,27,14,25,-$ $(16,1,29,5,4,3,2,8,6)$
Best distance (km)	18.0270	17.6570

Table 8. Results of the $n = 147$ problem.

$n = 147$	SGA without Seed	SGA with Seed
Generations	4000	4000
Population size	400	400
Crossover probability	0.8	0.8
Mutation probability	0.01	0.01
Best route	$(102, 103, 142, 113, 114, 125, 130, 134, 133, 132, -$	$(141, 140, 135, 136, 107, 138, 137, 139, 142, 147, -$
	$124, 117, 75, 61, 79, 82, 87, 21, 18, 19, -$	$41, 42, 40, 101, 44, 102, 104, 146, 105, 108, -$
	$53, 139, 141, 52, 36, 85, 8, 7, 10, 9, -$	$145, 109, 144, 143, 106, 103, 111, 112, 110, 113, -$
	$88, 76, 48, 46, 57, 50, 49, 47, 45, 121, -$	$115, 114, 116, 51, 126, 121, 125, 129, 130, 134, -$
	$120, 119, 116, 56, 64, 62, 99, 98, 69, 67, -$	$133, 131, 128, 127, 119, 132, 124, 117, 120, 122, -$
	$66, 122, 123, 129, 131, 128, 126, 55, 39, 71, -$	$123, 50, 118, 69, 64, 62, 68, 100, 99, 98, -$
	$73, 97, 68, 65, 63, 104, 146, 105, 137, 138, -$	$97, 67, 66, 49, 48, 47, 45, 46, 55, 43, -$
	$54, 20, 28, 32, 34, 41, 42, 33, 38, 31, -$	$56, 70, 96, 95, 94, 77, 78, 73, 75, 74, -$
	$29, 24, 25, 30, 35, 44, 106, 107, 143, 147, -$	$61, 60, 59, 39, 31, 37, 76, 84, 88, 86, -$
	$111, 112, 115, 144, 108, 127, 118, 95, 77, 83, -$	$90, 10, 9, 8, 7, 3, 2, 4, 1, 6, -$
	$3, 4, 2, 1, 14, 80, 110, 43, 51, 100, -$	$5, 12, 11, 13, 15, 17, 18, 16, 14, 89, -$
	$70, 74, 59, 58, 27, 26, 12, 17, 6, 5, -$	$91, 92, 93, 85, 87, 80, 79, 81, 19, 82, -$
	$22, 23, 72, 78, 86, 11, 13, 15, 16, 81, -$	$83, 71, 29, 27, 25, 26, 23, 22, 21, 20, -$
	$37, 96, 94, 84, 89, 90, 91, 93, 92, 60, -$	$24, 28, 72, 30, 32, 35, 34, 33, 38, 57, -$
	$40, 101, 145, 140, 135, 136, 109)$	$58, 65, 63, 54, 53, 52, 36)$
Best distance (km)	48.3510	30.4150

The longest, the shortest, and the average-distance itineraries for $n = 7$ can be seen in Figure 2. The limit of the municipality of Sopelana is marked using a thick red line.

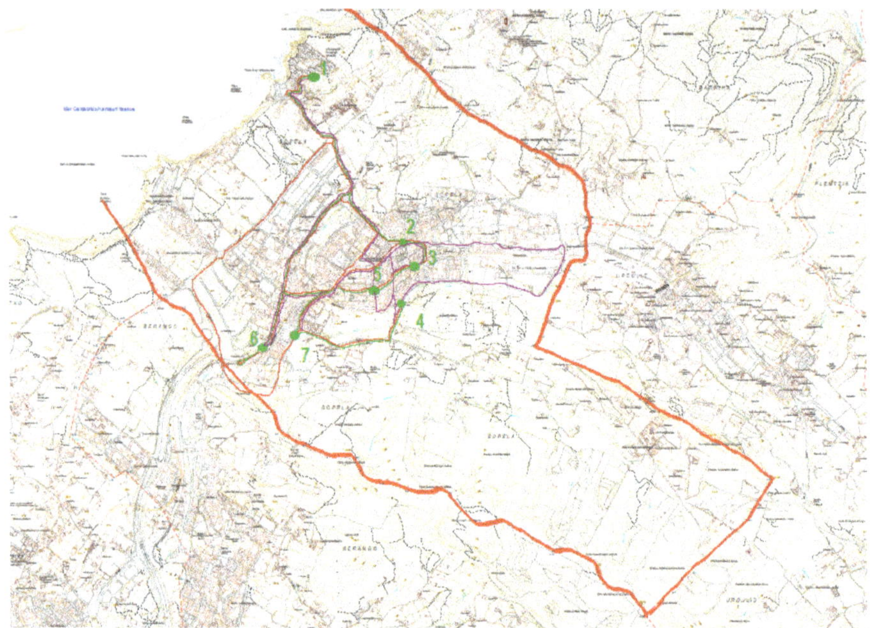

Figure 2. Routes of problem $n = 7$, longest in red, shortest in green, "average" in purple.

Additionally, several executions have been performed for each of the problems ($n = 7, 29, 147$). The best, the worst, the mean, and the median distance of this performance are presented in Tables 9–11. The values of these tables are acquired after performing 10 executions for each version of each problem.

Table 9. Results of 10 executions $n = 7$ problem, SGA.

$n = 7$	SGA without Seed	SGA with Seed
Generations	20	20
Population size	30	30
Crossover probability	0.8	0.8
Mutation probability	0.01	0.01
Best distance (km)	7.6700	7.6700
Worst distance (km)	8.0200	7.7700
Mean distance (km)	7.7050	7.6900
Median distance (km)	7.6700	7.6700

Table 10. Results of 10 executions $n = 29$ problem, SGA.

$n = 29$	SGA without Seed	SGA with Seed
Generations	4000	4000
Population size	200	200
Crossover probability	0.8	0.8
Mutation probability	0.01	0.01
Best distance (km)	17.3770	17.1870
Worst distance (km)	19.0170	18.3670
Mean distance (km)	18.1910	17.8530
Median distance (km)	17.9420	17.7420

Table 11. Results of 10 executions $n = 147$ problem, SGA.

$n = 147$	SGA without Seed	SGA with Seed
Generations	4000	4000
Population size	400	400
Crossover probability	0.8	0.8
Mutation probability	0.01	0.01
Best distance (km)	48.4740	29.6310
Worst distance (km)	54.6080	32.7670
Mean distance (km)	50.9096	31.7165
Median distance (km)	50.3485	31.9552

For the largest problems, $n = 29$ and $n = 147$, computation has been repeated choosing different parameters in order to obtain better results. The most important change is that a larger population size has been used, i.e., 4000 individuals. Execution data and results can be seen in Tables 12 and 13.

Table 12. Results of the $n = 29$ problem.

$n = 29$	SGA without Seed	SGA with Seed
Generations	800	800
Population size	4000	4000
Crossover probability	0.8	0.8
Mutation probability	0.8	0.8
Best route	$(16, 17, 23, 24, 26, 22, 19, 21, 20, 18, -$ $15, 11, 10, 12, 2, 1, 29, 5, 4, 3-$ $8, 6, 9, 7, 28, 27, 14)$	$(15, 11, 10, 13, 16, 17, 23, 24, 26, 22, -$ $19, 21, 20, 18, 14, 25, 28, 27, 12, 9, -$ $(-, 7, 4, 8, 6, 3, 5, 2, 1, 29)$
Best distance (km)	17.1700	16.9370

Table 13. Results of the $n = 147$ problem.

$n = 147$	SGA without Seed	SGA with Seed
Generations	1800	1800
Population size	4000	4000
Crossover probability	0.8	0.8
Mutation probability	0.8	0.8
Best route	$(107, 142, 53, 37, 73, 60, 59, 50, 58, 67, -$	$(141, 140, 135, 136, 107, 138, 137, 139, 142, 112, -$
	$66, 29, 26, 25, 24, 28, 30, 51, 121, 116, -$	$41, 42, 40, 101, 44, 102, 104, 146, 105, 106, -$
	$114, 115, 56, 36, 35, 33, 34, 46, 122, 124, -$	$103, 108, 144, 143, 145, 109, 147, 110, 43, 113, -$
	$117, 120, 126, 127, 132, 119, 19, 80, 17, 16, -$	$115, 114, 116, 123, 127, 126, 129, 130, 134, 133, -$
	$77, 78, 71, 76, 94, 82, 88, 85, 84, 39, -$	$132, 125, 128, 131, 121, 122, 124, 117, 120, 119, -$
	$31, 32, 38, 57, 143, 105, 146, 104, 103, 108, -$	$52, 118, 100, 69, 70, 94, 73, 97, 99, 98, -$
	$147, 136, 135, 112, 111, 79, 87, 81, 22, 23, -$	$68, 67, 66, 49, 48, 47, 45, 46, 55, 56, -$
	$72, 48, 45, 145, 144, 139, 138, 137, 123, 128, -$	$58, 64, 62, 96, 95, 77, 78, 71, 61, 75, -$
	$134, 133, 110, 43, 113, 41, 40, 42, 141, 140, -$	$74, 60, 59, 36, 35, 37, 76, 84, 88, 86, -$
	$64, 62, 68, 69, 70, 96, 95, 93, 92, 91, -$	$90, 10, 9, 8, 7, 3, 1, 2, 4, 6, -$
	$10, 3, 1, 2, 4, 83, 97, 65, 63, 61, -$	$5, 11, 12, 13, 15, 17, 16, 18, 14, 89, -$
	$75, 74, 118, 100, 99, 98, 49, 47, 55, 125, -$	$91, 92, 93, 85, 79, 87, 80, 81, 19, 83, -$
	$129, 130, 131, 54, 27, 20, 21, 18, 12, 11, -$	$82, 30, 29, 27, 25, 26, 23, 22, 21, 20, -$
	$6, 5, 7, 8, 9, 89, 90, 86, 14, 13, -$	$24, 28, 72, 39, 31, 32, 34, 33, 38, 57, -$
	$15, 52, 109, 44, 101, 102, 106)$	$50, 65, 63, 54, 53, 51, 111)$
Best distance (km)	40.3460	27.0950

The actual itineraries are presented in Figures 3 and 4, and the improved itineraries of Tables 12 and 13 obtained with the new parameters in Figures 5 and 6. In the case of problem $n = 147$, Figures 4 and 6, the first and the last location of the route are marked in red.

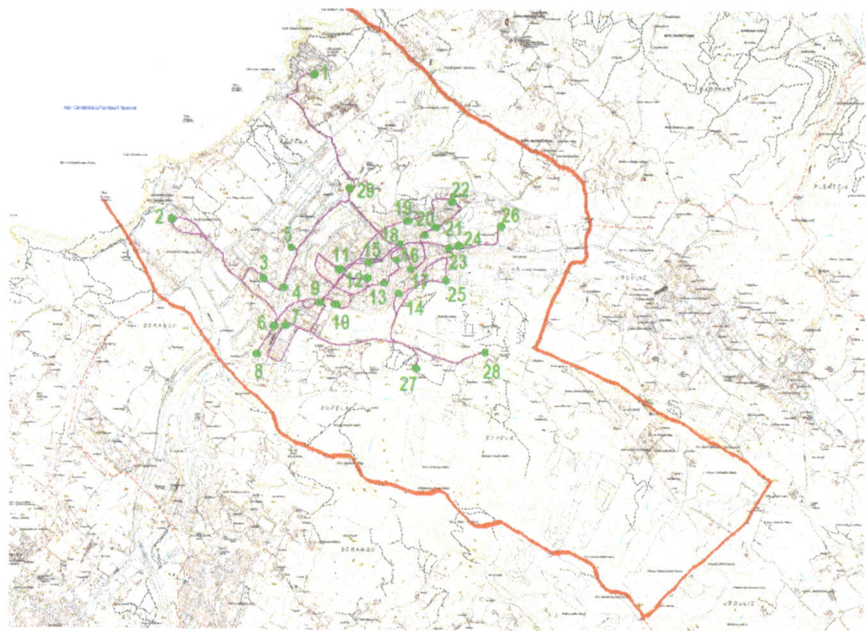

Figure 3. Actual route of problem $n = 29$.

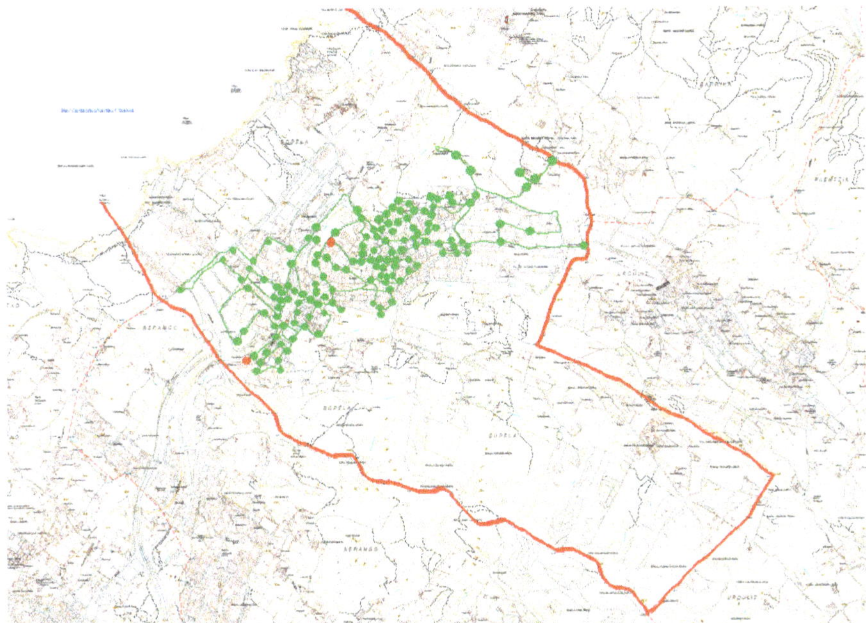

Figure 4. Actual route of problem $n = 147$.

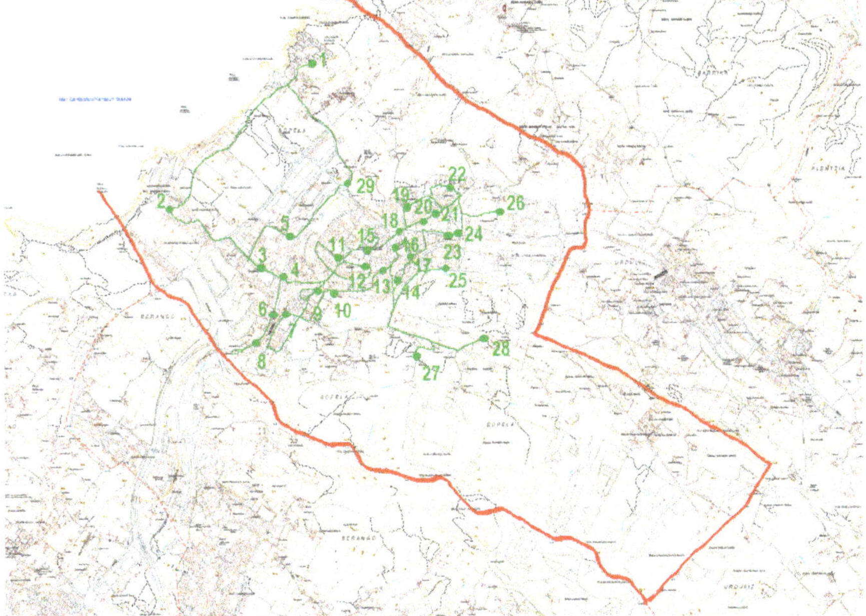

Figure 5. Smallest route obtained for problem $n = 29$.

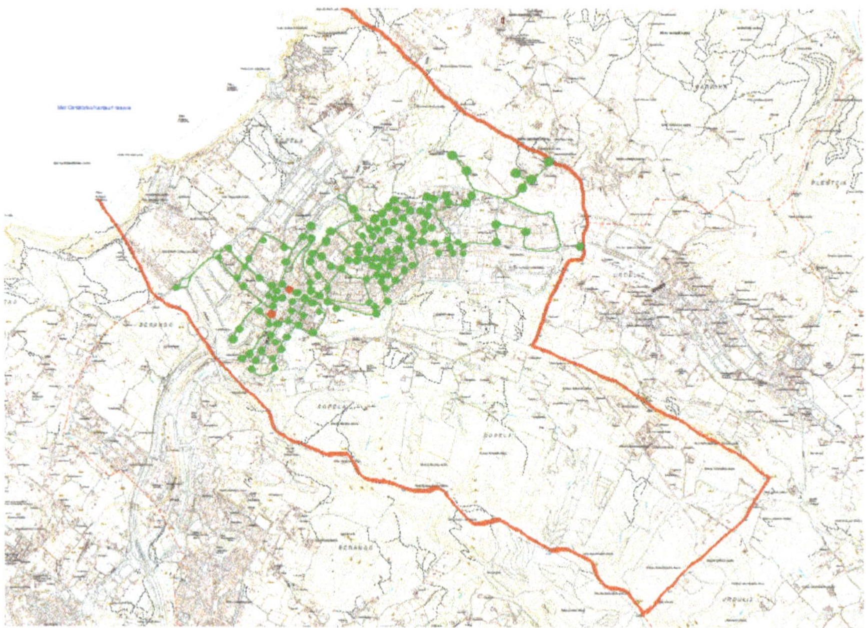

Figure 6. Smallest route obtained for problem $n = 147$.

6. Conclusions

In this work, three waste collection itineraries have been improved in a municipality of Biscay (Spain). The average itinerary of the reusable waste ($n = 7$ problem) has had a reduction of 2.78 km; and the actual itineraries of organic waste ($n = 29$ problem) and restwaste ($n = 147$ problem) have been reduced 5.98 km and 19.194 km, respectively. Taking into account the collection frequencies of these three itineraries, this makes a total reduction of 7400 km per year, that is to say, a reduction of the 40% of the total actual itineraries.

The truck has a continent of 13 tons and it has a compaction mechanism. Considering the following average data for the truck: vehicle of 26 Tn, speed limit 30 km/h, with 270 CV minimum engine power (1 CV = 735.39, 875 W= 0.986 HP) and diesel fuel type and 29 L/100 km consumption [30]. This implies a reduction of 5.58 Tn of CO_2 emissions, 0.43 Kg of CO, and 13.95 Kg of NOx per year. In addition to these improvements, a direct cost savings of 7294€ was obtained (considering the direct cost per kilometer calculation model recommended by the Ministry of Transportation, Logistics and Urban Agenda of Spain [31]).

It is worth mentioning that the research team performed an additional validation for the developed model. Specifically, the actual consumptions versus the ones proposed by the model were analyzed at the two comparable routes (the ones the truck made before and after the optimization), obtaining negligible differences. This double check on the results and the easiness of the solution have raised the interest of other commonwealths, such as the one of Lea-Artibai. Thus, the short and midterm future steps would be oriented to the application of the same procedure to other local communities, incorporating other parameters such as the elevation information of the routes.

Author Contributions: E.A., writing, conceptualization, and investigation; L.U., conceptualization and investigation; A.G., investigation, writing and funding acquisition and A.O.-Z., writing and funding acquisition. All authors have read and agreed to the published version of the manuscript.

Funding: This research was funded by Fundación BBK, partner of the Deusto Digital Industry Chair.

Acknowledgments: We would like to thank the partners of the Deusto Digital Industry Chair (Etxe-Tar, General Electric, Idom, Accenture, Fundación Telefónica, Fundación BBK) the interest and support shown during this research. We would like also to thank the municipality of Sopelana and the commonwealth of the regions of Lea-Artibai for placing their trust and confidence in our organization.

Conflicts of Interest: The authors declare no conflict of interest.

Abbreviations

The following abbreviations are used in this manuscript:

GA	Genetic Algorithm
VNS	Variable Neighborhood Search
SDG	Sustainable Development Goal
WCVRP	Waste Collection Vehicle Routing Problem
VRP	Vehicle Routing Problem
TSP	Travelling Salesman Problem
SA	Simulated Annealing
TWL	Threshold Waste Level
RFID	Radio Frequency Identification
NP	Nondeterministic Polynomial Time
SGA	Simple Genetic Algorithm

Appendix A. Data

Table A1. Coordinates of the seven locations.

Location	Latitude (Decimal Degrees)	Longitude (Decimal Degrees)
1	43.391464	−2.987950
2	43.381466	−2.980558
3	43.380186	−2.979504
4	43.377695	−2.980651
5	43.378805	−2.982968
6	43.375206	−2.992354
7	43.374066	−2.990935

Table A2. Distances of the seven locations in kilometers.

From ↓ To →	1	2	3	4	5	6	7
1	0 1.7	1.9	2.2	2.2	2.3	2.4	
2	1.7	0	0.23	0.55	0.55	1.7	1.1
3	1.9	0.27	0	0.8	0.8	2	1.3
4	3.4	1.8	0.5	0	0.4	1.9	0.95
5	2	0.4	0.4	0.4	0	1.2	0.8
6	2.4	1.7	1.9	1.9	1.2	0	1
7	3.8	1.7	2	1.6	1.6	1.7	0

Table A3. Coordinates of the 29 locations.

Location	Latitude (Decimal Degrees)	Longitude (Decimal Degrees)
1	43.391326	−2.988336
2	43.382808	−2.999907
3	43.379497	−2.992421
4	43.378769	−2.990454
5	43.381233	−2.989976
6	43.376588	−2.991162
7	43.376443	−2.990605
8	43.374686	−2.993086
9	43.377947	−2.987348
10	43.377956	−2.986103
11	43.379873	−2.985665
12	43.379364	−2.983446
13	43.378979	−2.982283
14	43.378942	−2.980098
15	43.380483	−2.982809
16	43.380490	−2.980981
17	43.380161	−2.979534
18	43.381520	−2.980494
19	43.382913	−2.979929
20	43.382040	−2.978645
21	43.382416	−2.977694
22	43.384166	−2.977144
23	43.381216	−2.976394
24	43.381311	−2.975217
25	43.379194	−2.977072
26	43.382615	−2.972005
27	43.374091	−2.979216
28	43.375154	−2.973282
29	43.384723	−2.984886

Table A4. Distances of the 29 locations in kilometers.

From ↓ To →	1	2	3	4	5	6	7	8	9	10	11	12	13	14	15	16	17	18	19	20	21	22	23	24	25	26	27	28	29
1	0	2.2	1.9	1.9	1.7	3	2.5	2.4	2.4	2.1	1.9	2.3	2.2	2.1	1.8	2	1.9	1.7	1.7	1.9	2	2.4	2	2.2	2.5	2.5	2.6	3	1.1
2	1.5	0	0.9	1.2	1.1	2	1.9	1.4	1.8	1.6	1.7	1.9	1.9	2.8	2.4	2.3	2.3	2.2	2.3	2.3	2.4	2.9	2.5	2.6	2.8	3	2.8	3.2	1.9
3	2	0.85	0	0.7	0.5	1.3	1.4	0.65	1.3	1.1	1.2	1.4	1.4	1.9	1.9	1.8	1.8	1.6	1.8	1.8	1.9	2.4	2	2.1	2.3	2.5	2.3	2.7	1.6
4	1.9	1.2	0.2	0	1.3	1.1	2	0.5	2	1.7	1.5	1.8	1.7	1.7	1.4	2.6	1.4	1.3	1.3	1.4	1.5	1.9	1.6	1.7	2.1	2	2.2	2.6	1.7
5	1.8	1.1	0.5	0.35	0	1.5	2	0.85	2	1.7	1.8	2.1	2	2.4	2.2	2.3	2.2	2	2.1	2.2	2.3	2.7	2.4	2.5	2.8	2.8	2.9	3.3	1.4
6	2.2	1.2	0.4	0.25	1.2	0	0.95	1.6	0.9	0.65	0.75	1	0.95	1.5	1.4	1.4	1.4	1.2	1.3	1.4	1.5	2	1.5	1.7	1.9	2.1	1.8	2.2	1.9
7	3.1	2.4	1.4	1.2	2.4	2.3	0	1.7	0.95	1	1.1	1.4	1.3	1.7	1.8	1.8	1.9	1.6	1.7	1.8	1.8	2.3	1.9	2.1	1.9	3.5	1.6	2	2.9
8	2.4	1.4	0.65	0.5	1.5	0.26	1.2	0	1.2	0.9	1	1.2	1.2	1.7	1.7	1.6	1.7	1.5	1.6	1.6	1.7	2.2	1.8	1.9	2.1	2.3	2.1	2.5	2.2
9	2.8	2.1	1.1	0.9	2.1	2	0.45	1.4	0	0.75	0.85	1.1	1	1.2	1.5	1.3	1.4	1.3	1.4	1.5	1.6	2	1.6	1.8	1.4	2.2	1.1	1.5	2.4
10	2.3	2.3	1.3	1.1	1.7	2.2	1.2	1.6	1.2	0	0.65	0.45	0.45	0.75	1.1	0.7	0.7	0.75	1	1	1	1.4	1.1	1.2	1.1	1.6	1.3	1.6	2
11	2.4	2	1	0.85	1.8	1.9	0.95	1.3	0.9	0.6	0	0.8	0.75	1.3	1.3	1	1.2	1	1.2	1.2	1.3	1.7	1.4	1.5	1.7	1.9	1.8	2.2	2.1
12	2	1.8	0.85	0.65	1.5	1.8	0.8	1.1	0.75	0.45	0.2	0	0.6	0.9	0.9	0.8	0.85	0.65	0.8	0.85	0.95	1.3	1	1.1	1.3	1.5	1.5	1.8	1.8
13	2.1	2.2	1.1	1	1.6	2.1	1.1	1.5	1	0.75	0.5	0.3	0	0.3	1	0.25	0.28	0.65	0.9	0.8	0.9	1.3	1	1.1	0.7	1.4	0.85	1.2	1.9
14	3.9	3.8	2.1	2	3.3	3.1	1.5	2.5	1.7	1.4	1.2	1	0.9	0	2.6	0.95	1	2.4	2.5	2.5	2.5	2.6	1.4	2.1	0.4	2.1	1.1	1.5	3.6
15	1.7	2.1	1.1	0.95	1.2	2	1.1	1.4	1	0.7	0.45	0.9	0.9	0.9	0	0.8	0.65	0.45	0.5	0.6	0.7	1.1	0.75	0.9	1.3	1.2	1.4	1.8	1.5
16	1.8	2.1	1.1	1	1.2	2	1.1	1.4	1	0.7	0.5	0.6	0.5	0.4	0.19	0	0.35	0.21	0.5	0.4	0.5	0.85	0.55	0.7	0.8	1	1	1.3	1.5
17	1.9	2.4	1.5	1.3	1.4	2.7	1.4	1.8	1.4	1	0.8	1	0.85	0.8	0.5	0.7	0	0.28	0.45	0.27	0.35	0.75	0.4	0.55	1.2	0.85	1.3	1.7	1.7
18	1.7	2.2	1.2	1	1.1	2.4	1.2	1.5	1.1	0.75	0.55	0.7	0.55	0.5	0.24	0.4	0.19	0	0.4	0.19	0.27	0.65	0.35	0.5	0.9	0.8	1	1.4	1.4
19	1.7	2.2	1.3	1.1	1.1	2.5	1.2	1.6	1.2	0.85	0.6	1	0.9	0.85	0.55	0.75	0.6	0.4	0	0.4	0.3	0.6	0.75	0.5	1.2	0.75	1.4	1.7	1.4
20	1.8	2.3	1.4	1.2	1.3	2.6	1.3	1.7	1.3	0.95	0.7	0.85	0.75	0.7	0.4	0.6	0.35	0.17	0.19	0	0.087	0.5	0.35	0.4	1.1	0.6	1.2	1.6	1.6
21	1.9	2.4	1.5	1.3	1.4	2.7	1.4	1.8	1.4	1	0.8	0.95	0.85	0.75	0.5	0.65	0.4	0.26	0.27	0.087	0	0.4	0.24	0.3	1.2	0.5	1.3	1.7	1.7
22	2	2.5	1.6	1.4	1.5	2.8	1.5	1.9	1.5	1.2	1	1.3	1.2	1.2	0.8	1.1	0.95	0.55	0.35	0.4	0.3	0	0.45	0.5	1.6	0.7	1.7	2.1	1.8
23	2	2.5	1.6	1.4	1.4	2.8	1.5	1.9	1.5	1.2	0.85	1	0.9	0.85	0.55	0.75	0.5	0.35	0.75	0.35	0.26	0.5	0	0.17	1.8	0.5	1.4	1.8	1.7
24	2.1	2.6	1.7	1.5	1.5	2.9	1.6	2	1.6	1.3	1	1.1	1	1	0.7	0.85	0.6	0.45	0.55	0.4	0.3	0.45	0.11	0	1.4	0.4	1.5	1.9	1.8
25	3.8	3.6	2	1.8	3.2	2.9	1.4	2.3	1.6	1.6	1.7	0.85	0.75	0.6	2.4	0.8	0.8	2.2	2.4	2.2	2.1	2.2	1.9	1.8	0	1.7	0.95	1.3	3.5
26	2.4	3	2	1.9	1.9	3.2	2	2.3	1.9	1.6	1.3	1.5	1.4	1.3	1	1.2	0.95	0.75	0.8	0.6	0.5	0.6	0.5	0.4	1.7	0	1.8	2.2	2.2
27	3.5	3.4	1.7	1.6	2.9	2.7	1.1	2	1.3	1.4	1.2	1	0.9	0.75	2.2	0.9	0.95	1.9	2.1	2.1	2.2	2.7	2.3	2.4	0.95	2.6	0	0.75	3.2
28	3.9	3.1	2.1	2	3.3	3	1.5	2.4	1.7	1.8	1.6	1.4	1.3	1.1	2.6	1.3	1.3	2.3	2.5	2.5	2.6	3.1	2.7	2.8	1.3	2.3	0.65	0	3.6
29	1.3	1.7	1	0.85	0.6	2	1.4	1.3	1.4	1.1	0.85	1.2	1.1	1	0.75	0.95	0.8	0.65	0.65	0.8	0.9	1.3	1	1.1	1.4	1.4	1.6	2	0

Table A5. Coordinates of the 147 locations.

Location	Latitude (Decimal Degrees)	Longitude (Decimal Degrees)
1	43.386866	−2.967695
2	43.385065	−2.969335
3	43.384389	−2.970417
4	43.385533	−2.970361
5	43.386875	−2.976296
6	43.385967	−2.975167
7	43.381338	−2.967009
8	43.382074	−2.969934
9	43.382585	−2.972301
10	43.381528	−2.972474
11	43.384161	−2.974454
12	43.383708	−2.975088
13	43.384144	−2.976702
14	43.384005	−2.976627
15	43.384120	−2.977955
16	43.383985	−2.977985
17	43.384236	−2.978082
18	43.383631	−2.978442
19	43.383199	−2.978524
20	43.383740	−2.979749
21	43.383519	−2.979418
22	43.383012	−2.979399
23	43.382807	−2.980273
24	43.383741	−2.980453
25	43.383798	−2.980842
26	43.383568	−2.980869
27	43.383510	−2.981054
28	43.382908	−2.980460
29	43.383270	−2.981680
30	43.382436	−2.981768
31	43.382340	−2.982082
32	43.382122	−2.982337
33	43.383011	−2.983951
34	43.382812	−2.983972
35	43.381533	−2.983377
36	43.381334	−2.983270
37	43.381829	−2.982122
38	43.382067	−2.986113
39	43.381568	−2.986365
40	43.381784	−2.988524
41	43.380237	−2.989935
42	43.381140	−2.989132
43	43.381249	−2.988139
44	43.381235	−2.989827
45	43.380718	−2.987577
46	43.380491	−2.987676
47	43.380686	−2.987247

Table A5. *Cont.*

Location	Latitude (Decimal Degrees)	Longitude (Decimal Degrees)
48	43.380298	−2.986529
49	43.379818	−2.985593
50	43.379931	−2.985285
51	43.380432	−2.984580
52	43.380354	−2.984325
53	43.380485	−2.983974
54	43.380450	−2.983486
55	43.379426	−2.987135
56	43.379123	−2.986105
57	43.378931	−2.986235
58	43.378462	−2.986664
59	43.380934	−2.982889
60	43.380474	−2.982763
61	43.380781	−2.981613
62	43.379854	−2.981971
63	43.380253	−2.982916
64	43.379868	−2.982576
65	43.379837	−2.982973
66	43.379389	−2.983617
67	43.379319	−2.983048
68	43.378805	−2.982968
69	43.378960	−2.982608
70	43.379314	−2.981559
71	43.381485	−2.980560
72	43.382740	−2.980696
73	43.380752	−2.980763
74	43.380498	−2.981023
75	43.380236	−2.980826
76	43.380501	−2.979872
77	43.381449	−2.979324
78	43.381796	−2.979530
79	43.382100	−2.978779
80	43.382583	−2.978996
81	43.382971	−2.979162
82	43.382396	−2.977782
83	43.382645	−2.977287
84	43.381542	−2.978895
85	43.381407	−2.977533
86	43.381706	−2.976436
87	43.382816	−2.978206
88	43.381285	−2.976771
89	43.381425	−2.975376
90	43.381242	−2.975296
91	43.380864	−2.975344
92	43.380848	−2.976503
93	43.380832	−2.977200
94	43.380186	−2.979504
95	43.380019	−2.980074
96	43.379958	−2.980587
97	43.378765	−2.981322
98	43.377759	−2.982722
99	43.377512	−2.982654
100	43.377344	−2.982785
101	43.381321	−2.992360
102	43.380362	−2.993327
103	43.380051	−2.993379
104	43.381045	−2.994875

Table A5. *Cont.*

Location	Latitude (Decimal Degrees)	Longitude (Decimal Degrees)
105	43.380561	−2.995728
106	43.379430	−2.993877
107	43.379093	−2.993990
108	43.379339	−2.992198
109	43.378773	−2.991203
110	43.378845	−2.990499
111	43.378604	−2.990441
112	43.378568	−2.990115
113	43.378646	−2.988672
114	43.378464	−2.987841
115	43.378462	−2.988852
116	43.378328	−2.987209
117	43.377763	−2.987006
118	43.377847	−2.986469
119	43.376388	−2.988905
120	43.376071	−2.988053
121	43.377174	−2.988653
122	43.377572	−2.988626
123	43.377533	−2.988873
124	43.377958	−2.988862
125	43.376747	−2.990108
126	43.375579	−2.989862
127	43.375428	−2.990021
128	43.375322	−2.990450
129	43.375529	−2.991228
130	43.374831	−2.991743
131	43.374468	−2.991131
132	43.373440	−2.991881
133	43.374343	−2.992137
134	43.378307	−2.991586
135	43.378739	−2.991224
136	43.378672	−2.990759
137	43.378758	−2.999239
138	43.378467	−2.990613
139	43.374214	−2.992418
140	43.374757	−2.993032
141	43.375189	−2.992432
142	43.375826	−2.994840
143	43.376289	−2.993976
144	43.377369	−2.992465
145	43.376535	−2.991381
146	43.376925	−2.991370
147	43.377435	−2.991360

The distances between the 147 locations can be found in [32].

References

1. European Commission. The Sustainable Development Goals—European Commission. Available online: https://ec.europa.eu/info/strategy/international-strategies/sustainable-development-goals_en (accessed on 7 February 2020).
2. Jovanovic, S. Expert Views (Part 2): What Are the Biggest Challenges in Municipal Waste Management in Serbia? Available online: https://balkangreenenergynews.com/expert-views-part-2-what-are-the-biggest-challenges-in-municipal-waste-management-serbia/ (accessed on 7 February 2020).
3. Caria, M.; Todde, G.; Pazzona, A. Modelling the Collection and Delivery of Sheep Milk: A Tool to Optimise the Logistics Costs of Cheese Factories. *Agriculture* **2018**, *8*, 5. [CrossRef]
4. Lin, S. Computer Solutions of the Traveling Salesman Problem. *Bell Syst. Tech. J.* **1965**, *44*, 2245–2269. [CrossRef]
5. Beltrami, E.J.; Bodin, L.D. Networks and vehicle routing for municipal waste collection. *Networks* **1974**, *4*, 65–94. [CrossRef]

6. Buhrkal, K.; Larsen, A.; Ropke, S. The Seventh International Conference on City Logistics The waste collection vehicle routing problem with time windows in a city logistics context peer-review under responsibility of the 7th International Conference on City Logistics. *Procedia Soc. Behav. Sci.* **2012**, *39*, 241–254. [CrossRef]

7. Tirkolaee, E.B.; Mahdavi, I.; Mehdi Seyyed Esfahani, M. A robust periodic capacitated arc routing problem for urban waste collection considering drivers and crew's working time. *Waste Manag.* **2018**, *76*, 138–146. [CrossRef] [PubMed]

8. Hannan, M.A.; Akhtar, M.; Begum, R.A.; Basri, H.; Hussain, A.; Scavino, E. Capacitated vehicle-routing problem model for scheduled solid waste collection and route optimization using PSO algorithm. *Waste Manag.* **2018**, *71*, 31–41. [CrossRef] [PubMed]

9. De Bruecker, P.; Beliën, J.; De Boeck, L.; De Jaeger, S.; Demeulemeester, E. A model enhancement approach for optimizing the integrated shift scheduling and vehicle routing problem in waste collection. *Eur. J. Oper. Res.* **2018**, *266*, 278–290. [CrossRef]

10. Ramos, T.R.P.; de Morais, C.S.; Barbosa-Póvoa, A.P. The smart waste collection routing problem: Alternative operational management approaches. *Expert Syst. Appl.* **2018**, *103*, 146–158. [CrossRef]

11. Benjamin, A.M.; Beasley, J.E. Metaheuristics for the waste collection vehicle routing problem with time windows, driver rest period and multiple disposal facilities. *Comput. Oper. Res.* **2010**, *37*, 2270–2280. [CrossRef]

12. Cervera Paz, A.; Lopez Molina, L.; Rodriguez Cornejo, V. Removing the pillars of logistics: The physical Internet. *Dyna* **2018**, *93*, 370–375. [CrossRef]

13. Route4me. Why You Shouldn't Rely On Free Mapping Tools For Planning Routes. Available online: https://blog.route4me.com/2017/07/google-maps-planning-routes/ (accessed on 7 February 2020).

14. OptimoRoute. OptimoRoute—Route and Schedule Planning and Optimization For Delivery and Field Service. Available online: https://optimoroute.com/ (accessed on 7 February 2020).

15. Altexsoft. How to Solve Vehicle Routing Problems: Route Optimization Software and their APIs. Available online: https://www.altexsoft.com/blog/business/how-to-solve-vehicle-routing-problems-route-optimization-software-and-their-apis/ (accessed on 7 February 2020).

16. Goti, A.; Sanchez, A.; Oyarbide-Zubillaga, A. Money based maintenance model constrained multi-objective optimization. In Proceedings of the IADIS International Conference Applied Computing (IADIS 2006), San Sebastian, Spain, 25–28 February 2006; pp. 49–56.

17. Bac, F.Q.; Perov, V.L. New evolutionary genetic algorithms for NP-complete combinatorial optimization problems. *Biol. Cybern.* **1993**, *69*, 229–234. [CrossRef]

18. Holland, J. *Adaptation in Natural and Artificial Systems*; University of Michigan Press: Ann Arbor, MI, USA, 1975.

19. Goldberg, D.E. *Genetic Algorithms in Search, Optimization and Machine Learning*; Addison-Wesley: Boston, FL, USA, 1989.

20. Davis, L. Applying adaptive algorithms to epistatic domains. *Proc. Int. Jt. Conf. Artif. Intell.* **1985**, *85*, 162–164.

21. Michalewicz, Z. *Genetic Algorithms + Data Structures = Evolution Programs*; Springer: Berlin, Germany, 1992.

22. Reeves, C. *Modern Heuristic Techniques for Combinatorial Problems*; Blackwell Scientific Publications: Oxford, UK, 1993.

23. Larrañaga, P.; Kuijpers, C.; Murga, R.; Inza, I.; Dizdarevich, S. Evolutionary algorithms for the traveling salesman problem: A review of representations and operators. *Artif. Intell. Rev.* **1999**, *13*, 129–170. [CrossRef]

24. Malhotra, R.; Chug, A. Software Maintainability: Systematic Literature Review and Current Trends. *Int. J. Softw. Eng. Knowl. Eng.* **2016**, *26*, 1221–1253. [CrossRef]

25. Europa Press. Spain Needs 10,000 Programmers Due to The Lack of Computer Engineers, According to the JBCNConf (España Necesita 10.000 Programadores por la Falta de Ingenieros InformáTicos, Según El JBCNConf). *La Vanguard*, 26 March 2019.

26. Mladenović, N.; Hansen, P. Variable neighborhood search. *Comput. Oper. Res.* **1997**, *24*, 1097–1100. [CrossRef]

27. Lin, S.; Kernighan, B.W. An effective heuristic algorithm for the Traveling Salesman Problem. *Oper. Res.* **1973**, *21*, 498–516. [CrossRef]

28. Ambati, B.K.; Ambati, J.; Mokhtar, M.M. Heuristic Combinatorial Optimization by Simulated Darwinian Evolution: A Polynomial Time Algorithm for the Traveling Salesman Problem. *Biol. Cybern.* **1991**, *65*, 31–35. [CrossRef]

29. De Jong, K.; Spears, W.M. Using Genetic Algorithms to Solve NP Complete Problems. In Proceedings of the Third International Conference on Genetic Algorithm, Los Altos, CA, USA, 4–7 June 1989; pp. 124–132.
30. Huella Medioambiental (Environmental Footprint)—Eco Calculator. Available online: https://ecocalculator.renault-trucks.com/es/huella-medioambiental (accessed on 8 January 2020).
31. Observatorio de Costes del Transporte de Mercancías por Carretera (Observatory of Road Transport Freight Costs). Available online: https://www.fomento.gob.es/CVP/handlers/pdfhandler.ashx?idpub=TTW103 (accessed on 7 April 2020).
32. Distances of the 147 Locations. Available online: https://docs.google.com/spreadsheets/d/1ZkiBnn-tGtyWV6pdxGEnqTBKIlPHJO3JqHkV5K5Nb6E/edit?usp=sharing (accessed on 8 January 2020).

Article

Fabrication of Ultrathin MoS₂ Nanosheets and Application on Adsorption of Organic Pollutants and Heavy Metals

Siyi Huang [1], Ziyun You [1], Yanting Jiang [1], Fuxiang Zhang [1], Kaiyang Liu [1], Yifan Liu [1], Xiaochen Chen [1] and Yuancai Lv [1,2,*]

[1]   Fujian Provincial Engineering Research Center of Rural Waste Recycling Technology, College of Environment & Resources, Fuzhou University, Fuzhou 350116, China; N170620033@fzu.edu.cn (S.H.); 061700527@fzu.deu.cn (Z.Y.); 061700506@fzu.deu.cn (Y.J.); 061700529@fzu.deu.cn (F.Z.); liukaiyang123@hotmail.com (K.L.); yfanym@fzu.edu.cn (Y.L.); chenxiaochen@fzu.edu.cn (X.C.)

[2]   Research Institute of Photocatalysis, Fuzhou University, Fuzhou 350116, China

*   Correspondence: yclv@fzu.edu.cn; Tel.: +86-135-6003-1596

Received: 10 April 2020; Accepted: 24 April 2020; Published: 26 April 2020

Abstract: Owing to their peculiar structural characteristics and potential applications in various fields, the ultrathin MoS₂ nanosheets, a typical two-dimensional material, have attracted numerous attentions. In this paper, a hybrid strategy with combination of quenching process and liquid-based exfoliation was employed to fabricate the ultrathin MoS₂ nanosheets (MoS₂ NS). The obtained MoS₂ NS still maintained hexagonal phase (2H-MoS₂) and exhibited evident thin layer-structure (1–2 layers) with inconspicuous wrinkle. Besides, the MoS₂ NS dispersion showed excellent stability (over 60 days) and high concentration (0.65 ± 0.04 mg mL⁻¹). The MoS₂ NS dispersion also displayed evident optical properties, with two characteristic peaks at 615 and 670 nm, and could be quantitatively analyzed with the absorbance at 615 nm in the range of 0.01–0.5 mg mL⁻¹. The adsorption experiments showed that the as-prepared MoS₂ NS also exhibited remarkable adsorption performance on the dyes (344.8 and 123.5 mg g⁻¹ of q_m for methylene blue and methyl orange, respectively) and heavy metals (185.2, 169.5, and 70.4 mg g⁻¹ of q_m for Cd²⁺, Cu²⁺, and Ag⁺). During the adsorption, the main adsorption mechanisms involved the synergism of physical hole-filling effects and electrostatic interactions. This work provided an effective way for the large-scale fabrication of the two-dimensional nanosheets of transition metal dichalcogenides (TMDs) by liquid exfoliation.

Keywords: transition metal dichalcogenides; liquid exfoliation; adsorption; quenching

---

1. Introduction

Given the special structure and potential applications, two-dimensional materials have drawn plenty of concerns [1,2], such as graphene, boron nitride, and molybdenum disulfide. Among them, the ultrathin molybdenum disulfide (MoS₂) nanosheets, which exhibit an evident layered structure, have attracted ample attentions because of their excellent performance on several fields, such as catalysis, sensors, and pollution remediation [1,3]. Recently, the ultrathin MoS₂ nanosheets were reported to show excellent prospects in pollution control [3,4]. Therefore, it was urgent to explore an effective method to produce ultrathin MoS₂ nanosheets.

To date, a few methods have been reported for efficient preparation of ultrathin MoS₂ nanosheets [5–10], for example, mechanical exfoliation, sputtering, atomic layer deposition, chemical methods, and liquid-based exfoliation. In spite of the excellent performance of the prepared monolayer or few-layer MoS₂ nanosheets through mechanical methods, the production efficiency was rather low, which severely limited the large-scale applications. Meanwhile, although most of the chemical methods like hydrothermal and solvent thermal routes could produce large-scale few-layer MoS_2 nanosheets, they generally needed strict reaction control, such as high temperature and pressure. Instead, due to the controllable operation and high production, liquid-based exfoliation was regarded as the most promising way for the production of ultrathin MoS_2 in large scale. According to previous studies [7,8], the solvent showed a significant impact on the exfoliation of MoS_2. Among which, pyrrolidone-based solvents like *N*-methyl-2-pyrrolidone displayed excellent MoS_2 exfoliation efficiency with 0.3 mg mL^{-1} of MoS_2 nanosheets concentration, because they showed similar surface energy with MoS_2. Nevertheless, considering their significant environmental risk, high toxicity and high-boiling points of the pyrrolidone-based solvents probably limited the large-scale application. To replace these toxic solvents, a number of polar solvents that own low boiling point and molecular weight were tested, such as water, methanol, ethanol, and isopropanol [11,12]. Unfortunately, given the different surface energy between the MoS_2 and polar solvents, most of the polar solvents showed dissatisfactory exfoliation efficiency [11]. Interestingly, it was reported that the MoS_2 exfoliation efficiency in the mixed solution with two of the polar micromolecular solvents was much better than those in the single solvent [13,14]. Meanwhile, it was worth noting that the MoS_2 exfoliation efficiency could be significantly improved when some organic small molecules, surfactants, or polymers were added in the polar micromolecular solvents, such as sodium cholate, Tween 80, Tween 85, sodium naphthalenide, Brij 30, Brij 700, Triton X-100, and so on [15–17]. Nevertheless, the strong van der Waals interaction between the MoS_2 layers still limited the MoS_2 nanosheets production.

Recently, owing to the effective break of the van der Waals force between the MoS_2 layers, quenching was found to be an effective way to exfoliate the graphene analogues [18–21]. Previous investigation showed that the high-quality ultrathin graphene sheets were fabricated by rapidly cooling the hot bulk graphite and pre-expanded graphite in aqueous solutions of NH_4HCO_3 and hydrazine hydrate, respectively [19,20]. Meanwhile, the boron nitride and MoS_2 nanosheets were also synthesized via the rapid quenching of hot bulk boron nitride and MoS_2 in the liquid N_2 [18,21]. Although the production efficiency of the nanosheets was dissatisfactory, we suspect that if the bulk MoS_2 was pretreated with the quenching process, the exfoliation efficiency of MoS_2 nanosheets in the polar micromolecular solvents could be significantly improved.

Herein, a hybrid strategy with the combination of quenching process and liquid-based exfoliation was employed to fabricate the ultrathin MoS_2 nanosheets. The microstructures, morphology, and optical properties were analyzed. In addition, the adsorption performance of dyes and heavy metals was also discussed.

2. Material and Methods

2.1. Materials

Ammonium tetrathiomolybdate ((NH_4)$_2MoS_4$) was provided by Sam Chemical Technology Co., Ltd. (Shanghai, China). Hydrazine monohydrate ($N_2H_4 \cdot H_2O$) was provided by Aladdin Reagent Co., Ltd. (Shanghai, China). Sodium hydroxide (NaOH), sulfuric acid (H_2SO_4), nitric acid (HNO_3), methylene blue (MB), methyl orange (MO), $AgNO_3$, $CuSO_4$, and $CdCl_2$ were obtained from Sinopharm Chemical Reagent Co. Ltd. (Shanghai, China). The double-distilled water was prepared with a Milli-Q water purification system (Milli-Q®Reference, Millipore, Billerica, Massachusetts, USA).

2.2. Fabrication of the MoS$_2$ Nanosheets

The MoS_2 nanosheets (MoS_2-NS) were prepared through a novel combined method including calcination at high temperature, quenching with liquid nitrogen, and ultrasonic-assisted peeling with hydrazine hydrate. Firstly, 4.000 g of (NH_4)$_2MoS_4$ was calcinated under nitrogen atmosphere at 800 °C for 5 h with a rate of 5 °C/min and then a black powder (named bulk MoS_2) was obtained. After that, the high-temperature bulk MoS_2 was quickly transferred into a Dewar bottle containing liquid nitrogen until the liquid nitrogen gasified completely. Subsequently, the pre-expanded bulk

MoS$_2$ was transferred into a serum bottle with 100 mL of hydrazine hydrate and the bottle was sonicated at a frequency of 40 kHz for 24 h. After centrifugation, the residual MoS$_2$ powders were added into another serum bottle with deionized water and sonicated for 12 h (In addition, the recycled hydrazine hydrate can be reused in a new procedure.). Finally, the resulting suspensions were centrifuged at 3000 rpm for 2 h and then the dark green MoS$_2$-NS dispersions were obtained. After dialyzed with dialysis tubing with 3000 dalton of molecular weight cut off, the obtained ultimate green dispersions were close to 7 of pH and stored in the fridge at 4 °C.

2.3. Adsorption Batch Experiments

Adsorption isotherm batch experiments were carried out in a 40-mL serum bottle containing 10 mL liquid with 0.1 g L^{-1} of the adsorbent concentration. The adsorption isotherm for MB and MO was conducted under 25 °C in the range of 0.5 to 50 mg L^{-1} of the MB and MO concentration and the pH was adjusted to 6.0 ± 0.1 with 1 M H$_2$SO$_4$ solution, while the experiments for heavy metals were conducted under 25 °C with metal concentration from 0.5 to 30 mg L^{-1} and the pH was adjusted to 5.0 ± 0.1 with 1 M H$_2$SO$_4$ solution. After sealed with polytetrafluoroethylene (PTFE) caps, all the bottles were shaken at 250 rpm for 6 h. At sampling points, one bottle was taken out. After filtered with 0.22 µm glass fiber filters (Tianjin Branch billion Lung Experimental Equipment Co., Ltd., Tianjin, China), the MB/MO concentrations were determined by UV-vis spectroscopy (UV-1780, SHIMADZU, Japan) at 664/464 nm, while the residual Cu^{2+}/Cd^{2+}/Ag$^+$ concentrations were analyzed with ICP-MS (XSERIES 2, Thermo). The adsorption isotherms data were treated with Langmuir and Freundlich models [22,23]. The experiments for the adsorption kinetics study were operated at 25 °C and 6.0 ± 0.1/5.0 ± 0.1 of pH (adjusted with 1 M H$_2$SO$_4$ solution) in 300 mL dyes/heavy metals solution (20 mg L^{-1} for dyes and 15 mg L^{-1} for heavy metals). All the samples were shaken at 250 rpm for 6 h. At sampling points, 1.5 mL of the solution was taken out and then filtered through the filters. The residual dyes/heavy metals concentrations were determined with ICP-MS. The adsorption kinetics data were treated with the pseudo-first order kinetic and pseudo-second-order non-linear kinetic models.

To study the effects of pH values (2–10)/(3–7) on dyes/heavy metals adsorption, the batch experiments were conducted at 25 °C in a serum bottle with 20 mg L^{-1}/15 mg L^{-1} of dyes/heavy metals concentration and 0.10 g L^{-1} of adsorbents.

2.4. Characterization

The X-ray powder diffraction (XRD) data of the bulk MoS$_2$ and MoS$_2$-NS were tested with X-ray powder diffractometer (MiniFlex600, Rigaku, Milwaukee, Wisconsin, USA) coupled with a Cu *Kα* line at 40 kV and 40 mA.

The microstructural features of prepared bulk MoS$_2$ and MoS$_2$-NS were observed with field emission scanning electron microscope (FESEM, Nova NanoSEM 230, FEI, Hillsboro, Oregon, USA), atomic force microscope (AFM, 5500, Agilent USA), and transmission electron microscope (TEM, TECNAI G2F20, FEI, Hillsboro, Oregon, USA).

The Raman data of bulk MoS$_2$ and MoS$_2$-NS were recorded by a confocal laser Raman microscopy (Invia Reflex, Renishaw, UK) with 532 nm of laser wavelength and 0.6 mW of laser energy.

The X-ray photoelectron spectroscopy (XPS) data were recorded with X-ray photoelectron spectrometer (ESCALAB 250, Thermo Scientific, Waltham, Massachusetts, USA) coupled with the Al *Kα* radiation at 15 kV and 51 W. The binding energies were confirmed by using the C1s component as the reference and the binding energy of C-C/H bonds were set at 284.5 eV.

The concentrations of MoS$_2$-NS dispersions were analyzed using UV-vis spectrophotometer (UV-2500, Shimadzu, Japan).

The Brunauer–Emmett–Teller (BET) surface areas of the bulk MoS$_2$ and MoS$_2$-NS were obtained from the analysis of N$_2$-adsorption isotherms at 77 K using the N$_2$ physisorption analyzer (ASAP2020, Micromeritics, Norcross, Georgia, USA).

3. Results and Discussion

3.1. Characterization of MoS$_2$ Nanosheets

3.1.1. Microstructures and Morphology

To study the variation of the crystal structure during the MoS$_2$-NS preparation, the precursor ((NH$_4$)$_2$MoS$_4$), bulk MoS$_2$ and MoS$_2$-NS were analyzed by XRD and the results were illustrated in Figure 1. As shown in Figure 1, after calcination under N$_2$, the characteristic peaks of (NH$_4$)$_2$MoS$_4$, located at 2θ = 17.2°, 18.44°, and 29.08°, fully vanished, suggesting the evident change of crystal structure. Instead, the peaks at 2θ = 14.6°, 33.48°, 39.82°, and 58.94° were assigned to the (100), (103), (105), and (110) plane of hexagonal MoS$_2$ phase (JCPDS card No. 65-0160), indicating that after calcination, the obtained bulk MoS$_2$ was hexagonal phase. The results were similar to Zhang et al.'s findings [24,25]. In addition, after exfoliation by sonication, the resulting MoS$_2$-NS still kept the same peaks with bulk MoS$_2$, manifesting that the 2H-MoS$_2$-NS was successfully obtained. However, compared to the bulk MoS$_2$, the (002) plane peak of MoS$_2$-NS became broadened and lower, suggesting an increase of the d spacing between MoS$_2$ layers [26]. Based on the full width at half maximum (FHWM) of the (002) plane, the layer number of the prepared MoS$_2$-NS could be calculated to be about ~2 layers through Scherrer's equation. To further confirm the structure of the exfoliated MoS$_2$-NS, Raman spectroscopy (Figure 1b) was employed to characterize the bulk MoS$_2$ and MoS$_2$-NS. Both the bulk MoS$_2$ and MoS$_2$-NS exhibited two dominant peaks ranging from 340 to 450 cm^{-1}, corresponding to the E$^1_{2g}$ and A$_{1g}$ mode of the hexagonal MoS$_2$, respectively [26,27], which convincingly proved the successful exfoliation of MoS$_2$-NS. Among which, the E$^1_{2g}$ mode peak at 381.6 cm^{-1} involved the in-layer displacements of Mo and S atoms, whereas the A$_{1g}$ mode peak at 407.9 cm^{-1} represented the out-of-layer symmetric displacements of S atoms along the c-axis [28]. Noticeably, compared to the bulk MoS$_2$, the E$^1_{2g}$ (377.8 cm^{-1}) and A$_{1g}$ (402.2 cm^{-1}) mode peaks displayed evident blue shift, and the interval (Δ = 24.4 cm^{-1}) between E$^1_{2g}$ and A$_{1g}$ peaks was lower than that of bulk MoS$_2$ (Δ = 26.3 cm^{-1}), which was ascribed to the decrease of the MoS$_2$ thickness. According to the previous literature [4,25,29], it was found that both E$^1_{2g}$ and A$_{1g}$ peaks were the characteristic peaks of MoS$_2$ and their frequencies would vary with the layer number. When the layer number increases, the interlayer van der Waals force in MoS$_2$ suppressed atom vibration, resulting in higher force constants [30]. On the contrast, the force constants between the layers would weaken with the layer number decreases. Thus, both E$^1_{2g}$ and A$_{1g}$ modes were supposed to stiffen (blue-shift) along with the reduction of MoS$_2$ layers.

Figure 2a–d presented the FESEM images of the bulk MoS$_2$ and MoS$_2$-NS. As seen in Figure 2a,b, the bulk MoS$_2$ displayed varisized particle-like morphology but clearly thick layer-structure. After exfoliation, the MoS$_2$-NS exhibited evident thin layer-structure with inconspicuous wrinkle (Figure 2c,d), which suggested the successful exfoliation of MoS$_2$-NS. Similar to the bulk MoS$_2$, the size of prepared MoS$_2$-NS still differed widely. In addition, the thickness of the prepared MoS$_2$-NS was analyzed by atomic force microscopy (AFM) (Figure 2). As seen in the AFM images (Figure 2e,f), the height profile of the two selected regions displayed a height of ~1.20 nm (±0.03 nm) for MoS$_2$-NS, which was about 2 times as thick as the theoretical thickness of monolayer MoS$_2$ (~0.65 nm) [27]. The evident platform of the height curves of the selected MoS$_2$-NS revealed the smooth surface for MoS$_2$-NS. Low-resolution TEM image (Figure 2g) also clearly depicted well-stacked layered structures (~2 layers) of the MoS$_2$-NS, which strongly confirmed the results of XRD and AFM. In the high-resolution TEM (HRTEM) images (Figure 2h), the lattice spacing of 0.27 and 0.16 nm between two adjacent lattice planes could be resolved, which were assigned to the (100) and (110) plane of MoS$_2$. In addition, the HRTEM image (Figure 2i) and associated fast Fourier transforms (FFT) (Figure 2j) from the center of the MoS$_2$-NS evidently exhibited hexagonally symmetric structure, which was consistent with the results of XRD analysis. All the above results manifested that the obtained MoS$_2$-NS still retained hexagonal single crystalline nature during pre-expansion and sonication treatments, which agreed with the previous findings [25,31,32]. However, the BET analysis (Figure 3) showed that the specific surface areas of bulk MoS_2 and MoS_2-NS were 5.6 and 26.6 m^2 g^{-1}, respectively, indicating that the exfoliation greatly changed the specific surface area of the MoS_2 materials. With the decreasing of the layers, more and more MoS_2 was exposed, resulting in a promotion of the BET surface.

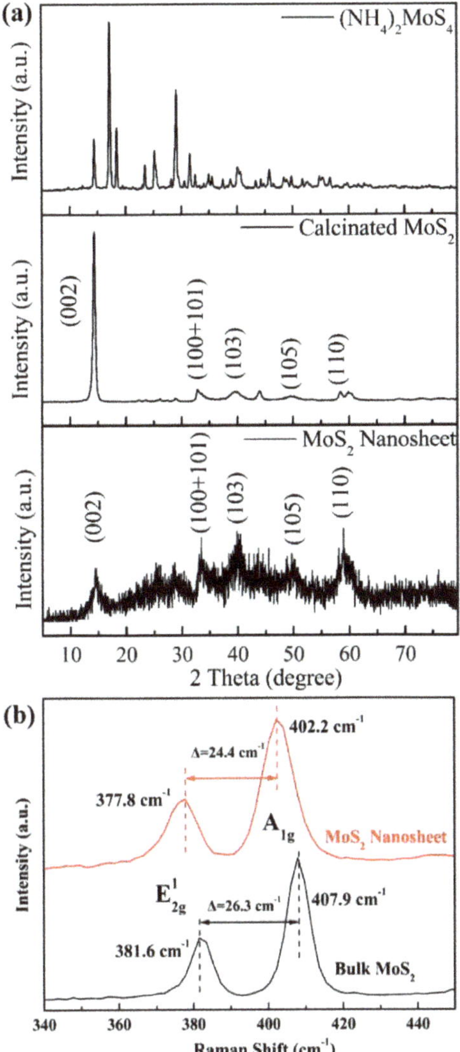

Figure 1. (**a**) XRD patterns of $(NH_4)_2MoS_2$, bulk MoS_2, and MoS_2-NS and (**b**) Raman spectra of bulk MoS_2 and MoS_2 nanosheets.

The chemical composition and element valence on the surface of MoS_2 NS were analyzed with XPS (Figure 4). As depicted in Figure 4a, the survey spectra clearly confirmed the presentence of C, O, NS, and Mo elements. The weak C1s (~284 eV) peak was attributed into the calibration of binding energy with carbon, while the N1s (~400 eV) peak was ascribed into the adsorbed hydrazine hydrate during the sonication. In Mo3d core-level spectra (Figure 4b), the appearance of Mo3d5/2 (233.5 eV) and Mo3d3/2 (232.6 eV) peaks for Mo3d doublet indicated the characteristic +4 oxidation state [1]. Besides, two weak Mo^{6+} 3d peaks (3d5/2 peak at 233.5 eV and 3d3/2 peak at 235.9 eV) were ascribed to the slight oxidation of MoS_2 NS edge during the MoS_2 transfer under high temperature [25]. In the high-resolution scans of S2p (Figure 4c), two feature peaks (S2p1/2 and S2p3/2) were observed at 162.0 and 163.3 eV, respectively, which greatly matched the binding energy of S^{2-} ions in 2H-MoS_2 [2]. In addition, the appearance of O1s also confirmed the oxidation of MoS_2. In the high-resolution spectra of O1s (Figure 4d), the peak of O^{2-} species located at 532.0 eV and the peak at 533.5 eV was attributed into the absorbed oxygen-containing material like H_2O [3].

Figure 2. Field emission scanning electron microscope (FESEM) images (**a–d**) of the bulk MoS_2 and MoS_2-NS, atomic force microscope (AFM) images (**e,f**), height profiles (inset), (transmission electron microscope) TEM and high-resolution transmission electron microscope (HRTEM) images of MoS_2 nanosheets (**g–i**), and the fast Fourier transforms (FFT) pattern of MoS_2 NS (**j**).

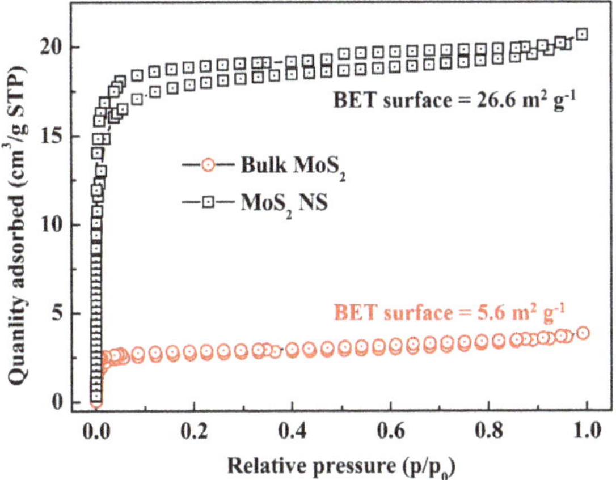

Figure 3. Brunauer–Emmett–Teller (BET) N_2 isotherms of the bulk MoS_2 and MoS_2 NS.

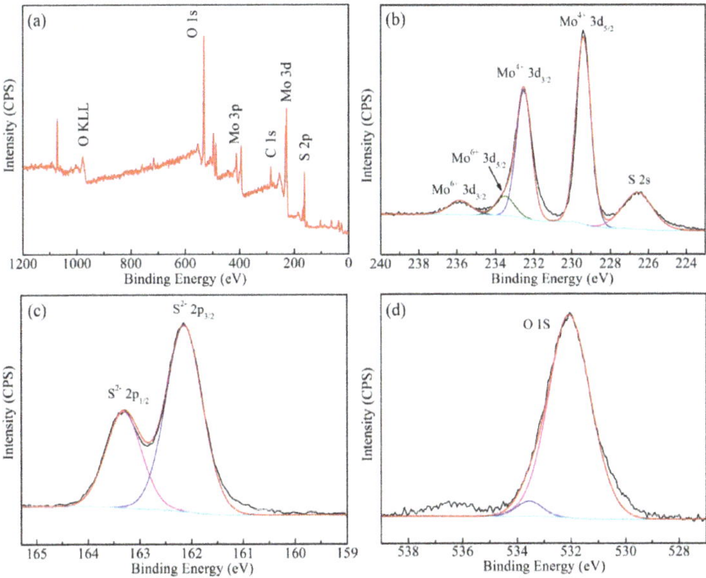

Figure 4. XPS survey spectra of MoS_2 (**a**) and high-resolution scans of Mo3d (**b**), S2p (**c**), O1s (**d**).

3.1.2. Optical Properties of MoS_2 Nanosheets Dispersion

To obtain pure few-layer MoS_2 NS, the suspensions were first centrifuged at 3000 rpm for 2 h to remove the no exfoliated precipitate. Figure 5a displayed the photographs of the as-prepared MoS_2 NS in water. As shown in the Figure 5a, the evident Tyndall phenomenon was observed both of the fresh MoS_2 NS dispersions and the dispersions after 60 days. Meanwhile, the UV-vis absorption spectra (Figure 5b,c) also exhibited no evident change during 60 days. All the results suggested the excellent stability (stable for over 60 days) of as-prepared MoS_2 NS dispersions. In addition, the UV-vis absorption spectra (Figure 5e) of the resulting MoS_2 NS dispersions with different concentrations (Figure 5d) displayed two distinctly characteristic peaks for 2H-MoS_2 [33]. The two peaks located at 615 (B-exciton) and 670 nm (A-exciton) were attributed to the direct excitonic transitions of MoS_2 at the

K point of the Brillouin zone [34,35]. According to Hai et al.'s study [25], the relationship between the concentrations of MoS$_2$ NS dispersions and the measured absorbance at a given wavelength (615 or 670 nm) were estimated by using the Beer–Lambert law. The fitting results (Figure 5f) proved that the concentrations of the dispersions showed good linear relationship (R^2 = 0.9996) with the absorbance at 615 nm in the range of 0.01–0.5 mg L^{-1}, which meant that the quantitative analysis of the MoS$_2$ NS dispersions was available. Based on the above relationship, the concentration of the as-prepared MoS$_2$ NS dispersions was 0.65 ± 0.04 mg mL^{-1}, which was much higher than previous findings [7,25]. The initial concentration of the bulk MoS$_2$ (2.510 g of bulk MoS$_2$ were obtained after the calcination of (NH$_4$)$_2$MoS$_4$) was 2.51 mg mL^{-1}, and the corresponding few-layer MoS$_2$ NS yield was calculated to be as high as 25.9% in water.

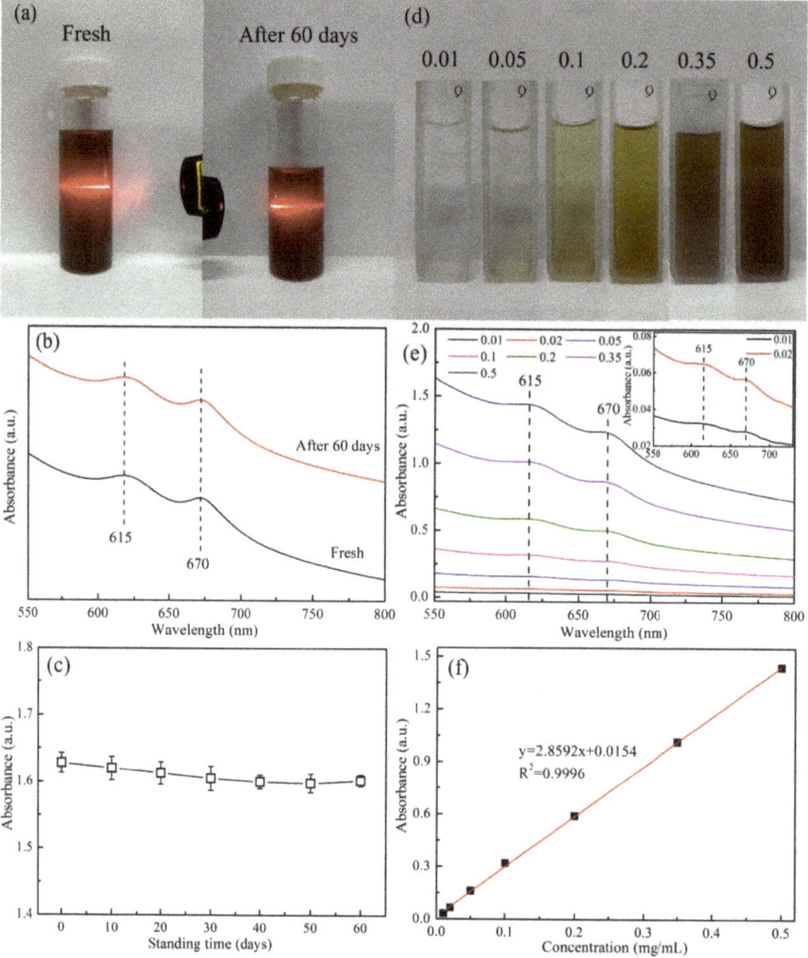

Figure 5. Photographs (**a**), UV-vis absorption spectra (**b**), absorbance change with standing time (**c**) of the prepared MoS$_2$ NS dispersions and photographs (**d**), UV-vis absorption spectra (**e**) and standard curve (**f**) of MoS$_2$ NS dispersion in different concentrations.

3.2. Adsorption Behavior of MoS$_2$-NS Towards Dyes and Heavy Metals

3.2.1. Adsorption Isotherms and Kinetics

The adsorption performance of the MoS$_2$ NS was tested by selecting two dyes (methylene blue, MB and methyl orange, MO) and three heavy metal ions (Cu^{2+}, Cd^{2+}, and Ag$^+$) as the targets. As seen in Figure 6a, in the bulk MoS$_2$ systems, the equilibrium adsorption capacities of the two dyes only slightly increased with the increasing of the dye concentrations, manifesting that the bulk MoS$_2$ exhibited unsatisfactory adsorption performance of MB and MO. Instead, the equilibrium adsorption capacities of MoS$_2$ NS for MB and MO significantly increased under high concentration of dyes, which were much larger than those of bulk MoS$_2$. Meanwhile, the as-prepared MoS$_2$ NS also displayed more excellent adsorption performance on heavy metals than the bulk MoS$_2$. All the results indicated that the exfoliation was beneficial to improve the adsorption performance of MoS$_2$, which was in accordance with previous studies [3,36,37].

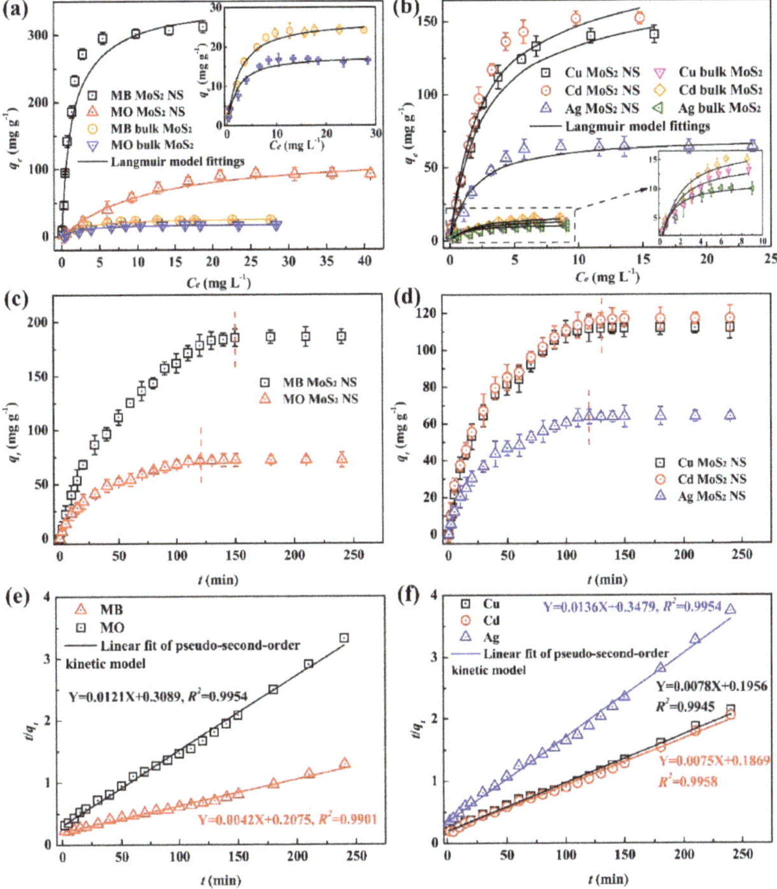

Figure 6. Adsorption isotherms (**a**) and kinetics (**c,e**) of MB, MO for bulk MoS$_2$ and MoS$_2$ NS at 20 mg L^{-1}; adsorption isotherms (**b**) and kinetics (**d,f**) of Cu^{2+}, Cd^{2+} and Ag$^+$ for bulk MoS$_2$ and MoS$_2$ NS at 15 mg L^{-1}.

In addition, to well study the adsorption behavior, the Langmuir and Freundlich models were employed to fit the experimental data (Figure 6a,b, Figure S1). As listed in Table 1, the high R^2 values suggested that the Langmuir model better described the adsorption of dyes and heavy metals onto MoS$_2$ NS and bulk MoS$_2$ than the Freundlich model. Based on the Langmuir model fitting, the relative parameters like the maximum adsorption capacity (q_m) and affinity constant (K_L) for dyes and heavy metals were obtained and listed in Table 1. The q_m values of MB and MO for MoS$_2$ NS were 344.8 and 123.5 mg g^{-1}, respectively, which were 12.77 and 6.94 larger than those (27.0 and 17.8 mg g^{-1} for MB and MO, respectively) of bulk MoS$_2$. Meanwhile, the similar results were observed in the heavy metal adsorption, indicating that the MoS$_2$ NS exhibited much more excellent adsorption performance than the bulk MoS$_2$. In addition, for the dyes, the higher q_m and K_L values of MB implied that MoS$_2$ materials exhibited better adsorption capacity and affinity to MB. For heavy metals, the highest q_m value occurred to Cd^{2+} (185.2 mg g^{-1}), following Cu^{2+} (169.5 mg g^{-1}) and Ag$^+$ (70.4 mg g^{-1}), indicating that MoS$_2$ NS were more beneficial to Cd^{2+} and Cu^{2+} adsorption than Ag$^+$.

Table 1. Fitted parameters for the adsorption of dyes and heavy metals on MoS$_2$ NS and bulk MoS$_2$.

Samples	Targets	Langmuir Model			Freundlich Model		
		q_m (mg g^{-1})	K_L (L mg^{-1})	R^2	$1/n$	K_f (L g^{-1})	R^2
MoS$_2$ NS	MB	344.8	0.725	0.994	0.648	92.796	0.875
	MO	123.5	0.393	0.980	0.724	9.023	0.969
Bulk MoS$_2$	MB	27.0	0.238	0.996	0.504	6.381	0.915
	MO	17.8	0.187	0.992	0.449	4.988	0.904
MoS$_2$ NS	Cd^{2+}	185.2	0.606	0.985	0.708	40.496	0.943
	Cu^{2+}	169.5	0.588	0.989	0.693	35.848	0.938
	Ag$^+$	70.4	0.339	0.993	0.533	18.326	0.914
Bulk MoS$_2$	Cd^{2+}	18.7	0.297	0.952	0.463	6.069	0.979
	Cu^{2+}	16.3	0.285	0.963	0.459	5.321	0.991
	Ag$^+$	11.7	0.134	0.987	0.389	4.913	0.968

Figure 6c,d displayed adsorption kinetics data for dyes and heavy metals over MoS$_2$ NS. As revealed in Figure 6c, both MO and MB adsorption increased rapidly at the beginning, then proceeded at a slower rate, and tended to equilibrium at the end. The similar results occurred to the adsorption of heavy metals. Besides, to further analyze the time-dependent variation during the adsorption process, pseudo-first-order and pseudo-second-order kinetic models were employed to fit the dyes and heavy metals adsorption on MoS$_2$ NS (Figure 6e,f). As shown in Table S1, the higher R^2 values suggested that the pseudo-second-order model better described both dyes and heavy metals adsorption than the pseudo-first-order model, suggesting that the electron transfer between MoS$_2$ NS and dye molecule or metal ions played a controlling role during the adsorption [38].

3.2.2. Adsorption Mechanism

Based on the results, the MoS$_2$ NS showed much better dye or metal adsorption performance than bulk MoS$_2$. According to the previous studies [3,37,39,40], the main mechanisms reported during the adsorption of dyes or metal by the inorganic materials involved physical hole-filling effects, electrostatic interactions, and ion exchange.

Physical Hole-Filling Effects

The specific surface area often displayed significant effect on the adsorption of the pollutants [41–43]. The adsorbents with large specific surface area usually owned abundant pores, which greatly provided a sufficient adsorption site to capture the pollutants, resulting in the promotion of their adsorption performance. For nano materials, the physical hole-filling effect was considered as one of the important adsorption mechanisms [41]. According to above-mentioned results, the obtained MoS$_2$ NS owned much larger specific surface area than bulk MoS$_2$, while the MoS$_2$ NS also exhibited more excellent adsorption performance on dyes and heavy metals. Thus, it could be inferred that the physical hole-filling effect probably played a vital role in the promotion of dyes or heavy metal adsorption. Herein, to verify the role of specific surface area during the dyes or heavy metal adsorption over MoS$_2$ NS and bulk MoS$_2$, the obtained q_e data were standardized with the BET surface area and the results were showed in Figure 7. As shown in Figure 7a, for dyes, the equilibrium adsorption capacities of MoS$_2$ NS for MB and MO were 312.0 and 92.6 mg g^{-1}, which were 12.89 and 5.61 times larger than those of bulk MoS$_2$, respectively. Meanwhile, the as-prepared MoS$_2$ NS also displayed excellent adsorption performance on heavy metals (Figure 7c), with 141.0, 152.8, and 64.2 mg g^{-1} for Cu^{2+}, Cd^{2+}, and Ag$^+$, respectively, which were 10.68, 10.12, and 6.42 folds larger than those of bulk MoS$_2$ (13.2, 15.1, and 10.0 mg g^{-1} for Cu^{2+}, Cd^{2+}, and Ag$^+$, respectively). After standardization (Figure 7b,d), all of the q_e ratios between the MoS$_2$ NS and bulk MoS$_2$ significantly decreased from 12.89 (MB), 5.61 (MO), 10.12 (Cu^{2+}), 10.68 (Cd^{2+}), and 6.42 (Ag$^+$) to 2.72, 1.12, 2.24, 2.11, and 1.33, respectively, suggesting that the physical hole-filling effect played positive role in the promotion of dyes or heavy metal adsorption over MoS$_2$.

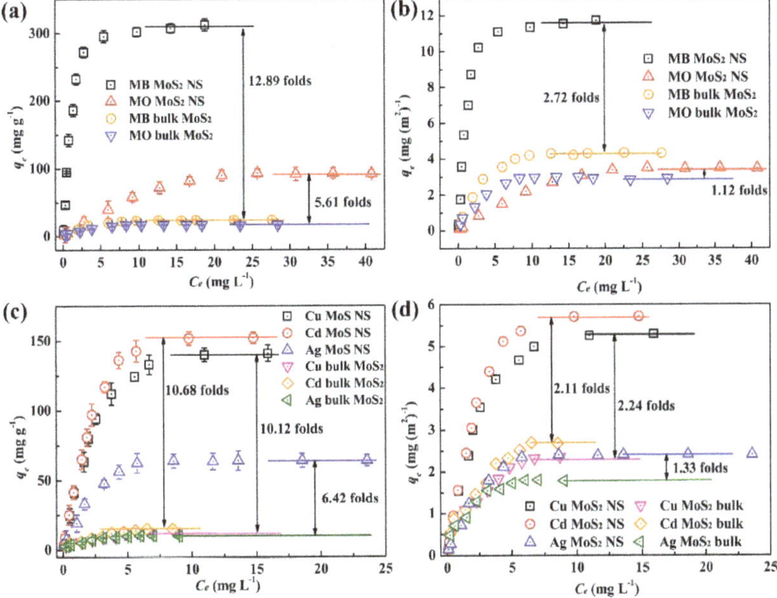

Figure 7. Equilibrium adsorption capacity (**a,c**) and standardized equilibrium adsorption capacity (**b,d**) of dyes (MO and MB) and heavy metals (Cu^{2+}, Cd^{2+}, and Ag$^+$) for bulk MoS$_2$ and MoS$_2$ NS.

In addition, no evident variation was observed between the standardized q_e values of MoS$_2$ NS and bulk MoS$_2$ (Figure 7b), meaning that the physical hole-filling effect was the sole mechanism during MO adsorption over MoS$_2$ NS. However, the significant enhancement between the standardized q_e values of MoS$_2$ NS and bulk MoS$_2$ (Figure 7b,d) suggested that besides the physical hole-filling effect, some other mechanisms were involved during the adsorption of MB and heavy metals over MoS$_2$ NS.

Electrostatic Interactions

Electrostatic interaction was often considered as a possible mechanism to explain the adsorption of dyes and heavy metals [37,40,44]. To confirm the role of electrostatic interaction during dyes and heavy metals adsorption over MoS$_2$ NS, the adsorption efficiency in various pH values were conducted.

As depicted in Figure 8a, the slight fluctuation among the q_e values for MO suggested that the MB adsorption over MoS$_2$ NS was not controlled by the pH values. Instead, the MB adsorption was notably influenced by the pH values. At low pH (<6) conditions, the q_e values increased with the pH value and reached a peak (186.2 mg g^{-1}) at pH = 6.0, and then gradually declined when pH > 6. Meanwhile, Zeta potential results (Figure 8c) showed that the isoelectric point of MoS$_2$ NS was about 3.8. This meant that the surface of MoS$_2$ NS displayed a positive charge when the pH value was below 3.8, while a negative charge above 3.8. As a typical cationic dye, MB molecules could strongly adhere to the MoS$_2$ NS through the electrostatic interaction once the surface charge of MoS$_2$ NS turned to negative, leading to an increasing of the q_e values.

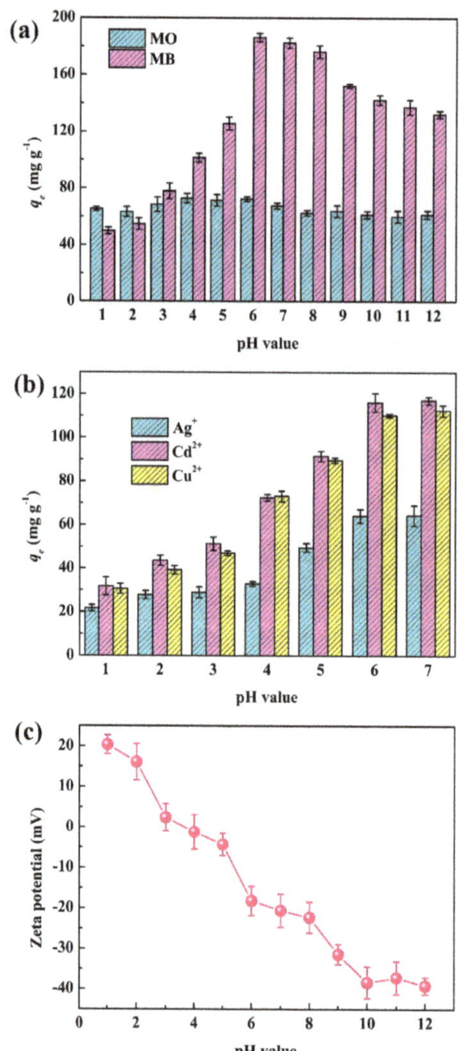

Figure 8. Effects of pH on dyes (**a**) and heavy metals (**b**) over MoS$_2$ NS, and the Zeta potential (**c**) of MoS$_2$ NS at different pH values. For dyes: 20 mg L^{-1} of the initial concentration, for heavy metals: 15 mg L^{-1} of the initial concentration.

Similarly, the pH also markedly influenced the adsorption of heavy metals over MoS_2 NS (Figure 8b). The q_e values of Cu^{2+}, Cd^{2+}, and Ag^+ evidently increased with an increasing pH, and stabilized at about 112.4, 117.0, and 64.4 mg g^{-1}, respectively. When the pH increased, the surface charge of MoS_2 NS turned to negative and the values gradually increased, which meant that stronger electrostatic interaction occurred between the heavy metal ions and MoS_2 NS at higher pH, resulting in improvement of the adsorption performance. In addition, the charge values of the heavy metal ions also showed visible effects on the adsorption capacity. Due to the lower value of the charge for Ag^+, the q_e value of Ag^+ was much lower than those of Cu^{2+} and Cd^{2+}, which was ascribed into the weaker electrostatic interaction between Ag^+ and MoS_2 NS. According to the Coulomb law, electrostatic interaction was in direct proportion to the value of the surface charge. The similar results were also found in Yang at al.'s studies [45].

Ion Exchange

According to previous studies [43,46], the ion exchange only occurred with heavy metals adsorption. It was well known that the affinity to the metal ions in the ion exchange process increased with the ion radius and the ion radius of Cd^{2+} and Cu^{2+} were 0.97 Å and 0.73 Å, respectively. If the ion exchange was the main adsorption mechanism, the number of the adsorbed Cd^{2+} should be larger than that of Cu^{2+}. Actually, in the system of 15 mg^{-1} L (Figure 8b), the molar adsorption capacity of Cd^{2+} (1.04 mmol g^{-1}, 117.0 mg g^{-1}) was visibly lower than that of Cu^{2+} (1.75 mmol g^{-1}, 112.4 mg g^{-1}), which indicated that the ion exchange was not the main mechanism during the heavy metals over MoS_2 NS. Similarly, Nguyen et al. also found that the ion exchange played a negligible role during the Cd^{2+} and Cu^{2+} adsorption over the activated carbon [43].

4. Conclusions

In summary, the ultrathin 2H-MoS_2 nanosheets with 1–2 layers were successfully obtained via a hybrid stagey with combination of quenching process and liquid-based exfoliation. The as-prepared 2H-MoS_2 nanosheets exhibited evident optical properties and could be accurately quantified with the absorbance at 615 nm in the range of 0.01–0.5 mg L^{-1}. Besides, the obtained 2H-MoS_2 nanosheets also showed a promising application in pollution control. It could be a candidate absorbent for the removal of dyes and heavy metals. This work provided an effective way for the large-scale fabrication of the two-dimensional nanosheets of transition metal dichalcogenides (TMDs) by liquid exfoliation.

Supplementary Materials: The following are available online at http://www.mdpi.com/2227-9717/8/5/504/s1, Figure S1. Linear fittings of dyes adsorption (a) and heavy metals (b and c) over bulk MoS2 and MoS2 NS with the Freundlich model, Table S1 Adsorption kinetics parameters of dyes and heavy metals adsorption over MoS2 NS.

Author Contributions: S.H. and Y.L. (Yifan Liu) conceived the study, designed the experiments and wrote the manuscript. S.H., Z.Y. and Y.J. performed the experiments. F.Z. and K.L. finished the characterization and data analysis. Y.L. (Yuancai Lv) and X.C. edited the manuscript. All authors have read and agreed to the published version of the manuscript.

Funding: This research was funded by by the Open Project Program of National Engineering Research Center for Environmental Photocatalysis (Grant No. 201904), Fuzhou University.

Conflicts of Interest: The authors declare no conflict of interest.

References

1. Kibsgaard, J.; Chen, Z.; Reinecke, B.N.; Jaramillo, T.F. Engineering the surface structure of MoS_2 to preferentially expose active edge sites for electrocatalysis. *Nat. Mater.* **2012**, *11*, 963–969. [CrossRef] [PubMed]
2. Oakes, L.; Carter, R.; Hanken, T.; Cohn, A.P.; Share, K.; Schmidt, B.; Pint, C.L. Interface strain in vertically stacked two-dimensional heterostructured carbon-MoS_2 nanosheets controls electrochemical reactivity. *Nat. Commun.* **2016**, *7*, 11796. [CrossRef] [PubMed]

3. Qiao, X.-Q.; Tian, F.-Y.; Hou, D.-F.; Hu, F.-C.; Li, D.-S. Equilibrium and kinetic studies on MB adsorption by ultrathin 2D MoS$_2$ nanosheets. *RSC Adv.* **2016**, *6*, 11631–11636. [CrossRef]

4. Ye, L.; Xu, H.; Zhang, D.; Chen, S. Synthesis of bilayer MoS$_2$ nanosheets by a facile hydrothermal method and their methyl orange adsorption capacity. *Mater. Res. Bull.* **2014**, *55*, 221–228. [CrossRef]

5. Lee, C.; Yan, H.; Brus, L.E.; Heinz, T.F.; Hone, J.; Ryu, S. Anomalous Lattice Vibrations of Single- and Few-Layer MoS$_2$. *ACS Nano* **2010**, *4*, 2695–2700. [CrossRef] [PubMed]

6. Zhang, Y.; Ye, J.; Matsuhashi, Y.; Iwasa, Y. Ambipolar MoS$_2$ Thin Flake Transistors. *Nano Lett.* **2012**, *12*, 1136–1140. [CrossRef]

7. Coleman, J.N.; Lotya, M.; O'Neill, A.; Bergin, S.D.; King, P.J.; Khan, U.; Young, K.; Gaucher, A.; De, S.; Smith, R.J.; et al. Two-Dimensional Nanosheets Produced by Liquid Exfoliation of Layered Materials. *Science* **2011**, *331*, 568–571. [CrossRef]

8. O'Neill, A.; Khan, U.; Coleman, J.N. Preparation of High Concentration Dispersions of Exfoliated MoS$_2$ with Increased Flake Size. *Chem. Mater.* **2012**, *24*, 2414–2421. [CrossRef]

9. Tao, J.; Chai, J.W.; Lü, X.; Wong, L.M.; Wong, T.I.; Pan, J.; Xiong, Q.; Chi, D.; Wang, S. Growth of wafer-scale MoS$_2$ monolayer by magnetron sputtering. *Nanoscale* **2015**, *7*, 2497–2503. [CrossRef]

10. Motola, M.; Baudys, M.; Zazpe, R.; Krbal, M.; Michalicka, J.; Rodriguez-Pereira, J.; Pavlinak, D.; Prikryl, J.; Hromadko, L.; Sopha, H.I.; et al. 2D MoS2 nanosheets on 1D anodic TiO$_2$ nanotube layers: An efficient co-catalyst for liquid and gas phase photocatalysis. *Nanoscale* **2019**, *11*, 23126–23131. [CrossRef] [PubMed]

11. Shen, J.; He, Y.; Wu, J.; Gao, C.; Keyshar, K.; Zhang, X.; Yang, Y.; Ye, M.; Vajtai, R.; Lou, J.; et al. Liquid Phase Exfoliation of Two-Dimensional Materials by Directly Probing and Matching Surface Tension Components. *Nano Lett.* **2015**, *15*, 5449–5454. [CrossRef] [PubMed]

12. Kim, J.; Kwon, S.; Cho, D.-H.; Kang, B.; Kwon, H.; Kim, Y.; Park, S.O.; Jung, G.Y.; Shin, E.; Kim, W.-G.; et al. Direct exfoliation and dispersion of two-dimensional materials in pure water via temperature control. *Nat. Commun.* **2015**, *6*, 8294. [CrossRef] [PubMed]

13. Dong, L.; Lin, S.; Yang, L.; Zhang, J.; Yang, C.; Yang, D.; Lu, H. Spontaneous exfoliation and tailoring of MoS$_2$ in mixed solvents. *Chem. Commun.* **2014**, *50*, 15936–15939. [CrossRef] [PubMed]

14. Zhou, K.-G.; Mao, N.-N.; Wang, H.; Peng, Y.; Zhang, H.-L. A Mixed-Solvent Strategy for Efficient Exfoliation of Inorganic Graphene Analogues. *Angew. Chem. Int. Ed.* **2011**, *50*, 10839–10842. [CrossRef] [PubMed]

15. Guardia, L.; Paredes, J.I.; Rozada, R.; Villar-Rodil, S.; Martinez-Alonso, A.; Tascón, J.D. Production of aqueous dispersions of inorganic graphene analogues by exfoliation and stabilization with non-ionic surfactants. *RSC Adv.* **2014**, *4*, 14115–14127. [CrossRef]

16. Smith, R.J.; King, P.J.; Lotya, M.; Wirtz, C.; Khan, U.; De, S.; O'Neill, A.; Duesberg, G.S.; Grunlan, J.C.; Moriarty, G.; et al. Large-Scale Exfoliation of Inorganic Layered Compounds in Aqueous Surfactant Solutions. *Adv. Mater.* **2011**, *23*, 3944–3948. [CrossRef]

17. Zheng, J.; Zhang, H.; Dong, S.; Liu, Y.; Nai, C.T.; Shin, H.S.; Jeong, H.Y.; Liu, B.; Loh, K.P. High yield exfoliation of two-dimensional chalcogenides using sodium naphthalenide. *Nat. Commun.* **2014**, *5*, 2995. [CrossRef]

18. Zhuc, W.; Gao, X.; Li, Q.; Li, H.; Chao, Y.; Li, M.; Mahurin, S.M.; Li, H.; Zhu, H.; Dai, S. Controlled Gas Exfoliation of Boron Nitride into Few-Layered Nanosheets. *Angew. Chem. Int. Ed.* **2016**, *55*, 10766–10770. [CrossRef]

19. Jiang, B.; Tian, C.; Wang, L.; Xu, Y.; Wang, R.; Qiao, Y.; Ma, Y.; Fu, H. Facile fabrication of high quality graphene from expandable graphite: Simultaneous exfoliation and reduction. *Chem. Commun.* **2010**, *46*, 4920. [CrossRef]

20. Tang, Y.B.; Lee, C.-S.; Chen, Z.; Yuan, G.D.; Kang, Z.H.; Luo, L.B.; Song, H.S.; Liu, Y.; He, Z.; Zhang, W.; et al. High-Quality Graphenes via a Facile Quenching Method for Field-Effect Transistors. *Nano Lett.* **2009**, *9*, 1374–1377. [CrossRef]

21. Van Thanh, D.; Pan, C.-C.; Chu, C.-W.; Wei, K.-H. Production of few-layer MoS2nanosheets through exfoliation of liquid N2–quenched bulk MoS$_2$. *RSC Adv.* **2014**, *4*, 15586–15589. [CrossRef]

22. Zhao, J.; Wang, Z.; Zhao, Q.; Xing, B. Adsorption of Phenanthrene on Multilayer Graphene as Affected by Surfactant and Exfoliation. *Environ. Sci. Technol.* **2013**, *48*, 331–339. [CrossRef] [PubMed]

23. Cao, Q.; Huang, F.; Zhuang, Z.; Lin, Z. A study of the potential application of nano-Mg(OH)$_2$ in adsorbing low concentrations of uranyl tricarbonate from water. *Nanoscale* **2012**, *4*, 2423. [CrossRef] [PubMed]

24. Zhang, D.-Q.; Chai, J.-X.; Jia, Y.-X.; Wang, L.-J.; Zhao, Z.-L.; Cao, M. Facile Preparation of Few-layer MoS_2-NS by Liquid-Phase Ultrasonic Exfoliation. In Proceedings of the 2017 International Conference on Manufacturing Engineering and Intelligent Materials (ICMEIM 2017), Guangzhou, China, 25–26 February 2017.

25. Hai, X.; Chang, K.; Pang, H.; Li, M.; Li, P.; Liu, H.; Shi, L.; Ye, J. Engineering the Edges of MoS_2 (WS_2) Crystals for Direct Exfoliation into Monolayers in Polar Micromolecular Solvents. *J. Am. Chem. Soc.* **2016**, *138*, 14962–14969. [CrossRef]

26. Ma, L.; Huang, G.; Chen, W.; Wang, Z.; Ye, J.; Li, H.; Chen, D.; Lee, J.Y. Cationic surfactant-assisted hydrothermal synthesis of few-layer molybdenum disulfide/graphene composites: Microstructure and electrochemical lithium storage. *J. Power Sources* **2014**, *264*, 262–271. [CrossRef]

27. Li, H.; Zhang, Q.; Yap, C.C.R.; Tay, B.K.; Edwin, T.H.T.; Olivier, A.; Baillargeat, D. From Bulk to Monolayer MoS_2: Evolution of Raman Scattering. *Adv. Funct. Mater.* **2012**, *22*, 1385–1390. [CrossRef]

28. Korn, T.; Heydrich, S.; Hirmer, M.; Schmutzler, J.; Schüller, C. Low-temperature photocarrier dynamics in monolayer MoS_2. *Appl. Phys. Lett.* **2011**, *99*, 102109. [CrossRef]

29. Chakraborty, B.; Matte, H.R.; Sood, A.K.; Ra, C.N.R. Layer-dependent resonant Raman scattering of a few layer MoS_2. *J. Raman Spectrosc.* **2013**, *44*, 92–96. [CrossRef]

30. Bagnall, A.; Liang, W.; Marseglia, E.; Welber, B. Raman studies of MoS_2 at high pressure. *Phys. B+C* **1980**, *99*, 343–346. [CrossRef]

31. Chou, S.S.; De, I.; Kim, J.; Byun, S.; Dykstra, C.; Yu, J.; Huang, J.; Dravid, V.P. Ligand Conjugation of Chemically Exfoliated MoS_2. *J. Am. Chem. Soc.* **2013**, *135*, 4584–4587. [CrossRef]

32. Wang, L.; Xu, Z.; Wang, W.; Bai, X. Atomic Mechanism of Dynamic Electrochemical Lithiation Processes of MoS_2 Nanosheets. *J. Am. Chem. Soc.* **2014**, *136*, 6693–6697. [CrossRef]

33. Wilcoxon, J.P.; Newcomer, P.P.; Samara, G.A. Synthesis and optical properties of MoS_2 and isomorphous nanoclusters in the quantum confinement regime. *J. Appl. Phys.* **1997**, *81*, 7934–7944. [CrossRef]

34. Guan, G.; Zhang, S.; Liu, S.; Cai, Y.; Low, M.; Teng, C.P.; Phang, I.Y.; Cheng, Y.; Duei, K.L.; Srinivasan, B.M.; et al. Protein Induces Layer-by-Layer Exfoliation of Transition Metal Dichalcogenides. *J. Am. Chem. Soc.* **2015**, *137*, 6152–6155. [CrossRef] [PubMed]

35. Mak, K.F.; Lee, C.; Hone, J.; Shan, J.; Heinz, T.F. Atomically Thin MoS_2: A New Direct-Gap Semiconductor. *Phys. Rev. Lett.* **2010**, *105*, 1–4. [CrossRef] [PubMed]

36. Wu, P.; Yin, N.; Li, P.; Cheng, W.; Huang, M. The adsorption and diffusion behavior of noble metal adatoms (Pd, Pt, Cu, Ag and Au) on a MoS_2 monolayer: A first-principles study. *Phys. Chem. Chem. Phys.* **2017**, *19*, 20713–20722. [CrossRef]

37. Zhan, W.; Jia, F.; Yuan, Y.; Liu, C.; Sun, K.; Yang, B.; Song, S. Controllable incorporation of oxygen in MoS_2 for efficient adsorption of Hg^{2+} in aqueous solutions. *J. Hazard. Mater.* **2020**, *384*, 121382. [CrossRef]

38. Lv, Y.; Zhang, R.; Zeng, S.; Liu, K.; Huang, S.; Liu, Y.; Xu, P.; Lin, C.; Cheng, Y.; Liu, M. Removal of p-arsanilic acid by an amino-functionalized indium-based metal–organic framework: Adsorption behavior and synergetic mechanism. *Chem. Eng. J.* **2018**, *339*, 359–368. [CrossRef]

39. Yang, H.; Yuan, H.; Hu, Q.; Liu, W.; Zhang, D. Synthesis of mesoporous C/MoS2 for adsorption of methyl orange and photo-catalytic sterilization. *Appl. Surf. Sci.* **2020**, *504*, 144445. [CrossRef]

40. Luo, J.; Fu, K.; Sun, M.; Yin, K.; Wang, D.; Liu, X.; Crittenden, J.C. Phase-Mediated Heavy Metal Adsorption from Aqueous Solutions Using Two-Dimensional Layered MoS_2. *ACS Appl. Mater. Interfaces* **2019**, *11*, 38789–38797. [CrossRef]

41. Choi, M.; Jang, J. Heavy metal ion adsorption onto polypyrrole-impregnated porous carbon. *J. Colloid Interface Sci.* **2008**, *325*, 287–289. [CrossRef]

42. Li, H.; Xie, F.; Li, W.; Fahlman, B.D.; Chen, M.; Li, W. Preparation and adsorption capacity of porous MoS_2 nanosheets. *RSC Adv.* **2016**, *6*, 105222–105230. [CrossRef]

43. Nguyen, H.D.; Tran, H.N.; Chao, H.-P.; Lin, C.-C. Activated Carbons Derived from Teak Sawdust-Hydrochars for Efficient Removal of Methylene Blue, Copper, and Cadmium from Aqueous Solution. *Water* **2019**, *11*, 2581. [CrossRef]

44. Li, R.; Deng, H.; Zhang, X.; Wang, J.J.; Awasthi, M.K.; Wang, Q.; Xiao, R.; Zhou, B.; Du, J.; Zhang, Z. High-efficiency removal of Pb(II) and humate by a CeO_2–MoS_2 hybrid magnetic biochar. *Bioresour. Technol.* **2019**, *273*, 335–340. [CrossRef] [PubMed]

45. Yang, K.; Chen, B.; Zhu, X.; Xing, B. Aggregation, Adsorption, and Morphological Transformation of Graphene Oxide in Aqueous Solutions Containing Different Metal Cations. *Environ. Sci. Technol.* **2016**, *50*, 11066–11075. [CrossRef] [PubMed]

46. Wang, C.; Yang, R.; Wang, H. Synthesis of ZIF-8/Fly Ash Composite for Adsorption of Cu^{2+}, Zn^{2+} and Ni^{2+} from Aqueous Solutions. *Materials* **2020**, *13*, 214. [CrossRef] [PubMed]

Article

Evaluation of Calcium Oxide Nanoparticles from Industrial Waste on the Performance of Hardened Cement Pastes: Physicochemical Study

Youssef Abdelatif [1,2,*], Abdel-Aal M. Gaber [3], Abd El-Aziz S. Fouda [1] and Tarek Alsoukarry [4]

[1] Chemistry Department, Faculty of Science, Mansoura University, Mansoura 35516, Egypt; asfouda@hotmail.com
[2] Production Manager, Sugar KSA, Savola Group, Jeddah 22337, Saudi Arabia
[3] Faculty of Sugar and Integrated Industries Technology and chemistry Department, Assiut University , Assiut 71516, Egypt; amabdelaal@hotmail.com
[4] Raw Building Materials Technology and Processing Research Institute, Housing & Building National Research Center, HBRC, Cairo 1770, Egypt; tarek_elsokkary@yahoo.com
* Correspondence: ayoussef104@yahoo.com

Received: 27 February 2020; Accepted: 23 March 2020; Published: 30 March 2020

Abstract: Large amounts of carbonated mud waste (CMW) require disposal during sugar manufacturing after the carbonation process. The lightweight of CMW enables its utilization as a partial replacement for the cement to reduce costs and CO_2 emissions. Here, various levels of CMW, namely, 0, 5, 10, 15, 20, and 25 wt.% were applied to produce composite cement samples with ordinary Portland cement (OPC) as a regular mix design series. Pure calcium oxide (CaO) nanoparticles were obtained after the calcination of CMW. The techniques of X-ray fluorescence spectrometers (XRF), Transmission electron microscope (TEM), Selected area diffraction (SAED), Scanning electron microscope (SEM), energy dixpersive X-ray (EDX), and dynamic light scattering (DLS) were used to characterize the obtained CaO nanoparticles. According to the compressive strength and bulk density results, 15 wt.% CMW was optimal for the mix design. The specific surface area increased from 27.8 to 134.8 m^2/g when the CMW was calcined to 600 °C. The compressive strength of the sample containing 15% CMW was lower than the values of the other pastes containing 5% and 10% CMW at all of the curing times. The porosity factor of the hardened cement pastes released with a curing time of up to 28 days. Excessive CMW of up to 25 wt.% reduced the properties of OPC.

Keywords: calcium oxide nanoparticles; calcination; blended cement paste; mix design; compressive strength; bulk density

1. Introduction

Solid waste disposal in many industries is proposed for cement partial replacement, owing to the fact that these types of wastes contain some of the pozzolanic behaviour [1]. However, the improvement of the cementations characteristic of cement through partial replacement is still required. The pozzolans include different materials with high levels of silicon dioxide, which activate the hydration process [2]. Other industrial waste materials, such as blast furnace slag, fly ash, and cellulosic paper pulp, have been known to strengthen their properties [3–5]. On the other side, many studies have examined the effects of hazardous constituents such as bacteria, heavy metals, and uncontrolled organic substances [6], which have a bad effect on the ecosphere and general health. Therefore, the creation of an innovative process to maximize the recovery of beneficial materials and/or energy in a renewable way is of interest in order to protect all sides from potential threats.

In the sugar industry, sugarcane press mud is produced after the juice filtration process [7,8]. Lime milk is added into sugar juice to coagulate colloidal substances, and is settled in the form of soluble and non-soluble substances [9,10]. Subsequently, CO_2 was used as a lime surplus precipitant, and the carbonated slurry, namely carbonated mud waste (CMW), was filtered off. CMW mainly contains pure $CaCO_3$, as well as other minerals, salts, and organic compounds with coloured components [11,12]. Huge amounts of CMW disposal are generated per day during sugar manufacturing, and it is a major issue especially in eutrophication [13]. Often, CMW is used carefully or is sold as immature compost to farmers, mainly for use as a soil conditioner [14], fertilizer [15], and for wax production [16]. The richness if the micronutrient content inside the CMW enables it to be used as a fertilizer in crops and horticulture applications [17].

Ordinary Portland cement (OPC)-based concrete is the first material in construction, and the comprehensive utilization of concrete is second only to water, comprising 70% of all building and construction materials [18]. However, OPC has many advantages over geopolymers, such as the wide availability of raw materials worldwide and the ease of application, but these processes release a huge number of greenhouse gases. Statistically, one ton of cement clinker releases about 0.1 tons of CO_2, and the cement industry largely accounts for 5% of global CO_2 emissions [19].

To combat global climate change, the carbon footprint of OPC-based concrete should be reduced. To this end, the amount of OPC used in concrete needs to be reduced, as OPC is the major contributor to the carbon footprint of concrete. This can be realized by partially replacing OPC with minerals or fully replacing OPC with alternative non-OPC binders that have a lower carbon footprint [20–22]. Almost all of the published studies reported an enhancement in the mechanical properties of concrete, with up to a 10% replacement of OPC by different wastes [23–25], while a few papers also found an enhancement by splitting the tensile strength at a 15% replacement level [26,27]. Considering the literature, some authors replaced OPC with 10% and 15% bagasse ash to optimize the replacement level of OPC, but these papers did not focus on the particle size when the waste was calcined to produce ultrafine nanopowder. Nanotechnology is a trend in many applications [28–31]. Pure CaO is used in many applications, such as the cement industry, biodiesel production, the petroleum industry, electric lighting, paper industry, and power production [32–38]. There are many sources available to prepare CaO nanoparticles, such as using chicken eggshells (with very low quantities) or any source rich in calcium carbonate, such as CMW sugar cane waste, before treatment. The rheological properties of cement paste containing CaO nanoparticle contents from origin waste have not been investigated. In the present work, an attempt has been made to investigate the effect of CaO nanoparticles on the rheological behavior of the blended cement pastes. X-ray fluorescence spectrometers (XRF), Transmission electron microscope (TEM), Selected area diffraction (SAED), Scanning electron microscope (SEM), energy dixpersive X-ray (EDX), and dynamic light scattering (DLS) tools were used to analyze the properties of CaO nanoparticles. To compare the results, plain cement was also considered in the experimental program described ahead.

Herein, CMW samples were collected from different batches during the sugar production process. The physical and chemical behaviours and engineering properties were conducted in order to study the feasibility of utilizing CMW and OPC blends with various ratios. The utilization of CMW with OPC at the nanoscale seems to be a sustainable approach and may add value with many benefits, such as minimizing the burden of hauling waste to landfills and economize free spaces for disposal or to other landfills, which reduces the carbon footprint. Thereat, an environmentally and green approach would conserve precious natural aggregates and cement for future generations. On-site utilization of CMW would curtail high transportation costs, which would be highly beneficial for sites situated far from markets (high altitude regions). Finally, an aesthetic benefit includes removing waste from roadside farms.

2. Experimental Techniques

2.1. Materials, Mix Design, and Procedures

The CMW batches were collected from the Saudi Sugar Company (SGC) as semi-batches, and the samples were washed with an ethanol/double distilled water (DDI) mixture, dried, and kept in an air atmosphere to remove unfavourable odours and impurities. The samples were completely dried and crushed to reach the nominal minimum size. The specific gravity was determined to be 2.96% and the water absorption was 0.84%. The collected samples were dried at 80 °C overnight to remove the moisture content. Then, calcination at different temperatures in a muffle furnace ranging from 200 to 600 °C was applied so as to obtain nanostructured materials. In this investigation, the main hydration characteristics of the various OPC–CMW blended cement pastes consisting of mixes of M 0.0, M 0.5 CMW, M 0.10 CMW, M 0.15 CMW, M 0.20 CMW, and M 0.25 CMW were studied at various hydration time intervals of 3, 7, and 28 days. These mixes were designated as follows, based on the weight percent:

M 0.0: (100% OPC);

M 0.5: (95% OPC and 5% CMW);

M 0.10: 90% OPC and 10% CMW;

M 0.15: (85% OPC and 15% CMW);

M 0.20: (80% OPC and 20% CMW) and

M 0.25: (75% OPC and 25% CMW).

The investigated parameters were the water content of hardened cement pastes, setting times, bulk density, total porosity, and compressive strength.

2.2. Preparation of Combined Cement Pastes

A pre-determined amount of OPC was placed on a smooth, hydrophobic surface and a crater was centered. A definite amount of pure water related to cement was poured into the crater using a trowel tool. The dry OPC took for approximately five minutes in order to absorb water, and then had vigorous stirring for another five minutes in order to complete the mixing. The designed moulds had cubic dimensions of 2.5 × 2.5 × 2.5 cm, and the reservoir of paste was manually pressed into the corners of moulds until a homogeneous surface was obtained. Afterwards the top layered was compacted using a thin-edged trowel to smooth the surface of the past. After moulding, the specimens were speedily cured against moisture at room temperature in a humidity reactor (100%) for at least one day. At the end of the curing period for moisturizing, the cubes were remoulded, and the curing was done hourly in tap water for 3, 7, and 28 days (ASTM: C191, 2008).

2.3. Methods of Investigation

As a preliminary measure, vicat apparatus (according to ASTM: C191, 2008) was used for determining the consistency and setting times (initial and final) of the cement pastes with water. The bulk density measurements were determined according to the following formulae:

$$\text{Bulk Density} = \frac{\text{Saturated weight}}{\text{Volume of sample}} \; (\text{g}/\text{cm}^3) \tag{1}$$

$$\text{The sample volume} = \frac{\text{Saturated weight} - \text{Suspended weight}}{\text{Density of water}} \; (\text{cm}^3) \tag{2}$$

$$\text{Bulk density} = \frac{\text{Saturated weight}}{\text{Saturated weight} - \text{Suspended weight}} \; (\text{g}/\text{cm}^3) \tag{3}$$

After that, the water content of the specimens and the total porosity were calculated according to the following well-known equation:

$$\text{Total porosity } (\varepsilon) = \frac{0.99 W_e \times d_p}{1 + W_t} \times 100 \tag{4}$$

where 0.99 is the free water-specific volume (cm^3/g), W_e is the free water content of the paste, d_p is the paste bulk density (g/cm^3), and W_t is the saturated hardened paste of the water content.

The total porosity of the hardened cement paste is equal to the volume of the pores/the volume of the pastes. According to ASTM: C-150, 2007, the compressive strength was tested using the three cubic specimens for the same cement pastes, and the curing time was used to determine the compressive strength of the hardened paste (MPa).

2.4. Characterization

X-ray fluorescence (XRF) micro XRF analysis (Mahwah, NJ, USA) was also used to detect the included oxide after burning. Surface area measurements of nitrogen sorption/desorption were made using Brunauer–Emmett–Teller (BET) at 77 K (Gemini, GA, USA). A high-resolution transmission electron microscopy (HR-TEM) JEOL (JEM-2100, Hitachi, Japan) microscope with a high voltage of 200 kV was used. The size distribution by intensity was measured by Malvern Instruments Ltd. (Zetasizer Ver.6.32, Malvern Instruments Ltd, UK). The surface morphology was obtained by scanning electron microscopy (SEM, Sirion, MA, USA) with an acceleration voltage of 20 kV. SEM equipped with a pendent energy-dispersive X-ray spectroscopy (EDX) detector (S-3400 N II, Hitachi, Japan) was used. The thermal stability was measured by TGA thermograms in the range of 0 to 800 °C, with a constant heating rate of 5 min^{-1}.

3. Results and Discussion

3.1. Structural and Morphological Analysis of Carbonated Mud Waste (CMW) Sample

Table 2 shows the surface characteristics of the sample before and after calcination. The specific surface area (S$_{\text{BET}}$) before treatment was 27.8 m^2/g, then, it gradually increased to 138.8 m^2/g when subjected to a temperature of 600 °C. The pore volume was lesser in the case of the treated CMW (0.0172 cc/g), and an almost similar pore size was observed owing to the elimination of different contaminants and chemically bonded water during the formation of CaO nanoparticles [40].

Table 1. Chemical constituent of the oxide composition in the carbonated mud waste CMW using X-ray fluorescence spectrometers (XRF) analysis.

Sample (Lime Mud)	Main Constituents in (wt.%)
3.82	SiO$_2$
0.03	TiO$_2$
0.90	Al$_2$O$_3$
0.38	Fe$_2$O$_3$$^{\text{tot.}}$
0.61	MgO
64.73	CaO
0.22	Na$_2$O
0.20	K$_2$O
0.40	P$_2$O$_5$
0.96	SO$_3$
0.03	Cl
12.37	Limited oxygen index (LOI) at 550 °C

Table 1. *Cont.*

Sample (Lime Mud)	Main Constituents in (wt.%)
27.66	LOI at1000 °C
0.001	Rb_2O
0.019	MnO
0.005	NiO
0.012	CuO
0.004	ZnO
0.013	SrO

Table 1 shows the XRF of the CMW sample for recognizing the chemical content in the form of oxides after burning the sample. The chemical compositions of the CMW sample mainly contained the CaO phase formed after burning $CaCO_3$, and at a temperature of nearly 255 °C, the organic matter disappeared. Furthermore, the table lists specific contaminants associated with the process through which they are generated [33,39]. Most of the problems regarding the utilization of this sludge were attributed to the presence of these contaminants, with most of them present in soluble form. However, after treatment, most of these contaminants disappeared, which provided an opportunity for them to be blended with cement to form a paste.

Table 2. Surface behavior comparison before and after the treatment of carbonated mud waste (CMW).

Sample	Barrett, Joyner, and Halenda (BJH) Desorption Summary		
	S_{BET} (m^2/g)	Pore Volume (cc/g)	Pore Radius (nm)
CMW before	27.8	0.0982	1.983
CMW after	138.8	0.0172	1.714

Figure 1 shows the results of the HR-TEM study of CMW after calcination. Figure 1a shows a marked irregularity in both the size and shape, with a high degree of aggregation-like structures. On the other side, there are more regular structures with polygonal plate-shaped nanoparticles of five or more sides, with a mean diameter of approximately 180 nm (red cycle) [41,42]. Furthermore, the CaO nanopowder consists of roughly spherical, slightly agglomerated nanoparticles with a particle size of 90 nm (blue circles). Figure 1b reveals the pure crystal phase of the CaO nanopowder [43]. Figure 1c shows the selected area electron diffraction (SAED) patterns of the CMW nanocomposites, which confirms the successful formation of ultra-crystalline calcium nanoparticles. Figure 1d shows the SEM morphology of the calcined CMW sample. The surface texture is almost aggregated crystal grains with distinct and smooth edges connected directly with each other. In Figure 1e, as agreed before treatment and calcination, there are a lot of undesirable contaminants, as shown in EDX analysis. The treated CMW shows a strong peak was obtained at 3.69 keV because of the effect of the presence of Ca (Figure 1e inset), as revealed by EDX. The particulate amount of Ca in the final CMW was 60.67%, indicating that most of the $CaCO_3$ precursors had been converted to CaO nanoparticles. Therefore, we can conclude that the prepared CaO nanoparticles were the major component in the prepared sample.

Figure 2 shows the DLS technique for the particle size distribution analyzer for the OPC and calcined CMW. The particle size was in the range of 50–100 nm, with a narrow size distribution in the case of the CMW nanostructure; this feature of the curve indicates the suitability of the CMW fine substitution. In some cases, the grain size, particle shape, crystalline pattern, and surface chemistry can be controlled by using calcination technique through the modulation of the solution composition, solvent properties, reaction temperature, and aging time [44].

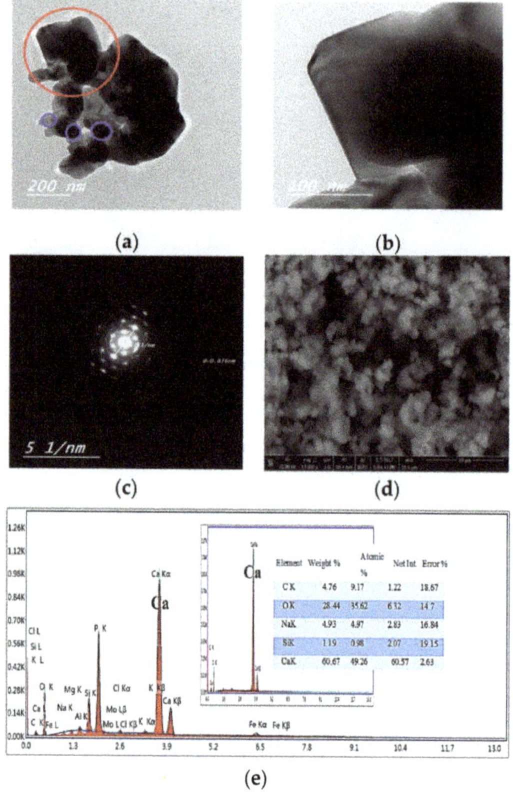

Figure 1. (**a**,**b**) TEM images of CMW with different magnification, (**c**) selected area electron diffraction (SAED) pattern for CMW, (**d**) SEM of CMW, and (**e**) EDX chart of CMW before and after calcination.

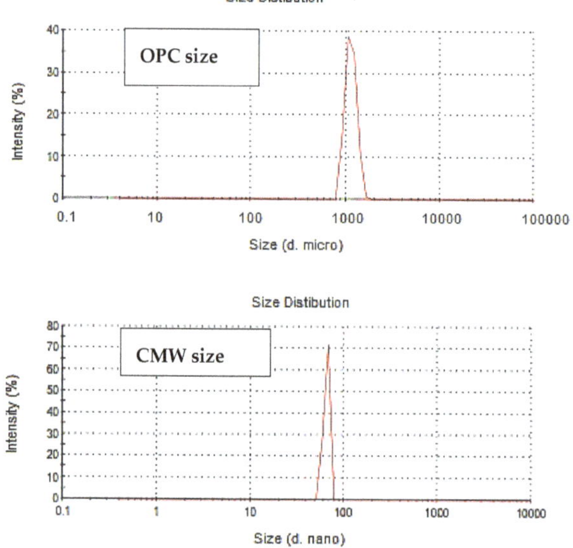

Figure 2. Particle size distribution of both cement and CMW after calcination.

3.2. Engineering Parameters

3.2.1. Compressive Strength Property

The compressive strength of OPC, including CMW, is presented in Table 3. The decrease in the hardened strength of the cement pastes containing more than 5% CMW is lower than that of the OPC at the early ages of hydration (3 days). This reinforced that the presence of the required amount of Ca(OH)$_2$ product of the hydration of OPC did not exist [45]. Also, the blended cement paste containing 15% CMW had compressive strength values lower than those of the other pastes containing 5% and 10% CMW at all of the curing ages. This result is attributed to the increase in the CMW content at the expense of OPC, which leads to the dilution of cement pastes and increased the void content, and thus resulted in decreases in compressive strength [46,47].

Table 3. Compressive strength of the ordinary Portland cement (OPC) blended cement pastes containing different ratios of calcined CMW for 3, 7, and 28 days.

Mixes	Compressive Strength (MPa)		
	3 days	7 days	28 days
OPC	52.56	69.23	71.88
OPC + 5%	34.91	61.39	69.23
OPC + 10%	31.48	53.25	67.76
OPC + 15%	30.5	47.07	51.68
OPC + 20%	30.01	42.27	47.17
OPC+ 25%	22.65	33.44	39.42

3.2.2. Bulk Density

Figure 3 shows that the bulk density nearly decreased with increasing the CMW content up to 25%, compared with the control OPC paste, which can be used as a lightweight material. In general, the results also indicate that the bulk density decreased with the increasing carbonated mud content of the OPC–CMW pastes, up to 15% (1.91 g/cm^3), in the early stage compared with the neat OPC paste (1.99 g/cm^3), owing to adding some porous calcined CMW, which had less density than the neat one [48].

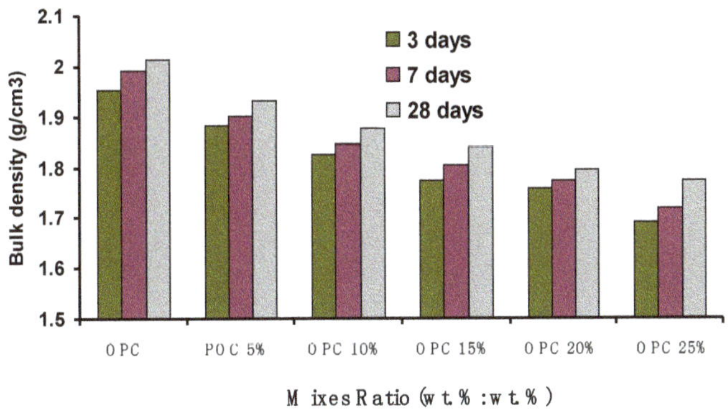

Figure 3. Bulk density of OPC and their mixes ratio with CMW.

3.2.3. Total Porosity

Figure 4 provides the total porosity of the hardened cement pastes. It has been validated that hardened cement paste is a kind of porous material [49]. The total porosity decreased with increasing the curing age to 28 days. This phenomenon is mainly because of the continuous cement hydration and accumulation of hydration products that fill the available pores within the cement matrix, resulting in pore refinement and a reduction in the total porosity. The blended cement pastes containing 25% carbonated mud exhibited higher total porosity values at all of the curing ages than those containing 5%, 10%, 15%, and 20% carbonated mud. Increasing the carbonated mud content over the OPC led to the dilution of cement pastes and an increase in the initial porosity to a relatively high value, which is controlled by the initial water/ cement ratios (W/C), hence increasing the total porosity [50].

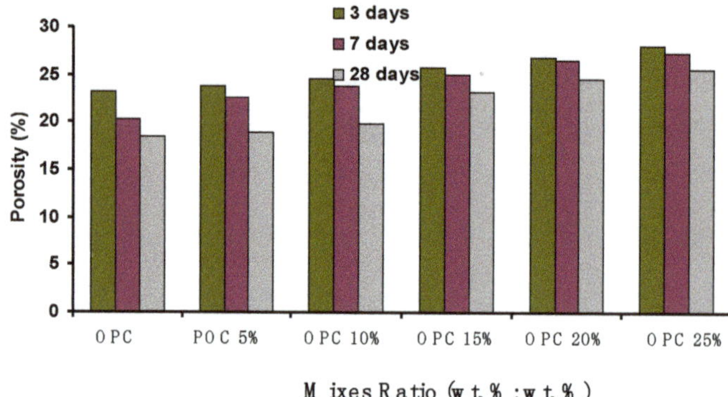

Figure 4. The total porosity of OPC and their mixes ratio with CMW.

3.2.4. Water Absorption

Water absorption was conducted to measure the amount of change in the water absorption process of CMW and the blended pastes, and the results are provided in Figure 5. The cement pastes containing carbonated mud up to 25% (20.4%) exhibited higher water absorption values at all of the curing times than the pure OPC paste. The porosity of the hardened cement paste shows that the total porosity decreased for the blended pastes, indicating that CMW-blended paste possesses a denser microstructure [51]. A high content of CMW mixed with OPC ensures the dilution of the cement paste and has a relatively high initial porosity, as controlled by the initial water/ cement ratios (W/C), hence increasing the water absorption. Furthermore, a dense microstructure is related to better anti-permeability properties, which means that blended pastes can withstand outside medium ingress, including water, so it is consistent with the water absorption test result.

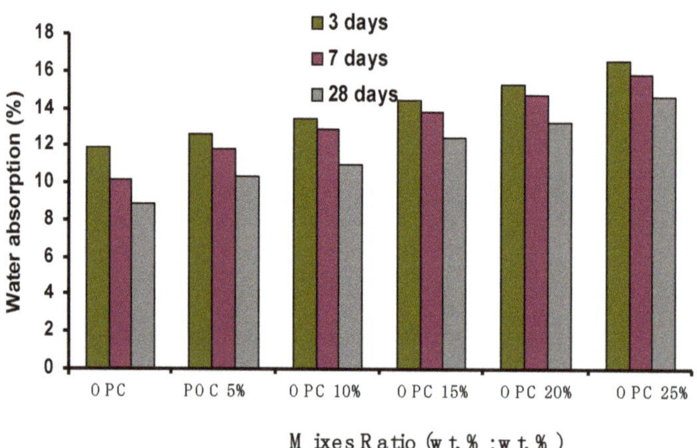

Figure 5. Water absorption of OPC and their mixes ratio with CMW.

3.2.5. Microstructure Analysis After a Fracture

Figure 6 provides the morphology of hydrated pristine OPC paste after 28 days. The SEM micrograph shows the formation of amorphous and microcrystalline phases of tobermorite-like hydrated products such as calcium silicate hydrate (CSH), along with calcium aluminate hydrates, and hexagonal calcium hydroxide; these hydrates are deposited in the originally water-filled spaces as well as in the anhydrate parts of the cement grains [52,53]. It was found that the microstructure of OPC and 5% and 10% CMW mix after 7 and 28 days of hydration formed poor crystalline particles of tobermorite-like CSH phases that were indulged, as well as the remaining anhydrate parts of cement and CMW grains. The pores that appeared in the neat OPC paste (without CMW) disappeared in this paste as a result of the filler effect of calcined CMW. The filling effect of CMW caused a notable increase in the values of compressive strength for these hardened pastes made of mixes of 5% and 10%. From the 15%, 20%, and 25% blended CM cement, partial voids could be observed, which reduced the durability.

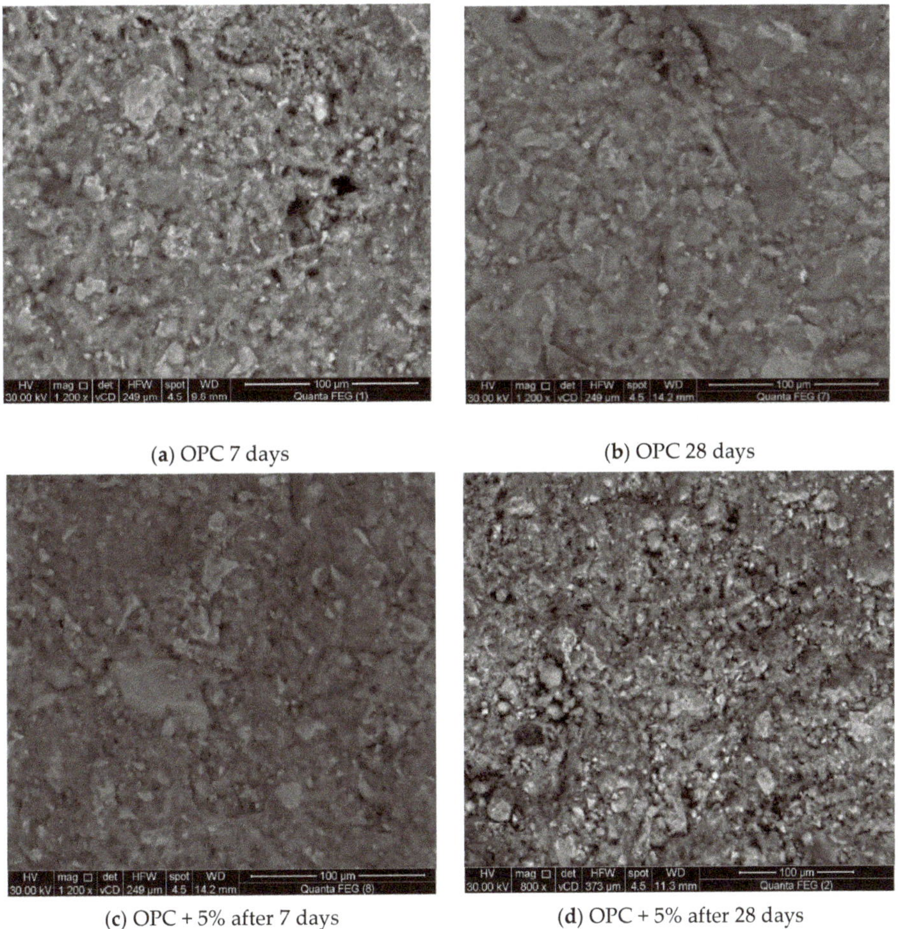

(**a**) OPC 7 days

(**b**) OPC 28 days

(**c**) OPC + 5% after 7 days

(**d**) OPC + 5% after 28 days

Figure 6. *Cont.*

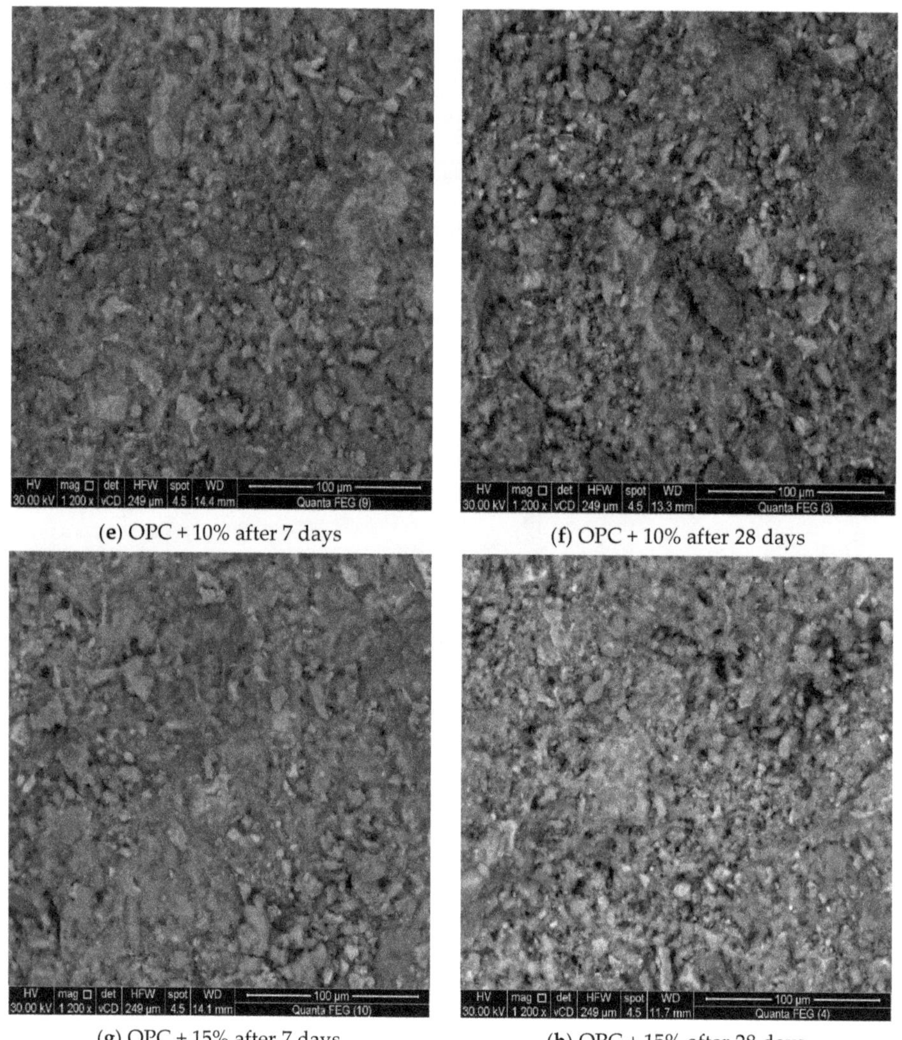

(**e**) OPC + 10% after 7 days

(**f**) OPC + 10% after 28 days

(**g**) OPC + 15% after 7 days

(**h**) OPC + 15% after 28 days

Figure 6. *Cont.*

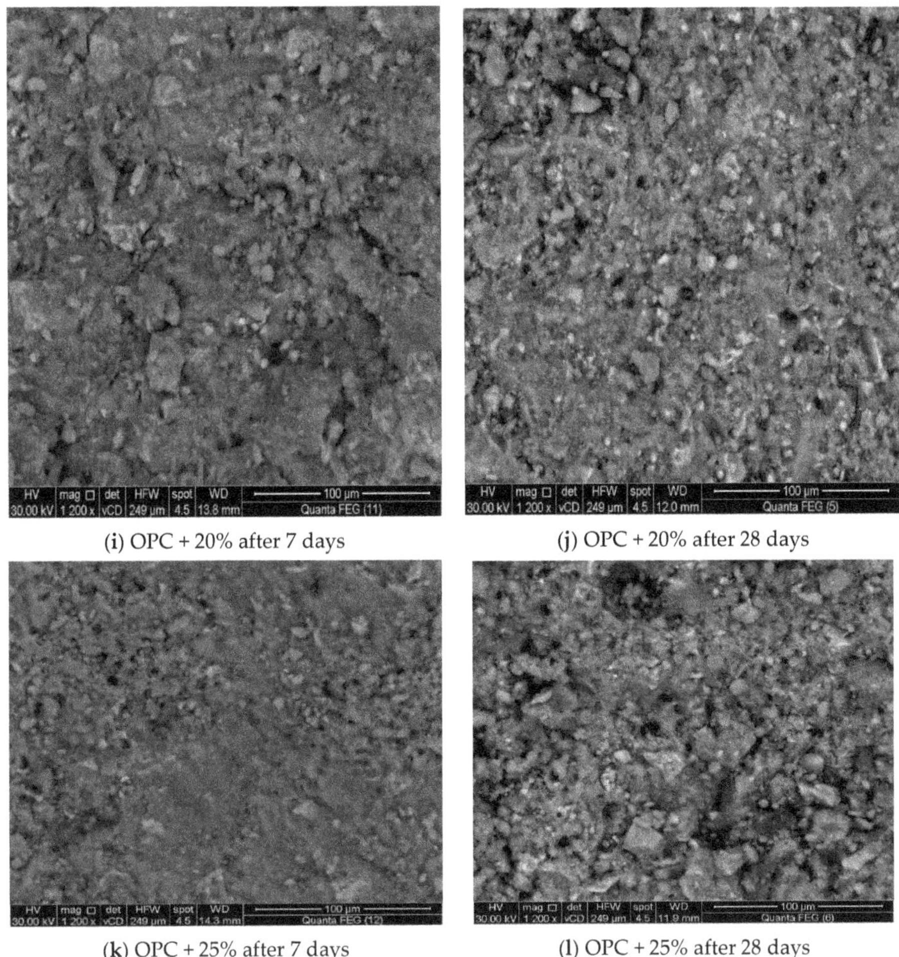

(**i**) OPC + 20% after 7 days   (**j**) OPC + 20% after 28 days

(**k**) OPC + 25% after 7 days   (**l**) OPC + 25% after 28 days

Figure 6. SEM micrographs of the OPC–CMW blended cement pastes at all curing times.

3.2.6. Thermal Stability Analysis

The thermal degradation stability of the mixed cement pastes containing 0%, 5%, 10%, 15%, and 20% CMW immersed in tap water for 7 days are displayed in Figure 7. The main first endothermic peak appeared at a temperature below 100 °C, because of the removal of the physically adsorbed water. The second endothermic peak was observed at approximately 100–180 °C due to the decomposition of CSH overlapping with calcium sulfoaluminate hydrates (ettringite and mono-sulfate hydrates). The third endothermic peak, located at approximately 475–500 °C, is characterized by the decomposition of portlandite Ca(OH)$_2$. The two ends of the overlapping endothermic peaks at 670–700 and 720–750 °C are related to the calcination of calcite (CaCO$_3$), with different degrees of crystallinity. The thermograms obtained for the hardened bare OPC paste display the areas and intensities of the peaks characteristic for CSH and Ca(OH)$_2$, which included the main hydration products. Beyond these areas, the peaks increase with increasing the hydration age. The effects of the partial replacement of the OPC by CMW cause a decrease in the peak area characteristic for the CSH, with increasing slag compared with that of the neat OPC paste. Furthermore, the peak area of Ca(OH)$_2$ observed in the thermograms decreased with the increasing CMW, due to the increase in their additional amounts. The thermograms obtained for all of the hardened cement pastes show an intensive endotherm characteristic for $CaCO_3$ and an increase with an increase in the amounts of mud.

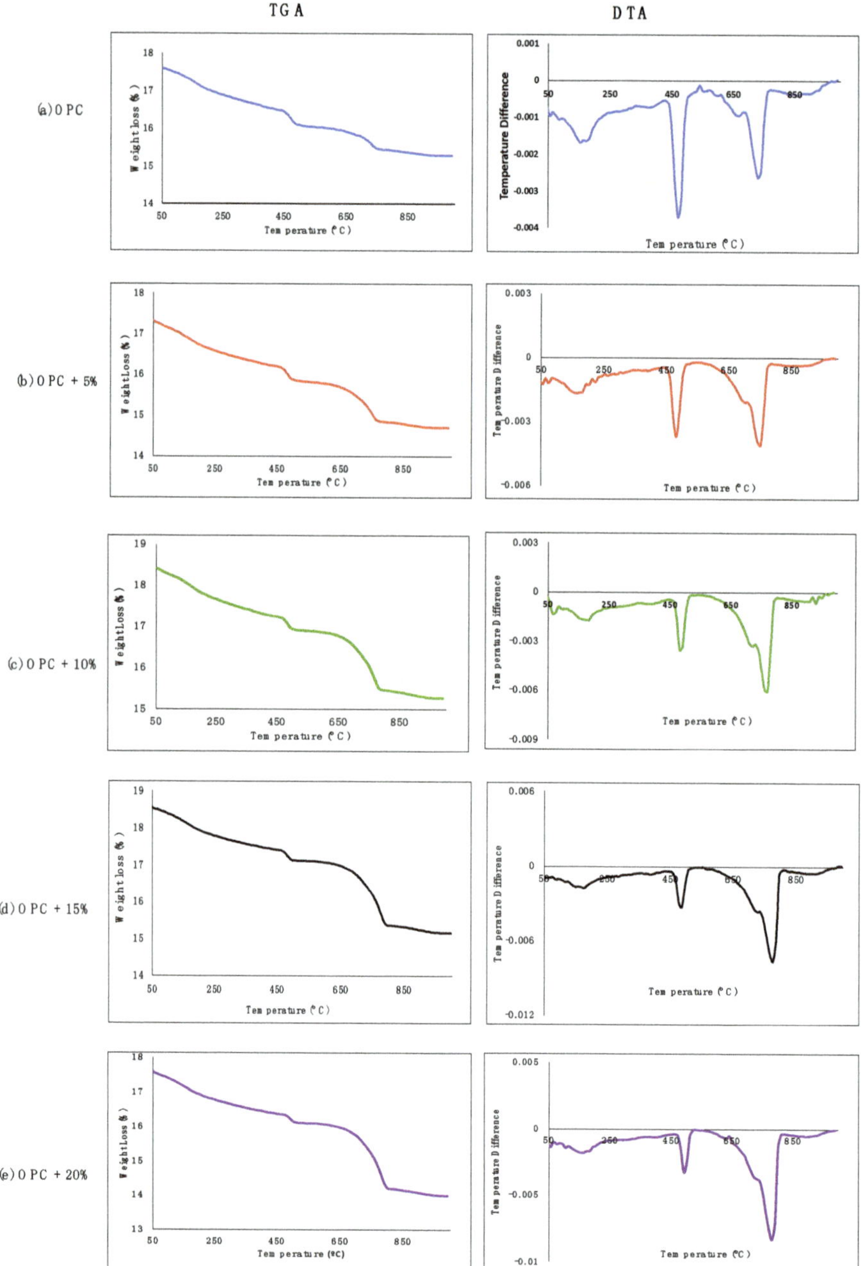

Figure 7. Thermogravimetric analysis (TGA) and differential thermal analysis (DTA) analysis of OPC–CMW blended cement pastes containing 5, 10, 15, and 20 wt. % calcined CMW nanoparticles.

3.3. Production Costs

The major goal of the present study is the argument of CMW from sugar factories, a waste originating after the carbonation process. In particular, waste materials as blended cement pastes or in the construction industry are improving the production process from economic aspects, and present an eco-friendly, viable option. Many published papers used various wastes, including palm oil mill waste [54], sewage sludge ash [55], rice husk, limestone, and the waste of activated mining coal [56]. All of the studies found that applying this disposal at a certain ratio presents a great advantage from economic, sustainable, and environmental viewpoints. Li and Yang [2] reported the utilization of Portland cement with sugar filter cake as the source of lime-based raw material, and the optimum replacement level was less than 20%. From an energy efficiency viewpoint, CMW will help reduce the need for excessive energy requirements during the stages of production, which may be beneficial for certain applications for which the energy and cost savings are vital, such as CO_2 reduction and improving the impact of climate change [57]. The incorporation of CMW not only lowered the needed energy, but also minimized the electricity consumption and saved time compared with the materials needed from natural resources. Moreover, the cost of transportation is ultimately diminished, and the economic implications of utilizing CMW instead of other costly materials, such as ZnO or TiO_2, is encouraging [58]. Mathematically, at normal operation, 60 tons of CMW are discharged per day, which requires at least six trucks. The cost of each truck is 40 SR, which is equal to \$10.66. In this manner, the total annual cost equivalent was 870.000 SR (\$232.222) in 2018, for example. Therefore, incorporating CMW into OPC enhances the cost-saving criteria and inserts green building infrastructure components to improve the engineering economy factor.

4. Conclusions

In summary, important considerations and informative data were presented to evaluate the feasibility of utilizing calcined CMW (5–25 wt. %) from a sugar refining company in the cement industry. The expected outcomes from this study provide guidelines regarding the characteristics of CMW, its engineering performance, and its sustainability. According to the overall methodology and characterization tests, the variation in the results for the different samples is not substantial, but a level of 15 wt.% is nearly acceptable and suitable for all of the parameters of hardened cement paste, as it contributes to a better waste management system, improves the sustainability in the cement industry, and prevents pathogenic effects during transportation. The small particle size of CaO with an average of ~50–100 nm is satisfactory for OPC compatibility. The compressive strength values (kg/cm^2) were studied for all of the curing ages. The engineering and environmental impacts were also affirmative concerning the amount of greenhouse gas emissions, daily energy consumption, and economic factors. Furthermore, CMW can reduce the cost of construction by up to \$230 per year.

Author Contributions: Conceptualization, Y.A.; Data curation, A.-A.M.G. and A.E.-A.S.F.; Investigation, T.A. All authors have read and agreed to the published version of the manuscript.

Funding: This research received no external funding.

Conflicts of Interest: The authors declare no conflict of interest.

References

1. Saeli, M.; Tobaldi, D.M.; Seabra, M.P.; Labrincha, J.A. Mix design and mechanical performance of geopolymeric binders and mortars using biomass fly ash and alkaline effluent from paper-pulp industry. *J. Clean. Prod.* **2019**, *208*, 1188–1197. [CrossRef]
2. Li, H.; Xu, W.; Yang, X.; Wu, J. Preparation of Portland cement with sugar filter mud as lime-based raw material. *J. Clean. Prod.* **2014**, *66*, 107–112. [CrossRef]
3. Huang, H.; Ye, G.; Damidot, D. Effect of blast furnace slag on self-healing of microcracks in cementitious materials. *Cem. Concr. Res.* **2014**, *60*, 68–82. [CrossRef]
4. Naidu, P.V.; Pandey, P.K. Replacement of cement in concrete. *Int. J. Environ. Res. Dev.* **2014**, *4*, 91–98.

5.   Balwaik, S.A.; Raut, S. Utilization of waste paper pulp by partial replacement of cement in concrete. *Int. J. Eng. Res. Appl.* **2011**, *1*, 300–309.

6.   Reddy, K.R.; Gopakumar, A.; Chetri, J.K. Critical review of applications of iron and steel slags for carbon sequestration and environmental remediation. *Rev. Environ. Sci. Bio/Technol.* **2019**, *18*, 127–152. [CrossRef]

7.   Payá, J.; Monzó, J.; Borrachero, M.V.; Díaz-Pinzón, L.; Ordonez, L.M. Sugar-cane bagasse ash (SCBA): Studies on its properties for reusing in concrete production. *J. Chem. Technol. Biotechnol. Int. Res. Process Environ. Clean Technol.* **2002**, *77*, 321–325. [CrossRef]

8.   Arif, E.; Clark, M.W.; Lake, N. Sugar cane bagasse ash from a high efficiency co-generation boiler: Applications in cement and mortar production. *Constr. Build. Mater.* **2016**, *128*, 287–297. [CrossRef]

9.   Fairbairn, E.M.; Americano, B.B.; Cordeiro, G.C.; Paula, T.P.; Filho, R.D.T.; Silvoso, M.M. Cement replacement by sugar cane bagasse ash: CO_2 emissions reduction and potential for carbon credits. *J. Environ. Manag.* **2010**, *91*, 1864–1871. [CrossRef]

10.   Atta, A.M.; Abdel-Bary, E.; Rezk, K.; Abdel-Azim, A. Fast responsive poly (acrylic acid-co-N-isopropyl acrylamide) hydrogels based on new crosslinker. *J. Appl. Polym. Sci.* **2009**, *112*, 114–122. [CrossRef]

11.   Sua-Iam, G.; Makul, N. Use of increasing amounts of bagasse ash waste to produce self-compacting concrete by adding limestone powder waste. *J. Clean. Prod.* **2013**, *57*, 308–319. [CrossRef]

12.   El-Shabasy, R.; Yosri, N.; El-Seedi, H.; Shoueir, K.; El-Kemary, M. A green synthetic approach using chili plant supported Ag/Ag2O@ P25 heterostructure with enhanced photocatalytic properties under solar irradiation. *Optik* **2019**, *192*, 162943. [CrossRef]

13.   Li, H.; Xu, J.; Wu, J.; Xu, W.; Xu, Y. Influence of sugar filter mud on formation of portland cement clinker. *J. Wuhan Univ. Technol. Mater. Sci. Ed.* **2013**, *28*, 746–750. [CrossRef]

14.   Madejón, E.; López, R.; Murillo, J.M.; Cabrera, F. Agricultural use of three (sugar-beet) vinasse composts: Effect on crops and chemical properties of a Cambisol soil in the Guadalquivir river valley (SW Spain). *Agric. Ecosyst. Environ.* **2001**, *84*, 55–65. [CrossRef]

15.   Anacleto, L.R.; Roberto, M.M.; Marin-Morales, M.A. Toxicological effects of the waste of the sugarcane industry, used as agricultural fertilizer, on the test system Allium cepa. *Chemosphere* **2017**, *173*, 31–42. [CrossRef] [PubMed]

16.   Burman, N.W.; Harding, K.G.; Sheridan, C.M.; van Dyk, L. Evaluation of a combined lignocellulosic/waste water bio-refinery for the simultaneous production of valuable biochemical products and the remediation of acid mine drainage. *Biofuels Bioprod. Biorefin.* **2018**, *12*, 649–664. [CrossRef]

17.   Kapur, M.; Kanwar, R. Influence of cane filter cakes and cattle manure on micronutrients content in sugar-beet and their availability in alkaline sandy loam soil. *Biol. Wastes* **1989**, *29*, 233–238. [CrossRef]

18.   Xu, Q.; Ji, T.; Gao, S.-J.; Yang, Z.; Wu, N. Characteristics and Applications of Sugar Cane Bagasse Ash Waste in Cementitious Materials. *Materials* **2019**, *12*, 39. [CrossRef]

19.   Paris, J.M.; Roessler, J.G.; Ferraro, C.C.; DeFord, H.D.; Townsend, T.G. A review of waste products utilized as supplements to Portland cement in concrete. *J. Clean. Prod.* **2016**, *121*, 1–18. [CrossRef]

20.   Moretti, J.P.; Sales, A.; Almeida, F.C.; Rezende, M.A.; Gromboni, P.P. Joint use of construction waste (CW) and sugarcane bagasse ash sand (SBAS) in concrete. *Constr. Build. Mater.* **2016**, *113*, 317–323. [CrossRef]

21.   Gopinath, A.; Bahurudeen, A.; Appari, S.; Nanthagopalan, P. A circular framework for the valorisation of sugar industry wastes: Review on the industrial symbiosis between sugar, construction and energy industries. *J. Clean. Prod.* **2018**, *203*, 89–108. [CrossRef]

22.   Singh, S.; Ransinchung, G.; Debbarma, S.; Kumar, P. Utilization of reclaimed asphalt pavement aggregates containing waste from Sugarcane Mill for production of concrete mixes. *J. Clean. Prod.* **2018**, *174*, 42–52. [CrossRef]

23.   Singh, S.; Ransinchung, G.; Kumar, P. An economical processing technique to improve RAP inclusive concrete properties. *Constr. Build. Mater.* **2017**, *148*, 734–747. [CrossRef]

24.   Bahurudeen, A.; Kanraj, D.; Dev, V.G.; Santhanam, M. Performance evaluation of sugarcane bagasse ash blended cement in concrete. *Cem. Concr. Compos.* **2015**, *59*, 77–88. [CrossRef]

25.   Gar, P.S.; Suresh, N.; Bindiganavile, V. Sugar cane bagasse ash as a pozzolanic admixture in concrete for resistance to sustained elevated temperatures. *Constr. Build. Mater.* **2017**, *153*, 929–936.

26.   Modani, P.O.; Vyawahare, M. Utilization of bagasse ash as a partial replacement of fine aggregate in concrete. *Procedia Eng.* **2013**, *51*, 25–29. [CrossRef]

27. Loh, Y.; Sujan, D.; Rahman, M.E.; Das, C.A. Sugarcane bagasse—The future composite material: A literature review. *Resour. Conserv. Recycl.* **2013**, *75*, 14–22. [CrossRef]

28. Shaban, N.Z.; Yehia, S.A.; Shoueir, K.R.; Saleh, S.R.; Awad, D.; Shaban, S.Y. Design, DNA binding and kinetic studies, antibacterial and cytotoxic activities of stable dithiophenolato titanium (IV)-chitosan Nanocomposite. *J. Mol. Liq.* **2019**, *287*, 111002. [CrossRef]

29. Shoueir, K.; Ahmed, M.; Gaber, S.A.A.; El-Kemary, M. Thallium and selenite doped carbonated hydroxyapatite: Microstructural features and anticancer activity assessment against human lung carcinoma. *Ceram. Int.* **2020**, *46*, 5201–5212. [CrossRef]

30. Abdelbar, M.F.; El-Sheshtawy, H.S.; Shoueir, K.R.; El-Mehasseb, I.; Ebeid, E.-Z.M.; El-Kemary, M. Halogen bond triggered aggregation induced emission in an iodinated cyanine dye for ultra sensitive detection of Ag nanoparticles in tap water and agricultural wastewater. *RSC Adv.* **2018**, *8*, 24617–24626. [CrossRef]

31. Salama, A.; Diab, M.A.; Abou-Zeid, R.E.; Aljohani, H.A.; Shoueir, K.R. Crosslinked alginate/silica/zinc oxide nanocomposite: A sustainable material with antibacterial properties. *Compos. Commun.* **2018**, *7*, 7–11. [CrossRef]

32. Oualha, M.A.; Omri, N.; Oualha, R.; Nouioui, M.A.; Abderrabba, M.; Amdouni, N.; Laoutid, F. Development of metal hydroxide nanoparticles from eggshell waste and seawater and their application as flame retardants for ethylene-vinyl acetate copolymer (EVA). *Int. J. Biol. Macromol.* **2019**, *128*, 994–1001. [CrossRef] [PubMed]

33. El-Kemary, M.A.; El-mehasseb, I.M.; Shoueir, K.R.; El-Shafey, S.E.; El-Shafey, O.I.; Aljohani, H.A.; Fouad, R.R. Sol-gel TiO_2 decorated on eggshell nanocrystal as engineered adsorbents for removal of acid dye. *J. Dispers. Sci. Technol.* **2018**, *39*, 911–921. [CrossRef]

34. Naveenkumar, R.; Baskar, G. Biodiesel production from Calophyllum inophyllum oil using Zinc doped Calcium oxide (Plaster of Paris) nanocatalyst. *Bioresour. Technol.* **2019**, *280*, 493–496. [CrossRef] [PubMed]

35. Benzennou, S.; Laviolette, J.P.; Chaouki, J. Kinetic study of microwave pyrolysis of paper cups and comparison with calcium oxide catalyzed reaction. *AIChE J.* **2018**, *65*, 684–690. [CrossRef]

36. Jiménez, P.E.S.; Perejón, A.; Guerrero, M.B.; Valverde, J.M.; Ortiz, C.; Maqueda, L.A.P. High-performance and low-cost macroporous calcium oxide based materials for thermochemical energy storage in concentrated solar power plants. *Appl. Energy* **2019**, *235*, 543–552. [CrossRef]

37. Rieger, J.; Kellermeier, M.; Nicoleau, L. Formation of nanoparticles and nanostructures—An industrial perspective on $CaCO_3$, cement, and polymers. *Angew. Chem. Int. Ed.* **2014**, *53*, 12380–12396. [CrossRef]

38. Atta, A.M.; El-Mahdy, G.A.; Al-Lohedan, H.A.; Shoueir, K.R. Electrochemical behavior of smart N-isopropyl acrylamide copolymer nanogel on steel for corrosion protection in acidic solution. *Int. J. Electrochem. Sci.* **2015**, *10*, 870–882.

39. Gunning, P.J.; Hills, C.D.; Carey, P.J. Accelerated carbonation treatment of industrial wastes. *Waste Manag.* **2010**, *30*, 1081–1090. [CrossRef]

40. Kočí, K.; Matějka, V.; Kovář, P.; Lacný, Z.; Obalová, L. Comparison of the pure TiO_2 and kaolinite/TiO_2 composite as catalyst for CO_2 photocatalytic reduction. *Catal. Today* **2011**, *161*, 105–109. [CrossRef]

41. Shoueir, K.; El-Sheshtawy, H.; Misbah, M.; El-Hosainy, H.; El-Mehasseb, I.; El-Kemary, M. Fenton-like nanocatalyst for photodegradation of methylene blue under visible light activated by hybrid green DNSA@ Chitosan@ $MnFe_2O_4$. *Carbohydr. Polym.* **2018**, *197*, 17–28. [CrossRef] [PubMed]

42. Shoueir, K.; Kandil, S.; El-hosainy, H.; El-Kemary, M. Tailoring the surface reactivity of plasmonic Au@ TiO_2 photocatalyst bio-based chitosan fiber towards cleaner of harmful water pollutants under visible-light irradiation. *J. Clean. Prod.* **2019**, *230*, 383–393. [CrossRef]

43. Abdel-Gawwad, H.; Heikal, M.; Mohammed, M.S.; El-Aleem, S.A.; Hassan, H.S.; García, S.V.; Rashad, A.M. Evaluating the impact of nano-magnesium calcite waste on the performance of cement mortar in normal and sulfate-rich media. *Constr. Build. Mater.* **2019**, *203*, 392–400. [CrossRef]

44. Habte, L.; Shiferaw, N.; Mulatu, D.; Thenepalli, T.; Chilakala, R.; Ahn, J.W. Synthesis of Nano-Calcium Oxide from Waste Eggshell by Sol-Gel Method. *Sustainability* **2019**, *11*, 3196. [CrossRef]

45. Rashad, A.M. An exploratory study on high-volume fly ash concrete incorporating silica fume subjected to thermal loads. *J. Clean. Prod.* **2015**, *87*, 735–744. [CrossRef]

46. Nguyen, H.-A.; Chang, T.-P.; Shih, J.-Y.; Chen, C.-T. Influence of low calcium fly ash on compressive strength and hydration product of low energy super sulfated cement paste. *Cem. Concr. Compos.* **2019**, *99*, 40–48. [CrossRef]

47. Fouad, R.R.; Aljohani, H.A.; Shoueir, K.R. Biocompatible poly (vinyl alcohol) nanoparticle-based binary blends for oil spill control. *Mar. Pollut. Bull.* **2016**, *112*, 46–52. [CrossRef]

48. Kroehong, W.; Jaturapitakkul, C.; Pothisiri, T.; Chindaprasirt, P. Effect of Oil Palm Fiber Content on the Physical and Mechanical Properties and Microstructure of High-Calcium Fly Ash Geopolymer Paste. *Arab. J. Sci. Eng.* **2018**, *43*, 5215–5224. [CrossRef]

49. Seleem, H.E.D.H.; Rashad, A.M.; Elsokary, T. Effect of elevated temperature on physico-mechanical properties of blended cement concrete. *Constr. Build. Mater.* **2011**, *25*, 1009–1017. [CrossRef]

50. Seleem, H.E.-D.H.; Rashad, A.M.; Elsokary, T. The Impact of Pozzolana in Securing Fire-resistive Properties of Concrete. *Silic. Ind.* **2009**, *9*, 245–254.

51. Wang, W.; Liu, X.; Guo, L.; Duan, P. Evaluation of Properties and Microstructure of Cement Paste Blended with Metakaolin Subjected to High Temperatures. *Materials* **2019**, *12*, 941. [CrossRef] [PubMed]

52. Amin, M.; El-Gamal, S.; Abo-El-Enein, S.; El-Hosiny, F.; Ramadan, M. Physico-chemical characteristics of blended cement pastes containing electric arc furnace slag with and without silica fume. *HBRC J.* **2015**, *11*, 321–327. [CrossRef]

53. Polat, R.; Demirboğa, R.; Karagöl, F. Mechanical and physical behavior of cement paste and mortar incorporating nano-CaO. *Struct. Concr.* **2019**, *20*, 361–370. [CrossRef]

54. Kanadasan, J.; Fauzi, A.; Razak, H.; Selliah, P.; Subramaniam, V.; Yusoff, S. Feasibility studies of palm oil mill waste aggregates for the construction industry. *Materials* **2015**, *8*, 6508–6530. [CrossRef] [PubMed]

55. Baeza-Brotons, F.; Garcés, P.; Payá, J.; Saval, J.M. Portland cement systems with addition of sewage sludge ash. Application in concretes for the manufacture of blocks. *J. Clean. Prod.* **2014**, *82*, 112–124. [CrossRef]

56. Mejia-Ballesteros, J.E.; Savastano, H., Jr.; Fiorelli, J.; Rojas, M.F. Effect of mineral additions on the microstructure and properties of blended cement matrices for fibre-cement applications. *Cem. Concr. Compos.* **2019**, *98*, 49–60. [CrossRef]

57. Gartner, E.; Hirao, H. A review of alternative approaches to the reduction of CO_2 emissions associated with the manufacture of the binder phase in concrete. *Cem. Concr. Res.* **2015**, *78*, 126–142. [CrossRef]

58. Senff, L.; Tobaldi, D.; Lemes-Rachadel, P.; Labrincha, J.; Hotza, D. The influence of TiO_2 and ZnO powder mixtures on photocatalytic activity and rheological behavior of cement pastes. *Constr. Build. Mater.* **2014**, *65*, 191–200. [CrossRef]

Article

Techno-Economic Analysis of CO_2 Capture Technologies in Offshore Natural Gas Field: Implications to Carbon Capture and Storage in Malaysia

Norhasyima Rahmad Sukor [1,2,*], Abd Halim Shamsuddin [1], Teuku Meurah Indra Mahlia [1,3] and Md Faudzi Mat Isa [2]

[1]   Department of Mechanical Engineering, Universiti Tenaga Nasional, Kajang 43000, Selangor, Malaysia; abdhalim@uniten.edu.my (A.H.S.); tmindra.mahlia@uts.edu.au (T.M.I.M.)

[2]   PETRONAS Research Sdn Bhd, Kawasan Institusi Bangi, Kajang 43000, Selangor, Malaysia; faudzi@petronas.com.my

[3]   School of Information, Systems and Modelling, Faculty of Engineering and IT, University of Technology Sydney, NSW 2007, Australia

*   Correspondence: pe20454@utn.edu.my; Tel.: +60-12-263-8532

Received: 18 November 2019; Accepted: 15 December 2019; Published: 19 March 2020

Abstract: Growing concern on global warming directly related to CO_2 emissions is steering the implementation of carbon capture and storage (CCS). With Malaysia having an estimated 37 Tscfd (Trillion standard cubic feet) of natural gas remains undeveloped in CO_2 containing natural gas fields, there is a need to assess the viability of CCS implementation. This study performs a techno-economic analysis for CCS at an offshore natural gas field in Malaysia. The framework includes a gas field model, revenue model, and cost model. A techno-economic spreadsheet consisting of Net Present Value (NPV), Payback Period (PBP), and Internal Rate of Return (IRR) is developed over the gas field's production life of 15 years for four distinctive CO_2 capture technologies, which are membrane, chemical absorption, physical absorption, and cryogenics. Results predict that physical absorption solvent (Selexol) as CO_2 capture technology is most feasible with IRR of 15% and PBP of 7.94 years. The output from the techno-economic model and associated risks of the CCS project are quantified by employing sensitivity analysis (SA), which indicated that the project NPV is exceptionally sensitive to gas price. On this basis, the economic performance of the project is reliant on revenues from gas sales, which is dictated by gas market price uncertainties.

Keywords: CO_2 capture; carbon capture and storage (CCS); offshore gas field; techno-economic analysis

1. Introduction

Anthropogenic carbon dioxide (CO_2) emissions in the atmosphere are significantly increasing and critically impacting global climate. Thus, there is an urgent need to gauge and combat its effect on global climate change. In general, the source of CO_2 comes mostly from combustion of fossil fuel, natural gas stream, and industrial processes [1–5]. As stated in BP energy statistics in 2018, there were 33,890.8 MT (metric tonne) of CO_2 emission worldwide and the value was driven by the high energy demand. Moreover, recently in the year 2018, the global energy consumption grew at a rate of 2.9%, which is almost twice its 10-year average of 1.5% per year that is the fastest trend since the year 2010. For natural gas, its production increased by 5.2% to 136,594 Tscf (Trillion standard cubic feet) while its consumption rose by 5.3% to 135.92 Tscf, which is one of the fastest rates of increase since the year 1984. This high demand is mainly driven by the booming Liquefied Natural Gas (LNG) industry in

China, Australia, the United States of America (USA), and Russia, accounting for the vast majority of the gains in both consumption and production [6].

Malaysia is an oil and natural gas producer and is strategically situated within the significant seaborne energy trade route and one of the largest energy consumers in Southeast Asia. Natural gas is indeed a very important commodity in Malaysia, as its effect extends far beyond the national power sector to further downstream products, the national economy, and international relationships. According to the BP energy statistics, natural gas production in Malaysia is steady at 2.56 Tscf per year, while the consumption is 1.46 Tscf in 2018 [6]. As LNG exporter, Malaysia is placed second after Qatar and has one of the world's largest LNG production facilities; Malaysia Liquefied Natural Gas (MLNG) complex that is aimed to supply the demand of their primary customer from Japan, Korea, and Taiwan [7]. There is an estimated 37 Tscf of natural gas that remains undeveloped in Malaysia's gas fields, whereby its CO_2 compositions exceed 10% volumetric of the produced acid gas [8]. Most of these gas fields were not economically viable in the past due to the presence of large capacities of CO_2 and are always associated with potentially high corrosion risks to the topside facilities and pipelines. The development of offshore high CO_2 gas fields requires prudent strategies of CO_2 separation technology in order to optimize both the capital and the operating expenses for CCS [9].

Nonetheless, Malaysia is committed to mitigating CO_2 emissions and has assigned its Ministry of Energy, Science, Technology, Environment, and Climate Change (MESTECC), with the responsibility of developing the national emissions reduction plan. The commitment was to voluntary reduce the economy's carbon intensity by 40% by 2020 during the Copenhagen 2009 Climate Summit and has repledged to attain 45% CO_2 emission by 2030 [10–12]. The repledge was motivated by promising results with a reduction of 33% of CO_2 achieved between 2005 and 2015 [11,13]. The Government of Malaysia is gauging various mitigation plans and energy efficiency alternatives, including joint CCS feasibility study with International Energy Agency (IEA) as well as future implementation of the Energy Efficiency and Conservation policy to be tabled estimated by the end of 2019 or early 2020 [14,15]. Since its inception in 1974, Petroliam Nasional Berhad (PETRONAS), solely owned by the Government of Malaysia, is the national oil and gas company and has been entrusted with the responsibility of developing the nation's oil and gas resources. To address the CO_2 emission, PETRONAS has developed its Carbon Commitment pledge to support the initiatives by promoting natural gas as low-carbon fuel and application in the power and transportation industry. This is supported by their success in building the world's first pioneer floating Liquefied Natural Gas (LNG) facility. In their enhanced Carbon Commitment 2017, all future PETRONAS's projects for CO_2 containing gas field development in upstream shall incorporate CCS technologies during conceptual design stage [16].

Carbon capture and storage (CCS) is a methodology to separate CO_2, then to store the CO_2, commonly originating from power generation, industrial processes, and from CO_2 gas fields. CCS is an offset for continued fossil fuel exploration, while at the same time achieving the targeted reduction of carbon emissions. Hence, CCS has high prospects to be one of the solutions for CO_2 emission mitigation technology in Malaysia, with added advantage especially if a particular technology is suitable for deployment, has competitive cost, and has nearby storage capacity availability.

The Intergovernmental Panel on Climate Change (IPCC) report [17] has indicated that in the absence of CCS implementation, the required total cost to mitigate global climate change may escalate up to 138%. Furthermore, it is a real challenge to reach the targeted limit of temperature increase to 2 °C scenario and an even greater one to achieve to 1.5 °C based on the Paris Agreement enacted in 2015 [18–20]. To achieve this ambitious goals, various international bodies such as the Global Carbon Capture Storage Institute (GCCSI), Carbon Capture & Storage Association (CCSA), Cooperative Research Centre for Greenhouse Gas Technologies (CO2CRC), and many other prominent research centers has been established to innovate CCS technologies solutions for economically viable path for CO_2 emissions reduction [21]. CCS has been recently known to play a vital role in global climate change mitigation, however, there are certain challenges that need to be addressed before its implementation. One of the main challenges recognized is the fairly high cost of the CCS technology for integration with the existing facilities as well as for widespread replicated implementation [17,22]. Currently, the high-cost driver is highly dependent on technology maturity as well as the availability of the storage sites [23]. In terms of the CCS process, the capture stage is certainly the highest cost in the CCS chain [24]. The capture stage is generally estimated to represent 70–80% of the overall total cost for CCS chain [25–27]. Reducing the CO_2 capture cost is the utmost essential phase in the CCS chain in order to be economically viable in the energy industry [27]. In this study, it focuses on the techno-economic of CO_2 capture technologies suitable for natural gas field applications.

2. Materials and Method

This techno-economic analysis of CCS includes a gas field model, a revenue model, and a cost model as shown in Figure 1. This analysis framework is adopted from previous work by GCCSI [28] and CO2CRC [29]. These previous studies are largely recognized and the method is used regularly for estimating CCS costs. In this study, calculation methods were adapted from the existing studies and tabulated to describe the parameters of natural gas field economic performance for different cases of CO_2 capture technology. In the instance where no specific data on the subject is available, assumptions were made using comparative data. This study also focuses on the capture step as it contributes to the highest cost in CCS chain.

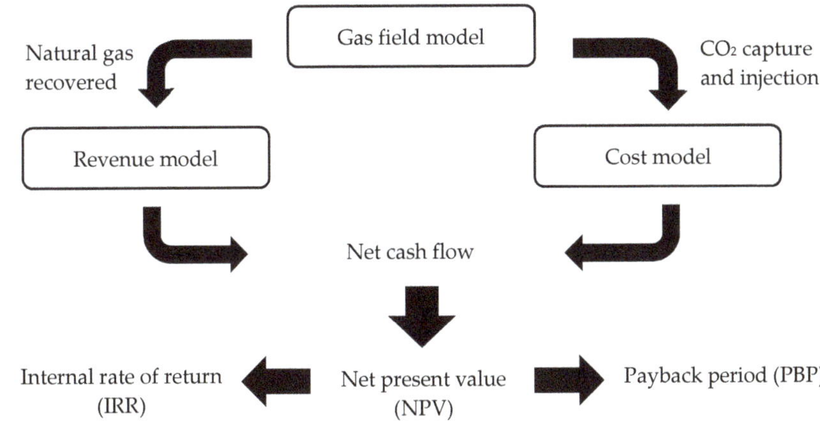

Figure 1. Schematic of the techno-economic analysis framework.

The CCS performance results for the natural gas field models are incorporated into two types of models, which are the revenue model and cost model. The revenue model comes from the value of the recovered natural gas, while the cost model is the cost of CCS, including CO_2 capture, transportation, and injection for storage. The output of both models is net cash flow that is then discounted to obtain fiscal precursors comprising of Net Present Value (NPV), Internal Rate of Return (IRR), and Pay Back Period (PBP) of the project as shown in Figure 1.

Since Malaysia has not yet regulated CCS implementation in its national regulatory setting, the cost model does not take into consideration the value of incentives and credits that could be obtained for CO_2 storage, which would be a potentially added economic benefit to offset cost framework [30,31]. In this study, the economic analysis evaluates a CCS project's impact by evaluating its costs and benefits to the overall economy. This techno-economic analysis compares four scenarios where the project is implemented using various CO_2 capture technology options, assessing the effects, including quantifying parameters in quantitative terms.

2.1. Gas Field Model

This paper investigates the cost of carbon capture from the natural gas field through a case study. In this paper, Tangga Barat Cluster situated in the PM (Peninsular Malaysia) 313 Block, as the referred case study is a typical natural gas field at water depths in the range of 60 m to 71 m, about 150 km north-east of Kertih, Terengganu offshore Peninsular Malaysia. Tangga Barat Cluster is presently operated by PETRONAS Carigali Sdn Bhd (PCSB), the upstream arm for PETRONAS that has 100% working interest. The content level of CO_2 at Tangga Barat is above the gas specifications required for sales gas. Tangga Barat Cluster consists of a total of 3 producing platforms, which are Laho (LHDP-A), Melor (MLDP-A), and Tangga (TGDP-A), and a central processing platform, which is Tangga Barat (TBCP-A), while a separate riser and wellhead platform TBDR-A bridge-linked to TBCP-A will accommodate the development wells for the Tangga Barat field as shown in Figure 2. Existing gas facilities nearest to the Tangga Barat Cluster fields are the gas production and pipeline facilities of the Resak field, which is located 52 km away. The main power generation, processing facilities, natural gas compression, utilities, and living quarters is located at the central processing platform (TBCP-A). The Tangga Barat field was chosen as the gathering point for gas from the Tangga, Melor, and Laho fields based on the fact that the Tangga Barat gas volumes remain the largest along the four fields and distance is the shortest to the Resak complex. It is designed for a peak capacity of 440 MMscf/d of raw gas with an initial blended CO_2 level of 37 mole% (Tangga Barat CO_2 content) prior to CO_2 removal. The raw gas is then processed, pretreated, reducing the CO_2 content to 8 mole% to meet the specifications of the downstream sales gas process requirement from the PETRONAS gas processing plant at Kertih Terengganu [32].

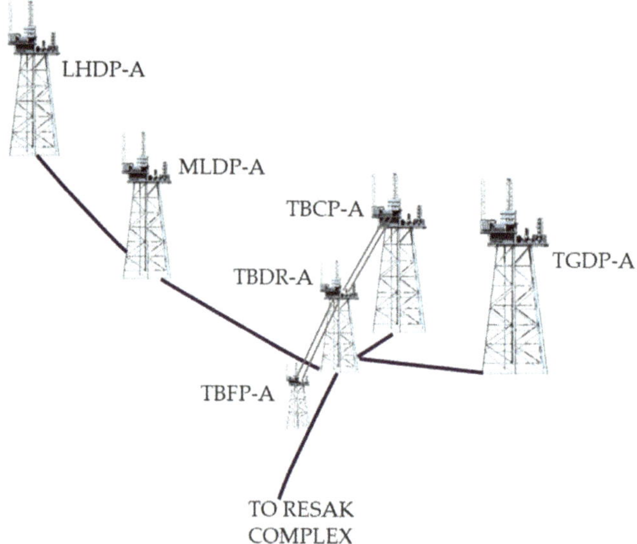

Figure 2. Tangga Barat Development Project.

In this case study, an annual average sales gas of 220 MMscf/d with 15 years of production life is estimated [33]. CO_2 storage identified for this study is situated at an underground geological storage site approximately 20 km away from the Tangga Barat fields (near the Laho and Tangga fields) whereby the injected CO_2 is to be stored while ensuring there is no increase of the reservoir pressure above the fracture pressure of the storage formation [34,35]. The facilities required will be a dedicated CO_2 compression platform and its associated pipelines and injection wells at the Tangga wellhead platform [34,35].

2.1.1. CO_2 Capture Technologies

There are several routes available for CO_2 capture technologies from the gas stream. The technologies are generally based on different physical and chemical processes comprising of absorption, cryogenic distillation, and membrane. The selection of technology route mainly relies on the type of plant, gas type, and CO_2 separation level. In this study, to meet the Tangga Barat gas field sales gas specifications, suitable CO_2 capture technologies that have been widely been used for CO_2 separation from natural gas were evaluated. Four technologies have been selected for this techno-economic analysis study, which are:

1. Polymeric membrane
2. Chemical absorption amine (MEA)
3. Physical absorption solvent (Selexol) (UOP LLC, IL, USA)
4. Cryogenic distillation

Membrane is a type of semipermeable barrier that has the ability to separate components using various separation mechanisms such as sorption/diffusion, adsorption/diffusion, ion-conducting, and molecular sieving, with selection of organic material (polymeric) and inorganic material (metallic, zeolite, ceramic). Figure 3a shows the basic method for membrane separation for CO_2 capture. Among all the available mechanisms, polymeric membrane with solution-diffusion has been commercially applied in the upstream business for natural gas separation at offshore CO_2 containing gas fields. Natural gas separation utilizing polymeric hollow fiber cellulose triacetate membrane has been installed at an offshore platform in Thailand [36]. In this solution-diffusion process, it starts with gas molecules from the feed side being absorbed by the membrane, which is then diffused across the membrane matrix and finally desorbed through to the other permeate side. The membrane selectivity is dependent on the polymer molecular structure that allows preferential passing of selected gas based on their molecule sizes, while the membrane permeability is highly dependent on the gas solubility [37–39].

Chemical absorption technology for CO_2 capture has been commonly applied in the petroleum and natural gas industry. CO_2 is an acidic gas; therefore, chemical absorption of CO_2 is a method that utilizes necessary solvents for acid-based neutralization reactions in gaseous streams/flue gas. In this process, CO_2 reacts with chemical solvents to produce an intermediate compound that is weakly bonded, which is further broken down using heating, then regenerating back into the original solvent to produce pure CO_2 stream. For more than 60 years, chemical absorption process using amine chemical solution, such as monoethanolamine (MEA), has been commercially applied in the natural gas industry and is considered as one of the most well-known chemical absorption technologies to absorb CO_2 from natural gas [27]. In this process, in order to capture CO_2, in a packed absorption column, the flue gas is bubbled through the solvent during which preferentially separates the CO_2 from the flue gas stream [40,41]. Next, the solvent flows through a regenerator unit, whereby the absorbed CO_2 is separated from the solvent by counterflowing the steam at temperature between 100–200 °C. This results in condensation of water vapor that leaves a high concentration of more than 99% of CO_2 stream, which would then be compressed and used for storage or commercial. Finally, the lean solvent is cooled down to the temperature of 40–65 °C and is recycled back into the absorption column. In fundamental, the chemical reaction is shown as follows:

$$C_2H_4OHNH_2 \ (MEA) + H_2O + CO_2 \leftrightarrow C_2H_4OHNH_3^+ + HCO_3^- \tag{1}$$

During the absorption process using MEA, the chemical reaction starts from left to right while during regeneration, the reaction is from the opposite direction, right to left [40,42].

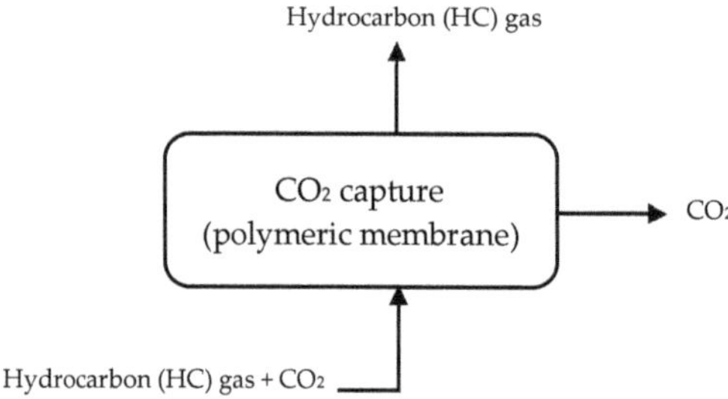

3a. Separation with membrane.

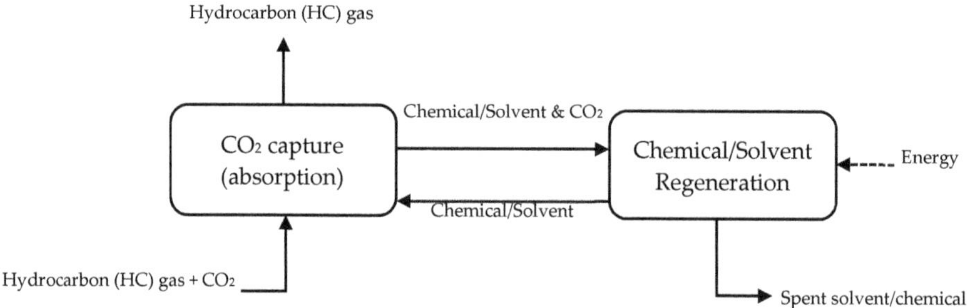

3b. Separation with chemical/physical absorption.

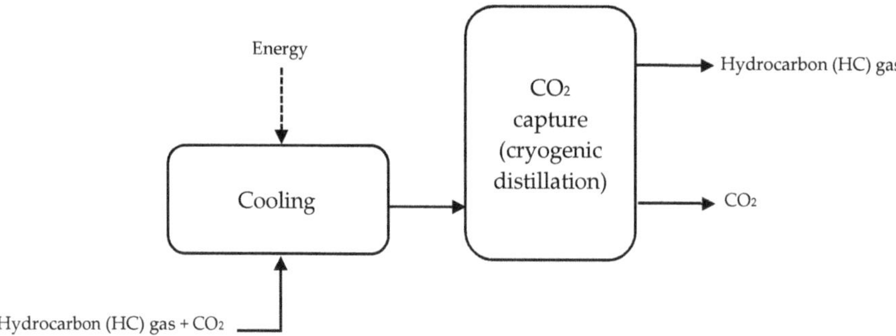

3c. Separation with cryogenic distillation.

Figure 3. (a–c) Basic methods for CO_2 capture from natural gas.

Physical absorption is also a commercially available CO_2 capture method in the petroleum and natural gas industry. In this physical solvent process, it employs organic solvents to absorb gas components physically rather than reacting chemically. The solubility of CO_2 of solvents determines the separation rate of CO_2 by physical absorption processes, in which the solubility hinge on the temperature and the partial pressure and of the feed gas. Low temperature and high CO_2 partial pressure will increase the CO_2 in the solvent solubility. Selexol (dimethylether of polyethylene glycol) is a liquid glycol-based solvent and has since been used to process natural gas especially for bulk CO_2 removal. In the Selexol process, the flue gas is first dehydrated. After that, in an absorption column at a pressure about 450 psi with a low temperature of 0–5 °C, the flue gas is contacted with the solvent to produce a loaded CO_2 solvent. Then, flash desorption or separation of the CO_2 loaded solvent will regenerate the original solvent. Finally, the produced CO_2 gas stream is compressed and stored while the regenerated solvent is recycled back to the column [25,40,43]. Figure 3b shows the basic method for CO_2 capture through absorption.

In the cryogenic distillation CO_2 separation method, CO_2 is purified by separating the gas mixtures using distillation at low temperature and fractional condensation, leveraging on their de-sublimation properties and different condensation level as shown in Figure 3c. The low-temperature distillation is a mature process used to liquefy and purify CO_2 from relatively high purity (>90%) sources. The process starts with the cooling of the gas to a very low temperature of $<-73.3\ °C$, to freeze out/liquefied and separate the CO_2. During which, cryogenic air separation unit (ASU) was used to supply high purity oxygen to a boiler. This step condenses most of the moisture and removes any carried over particles. The produced high purity oxygen is then blended with the recycled flue gas before combusted in the boiler to maintain combustion conditions comparable to air fired configuration. This is crucial as the presently existing materials for construction could not survive the high temperatures resulting from combustion in pure oxygen [44]. Table 1 shows the advantages and disadvantages of the selected CO_2 capture technologies.

Table 1. Advantages and disadvantages of the selected CO_2 capture technologies.

CO_2 Capture Technology	Advantages	Disadvantages
Membrane	• High recovery rate (up to 95%) • Regeneration energy not required • Simple modular system, small footprint • No waste stream • No chemical process [36] • Ability to adapt to changing gas compositions • Highly require pretreatment [36]	• Could be clogged by impurities in the gas stream • Membrane wetting prevention is a challenge • Plasticization of the membrane (polymeric membrane) • Sensitive to trace elements, e.g., sulfur compound [38]
Chemical absorption using amine (MEA)	• High recovery rate (up to 95%) and product purity >99 vol.% achievable [42]	• Process consumes considerable energy • In presence of O_2, occurrence of solvent degradation and equipment corrosion • Concentration of SO_x and NO_x existence in the gas stream combined with amine produces heat-stable salt • High energy for solvent regeneration is required [45]
Physical absorption solvent (Selexol)	• Low utility consumption • Higher capacity to absorb gases than amines • Able to remove H_2S and organic sulfur compound • Able to simultaneous gas stream dehydration [38]	• Hydrocarbon losses, resulting in reduced product output and often require recycle compression • High energy for solvent regeneration is required [45]
Cryogenic distillation	• High recovery rate (>90%) • Eliminates water consumption and chemicals, hence no corrosion issues • Allows pure CO_2 recovery (liquid), and transportable or pumpable to injection site conveniently [38]	• Process consumes high energy

2.1.2. CO_2 Storage and Abandonment

The storing of the produced CO_2 underground could benefit in addressing climate change by ensuring it is not released into the atmosphere. The prospective opportunities for CO_2 storage are at geological storage saline aquifers and depleted oil and gas reservoirs. These approaches are considered as the most feasible storage as there is great potential capacity for storage at an estimation of 236 Gt, worldwide [46,47].

In CCS chain, after the capture process, the CO_2 is compressed into a 'supercritical phase' whereby it becomes fluid almost as dense as water and is transported by pumping it along the pipeline to the storage site and subsequently into an underground geological formation that is typically situated several kilometers under the earth's surface. As the injected CO_2 is slightly more buoyant than the salty water that coexists within the storage reservoir, a portion of the injected CO_2 will migrate to the top of the formation and acts as a seal as it is trapped underneath the impermeable caprock structure. Some of the trapped CO_2 will dissolve slowly into the saline water and become trapped permanently (solution trapping), whilst others become trapped in tiny pore spaces (residual trapping). The ultimate trapping process involves dissolved CO_2 reacting with the reservoir rocks, which forms into a new mineral. This process called mineral trapping effectively locks the CO_2 into a solid mineral. As the storage mechanisms change over time from solution to residual then to mineral, the CO_2 becomes less and less mobile. Therefore, the longer the CO_2 is stored, the lower the risk of any leakage. Typically, the following geologic characteristics are associated with effective storage sites, whether at onshore or offshore site [48]:

- rock formations have adequate millimeter-sized voids or pores; this is to ensure available capacity to store the CO_2;
- pores in the rock are efficiently linked, which is a feature called permeability for injectivity; this is to receive the amount of CO_2 at the rate it is injected, allowing the CO_2 to move and spread out within the underground storage formation;
- extensive cap rock or barrier at the top of the storage formation; this is to contain the CO_2 permanently;
- stable geological environment; this is to avoid any potential geological effect that could compromise the storage site integrity.

The saline aquifers can be found at 700–1000 m below ground level and it consists of porous rocks with high salinity formation brine. These saline aquifers have no commercial value but it can be used to store the captured CO_2. These aquifers can be found at onshore/offshore and are usually estimated to have large storage potential volume. However, from an economic view, it is less desirable as storage option due to the unavailability of required infrastructure such as injection wells, surface equipment, and pipelines that would need a brand new capital investment altogether as compared to depleted oil and gas reservoirs. Depleted oil and gas reservoirs are more likely to be used for projects as extensive information from geological and hydrodynamic assessments is already available and equipped with readily available infrastructures. On average, about 40% of the residual is left after a typical oil and gas field production. The CO_2 can be injected in the depleted (or almost depleted) oil and gas reservoirs to increase their pressure, which in turn could provide the driving force to extract residual oil and gas while the CO_2 can then be stored inside permanently. This makes it more economical than storage in saline aquifers [46].

After the geological storage has reached its capacity, it will be retained for an extremely long period with an estimated minimum residence time of 1000 years without seepage/leakage back to the surface, also referred to as abandonment phase [48]. During which, the migration of the injected CO_2 in the storage volume will be monitored to ensure that it will not have any negative effect on the surrounding environment, e.g., groundwater pollution. In the monitoring stage, there are a few available methods that are commonly used, seismic monitoring, geo-electrical methods, temperature logs, gravimetry methods, remote sensing, geochemical sampling, atmospheric monitoring, tracers, soil gas monitoring, and microbiology [46]. These methods have since been applied at offshore CO_2 storage sites worldwide, such as at the Sleipner and Snohvit project. For example, at Sleipner, since the start of injection, it was found that the CO_2 plume has risen through shale rock layers nearing the caprock using seismic monitoring method. The project is currently under post-injection process and no leakage has been reported to date [48,49].

2.1.3. Carbon Capture and Storage (CCS) at Tangga Barat

The Tangga Barat CCS case study comprising of different CO_2 capture technologies units, transportation, and storage is shown in Figure 4. It begins with the central processing platform TBCP-A that will treat all produced gas originating from the producing platforms, TGDP-A, LHDP-A, and MLDP-A. The selected CO_2 capture technologies options consist of polymeric membrane, chemical absorption amine, physical absorption solvent (Selexol), and cryogenic distillation. Among the main criteria used to shortlist the selected CO_2 capture technologies is that the technologies have already been proven and suitable for large applications (>200 MMscfd) and capable of bulk removal of CO_2 (37% mole to 8% mole) as per Tangga Barat sales gas specification requirement. Following the CO_2 capture, the treated gas is transported via offshore pipeline to Resak platform for compression followed by pumping for sales export. Then, the captured CO_2 will be stored at the identified geological site near Tangga Barat. Moreover, please note that the intention of this study is not to endorse nor discriminate any of the CO_2 capture technologies presented, but rather to gauge the CCS techno-economic impact of using different method of CO_2 capture.

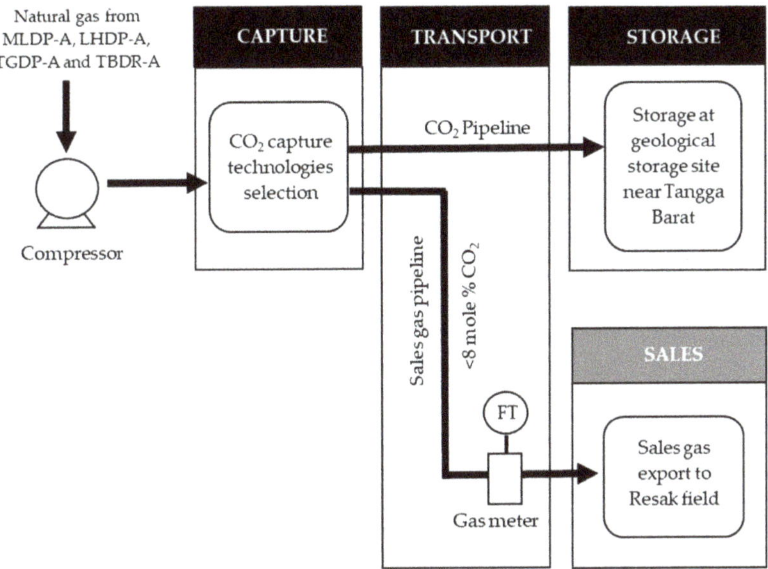

Figure 4. Carbon capture and storage (CCS) concept at Tangga Barat gas field.

2.2. Revenue Model

Based on the annual gas production estimates, the annual revenue estimation could be generated. In the US Energy Information Administration (EIA) projections reference case, natural gas prices remain comparatively low as compared to average historical prices. To estimate the economic, an average gas price between the year 2000 and projection up to 2050 whereby the estimated gas price for the year 2020 was used for this analysis [50]. The gas produced from Tangga Barat data is based on estimated studies from PETRONAS [33]. The revenue model calculates net revenue from the recovered gas production and then sold at a specified gas price, with a 10% deduction in royalty and 38% tax deduction based on Malaysian Petroleum Income Tax Act [34]. This revenue model structure will generate a time series of incoming net cash flows in which the resultant net revenue of the project could be assessed incrementally during production lifetime period and also cumulatively during the end of the project life. The tax payment is based on the estimation of capital cost depreciation of 5 years. There are also tax incentives being introduced by some countries, for example in the United States, the federal government has sanctioned supplementary tax credit for CO_2 sequestration activities that provide USD 10 per ton CO_2 stored, 15% tax credit applied to all costs associated with CO_2 purchase cost as well as CO_2 operation cost for injection [51]. For this study, since there is no tax incentive for CCS projects implemented in Malaysia, hence no additional tax incentives have been calculated. Table 2 shows the associated parameters assessed in this revenue model.

Table 2. Input and definition for parameters assessed in the revenue model.

Parameter	Input/Definition
Gas produced (Million British Thermal Unit (MMBtu))	Incremental or cumulative natural gas production volume simulated for reservoir performance estimation
Gas price (USD)	Average market price of natural gas
CO_2 produced (tonne)	Incremental or cumulative CO_2 production volume simulated for reservoir performance estimation
CO_2 price (USD)	Average market price of CO_2
Gross revenue (USD)	Revenue from natural gas recovered and cost and sold at specified gas price
Royalty (%)	10% of gross revenue
Tax (%)	38% of (revenue- royalty - maintenance cost- abandonment cost - capital cost depreciation)
Net revenue (USD)	Revenue after deduction of royalty and tax (Gross revenue (\$) − Royalty (\$) + Tax (\$))

2.3. Cost Model

In this study, the CCS project cost model could be categorized into four main types: capital cost, CO_2 capture cost, operation and maintenance (O&M) cost, and abandonment cost. In the cost model, the capital costs are accrued at the start of the project. Capital cost accounts for the gas field, compressor, pipeline, wells, and injection platforms. These capital costs are amortized over the lifetime of the field using the project discount rate [52]. For CO_2 capture cost, it is estimated based on the requirement of the gas field, including the processing equipment, production equipment, injection of platforms, and wells along with transportation pipeline. Furthermore, it includes 'on-cost' which is the insurance, obtaining rights-of-way, and legal and regulatory cost. CO_2 capture cost is the capital cost to implement the selected CO_2 capture technologies for the CCS project. O&M costs comprise of services, consumables, labor for site operation and maintenance (surface & subsurface), as well as general administration purposes [53]. Abandonment cost is the decommissioning cost after the injection period. These thorough analyses of CCS cost have been conducted by prominent research centers in this field. The costs in this study were estimated based on studies conducted by established studies by reputable bodies such as Asia Pacific Economic Cooperation (APEC), CO2CRC, IEAGHG, and GCCSI as well as previous research reports [22].

All cost data were updated to represent constant 2020 US Dollar (USD):

$$PV = \frac{FV}{(1 + i\%)^n} \qquad (2)$$

Whereby, PV is present value, FV is future value, i is discount rate, and n is the project lifetime number of years. Table 3 shows the parameters included in the cost model.

Table 3. Input and definition of the cost model.

Cost Model Parameter	Input/Definition	Study Referenced
Capital cost	Estimated unit cost for Tangga Barat	Asia-Pacific Economic Cooperation (APEC) and Asia Cooperative Research Centre for Greenhouse Gas Technologies (CO2CRC) [34]
CO_2 capture cost	Mean percentage increase in capital cost (over reference plant)	Rubin et al. [22]
	Cost data	Muhamad et al. [33]
Operation and Maintenance (O&M) costs	2.5% from plant cost	Klemes et al. [53]
Abandonment cost	Estimated unit cost for Tangga Barat	APEC and CO2CRC [34]

2.4. Cost Evaluation Metrics

2.4.1. Net Present Value (NPV)

NPV represents the value of future cash flows that is accrued incrementally and also cumulatively over the CCS project lifetime duration. It is the total value of cash inflows and outflows that are discounted to account for the time value of money and the risks/uncertainties associated with future cash flows. Net revenues (cash inflows) obtained from the revenue model was incorporated with the total costs (cash outflows) calculated by the cost model to develop NPV estimates in time series (years). At the end of the project lifetime, the cumulative NPV calculated is used as input to estimate the CCS project profitability, whereby positive values represent scenarios of profit while negative values represent fiscal losses sustained by investors/stakeholders in the project. The project is considered feasible if NPV > 0 and vice versa. NPV calculation is as presented [54]:

$$\mathrm{NPV}\,(i,N) = \sum_{t=0}^{N} \frac{C_t}{(1+i)^t} \tag{3}$$

Whereby; i is the discount rate, C_t the annual cash flow in the t th year, and N the total number of years. Time period $t = 0$ relates to the investment during the project lifetime.

2.4.2. Internal Rate of Return (IRR)

In the CCS project, the overall investment along with its profit gain could be gauged using IRR. From the calculations, as the IRR value increases, the greater the project gain will be and vice versa. Moreover, it is the discount rate value that is resulting the project cash flow NPV to become equal to zero (0), hence defining the least rate of return required in order to achieve project viability. If the IRR value obtained is higher than the defined discount rate, hence the CCS project is considered viable. The IRR is calculated and shown as [54,55]:

$$\mathrm{NPV}\,(IRR,N) = \sum_{t=0}^{N} \frac{C_t}{(1+IRR)^t} = 0 \tag{4}$$

Whereby; i is the discount rate, C_t the annual cash flow in the t th year, and N is the total number of years. Time period $t = 0$ relates to the investment during the project lifetime.

2.4.3. Payback Period (PBP)

The duration of time needed to recover the initially invested capital cost in a CCS project is represented by PBP. From the calculations, as the PBP value becomes shorter, the project fiscal viability strength will increase and vice versa. Moreover, PBP is defined by identifying the specific year when the calculated cumulative cash flow becomes positive and achieves breakeven. The PBP is expressed as below [54]:

$$\mathrm{PBP} = 1 + \mathrm{A} - \frac{\mathrm{B}}{\mathrm{C}} \tag{5}$$

Whereby; A is the final year with negative cumulative cash flow, B is the value of cumulative cash flow at the end of year A, and C is the total annual cash flow during the year after A.

2.4.4. Sensitivity Analysis (SA)

SA is an assessment of the projected performance with key assumptions and projections based on the base values. There are several important key variables that could impact CCS project, which are the gas price, discount price, initial capital cost, and CO_2 capture cost and tax. Key economic assumptions and associated references used as input in this study are listed in Table 4.

Table 4. Key economic assumptions and references in this study.

Input Data	Data	Remarks
Year Enacted	2020	
Project lifetime (N, year)	15	from Reference [34]
Interest rate (%)	3	from Reference [56]
Discount rate (R, %)	8	from Reference [57]
Plant capacity (MMscf/d)	440	from Reference [32]
Average gas sales (MMscf/d)	220	from Reference [34]
Gas price ($/MMBtu)	2.8	from Reference [58]
CO_2 price ($/tonne)	23	from Reference [59,60]
Volume of CO_2 (tonne/d)	1646	equivalent to (81 MMScf/d) from Reference [34]
Capital cost USD Million	427.9	estimate basis constant year 2020 with reference location Malaysia, modified from Reference [34]
CO_2 capture technologies cost (USD Million):		
– polymer membrane	212	modified from Reference [33]
– chemical absorption (Amine)	410.79	96% increase in capital cost [22]
– physical absorption solvent (Selexol)	162.61	38% increase in capital cost [22]
– cryogenic distillation	289.39	91% increase in capital cost [22]
Tax (%)	38	from Reference [61]
Operation and maintenance cost (%)	2.5	from Reference [53]
Abandonment cost (USD Million)	50	from Reference [34]

3. Techno-Economic Model Results and Discussion

This study on techno-economic analysis provides the fiscal precursor and its outlook for Tangga Barat natural gas field for CCS implementation starting from the year 2020 with a projection of 15 years project lifetime. The calculation outputs of project cash flow and total annual cash outflows, as well as its inflows throughout its lifetime, is as shown in Figure 5a–d. It can be seen that as the project extends into the exploration stage, the annual project cash flow will begin to increase in a positive note. During operation, inflow arrives from the natural gas sales revenue and outflows derive from the O&M, royalty, and tax costs.

For base case (without any CO_2 capture technology implementation), during project investment in the year 2020, the initial sum proceeding from equity and capital cost is USD 427.9 Million [34]. It reflects the extent of the total cost expenses required to start a project indeed. The initial investment increases as it includes the implementation of CO_2 capture technologies. The total capital cost required including the CO_2 capture technologies implementation for the polymeric membrane is USD 639.9 Million, chemical absorption (Amine) is USD 838.69 Million, using physical absorption solvent (Selexol) is USD 590.51 Million, and using cryogenic distillation is USD 817.29 Million (refer Table 5). These costs are investments to achieve the targeted natural gas production output of 83,512,000 MMBtu per year for the overall project lifetime of 15 years.

For polymeric membrane, the case is shown in Figure 5a, where the project annual cash flow will gradually surpass its values in the succeeding years, which is the result of large investments for the capital cost being subjected to reduction from financial amortizations. This trend is reflected by the slope of the curve ranging between the year 2020 and 2027. Only by 2028 onwards, the gas field is anticipated to operate at a fairly constant speed and ultimately reaches breakeven. After that, the project cash flow is inclined to increase continuously as the revenue exceeds the cost expenditures during which its inflows and outflows also increase based on the specified discount rates over the project lifetime. From which, the revenue demonstrates a relatively optimistic cash flow increment over the years, which in turn highlights the significance of the profit obtained from natural gas sales price toward the overall feasibility of the CCS project. The entire field system and equipment will free from debt without further amortization by the end of the field production lifetime. Case using chemical absorption (Amine) is shown in Figure 5b, whereby a larger capital investment than the previous case (polymeric membrane) was made causing the project annual cash flow exceeds even higher. This is shown by the curve slope between 2020 and 2029. By 2030 onwards, the gas field reaches breakeven. For physical absorption solvent (Selexol) case shown in Figure 5c, the initial capital cost is the lowest of all the CO_2 capture technologies analyzed. The project is projected to breakeven in 2027 with payback obtained 7.94 years, which is about half of the project lifetime. In the case of using cryogenic distillation, the initial capital cost is slightly lower as compared to using Amine. As a result, the project is projected to breakeven one year earlier in 2029 (Figure 5d).

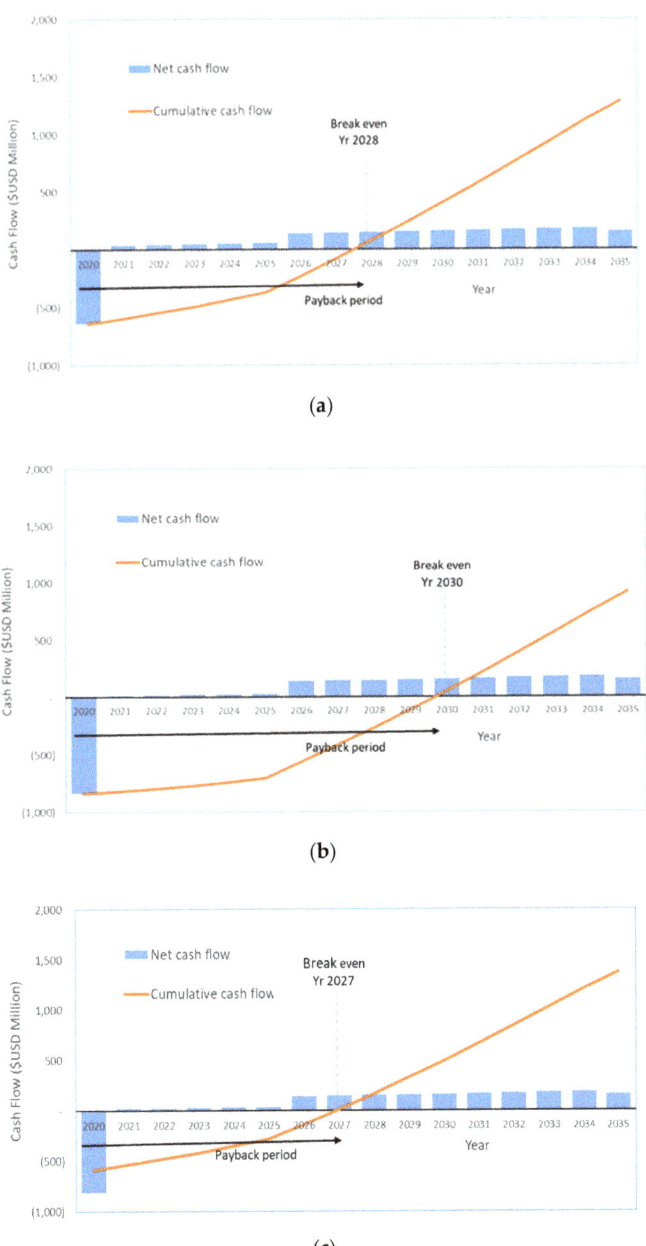

(a)

(b)

(c)

Figure 5. *Cont.*

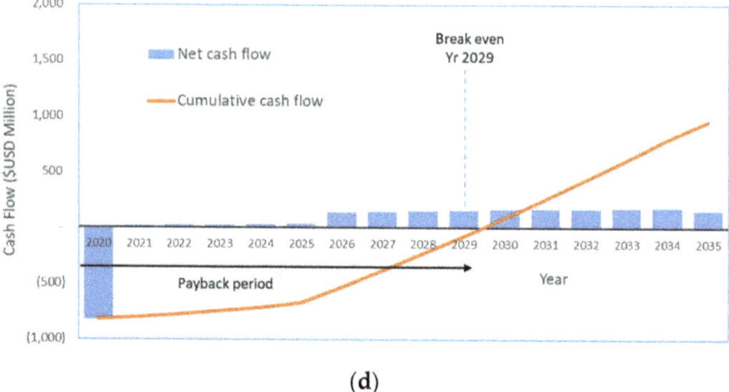

(**d**)

Figure 5. (**a**) Project cash flow base case using polymeric membrane. (**b**) Project cash flow using chemical absorption (Amine). (**c**) Project cash flow using physical absorption solvent (Selexol). (**d**) Project cash flow using cryogenic distillation.

Table 5. Economic results of the base case and the four CO_2 capture technologies.

Parameter	Polymeric Membrane	Chemical Absorption (Amine)	Physical Absorption (Selexol)	Cryogenic Distillation
Capital cost (USD Million)	639.90	838.69	590.51	817.29
O&M cost including abandonment cost (USD Million)	289.96	364.51	271.44	356.49
Net revenue (USD Million)	1923.06	1753.59	1965.17	1771.83
Payback period (PBP)	8.55	10.68	7.94	10.45
Net present value (NPV)	292.94	(6.67)	367.39	25.58
Internal rate of return (IRR)	13%	8%	15%	8%

Table 5 displays the main outputs from the model for the four CO_2 capture technologies scenarios. Generally, a CCS project is considered to be techno-economically viable when it achieves the trio combination fiscal precursor of NPV > 0, followed by IRR value higher than the specified discount rate along with PBP value being lower than the gas field production lifetime. The physical absorption solvent using Selexol result shows the most favorable profitability with a fairly positive NPV equal to USD 367.37 Million with an IRR equal to 15% with the shortest PBP as compared to the other capture technologies. This is followed by CO_2 capture using a polymeric membrane with just a slight difference with NPV of USD 292.94, IRR at 13%, which is higher than the specified discount rate, and PBP of 8.55 years. Cryogenic distillation comes third at NPV of USD 25.58, IRR at 8%, which is the same as the specified discount rate, and PBP of 10.45 years. And finally, the least feasible CO_2 capture is by using chemical absorption (Amine) as the NPV has negative value at −USD 6.67, IRR same as the specified discount rate at 8%, and the longest PBP at 10.68 years. Even so, although the obvious viability is by physical absorption solvent using Selexol, oil and gas producers/investors may not find it is as economically attractive since there are many other factors that may directly influence project decision making. External factors such as market price and financial support also are very likely to have either a negative or positive impact. On that account, it is imperative that consideration of associated risks are analyzed when conducting investment analysis. For that reason, in order to decide which considered variables are the most significant to the CCS project, SA would be the best path to reflect its worthiness.

3.1. Sensitivity Analysis (SA) Results and Discussion

Sensitivity Analysis (SA) examines the effect of input variables variation on the model's conclusion to gauge and identify the most significant variables affecting the CCS project performance. By quantifying their elasticity range and assessing the project performance effect when subjected to different scenarios impacted either by the favorable or unfavorable perspective of the variables concurrently, the feasibility of the project fiscal precursor could be defined. In this study, these variables are considered to be significantly affecting the project viability in terms of economic as well as performance risks:

- Gas price provides input to the project revenue and has long-term uncertainty as it is highly dictated by supply and demand, global market fluctuation, and political–economic decisions;
- Tax is dependent on the country tax requirement as well as its government political decision;
- Discount rate directly affects the present value of future costs and its profits;
- Initial capital cost is highly reliant on the capital outflow investment ability by the oil and gas producers and investors.
- CO_2 capture technology cost is highly reliant on the technology provider profit margin and its maturity status;

Figure 6a–d depicts variation scenarios and its impact change to the five input variables, gas price, discount rate, tax, CO_2 capture technology cost, initial capital cost, and its effect on the risks affecting the fiscal precursor that is NPV and its sensitivity range. The sensitivity bound is defined as favorable/unfavorable by changing the baseline value, testing the different scenarios of that certain investment based on the specified five input variables. A ±10% range was implemented for the initial capital cost, tax, and CO_2 capture technologies and this is as supported by the literature [54]. For the gas price, the range of ±USD 1.0 Million/MMBtu was selected as it reflects the gas price trend for the past 5 years [58]. The discount rate of ±2% was estimated based on previous studies conducted by established bodies such as the US Department of Energy, International Energy Agency Greenhouse Gas (IEAGHG), Electric Power Research Institute (EPRI), and UK Department of Energy and Climate Change (DECC) [62].

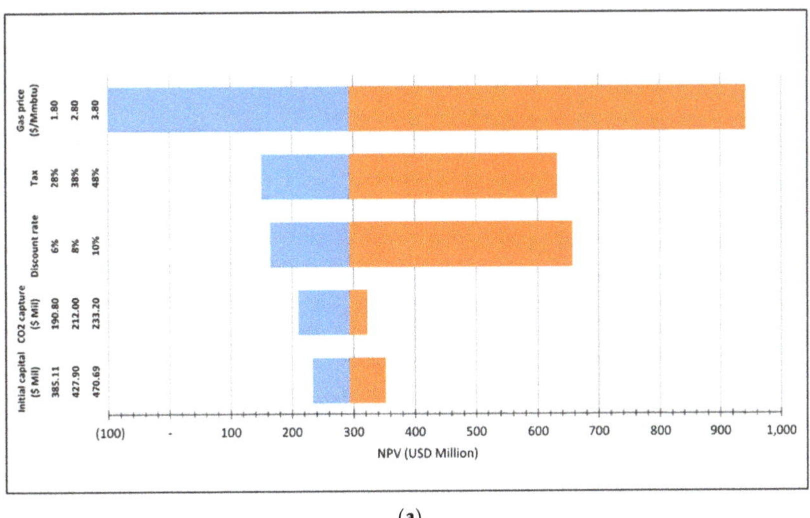

(**a**)

Figure 6. *Cont.*

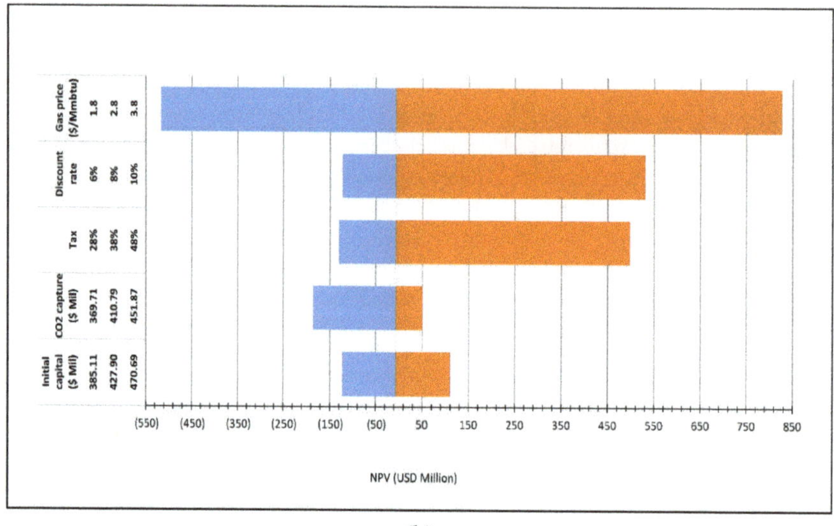

(**b**)

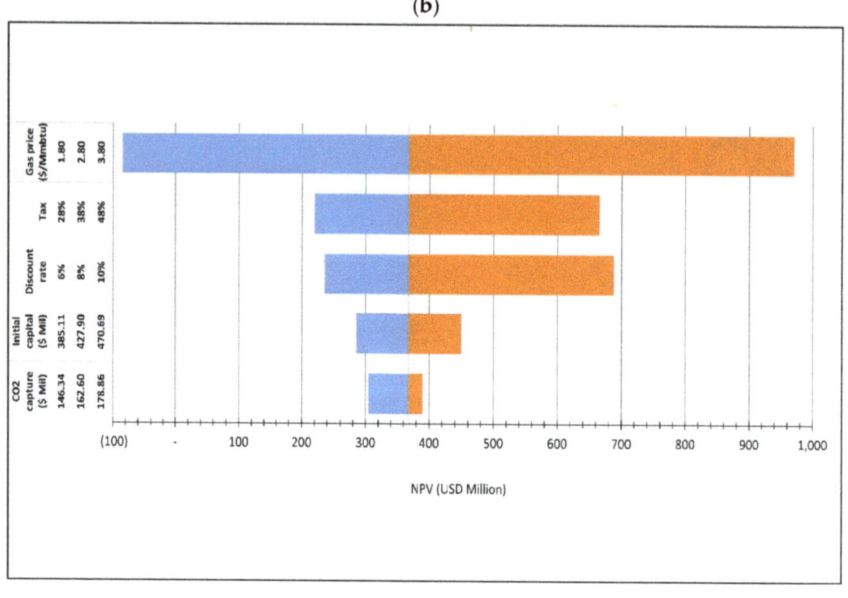

(**c**)

Figure 6. *Cont.*

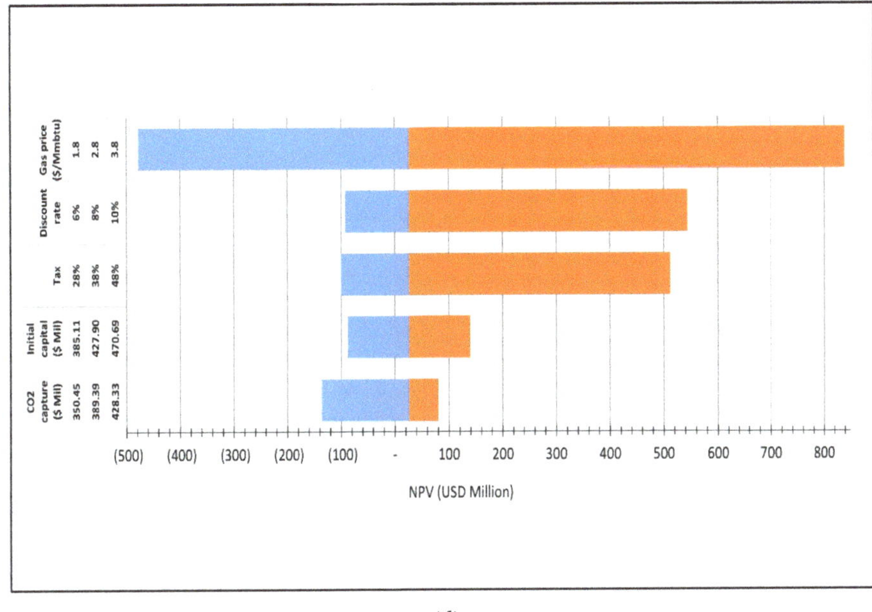

(**d**)

Figure 6. (**a**) Sensitivity analysis (SA) of Net Present Value (NPV) for CO_2 capture using membrane. (**b**) SA of NPV for CO_2 capture using chemical absorption (Amine). (**c**) SA of NPV for CO_2 capture using physical absorption solvent (Selexol). (**d**) SA of NPV for CO_2 capture using cryogenic distillation.

The SA result for all four scenarios of CCS with different CO_2 capture technologies is reflected by NPV, which visualizes the gas price as the main key influencer affecting the viability of the project. In comparison to the other variables, gas price demonstrates the most substantial NPV impact change. The output from discount rate, tax, CO_2 capture cost, and initial capital cost (from the most affecting) variables are found to less influencing as compared to gas price. The feasibility of the project could significantly be affected by gas price during all four scenarios that are using polymeric membrane, chemical absorption (Amine), physical absorption solvent (Selexol), and cryogenic distillation in an unfavorable scenario. For example, in the case of gas price of USD 1.8/MMBtu, it predicts a negative NPV of up to −USD 161 Million, −USD 517.15 Million, −USD 84.91 Million, and −USD 487.14 Million, respectively. On the other hand, when the gas price increases to USD 3.8/MMBtu, it considerably increases the projects NPV up to USD 941.54 Million, USD 826 Million, USD 970.25 Million, and USD 838.42 Million also, respectively. High gas prices would, therefore, be beneficial for the CCS project's economic feasibility. It also could be seen that, since the uncertainties of the gas price are heavily subjected to the global supply and demand, market fluctuation, political and economic decisions, and government subsidies, henceforth making it one of the most vital factors to deal with to ensure the project success.

In the case of discount rate and tax, both variables share a nearly similar impact on the NPV. The increase of the discount rate to 10% per annum results in a decrease in NPV to USD 163.90 Million for membrane, −USD 123.06 Million for chemical absorption (Amine), USD 235.21 for physical absorption solvent (Selexol), and −USD 92.18 Million for cryogenic distillation, meanwhile a decrease to 6% per annum resulting in an increase of NPV to USD 656.73 Million, USD 530.54 Million, USD 688.08 Million, and USD 544.12 Million, respectively. Increasing the tax to 48% decreases the NPV to USD 150.13 Million for membrane, −USD 130.85 Million for chemical absorption (Amine), USD 219.94 for physical absorption solvent (Selexol), and −USD 100.61 Million for cryogenic distillation, whilst a decrease to 28% results in an increase of NPV to USD 632.05 Million, USD 497.86 Million, USD 665.39 Million, and USD 512.31 Million, respectively. Both variables are effective, whereby the specified discount rate exhibits high impact on the calculation for the discounted cash flow of NPV whilst tax is highly dependent on local government decision and profitability.

Both CO_2 capture technology cost and initial capital cost results variable are also almost similar and has the least effect on NPV. An increase of the capture technology by 10% results in a reduction in NPV to USD 210.14 Million for membrane, −USD 185.80 Million for chemical absorption (Amine), USD 303.88 for physical absorption solvent (Selexol), and −USD 136.95 Million for cryogenic distillation, to whilst a decrease by 10% results in an increase of NPV to USD 322.29 Million, USD 50.20 Million, USD 389.90 Million, and USD 79.48 Million, respectively. Meanwhile, when the capital cost is also increased by 10%, it also results in a reduction in NPV to USD 233.70 Million, −USD 122.78 Million for chemical absorption (Amine), USD 285.64 for physical absorption solvent (Selexol), and −USD 87.57 Million for cryogenic distillation, to whilst a decrease by 10% results in an increase of NPV to USD 352.18 Million, USD 109.43 Million, USD 449.14 Million, and USD 138.72 Million, respectively. This shows that the CO_2 capture technologies cost has the least compromising effect on the NPV performance as it has the least negative range scenario as compared to the other input variables.

3.2. Limitation of This Techno-Economic Analysis

The techno-economic analysis conducted in this study achieves the targeted evaluation of CCS prospect with four different CO_2 capture technologies scenarios at Tangga Barat gas field could be used as an early decision making indicator for investors and enabling other natural gas fields to be rapidly evaluated. However, this study was based on limited and readily available statistical economic parameters and storage reservoir properties. A more detailed storage site characterization and equipment and material cost parameters inventory for the Tangga Barat CCS during detailed engineering design stage could assist in obtaining a higher precision project techno-economics.

4. Conclusions

In this study, it presents a techno-economic analysis of an actual gas field Tangga Barat in Peninsular Malaysia and assesses its potential for CCS to be implemented in Malaysia as a means of achieving targeted CO_2 reduction aim. The deployment of this CCS project will contribute to a significant decrease of the environmental impact, by storing the captured CO_2 at an estimation of 1646 tonnes/day. The result of this paper focuses on the possibilities for CCS project comprising of 4 cases of CO_2 capture for offshore gas field along with a fixed CO_2 transport and storage infrastructure development. In this study analysis, a techno-economic model spreadsheet is established to obtain the fiscal precursors, which include NPV, IRR, and PBP. In the cost model, the considered cost factor is capital cost that includes CO_2 capture technology cost, O&M cost, and abandonment cost incurred. While for the revenue model, the estimation is based on the profit from the produced natural gas sales. The techno-economic analysis is conducted based on previous literature and then benchmarked and updated catering to suit market gas prices, plant equipment, revenues, operating cost, tax rates, and interest rates. Then, a SA is conducted to evaluate the risks and measures associated with the project investment uncertainties. Favorable and unfavorable scenarios are analyzed within a defined elasticity range basing from the baseline value input. The selected variables include gas price, discount rate, tax, CO_2 capture technology cost, and initial capital cost were used to evaluate CCS project feasibility potential.

From the techno-economic analysis model, it presents a positive outlook stating its achievability to deploy the project based on Malaysia's current market and economic setting. The analysis shows two prospective CO_2 capture technology with high positive NPVs by using physical absorption solvent (Selexol) followed by the polymeric membrane. The most prospective CO_2 capture method is using Selexol with IRR at 15% and PBP of 7.94 years. This is closely followed by CO_2 capture using membrane with just a slight difference with IRR at 13% and PBP of 8.55 years. From this analysis, it could be concluded that these two variants of CO_2 capture have a huge economic prospect for deployment at Tangga Barat. Based on the SA, the NPV exhibits to be significantly most sensitive to gas price, followed by discount rate and tax. It also shows that both CO_2 capture technology and initial capital cost has minimal influence on NPV. Although the project feasibility and acceptable risks are provided by all four CCS scenarios from the economic model calculations, two distinct capture technologies, which are using membrane and physical absorption solvent (Selexol), give high IRR with an acceptable time of return period (less than half the production years). This study shows that either one of the CO_2 technologies could be implemented at Tangga Barat, which could further be influenced by the oil and gas producer or investor inclinations, starting with this economics and extending to political and geological factors. Furthermore, the SA illustrates that the gas price has the most significant impact over NPV as compared among all the considered variables. This scenario portrays that the gas sales revenues that is controlled by the uncertain market, tariffs, and reimbursements will eventually impact the overall economic viability of the CCS project.

In order to achieve successful deployment of CCS projects, collaborations between governments, international organizations, and the private sector would be essential and, of course, international financing will be critical to project success. In the scenario, for the case of Malaysia, to adopt and extend its current reduction of tax incentives for marginal field from 38% to 25% for CCS projects, this incentive would positively influence the key decision factor for oil and gas producers/investors to implement CCS. Furthermore, it is hopeful that the upcoming future implementation of the Energy Efficiency and Conservation act in Malaysia would be an appropriate avenue for the climate policy to manage and positively impact the control of CO_2 emissions. The accessibility to sufficient storage capacity could be another restrictive element to reduce CO_2 emissions through CCS, even though there are some fundamental studies that believe there are available geologic storage and also studies on storing back the CO_2 into the origin source well/depleted field. These uncertainties and lack of storage capacity could be overcome by implementing CO_2 utilization method, by transforming CO_2 into valuable products that could serve as project revenue and eventually offset the initial investment cost altogether. Finding creative, cost-effective ways to capture and profit from this CO_2 for use in CCS can result in a vast improvement in the long-term commercial viability of natural gas production in Malaysia, along with increasing the productive life of the economy's natural gas resources. This study on techno-economic analysis could be a reference point in terms of economic viability for implementation of CCS in other CO_2 gas fields in Malaysia and globally.

Author Contributions: N.R.S. conducted the techno-economic analysis and wrote the paper; T.M.I.M. reviewed the paper preparation; A.H.S. checked and proofread the final version of the paper; and M.F.M.I. provided the resources for the analysis and reviewed the paper. All authors have read and agreed to the published version of the manuscript.

Funding: This research was funded by the Centre for Advanced Modeling and Geospatial Information Systems (CAMGIS) [Grant no. 321,740.2232397] and research development fund; School of Information, Systems and Modelling, University of Technology Sydney, Australia. This research also received financial support from AAIBE Chair of Renewable Energy (Grant no: 201801 KETTHA).

Acknowledgments: The authors would like to acknowledge this research as supported by MyBrain15, Ministry of Higher Education Malaysia.

Conflicts of Interest: There is no conflict of interest.

References

1. Kostowski, W.J.; Usón, S. Thermoeconomic assessment of a natural gas expansion system integrated with a co-generation unit. *Appl. Energy* **2013**, *101*, 58–66. [CrossRef]
2. Silitonga, A.S.; Masjuki, H.H.; Ong, H.C.; Sebayang, A.H.; Dharma, S.; Kusumo, F.; Siswantoro, J.; Milano, J.; Daud, K.; Mahlia, T.M.I.; et al. Evaluation of the engine performance and exhaust emissions of biodiesel-bioethanol-diesel blends using kernel-based extreme learning machine. *Energy* **2018**, *159*, 1075–1087. [CrossRef]

3. Ong, H.C.; Masjuki, H.H.; Mahlia, T.M.I.; Silitonga, A.S.; Chong, W.T.; Yusaf, T. Engine performance and emissions using Jatropha curcas, Ceiba pentandra and Calophyllum inophyllum biodiesel in a CI diesel engine. *Energy* **2014**, *69*, 427–445. [CrossRef]

4. Silitonga, A.; Shamsuddin, A.; Mahlia, T.; Milano, J.; Kusumo, F.; Siswantoro, J.; Dharma, S.; Sebayang, A.; Masjuki, H.; Ong, H.C. Biodiesel synthesis from Ceiba pentandra oil by microwave irradiation-assisted transesterification: ELM modeling and optimization. *Renew. Energy* **2020**, *146*, 1278–1291. [CrossRef]

5. Ong, H.C.; Milano, J.; Silitonga, A.S.; Hassan, M.H.; Shamsuddin, A.H.; Wang, C.T.; Mahlia, T.M.I.; Siswantoro, J.; Kusumo, F.; Sutrisno, J. Biodiesel production from Calophyllum inophyllum-Ceiba pentandra oil mixture: Optimization and characterization. *J. Clean. Prod.* **2019**, *219*, 183–198. [CrossRef]

6. BP. *BP Statistical Review of World Energy*; BP: London, UK, 2018.

7. Calabrese, J. Positioning Malaysia within the Global Energy Landscape. Available online: https://www.mei.edu/publications/positioning-malaysia-within-global-energy-landscape#_ftn28 (accessed on 8 July 2019).

8. Rahman, F. Development of innovative membrane for offshore high CO_2 separation. In Proceedings of the World gas conference 2012, Kuala Lumpur, Malaysia, 4–8 June 2012.

9. Harun, N.D.A.R. Technical challenges and solutions on natural gas development in Malaysia. In Proceedings of the Petroleum Policy and Management (PPM) Project 4th Workshop of the China, Beijing, China, 30 May–3 June 2006.

10. Oh, T.H. Carbon capture and storage potential in coal-fired plant in Malaysia—A review. *Renew. Sustain. Energy Rev.* **2010**, *14*, 2697–2709. [CrossRef]

11. Bernama. Malaysia re-pledges to achieve 45 per cent CO_2 emission by 2030. *New Straits Times*, 22 April 2016.

12. Silitonga, A.S.; Masjuki, H.H.; Mahlia, T.M.I.; Ong, H.C.; Chong, W.T. Experimental study on performance and exhaust emissions of a diesel engine fuelled with Ceiba pentandra biodiesel blends. *Energy Convers. Manag.* **2013**, *76*, 828–836. [CrossRef]

13. Lokman, T. PM: Malaysia on Course to Reduce Carbon Emissions by 40 Pct by 2020. Available online: https://www.nst.com.my/news/nation/2017/12/310231/pm-malaysia-course-reduce-carbon-emissions-40-pct-2020 (accessed on 5 December 2019).

14. Energy Efficiency and Conservation Bill to be tabled end of this year, says minister. *Malaymail*, 4 July 2019.

15. International Energy Agency. *KETTHA/IEA CCS Roundtable*; International Energy Agency: Paris, France, March 2011.

16. PETRONAS. *PETRONAS Sustainability Report 2017*; PETRONAS: Kuala Lumpur, Malaysia, 2017; p. 82. Available online: https://www.petronas.com/ws/sites/default/files/2018-07/sustainability-report-2017.pdf (accessed on 17 October 2019).

17. IPCC. *Climate Change 2014: Synthesis Report. Contribution of Working Groups I, II and III to the Fifth Assessment Report of Intergovernmental Panel on Climate Change*; IPCC: Geneva, Switzerland, 2014.

18. Nations, U. Summary of the Kyoto Protocol. Available online: http://bigpicture.unfccc.int/#content-the-paris-agreemen (accessed on 17 October 2019).

19. Hulme, M. 1.5 degrees C and climate research after the Paris Agreement. *Nat. Clim. Chang.* **2016**, *6*, 222–224. [CrossRef]

20. Rogelj, J.; den Elzen, M.; Hohne, N.; Fransen, T.; Fekete, H.; Winkler, H.; Chaeffer, R.S.; Ha, F.; Riahi, K.; Meinshausen, M. Paris Agreement climate proposals need a boost to keep warming well below 2 degrees C. *Nature* **2016**, *534*, 631–639. [CrossRef]

21. United Nations Economic Commission for Europe. The Role of Fossil Fuels in Delivering a Sustainable Energy Future. Available online: https://energy.gov/sites/prod/files/2016/09/f33/DOE%20-%20Carbon%20Capture%20Utilization%20and%20Storage_2016-09-07.pdf (accessed on 17 October 2019).

22. Rubin, E.S.; Davison, J.E.; Herzog, H.J. The cost of CO_2 capture and storage. *Int. J. Greenh. Gas Control* **2015**, *40*, 378–400. [CrossRef]

23. Durmaz, T. The economics of CCS: Why have CCS technologies not had an international breakthrough? *Renew. Sustain. Energy Rev.* **2018**, *95*, 328–340. [CrossRef]

24. Budinis, S.; Krevor, S.; Dowell, N.M.; Brandon, N.; Hawkes, A. An assessment of CCS costs, barriers and potential. *Energy Strategy Rev.* **2018**, *22*, 61–81. [CrossRef]

25. Blomen, E.; Hendriks, C.; Neele, F. Capture technologies: Improvements and promising developments. *Energy Procedia* **2009**, *1*, 1505–1512. [CrossRef]

26. Mondal, M.K.; Balsora, H.K.; Varshney, P. Progress and trends in CO_2 capture/separation technologies: A review. *Energy* **2012**, *46*, 431–441. [CrossRef]

27. Yang, H.; Xu, Z.; Fan, M.; Gupta, R.; Slimane, R.B.; Bland, A.E.; Wright, I. Progress in carbon dioxide separation and capture: A review. *J. Environ. Sci.* **2008**, *20*, 14–27. [CrossRef]

28. Rubin, E.; Booras, G.; Davison, J.; Eksrom, C.; Matuszewski, M.; McCoy, S.; Short, C. *Toward a Common Method for Cost-Estimation for CO_2 Capture and Storage at Foosil Fuelower Plant*; Global CCS Institute: Melbourne, Australia, 2013.

29. Allinson, G.; Neal, P.; Ho, M.; Wiley, D.; McKee, G. *CCS Economics Methodology and Assumptions*; RPT06-0080; School of Petroleum Engineering, The University of New South Wales: Sydney, Australia, 2006.

30. *Malaysian CCS Legal and Regulatory Workshop*; Global CCS Institute: Melbourne, Australia, 2013.

31. Lai, N.Y.G.; Yap, E.H.; Lee, C.W. Viability of CCS: A broad-based assessment for Malaysia. *Renew. Sustain. Energy Rev.* **2011**, *15*, 3608–3616. [CrossRef]

32. Teh, Y.H.; Theseira, K.; Abdul Karim, A.H.; Hashim, N.S.; Yakob, A.R.; Musa, A.S.; Borhan, N.A.; Muhamad, S.; Sykahar, M.W. Preparing a Gas Field Development Plan: Tangga Barat Cluster Gas Project. In Proceedings of the International Petroleum Technology Conference, Kuala Lumpur, Malaysia, 1 January 2008; p. 10.

33. Muhamad, S.; Teh, Y.H.; Hassan, N.H.; Sabri, H.A.R.; Arif, I.A.M.; A. Karim, A.H. Acid Gas Removal System for Tangga Barat Cluster Gas Development—Case Study. In Proceedings of the SPE Asia Pacific Oil and Gas Conference and Exhibition, Perth, Australia, 1 January 2008; p. 24.

34. APEC Energy Working Group. *Assessment of the Capture and Storage Potential of CO_2 Co-Produced with Natural Gas in South-East Asia*; Asia-Pacific Economic Cooperation: Western Australia, Australia, 2010.

35. Hong Teh, Y.; Theseira, K.; Karim, A.; Shima Hashim, N.; Razak Yakob, A.; Syrhan Musa, A.; Aisyah, N.; Muhamad, S.; Wakif Sykahar, M. Preparing a Gas Field Development Plan: Tangga Barat Cluster Gas Project. In Proceedings of the International Petroleum Technology Conference, Kuala Lumpur, Malaysia, 3–5 December 2008. [CrossRef]

36. Siagian, U.W.R.; Raksajati, A.; Himma, N.F.; Khoiruddin, K.; Wenten, I.G. Membrane-based carbon capture technologies: Membrane gas separation vs. membrane contactor. *J. Nat. Gas Sci. Eng.* **2019**, *67*, 172–195. [CrossRef]

37. Hashemifard, S.A.; Ahmadi, H.; Ismail, A.F.; Moarefian, A.; Abdullah, M.S. The effect of heat treatment on hollow fiber membrane contactor for CO_2 stripping. *Sep. Purif. Technol.* **2019**, *223*, 186–195. [CrossRef]

38. Olajire, A.A. CO_2 capture and separation technologies for end-of-pipe applications—A review. *Energy* **2010**, *35*, 2610–2628. [CrossRef]

39. Zhang, Y.; Sunarso, J.; Liu, S.; Wang, R. Current status and development of membranes for CO2/CH4 separation: A review. *Int. J. Greenh. Gas Control* **2013**, *12*, 84–107. [CrossRef]

40. Spigarelli, B.P.; Kawatra, S.K. Opportunities and challenges in carbon dioxide capture. *J. CO_2 Util.* **2013**, *1*, 69–87. [CrossRef]

41. Ağralı, S.; Üçtuğ, F.G.; Türkmen, B.A. An optimization model for carbon capture & storage/utilization vs. carbon trading: A case study of fossil-fired power plants in Turkey. *J. Environ. Manag.* **2018**, *215*, 305–315. [CrossRef]

42. Figueroa, J.D.; Fout, T.; Plasynski, S.; McIlvried, H.; Srivastava, R.D. Advances in CO_2 capture technology—The U.S. Department of Energy's Carbon Sequestration Program. *Int. J. Greenh. Gas Control* **2008**, *2*, 9–20. [CrossRef]

43. Demirel, Y.; Matzen, M.; Winters, C.; Gao, X. Capturing and using CO_2 as feedstock with chemical looping and hydrothermal technologies. *Int. J. Energy Res.* **2015**, *39*, 1011–1047. [CrossRef]

44. Koytsoumpa, E.I.; Bergins, C.; Kakaras, E. The CO_2 economy: Review of CO_2 capture and reuse technologies. *J. Supercrit. Fluids* **2017**. [CrossRef]

45. Rubin, E.S.; Mantripragada, H.; Marks, A.; Versteeg, P.; Kitchin, J. The outlook for improved carbon capture technology. *Prog. Energy Combust. Sci.* **2012**, *38*, 630–671. [CrossRef]

46. Leung, D.Y.C.; Caramanna, G.; Maroto-Valer, M.M. An overview of current status of carbon dioxide capture and storage technologies. *Renew. Sustain. Energy Rev.* **2014**, *39*, 426–443. [CrossRef]

47. Stangeland, A. A model for the CO_2 capture potential. *Int. J. Greenh. Gas Control* **2007**, *1*, 418–429. [CrossRef]

48. Aminu, M.D.; Nabavi, S.A.; Rochelle, C.A.; Manovic, V. A review of developments in carbon dioxide storage. *Appl. Energy* **2017**, *208*, 1389–1419. [CrossRef]

49. Bachu, S. Review of CO_2 storage efficiency in deep saline aquifers. *Int. J. Greenh. Gas Control* **2015**, *40*, 188–202. [CrossRef]

50. *Annual Energy Outlook 2019—With Projections to 2050*; US Energy Information Administration: Washington, DC, USA, 2019.

51. NEORI. Carbon Dioxide Enhanced Oil Recovery: A Critical Domestic Energy, Economic, and Environmental Opportunity. 2012. Available online: https://www.ourenergypolicy.org/ (accessed on 17 October 2019).

52. Rubin, E.S.; Short, C.; Booras, G.; Davison, J.; Ekstrom, C.; Matuszewski, M.; McCoy, S. A proposed methodology for CO_2 capture and storage cost estimates. *Int. J. Greenh. Gas Control* **2013**, *17*, 488–503. [CrossRef]

53. Klemeš, J.; Bulatov, I.; Cockerill, T. Techno-economic modelling and cost functions of CO_2 capture processes. *Comput. Chem. Eng.* **2007**, *31*, 445–455. [CrossRef]

54. Cardoso, J.; Silva, V.; Eusébio, D. Techno-economic analysis of a biomass gasification power plant dealing with forestry residues blends for electricity production in Portugal. *J. Clean. Prod.* **2019**, *212*, 741–753. [CrossRef]

55. Zhu, Y.; Zhai, R.; Yang, Y.; Reyes Belmonte, M. Techno-Economic Analysis of Solar Tower Aided Coal-Fired Power Generation System. *Energies* **2017**, *10*, 1392. [CrossRef]

56. Malaysia Policy Rate. Available online: https://www.ceicdata.com/en/indicator/malaysia/policy-rate (accessed on 17 November 2019).

57. Irlam, L. *The Costs of CCS and Other Low-Carbon Technologies in the United States- 2015 Update*; Global CCS Institute: Melbourne, Australia, 2015.

58. Natural Gas Price. Available online: https://oilprice.com/Energy/Natural-Gas (accessed on 10 March 2019).

59. Price of CO2. Available online: https://www.iberdrola.com/about-us/utility-of-the-future/regulation-our-vision/price-co2 (accessed on 1 February 2019).

60. Khorshidi, Z.; Ho, M.T.; Wiley, D.E. Techno-Economic Study of Biomass Co-Firing with and without CO_2 Capture in an Australian Black Coal-Fired Power Plant. *Energy Procedia* **2013**, *37*, 6035–6042. [CrossRef]

61. Pamela Tomski, C.H. *Feasibility of Accelerating the Deployment of Carbon Capture, Utilization and Storage in Developing APEC Economies*; Asia-Pacific Economic Cooperation: Singapore, 2014; p. 124.

62. Rubin, E.S. *Methods and Measures for CCS Cost*; Carnegie Mellon University: Paris, France, 2011.

Article

Synthesis and Characterization of New Schiff Base/Thiol-Functionalized Mesoporous Silica: An Efficient Sorbent for the Removal of Pb(II) from Aqueous Solutions

Moawia O. Ahmed *, Ameen Shrpip and Muhammad Mansoor

Chemistry Department, College of Science, Taibah University, P.O. Box 30003, Al Madinah Al Munawarah 41477, Saudi Arabia; shrpip@hotmail.com (A.S.); mansoorishaque@hotmail.com (M.M.)
* Correspondence: moahmed@taibahu.edu.sa; Tel.: +966-5655-48322

Received: 25 January 2020; Accepted: 18 February 2020; Published: 21 February 2020

Abstract: A new type of silica hybrid material functionalized with Schiff base-propyl-thiol and propyl-thiol groups (adsorbents **1** and **2**, respectively) was synthesized using a co-condensation method. The synthesized materials and their starting materials were successfully characterized using a variety of techniques such as Fourier transform infrared spectroscopy (FTIR), scanning electron microscopy (SEM), X-ray diffraction (XRD), nitrogen adsorption–desorption isotherms, the Brunauer–Emmett–Teller (BET) surface area calculation method, the Barrett, Joyner, and Halenda (BJH) pore size calculation method, thermogravimetry analysis (TGA), and 1H and 13C nuclear magnetic resonance (NMR) spectra. The results indicate that the new material (adsorbent 1) has a large surface and possesses different functional groups (-SH, -OH, -C=O and –N=C). The newly synthesized hybrid materials (**1** and **2**) were investigated as potential adsorbents for removal of toxic heavy metals, such as Pb(II) from aqueous solutions. The adsorption results show that materials **1** and **2** have different sorption properties and were found to be effective adsorbents for Pb(II) removal from aqueous solutions. In addition, compound **1** exhibited a higher adsorption capacity for Pb(II) compared to compound **2**. The results showed that the optimum pH for Pb(II) sorption was 6.5. Contact time was observed to occur after 30 min for 25 mg L^{-1} Pb(II) concentration and adsorbent dose of 0.4 g L^{-1} at 25 °C.

Keywords: heavy metals; hybrid materials; functionalized; Schiff base; lead; Langmuir and Freundlich

1. Introduction

Excessive heavy metal concentrations in drinking water pose significant threats to human health and the environment. For example, lead (Pb) is a very commonly used metal with a relatively high atomic mass, atomic number, and density of 11.34 g/cm^3. Pb is widely used in battery production, paints, and fertilizers. However, exposure to high Pb levels can cause major problems in the biosynthesis of hemoglobin (Hb), which is important for oxygen transport in humans and other vertebrates. In addition, Pb can cause serious damage to the liver, kidneys, and nervous system. Elimination of highly toxic heavy metal ions, such as Pb, mercury (Hg), chromium (Cr), copper (Cu), cobalt (Co), and zinc (Zn), from water is necessary and recently has been the subject of extensive research. Polluted water can cause serious environmental problems and can cause harm to both humans and other animal life [1–4].

A number of techniques are used for treatment of wastewater discharge. Some of the primary methods include precipitation, flocculation, flotation, and chelation reactions. Other methods include ion-exchange columns, solvent extraction, reverse osmosis, and electrolytic plating on an anode. Nonetheless, these methods suffer from being costly and inappropriate for removing a lot of contaminants; for instance, large amounts of chemicals are needed for lime precipitation, and its electrolytic recovery suffers from corrosion [5–13]. Recently, low-cost adsorbents derived from functionalized inorganic–organic hybrid materials have been extensively used as adsorbents for removal of heavy metal from wastewater because of their notable properties, such as strong binding affinities and high adsorption capacities towards heavy metal ions; in addition, they have large surface areas and good chemical, thermal, and mechanical stabilities [14–24].

Among the different types of adsorbents is mesoporous silica material, which is widely used and a well-known adsorbent for removal of toxic heavy metals. This material consists of a porous structure, and there are two types: (1) hexagonal (MCM-41), which is a silica-based mesoporous material with size between 50 and 120 nm, with different morphologies and hollow and core-shell spheres and (2) three-dimensional cube (MCM-48), which has four different structures (worm-like, helical, radial, and lamellar). MCM-41 has been extensively used in different applications, such as catalysis, adsorption, and separation processes, and as drug carriers [25]. Recently, modifications of MCM-41 with organic compounds in order to enhance the properties of MCM-41 and to improve its adsorption capacities towards toxic heavy metals has attracted more interest [26–29]. For example, Liu and co-workers [30] prepared novel zwitterion hybrid polymer adsorbents for adsorption of Pb(II) and Cu(II) ions from aqueous solution. Pavan et al. [31] prepared a stable mesoporous aniline/silica sorbent material for adsorption of Co, Zn, and cadmium chlorides ($CoCl_2$, $ZnCl_2$, and $CdCl_2$, respectively) from aqueous and ethanol solutions.

On the other hand, the stable azomethine group (-C=N-R, in which R is either an alkyl or aryl), also known as an imine or Schiff base, has been shown to play an important role in stabilizing metal complexes because it can act as a good binding site for many transition metals and therefore can form stable coordination complexes. Recent studies have demonstrated that mesoporous silica materials (MCM-41) can be immobilized with organic groups such as thiols, anilines, and Schiff bases and then used successfully for removal of toxic heavy metals from contaminated water. For example, Radi et al. [32] anchored a Schiff-base on silica gel for adsorption of Cd(II), Cu(II), and Zn(II) from aqueous solutions. The author concluded that the adsorbent can be regenerated and used several times without loss of its activity. However, the disadvantage of their adsorbent was the maximum sorption of Cu(II), Zn(II), and Cd(II) occurred at a pH of >8. Therefore, at this pH, hydrolysis of these metal ions may occur, and this hydrolysis makes it difficult to distinguish between hydrolyzed and adsorbed metals.

The goal of this present study was to prepare new modified mesoporous silica materials as adsorbents with large surface areas, regular pores, and high adsorption capacities for removal of toxic heavy metal ions from contaminated water. For this purpose, we have fabricated two hybrid materials, namely, Schiff base-modified and propyl-thiol-modified silicas (adsorbents **1** and **2**, respectively) by a direct co-condensation method using (3-mercaptopropyl) trimethoxysilane as the silica source. The removal of heavy metal ions such as Pb(II) from aqueous solutions by adsorbents **1** and **2** was examined by considering the effect of different factors, such as pH and contact time, between the absorbent and surrounding solution.

2. Materials and Methods (Experimental Section)

2.1. Materials

2-amino-4,5-dimethoxybenzoic acid, 4-hydroxybenzaldehyde, (3-mercaptopropyl) trimethoxysilane, and the 1000 mg/L standard solution with Pb(II) were purchased from Sigma Aldrich and used without modifications. HCl 0.5 M and NaOH 0.5 M were used to adjust the required pH. All reagents and solvents (analytical grade) were used without any further purification. All solutions were prepared with fresh deionized water obtained from a Milli-Q water system.

2.2. Instrumentation

The residual Pb (II) concentration was determined by inductively coupled plasma mass spectrometry (ICP-MS; Agilent 7500). A digital pH meter was used to determine the pH of water samples. Fourier-transform infrared (FTIR) spectra were obtained with a Perkin Elmer System 2000. Scanning electron microscopy (SEM) images were obtained on an FEI-Quanta 200. The ^{1}H and ^{13}C nuclear magnetic resonance (NMR) spectra of the Schiff base were obtained with a CP MAX CXP 300 MHz. A specific area of modified silica was determined by using the Brunauer–Emmett–Teller (BET) equation. Nitrogen adsorption–desorption was obtained by means of a Thermoquest Sorpsomatic 1990 analyzer after the material had been purged in a stream of dry nitrogen. Thermal gravimetric analysis (TGA) was conducted on a SDT Q 600 analyzer. Powder X-ray diffraction (XRD) patterns for determining phase purity and structure of the end product were performed on an XRD P-6000-Shimadzu X-ray diffractometer (40 kV/20 mA) using a conventional θ–2θ reflection geometry and Cu K α radiation (λ = 1.5406 A°).

2.3. Preparation of the Schiff Base 2-(4-Hydroxybenzylideneamino)-4,5-Dimethoxybenzoic Acid

2-amino-4,5-dimethoxybenzoic acid (1 mmol) and 4-hydroxybenzaldehyde (1 mmol) were mixed and heated to reflux for 4 h in ethanol (40 mL). The orange precipitate of the Schiff base was filtered, washed with cooled ethanol, and dried. The yield was 80% (Scheme 1). Analytical conditions were set for several different methods: (1) FTIR (KBr, cm^{-1}): 1521 (vC=N Schiff base), 1601 (vC=O), 1452–1354 (vC=C), 3442 phenolic hydroxide v (OH); (2) 1HNMR (DMSO, ppm, 400 MHz): 3.62 (s 3H), 3.72 (s, 3H), 6.32 (s H), 6.91–0.6.93 (d, 2H), 712 (s, H), 7.75–7.77 (d, 2H) and 9.78 (s, H); and (3) 13C NMR (dimethyl sulphoxide, DMSO, ppm, 400 MHz): 55.64, 55.38, 99.49, 101.05, 113.59, 116.31(2C), 128.84, 132.61(2C), 139.55, 148.71, 154.94, 163.80, 169.62, and 191.50.

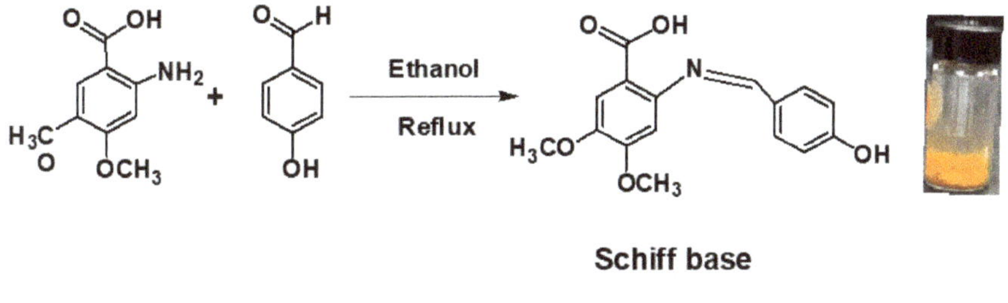

Scheme 1. Schiff base synthesis.

2.4. Preparation of Schiff Base-Thiol Silica (Adsorbent 1)

The Schiff base 2-(4-hydroxybenzylideneamino)-4,5-dimethoxybenzoic acid) (3) (first reaction) was added to a solution of deionized water/EtOH (25.0 mL, 50% *v/v*). The resulting suspension was stirred for 30 min. To this suspension, 3-(triethoxysilyl) propane-1-thiol (two equivalents) in deionized water/EtOH (25.0 mL, 50% *v/v*) was added. A few drops of base were added as catalyst. The mixture was stirred for 72 h at 60°C. The resulting yellowish solid (adsorbent **1**) was filtered and washed several times with distilled water and ethanol. The solid was then dried under vacuum overnight at 105 °C before characterization (Scheme 2).

The propyl-thiol functionalized silica material (adsorbent **2**) was prepared by the sol-gel method (direct synthesis) after stirring (3-mercaptopropyl) trimethoxysilane in an ethanol/water mixture (50% *v/v*) using base as catalyst (Scheme 3). The mixture was stirred for 72 h at 60 °C. The resulting white solid was filtered and washed several times with distilled water and ethanol. The solid was then dried under vacuum overnight at 105 °C (Scheme 3).

Scheme 2. The synthesis of adsorbent 1.

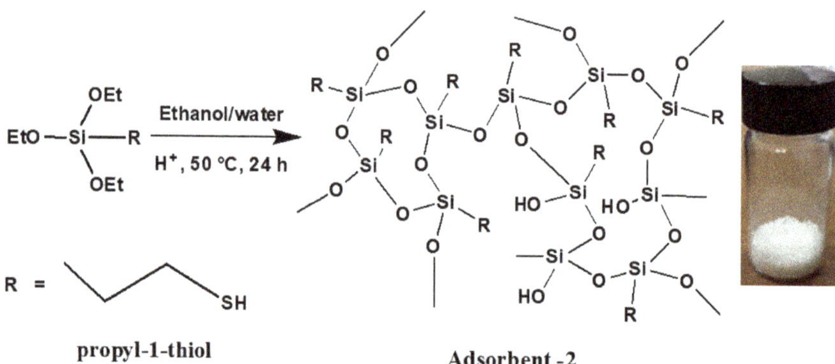

Scheme 3. Preparation of adsorbent 2.

2.5. Sorption Experiment

2.5.1. pH Optimization

The effects of pH on the adsorption of Pb(II) by adsorbents **1** and **2** was investigated at room temperature. Adsorption checks were carried out by adding 10 mg of adsorbent **1** or **2** to each flask in a series of 100 mL Erlenmeyer flasks. To each flask, 50 mL of Pb(II) solution (25.52 mg/L) was added. Hydrochloride acid (HCl; 0.5 M) and sodium hydroxide (NaOH; 0.5 M) were used to adjust the initial pH of the solutions between 2 and 7. A pH meter was used to measure the pH of the solutions. Each suspension was stirred at room temperature for 2 h at 25 °C over various pH ranges (2.0, 4.0, 6.0, and 7.0). The solution was then filtered, and the amount of Pb(II) ions in the filtrate was analyzed for residual Pb(II) ions using inductively coupled plasma mass spectrometry (ICP-MS) in triplicate.

2.5.2. Effect of Contact Time for Pb(II)

The 1000 mg/L standard solution of lead Pb(II) ions was diluted with deionized water to obtain 50 mL of 40 mg/L. Adsorption checks were carried out by adding 20 mg of each adsorbents **1** and **2** to three Erlenmeyer flasks (100 mL). To each flask, 50 mL of Pb(II) solution (of 40 mgL^{-1}) was added. The pH of solutions was adjusted to the optimum values (6.5 ± 0.2). The mixture was stirred constantly at 300 rpm in order to allow contact at different contact times. Five milliliter portions from each solution were filtered at measured time intervals (0 to 340 min), and the residual Pb(II) ion concentrations were measured using ICP-MS in triplicate. Results expressed as ppm for adsorbents **1** and **2**, respectively. The sorption capacities at equilibrium (mg g^{-1}) and the percentage adsorption (% ads) of adsorbents **1** and **2** were calculated using the following Equations (1) or (2) [4]:

$$qe = [C_0 - C_e]/m \times V \tag{1}$$

$$\% \ ads = [C_0 - C_e]/C_0 \times 100 \tag{2}$$

where q_e is the equilibrium uptake capacity (mg g^{-1}), C_0 is the initial concentration of Pb(II) (mg L^{-1}), C_e is the equilibrium concentration of Pb(II) expressed in mg L^{-1}, V is the volume of aqueous solutions (L), and m is the weight of the adsorbent **1** or **2** (g).

3. Results and Discussion

3.1. Synthesis of Adsorbents 1 and 2

The aim of this work was to improve the sorption efficiency of silica for removal of toxic heavy metals from polluted water. For this purpose, two mesoporous silica materials (adsorbents) were prepared. Adsorbent **1** was silica functionalized with Schiff base-propyl-thiol and adsorbent **2** was silica functionalized with propyl-thiol. The Schiff base was synthesized from 2-amino-4,5-dimethoxybenzoic acid and 4-hydroxybenzaldehyde via a condensation reaction in ethanol using acetic acid as a catalyst (Scheme 1). The Schiff base purity was confirmed by FTIR (Figure 1), ^{1}H and ^{13}C NMR spectroscopy (See Supporting Information; Supplementary Materials, Figures S1 and S2). The synthetic route of adsorbent **1** is shown in Scheme 2, the organic precursor 3-(triethoxysilyl) propane-1-thiol was added to a suspension containing Schiff base 2-(4-hydroxybenzylideneamino)-4,5-dimethoxybenzoic acid), ethanol, water and ammonia as a catalyst to obtain Schiff base-thiol silica gel (adsorbent **1**) through the sol-gel reaction. The silica functionalized with propyl-thiol (adsorbent **2**) was prepared by co-condensation (direct synthesis) of the organic precursor 3-(triethoxysilyl) propane-1-thiol in the presence of ammonia as a catalyst (Scheme 3).

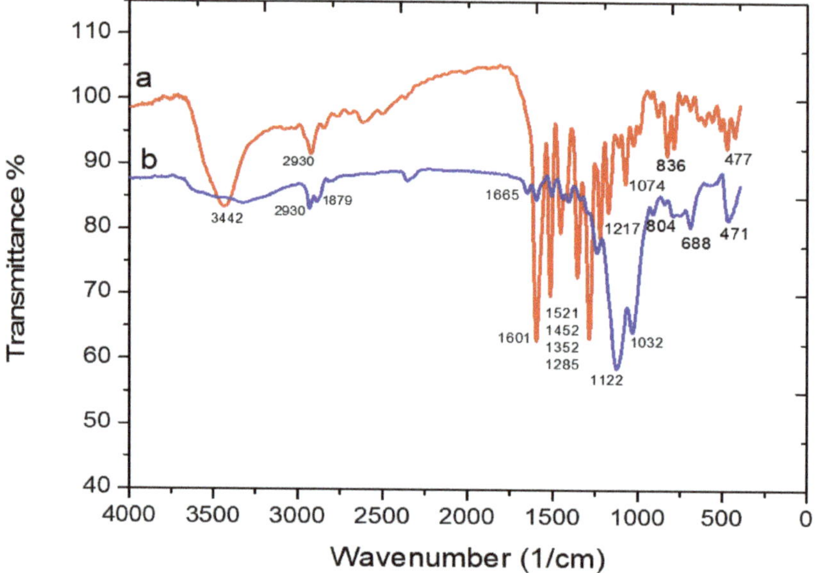

Figure 1. FTIR spectra of (a) free Schiff base and (b) adsorbent **1**.

The two adsorbents (**1** and **2**) were isolated as yellowish and white powders, respectively. The hybrid materials were washed several times with water, ethanol, and dried under vacuum before characterization. The synthesized adsorbents were completely characterized by FTIR, XRD, SEM, energy-dispersive X-ray spectroscopy (EDX), TGA, and BET analyses.

3.2. Characterization

3.2.1. The FT-IR Spectra of Free Schiff Base and the Schiff Base Functionalized Silica (Adsorbent **1**)

The FT-IR spectra of free Schiff base and the Schiff base functionalized silica (adsorbent **1**) are shown in Figure 1. Figure 1a displays the characteristic peaks of free Schiff base. The broad band at 3442 cm^{-1} was assigned to the vibration of the phenolic group υ(OH). In addition, the absorptions at 1601 and 1521 cm^{-1} in the free Schiff base were due to carbonyl υ(C=O) and azomethine υ(C=N), respectively. These vibrations appeared at lower frequencies due to strong intramolecular hydrogen bonds in the solid state of the free Schiff base. The weak peaks observed at 2921 cm^{-1} belonged to the stretching modes of the aliphatic -C-H bond. Figure 1b, shows the FT-IR spectrum of adsorbent **1**. The characteristic vibration of the phenolic group υ(OH) at 3442 cm^{-1} in free Schiff base (Figure 1a) was decreased indicating successful functionalization of Schiff base on the silica surface. However, adsorbent **1** displayed broad bands in the region of 3250 to 3600 cm^{-1}, which were assigned to the silanol stretching modes ν(Si-OH). The intensive absorption peaks observed at 1122–1032 cm^{-1} are due to asymmetric ν(Si–O-Si) and the band at 804 cm^{-1} is due to symmetric stretching vibrations of Si-O-Si [33]. The weak peaks observed at 2930 cm^{-1} belong to the stretching modes of the aliphatic -C-H bond. The appearance of bands around 1665–1402 cm^{-1} were caused by C=O, C=N, and C=C vibrations, confirming the anchoring of the organic molecule (Schiff base) onto the silica surface [34]. Moreover, the characterization features of adsorbent **1** compared with the free Schiff base were the shift of the carbonyl group from 1601 cm^{-1} (in free Schiff base) to 1665 cm^{-1} (in adsorbent **1**) and the disappearance of the thiol S-H peak at 2556 cm^{-1} compared to adsorbent **2**.

3.2.2. The FT-IR Spectra of Adsorbent 1 before and after Loading Pb(II) Ions

The FT-IR spectra of adsorbent **1** before and after loading Pb(II) ions are shown in Figure 2. It can be seen that the results for the Pb(II) ions loaded sample were almost same as the pre-adsorption sample, However, some of the bands after lead ions were loaded shifted to a lower value than that of the pre-adsorption sample, confirming the adsorption of the Pb(II) ions on the silica surface. This suggests the involvement of the oxygen atom of the carboxylate anion with the azomethine nitrogen and the Pb(II) ion [35]. The infrared spectrum in Figure 2b revealed the asymmetric stretching vibration of $v_{as}(COO^-)$ of adsorbent **1** at 1665 cm^{-1} was shifted to a lower wave number (1642 cm^{-1}) after the Pb(II) ion was loaded, indicating that the lead ion coordination took place via the oxygen atom of the carboxylate anion [36].

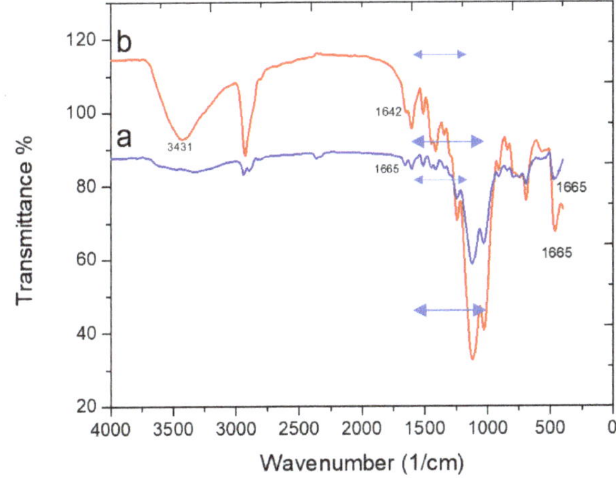

Figure 2. FTIR absorption spectrum of adsorbent **1**, (a) before absorption and (b) after adsorption.

3.2.3. The FT-IR Spectra of Adsorbent 2 before and after Loading Pb(II) Ions

Figure 3 shows the FT-IR spectroscopy results of adsorbent **2** samples taken before and after Pb(II) ion adsorption studies. The spectrum of synthesized adsorbent **2** before adsorption studies (Figure 3a) presents a broad band centered at 3495 cm^{-1}, which was assigned to the OH vibrations in the silica framework, while the strong absorption peak located at 1128 cm^{-1} was due to asymmetric v(Si–O–Si), and the band at 804 cm^{-1} was due to symmetric v(Si–O) stretching vibrations [33]. The weak peaks observed at 2921 cm^{-1} belong to the stretching modes of the aliphatic -C-H bond. The characteristic thiol S-H functional group peak was detected at 2556 cm^{-1} confirming the anchoring of the organic molecule (propyl-thiol) onto the silica surface. Compared with the spectrum of adsorbent **2** after Pb(II) ion loading (Figure 3b), it can be seen that the results for the lead ion loaded sample were similar to that before adsorption, except the intense broad peak centered at 3431 cm^{-1} was attributed to coordinative water in the coordination sphere of Pb(II) Schiff base complex. The characteristic peaks at 2556 cm^{-1} has diminished, suggesting that adsorbent **2** interacted with soft Lewis acid Pb(II) ions through the thiol groups.

3.2.4. XRD of Adsorbents 1 and 2

The XRD patterns of Schiff base-functionalized silica (adsorbent **1**) and propyl-1-thiol functionalized silica (adsorbent **2**) are illustrated in Figure 4. The XRD patterns of both adsorbents **1** and **2** showed weak broad peaks at a low 2θ angle (7.47°) and broad peaks at a high 2θ angle (20.21°) for adsorbent **1** and at 23.57° for adsorbent **2**, indicating the amorphous nature of the silica in both adsorbents [37]. Furthermore, the shift of the broad peak at a high 2θ angle (23.57°) for adsorbent

2 to 20.21° for adsorbent **1** suggests structural perturbations resulting from the incorporation of the Schiff base within the silica framework [34]. The XRD patterns of adsorbent **1** show sharp peaks at 37.73, 43.79, 64.29, 77.47, 81.81, and 98.37 of 2θ. These sharp peaks were due to the ligand (Schiff base) immobilized on the surface of silica [38].

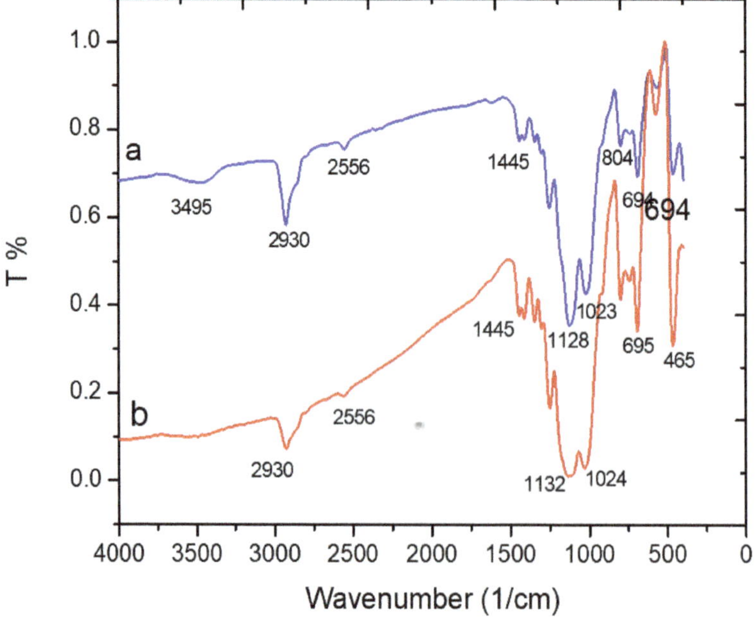

Figure 3. FTIR absorption spectrum of adsorbent (2), (a) before absorption and (b) after adsorption.

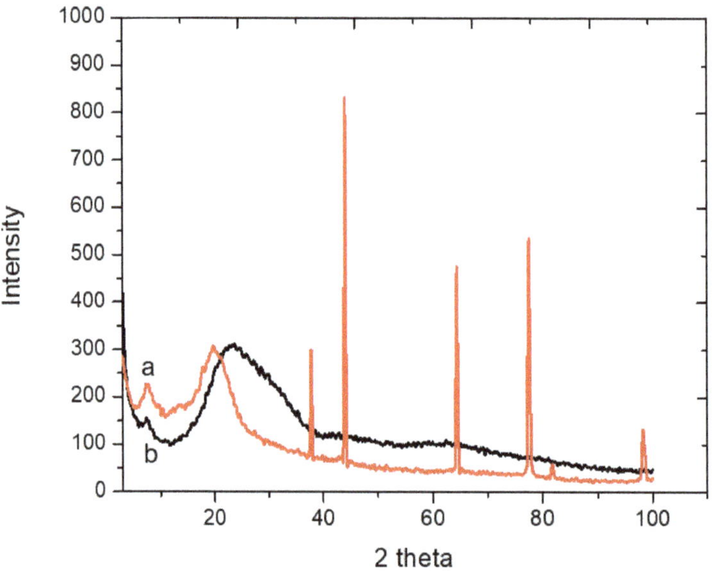

Figure 4. The X-ray diffraction (XRD) patterns of (a) adsorbent **1** and (b) adsorbent **2**.

3.2.5. SEM of Adsorbents **1**

The SEM images of adsorbent **1** are shown in Figure 5. As revealed in Figure 6, adsorbent **1** showed uniform distribution of spherical particles with different sizes. The presence of the organic components (Schiff base and propyl thiol groups) on the overall surface caused the surface of adsorbent **1** to be rough as a result of the microporous structure. Furthermore, the EDX pattern of adsorbent **1** is shown in Figure 6. EDX data in Table 1 confirm the formation of adsorbent **1,** as shown by the Si, S, N, C, and O peaks, thus confirming the presence of organic components (Schiff base and propyl thiol groups) on the silica surface.

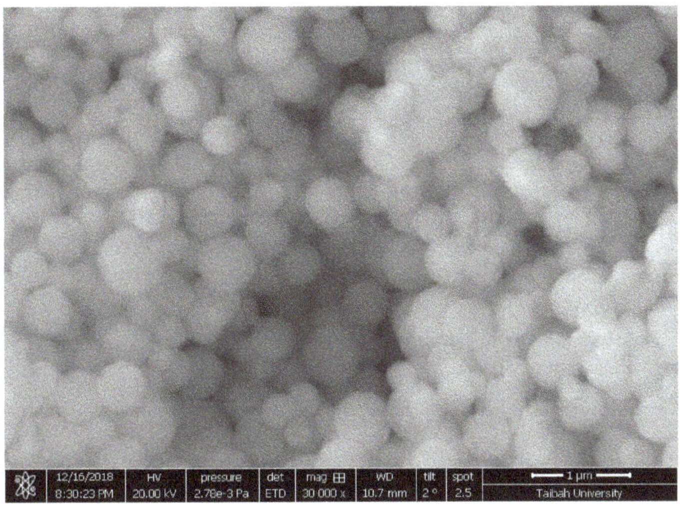

Figure 5. Scanning electron microscopy (SEM) images of pure adsorbent **1** before adsorption.

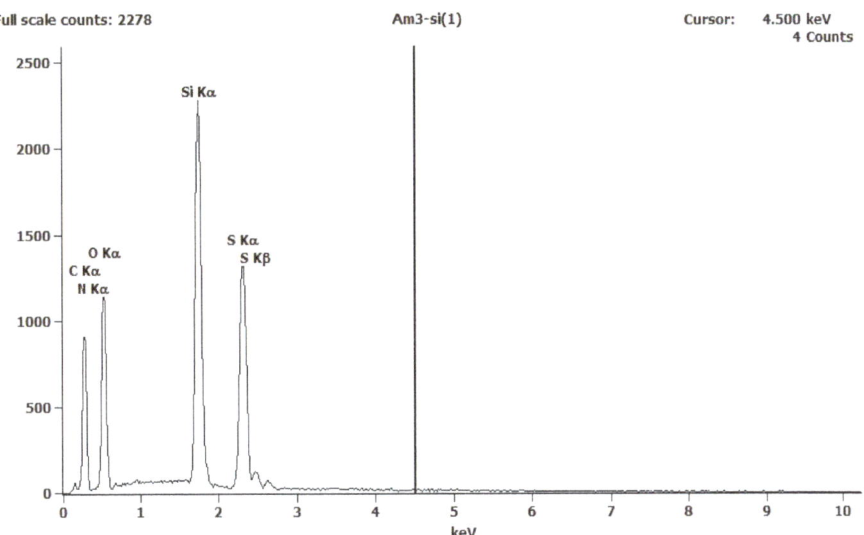

Figure 6. Energy-dispersive X-ray spectroscopy (EDX) spectrum of adsorbent **1** before adsorption.

Table 1. (EDX) data of adsorbent **1** before adsorption.

Element	Weight %	Atom %
C	20.99	34.36
N	1.19	1.66
O	22.46	27.59
Si	28.09	19.66
S	27.28	16.73
Total	100.00	100.00

Figure 7 presents an SEM micrograph of Pb(II) ions loaded onto adsorbent **1**. SEM analysis revealed that adsorbent **1** displayed agglomerated spherical particles with different sizes and granular morphologies. The accompanying EDX spectrum of the silica-functionalized Schiff base shown in Figure 8 confirms the presence of Pb(II) ions on the surface of adsorbent **1**. The EDX data in Table 2 confirmed that adsorbent **1** was very effective in removing Pb(II) ions from aqueous solutions. The EDX results were in good agreement with the results obtained by inductively coupled plasma mass spectrometry (ICP-MS).

Figure 7. SEM images of adsorbent **1** after adsorption.

Table 2. (EDX) data of adsorbent **1** after adsorption.

Element	Weight %	Atom %
C	24.07	36.73
N	1.78	2.33
O	29.81	34.16
Si	21.60	14.10
S	22.08	12.63
As	0.00	0.00
Pb	0.66	0.06
Total	100.00	100.00

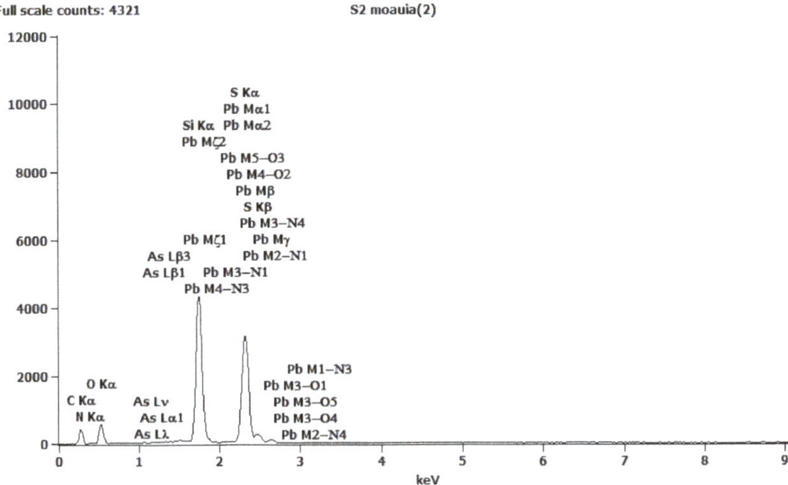

Figure 8. (EDX) spectrum of adsorbent **1** after adsorption.

3.2.6. Field Emission Scanning Microscope (FESEM) of Adsorbent **2**

Figure 9 shows the field emission scanning microscope (FESEM) image of adsorbent **2**. As shown in Figure 10, the hybrid is spherical with a rough surface. The SEM image showed a homogeneous material with pores covering the entire structure. The pore size varied between 2 and 8 µm. The presence of the overall surface of the propyl thiol groups caused the surface of adsorbent **2** to be rough with a microporous structure. Moreover, the EDX pattern of adsorbent **2** (Figure 10) confirms the purity of adsorbent **2** and also confirms the presence of propyl thiol groups on the silica surface. The EDX spectrum also presents a minor peak of Au and Cl that resulted from the thin film of gold deposited on the insulating sample in order to make it conductive and, hence, easy to visualize using SEM.

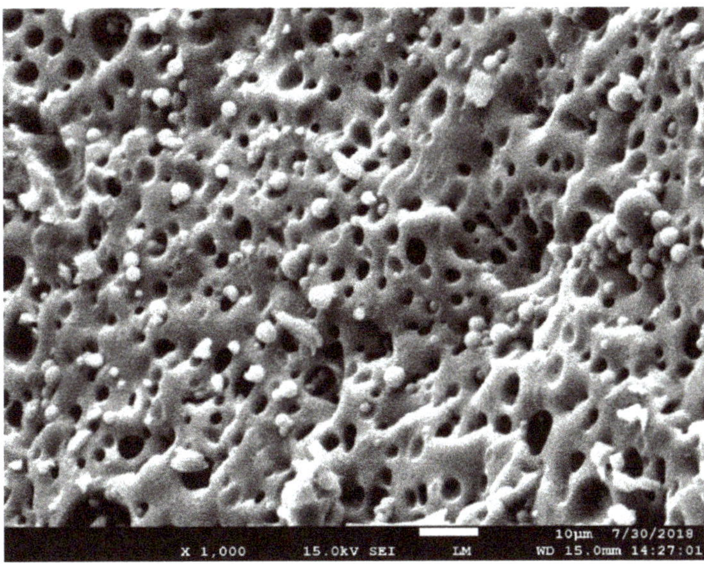

Figure 9. Field emission scanning electron microscopy (FESEM) images of adsorbent **2** before adsorption.

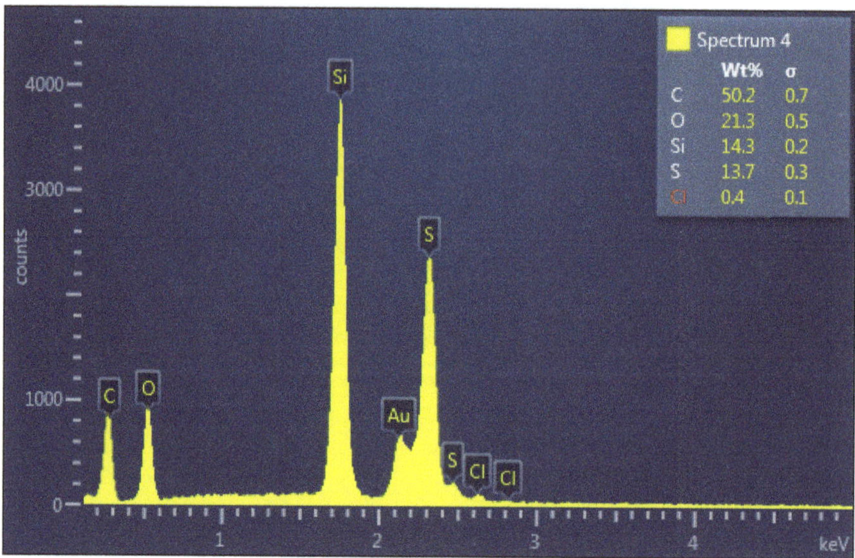

Figure 10. EDX images of adsorbent **2** (inset is the EDX data of adsorbent **1** before adsorption).

Figure 11 presents an SEM micrograph of Pb(II) ions loaded onto adsorbent **2**. SEM analysis revealed that adsorbent **2** displayed spherical particles with different sizes and rough morphology. The accompanying EDX spectrum of the silica-functionalized Schiff base, shown in Figure 12, confirms the presence of Pb(II) ions on the surface of adsorbent **2** (see Table 3). However, the EDX spectrum confirmed that adsorbent **2** was also effective in removing Pb(II) ions from aqueous solutions. EDX studies have shown that both adsorbents (**1** and **2**) have strong affinities to divalent metal cations (Pb(II)).

Figure 11. SEM images of adsorbent **2** after adsorption.

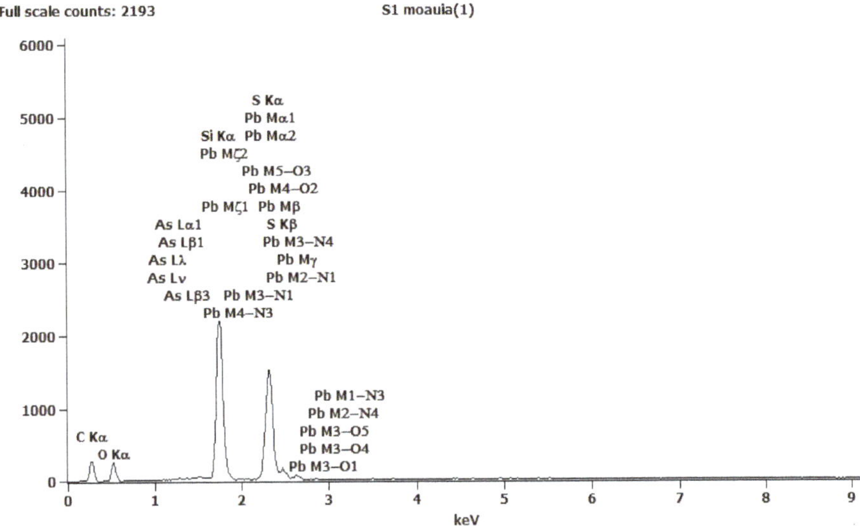

Figure 12. EDX spectrum of adsorbent **2** after adsorption.

Table 3. EDX data of adsorbent **2** after adsorption.

Element	Weight %	Atom %
C	30.19	46.20
O	24.74	28.42
Si	21.87	14.32
S	18.54	10.63
As	0.10	0.02
Pb	4.55	0.40
Total	100.00	100.00

3.2.7. TGA of Adsorbent **1** and **2**

The TGA thermal details of adsorbents **1** and **2** are presented in Figure 13. The thermogram of adsorbent **1** (Figure 13a) shows that there are essentially two regions of weight loss; in the first region, a slight weight lost in the range of 45 to 285 °C due to the loss of physiosorbed solvents used in the preparation (2.85%) can be seen. The second region appears in the range of 285 to 475 °C (with lost weight equal to 28.1% and with DTG_{max} at 346 °C) due to decomposition of the Schiff base and propyl-thiol groups immobilized on the silica surface [39]. The remaining weight loss in the range of 475 to 885 °C (with total weight loss equal to 26.23%) was due to the condensation of the hydroxyl group silanol (Si-O-OH) to yield siloxane bonds (Si-O-Si) [38]. At 885 °C, adsorbent **1** around 57.18% was degraded, leaving 42.82% behind as residue.

3.2.8. Surface Properties of Adsorbents **1** and **2**

Using nitrogen adsorption isotherms, the surface area, pore volume, and pore size of adsorbents **1** and **2** were measured. It is noted that adsorbent **1** exhibited a BET surface area of 22.0452 m^2/g, the pore volume (BJH) of adsorbent **1** was 0.120986 cm^{3}g^{-1} with pore sizes of 15.7644 nm (BET) and 22.7447 nm (BJH). On the other hand, adsorbent **2**, which was prepared by direct hydrolysis and condensation of 3-(triethoxysilyl) propane-1-thiol, exhibited a BET surface area of 9.3475 m^{2}g^{-1}. The pore volume (BJH) of adsorbent **2** was 0.049107 cm^{3}g^{-1} with pore sizes of 9.68719 nm (BET) and 25.0027 nm (BJH). The reduced surface area and pore volume of adsorbent **2** were due to the smaller size of the propyl-thiol group. The noticeable increase in the BET surface area of adsorbent **1** was attributed to the Schiff base and propyl-thiol groups anchored on the silica framework. This indicates that adding more organic moieties to silica framework results in a large surface area, which increases the possibility of extracting metal ions [40].

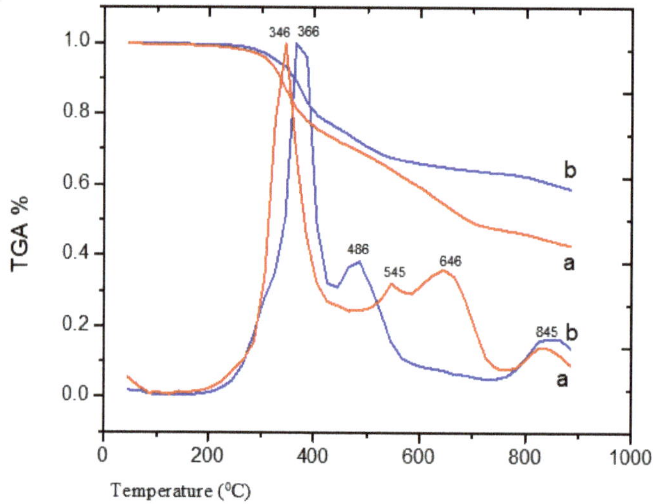

Figure 13. Thermogravimetric (TGA) curve of (a) adsorbent **1** and (b) adsorbent **2**.

3.3. Adsorption Study

3.3.1. Effects of pH on the Adsorption of Pb(II) by Adsorbent **1**

The effects of the pH value on the adsorption of Pb(II) ions by adsorbents **1** and **2** are shown in Figure 14. The results show an increase in percent adsorption (% ads) for sorption of Pb(II) ions by both adsorbents with increasing pH reaching a maximum of 91.30% and 73.54% at pH 7.0 for adsorbents **1** and **2** respectively.

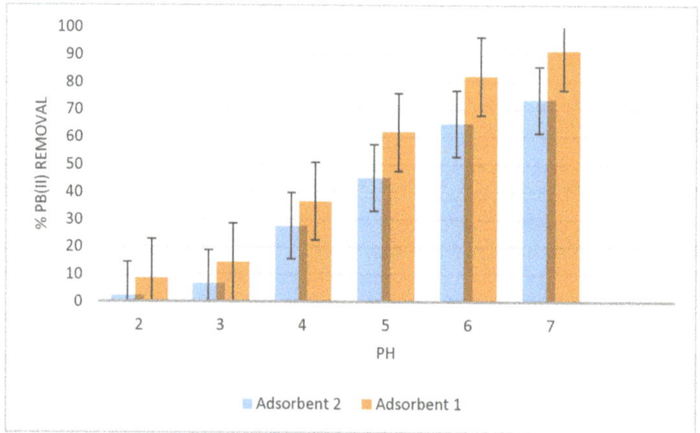

Figure 14. Effects of pH on the adsorption of Pb(II) (mean ± RSD) by adsorbent **1** and adsorbent **2**. (Experimental conditions: C_o = 25.52 mg/L, dosage = 0.01 g per 50 mL, shaking time 2 h, mixing rate = 300 rpm; T = 25 °C). Each pH measurement was done in triplicate.

From the plot in Figure 14, it can be observed that the Pb(II) ion uptake sharply increased from pH 2.0 to 7.0. At low pH values, the interaction of Pb(II) with adsorbents **1** and **2** decreases because the surfaces of the adsorbents were fully covered by hydronium (H_3O^+) ions, resulting in minimum adsorption. As pH increased, the interaction of Pb(II) with adsorbents **1** and **2** increased, and the adsorption also increased. Hence, the optimum pH for the maximum sorption of Pb(II) was pH = 7.0. Above a pH of 7.0, we observed precipitation of Pb(II) ions in the form of $Pb(OH)_2$ in a control sample (without adsorbents).

3.3.2. Effect of Contact Time and Kinetics for the Adsorption of Pb(II) by Adsorbents **1** and **2**

The kinetics of Pb(II) removal were determined in order to understand the adsorption behavior of adsorbents **1** and **2**. Figure 15 illustrates the Pb(II) ions adsorption on adsorbents **1** and **2** as a function of contact time (0–340 min). As can be seen from Figure 15, the rate of adsorption of Pb(II) ions was fairly rapid in the initial stages (in the first 50 min). After 60 min, the rate of adsorption slowed down as time progressed and reached a constant value around 60 min (equilibrium time). The initial rapid adsorption rate before 50 min could be explained by the fact that at the beginning of the adsorption process, the sites for adsorption were available and open for Pb(II) ions leading to a higher adsorption rate. As time progressed to greater than 60 min, the adsorption of Pb(II) ions slowed down because all binding sites were occupied by Pb(II) ions.

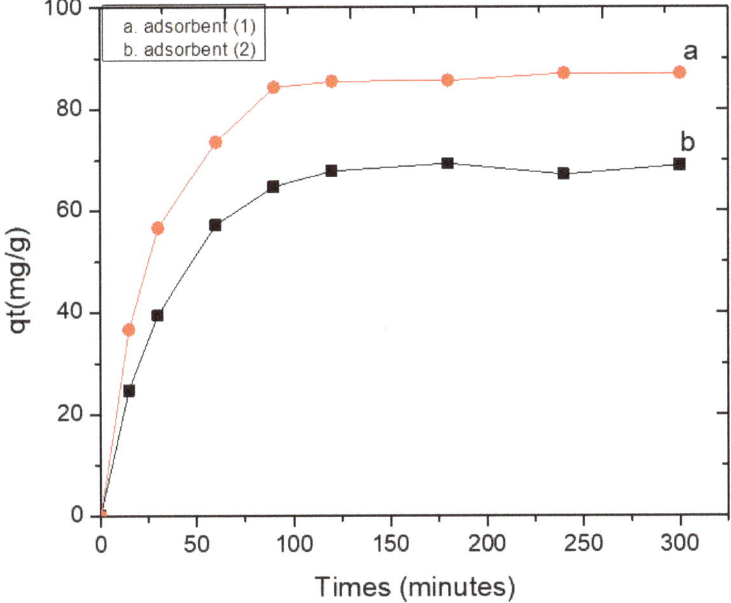

Figure 15. Kinetic adsorption of Pb(II) on adsorbent **1** (a) and adsorbent **2** (b) as a function of contact time (pH 6.5 ± 0.2, initial metal concentration = 20 mg/L, dosage = 50 mg/50 mL, mixing rate = 50 rpm, T = 25 °C).

3.3.3. Adsorption Kinetics for the Adsorption of Pb(II) by Adsorbents **1** and **2**

The dynamics of the adsorption by the adsorbents **1** and **2** were evaluated using the Lagergren's pseudo-first-order and the McKay and Ho's pseudo-second-order models as defined in Equations (3) and (4), respectively [41,42]:

$$Log\ (q_e - q_t) = -K_1\ t/2.303 + Log\ (q_e) \tag{3}$$

$$t/q_t = 1/K_2 q e^2 + t/q_e \tag{4}$$

where q_e is the adsorption capacity (mg g^{-1}) at equilibrium, q_t is the adsorption capacity (mg g^{-1}) at any time t, k_1 is the rate constants of first-order sorption (min^{-1}), and k_2 (g mg^{-1} min^{-1}) is the rate constant for the pseudo-second-order adsorption.

Figure 16 shows the linear plot of t/q_t versus t for the Lagergren's pseudo-second-order model for the adsorption of Pb(II) for adsorbents **1** and **2**, respectively. The equilibrium rate constants of the pseudo-second-order model (k_2) were 1.23×10^{-4} and 1.03×10^{-3} g mg^{-1} min^{-1} for Pb(II) for adsorbents **1** and **2,** respectively (Table 4). The pseudo-second-order model (Equation (4)) fit the experimental data very well with a correlation coefficient (R^2) of 0.99, which was close to unity as shown in Table 4 and Figure 16. On the other hand, our calculations demonstrated that the resulting experimental data were not fitted to the pseudo-first-order model. Therefore, the adsorption kinetics of Pb(II) ions on the three adsorbents showed that the Pb(II) adsorption followed a second-order reaction. Moreover, the experimental values of qe (Exp.) for both adsorbents were very close to the calculated values qe (Cal.) measured from the pseudo-second-order equation.

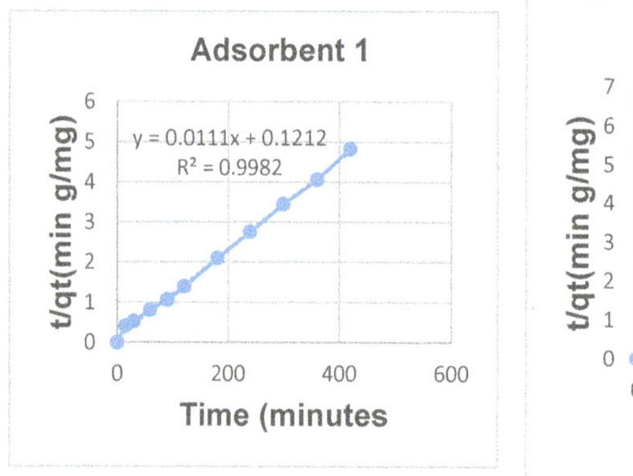
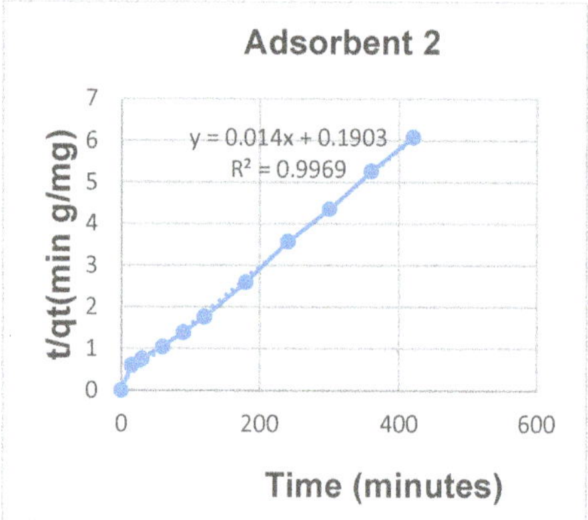

Figure 16. Lagergren pseudo-second-order kinetics adsorption for Pb(II) onto adsorbents **1** and **2** (Experimental conditions: pH 6.5 ± 0.2, initial metal concentration = 20 mg/L, dosage = 50 mg/50 mL, mixing rate = 50 rpm, T = 25 °C).

Table 4. Coefficients of sorption kinetics (Lagergren's pseudo-second-order model) for Pb (II) removal by adsorbents **1** and **2** and their correlation coefficient (R^2). (Experimental conditions: pH 6.5 ± 0.2, initial metal concentration = 20 mg/L, 50 mg/50 mL, mixing rate = 50 rpm, T = 25 °C).

Parameters	Adsorbent 1	Adsorbent 2
Calculated equilb. uptake q_e (mg/g)	90.09	71.43
K_2 (g mg^{-1} min^{-1})	1.23×10^{-4}	1.03×10^{-3}
R^2	0.9982	0.9969

Furthermore, the effect of temperature on the adsorption behavior for Pb(II) ions was also investigated. The effect of temperature on Pb(II) ions adsorption by adsorbents **1** and **2** is shown in Figure 17. The removal of Pb(II) ions by both adsorbents increased slightly with increasing solution temperature from 25 to 50 °C. This result suggested that the adsorption mechanism associated with Pb(II) ions on both adsorbents involved a temperature-dependent process to some extent. The adsorbent's weak sensitivity to temperature is essential to practical applications, thereby enabling both synthesized adsorbents to be potentially applied to the practical treatment of Pb(II) ions at room temperature.

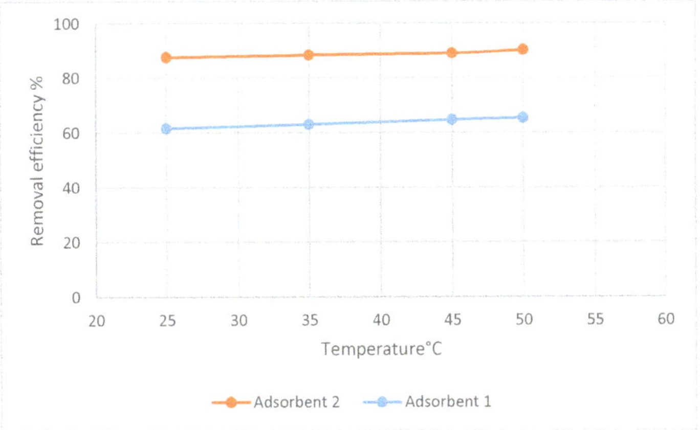

Figure 17. Effect of the temperature on Pb (II) ion adsorbtion (mean ± RSD) on adsorbent **1** and adsorbent **2**. (pH 6.5 ± 0.2, initial metal concentration = 20 mg/L Pb(II), dosage = 50 mg/50 mL, at different temperatures for 60 min). Each measurement has been done in triplicate.

3.3.4. Adsorption Isotherms

The adsorption isotherms were used to determine the affinity of adsorbents **1** and **2** to Pb(II) ions. The amount of Pb(II) ion per gram of adsorbent (q_e) was defined as shown in Equation (1) above. The sorption of Pb(II) into adsorbents **1** and **2** was described by Langmuir [43,44] and Freundlich [45] models:

- The Langmuir Isotherm Model [43]

$$q_e = q_m \, K_L \, C_e/1 + K_L \, C_e \tag{5}$$

where q_e is the is the lead concentration adsorbed per specific amount of adsorbent (mg g^{-1}), C_e is the equilibrium concentration of Pb(II) expressed in mg L^{-1}, and q_m is the maximum amount of lead ions required to form a monolayer (mg g^{-1}). The Langmuir equation can be rearranged to a linear form (Equation (6)) for the convenience of plotting and determination of the Langmuir constant (K_L, the first coefficient related to the energy of adsorption) as below. The values of q_m and K_L can be determined from the linear plot of 1/C_e versus 1/q_e:

$$1/q_e = 1/q_m + 1/K_L \, q_m \times 1/C_e \tag{6}$$

The dimensionless equilibrium parameter or separation factor (R_L) could be expressed as in the following equation [44]:

$$R_L = 1/1 + K_L C_0 \tag{7}$$

where C_0 is the initial concentration of the lead ion, and K_L is the Langmuir constant. The value of R_L is between 0 and 1 for favourable adsorption, while $R_L > 1$ represents unfavorable adsorption, and $R_L = 1$ represents linear adsorption. The adsorption process is irreversible if $R_L = 0$.

- The Freundlich Isotherm Model [45]

$$q_e = K_F \, C_e^{1/n} \tag{8}$$

$$\text{Log } q_e = \text{Log} K_F + (1/n) \text{ Log } C_e \tag{9}$$

where, q_e is the equilibrium uptake capacity (mg g^{-1}) and C_e is the equilibrium concentration of Pb(II)) expressed in mg L^{-1}. k_F is the first coefficient related to the energy of adsorption, and n is a coefficient. Both constants k_F and n can be calculated from the plot of log q_e vs. log Ce.

* The coefficients were calculated from linearized forms of sorption isotherms models Equations (6) and (9).

3.3.5. Linear Fitting of the Isotherm Models

Linear fitting of the Langmuir and Freundlich isotherm models is shown in Figures 18 and 19 respectively; the corresponding parameters of sorption isotherm models calculated are illustrated in Table 5. From the results shown in Table 5, the adsorption process of Pb(II) on both adsorbents can be described by the linear form of the Langmuir Isotherm Model, which produced higher R^2 values of 0.959 and 0.983 for Pb(II) onto adsorbents **1** and **2**, respectively, compared to the Freundlich isotherm model, which produced low R^2 values of 0.902 and 0.871 for Pb(II) onto adsorbents **1** and **2** respectively. Therefore, the adsorption isotherms for both adsorbents fitted well with the Langmuir model more than with Freundlich model; hence, the sorption process of Pb(II) ions involved in both adsorbents occurred by chemical complexation (chemisorption). The separation factor R_L (Equation (7)) was found to be between 0 and 1 (Table 5) indicating a favorable adsorption process [44].

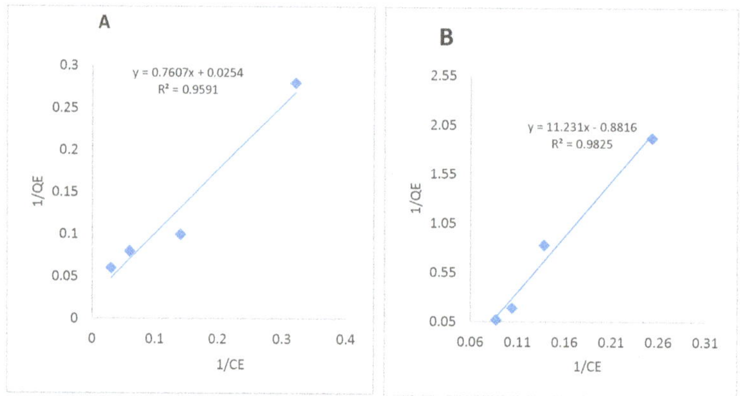

Figure 18. The Langmuir Isotherm Model adsorption for Pb(II) onto (**A**) adsorbent **1** and (**B**) adsorbent **2**. (Experimental conditions: dosage = 0.015 g (adsorbent **1**) and 0.1 (adsorbent **2**) per 50 mL; T = 25 ± 1 °C; contact time = 60 min.; pH = 6.5 ± for Pb(II).

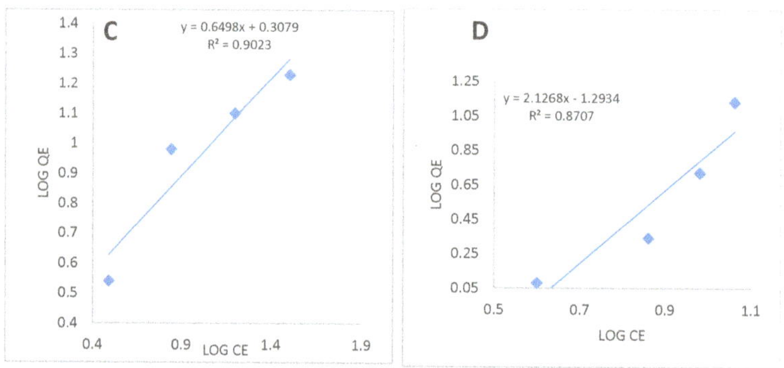

Figure 19. The Freundlich Isotherm Model adsorption for Pb(II) onto (**C**) adsorbent **1** and (**D**) adsorbent **2**. (Experimental conditions: dosage = 0.015 g (adsorbent **1**) and 0.1 (adsorbent **2**) per 50 mL; T = 25 ± 1 °C; contact time = 60 min.; pH = 6.5 ± for Pb(II).

Table 5. Coefficients of two different sorption isotherm models for Pb(II) removal by adsorbents **1** and **2** and their correlation coefficient (R^2). (Experimental conditions: dosage = 0.015 g (adsorbent **1**) and 0.1 (adsorbent **2**) per 50 mL; T = 25 ± 1 °C; contact time = 60 min.; pH = 6.5 ± 0.2 for Pb(II).

| Adsorbent | Langmuir Isotherm | | | | | | |
	Freundlich Isotherm						
	qm (mg/g)	K (L/mg)	RL	R2	KF	1/n	R2
1	39.37	0.0333	0.601	0.956	2.032	1.54	0.902
2	1.134	0.079	0.717	0.983	0.051	0.470	0.871

a.   The Langmuir Isotherm Model

b.   The Freundlich Isotherm Model

3.3.6. Complexation of Pb(II) by Adsorbent 1

Adsorbent **1** showed adsorption of Pb(II) ions due to the presence of Schiff base functionality (azomethine (C=N), carboxylate oxygen, and active thiol and silanol groups on the surface), whereas adsorbent **2** adsorbs Pb(II) ions due to the presence of active thiol and silanol groups only on the surface. Adsorbent **1** showed better adsorption of Pb(II) ions, which may be attributed to the presence of potentially good coordinating Schiff base sites and active carboxylate oxygen, in addition to thiol and hydroxyl groups. The shift of asymmetric stretching vibration of $\nu_{as}(COO^-)$ of adsorbent **1** at 1665 cm^{-1} to a lower wave number 1642 cm^{-1} and the disappearance of the thiol S-H peak at 2556 cm^{-1} after Pb(II) ion was loaded confirmed the chemical complexation of Pb(II) by adsorbent **1** (Figure 2b). The FTIR results (Figure 2b) support the structures proposed for Pb(II) ion interaction with adsorbent **1** (complex) during the adsorption process [46]. The Pb (II) ions interaction with adsorbent **1** is shown in Figure 20 below.

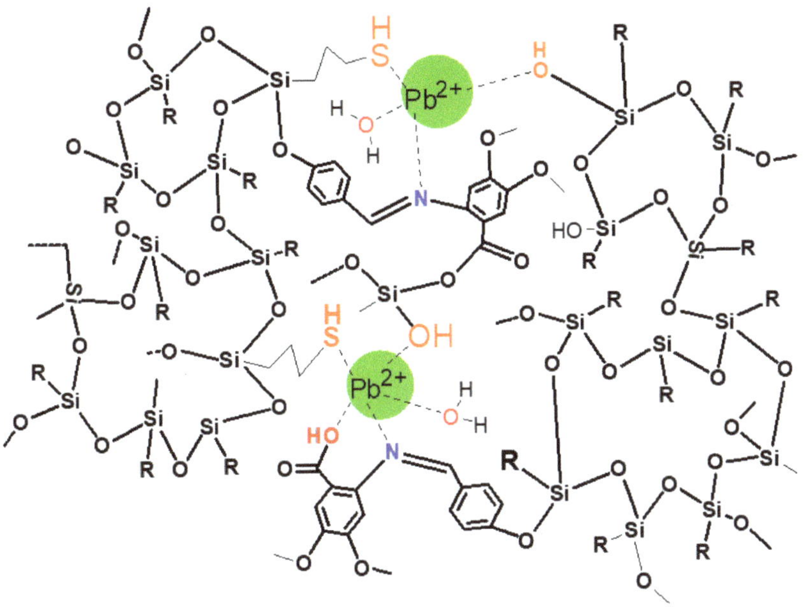

Figure 20. Pb(II) ion adsorbent **1** complex formation.

4. Conclusions

In this study, novel silica functionalized with Schiff-base (adsorbent **1**) and propyl-thiol (adsorbent **2**) were successfully prepared. Direct hydrolysis of 3-(triethoxysilyl) propane-1-thiol gave adsorbent **2**, and mixing 3-(triethoxysilyl) propane-1-thiol with a Schiff base produced a good yield of adsorbent **1**. The new materials were well characterized by FTIR, SEM, XRD, nitrogen adsorption–desorption isotherms, BET surface area, BJH pore sizes, TGA, and ^{1}H and ^{13}C NMR spectra. The newly synthesized adsorbents **1** and **2** displayed good thermal stability and adsorption capacity from aqueous solutions towards Pb(II) ions. The higher removal efficiency for Pb(II) by adsorbent **1** could be attributed to the higher surface area (20.1656 m^2/g), which facilitated the interaction of free Pb(II) ions present in solution with the surface of adsorbent **1**. The adsorption isotherms of Pb(II) fitted well with the Langmuir Isotherm Model, indicating that the adsorption mechanism followed Langmuir monolayer adsorption, and the adsorption kinetics of Pb(II) followed the pseudo-second-order kinetic models. In addition, the newly synthesized materials were proven to be cost-effective adsorbents and could be used for the selective extraction of Pb(II) from contaminated water. Overall, our studies suggested that adsorbent **1** could be considered as a good candidate for removing Pb(II) from aqueous solutions.

Supplementary Materials: The following are available online at http://www.mdpi.com/2227-9717/8/2/246/s1, Figure S1. ^{1}H-nuclear magnetic resonance (NMR) absorption spectrum of the Schiff base. Figure S2. ^{13}C NMR spectra of the Schiff base. Figure S3. Thermogravimetric (TGA) curves of the Schiff base.

Author Contributions: Methodology: M.O.A., A.S., and M.M.; Software: M.O.A., A.S.; Writing—Original Draft Preparation: A.S.; Writing—Review & Editing: M.O.A.; Project Administration: M.O.A. All authors have read and agreed to the published version of the manuscript.

Funding: This research received no external funding.

Conflicts of Interest: The authors declare no conflicts of interest.

References

1. Huuha, T.S.; Kurniawan, T.A.; Sillanpaa, M.E.T. Removal of silicon from pulping whitewater using integrated treatment of chemical precipitation and evaporation. *Chem. Eng. J.* **2010**, *158*, 584–592. [CrossRef]
2. El-Nasser, A.; Parish, R. Solid polysiloxane ligands containing glycine- or iminodiacetate-groups: Synthesis and application to binding and separation of metal ions. *J. Chem. Soc. Dalton Trans.* **1999**, *19*, 3463–3466. [CrossRef]
3. El-Nahhal, I.M.; El-Shetary, B.A.; Salib, K.A.R.; El-Ashgar, N.M.; El-Hashash, A.M. Uptake of divalent metal ions (Cu^{2+}, Ni^{2+}, and Co^{2+}) by polysiloxane immobilized triamine-thiol- and thiocyanate ligand system. *Anal. Lett.* **2001**, *34*, 2189–2202. [CrossRef]
4. Najafi, M.; Rostamian, R.; Rafati, A.A. Chemically modified silica gel with thiol group as an adsorbent for retention of some toxic soft metal ions from water and industrial effluent. *Chem. Eng. J.* **2011**, *168*, 426–432. [CrossRef]
5. Esalah, O.; Weber, M.; Vera, J. Removal of lead, cadmium and zinc from aqueous solutions by precipitation with sodium Di-(*n-octyl*) phosphinate. *Can. J. Chem. Eng.* **2000**, *78*, 948–954. [CrossRef]
6. Lertlapwasin, R.; Bhawawet, N.; Imyim, A.; Fuangswasdi, S. Ionic liquid extraction of heavy metal ions by 2-aminothiophenol in 1-butyl-3-methylimidazolium hexafluorophosphate and their association constants. *Sep. Purif. Technol.* **2010**, *72*, 70–76. [CrossRef]
7. Emamjomeh, M.M.; Sivakumar, M. Review of pollutants removed by electrocoagulation and electrocoagulation/flotation processes. *J. Environ. Manag.* **2009**, *90*, 1663–1679. [CrossRef]
8. Mahmoud, A.; Hoadley, A. An evaluation of a hybrid ion exchange electrodialysis process in the recovery of heavy metals from simulated dilute industrial wastewater. *Water Res.* **2012**, *46*, 3364–3376. [CrossRef]
9. Yurloval, L.; Kryvoruchko, A.; Kornilovich, B. Removal of Ni (II) ions from wastewater by micellar-enhanced ultrafiltration. *Desalination* **2002**, *144*, 255–260. [CrossRef]
10. Greenlee, L.; Lawler, D.; Freeman, B.; Marrot, B.; Moulin, P. Reverse Osmosis Desalination: Water Sources, Technology and Today's Challenges. *Water Res.* **2009**, *43*, 2317–2348. [CrossRef]
11. Benit, Y.; Ruiz, M. Reverse osmosis applied to metal finishing wastewater. *Desalination* **2002**, *142*, 229–234. [CrossRef]

12. Rubio, J.; Souza, M.; Smith, R.W. Overview of flotation as a wastewater treatment technique. *Miner. Eng.* **2002**, *15*, 139–155. [CrossRef]

13. Ghurye, G.; Clifford, D.; Tripp, A. Iron coagulation and direct microfiltration to remove arsenic from groundwater. *Am. Water Work. Assoc.* **2004**, *96*, 143–152. [CrossRef]

14. Wang, L.; Wu, X.; Xu, W.; Huang, X.; Liu, J.H. Stable Organic–Inorganic Hybrid of Polyaniline/α-Zirconium Phosphate for Efficient Removal of Organic Pollutants in Water Environment. *Appl. Mater. Interfaces* **2012**, *4*, 2686–2692. [CrossRef]

15. Gao, B.; Gao, Y.; Li, Y. Preparation and chelation adsorption property of composite chelating material poly (amidoxime)/SiO$_2$ towards heavy metal ions. *Chem. Eng. J.* **2010**, *158*, 542–549. [CrossRef]

16. Zaitseva, N.; Zaitsev, V.; Walcarius, A. Chromium (VI) removal via reduction–sorption on bi-functional silica adsorbents. *J. Hazard. Mater.* **2013**, *250*, 454–461. [CrossRef]

17. Simsek, E.; Duranoglu, D.; Beker, U. Heavy Metal Adsorption by Magnetic Hybrid-Sorbent: An Experimental and Theoretical Approach. *Sep. Sci. Technol.* **2012**, *47*, 1334–1340. [CrossRef]

18. Suchithra, P.; Vazhayal, L.; Mohamed, A.; Ananthakumar, S. Mesoporous organic–inorganic hybrid aerogels through ultrasonic assisted sol–gel intercalation of silica–PEG in bentonite for effective removal of dyes, volatile organic pollutants and petroleum products from aqueous solution. *Chem. Eng. J.* **2012**, *200*, 589–600. [CrossRef]

19. Repo, E.; Warchoł, J.; Bhatnagar, A.; Sillanpää, M. Heavy metals adsorption by novel EDTA-modified chitosan–silica hybrid materials. *J. Colloid Interface Sci.* **2011**, *358*, 261–267. [CrossRef]

20. Ge, P.; Li, F.; Zhang, B. Synthesis of modified mesoporous materials and comparative studies of removal of heavy metal from aqueous solutions. *Pollut. J. Environ. Stud.* **2010**, *19*, 301–308.

21. Pang, Y.; Zeng, G.; Tang, L.; Zhang, Y.; Liu, Y.; Lei, X.; Li, Z.; Zhang, J.; Xie, G. PEI-grafted magnetic porous powder for highly effective adsorption of heavy metal ions. *Desalination* **2011**, *281*, 278–284. [CrossRef]

22. Wang, L.; Zhang, J.; Wang, A. Fast removal of methylene blue from aqueous solution by adsorption onto chitosan-g-poly (acrylic acid)/attapulgite composite. *Desalination* **2011**, *266*, 33–39. [CrossRef]

23. Mercier, L.; Pinnavaia, T. Heavy Metal Ion Adsorbents Formed by the Grafting of a Thiol Functionality to Mesoporous Silica Molecular Sieves: Factors Affecting Hg (II) Uptake. *Environ. Sci. Technol.* **1998**, *32*, 2749–2754. [CrossRef]

24. Sanchez, C.; Julián, B.; Belleville, P.; Popall, M. Applications of hybrid organic–inorganic nanocomposites. *J. Mater. Chem.* **2005**, *15*, 3559–3592. [CrossRef]

25. Yanagisawa, T.; Schhimizu, T.; Kuroda, K.; Kato, C. The preparation of alkyltrimethylammonium-kanemite complexes and their conversion to microporous material. *Bull. Chem. Soc.* **1990**, *63*, 988–992. [CrossRef]

26. Kubota, Y.; Nishizaki, Y.; Sugi, Y. High Catalytic Activity of as-Synthesized, Ordered Porous Silicate–Quaternary Ammonium Composite for Knoevenagel Condensation. *Chem. Lett.* **2000**, *29*, 998–999. [CrossRef]

27. Wu, J.; Liu, X.; Tolbert, S. High-Pressure Stability in Ordered Mesoporous Silicas: Rigidity and Elasticity through Nanometer Scale Arches. *J. Phys. Chem.* **2000**, *104*, 11837–11841. [CrossRef]

28. Wu, J.; Abu-Omar, M.; Tolbert, S. Fluorescent Probes of the Molecular Environment within Mesostructured Silica/Surfactant Composites under High Pressure. *Nanoletters* **2001**, *1*, 27–31. [CrossRef]

29. Diaz, J.; Balkus, J.; Bedioui, F.; Kurshev, V.; Kevan, L. Synthesis and Charact-erization of Cobalt-Complex Functionalized MCM-41. *Chem. Mater.* **1997**, *9*, 61–67. [CrossRef]

30. Liu, J.; Song, L.; Shao, G. Novel Zwitterionic Inorganic—Organic Hybrids: Kinetic and Equilibrium Model Studies on Pb^{2+} Removal from Aqueous Solution. *J. Chem. Eng. Data* **2011**, *56*, 2119–2127. [CrossRef]

31. Pavan, F.; Costa, T.; Benvenutti, E. Adsorption of CoCl$_2$, ZnCl$_2$ and CdCl$_2$ on aniline/silica hybrid material obtained by sol-gel method. *Colloids Surf. A Physicochem. Eng. Asp.* **2003**, *226*, 95–100. [CrossRef]

32. Radi, S.; Tighadouini, S.; Bacquet, M.; Degoutin, S.; Cazier, F.; Zaghrioui, M.; Mabkhot, Y.N. Organically Modified Silica with Pyrazole-3-carbaldehyde as a New Sorbent for Solid-Liquid Extraction of Heavy Metals. *Molecules* **2014**, *19*, 247–262. [CrossRef]

33. Qu, R.; Zhang, Y.; Qu, W.; Sun, C.; Chen, J.; Ping, Y.; Chen, H.; Niu, Y. Mercury adsorption by sulfur- and amidoxime-containing bifunctional silica gel based hybrid materials. *Chem. Eng. J.* **2013**, *219*, 51–61. [CrossRef]

34. Anbarasu, G.; Malathy, M.; Karthikeyan, P.; Rajavel, R. Silica functionalized Cu (II) acetylacetonate Schiff base complex: An efficient catalyst for the oxidative condensation reaction of benzyl alcohol with amines. *J. Solid State Chem.* **2017**, *253*, 305–312. [CrossRef]

35. Singh, H.; Singh, J.B. Synthesis and Characterization of New Lead (II) and Organotin (IV) Complexes of Schiff Bases Derived from Histidine and Methionine. *Int. J. Inorg. Chem.* **2012**, *2012*. [CrossRef]

36. Sandhu, G.K.; Verma, S.P. Triorganotin (IV) derivatives of five membered heterocyclic 2-carboxylic acids. *Polyhedron* **1987**, *6*, 587–592. [CrossRef]

37. Miao, J.; Qian, J.; Wang, X.; Zhang, Y.; Yang, H.; He, P. Synthesis and characterization of ordered mesoporous silica by using polystyrene microemulsion as templates. *Mater. Lett.* **2009**, *63*, 989–990. [CrossRef]

38. Helloa, K.; Ibrahimb, A.; Jawad, K.; Shneineb, J. Simple method for functionalization of silica with alkyl silane and organic ligand. *South Afr. J. Chem. Eng.* **2018**, *25*, 159–168. [CrossRef]

39. Naghmeh, D.; Niaz, M.; Mahmood, T. Preparation and Characterization of a Molybdenum (VI) Schiff Base Complex as Magnetic Nanocatalyst for Synthesis of 2-Amino-4H-benzo[h]chromenes. *J. Nanostruct.* **2016**, *6*, 312–321. [CrossRef]

40. Xue, X.; Li, F. Removal of Cu (II) from aqueous solution by adsorption onto functionalized SBA-16 mesoporous silica. *Microporous Mesoporous Mater.* **2008**, *116*, 116–122. [CrossRef]

41. Ramesh, A.; Hasegawa, H.; Maki, T.; Ueda, K. Adsorption of inorganic and organic arsenic from aqueous solutions by polymeric Al/Fe modified montmorillonite. *Sep. Purif. Technol.* **2007**, *56*, 90–100. [CrossRef]

42. Ho, Y.; McKay, G. Pseudo-second order model for sorption processes. *Process Biochem.* **1999**, *34*, 451–465. [CrossRef]

43. Langmuir, I. The adsorption of gases on plain surface of glass, mica and platinum. *J. Am. Chem. Soc.* **1918**, *40*, 1361–1368. [CrossRef]

44. Weber, T.W.; Chakravorti, R.K. Pore and solid diffusion models for fixed bed adsorbers. *AIChE J.* **1974**, *20*, 228–238. [CrossRef]

45. Freundlich, H. Über die adsorption in lösungen. Zeitschrift für Physikalische. *Chemie* **1907**, *57*, 385–470. [CrossRef]

46. Karthik, R.; Meenakshi, S. Removal of Pb (II) and Cd (II) ions from aqueous solution using polyaniline grafted chitosan. *Chem. Eng. J.* **2015**, *263*, 168–177. [CrossRef]

 *processes*

Article

Quantifying the Benefit of a Dynamic Performance Assessment of WWTP

Silvana Revollar [1,*], Montse Meneses [2], Ramón Vilanova [2], Pastora Vega [1] and Mario Francisco [1]

[1] Department of Computer Science and Automatic, University of Salamanca, 37008 Salamanca, Spain; pvega@usal.es (P.V.); mfs@usal.es (M.F.)

[2] Department Telecommunications and Systems Engineering, School of Engineering, Universitat Autonoma de Barcelona, 08193 Bellaterra, Spain; montse.meneses@uab.cat (M.M.); Ramon.Vilanova@uab.cat (R.V.)

* Correspondence: srevolla@usal.es

Received: 31 December 2019; Accepted: 5 February 2020; Published: 7 February 2020

Abstract: In this work a comprehensive analysis of the environmental impact of the operation of a wastewater treatment plant (WWTP) using different control strategies is carried out considering the dynamic evolution of some environmental indicators and average operation costs. The selected strategies are PI (proportional integral) control schemes such as dissolved oxygen control in the aerobic zone (DO control), DO control and nitrates control in the anoxic zone (DO + NO control) and regulation of ammonium control at the end of aerobic zone (Cascade S_{NHSP}) commonly used in WWTPs to maintain the conditions that ensure the desired effluent quality in a variable influent scenario. The main novelty of the work is the integration of potential insights into environmental impact from the analysis of dynamic evolution of environmental indicators at different time scales. The consideration of annual, bimonthly and weekly temporal windows to evaluate performance indicators makes it possible to capture seasonal effects of influent disturbances and control actions on environmental costs of wastewater treatment that are unnoticed in the annual-based performance evaluation. Then, in the case of periodic events, it is possible to find solutions to improve operation by the adjustment of the control variables in specific periods of time along the operation horizon. The analysis of the annual average and dynamic profiles (weekly and bimonthly) of environmental indicators showed that ammonium-based control (Cascade S_{NHSP}) produce the best compromise solution between environmental and operation costs compared with DO control and DO + NO control. An alternative control strategy, named $S_{NHSP\ var}\ Q_{carb\ var}$, has been defined considering a sequence of changes on ammonium set-point (S_{NHSP}) and carbon dosage (Q_{carb}) on different temporal windows. It is compared with DO control considering weekly and bimonthly profiles and annual average values leading to the conclusion that both strategies, Cascade S_{NHSP} and $S_{NHSP\ var}\ Q_{carb\ var}$, produce an improvement of dynamic and annual average environmental performance and operation costs, but benefits of Cascade S_{NHSP} strategy are associated with reduction of electricity consumption and emissions to water, while $S_{NHSP\ var}\ Q_{carb\ var}$ strategy reduces electricity consumption, use of chemicals (reducing external carbon dosage) and operation costs.

Keywords: wastewater treatment plants; environmental costs; PID control; dynamic assessment of performance

---

1. Introduction

Historically, the primary objective for collecting wastewater was sanitation to prevent the spread of waterborne diseases. Nowadays, wastewater treatment continuously evolves as the awareness of emerging environmental problems grows. The knowledge about the influence of human activities on climate change has widened the scope for treatment plants beyond only effluent water quality and cost. Today greenhouse gas emissions, energy efficiency and resource recovery also need to be considered when evaluating operational strategies by also minimizing the operational costs in order to achieve sustainable treatments. The optimization of the operations of a wastewater treatment plant (WWTP) is not an easy task. The influent load is constantly varying in flow and concentration, is naturally uncontrolled and arrives every hour of the day, all year round. Rainfall events affect wastewater composition because, in combined sewers, these events increase the flow and pollutants stored in sewer sediments and/or deposited on impervious surfaces are washed out [1,2]. A wastewater treatment plant cannot shut down for review and maintenance. Moreover, the construction with sequential unit processes in combination with multiple return feeds create numerous feed-back effects that makes the processes interconnected in an intricate manner. A WWTP should be considered as an integrated process, where primary/secondary clarifiers, activated sludge reactors, anaerobic digesters, thickener/flotation units, dewatering systems, storage tanks are interconnected and need to be operated and controlled not as individual unit operations, but taking into account all the interactions amongst the processes. Models should describe the processes and their interactions in detail considering the ambient conditions. Thereby, the plant-wide effects are captured so that the overall result can be surveyed, analyzed and sub-optimization avoided. In this complex scenario mathematical modelling and simulation provide a solid base for decision support when evaluating WWTP operations.

Researchers and design engineers in wastewater treatment (WWT) are aware of alternative modelling approaches that can be used to evaluate the appropriateness of control strategies to ensure the quality of the treated water with respect to the regulations in the presence of frequent and large disturbances and variable influent characteristics. The control of the activated sludge process (ASP) is crucial for the appropriate operation of WWTPs. ASP is a commonly used biological treatment, especially in large wastewater catchments. In this biological process, the control of aeration is particularly demanding: its optimization is linked to the minimization of the energy used in a plant [3]. Through modelling and simulation studies, not only can the present operations be evaluated but also future scenarios investigated, for example: load forecasts, plant expansions or alternative operational strategies. Some recent works [4,5] have demonstrated how model-based tools can be used in practice to improve the performance of WWTPs. A scenario-based optimization approach that connects effluent quality variables and energy demand and production, by a simulation procedure is proposed in [4] to improve the energy efficiency of an Italian WWTP using the model developed and calibrated in [6]. Potential savings on annual average energy consumption are reported and effluent quality is improved by operational changes, furthermore, the results showed that modifications in design could affect positively the energy and greenhouse gas balance of the plant. In [5], mass balances have been used to evaluate the impact of operation and plant parameters on nitrogen and organic matter removal efficiencies in another Italian WWTP.

An appropriate management of WWTP can produce significant economic and environmental benefits. A holistic assessment procedure that considers the environmental costs of wastewater treatment is necessary to attain a sustainable operation, minimizing energy consumption and greenhouse emissions. Previous works [7–11] address the integration of the analysis of environmental impact on the evaluation of performance of control strategies applied to WWTP. Specifically, annual-based life-cycle assessment (LCA) is used for the evaluation of economic and environmental performance of a WWTP in [8,9], LCA is applied considering annual average inventory. Global performance indicators are proposed in [10,11]. In [10] an integral performance index that quantifies the effect of the main control actions on water quality, operational cost and greenhouse gas emissions is used to measure the global positive effect of control systems on the plant operation. In [11], an overall efficiency index is used as the controlled variable of a holistic optimizing proportional integral (PI)-control strategy that introduce plantwide considerations.

Nevertheless, there are few studies discussing the additional benefit of adding a new dimension related to dynamic analysis within the performance evaluation procedures [12]. Regarding the LCA methodology that is typically used to evaluate environmental impact of production systems, several authors have critic the lack of a temporal dimension, even though inputs and environmental mechanisms are time varying [12]. Few works can be found that consider the time dependency of indicators of environmental performance of WWTPs. In [12] a dynamic LCA methodology is proposed and a WWTP is used as case study to evaluate the sensitivity of LCA results to temporal parameters. In [11], the evolution of environmental performance indicators is considered in [11] to evaluate the impact of control strategies. In [13], a dynamic model of activated sludge reactors working under an intermittent aeration regime is developed to evaluate the link between aeration and effluent quality, the analysis of airflow rate influence on performance allow to increase process efficiency, producing a reduction of 14.5% on power consumption. The selection of a time horizon is equivalent to giving a weight to time and is one of the most critical parts of the carbon accounting processes [14,15].

The Benchmark Simulation Model No. 2 (BSM2) is a standard simulation model developed as a reference scenario to implement and evaluate control strategies [7–11,16–19]. The BSM2 represents the water line and the sludge line of a municipal WWTP considering a dynamic influent that contains everything from short-term diurnal variations and weekend effects to long-term variations for temperature and holidays periods [14,17,18]. The BSM2 platform is selected in this work to represent a municipal WWTP, in order to demonstrate the benefit of adding this extra dynamic dimension to the simulation.

In this paper a comprehensive analysis of the environmental impact of control/operational strategies is performed through a dynamic perspective from a WWTP operation. The main novelty of the work is the introduction of potential insights into environmental impact from the analysis of the evolution of environmental indicators considering different time scales: annual, bimonthly and weekly. The consideration of different temporal windows makes possible to capture periodic seasonal effects associated with influent variations and interactions between control actions and environmental costs of wastewater treatment that are unnoticed in the traditional performance evaluation based on the analysis of annual average indicators. Indeed, the analysis of plant behavior in shorter time horizons makes possible to capture dynamic effects that are hidden by the evaluation using annual based indicators of performance.

The main objective is to show the benefits that result from adding dynamic perspective to plant performance evaluation criteria evaluation of control/operational strategies. The analysis makes possible to find solutions to improve operation by the adjustment of the control variables in specific periods of time along the operation horizon. It allows to improve the wastewater treatment in terms of energy efficiency, resources recovery and greenhouse gas emissions, while not compromising effluent quality and still maintaining control of the operational cost.

The remainder of this paper is organized as follows. Descriptions of a reference wastewater treatment plant (BSM2 model), control strategies and environmental performance indicators are presented in Section 2. Results of the evaluation of annual average and dynamic (weekly and bimonthly temporal widows) of environmental costs considering the selected control strategies and the results of the comparison with alternative operation strategy are presented in Section 3, closing with some conclusions.

2. Materials and Methods

2.1. Description of the Wastewater Treatment Plant and Control Strategies. Benchmark Simulation Model No. 2 (BSM2) Platform

The process lines commonly distinguished in a municipal WWTP are water line, where pollution removal is carried out, sludge line and gas line. In this work, the water and sludge lines of a municipal WWTP are represented using the BMS2. This recognized simulation platform describes the plant layout, the simulation model, the influent profile and the evaluation protocol [17–19]. The BSM2 plant comprises primary clarification and activated sludge process units in the water line and anaerobic digestion, thickening and dewatering operations in the sludge line (Figure 1). The plant is designed for an average influent dry weather flow rate of 20,648.36 m^3/d and an average biodegradable chemical oxygen demand (COD) in the influent of 592.53 g/m^3. Its hydraulic retention time, computed considering average dry weather flow rate and total tank volume of 18,900 m^3, is 22 h (total tank volume includes: primary clarifier (900 m^3) + biological reactor (12,000 m^3) + secondary clarifier (6000 m^3)) [18,19].

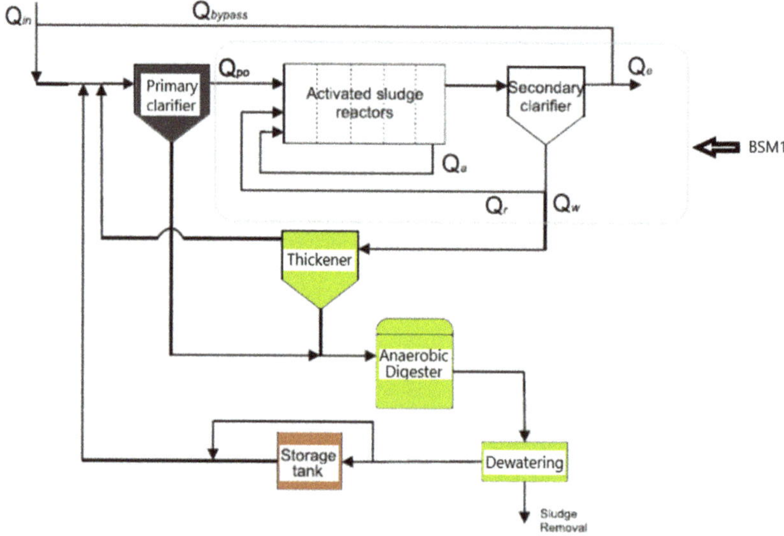

Figure 1. Benchmark Simulation Model No. 2 (BSM2) plant layout.

In the water line, a modification of the benchmark simulation model 1 (BSM1) is used to represent the activated sludge process (ASP) where biological nitrogen and organic matter removal take place by means of nitrification and denitrification processes [16]. In nitrification process, nitrogenated compounds (mostly in the form of ammonium NH$_4$) are sequentially oxidized to nitrite and to nitrate by autotroph bacteria that are strict aerobes, while heterotrophs transform nitrates in nitrogen gas (N$_2$) by denitrification. These biological processes are carried out in a system of 5 bioreactors in series. The first two reactors are anoxic and perfectly mixed to facilitate denitrification, the last three reactors are aerated to promote the nitrification step. Nitrate is recirculated from the aerobic to the anoxic zone (internal recycle flow Q$_a$). A secondary clarifier separates clean water from sludge. The clean effluent (Q$_e$) is discharged and the sludge is partly a wastage flow (Q$_w$) that is fed to the sludge line and partly recycled to the anoxic zone (external recycle flow Q$_r$). The Activated Sludge Model no. 1 (ASM1) [20,21] describes these biological processes and the effect of temperature in the biological kinetics considering eight biological processes and 13 state variables on each reactor. The settler (secondary clarifier) is described using the model of Takács et al. [22].

In the sludge line, a thickener prepares the sludge collected from the primary and secondary clarifiers for the anaerobic digestion. A dewatering unit is used to increase the concentration of the stabilized sludge. As shown in Figure 1, there is a storage tank before recycling the remaining sludge to the water line and the liquids collected in the thickening and dewatering steps are recycled to the primary settler [18]. The digester is modeled using the anaerobic digestion model (ADM1) of [23].

Significant Operating Variables and Control Strategies

Since the main objective of a BSM2 plant is nitrogen removal, the aim of the control strategies applied to the BSM2 plant is to ensure the appropriate conditions for nitrification/denitrification processes in the ASP. Dissolved oxygen (DO) concentration in the aerobic zone is a determining variable for oxidation mechanisms involved in nitrification process. On the other hand, carbon dosage is a key variable when external carbon source is required in the anoxic zone to provide readily biodegradable substrate to heterotrophs. The sludge age or solids retention time (SRT), which is a measure of the time that sludge (cells, microorganisms) remain in the system, and the food to microorganism ratio (F:M), that represents the balance between the quantity of substrate available and the quantity of microorganisms in the biological reactors, are other important factors for the appropriate course of biological reactions.

The available control handles in the ASP bioreactors are the airflow rate, the internal recirculation (Q_a), the sludge recirculation (Q_r), the sludge purge flow (Q_w) and the external carbon dosage (Q_{carb}). Then, DO control or ammonium-based supervisory control (regulation of ammonium concentration S_{NH}) in the aerobic basin is performed by manipulation of the airflow rate. Control of nitrates concentration (S_{NO}) in the anoxic zone is carried out by manipulation of internal recirculation flow (Q_a) that transport nitrates produced in the aerobic zone to anoxic zone. Moreover, carbon dosage (Q_{carb}) is a manipulated variable that affects nitrates concentration (S_{NO}) also. In practice, it is usual to maintain constant values of Q_a and Q_{carb} over long periods of time to regulate S_{NO}. Purge flow (Q_w) is used to regulate the sludge age (or SRT) and external recirculation (Q_r) regulates the F:M ratio. A detailed description can be found in [24,25].

In the sludge line, the loading rate, given in part by ASP purge flow (Q_w), and the composition of the input sludge flow affect the characteristics of the biogas and stabilized sludge in the anaerobic digestion. Temperature is another critical variable, that is maintained between 32–35 °C using energy from biogas to heat the sludge input flow. Other important operation variables are the solids retention time (more than 20 days) and pH (6.8–7.2).

- Operation strategies: BSM2 default strategy and modifications

The default operation strategy of BSM2 plant includes a DO control scheme in the activated sludge process, distinguished as a DO default control scheme, and open-loop actions to regulate the levels of nitrates in the system, sludge age, F:M ratio control and digester temperature:

(a) manipulation of Q_w that is fixed to 450 m³/d in the warmer season and changed to 300 m³/d in the colder season.

(b) constant carbon dosage to the first reactor in the anoxic zone: Q_{carb} = 2 m³/d with a concentration of 40,000 g/m³

(c) constant internal and external recirculation flowrate: Q_r = 20,648 m³/d and Q_a = 61,944 m³/d.

(d) keeping digester temperature (T) at 35 °C by using biogas for heating.

(e) PI feedback control of DO concentration using the oxygen transfer coefficient (K_{La}) as manipulated variable instead of a basic control loop to regulate airflow rate. DO concentration in the fourth aeration tank (S_{O4}) is regulated to a constant set-point of 2 g r/m³ manipulating the oxygen transfer coefficient of the fourth reactor (K_{La4}), while the transfer coefficients of the third and fifth reactors (K_{La3} and K_{La5}) are computed considering a gain of 1 and 0.5 (Figure 2a). This scheme is named the DO default control.

The objective of operation strategy is to maintain levels of pollutants in the effluent, such as Nitrogen (N_{tot}), ammonium (S_{NH}), nitrates (S_{NO}), total chemical oxygen demand (COD_t), total suspended solids (TSS) and biological oxygen demand (BOD5), bellow the limits given by effluent quality requirements. The indicators considered in this work are given by BSM2 platform [17–19]: $N_{tot} < 18$ gN/m³, $S_{NH} < 4$ gN/m³, $COD_t < 100$ gCOD/m³.

In this work, two alternatives to the DO control scheme of the default BSM2 strategy described above (Figure 2a) are considered (See Figure 2b,c):

- The addition of a closed loop for the control of nitrates concentration at the end of the anoxic zone using the internal recycle (Q_a). The combination of both loops, DO default and nitrates control, is similar to the default control strategy of BSM1 platform [13]. Here, it is named DO + NO control.

- Substituting direct DO control by ammonium-based control including an external PI controller to compute the DO set-point for the internal controller. The ammonium concentration in the fifth reactor (S_{NH5}) is the measured variable. This scheme is named Cascade S_{NHSP}.

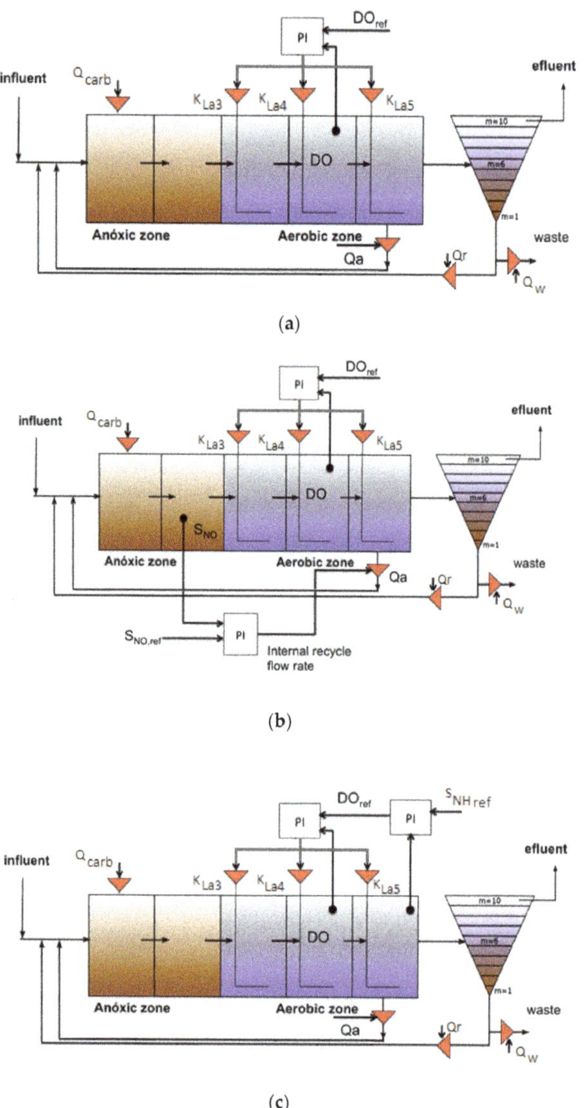

Figure 2. Operation strategies applied to BSM2 platform. (**a**) BSM2 Default operation strategy (dissolved oxygen (DO) default), (**b**) BSM2 Default operation strategy with nitrates control scheme (DO + nitrates control (NO)), (**c**) BSM2 Default operation strategy with ammonium-based control scheme (Cascade S_{NHSP}).

The three closed-loop control schemes (DO default, DO + NO control and Cascade S_{NHSP}) use PI controllers of the form: $u(t) = Kp \cdot \left(y(t) - y_{sp}\right) + \frac{Kp}{Ti} \int \left(y(t) - y_{sp}\right)dt + \frac{1}{Tt} \int \left(y_{lim} - y(t)\right)dt$ where u is the manipulated variable, y is the controlled variable, y_{sp} the desired set-point, y_{lim} the limit values of the controlled variables, Kp is the proportional gain and Ti the integral time and Tt the anti-windup constant. The tuning parameters of the controllers can be found in [19] for DO controller, [20] for nitrates controller and [10] for ammonium controller.

Table 1 summarizes the BSM2 operation strategy, including the three possible control schemes studied in this paper and the effluent quality objectives.

Table 1. BSM2 Default operation strategy with the alternative control schemes.

Common Conditions	Possible Control Schemes		
	Control Scheme	Short Denomination	Variables
Q_{carb} = 2 m^3/d Q_r = 20648 m^3/d Q_a = 61944 m^3/d In warm season, bimesters 1, 2 and 6: Q_w = 450 m^3/d In cold season bimesters 3–5: Q_w = 300 m^3/d	DO control scheme	DO Default	S_{O4SP} = 2 g/m^3 Kp = 25, Ti = 0.002, Tt = 0.001
	DO control scheme in the aerobic zone combined with nitrates control in the anoxic zone	DO + NO control	S_{O4SP} = 2 g/m^3 Kp = 25, Ti = 0.002, Tt = 0.001
			S_{NO2SP} = 1 g/m^3 Kp = 1000, Ti = 0.025, Tt = 0.015
	Ammonium-based control scheme	Cascade S_{NHSP}	S_{NHSP} = 1 g/m^3 Kp = −1, Ti = 1, Tt = 0.2
Effluent quality objectives			
Total nitrogen (N_{tote})		<18 gN/m^3	
Ammonium concentration (S_{NHe})		<4 gN/m^3	
Chemical oxygen demand (COD_e)		<100 gCOD/m^3	

S_{O4SP} is DO set-point in the 4th reactor, S_{NO2SP} is nitrates set-point in the 2nd reactor, S_{NHSP} is ammonium set-point in the 5th reactor.

- Performance evaluation protocol

The BSM2 platform provides an evaluation protocol for the control strategies tested in the plant [12]. These indicators are computed, using a code provided by BSM2 simulator for an evaluation period of one year, starting 1 July. The program computes the most important variables associated with the load and composition of influent, effluent, biogas and sludge in a given temporal window. The notation for the interesting variables (most of them ASM1 variables): S_I—soluble inert organic matter, S_S—readily biodegradable substrate, X_I—particulate inert organic matter, X_S—slowly biodegradable substrate, $X_{B,H}$—active heterotrophic biomass, $X_{B,A}$—active autotrophic biomass, X_P—particulate products arising from biomass decay, S_O—oxygen, S_{NO}—nitrate and nitrite nitrogen, S_{NH}—NH_4^+ + NH_3 nitrogen, S_{ND}—soluble biodegradable organic nitrogen, X_{ND}—particulate biodegradable organic nitrogen, Q_{in}—influent flowrate, Q_e—effluent flowrate, T_{in}—influent temperature, T—digester temperature A sub-index is used to indicate the number of the reactor associated with the variables when is necessary.

Some indicators given by BSM2 platform are the effluent quality index (EQI) that measures the effluent water quality as a weighted average of effluent COD, BOD, ammonia, nitrate and total solid loads, the overall cost index (OCI) [17–19] that provides a relative comparison for the operational cost including, power for mixing aeration and pumping, carbon source addition, heating of the digester, utilization of biogas and disposal of sludge. A modification is introduced in this work to obtain direct carbon dioxide (CO_2) emissions produced in ASP.

The effluent quality index [19,20]:

$$EQI = C_1 \int_{t_0}^{tf(days)} [2 \cdot SS_e + COD_e + 30 \cdot N_{tote} + 10 \cdot S_{NO,e} + 2 \cdot BOD_e] Q_e dt \left(\frac{kg\ pollution}{d} \right) \quad (1)$$

where $C_1 = \frac{1}{T \cdot 1000}$ and T is the evaluation period.

The BOD, COD, total nitrogen concentration (N_{tot}) and suspended solids (SS) are computed as [19,20]:

$$BOD_e = 0.25 \cdot (S_{Se} + X_{Se} + (1 - 0.08) \cdot (X_{B,Ae} + X_{B,He})) \left(g/m^3 \right) \quad (2)$$

$$COD_e = (S_{Se} + S_{Ie} + X_{Se} + X_{Ie} + X_{B,He} + X_{B,Ae} + X_{Pe})\,(g/m^3) \tag{3}$$

$$N_{tote} = S_{NOe} + S_{NHe} + X_{NDe} + i_{XB}(X_{B,He} + X_{B,Ae}) + i_{XP}(X_{Pe} + X_{Ie})\,(g/m^3) \tag{4}$$

$$SS_e = 0.75 \cdot (X_{S,e} + X_{I,e} + X_{B,H,e} + X_{B,A,e} + X_{P,e})\,(g/m^3) \tag{5}$$

where the subscript index: *e* is used to distinguish the variables in the effluent.

The influent quality index (IQI) has been defined to characterize the influent [19,20]:

$$IQI = C_1 \int_{t_0}^{tf(days)} [2 \cdot SS_i + COD_i + 30 \cdot N_{toti} + 10 \cdot S_{NO,i} + 2 \cdot BOD_i]Q_{in}dt\left(\frac{kg\ pollution}{d}\right) \tag{6}$$

where SS_i, COD_i, N_{toti}, BOD_i are analogous to SS_e, COD_e, N_{tote}, BOD_e but the subscript index: *i* is used to denote the variables in the influent.

The global operational cost (OCI) is:

$$OCI = AE + PE + 3 \cdot SP + 3 \cdot EC + ME - 6 \cdot MP + HE_{net}\ (EUR/d) \tag{7}$$

where AE represents the aeration energy in the activated sludge process, PE is the pumping energy in the full plant (involving all flows), ME is the mixing energy in the full plant, SP is the sludge production for disposal, EC is the external carbon addition and MP is the methane production and HE_{net} is [19,20]:

$$HE_{net} = \max\left(0, HE - 7 \cdot MET_{prod}\right)\ (kWh/d) \tag{8}$$

where HE is heating energy necessary to heat the sludge to the digester operating temperature and MET_{prod} is the methane production (kg/d).

Regarding greenhouse emissions, direct emissions from activated sludge process are calculated as in [26], considering the following equations:

$$2.57C_{2.43}H_{3.96}O + 2.5O_2 + NH_3 \rightarrow C_5H_7O_2N + 1.24CO_2 + 3.09H_2O \tag{9}$$

$$48.59NH_3 + 5CO_2 + 90.19O_2 \rightarrow C_5H_7O_2N + 47.59HNO_3 + 45.59H_2O \tag{10}$$

$$2.57C_{2.43}H_{3.96}O + 2HNO_3 + NH_3 \rightarrow 1.24CO_2 + C_5H_7O_2N + 4.09H_2O + N_2 \tag{11}$$

where $C_{2.43}H_{3.96}O$ represents readily biodegradable substrate (S_S) and $C_5H_7O_2N$ represents heterotroph and autotroph biomass ($X_{B,H}$ and $X_{B,A}$).

2.2. Evaluation of the Impact of Dynamic Behavior Actions on Environmental and Operating Costs

Several characteristics of WWTPs make the optimization of their operation a challenging problem:

1. The objective of a WWTP is to minimize the emissions to water; however, wastewater treatment implies environmental impacts associated with energy consumption, use of chemicals and emissions to soil and air (solids and greenhouse gases). The control actions carried out to ensure the desired elimination of water pollutants (i.e., emissions to water) affect those environmental costs as well as the operation costs. Then, for a sustainable operation of WWTPs control systems should lead to operating conditions that satisfy the compromise between environmental costs, operating costs and appropriated plant performance.

2. The influent is constantly varying in flow and concentration mainly due to seasonal and daily variations of population activities, which modifies continuously the load to be treated. Temperature and rainfall effects produce significant changes in the processes affecting its efficiency. Since biological processes are non-linear, the variability of the inputs produces a variable behavior,

which makes it difficult to find and maintain operating conditions that ensure the desired process performance with an optimal use of resources and minimum evitable emissions.

3. Due to the interactions and interconnections between the different units, the control actions performed in ASP have an impact in the whole plant effecting, sludge and biogas production which are emissions to soil and air, respectively.

Then, dynamic analysis of the effect of control actions on environmental and operating costs facilitate to detect dynamic effects on environmental indicators that are hidden in the annual based analysis of environmental impact [12,27] but can be relevant at smaller time scales. Identifying the effect of periodic variations and particular events in the influent in different temporal windows along the year provide the means to determine alternative control actions that can be applied to improve plant efficiency in such scenarios.

Conflicts regarding energy use, greenhouse gases emissions and use of chemicals are considered in the analysis:

1. Electricity used to perform control actions is the most important factor in the operating costs being aeration the main contribution [28] and it is also the major source of indirect carbon dioxide (CO_2) emissions, then, the minimization of electricity consumption in the control actions improve economics as well as environmental costs.

2. Heating requirements to keep the temperature of the anaerobic digestion close to 35 °C are covered using the biogas obtained as by-product. Biogas in excess can be used to produce electricity and reduce operating costs, but indirect emissions of CO_2 are produced when using biogas to obtain energy.

3. An external carbon source is required to provide enough organic matter to heterotrophs for denitrification. Carbon dosage can decrease the concentration of nitrates, reducing total nitrogen emissions to water but it is a chemical additive that increases also operating costs.

4. Biological processes in ASP produce greenhouse gases as CO_2. Biogas from anaerobic digestion contains methane, CO_2 and hydrogen. The sludge for disposal is a solid residue that affects soil.

2.2.1. Analysis Procedure

The general idea is to evaluate the impact of the control actions to find an operation strategy that produce a satisfactory trade-off between environmental and operating costs. It should be considered that wastewater treatment plants (WWTP) are subject to large disturbances related to variations in the flow and composition of the incoming wastewater. These variations are associated with human activity in the catchment or rainfall effects and seasonal effects due to the temperature changes along the year. The influent variations affect process behavior and produce a reaction of the control system to reject disturbances and maintain the appropriated operating conditions.

Annual, bimonthly and weekly periods are considered to capture such cause–effect relations in different operating windows. The annual average values of the environmental indicators and operating costs (OCI) measures the performance in the full operating horizon. The weekly profiles capture the effect of short time variations associated with rain events and human activities and bimonthly profile allows the observation of long-time effects in influent flow and temperature associated with the different seasons. This analysis makes it possible to identify the changes on operation variables that can be made in a specific temporal window to improve the plant behavior.

The criteria to select the environmental indicators considered in this study is the possibility of being affected by control actions, even though data provided by BSM2 protocol can be used to perform a more detailed environmental analysis. The selected indicators are:

For energy:

-   Electricity consumption (kW) that includes aeration energy in the ASP (AE), pumping energy (PE) and mixing energy (ME) used in the whole plant.

- Heating energy (HE) in kW h required to maintain the sludge fed to digester at 35 °C.
- For chemicals usage:
- The load of the external carbon source that is methanol with a concentration of 40,000 g/m^3 with a flowrate given by Q_{carb}.

For emissions to air:

- Amount in g of methane (Biogas CH_4) and carbon dioxide (CO_2-digester) produced in anaerobic digester.
- Amount in g of carbon dioxide produced in ASP (CO_2-ASP), computed from the relations given by Equations (9)–(11) introduced in the BSM2 evaluation program.
- Total CO_2 that combines CO_2 from digester and CO_2 from ASP.

For emissions to soil:

- Amount in kg of sludge for disposal.
- For emissions to water:
- Amount in g of ammonium in the effluent ($S_{NH\ Load}$ effluent)
- Amount in g of total nitrogen in the effluent ($N_{tot\ Load}$ effluent)
- Amount in g of COD in the effluent (COD effluent)
- Effluent quality index (EQI).

Operational costs are measured using the OCI in euros/day (EUR/d).

All indicators are computed for a given temporal window and are expressed with respect average influent flow Q_{in} (m^3/h) in such time period. Then, $N_{tot\ Load}/Q_{in}$ (g/m^3) and $S_{NH\ Load}/Q_{in}$ (g/m^3) are referred as concentrations N_{tot} and S_{NH}.

- Temporal characterization of BSM2 dynamic influent

The BSM2 model represents a plant located in the northern hemisphere. The available dynamic influent profile describes seasonal changes of temperature and influent flowrate, characteristic daily and weekly variations associated with population activities and precipitations [8,12]. Both daily and seasonal variations of temperature are modelled with a sinus function [8]. The evaluation period contemplated in the simulation platform is one year, starting 1 July in a plant located in the northern hemisphere.

The characterization of influent behavior is important to identify the significant events on influent behavior that affect operating conditions and the temporal window that capture such an event. Weekly and bimonthly profiles of the most important influent variables: temperature, influent flow Q_{in}, and influent concentrations of COD and N_{tot} (Equations (3) and (4)) are presented in Figure 3. It is observed that temperature profile (weekly or bimonthly) follows a senoidal function with a minimum in the colder (4th) bimester and a maximum in the warmer (1st) bimester. Weekly profiles of influent flow (Q_{in}), total nitrogen (N_{tot}) and COD observed in Figure 3 exhibit frequent disturbances with eventual minimums and peaks due to population activities and rain events, while bimonthly profiles show the seasonal effects as the period with the highest influent flow (Q_{in}) that is the 3rd bimester, the driest period (1st bimester), the period with lower load (3rd bimester) and the period with the lowest load (2nd bimester). Table 2 summarizes the annual averages of influent variables as well as the maximum and minimum values in the different time scales, quantifying the variations observed in Figure 3. The information provided by Table 2 allows us to demonstrate that WWTP influent exhibits variations of temperature of approximately 10 °C between the colder and warmer period that affect significantly biological processes kinetics. Moreover, the quantification of the differences between the maximum and minimum values of influent flow and load in the different time scales, shows how relevant are the changes in the influent that affect WWTP behavior. In order to maintain the desired WWTP performance, different control actions are executed to face these appreciable variations of influent characteristics detected in different time horizons. These actions affect environmental performance of WWTP; a dynamic analysis of environmental indicators is interesting to determine the impact of control actions considering the time varying characteristics of the process.

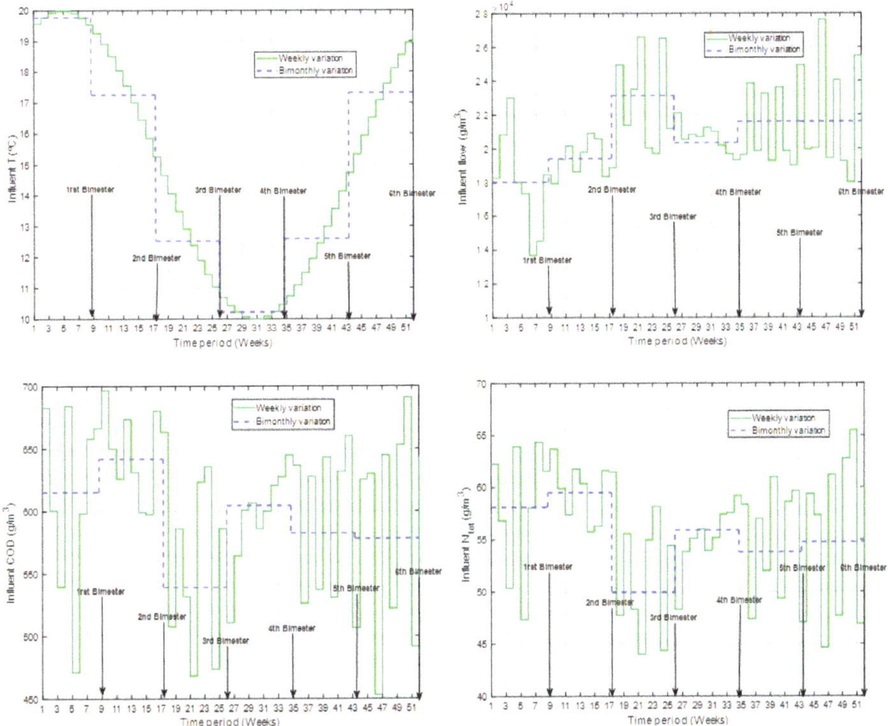

Figure 3. Weekly and bimonthly profile of influent temperature (°C), flow rate Q_{in} (m³/d), total nitrogen concentration N_{tot} concentration (g/m³) and COD concentration (g/m³).

Table 2. Characteristic values of the significant variables of the influent including weekly and bimonthly means (W. Av.: Weekly average, Bi-m. Av.: Bimonthly average).

Variable	Average	Maximum	Minimum	W. Av. max.	W. Av. min.	Bi-m. Av. max.	Bi-m. Av. min.
T (°C)	15	20.5	9.5	20	10	19.8	10.2
Q_{in} (m³/d)	20,668	85,841	5146	27,800	13,800	23,200	18,000
N_{toti} (g/m³)	55.2	114.2	7.7	65.5	44	59.6	50
COD influent (g/m³)	592.2	1213.0	36.5	695	454	615	540

3. Results

The evaluation of process behavior is performed for an operation cycle of one year using the control strategies described in Table 1 (DO default, DO + NO control and Cascade S_{NHSP}). The selected environmental indicators and operating costs are computed considering the different temporal windows to capture: (1) the impact of slow disturbances associated with seasonal behavior of influent, and (2) the impact of variations on influent flowrate and load detected in weekly and bimonthly periods.

Following the BSM2 protocol, a simulation of 609 days is carried out but only the last 365 days (one year) are considered to compute the performance indices and environmental indicators [11]. The.output data is stored with a sampling time of 15 min. These outputs are used for the calculation of the environmental indicators and the OCI. Thus, the annual, bimonthly and weekly mean of the selected environmental indicators (Section 2.2.1) are computed, and weekly and bimonthly dynamic profiles are obtained to show the effect of control actions and influent variations on different temporal windows.

This is a first step of the analysis, whereby different control strategies are compared and the control scheme that produce the best compromise between environmental and operational costs is selected. In a second stage, the effect of set-point changes and carbon dosage (Q_{carb}) on plant behavior with the selected control strategy is evaluated, in order to determine control movements that improves environmental and operating costs in a given operation window.

3.1. Analysis of the Effect of Control Actions and Influent Variations on Environmental Indicators Considering Different Temporal Windows. Different Activated Sludge Process (ASP) Control Strategies

- Analysis of behavior in the full operating period (one year)

Table 3 presents the annual average values of environmental indicators and operation costs computed with respect to the volume of treated wastewater.

Table 3. Annual values of environmental indicators and operating costs of BSM2 plant with respect to the volume of treated wastewater with different control schemes.

	Environmental Indicators	DO Default	DO + NO Control	Cascade S_{NHSP}
Energy	Electricity (AE + PE + ME) (kW h/m^3)	0.263	0.263	0.243
	Heating energy (kW h/m^3)	0.204	0.204	0.204
Chemicals	External carbon (kg COD/m^3)	0.039	0.039	0.039
Emissions to air	Biogas CH$_4$ (g/m^3)	52.51	52.51	52.53
	CO$_2$ (Digester) (g/m^3)	75.61	75.60	75.63
	CO$_2$ (ASP) (g/m^3)	91.52	95.52	87.20
Emissions to soil	Sludge for disposal (kg/m^3)	131.1	132.0	131.0
Emissions to water	S_{NH} effluent (g/m^3)	0.474	0.312	1.052
	N_{tot} effluent (g/m^3)	13.53	17.56	11.34
	COD effluent (g/m^3)	48.99	49.01	49.07
	EQI (kg/m^3)	0.270	0.307	0.260
Operation costs	OCI (EUR/m^3)	0.457	0.457	0.437

AE: Aeration energy, PE: Pumping energy, ME: Mixing energy, COD: Chemical oxygen demand, S_{NH}: ammonium concentration, N_{tot}: Total nitrogen concentration, EQI: Effluent quality index.

Energy consumption. The lowest consumption of electricity is attained with the Cascade S_{NHSP} scheme. This strategy varies DO set-point to regulate effluent ammonium concentration which reduces the consumption of energy for aeration, while the other schemes keep a constant DO set-point of 2 g/m^3 in the full operating period. Regarding heating requirement of digester, the heating energy (HE) is equal with the three control schemes.

Use of chemicals. A constant carbon dosage (Q_{carb} = 2 m^3/d) is applied in all cases.

Emissions to air and emissions to soil. The amount of CH$_4$, CO$_2$ (emissions to air) and sludge (emissions to soil) produced in anaerobic digester is the same with the three control schemes. Conversely, the amount of CO$_2$ produced by biological processes in ASP varies with the different control schemes, attaining the lowest levels with Cascade S_{NHSP} scheme.

Emissions to water. The lowest N_{tot} concentration in the effluent is obtained with Cascade S_{NHSP} scheme but also the highest concentration of ammonium (S_{NH}). The Cascade scheme exhibits the lowest EQI, i.e., the lowest pollution load in the effluent considering nitrogenated compounds, organic matter and biomass. The concentration of COD in the effluent is similar for the three schemes.

The OCI depends on electricity, heating energy, chemicals and sludge treatment. Electricity usage is the only factor of OCI that varies with the different control schemes, therefore, the lower OCI is obtained with Cascade S_{NHSP} scheme due to the reduction of energy requirements attained with this control scheme.

Table 4 present the variation of environmental and operating costs indicators observed with ammonium-based control (Cascade S_{NHSP}) and DO and nitrates control (DO + NO control) relative to the DO default control scheme. The major impact of Cascade S_{NHSP} scheme is observed on total nitrogen concentration with an improvement of 16.2% and S_{NH} levels with a significant increment of 121%. The Cascade control scheme has a positive effect on five of the six indicators (5/6) decreasing its annual average values and the indicators that are worsened, and the levels of S_{NH} (1.053 g/m^3) are still below the desired limits (4 g/m^3 in Table 1) with a back-off of 70%. Therefore, it can be concluded from the annual analysis that Cascade S_{NHSP} scheme produce the best trade-off between environmental and operating costs.

Table 4. Environmental and cost indicators of Cascade S_{NHSP} and DO + NO control relative to default DO control scheme.

	Environmental Indicators	**DO + NO Control**	**Cascade S_{NHSP} = 1**
Energy	Electricity (AE + PE + ME) (kW h/m^3)	-	−7.6%
Emissions to air	CO$_2$ (ASP) (g/m^3)	+4.3%	−4.7%
	S_{NH} effluent (g/m^3)	−34.2%	+121%
Emissions to water	N_{tot} effluent (g/m^3)	+22.9%	−16.2%
	EQI (kg/m^3)	+37.03%	−3.7%
Operation costs	OCI (EUR/m^3)	-	−4.4%

AE: Aeration energy, PE: Pumping energy, ME: Mixing energy, COD: Chemical oxygen demand, S_{NH}: ammonium concentration, N_{tot}: Total nitrogen concentration, EQI: Effluent quality index.

Now, dynamic analysis is carried out in to provide insight on the dynamic effect of control actions on environmental and operational costs that are hidden when annual average values of indicators are considered.

- Analysis of dynamic behavior considering weekly and bimonthly time scales

The dynamic evolution of the environmental indicators considering weekly and bimonthly temporal windows is presented here. For simplicity, a profile of total CO$_2$ emissions is presented instead of separated profiles of CO$_2$ from the digester and CO$_2$ from the ASP, and N_{tot}, S_{NH} and EQI profiles are considered to characterize emissions to water.

Figure 4 presents the bimonthly and weekly dynamic profiles of environmental indicators associated with energy usage: electricity consumption and heating energy (HE). The dynamic profile of electricity consumption with the three control schemes exhibits the combined effect of influent flow, COD and N_{tot} that affect the load to be treated. There are peaks on weeks 6, 7 and 50 and minimums on weeks 22, 24 and 46 corresponding to extreme values of influent variables in the weekly profiles (See Figure 2). A seasonal effect is detected in the bimonthly profile, electricity consumption is maximum in the first and second bimesters, that is the driest period of the year, and more energy is required to track DO set-point in these conditions and decreases significantly in the 3rd bimester, that is the period with the largest influent flow and lowest concentration of pollutants. The temperature effect is not evidenced in the evolution of this indicator.

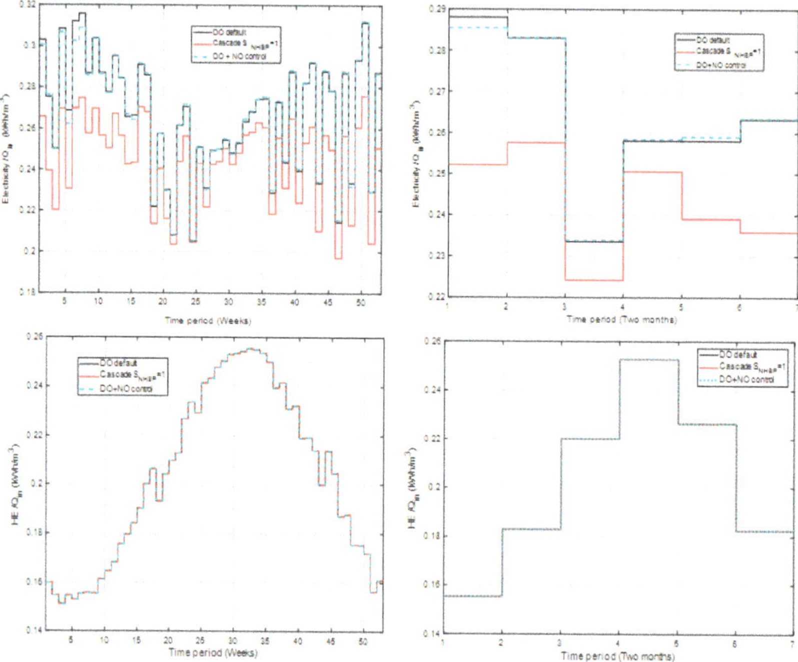

Figure 4. Weekly and bimonthly profiles of environmental indicators associated with energy use: electricity consumption, Electricity/Q_{in} (kW h/m^3); heating energy, HE/Q_{in} (kW h/m^3).

Electricity consumption is associated with control movements of the control schemes (i.e., aeration energy, pumping energy), then the frequent variations observed in the weekly profile of the control schemes that keep a constant DO set-point (DO default and DO + NO control) are associated with control actions that keep DO close to set-point in the presence of disturbances in the load. The similitude between DO default and DO + NO control profiles suggest that variations in electricity consumption are mainly produced by aeration. On the other hand, the Cascade S_{NHSP} scheme exhibits variations with load similar to the variations observed with DO-based control schemes, but electricity consumption is significantly reduced since it is not necessary to increase aeration to keep a fixed DO-set point. In the bimonthly profiles, the periods of maximum and minimum consumption coincide with DO-based schemes, but the variation pattern is completely different.

Regarding heating energy (HE) profiles, a clear effect of temperature with maximum heating requirements in the colder period and minimum requirements in the warmer season are observed in weekly and bimonthly profiles. The profiles with the three control schemes coincide exactly, which indicates that the effect of tested control schemes on energy requirements of digester is negligible.

Figure 5 shows the weekly and bimonthly profiles of the environmental indicators associated with emissions to air and soil: CH$_4$ content on biogas, sludge production and carbon dioxide emissions. There is no evidence of a significant temperature effect in the profiles of the three variables. The weekly and bimonthly profiles of CH$_4$ content on biogas reproduce the variation of COD concentration in the influent; it is recognized that COD content in the digester feed have a significant effect on biogas production [19], so influent variations of COD are reflected in the composition of Q_w that is fed to the digester. In the case of sludge for disposal, the dynamic pattern does not coincide with influent variations, except for the periods of minimum production in the third bimester that coincides with the maximum influent flow and minimum N_{tot} and COD concentration. Regarding the effect of control schemes, identical weekly and bimonthly profiles of CH$_4$ content in biogas and sludge for disposal are obtained with the three control schemes indicating a negligible influence of control actions on these variables. CO_2 production in ASP is governed by biological processes that are affected by load to be treated, so the effect of influent variables is evidenced in the profiles of CO_2 emissions. The frequent changes observed in the weekly profiles coincide with continuous influent variations and bimonthly profiles, allowing us to distinguish a period with the lowest emissions in the 3rd bimester and higher emissions in the driest period, the 1st and 2nd bimesters. The CO_2 emissions profiles with the three control schemes exhibit the same pattern of variation but different magnitudes, the Cascade S_{NHSP} scheme produces the lowest CO_2 emissions in the full operation horizon since treatment intensity is reduced due to reduction in aeration, but on the other hand DO + NO control produces the higher emissions since strict treatment requirements are imposed by simultaneous regulation of DO and NO set-point.

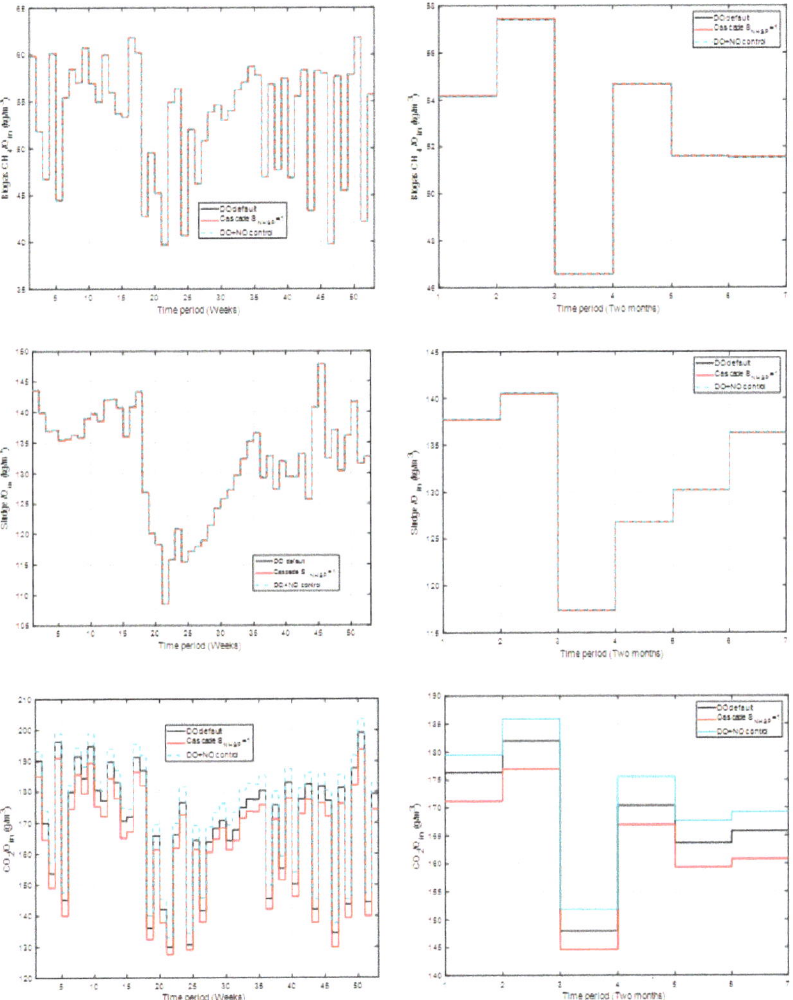

Figure 5. Weekly and bimonthly profiles of environmental indicators associated with emissions to air and soil: methane content in biogas, Biogas CH_4/Q_{in} (kg/m^3), sludge for disposal, Sludge/Q_{in} (kg/m^3), and total carbon dioxide emissions, CO_2/Q_{in} (g/m^3).

Figure 6 shows the weekly and bimonthly profiles of the environmental indicators associated with emissions to water: total nitrogen (N_{tot}), ammonium concentration (S_{NH}) and EQI. The influence of influent variation is evidenced by the valleys between weeks 5 and 7, in all indicators, and some peaks that coincide with extreme values of influent concentration and flowrate. Seasonal effect of temperature and load are observed in the bimonthly profiles, especially in the case of S_{NH} and EQI that exhibits higher values of the indicators in the period of lower temperature (4th bimester) and lower values in the warmer period (1st bimester). These variables determine the effluent quality attained with wastewater treatment, and then they are significantly affected by control actions. The weekly and bimonthly profiles of total nitrogen (N_{tot}) are completely different depending on control strategy. In the case of the DO control and DO + NO control schemes, it is evident that variations associated with changes in the influent load are attenuated by control actions that regulate nitrogen removal adjusting DO concentration to a constant DO set-point. The bimonthly profiles with DO-based control schemes do not suggest a seasonal effect. In the case of the Cascade S_{NHSP} scheme, larger variations associated with changes in influent load are observed in the weekly profile and, in the bimonthly profile, it is observed how N_{tot} increases in the colder period, where biological removal is slower, and decreases in the warmer bimesters, where microorganism activity increases. This control scheme varies the DO set-point with ammonium concentration in the last bioreactor, then pressure on biological nitrogen removal is reduced and the effect of other variables is more notorious. Regarding the S_{NH} and EQI profiles, they exhibit a similar variation pattern with the three control strategies with significant differences in the magnitudes of the indicators. The regulation of nitrates concentration with the DO + NO control scheme produce minimum levels of ammonium in the effluent in the full operation period while the highest values are attained with Cascade S_{NHSP} scheme. EQI measures pollution content in the effluent including COD, BOD, total nitrogen and ammonium, and this indicator exhibits the lower values with the Cascade S_{NHSP} scheme but, it is detected that in the periods of largest influent flow, lowest concentration of influent pollutants and lower temperature, the 3rd and 4th bimester, Cascade S_{NHSP} scheme and DO control attain the same values. The worst EQI profiles correspond to DO + NO control.

Summarizing, the analysis of the weekly and bimonthly profiles provide evidence that the effect of control strategies on environmental indicators associated with the sludge line as heating energy (HE), CH_4 content in biogas and sludge production is minimal. The dynamic profiles allow us to detect the significant effect of influent temperature for HE, and CH_4 content in biogas, while sludge production is affected by seasonal behavior of influent flow rate. The electricity consumption is associated with manipulated variables of the control schemes such as aeration energy and pumping energy, so the dynamics of electricity consumption depends on control actions performed to deal with frequent and seasonal changes in the influent load. Dynamic behavior of indicators of emissions to water and CO_2 emissions is determined by control actions performed to regulate the nitrogen removal process. However, analysis of those dynamic profiles allows to detect seasonal effects of influent load and temperature in CO_2 emissions, S_{NH}, EQI and N_{tot}, that cannot be observed in a study based on the evaluation of annual average environmental indicators. Then, analysis of dynamic performance considering different time scales provides insights into the effect of seasonal and periodic influent disturbances that can be useful to take adequate control decisions. Moreover, it allows us to capture the interactions between control actions and environmental impacts that can be addressed by the opportune adjustment of control variables.

This statement can be supported by comparing maximum difference between the values of indicators for the weekly and bimonthly profiles and the average annual value (DO control is used as a reference for comparison) shown in Table 5. Average values provide a quantification of performance in the full operational horizon, while dynamic profiles provide information of the changes experimented by indicators along the operation horizon that cannot be appreciated using annual based indicators. The possibility to detect such dynamic effects increases as the operation window decreases. In Table 5, the differences between the maximum and minimum values for the weekly time scales are attenuated in the bimonthly time scale.

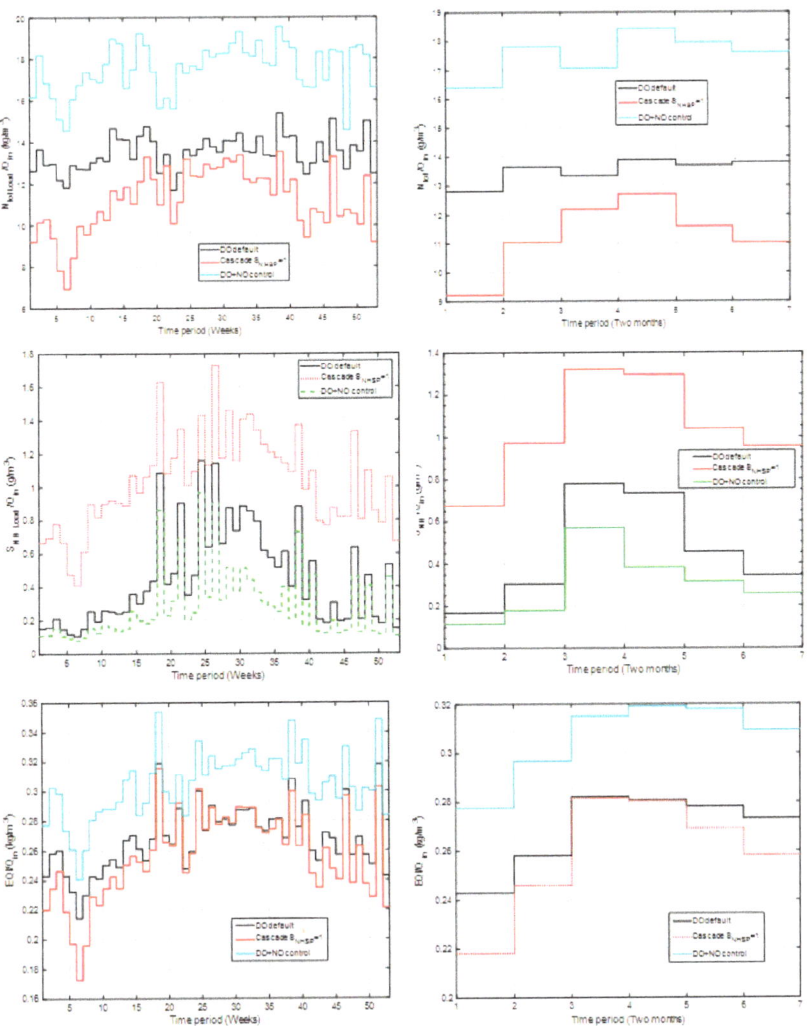

Figure 6. Weekly and bimonthly profiles of environmental indicators associated with emissions to water: $N_{tot\ Load}/Q_{in}$ (g/m^3), $S_{NH\ Load}/Q_{in}$ (g/m^3), and EQI/Q_{in} (kg/m^3).

In order to select the control strategy that exhibits the best dynamic performance, a quantitative comparison of the mean bimonthly values of environmental indicators significantly affected by control actions (Electricity, N_{tot} and EQI) is presented in Table 6. It presents the variation of indicators obtained with the Cascade S_{NHSP} scheme and DO + NO control with respect to the DO control scheme, since it is the typical strategy implemented in WWTPs. Weekly mean values of indicators are not presented, because a large amount of data had to be reported and qualitative information from Figures 4–6 have been sufficient to observe the dynamic effect in a shorter time horizon. The values reported in Table 5 evidence the improvement of electricity consumption, N_{tot} concentration and EQI attained with the Cascade S_{NHSP} scheme in the full operation period with respect to the DO control scheme. Effluent

S_{NH} concentration is worsened, but the highest value of S_{NH} in the weekly profile is still significantly below the desired limit (4 g/m^3), so it is admissible. The DO + NO control scheme improves only S_{NH} concentration and worsens N_{tot} concentration, attaining values that violates the desired limits in the weekly and bimonthly profiles. Thus, the analysis of the average annual indicators and the qualitative observation of dynamic profiles leads to the conclusion that the best performance in terms of environmental and operational costs is achieved with the Cascade S_{NHSP} scheme.

Table 5. Maximum and minimum values in the weekly and bimonthly profile, and average values of environmental indicators with respect to the volume of treated wastewater with DO control scheme (W. Av.: Weekly average, Bi-m Av.: Bimonthly average).

Environmental Indicators	Annual Average	W. Av. max.	W. Av. min.	Bi-m. Av. max.	Bi-m. Av. min.
Electricity (AE + PE + ME) (kW h/m^3)	0.263	0.315	0.204	0.288	0.224
Heating energy (kW h/m^3)	0.204	0254	0.150	0.252	0.155
Biogas CH$_4$ (g/m^3)	52.51	63	40	57.5	46.5
CO$_2$ (Total) (g/m^3)	167.1	200.0	130.0	182.0	148.0
Sludge for disposal (kg/m^3)	131.1	148.0	108.0	140.5	118.5
S_{NH} effluent (g/m^3)	0.474	1.150	0.100	0.800	0.180
N_{tot} effluent (g/m^3)	13.53	15.20	11.80	14.00	12.80
EQI (kg/m^3)	0.270	0.320	0.215	0.282	0.242

AE: Aeration energy, PE: Pumping energy, ME: Mixing energy, S_{NH}: ammonium concentration, N_{tot}: Total nitrogen concentration, EQI: Effluent quality index.

Table 6. Variation of bimonthly means of environmental indicators mainly affected by the Cascade S_{NHSP} and DO + NO control schemes with respect to default DO control scheme.

Cascade S_{NHSP} Scheme with Respect to DO Control						
Indicator/Bimester	1	2	3	4	5	6
Electricity %	−13	−9.0	−4.0	−2.0	−6.0	−11
N_{tot} %	−29	−20	−10	−9.0	−16	−21
S_{NH} %	230	227	68	73	131	174
EQI %	−8.3	−6.0	0	0	−3.5	−5.4
DO + NO Control with Respect to DO Control						
Indicator/Bimester	1	2	3	4	5	6
Electricity %	−1.0	0	0	1.0	1.0	0
N_{tot} %	28	30	26	32	30	26
S_{NH} %	−50	−33	−26	−47	−29	−23
EQI %	17	15	12	14	14	13

S_{NH}: ammonium concentration, N_{tot}: Total nitrogen concentration, EQI: Effluent quality index.

In order to determine the effect of other control actions on environmental indicators when using Cascade S_{NHSP} scheme, in the next sections the effect of ammonium set-point S_{NHSP} and external carbon dosage Q_{carb} variations is evaluated. For the sake of simplicity, the analysis is performed considering the annual average values of the indicators significantly affected by S_{NHSP} and Q_{carb} variations, the weekly dynamic profiles for ammonium set-point changes and bimonthly profiles for carbon dosage variation and the comparison of the bimonthly mean values. The idea is to detect this through observation of the dynamic effect of these control actions, when the changes with respect to default Cascade S_{NHSP} scheme with constant S_{NHSP} = 1 g/m^3 and Q_{carb} = 2 m^3/d can be favorable to environmental performance.

3.1.1. Different Set-Points for the Ammonium-Based Control Scheme (Cascade S_{NHSP})

Three possible set-points are considered for the Cascade S_{NHSP} scheme, default set-point 1 g/m^3, a set-point close to the admitted limit 4 g/m^3 and a relaxed set-point 6 g/m^3. The annual average values of the environmental indicators affected by control actions (Electricity consumption, CO_2 emissions and effluent variables: N_{tot}, S_{NH} and EQI) are presented in Table 6, together with the variations relative default Cascade scheme with $S_{NHSP} = 1$ g/m^3.

Increasing ammonium set-point implies increasing ammonium concentration in the effluent (S_{NH}) as is observed in the values reported in Table 7, it reflects also on EQI. However, other indicators as energy consumption, CO_2 emissions, N_{tot} in the effluent and operation costs (OCI) are improved when requirements on ammonium concentration in the effluent are reduced. Moreover, the negative impact of relaxing S_{NH} set-point can be tolerable if it is compensated by a significant improvement on other indicators, since S_{NH} in the effluent is still separated from its limit (4 g/m^3) with an approximated back off of 56% in the worst case ($S_{NHSP} = 6$) and the increment on EQI is small (1.9%). The analysis of dynamic behavior of the indices allows us to detect particular situations in a given temporal window where combination of the effects of influent variations, control actions and ammonium set-point variation produce a positive effect on environmental performance. Weekly profiles are considered since ammonium set-point changes affects biological processes that occur in a short time scale [8].

Table 7. Annual values of environmental indicators and operating costs of a BSM2 plant with respect to the volume of treated wastewater using the ammonium-based control scheme (Cascade S_{NHSP}) with different set-points and variation relative to 1 g/m^3 set-point.

	Environmental Indicators	$S_{NHSP} =$ 1 g/m^3	$S_{NHSP} =$ 4 g/m^3	$S_{NHSP} =$ 6 g/m^3	Relative Variation $S_{NHSP} = 4\%$	Relative Variation $S_{NHSP} = 6\%$
Energy	Electricity (AE + PE + ME) (kW h/m^3)	0.243	0.234	0.231	−3.7	−5.0
Emissions to air	CO_2 (ASP) (g/m^3)	87.20	84.32	83.29	−3.3	−3.7
Emissions to water	S_{NH} effluent (g/m^3)	1.052	1.562	1.765	48.5	67.8
	N_{tot} effluent (g/m^3)	11.34	10.54	10.36	−7.0	−8.64
	EQI (kg/m^3)	0.260	0.263	0.265	1.15	1.9
Operating costs	OCI (EUR/d)	0.437	0.428	0.425	−2.1	−2.7

AE: Aeration energy, PE: Pumping energy, ME: Mixing energy, S_{NH}: ammonium concentration, N_{tot}: Total nitrogen concentration, EQI: Effluent quality index.

Figure 7 presents weekly profiles of electricity consumption and CO_2 emissions and Figure 8 presents weekly profiles of the indicators associated with emissions to water: N_{tot}, S_{NH} and EQI. The variation of bimonthly means of electricity, N_{tot}, S_{NH} and EQI obtained with the Cascade scheme with $S_{NHSP} = 4$ g/m^3 and $S_{NHSP} = 6$ g/m^3 with respect to original $S_{NHSP} = 1$ g/m^3 is presented in Table 7.

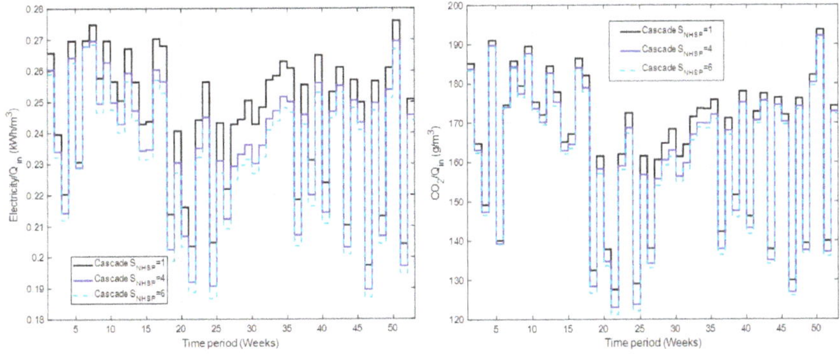

Figure 7. Weekly profiles for electricity consumption Electricity/Q_{in} (kW h/m^3) and total carbon dioxide emissions CO_2/Q_{in} (g/m^3) under ammonium-based control scheme (Cascade S_{NHSP}).

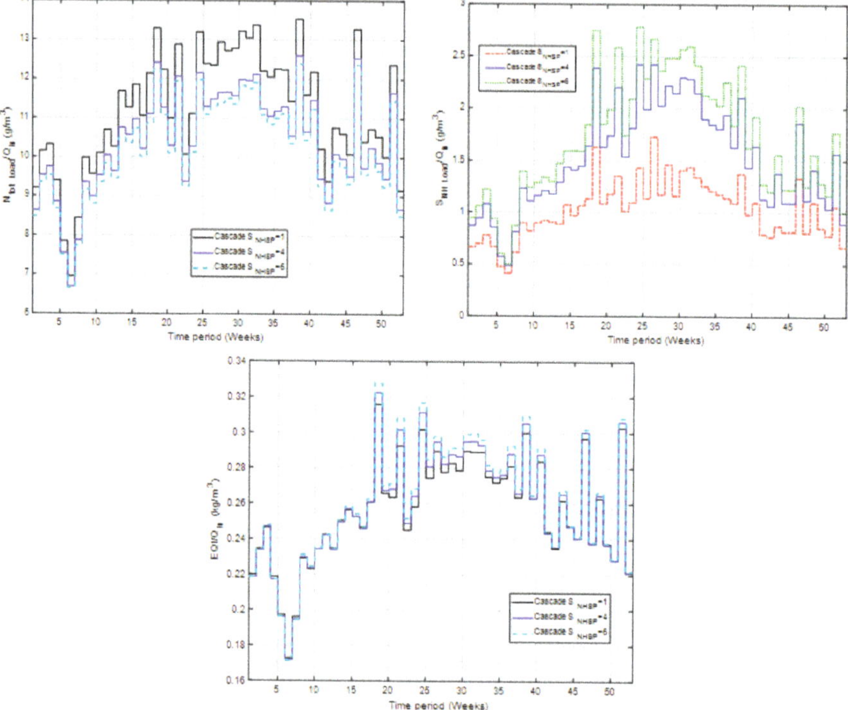

Figure 8. Weekly profiles of environmental indicators associated with emissions to water: $N_{tot\ Load}/Q_{in}$ (g/m^3), $S_{NH\ Load}/Q_{in}$ (g/m^3) and EQI/Q_{in} (kg/m^3) under the ammonium-based control scheme (Cascade S_{NHSP}).

In Figure 7 it can be observed that profiles of electricity consumption and CO_2 emissions decreases their magnitude as S_{NHSP} increases but exhibit the same pattern of variation. The variation of electricity consumption reported in Table 8 provide evidence that the changes are larger between the 3rd and 5th bimesters where load and temperature effects are significant. Regarding indicators of emissions to water, in Figure 8 it is observed that magnitude of N_{tot} profile decreases as ammonium set-point increases, but the effect is notorious in the colder weeks (20–40). The opposite effect is observed in S_{NH} and EQI profiles, ammonium concentration in the effluent and EQI increases as S_{NHSP} increases, but it is more notorious in the colder weeks. Moreover, S_{NH} values are significantly affected while the impact on EQI can be negligible in the warmer weeks. These observations are supported by the quantitative information reported in Table 8.

The solution to improve plant performance modifying ammonium set-point (S_{NHSP}) depends on different factors. Increasing the S_{NHSP} reduces electricity consumption and CO_2 emissions and minimizes N_{tot} but increases ammonium emissions (S_{NH}). The S_{NH} set-point could be increased in specific periods of time, where other factors compensate the deterioration of EQI and emissions of ammonium to water, to produce a positive effect on electricity consumption, CO_2 emissions and N_{tot}. The adjustment of carbon dosage, the effect of which is evaluated in the next section, can produce conditions favorable to increase ammonium set-points in particular temporal windows.

Table 8. Variation of bimonthly means of environmental indicators affected by ammonium set-point (S_{NHSP}) changes with respect to the Cascade S_{NHSP} control scheme with $S_{NHSP} = 1$ g/m^3.

$S_{NHSP} = 4$ g/m^3 Scheme with Respect to $S_{NHSP} = 1$ g/m^3						
Indicator/Bimester	**1**	**2**	**3**	**4**	**5**	**6**
Electricity %	−2.4	−3.5	−4.9	−4.8	−3.8	−3.0
N_{tot} %	−5.9	−4.5	−5.7	−9.0	−7.3	−5.5
S_{NH} %	39	35	52	62	52	42
EQI %	0	0	2.5	1.8	3.3	0.4
$S_{NHSP} = 6$ g/m^3 Scheme with Respect to $S_{NHSP} = 1$ g/m^3						
Indicator/Bimester	**1**	**2**	**3**	**4**	**5**	**6**
Electricity %	−2.8	−4.7	−6.7	−6.0	−5.0	−4.2
N_{tot} %	−8.1	−8.2	−8.2	−10.6	−8.6	−6.4
S_{NH} %	54	50	76	84	71	53
EQI %	0	0	3.6	3.2	1.9	0.8

S_{NH}: ammonium concentration, N_{tot}: Total nitrogen concentration, EQI: Effluent quality index.

3.1.2. Effect of Variation of External Carbon Dosage (Q_{carb}) with the Ammonium-Based Control Scheme (Cascade S_{NHSP})

In the default operation strategy, an external carbon source with a concentration of 40,000 g/m^3 is added to the first anoxic reactor at a constant flowrate $Q_{carb} = 2$ m^3/d. The effect of variations of Q_{carb} to lower values, including $Q_{carb} = 0$ is evaluated considering annual and bimonthly time scales since it affects biological processes in a medium time scale, and it is easier to appreciate this effect using bimonthly profiles.

Table 9 presents the annual average values of environmental indicators and operating costs computed with respect to the volume of treated wastewater, and the variations observed on the annual average values of the indicators relative to the default $Q_{carb} = 2$ m^3/d are presented in Table 10. In Tables 9 and 10 it is observed that variation of Q_{carb} affect, even slightly, all environmental indicators from the water and sludge line. Decreasing carbon dosage produce a slight positive effect ranging between 1% to 4% on electricity consumption, CO_2 emissions from the digester, sludge production, S_{NH} concentration in the effluent and heating energy (HE). On the other hand, a slight negative impact is observed in biogas production, CO_2 emissions from ASP and COD in the effluent.

Table 9. Annual values of environmental indicators and operating costs of the BSM2 plant with respect to the volume of treated wastewater using the ammonium-based control scheme (Cascade S_{NHSP}) with $S_{NHSP} = 1$ g/m^3 and different values of Q_{carb}.

	Environmental Indicators	$Q_{carb} = 2$ g/m^3	$Q_{carb} = 1$ g/m^3	$Q_{carb} = 0.5$ g/m^3	$Q_{carb} = 0$ g/m^3
Energy	Electricity (AE + PE + ME) (kW h/m^3)	0.243	0.237	0.235	0.234
	Heating energy (kW h/m^3)	0.204	0.203	0.202	0.201
Chemicals	External carbon (kg COD/m^3)	0.039	0.019	0.010	0
Emissions to air	Biogas CH$_4$ (g/m^3)	52.53	51.90	51.59	51.28
	CO$_2$ (Digester) (g/m^3)	75.63	74.77	74.35	73.92
	CO$_2$ (ASP) (g/m^3)	87.20	88.55	89.44	90.49
Emissions to soil	Sludge for disposal (kg/m^3)	131.01	128.40	127.08	125.77
Emissions to water	S_{NH} effluent (g/m^3)	1.052	1.028	1.019	1.011
	N_{tot} effluent (g/m^3)	11.34	12.83	13.86	15.09
	COD effluent (g/m^3)	49.07	48.71	48.54	48.38
	EQI (kg/m^3)	0.260	0.273	0.282	0.294
Operation costs	OCI (EUR/d)	0.437	0.369	0.336	0.303

AE: Aeration energy, PE: Pumping energy, ME: Mixing energy, COD: Chemical oxygen demand, S_{NH}: ammonium concentration, N_{tot}: Total nitrogen concentration, EQI: Effluent quality index.

Table 10. Comparison of the influence on environmental and cost indicators of carbon dosage variations relative to default $Q_{carb} = 2\ m^3/d$ with the ammonium-based control (Cascade S_{NHSP}) with $S_{NHSP} = 1\ g/m^3$.

	Environmental Indicators	$Q_{carb} = 1\ g/m^3$	$Q_{carb} = 0.5\ g/m^3$	$Q_{carb} = 0\ g/m^3$
Energy	Electricity (AE + PE + ME) (kW h/m^3)	−2.5%	−3.3%	−3.7%
Chemicals	External carbon (kg COD/m^3)	−51.2%	−74.4%	−100%
Emissions to air	CO_2 (ASP) (g/m^3)	1.6%	2.6%	3.8%
Emissions to soil	Sludge for disposal (kg/m^3)	−2%	−3%	−4%
	S_{NH} effluent (g/m^3)	−2.3%	−3.2%	−3.9%
Emissions to water	N_{tot} effluent (g/m^3)	13.1%	22.2%	33.1%
	EQI (kg/m^3)	5%	8.5%	13.8%
Operation costs	OCI (EUR/d)	−15.6%	−23.1%	−30.7%

AE: Aeration energy, PE: Pumping energy, ME: Mixing energy, S_{NH}: ammonium concentration, N_{tot}: Total nitrogen concentration, EQI: Effluent quality index.

The reduction in the use of chemicals, measured as the amount in kg COD of external carbon, is proportional to Q_{carb}. Carbon dosage affects directly the operation costs since OCI includes a term that accounts external carbon with a cost factor of 3 EUR/kg, then operation costs can be reduced 15.6% when half of the carbon dosage is used and can be reduced to 30% eliminating carbon dosage ($Q_{carb} = 0$).

The variables that are significantly affected by Q_{carb} are N_{tot} and EQI which vary up to 33.1% and 13.8% respectively. The amount of organic matter provided by Q_{carb} is used as substrate by heterotrophs for denitrification, then reducing available substrate to transform nitrates (S_{NO}) to N_2 gas increases the amount of nitrates in the effluent and consequently N_{tot} and EQI.

The dynamic behavior of total nitrogen in the effluent (N_{tot}) and EQI is observed using bimonthly profiles (Figure 9), since the effect of Q_{carb} variation on these variables is clearly observed considering this time scale. The N_{tot} and EQI profiles shown in Figure 9 exhibit the same variation patterns with different magnitudes for the different values of Q_{carb}, and the magnitude of the profiles increases proportionally to Q_{carb} reduction. There is only one exception in the case of EQI that exhibits a different trend in the 4th bimester with $Q_{carb} = 2\ g/m^3$. Table 11 presents the variation of bimonthly mean values of N_{tot} and EQI for the different Q_{carb} values relative to the default $Q_{carb} = 2\ g/m^3$.

Decreasing carbon dosage implies a significant reduction in the use of chemicals and operating costs but produces a negative impact on total nitrogen and effluent quality index. A comprehensive evaluation of Q_{carb} effect allows us to determine the temporal windows where other effects compensate the negative impact on N_{tot} and EQI of Q_{carb} reduction, to minimize operation costs and the use of chemicals.

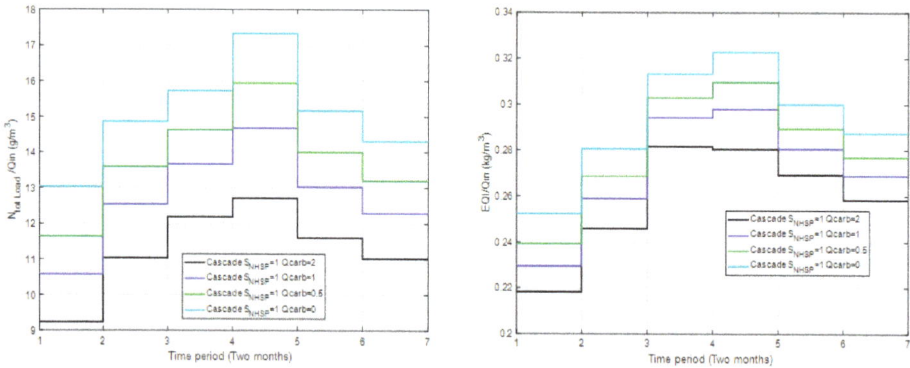

Figure 9. Bimonthly profiles of $N_{tot\ Load}/Q_{in}$ (g/m^3) and EQI/Q_{in} (kg/m^3) for Q_{carb} variations under ammonium-based control (Cascade S_{NHSP}).

Table 11. Bimonthly variations on N_{tot} and EQI with carbon dosage variations relative to default $Q_{carb} = 2$ m^3/d.

	Indicator/Bimester	1	2	3	4	5	6
$Q_{carb} = 1$ g/m^3	N_{tot}%	15	15	12	16	12	11
	EQI %	4.6	6.1	3.6	6.4	3.7	4.7
$Q_{carb} = 0.5$ g/m^3	N_{tot}%	28	24	23	26	21	20
	EQI %	9.1	8.0	7.1	11	7.4	5.8
$Q_{carb} = 0$	N_{tot}%	41	36	30	34	30	29
	EQI %	14.6	14.3	10.6	15.0	11.1	12.4

N_{tot}: Total nitrogen concentration, EQI: Effluent quality index.

3.2. Selection of the Alternative Strategy for the Best Trade-off Solution

The ammonium-based control scheme with constant $S_{NHSP} = 1$ g/m^3 and $Q_{carb} = 2$ m^3/d is selected as the best trade-off solution between environmental and operational costs compared with the DO default control and DO + NO control schemes. The analysis of the dynamic behavior in the weekly and bimonthly time scales, including the effect of variations of ammonium set-point S_{NHSP} and Q_{carb}, allows us to determine how control actions and influent variables affects environmental indicators in different temporal windows. The analysis makes it possible to determine the temporal windows where different control actions can be applied to improve the environmental indicators. Thus, different combinations of ammonium set-points and a fixed sequence of changes of carbon dosage Q_{carb} have been evaluated to find the combination of control actions in the operational period that produce a positive effect on environmental and operation costs, preserving the desired performance. The sequence of control movements on S_{NHSP} and Q_{carb} is presented in Figure 10.

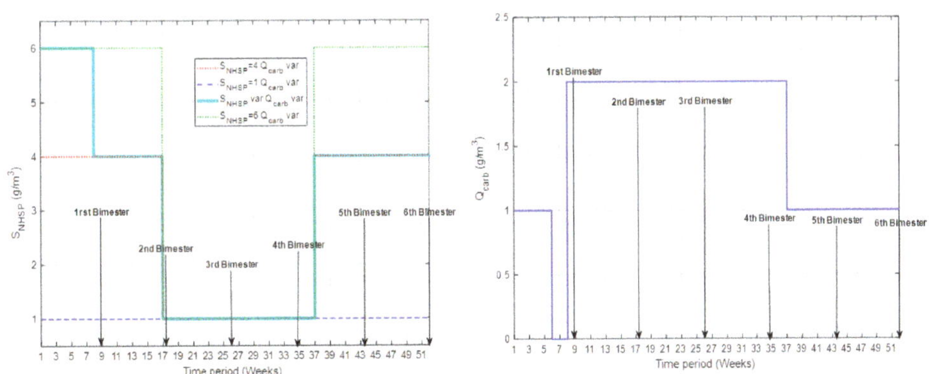

Figure 10. Ammonium set-point S_{NHSP} (g/m^3) and Q_{carb} (m^3/d) variations considered as alternative control actions.

From the analysis of dynamic behavior (weekly and bimonthly profiles), a period between the 3rd and 4th bimesters has been detected where influent conditions and low temperature affects negatively the indicators of emissions to water: N_{tot}, S_{NH} concentration and EQI. As shown in Figure 10, strict ammonium set-point $S_{NHSP} = 1$ g/m^3 and default $Q_{carb} = 2$ g/m^3 are maintained between weeks 17 and 37, where these effects are notorious, and the three different ammonium set-points (1, 4 and 6 g/m^3) are considered for the rest of the operation period. A fixed sequence of movements of Q_{carb} is applied, in the first 6 weeks where low N_{tot} and S_{NH} levels are observed with the different ammonium set-points (Figure 8) carbon dosage is reduced to $Q_{carb} = 1$ g/m^3, in between weeks 7 and 9 the minimum values of N_{tot} and S_{NH} are attained, then carbon dosage is cut, it is increased to $Q_{carb} = 2$ g/m^3 between weeks 17 and 37, and it is finally reduced to $Q_{carb} = 1$ g/m^3 in the last weeks when N_{tot} and S_{NH} levels decrease. The combination of the sequence of different ammonium set-points and given sequence for carbon dosage, produce three different strategies named: $S_{NHSP} = 1\ Q_{carb\ var}$, $S_{NHSP} = 4\ Q_{carb\ var}$ and $S_{NHSP} = 6\ Q_{carb\ var}$. The weekly and bimonthly dynamic profiles of the different environmental indicators affected by the aforementioned strategies are shown in Figures A1–A3 in Appendix A. From the observation of weekly and bimonthly profiles of environmental indicators, the changes that produce a positive effect detected in a specific temporal window are selected to produce a strategy named $S_{NHSP\ var}\ Q_{carb\ var}$, that combines the sequence of S_{NHSP} changes and Q_{carb} changes presented in Figure 10. Those changes are $S_{NHSP} = 6\ g/m^3$ between weeks 1 and 8 where S_{NH} levels are the minimum, improvement attained with stricter ammonium set-point is not significant, but electricity consumption can be reduced by increasing S_{NHSP}, $S_{NHSP} = 4\ g/m^3$ between weeks 8 and 16 and weeks 38 to 53 to reduce electricity consumption and attain acceptable levels of S_{NH} and $S_{NHSP} = 1\ g/m^3$ between weeks 17 to 37 where treatment is difficult due to load and temperature effect. The proposed strategies are summarized in Table 12.

Table 12. Alternative strategies proposed to improve environmental and operation costs.

Name	S_{NHSP}	Q_{carb}
Cascade S_{NHSP}	Constant 1 g/m³	Constant 2 g/m³
$S_{NHSP} = 1\ Q_{carb\ var}$	Constant 1 g/m³	Q_{carb} sequence shown in Figure 10
$S_{NHSP} = 4\ Q_{carb\ var}$	$S_{NHSP} = 1\ g/m^3$ between weeks 17 and 37, $S_{NHSP} = 4\ g/m^3$ in the rest of operational period	Q_{carb} sequence shown in Figure 10
$S_{NHSP} = 6\ Q_{carb\ var}$	$S_{NHSP} = 1\ g/m^3$ between weeks 17 and 37, $S_{NHSP} = 6\ g/m^3$ in the rest of operational period	Q_{carb} sequence shown in Figure 10
$S_{NHSP\ var}\ Q_{carb\ var}$	$S_{NHSP} = 6\ g/m^3$ between weeks 1 and 8, $S_{NHSP} = 4\ g/m^3$ between weeks 8 and 16, $S_{NHSP} = 1\ g/m^3$ between weeks 17 and 37, $S_{NHSP} = 4\ g/m^3$ from week 38 to the end of the operational period	Q_{carb} sequence shown in Figure 10

S_{NHSP}: Ammonium set-point, Q_{carb} carbon dosage.

It is important to mention that the control decisions described above have been motivated by the observation of situations on specific periods of time (weeks or bimesters) on dynamic profiles, that can be changed to improve environmental performance. These situations could not be detected by a traditional analysis of annual average environmental indicators.

The performance of $S_{NHSP\ var}\ Q_{carb\ var}$ strategy is compared with the Cascade S_{NHSP} scheme with $S_{NHSP} = 1\ g/m^3$. First, weekly and bimonthly profiles of environmental indicators are obtained and compared to observe the temporal windows where environmental indicators are affected by the proposed strategy. Afterwards, the two alternative strategies are compared with DO default scheme, that is the usual control strategy implemented in WWTPs, considering the annual average values of environmental indicators and operating costs to evaluate the global improvement of the control actions determined after the dynamic analysis of behavior.

Figure 11 shows the weekly and bimonthly profile for electricity consumption, Figure 12 shows the bimonthly profiles of the indicators of biogas and sludge production, and Figure 13 the weekly and bimonthly profile for CO_2 emissions. In Figure 11 it is observed that weekly and bimonthly profiles of the proposed $S_{NHSP\ var}\ Q_{carb\ var}$ strategy attains lower values than Cascade S_{NHSP} scheme in the full operation horizon except for the period between weeks 17 and 37, where identical control actions are applied, and profiles coincide. The reduction of electricity consumption obtained with the $S_{NHSP\ var}\ Q_{carb\ var}$ strategy is 4.7% and 3.1% in the 1st and 2nd bimesters and 3.8% and 4.6% in the 5th and 6th bimesters. A slight reduction of biogas (ranging between 1.4 and 2.2%) and sludge production (only 1%) is achieved with the proposed strategy as observed in Figure 12, where $S_{NHSP\ var}\ Q_{carb\ var}$ profile is below Cascade S_{NHSP} profile in the full operation period except for the period between weeks 17 and 37. A similar positive effect is observed in Figure 13 for CO_2 emissions that are slightly reduced by proposed strategy.

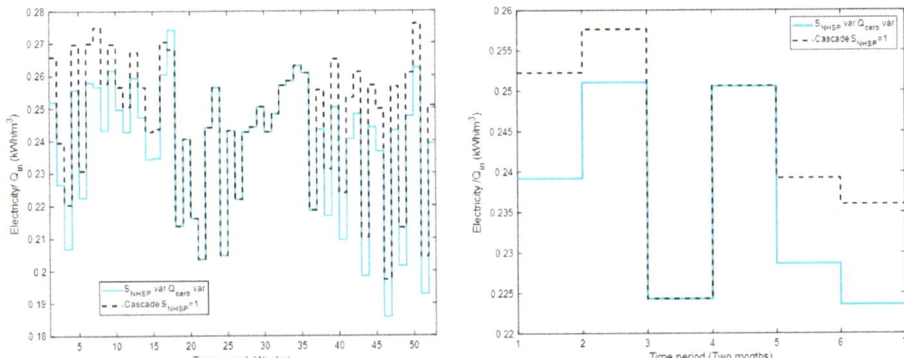

Figure 11. Weekly and bimonthly profile for electricity consumption indicator (kW h/m^3) with Cascade S_{NHSP} and the alternative strategy.

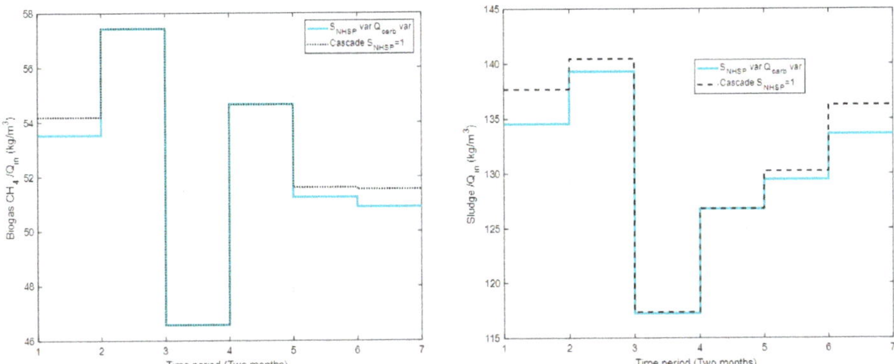

Figure 12. Bimonthly profile for biogas (kg/m^3) and sludge production (kg/m^3) indicators with Cascade S_{NHSP} and the alternative strategy.

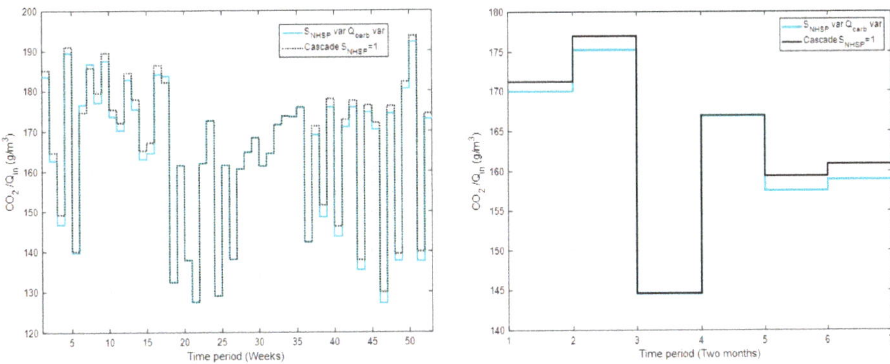

Figure 13. Weekly and bimonthly profile for total CO2 emissions (g/m3) indicators with Cascade S_{NHSP} and the alternative strategy.

The weekly and bimonthly profile of the indicators associated with emissions to water (effluent N_{tot}, S_{NH} and EQI) with the proposed $S_{NHSP\,var}$ $Q_{carb\,var}$ strategy and Cascade S_{NHSP} scheme are presented in Figure 14. The control actions of the proposed scheme produce a negative effect on N_{tot},

S_{NH} and EQI profiles that are worsened in most of the operation period with respect to the Cascade S_{NHSP} scheme. Considering the first bimester as the worst temporal period, N_{tot} is worsened up to 8.7% and EQI is increased up to 4.5%, while S_{NH} increases up to 28% in the 2nd bimesters even though the worst situation, identified in the week 18, is still distant from the limit value that is 4 g/m^3.

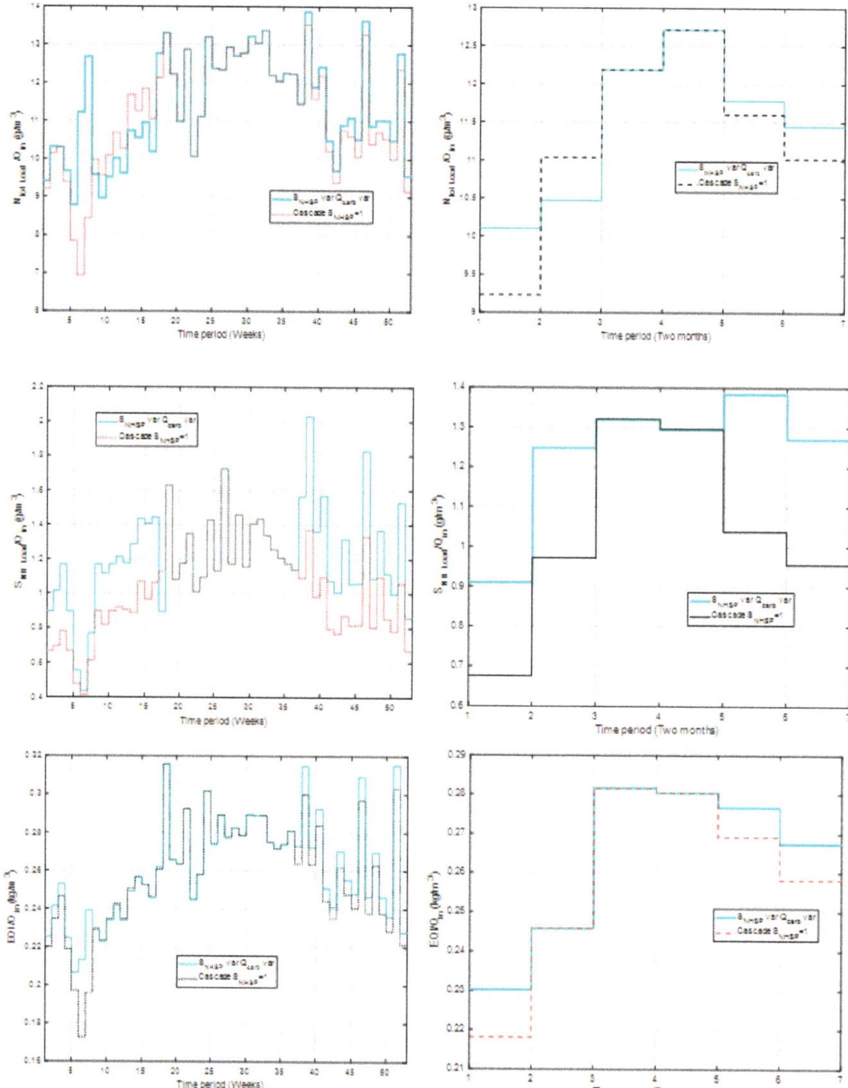

Figure 14. Weekly and bimonthly profile for emissions to water indicators $N_{tot\,Load}/Q_{in}$ (g/m^3), $S_{NH\,Load}/Q_{in}$ (g/m^3) and EQI/Q_{in} (kg/m^3) with Cascade S_{NHSP} and the alternative strategy.

The annual average values are presented in Table 13 and the comparison relative to the DO default scheme is presented in Table 14. Despite deterioration of the S_{NH} indicator, that is a consequence of the application of ammonium control with variable DO set-point, $S_{NHSP\,var}\,Q_{carb\,var\,strategy}$ and Cascade S_{NHSP} strategy with S_{NHSP} = 1 g/m^3 and Q_{carb} = 2 g/m^3 produce a significant improvement to the rest of environmental indicators and operation costs in comparison with DO default strategy. The variations of S_{NHSP} and Q_{carb} in the appropriated temporal windows reduce the use of chemicals, electricity consumption and consequently operation costs, which could compensate the increment in the levels of S_{NH} in the effluent.

Table 13. Annual values of environmental indicators and operating costs of BSM2 plant with respect to the volume of treated wastewater with the Cascade S_{NHSP} and proposed alternative strategies.

	Environmental Indicators	DO Default	Cascade $S_{NHSP} = 1$ g/m^3	Variable $S_{NHSP\,var}$-$Q_{carb\,var}$
Energy	Electricity (AE + PE + ME) (kW h/m^3)	0.263	0.243	0.236
	Heating energy (kW h/m^3)	0.204	0.204	0.204
Chemicals	External carbon (kg COD/m^3)	0.039	0.039	0.029
Emissions to air	Biogas CH$_4$ (g/m^3)	52.51	52.53	52.25
	CO$_2$ (Digester) (g/m^3)	75.61	75.63	75.25
	CO$_2$ (ASP) (g/m^3)	91.52	87.20	86.49
Emissions to soil	Sludge for disposal (kg/m^3)	131.1	131.0	129.8
Emissions to water	S_{NH} effluent (g/m^3)	0.474	1.052	1.244
	N$_{tot}$ effluent (g/m^3)	13.53	11.34	11.48
	COD effluent (g/m^3)	48.99	49.07	48.93
	EQI (kg/m^3)	0.270	0.260	0.265
Operation costs	OCI (EUR/d)	0.457	0.437	0.400

AE: Aeration energy, PE: Pumping energy, ME: Mixing energy, COD: Chemical oxygen demand, S_{NH}: ammonium concentration, N$_{tot}$: Total nitrogen concentration, EQI: Effluent quality index.

Table 14. Comparison of the influence on environmental and cost indicators of alternative strategies' control relative to the default DO control scheme.

	Environmental Indicators	Cascade $S_{NHSP} = 1$ g/m^3	Variable $S_{NHSP\,var}$-$Q_{carb\,var}$
Energy	Electricity (AE + PE + ME) (kW h/m^3)	−7.6%	−10.3%
Chemicals	External carbon (kg COD/m^3)	0%	−25.6%
	CO$_2$ (ASP) (g/m^3)	−4.7%	−5.5%
Emissions to water	S_{NH} effluent (g/m^3)	+121%	+162%
	N$_{tot}$ effluent (g/m^3)	−16.2%	−15.15%
	EQI (kg/m^3)	−3.7%	−1.81%
Operation costs	OCI (EUR/d)	−4.4%	−14.25%

AE: Aeration energy, PE: Pumping energy, ME: Mixing energy, S_{NH}: ammonium concentration, N$_{tot}$: Total nitrogen concentration, EQI: Effluent quality index.

In order to provide a condensed view of the advantages and disadvantages of the control schemes considered in this work: DO control, DO + NO control, Cascade S_{NHSP} and the proposed modification named $S_{NHSP\,var}$ $Q_{carb\,var}$, Table 15 summarizes the most important effects of control actions on environmental costs. The consideration of different temporal windows to observe WWTP behavior under different control schemes, on a dynamic influent scenario, have been a useful tool to detect seasonal effects and the influence of control actions performed to maintain the desired operating conditions on environmental indicators. The analysis of weekly and bimonthly dynamic profiles allows us to capture the interactions between control actions and environmental impacts that can be addressed by the opportune adjustment of control variables. The proposed methodology that combines the comprehensive analysis of annual average indicators and the qualitative observation of dynamic profiles allows us to determine the control scheme that produces the best compromise solution between environmental and operation costs. Moreover, the introduction of the analysis of dynamic profiles in the evaluation of the environmental impact of wastewater treatment makes it possible to determine the temporal windows where different control actions that can be applied to improve the environmental indicators. Thus, in this specific case study, the Cascade S_{NHSP} scheme was selected from existing control strategies, and its overall performance has been improved introducing different combinations of ammonium set-points and a sequence of changes of carbon dosage Q_{carb}.

Table 15. Summary of the effect on dynamic evolution of environmental indicators and average operation costs of the three evaluated proportional integral (PI) control schemes (DO control, DO+ NO control, Cascade S_{NHSP}).

Control Scheme	Effect on Environmental Indicators and Average Operation Costs
DO control	**Advantages:** - Affects positively all indicators of emissions to water, S_{NH} (g/m^3) and N_{tot} (g/m^3) levels in the effluent and EQI are simultaneously improved in the annual, bimonthly and weekly time periods. - Simpler control structure with respect to DO + NO control and Cascade S_{NHSP}. **Disadvantages:** - Increases consumption of electricity due to increments of aeration energy to keep the imposed DO set-point in the periods of higher load. - Higher annual average operation costs (OCI) due to larger consumption of energy.
DO + NO control	**Advantages:** - Minimizes the emissions of ammonium to effluent producing the lowest levels of S_{NH} (g/m^3) of the three strategies in the annual, bimonthly and weekly time periods. **Disadvantages:** - Increases consumption of electricity due to increments of aeration energy to keep the imposed DO set-point in the periods of higher load. - Increases emissions to air, producing larger CO_2 emissions from ASP in the full operation period due to higher pressure on denitrification. - Increases annual average operation costs (OCI) due to larger consumption of energy.
Cascade S_{NHSP}	**Advantages:** - Decreases aeration energy in the periods of lower load and higher temperature due to the possibility of varying the DO set-point, producing significant energy savings. - Affects positively emissions to air, reducing CO_2 emissions from ASP in the full operation period. - Minimizes the total nitrogen emissions to effluent, producing the lowest levels of N_{tot} (g/m^3) of the three strategies in the annual, bimonthly and weekly time periods. - Sensitivity to dynamic effect of temperature on N_{tot} (g/m^3) in the effluent can be exploited to apply other control actions to improve performance. - Minimizes annual average operation costs (OCI) due to reduction of energy use. **Disadvantages:** - Increases the S_{NH} (g/m^3) levels released in the effluent. - Complex control structure with respect to DO control.
S_{NHSP} var Q_{carb} var strategy	**Advantages:** - Reduce energy consumption due to the possibility of varying DO set-points, since the strategy is based on Cascade S_{NHSP} scheme. - Affects positively environmental and operation costs owing to the implementation of different S_{NH} (g/m^3) set-points and carbon dosage Q_{carb}, in selected operation windows. - Affects positively environmental indicators that are not affected by DO control, DO + NO control and cascade S_{NHSP} scheme, such as the use of chemicals, biogas production and sludge production. - Minimizes annual average operation costs (OCI) due to reduction of energy use and carbon dosage. **Disadvantages:** - Increases the S_{NH} (g/m^3) levels released in the effluent. - Could requires complex control structure to supervise control actions.

As can be observed from Table 15, the minimization of electricity consumption is an expected advantage of control strategies. Electricity consumption is strongly dependent on control actions associated with aeration and pumping performed to deal with frequent and seasonal changes in the influent load. Since, energy consumption is a crucial variable for improving WWTP efficiency, it affects simultaneously the operation costs and environmental costs. So, systematic analysis of its dynamic behavior can be helpful for the decision-making process on WWTPs management. For future work, the available tools that describe aeration system [13] and alternative renewable energy sources [29] can be useful to implement innovative operation strategies oriented to upgrade environmental performance of the plant by applying appropriate energy-management strategies.

On the other hand, behavior of indicators of CO_2 emissions and indicators of emissions to water is determined by control actions performed to regulate nitrogen removal process. Some control strategies such as DO control can affect positively all indicators of emissions to water, with the corresponding increase of electricity consumption as indicated in Table 15. The DO+ NO and Cascade S_{NHSP}-based strategies have to deal with the compromise of improving ammonium removal or total nitrogen concentration in the effluent. It is affected also by carbon dosages, that have a significant influence on operation costs. Then, dynamic analysis allows us to detect seasonal effects of influent load and temperature in CO_2 emissions, S_{NH}, EQI and N_{tot}, that cannot be observed in a study based on the evaluation of annual average environmental indicators, carbon dosage Q_{carb} can be regulated considering the operation periods where it is possible to reduce carbon dosage preserving the desired N_{tot} vs. S_{NH} compromise in the effluent load.

It is important to mention that the control decisions described above have been motivated by observation of situations during specific periods of time (weeks or bimesters) on dynamic profiles, that can be changed to improve environmental performance. These situations could not be detected by a traditional analysis of annual average environmental indicators. Moreover, the comparison of the annual average indicators provides a global perspective of environmental and economic performance of control strategies in the full operational period. Nevertheless, the analysis of the evolution of environmental indicators considering different temporal windows (weekly and bimonthly) allows us to determine which situations produce such overall result, when this situations occurs, in the case of seasonal variations of influent conditions and, in the case of the interactions between control actions and environmental costs of the treatment in the presence of influent variations.

4. Conclusions

In this paper the assessment of environmental costs of the operation of a WWTP employing three different control strategies (DO control scheme, DO + NO control, Cascade S_{NHSP}) integrating analysis of dynamic performance in different time scales (annual, bimonthly and weekly) has been carried out. The dynamic assessment has been based on environmental indicators classified into the following categories: energy indicators that measure electricity consumption and heating energy, indicators of emissions to air measuring CO_2 emissions from the activated sludge process and anaerobic digestion, emissions to soil associated with the production of sludge for disposal and emissions to water indicators associated with total nitrogen concentration in the effluent N_{tot}, ammonium concentration S_{NH} and pollution to effluent measured with the effluent quality index (EQI).

The analysis of dynamic profiles on different temporal windows makes it possible to identify operation periods where load, temperature effects and control actions have a significant impact on environmental indicators. These effects cannot be detected in a study based on the evaluation of annual average environmental indicators. The analysis of dynamic profiles of environmental indicators considering different time scales allows us to identify the seasonal influent disturbances and periodic variations that affect environmental performance such as seasonal changes of temperature and influent flow rate. This information is useful to take adequate control decisions that improve the environmental performance of the plant in these situations. Moreover, it allows us to capture interactions between control actions and environmental impacts occurring in specific periods of time that can be addressed by the opportune adjustment of control variables.

The observation of the annual average values of environmental indicators and operational costs showed that ammonium-based control (Cascade S_{NHSP}) produces the best compromise solution between environmental and operating cost compared with DO default control and DO + NO control. The analysis of dynamic profiles (weekly and bimonthly) showed that the Cascade S_{NHSP} perform better than the other control schemes in the periods where disturbances on load and seasonal effects of temperature and influent flow rate affect plant behavior. The ammonium-based control relaxes the requirements on ammonium concentration in the effluent, but reduces energy consumption, CO_2 emissions, total nitrogen concentration, and EQI. This is appreciated in the annual-based analysis of environmental performance, but also in the weekly and bimonthly dynamic profiles. The evaluation of the effect of S_{NH} set-point changes and carbon dosage on performance of the Cascade S_{NHSP} scheme allows us to determine the specific temporal windows where these actions produce a positive effect. Thus, a control strategy $S_{NHSP\ var}\ Q_{carb\ var}$ defined by a sequence of changes on S_{NHSP} and carbon dosage is proposed. The comparison of the proposed strategies with DO default control considering dynamic profiles and annual averages values leads to the conclusion that both alternatives improve environmental performance, but benefits of the Cascade S_{NHSP} scheme are associated with improvement of electricity consumption and emissions to water indicators N_{tot} and EQI, while the $S_{NHSP\ var}\ Q_{carb\ var}$ strategy reduces electricity consumption, use of chemicals (reducing external carbon dosage), and operational costs.

Author Contributions: Conceptualization, M.M. and R.V.; methodology, M.M.; software, S.R.; formal analysis, M.M. and S.R.; investigation, All authors; writing—original draft preparation, M.M. and S.R.; writing—review and editing, P.V. and M.F.; supervision, R.V. and P.V.; project administration, M.F. All authors have read and agreed to the published version of the manuscript.

Funding: The authors wish to thank the support of the Spanish Government through the Ministerio de 662 Economía y Competitividad (MINECO) projects DPI2015-67341-C2-1-R, DPI2016-77271-R also with FEDER funding.

Acknowledgments: To the WWTP of Salamanca (Aqualia) for allowing our research group visiting the plant and the IWA Task Group from the Department of Industrial Electrical Engineering and Automation (IEA), Lund University, Sweden (Ulf Jeppsson, Christian Rosen) for the BSM1 models.

Conflicts of Interest: The authors declare no conflict of interest.

Appendix A

As shown in Figure 10, strict ammonium set-point $S_{NHSP} = 1$ g/m^3 and default $Q_{carb} = 2$ g/m^3 are maintained between weeks 17 and 37, and the three different ammonium set-points (1, 4 and 6 g/m^3) are considered for the rest of the operational period. A fixed sequence of movements of Q_{carb} is applied, in the first 6 weeks $Q_{carb} = 1$ g/m^3, carbon dosage is cut between weeks 7 and 9 and then, it is increased to $Q_{carb} = 2$ g/m^3 between weeks 17 and 37, to be finally reduced to $Q_{carb} = 1$ g/m^3 in the last weeks. The combination of the sequence of ammonium set-point changes and carbon dosage variation, produce three different strategies named: $S_{NHSP} = 1\ Q_{carb\ var}$, $S_{NHSP} = 4\ Q_{carb\ var}$ and $S_{NHSP} = 6\ Q_{carb\ var}$. The weekly and bimonthly dynamic profiles of the different environmental indicators: electricity consumption, emissions of CO_2 and emissions to water affected by the mentioned strategies are shown in Figures A1–A3.

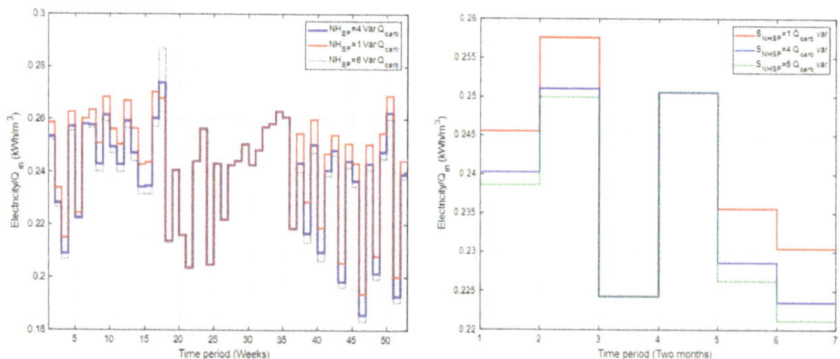

Figure A1. Weekly and bimonthly profile for electricity consumption indicator (kW h/m^3) with alternative control actions.

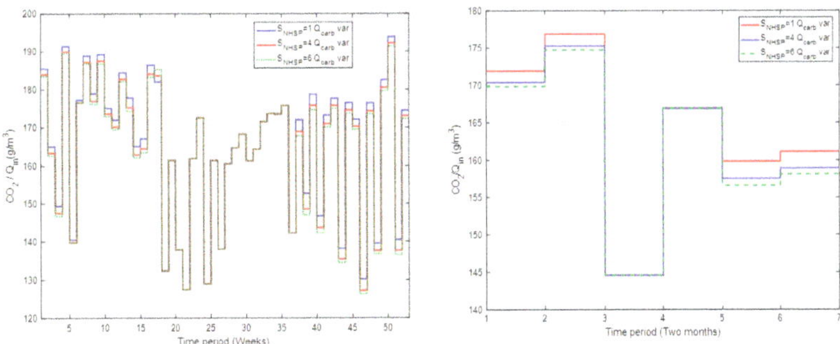

Figure A2. Weekly and bimonthly profile for total CO_2 emissions (g/m^3) indicators with alternative control actions.

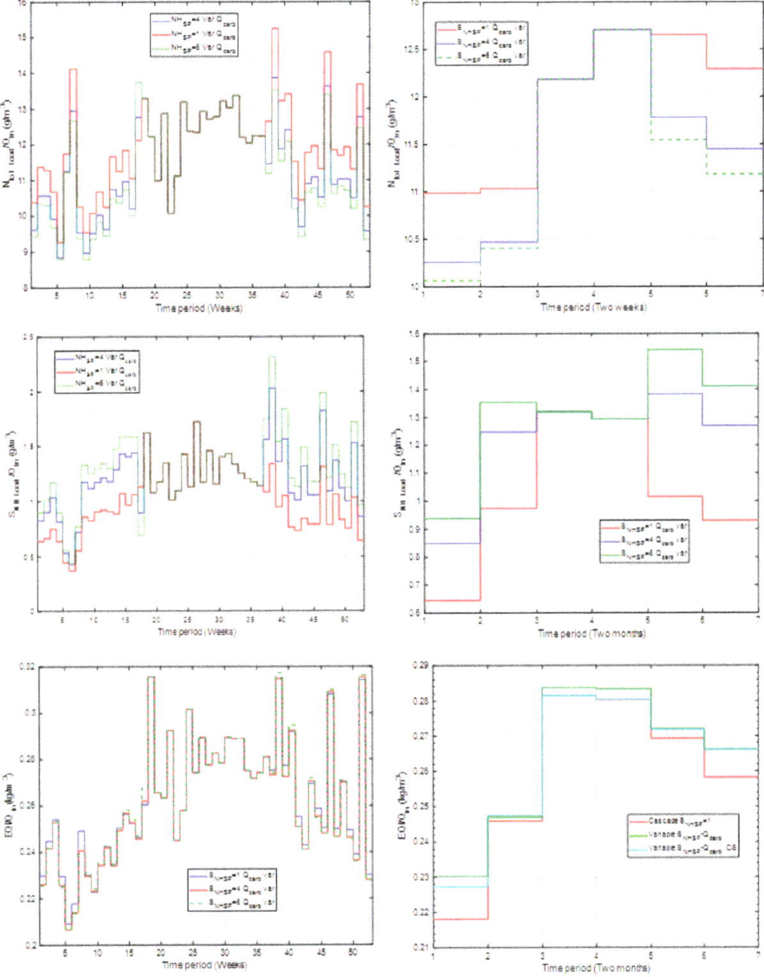

Figure A3. Weekly and bimonthly profile for emissions to water indicators $N_{tot\ Load}/Q_{in}$ (g/m^3), $S_{NH\ Load}/Q_{in}$ (g/m^3) and EQI/Q_{in} (kg/m^3) with alternative control actions.

Table A1. Annual values of environmental indicators and operating costs of BSM2 plant with respect to the volume of treated wastewater with alternative strategies.

	Environmental Indicators	Variable $S_{NHSP} = 1 \cdot Q_{carb\ var}$	Variable $S_{NHSP} = 4 \cdot Q_{carb\ var}$	Variable $S_{NHSP} = 6 \cdot Q_{carb\ var}$
Energy	Electricity (AE + PE + ME) (kW h/m^3)	0.240	0.236	0.235
	Heating energy (kW h/m3)	0.204	0.204	0.204
Chemicals	External carbon (kg COD/m^3)	0.029	0.029	0.029
Emissions to air	Biogas CH$_4$ (g/m^3)	52.25	52.25	52.25
	CO$_2$ (Digester) (g/m^3)	75.24	75.25	75.25
	CO$_2$ (ASP) (g/m^3)	87.80	86.54	86.08
Emissions to soil	Sludge for disposal (kg/m^3)	129.8	129.8	129.8
Emissions to water	S_{NH} effluent (g/m^3)	1.039	1.236	1.318
	N_{tot} effluent (g/m^3)	12.00	11.51	11.38
	COD effluent (g/m^3)	48.91	48.93	48.94
	EQI (kg/m^3)	0.260	0.265	0.265
Operating costs	OCI (EUR/d)	0.400	0.400	0.400

AE: Aeration energy, PE: Pumping energy, ME: Mixing energy, COD: Chemical oxygen demand, S_{NH}: ammonium concentration, N_{tot}: Total nitrogen concentration, EQI: Effluent quality index.

References

1. Gasperi, J.; Gromaire, M.C.; Kafi, M.; Moilleron, R.; Chebbo, G. Contributions of wastewater, runoff and sewer deposit erosion towet weather pollutant loads in combined sewer systems. *Water Res.* **2010**, *44*, 5875–5886. [CrossRef]
2. Gasperi, J.; Zgheib, S.; Cladière, M.; Rocher, V.; Moilleron, R.; Chebbo, G. Priority pollutants in urban stormwaters: part 2–Case of combined sewers. *Water Res.* **2012**, *46*, 6693–6703. [CrossRef] [PubMed]
3. Atinkpahoun, C.; Le, N.H.; Pontvianne, S.; Poirot, H.; Leclerc, J.P.; Pons, M.N.; Soclo, H.H. Population mobility and urban wastewater dynamics. *Sci. Total Environ.* **2018**, *622*, 1431–1437. [CrossRef] [PubMed]
4. Borzooei, S.; Campo, G.; Cerutti, A.; Meucci, L.; Panepinto, D.; Ravina, M.; Riggio, V.; Ruffino, B.; Scibilia, G.; Zanetti, M. Optimization of the wastewater treatment plant: From energy saving to environmental impact mitigation. *Sci. Total Environ.* **2019**, *691*, 1182–1189. [CrossRef] [PubMed]
5. Malinverni, F.; Genon, G. Development of a Practical Tool for the Assessment of Biological Performances in Wastewater Treatment Plants. *Clean Soil Air Water* **2016**, *44*, 1435–1443. [CrossRef]
6. Borzooei, S.; Amerlinck YAbolfathic, S.; Panepintoa, D.; Nopens, I.; Lorenzi, E.; Meucci, L.; Zanettia, M. Data scarcity in modelling and simulation of a large-scale WWTP: Stop sign or a challenge. *J. Water Process Eng.* **2019**, *28*, 10–20. [CrossRef]
7. Flores-Alsina, X.; Corominas, L.; Snip, L.; Vanrolleghem, P.A. Including greenhouse gas emissions during benchmarking of wastewater treatment plant control strategies. *Water Res.* **2011**, *45*, 4700–4710. [CrossRef]
8. Meneses, M.; Concepción, H.; Vrecko, D.; Vilanova, R. Life Cycle Assessment as an environmental evaluation tool for control strategies in wastewater treatment plants. *J. Clean. Prod.* **2015**, *107*, 653–661. [CrossRef]
9. Meneses, M.; Concepción, H.; Vilanova, R. Joint Environmental and Economical Analysis of Wastewater Treatment Plants Control Strategies: A Benchmark Scenario Analysis. *Sustainability* **2016**, *8*, 360. [CrossRef]
10. Barbu, M.; Vilanova, R.; Meneses, M.; Santin, I. On the evaluation of the global impact of control strategies applied to wastewater treatment plants. *J. Clean. Prod.* **2017**, *149*, 396–405. [CrossRef]
11. Revollar, S.; Vilanova, R.; Vega, P.; Francisco, M.; Meneses, M. Wastewater Treatment Plant Operation: Simple Control Schemes with a Holistic Perspective. *Sustainability* **2020**, *12*, 768. [CrossRef]
12. Shimako, A. Contribution to the development of a dynamic Life Cycle Assessment method. Ph.D Thesis, INSA de Toulouse, Toulouse, French, 2017.
13. Sánchez, F.; Rey, H.; Viedma, A.; Nicolás-Pérez, F.; Kaiser, A.S.; Martínez, M. CFD simulation of fluid dynamic and biokinetic processes within activated sludge reactors under intermittent aeration regime. *Water Res.* **2018**, *139*. [CrossRef] [PubMed]
14. Gernaey, K.; Flores Alsina, X.; Rosen, C.; Benedetti, L.; Jeppsson, U. Dynamic influent pollutant disturbance scenario generation using a phenomenological modelling approach. *Environ. Model. Softw.* **2011**, *26*, 1255–1267. [CrossRef]

15. Fearnside, P.M. Why a 100-year time horizon should be used for global warming mitigation calculations. *Mit. Adapt. Strat. Glob. Chang.* **2002**, *7*, 19–30. [CrossRef]

16. Gernaey, K.; Jeppsson, U.; Vanrolleghem, P.; Copp, J.; Steyer, J. *Benchmarking of Control Strategies for Wastewater Treatment Plants*; IWA Publishing: Colchester, UK, 2010.

17. Rieger, L.; Gillot, S.; Langergraber, G.; Ohtsuki, T.; Shaw, A.; Takács, I.; Winkler, S. *Guidelines for Using Activated Sludge Models*; IWA Scientific and Technical Report No. 22; IWA Publishing: London, UK, 2012.

18. Jeppsson, U.; Pons, M.N.; Nopens, I.; Alex, J.; Copp, J.B.; Gernaey, K.V.; Rosen, C.; Steyer, J.P.; Vanrolleghem, P.A. Benchmark simulation model no 2: general protocol and exploratory case studies. *Water Sci. Technol.* **2007**, *56*, 67–78. [CrossRef] [PubMed]

19. Alex, J.; Benedetti, L.; Copp, J.; Gernaey, K.; Jeppsson, U.; Nopens, I.; Pons, M.; Rosen, C.; Steyer, J.; Vanrolleghem, P.A. *Benchmark Simulation Model No. 2 (BSM2), Technical Report No 3. IWA Taskgroup on Benchmarking of Control Strategies for WWTPs*; IWA publishing: London, UK, 2018.

20. Alex, J.; Benedetti, L.; Copp, J.; Gernaey, K.; Jeppsson, U.; Nopens, I.; Pons, M.; Rieger, L.; Rosen, C.; Steyer, J.; et al. *Benchmark Simulation Model no. 1 (BSM1). In IWA Taskgroup on Benchmarking of Control Strategies for WWTPs. Dpt. of Industrial Electrical Engineering and Automation*; LUTEDX-TEIE 7229; Lund University Cod.: Lund, Sweden, 2008; pp. 1–62.

21. Henze, M.; Grady , C.P.L., Jr.; Gujer, W.; Marais, G.V.R.; Matsuo, T. *Activated Sludge Model n 1*; IAWQ Scientific and Technical Report n 1; IAWQ: London, UK, 1987.

22. Takács, I.; Patry, G.G.; Nolasco, D. A dynamic model of the clarification thickening process. *Water Res.* **1991**, *25*, 1263–1271. [CrossRef]

23. Batstone, D.J.; Keller, J.; Angelidaki, I.; Kalyuzhnyi, S.V.; Pavlostathis, S.G.; Rozzi, A.; Sanders, W.T.M.; Siegrist, H.; Vavilin, V.A. *Anaerobic Digestion Model No. 1*; IWA STR No. 13; IWA Publishing: London, UK, 2002.

24. Olsson, G. ICA and me a subjective review. *Water Res.* **2012**, *46*, 1585–1624. [CrossRef] [PubMed]

25. Åmand, L.; Olsson, G.; Carlsson, B. Aeration control—A review. *Water Sci. Technol.* **2013**, *67*, 2374–2398. [CrossRef]

26. Takács, I.; Vanrolleghem, P. *Elemental Balances in Activated Sludge Modelling*; IWA Publishing: London, UK, 2006.

27. Levasseur, A.; Lesage, P.; Margni, M.; Deschênes, L.; Samso, R. Considering Time in LCA: Dynamic LCA and Its Application to Global Warming Impact Assessments. *Environ. Sci. Technol.* **2010**, *44*, 3169–3174. [CrossRef]

28. Panepinto, D.; Fiore, S.; Zappone, M.; Genon, G.; Meucci, L. Evaluation of the energy efficiency of a large wastewater treatment plant in Italy. *Appl. Energy* **2016**, *161*, 404–411. [CrossRef]

29. Nguyen, H.; Safder, U.; Nguyen, X.; ChangKyoo, Y. Multi-objective decision-making and optimal sizing of a hybrid renewable energy system to meet the dynamic energy demands of a wastewater treatment plant. *Energy* **2019**, *2019*, 116570. [CrossRef]

Article

Stabilization/Solidification of Strontium Using Magnesium Silicate Hydrate Cement

Tingting Zhang [1], Jing Zou [1], Yimiao Li [1], Yuan Jia [2,*] and Christopher R. Cheeseman [3]

[1] Faculty of Infrastructure Engineering, Dalian University of Technology, Dalian 116024, China; tingtingzhang@dlut.edu.cn (T.Z.); zou_jing8@mail.dlut.edu.cn (J.Z.); lym_z@mail.dlut.edu.cn (Y.L.)

[2] Hebei Provincial Key Laboratory of Inorganic Nonmetallic Materials and Hebei Provincial Industrial Solid Waste Comprehensive Utilization Technology Innovation Center, College of Materials Science and Engineering, North China University of Science and Technology, Tangshan 063210, China

[3] Department of Civil and Environmental Engineering, Imperial College London, London SW7 2AZ, UK; c.cheeseman@ic.ac.uk

* Correspondence: jia132012@ncst.edu.cn; Tel.: +86-0315-8805020

Received: 17 November 2019; Accepted: 30 December 2019; Published: 1 February 2020

Abstract: Magnesium silicate hydrate (M–S–H) cement, formed by reacting MgO, SiO_2, and H_2O, was used to encapsulate strontium (Sr) radionuclide. Samples were prepared using light-burned magnesium oxide and silica fume, with sodium hexametaphosphate added to the mix water as a dispersant. The performance of the materials formed was evaluated by leach testing and the microstructure of the samples was also characterized. The stabilizing/solidifying effect on Sr radionuclide in the $MgO–SiO_2–H_2O$ system with low alkalinity is demonstrated in the study. The leaching rate in a standard 42-day test was 2.53×10^{-4} cm/d, and the cumulative 42-day leaching fraction was 0.06 cm. This meets the relevant national standard performance for leaching requirements. Sr^{2+} was effectively incorporated into the M–S–H hydration products and new phase formation resulted in low Sr leaching being observed.

Keywords: magnesium silicate hydrate; radioactive waste; stabilization/solidification; strontium; leaching

---

1. Introduction

The volume of low level radioactive waste (LLW) and intermediate level radioactive waste (ILW) accounts for ~95% of the total volume of nuclear waste. Strontium and cesium are important contaminants in LLW and ILW, because these radionuclides are most often present in the cooling water of nuclear reactors. These nuclides have long half-lives (^{90}Sr has a half-life of 28.8 years and ^{137}Cs has a half-life of 30.5 years), can readily migrate, and are known to cause carcinogenesis in contaminated living organisms [1–6]. Therefore, the management of low- and intermediate-level radioactive wastes containing Cs and Sr remains a major challenge to the nuclear energy industry.

Stabilization/solidification (S/S) includes a broad range of waste treatment technologies, which use asphalt or cementitious materials to convert wastes into a solid form that is more suitable for transportation and long-term storage [7]. Effective solidification transforms a radioactive waste liquid into a solidified material, reducing the potential for radionuclides to migrate into the biosphere and reducing the volume of material requiring storage/disposal. For low- and intermediate-level wastes, solidification in cementitious materials is the preferred option. This uses the hydration and hardening characteristics of Portland cement to achieve physical encapsulation, adsorption, and chemical bonding of the radionuclides. The processing is relatively simple and low cost [8]. However, there are issues with conventional cement stabilization/solidification. These include expansion of the solidified body and the potential for heavy metal salts to retard cement hydration reactions. Therefore, research on new types of solidification systems, such as alkali-activated cementitious materials, alkali–slag–clay mineral composite cementitious materials, phosphate cement, and phosphate cement–clay mineral composite systems, is ongoing [9–15].

Alkaline slag cement systems based on metakaolin have been used for the solidification of simulated radioactive wastes and produced low Cs^+ and Sr^{2+} leaching rates [11]. However, the pH of these systems is higher than 12, which is not conducive to long-term solidification because of the formation of mobile metal complexes. Phosphate cements are also being investigated for solidifying nuclear wastes. These are rapid-setting and hardening and the solidified wastes formed are reported to be resistant to leaching and damage from freeze–thaw cycling [16–18]. However, the setting of phosphate cements is difficult to control and a highly exothermic reaction limits the ability to solidify some types of nuclear waste [19].

Cements that hydrate to form magnesium silicate hydrate (M–S–H gel) are formed from reactions between MgO, SiO_2, and H_2O, and these can have a low porosity, large specific surface area, high strength, low heat of hydration, and low pH [20–25]. M–S–H gel was first discovered in sulfate-eroded marine concrete [26]. Natural magnesium-rich silicate minerals are reported to adsorb heavy metal ions and hazardous organic compounds [27,28]. Synthetic magnesium silicate has excellent adsorption of radioactive U and methylene blue [29,30]. Magnesium silicate systems have also been used to encapsulate nuclear wastes containing active metals such as Mg and Al. The low pH and pore solution composition resulted in effective Al encapsulation with minimal H generation [25,31–33]. Some nuclear wastes contain significant quantities of mixed wastes including the main nuclides Sr and Cs, and Mg and Al alloys. While the high pH in Portland cement-based binders passivates the corrosion of Mg alloys, Al alloys corrode under high pH conditions with the evolution of H_2 gas [26]. The pH of water in equilibrium with the M–S–H system is in the range from 9.5 to 10.5 and, therefore, the M–S–H system may be suitable for stabilizing this type of nuclear waste [34].

The aim of this research was to assess, for the first time, the effectiveness of M–S–H forming cement systems for the stabilization/solidification of simulated radioactive Sr. The samples formed were leached tested and the compressive strength were determined. In addition, the microstructure was characterized to understand the effect of Sr addition on hydration reactions and the phases formed.

2. Experimental

2.1. Materials

Light-burned MgO (Mag Chem 30, M.A.F. Magnesite B.V. Den Haag, The Netherlands) with an MgO content >98% and an activity of 82.16% was used (Table 1). The composition of the silica fume used had a SiO_2 content >97% (SiO_2, model 920U, Shanghai Elkem, Shanghai, China), as shown in Table 1. Sodium hexametaphosphate (Na–HMP) with a purity >95% was also used (Sinopharm Group, Shanghai, China). The radionuclide ^{90}Sr was replaced with the stable isotope ^{88}Sr and added to samples as strontium nitrate ($Sr(NO_3)_2$) with purity >95%.

Table 1. Chemical composition of light-burned magnesia and silica fume (wt.%).

Chemical Composition (wt.%)	MgO	CaO	SiO_2	Fe_2O_3	Al_2O_3	K_2O	MnO	SO_3
light-burned magnesia	98.45	0.62	0.49	0.18	0.04	–	0.04	–
silica fume	0.45	0.43	97.34	0.05	0.12	0.97	0.04	0.35

2.2. Methods

The mass ratio of the light-burned MgO to silica fume was 2:3. The amount of sodium hexametaphosphate added was 2 wt.% of the total mass of the MgO and silica fume [34]. The water-to-binder mass ratio varied in different samples between 0.6 and 0.8 and deionized water was used in all mixes. The mass fraction of Sr^{2+} added as $Sr(NO_3)_2$ varied from 0.8 wt.% to 3.2 wt.% of the total mass of the MgO and silica fume.

The raw materials including the required amount of $Sr(NO_3)_2$ were blended together, and then deionized water containing the Na–HMP was added and mixed to form cement paste samples. This was then poured into $20 \times 20 \times 20$ mm steel moulds to form cubic samples or $ø50 \times 50$ mm moulds to form cylindrical samples. The samples were removed from the moulds when the samples were sufficiently hard, and then they were stored under standard curing conditions of $20 \pm 2\,°C$ at a humidity of >95%.

Samples were tested in compression after curing for 3, 7, and 28 days in accordance with the relevant standard (GB/T 17671-1999) [35]. The leaching tests were completed in accordance with GB/T 7023-2011 [36], the standard test method for leachability of low and intermediate level solidified radioactive waste forms, and GB 14569.1-2011 [37] performance requirements for low- and intermediate-level radioactive waste form-cemented waste form.

The leaching tests used the $ø50 \times 50$ mm cylindrical samples cured for 28 days. The end faces were polished with sandpaper and the samples were suspended in a polyethylene container using nylon filament. Deionized water and simulated seawater were used as the leachants. The composition of simulated seawater is shown in Table 2. The leachates were replaced after 1, 3, 7, 10, 14, 21, 28, 35, and 42 days, and the concentration of Sr^{2+} in the leachate was measured by inductively coupled plasma optical emission spectrometry (ICP-OES). The leaching rate and cumulative amount of Sr leached were calculated using the following formulas [36]:

$$R_n = \frac{m_n/m_0}{(F/V)t_n} \tag{1}$$

$$P_t = \frac{\Sigma m_n/m_0}{F/V} \tag{2}$$

where R_n is the leaching rate of Sr^{2+} in the n-th leaching cycle (cm/d); m_n is the mass of Sr^{2+} leached in the n-th leaching cycle (g); m_0 is the initial mass of Sr^{2+} in the leaching test sample (g); F is the geometric surface area of the sample in contact with the leachate (cm^2); V is the volume of sample (cm^3); t_n is the number of days in the n-th leaching cycle; and P_t is the cumulative leaching fraction of Sr^{2+} at time t (cm).

Phase analysis of solidified M–S–H cement samples used X-ray diffraction (XRD; D8 Advance AXS, Brooklyn, Germany) with a Cu target, a working voltage of 40 kV, a working current of 40 mA, a scanning range of 5~80°, and a scanning speed of 0.02° 2θ/min. Scanning electron microscopy (SEM; Nova Nano SEM 450; FEI, Hillsboro, OR, USA) was used to examine the microstructure of the solidified samples at a resolution of 3.5 nm at 30 kV. The test samples examined were taken from a central thin section taken out after crushing the solidified body to a particle size of 3–5 mm. Energy dispersive spectrometry (EDS) was used to perform an elemental analysis of samples.

Table 2. Synthetic components of simulated seawater.

	NaCl	$MgCl_2$	Na_2SO_4	$CaCl_2$	KCl	$NaHCO_3$	KBr
Quality/g	23.50	4.98	3.92	1.10	0.66	0.19	0.09

Note: Dilute to 1000 g with water; the ionic strength should be 0.71 mol/kg.

3. Results

3.1. Compressive Strength

Figure 1 shows the effect of Sr^{2+} and the water–cement ratio on the 28-day compressive strengths of M–S–H cement-solidified Sr^{2+} samples. The compressive strength decreases with increasing water–cement ratio and increasing Sr^{2+} content. For samples with a water–cement ratio of 0.6, the compressive strength with a Sr^{2+} content of 3.2 wt.% decreased by 48.1% compared with the control samples. For samples with a water–cement ratio of 0.65–0.8, the corresponding reductions were 40.7%, 41.8%, 38.3%, and 49.8%, respectively. The presence of strontium significantly inhibits the development of compressive strength. However, even at the highest water–cement ratio of 0.8 and Sr^{2+} content of 3.2 wt.%, the 28-day compressive strength was ~10 MPa, which exceeds the 7 MPa required in the specification [36].

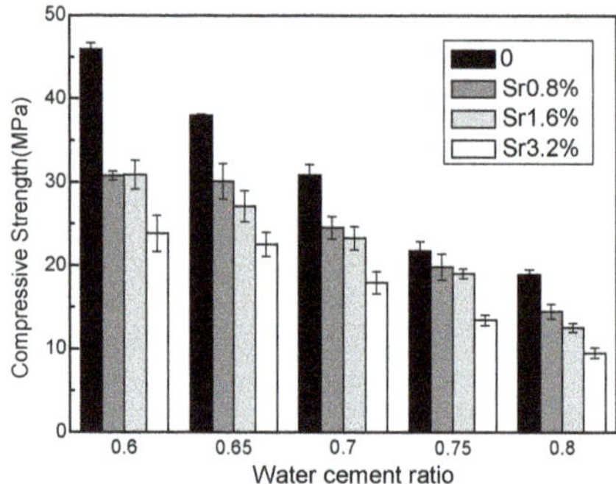

Figure 1. Influence of Sr content and water–cement ratio on 28-day compressive strength of the $MgO–SiO_2–H_2O$ solidified bodies.

3.2. Crystalline Phase Analysis

Figure 2 shows XRD data for M–S–H cements with 0 wt.% and 3.2 wt.% Sr^{2+} added. Figure 2a shows sharp $Mg(OH)_2$ peaks at 18.6°, 38.0°, 50.8°, and 58.6°. The two peaks at 42.9° and 62.3° correspond to MgO. The diffuse background appearing at 2θ between 32 and 38° and between 58 and 62° corresponds to M–S–H gel [34]. This shows that $Mg(OH)_2$ is a reaction product in the initial stages of hydration, accompanied by the formation of some M–S–H gel. The content of $Mg(OH)_2$ decreases as hydration proceeds.

Figure 2b shows that the addition of Sr retards hydration. Some $SrCO_3$ identified by the peak at 25° was present in the hydration products at all ages. Hydration was slower than the control samples without added Sr^{2+}. Some MgO had not hydrated after 28 days and peaks associated with $Mg(OH)_2$ remained strong and the amount of M–S–H gel was reduced. It is speculated that strontium ions promote the hydrolysis of MgO to form $Mg(OH)_2$ during the initial stage of hydration, but later compete for hydroxyl ions to form strontium hydroxide, which slows down the rate of M–S–H gel formation in the later stages and leads to the slow development of strength.

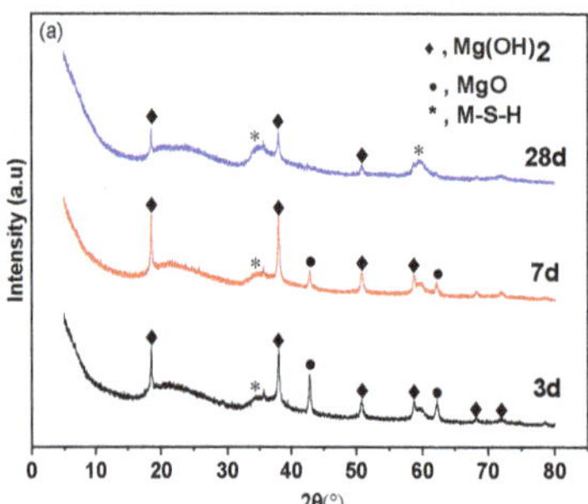
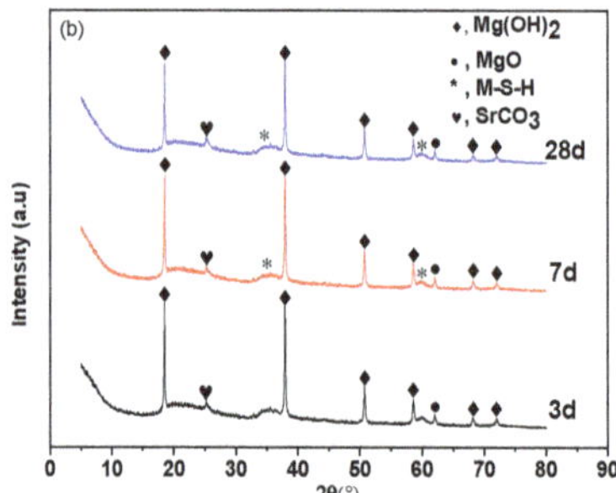

Figure 2. X-ray diffraction spectra of different curing periods of the $MgO–SiO_2–H_2O$ solidified bodies with or without added strontium. (**a**) No added strontium; (**b**) with 3.2 wt.% added strontium.

3.3. Microstructural Analysis

Figure 3 shows SEM images of samples with 0 wt.% and 3.2 wt.% Sr^{2+} hydrated for 3, 7, and 28 days. Figure 3a shows that the hydration products of magnesium silicate cement at the initial stage (three days) of hydration were mainly short rod-like magnesium hydroxide crystals. After seven days, the number of Mg(OH)$_2$ crystals decreased, and floc M–S–H gel overlapped the surface of silica fume particles. After hydration for 28 days, compact internal structures were formed, except for some unreacted spherical silica fume particles.

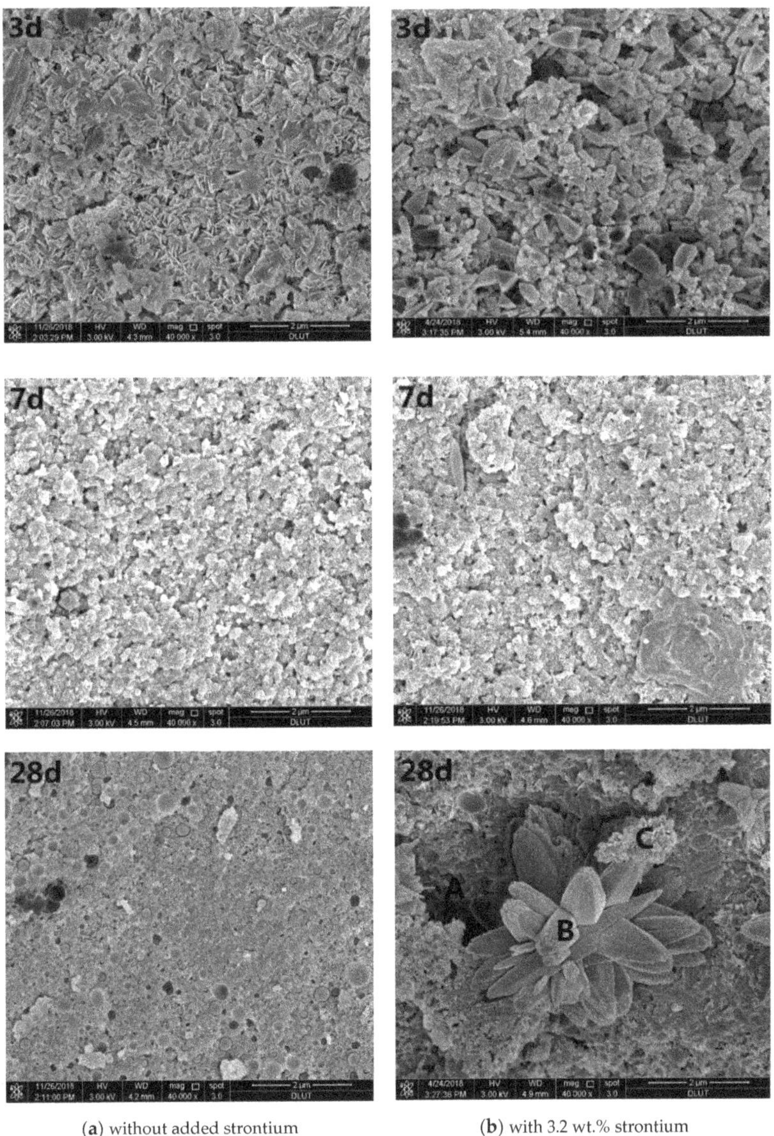

(**a**) without added strontium　　　　　(**b**) with 3.2 wt.% strontium

Figure 3. Scanning electron microscopic (SEM) images of different curing periods of the MgO–SiO$_2$–H$_2$O solidified bodies with or without strontium; (**a**) without added strontium; (**b**) with 3.2 wt.% strontium.

Figure 3b shows that solidified Sr in the magnesium silicate cement affected the MgO–SiO$_2$–H$_2$O hydration environment. At the beginning of hydration, a large number of pine nut-shaped crystalline Mg(OH)$_2$ structures can be seen, while a small amount of flocculent M–S–H gel was dispersed therein.

There was also a small number of cubic pyramidal $Mg(OH)_2$ crystals. As the hydration reaction progressed, the number of $Mg(OH)_2$ crystals decreased gradually. After seven days' hydration, the main hydration product was a flocculent M–S–H gel, although some $Mg(OH)_2$ crystals were still present. The crystals basically disappeared from the SEM images after 28 days, while the amount of flocculent substances continued to increase. Specially, a petal-like substance appeared in the sample when the hydration process lasted for 28 days.

EDS analysis of the MgO-SiO_2-H_2O solidified body with 3.2 wt.% Sr^{2+} after curing for 28 days produced the elemental distribution is shown in Table 3. Strontium ions were mainly concentrated in the petal-like features. The elements in the petals were mainly C, O, and Sr, and the atomic percentages were 32.30%, 54.82%, and 9.20%, respectively. This is characteristic of one type of strontium carbonate and is consistent with the XRD analysis. It can be seen that Sr^{2+} is chemically bound to the magnesium silicate cement matrix, which is conducive to improving the long-term leaching resistance of the cured body. In addition, the atomic molar ratios of Mg and Si in the regions A, B and C were 0.56, 0.86 and 0.76, respectively, indicative of M-S-H gel [38]. It also shows that the content of unreacted silica fume in the system was high, which indicates that the presence of strontium ions significantly reduces the rate of hydration in magnesium silicate cement.

Table 3. Energy dispersive spectrometry (EDS) analysis of the MgO–SiO$_2$–H$_2$O solidified body with 3.2 wt.% Sr^{2+}, cured for 28 days.

Element	A		B		C	
	Weight/%	Atom/%	Weight/%	Atom/%	Weight/%	Atom/%
C	5.61	10.50	17.85	32.30	14.25	26.32
O	32.52	45.72	40.36	54.82	38.52	53.43
Mg	15.25	14.11	1.39	1.24	4.63	4.23
Si	31.57	25.28	1.88	1.45	7.04	5.56
Sr	13.39	3.44	37.11	9.20	32.19	8.15

Note: The points A, B, and C are marked in Figure 3b.

3.4. Leach Test Results

Figure 4 shows the leaching rate and cumulative fraction leached from solidified bodies with different water–cement ratios doped with 1.6 wt.% Sr^{2+} at 25 °C under deionized water leaching conditions. This indicates that the leaching rate of different water–cement ratios of the magnesium silicate cement solidified body samples initially increased, but then decreased. At the beginning of the leaching test (before seven days), the Sr^{2+} leaching rate decreased rapidly, but this slowed after seven days. After 28 days, the leaching rate tended to be consistent and gradually reached a steady state. However, the difference in leaching results between the water–cement ratios was small. The effect of the water–cement ratio on leaching from the solidified body of magnesium silicate cement was not obvious. After 42 days of leaching, the maximum leaching rate of each solidified sample was 4.64×10^{-4} cm/d, and the minimum was 2.53×10^{-4} cm/d. This was one order of magnitude lower than the limit (1×10^{-3} cm/d) required by the National Standards. Cumulative leaching from the cumulative fraction leached score curves results during the first 10 days increased rapidly, and the growth rate gradually stabilized after 28 days, which was consistent with the trend in the leaching rate. The 42-day cumulative fraction that leached from all samples was 0.06~0.07 cm, which only accounted for about 35% of the specification limit in GB 14569.1-2011 [37], which stipulates that the 42-day cumulative fraction leached from the cement solidified body to Sr^{2+} is 0.17 cm.

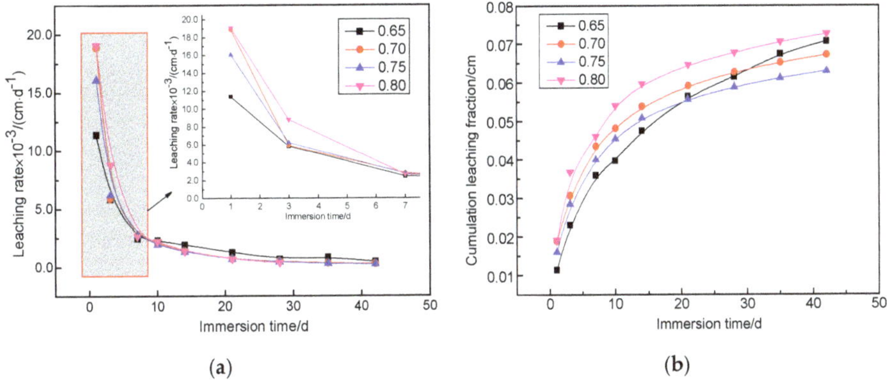

Figure 4. Leaching rate and cumulative fraction leached curves of different water–cement ratios for the MgO–SiO$_2$–H$_2$O solidified bodies with 1.6 wt.% Sr^{2+} under leaching conditions at an ambient temperature of 25 °C and using deionized water as leachant. (**a**) Leaching rate curves (**b**) Cumulative fraction leached curves

Figure 5 shows the leaching rate and cumulative fraction leached of magnesium silicate cement solidified samples with 1.6 wt.% strontium using a water–cement ratio of 0.65 under different leaching conditions. In the case of changes in temperature and leachant, the leaching rate and cumulative fraction leached of the solidified body during each leaching period were consistent with those tested at 25 °C using deionized water as leachant. In contrast, the effect of the leachant on the solidification of Sr^{2+} by the magnesium silicate cement was more significant than the ambient leaching temperature.

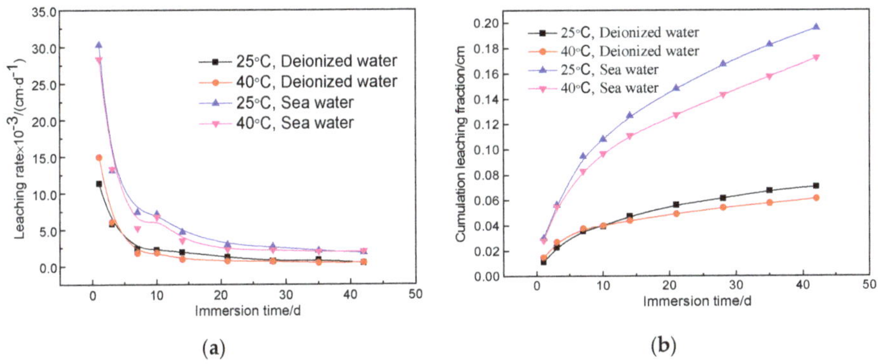

Figure 5. Leaching rate and cumulative fraction leached curves of the MgO–SiO$_2$–H$_2$O solidified bodies (water–cement ratio, 0.65; Sr^{2+} content, 1.6 wt.%) under different leaching conditions. (**a**) Leaching rate curves (**b**) Cumulative fraction leached curves

The leaching rate of the solidified body under simulated seawater conditions improved significantly compared with the results in deionized water. When the ambient leaching temperature was 25 °C, the 42-day leaching rates of the leachant for deionized water and simulated seawater were 4.64×10^{-4} cm/d and 1.89×10^{-3} cm/d, respectively; the corresponding values were 5.11×10^{-4} cm/d and 2.08×10^{-3} cm/d, respectively, when the temperature was increased to 40 °C. From the cumulative fraction leached curve, when the leaching temperature was 25 °C, the cumulative fraction leached after 42 days under simulated seawater conditions was 0.20 cm, which was 2.9 times that at the same temperature using deionized water (0.07 cm). When the leaching temperature was 40 °C, the 42-day cumulative fraction leached under simulated seawater was 0.17 cm, which was about twice that with deionized water (0.06 cm). These results show that the magnesium silicate cement reduces leaching of the radionuclide, particularly in the standard leaching environment (25 °C, deionized water). The 42-day leaching rate was <50% of the standard limit.

Mercury intrusion porosimetry (MIP) test results (Figure 6 and Table 4) show that most pores in the M–S–H solidified body are less than 10 nm and the total porosity is ~30.5%. This shows that the resistance of nuclide ion diffusion through the pore solution in the material is large, which reduces the leaching rate of the solidified body.

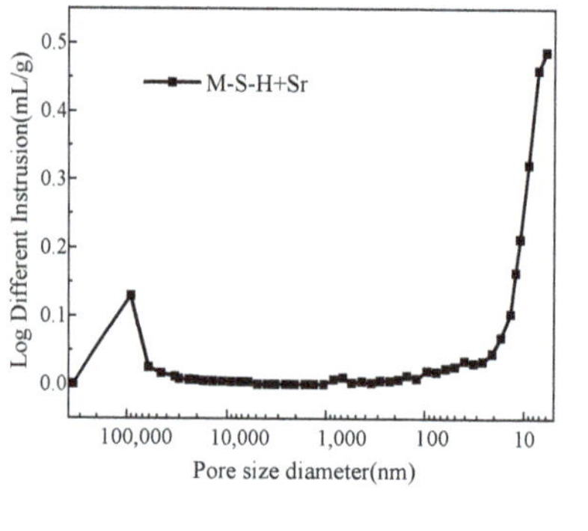
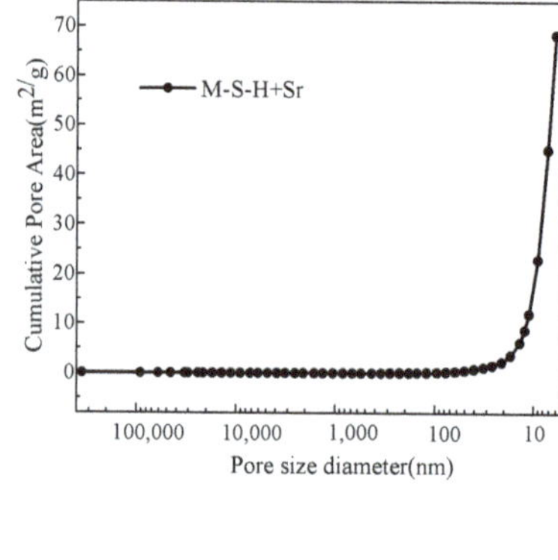

(a)   (b)

Figure 6. Mercury intrusion porosimetry (MIP) results of the MgO–SiO$_2$–H$_2$O solidified body (water–cement ratio, 0.65; Sr^{2+} content, 1.6 wt.%); (**a**) log different intrusion; (**b**) cumulative pore area.

Table 4. Mercury intrusion porosimetry (MIP) results of the MgO–SiO$_2$–H$_2$O solidified body with 1.6 wt.% Sr^{2+}.

Average Pore Diameter/nm	Median Pore Diameter/nm	Total Pore Area/m^2/g	Porosity/%
15.30	8.0	68.34	30.53

4. Discussion

Solidified bodies formed using cement as a substrate material for solidifying waste are a heterogeneous multiphase system composed of various solid products, including hydration products, residual clinkers, radioactive waste, and a small amount of liquid and air existing in the pores. The system solidifies radionuclide ions mainly by physical encapsulation, adsorption, and by forming a solid solution. Encapsulation relies on the density of the cement to retain the nuclide ions and spatially inhibits outward diffusion. Adsorption refers to the adsorption of a nuclide on the surface of the solidified body or hydration products. Forming a solid solution causes improved retention of the nuclide ion compared with the former methods. This is a chemical action that participates nuclides in the internal hydration products of the cement solidified body. According to the experimental results, the solidifying effect of MgO–SiO$_2$–H$_2$O on the radionuclide included the above ways, indicating that the system has good potential for encapsulating nuclear waste.

The encapsulation effect is mainly related to the pore size distribution of the solidified body, and the pore size distribution of the cement is mainly affected by the water–cement ratio, the admixture, and the curing age. These factors further affect physical properties of the solidified body, such as compressive strength and leaching rate. The presence of strontium changed the hydration environment of the magnesium silicate cement and the fluidity of the system decreased with increases in strontium content. The molding process becomes more difficult resulted in a decrease in the density of the solidified body, so the system does not have obvious advantages for physical inclusion of the nuclide.

This is the reason that the compressive strength of the solidified body shown in Figure 1 decreased as the water–cement ratio increased. According to the XRD and SEM results, MgO and SiO_2 reacted slowly during the initial stage of hydration and the reaction product was prismatic crystalline $Mg(OH)_2$. As the reaction progresses, $Mg(OH)_2$ hydrates to form M–S–H gel. Previous work has shown that the M–S–H gel is the main strength source in $MgO–SiO_2–H_2O$ cements [38] and the compressive strengths of magnesium silicate cement increase at later stages.

Adsorption involves chemical adsorption, which is caused by chemical bonds, mainly through the exchange of specific ions, and physical adsorption, which is generated by intermolecular forces, mainly through surface energy [39]. In addition, the molecular structure of M–S–H gel is close to the molecular structure of sepiolite with a specific surface area of ~200 m^2/g, which has advantages for adsorbing heavy metal ions [40]. The leaching results in Figures 4 and 5 are a good indication of the effect of $MgO–SiO_2–H_2O$ cements on limiting the leaching of radionuclide ions.

At the beginning of the leaching tests, part of the Sr^{2+} adhering to the surface of the solidified body contacting the leachant is dissolved. The Sr^{2+} present in the connected pore solution also enters the leachate by liquid–liquid diffusion. The solution in closed pores is in equilibrium with the solid phase. As the leaching period extends, part of the Sr^{2+} encapsulated inside the solidified body was released gradually to the outer surface of the solidified body by solid phase diffusion, and this can also enter the leaching solution. In addition, the solidified body continues to hydrate during the leaching process, and more hydration products are formed to fill the pores, which reduces the diffusion rate of Sr^{2+}. Therefore, the leaching rate of the magnesium silicate cement solidified body tends to be high during the early stage and low during the later stages. In addition, the SEM image in Figure 3 shows that Sr participates in the hydration process of the magnesium silicate cement, changing the final hydration product type, and existing in the solidified body in the form of strontium carbonate. Although this reduced the strength of the solidified body, the leaching resistance of the magnesium silicate cement solidified body was improved.

5. Conclusions

The effect of Sr^{2+} on the hydration process of $MgO–SiO_2–H_2O$ cement was evaluated by assessing the mechanical properties, the phases, and by microstructural analysis. From results of strength, leaching tests and pore diameter distribution analysis, the stabilization performance of the M–S–H cement on Sr^{2+} was revealed and the following conclusions were drawn:

(1) Incorporating the Sr nuclide reduced the working performance of the magnesium silicate cement and inhibited the development of compressive strength owing to the inhibition of hydration process.

(2) The leaching rate and cumulative leaching fraction of the solidified body after immersion for 42 days in the standard environment (4.64×10^{-4} cm/d) and high temperature seawater (1.89×10^{-3} cm/d) were one order of magnitude lower than the limit value in the National Standard.

(3) The presence of Sr affected the hydration reactions of the magnesium silicate cement and was encapsulated the interior of the matrix in the form of a strontium carbonate precipitate. This was beneficial to ensure low long-term leaching performance of the solidified body.

Author Contributions: T.Z. and Y.J. conceived and designed the experiments; J.Z. and Y.L. performed the experiments; Y.J. and J.Z. analyzed the data; T.Z. contributed reagents/materials/analysis tools; J.Z. wrote the paper; Y.J. and C.R.C. reviewed the paper. All authors have read and agreed to the published version of the manuscript.

Funding: This research was funded by [National Natural Science Foundation of China] grant number [No. 51778101; No. 61704017; No. 51808217], [National Key R&D program of China] grant number [2017YFE0107000], [State Key Laboratory of Coastal and Offshore engineering of Dalian University of Technology] grant number [LP1808], [Dalian High-level Talent Innovation Program] grant number [2017RQ051], [The fundamental Research Funds for the Central Universities] grant number [DUT19JC27], [Natural Science Foundation of Hebei Province] grant number [E2019209403], [Key R&D Program of Hebei Province] grant number [19273803D] and [Post-doctoral Research Project of Hebei Province] grant number [B2019003028].

Acknowledgments: The authors wish to express their gratitude and sincere appreciation for the financial support provided by the National Natural Science Foundation of China, Institute of Building Materials (Dalian University of Technology) and Hebei Provincial Key Laboratory of Inorganic Nonmetallic Materials (North China University of Science and Technology).

Conflicts of Interest: The authors declare no conflict of interest.

References

1. IAEA. *Classification of Radioactive Waste*; Safety Series GSG-1; IAEA: Vienna, Austria, 2009.
2. Hashimoto, S.; Matsuura, T.; Nanko, K.; Linkov, I.; Shaw, G.; Kaneko, S. Predicted spatio-temporal dynamics of radio cesium deposited onto forests following the Fukushima nuclear accident. *Sci. Rep.* **2013**, *3*, 2564. [CrossRef] [PubMed]
3. Koarashi, J.; Atarashi-Andoh, M.; Amano, H.; Matsunaga, T. Vertical distributions of global fallout ^{137}Cs and ^{14}C in a Japanese forest soil profile and their implications for the fate and migration processes of Fukushima-derived ^{137}Cs. *J. Radioanal. Nucl. Chem.* **2017**, *311*, 473–481. [CrossRef]
4. Ota, M.; Nagai, H.; Koarashi, J. Modeling dynamics of ^{137}Cs in forest surface environments: Application to a contaminated forest site near Fukushima and assessment of potential impacts of soil organic matter interactions. *Sci. Total Environ.* **2016**, *551*, 590–604. [CrossRef] [PubMed]
5. Buesseler, K.O.; Jayne, S.R.; Fisher, N.S.; Rypina, I.I.; Baumann, H.; Baumann, Z.; Breier, C.F.; Casacuberta, E.M.; Masqué, N.P.; Garcia-Orellana, J.; et al. ^{90}Sr and ^{89}Sr in seawater off Japan as a consequence of the Fukushima Dai-ichi nuclear accident. *Bioresources* **2013**, *10*, 3649–3659. [CrossRef]
6. Kubota, T.; Shibahara, Y.; Fukutani, S.; Fujii, T.; Ohta, T.; Kowatari, M.; Mizuno, S.; Takamiya, K.; Yamana, H. Cherenkov counting of ^{90}Sr and ^{90}Y in bark and leaf samples collected around Fukushima Daiichi Nuclear Power Plant. *J. Radioanal. Nucl. Chem.* **2015**, *303*, 39–46. [CrossRef]
7. Steinhauser, G.; Schauer, V.; Shozugawa, K. Concentration of Strontium-90 at selected hot spots in Japan. *PLoS ONE* **2013**, *8*, e57760. [CrossRef] [PubMed]
8. Toth, F.L.; Rogner, H.H. Oil and nuclear power: Past, present, and future. *Energy Econ.* **2006**, *28*, 1–25. [CrossRef]
9. Halim, C.E.; Amal, R.; Beydoun, D.; Scott, J.A.; Low, G. Low, Implications of the structure of cementitious wastes containing Pb (II), Cd (II), As (V), and Cr (VI) on the leaching of metals. *Cem. Concr. Res.* **2004**, *34*, 1093–1102. [CrossRef]
10. Shi, C.J.; Fernández-Jiménez, A. Stabilization/solidification of hazardous and radioactive wastes with alkali-activated cements. *J. Hazard. Mater.* **2006**, *137*, 1656–1663. [CrossRef]
11. Qian, G.G.; Sun, D.D.; Tay, J.H. New aluminium-rich alkali slag matrix with clay minerals for immobilizing simulated radioactive Sr and Cs waste. *J. Nucl. Mater.* **2001**, *299*, 199–204. [CrossRef]
12. Guangren, Q.; Yuxiang, L.; Facheng, Y.; Rongming, S. Improvement of metakaolin on radioactive Sr and Cs immobilization of alkali-activated slag matrix. *J. Hazard. Mater.* **2002**, *92*, 289–300. [CrossRef]
13. Plecas, I.; Pavlovic, R.; Paclovic, S. Leaching behavior of ^{60}Co and ^{137}Cs from spent ion exchange resins in cement–bentonite clay matrix. *J. Nucl. Mater.* **2004**, *327*, 171–174. [CrossRef]
14. Wei, M.L. Research on Mechanism and Long-Term Performance of a New Phosphate-Based Binder for Stabilization of Soils Contaminated with High Levels of Zinc and Lead. Ph.D. Thesis, Southeast University, Nanjing, China, 2017.
15. Zhang, H.; He, P.J.; Shao, L.M. Fate of heavy metals during municipal solid waste incineration in Shanghai. *J. Hazard. Mater.* **2008**, *156*, 365–373. [CrossRef] [PubMed]
16. Cao, X.; Wahbi, A.; Ma, L.; Li, B.; Yang, Y. Immobilization of Zn, Cu, and Pb in contaminated soils using phosphate rock and phosphoric acid. *J. Hazard. Mater.* **2009**, *164*, 555–564. [CrossRef] [PubMed]
17. You, C.; Qian, J.; Qin, J.; Wang, H.; Wang, Q.; Ye, Z. Effect of early hydration temperature on hydration product and strength development of magnesium phosphate cement (MPC). *Cem. Concr. Res.* **2015**, *78*, 179–189. [CrossRef]
18. Lai, Z.; Lai, X.; Shi, J.; Lu, Z. Effect of Zn^{2+} on the early hydration behavior of potassium phosphate based magnesium phosphate cement. *Constr. Build. Mater.* **2016**, *129*, 70–78. [CrossRef]
19. Lai, Z.Y.; Qian, J.S.; Liang, P. Study on the curing properties of ^{90}Sr by magnesium phosphate cement. *Acta Sci. Circumstantiae* **2011**, *31*, 2792–2797. [CrossRef]

20. Qiao, F.; Chau, C.K.; Li, Z. Property evaluation of magnesium phosphate cement mortar as patch repair material. *Constr. Build. Mater.* **2010**, *24*, 695–700. [CrossRef]

21. Fei, J.; Al-Tabbaa, A. Strength and hydration products of reactive MgO–silica pastes. *Cem. Concr. Compos.* **2014**, *52*, 27–33. [CrossRef]

22. Fei, J.; Gu, K.; Al-Tabbaa, A. Strength and drying shrinkage of reactive MgO modified alkali-activated slag paste. *Constr. Build. Mater.* **2014**, *51*, 395–404. [CrossRef]

23. Li, Z.; Zhang, T.; Hu, J.; Tang, Y.; Niu, Y.; Wei, J.; Yu, Q. Characterization of reaction products and reaction process of MgO-SiO2-H2O system at room temperature. *Constr. Build. Mater.* **2014**, *61*, 252–259. [CrossRef]

24. Wei, J.X. *Study on MgO-SiO$_2$-H$_2$O Cementitious System and Its Mechanism of Hydration and Hardenin*; China Academy of Building Materials Science: Beijing, China, 2004.

25. Wei, J.X.; Chen, Y.M.; Li, Y.X. The reaction mechanism between MgO and microsilica at room temperature. *J. Wuhan Univ. Technol. Mater. Sci. Ed.* **2006**, *21*, 88–91. [CrossRef]

26. Zhang, T.T.; Cheesemana, C.R.; Vandeperre, L.J. Development of low pH cement systems forming magnesium silicate hydrate (M-S-H). *Cem. Concr. Res.* **2011**, *41*, 439–442. [CrossRef]

27. Cole, W.F.; Demediuk, T. X-Ray, thermal, and Dehydration studies on Magnesium oxychlorides. *Aust. J. Chem.* **1955**, *8*, 234–251. [CrossRef]

28. Sevim, A.M.; Hojiyev, R.; Gül, A.; Çelik, M.S. An investigation of the kinetics and thermodynamics of the adsorption of a cationic cobalt porphyrazine onto sepiolite. *Dyes Pigments* **2011**, *88*, 25–38. [CrossRef]

29. Alexandre-Franco, M.; Albarrán-Liso, A.; Gómez-Serrano, V. An identification study of vermiculites and micas: Adsorption of metal ions in aqueous solution. *Fuel Process. Technol.* **2011**, *92*, 200–205. [CrossRef]

30. Franco, F. Adsorption of Methylene Blue on magnesium silicate: Kinetics, equilibria and comparison with other adsorbents. *J. Environ. Sci.* **2010**, *22*, 467–473. [CrossRef]

31. Fouad, H.K.; Bishay, A.F. Uranium uptake from acidic solutions using synthetic titanium and magnesium based adsorbents. *J. Radioanal. Nucl. Chem.* **2010**, *283*, 765–772. [CrossRef]

32. Zhang, T.T.; Vandeperre, L.J.; Cheeseman, C. Magnesium-silicate-hydrate cements for encapsulating problematic aluminium containing wastes. *J. Sustain. Cem.-Based Mater.* **2012**, *1*, 34–45. [CrossRef]

33. Zhang, T.T.; Vandeperre, L.J.; Christopher, R. Formation of magnesium silicate hydrate (M-S-H) cement pastes using sodium hexametaphosphate. *Cem. Concr. Res.* **2014**, *65*, 8–14. [CrossRef]

34. Jia, Y. *The Effect of Na-HMP and CaO on the Reaction Mechanism of MgO-SiO$_2$-H$_2$O System*; Dalian University of Technology: Dalian, China, 2017.

35. GB/T 17671-1999. Determinations for Isotopes of Lead, Strontium and Neodymium in Rock Samples. Available online: https://books.google.com.br/books?id=ghrVAwAAQBAJ&pg=PA477&lpg=PA477&dq= GB/T+17671-19992011 (accessed on 10 December 2019).

36. GB/T 7023-2011. Standard Test Method for Leachability of Low and Intermediate Level Solidified Radioactive Waste Forms. Available online: https://www.chinesestandard.net/PDF/English.aspx/GBT7023-2011 (accessed on 10 December 2019).

37. GB 14569.1-2011. Performance Requirements for Low and Intermediate Level Radioactive Waste Form—Cemented Waste Form. Available online: http://english.mee.gov.cn/Resources/standards/ Radioactivity/radiation/201111/t20111101_219413.shtml (accessed on 10 December 2019).

38. Li, Z.H. *Reaction Mechanisms and Application Study of MgO-SiO$_2$-H$_2$O Cementitious System*; South China University of Technology: Guangzhou, China, 2015.

39. Lai, Z.Y. *Study on Solidification of Low and Medium Radioactive Waste with Magnesium Phosphate Cement*; Chongqing University: Chongqing, China, 2012.

40. Brew, D.R.M.; Glasser, F.P. Synthesis and characterisation of magnesium silicate hydrate gels. *Cem. Concr. Res.* **2005**, *35*, 85–98. [CrossRef]

Article

Thermal Isolation of a Clean Alloy from Waste Slag and Polymeric Residue of Electronic Waste

Rasoul Khayyam Nekouei *, Samane Maroufi, Mohammad Assefi, Farshid Pahlevani and Veena Sahajwalla

Centre for Sustainable Materials Research and Technology (SMaRT), School of Materials Science and Engineering, University of New South Wales (UNSW), Sydney 2052, Australia; s.maroufi@unsw.edu.au (S.M.); m.assefi@unsw.edu.au (M.A.); f.pahlevani@unsw.edu.au (F.P.); veena@unsw.edu.au (V.S.)
* Correspondence: r.nekouei@unsw.edu.au

Received: 7 November 2019; Accepted: 20 December 2019; Published: 2 January 2020

Abstract: Unprecedented advances and innovation in technology and short lifespans of electronic devices have resulted in the generation of a considerable amount of electronic waste (e-waste). Polymeric components present in electronic waste contain a wide range of organic materials encompassing a significant portion of carbon (C). This source of carbon can be employed as a reducing agent in the reduction of oxides from another waste stream, i.e., steelmaking slag, which contains ≈20 wt%–40 wt% iron oxide. This waste slag is produced on a very large scale by the steel industry due to the nature of the process. In this research, the polymeric residue leftover from waste printed circuit boards (PCBs) after a physical-chemical recycling process was used as the source of carbon in the reduction of iron oxide from electric arc furnace (EAF) slag. Prior to the recycling tests, the polymer content of e-waste was characterized in terms of composition, morphology, thermal behavior, molecular structure, hazardous elements such as Br, the volatile portion, and the fixed carbon content. After the optimization of the ratio between the waste slag (Fe source) and the waste polymer (the carbon source), the microstructure of the recycled alloy showed no Br, Cl, S, or other contamination. Hence, two problematic and complex waste streams were successfully converted to a clean alloy with 4 wt% C, 4% Cr, 2% Si, 1% Mn, and 89% Fe.

Keywords: solid waste; cleaner production; electronic waste; recycling; waste printed circuit boards; waste EAF slag

---

1. Introduction

In the early 21st century, rapid advances in technology and short lifespans of devices, machines, and equipment caused the generation of a large volume of waste in different industries, whether in the well-established steelmaking [1] or in the modern electronics sectors [2]. Management of these waste resources is a critical issue for researchers, and approaches to tackle these problems are chiefly dependent on the content and types of valuable materials/elements in the waste and the environmental impacts of the waste [3]. Due to the large-scale production of steelmaking, the amount of generated waste from this process can be more than thousands of tons annually. Such slag byproducts typically contain a significant amount of iron oxide [4], implying that a significant portion of iron is lost along with the disposal of the slag. On the other hand, electronic waste (e-waste) comes from everywhere that humans live. The main issue is the hazardous effect on the environment and humans' health due to the presence of hazardous elements, such as Cd, Hg, and Pb [5]. Hence, there are clearly various purposes for somehow recycling various wastes, but in all cases, the final product must be worthwhile from an economic viewpoint.

E-waste consists of discarded electronics that contain a variety of materials, including metals, ceramics, and polymers entangled with each other [5]. Many studies successfully recycled e-waste via physical, mechanical, and chemical approaches [6]. Some researchers focused on the metallic content [7] while others worked on the ceramic [8] or polymer portions [9]. However, less research has been performed regarding how to use these separated/recycled wastes for new purposes. In other words, because of the rapid growth in e-waste generation, recycling industries need to obtain potential customers for their recycled products, such as polymeric components recovered from e-waste.

Polymeric components from e-waste contain many different organic materials, including phenoxy resins, polyvinyl acetate, and vinyl chloride [5]. These are thermoset types of polymers, which have limitations in terms of their applications that are hard to degrade compared to thermoplastics [10]. Also, this e-waste emits hazardous dioxins and elements, for instance, Br and Cl [11]. However, a significant portion of these waste polymers is carbon (C) [12], which can be used as a potential source of carbon in the recycling process of other types of waste, i.e., the discarded slag from the steelmaking industry.

Over the last few decades, traditional steelmaking furnaces were replaced by electric-arc furnaces (EAFs) [1]. The byproduct slag from EAF contains $\approx$15 wt%–20 wt% Fe in the form of Fe_2O_3 and FeO. Depending on the conditions and efficiency of the process, the Fe [4] and other heavy metal contents, such as Cr [13], can change. Several studies focused on practical applications of EAF slag, such as fiber-reinforced concrete for pavements [14], aggregate in asphalt mixes [13], geo-fill materials [1], and P-removal agents [15]. These applications are innovative and economical; however, original iron ores contain around 60 wt% iron oxide [16] compared to the 20 wt%–40 wt% iron oxide content found in EAF slag. Thus, in order to avoid the continuous depletion of deposit ores (i.e., Fe, Cr, Si, and Mn), the recycling of these waste slags seems to be profitable. Nevertheless, due to the lower Fe content in EAF slag, the activity of C must be higher and traditional methods of using conventional carbonaceous resources (i.e., coke and coal) may not be efficient. Since the usual coals are not very promising when it comes to providing a high activity for the reduction of EAF waste slag, the carbon sourced from the polymer content of e-waste may be an alternative reducing agent. This carbon is much more reactive than coal carbon and the final surface area of the carbon is much higher, which enhances the rate of the reaction [17]. Using high surface area carbon or a gaseous atmosphere, e.g., CO and/or CH_4, could efficiently enhance reduction [18]. Another advantage of using polymeric material for reduction is the presence of hydrogen (H) in addition to carbon, which increases the reduction speed and makes it more environmentally friendly, as it ill produces H_2O after the process instead of greenhouse gasses.

In this research, the leftover polymeric residue waste of printed circuit boards (PCBs) as a result of the physical-chemical process was used as a source of activated C in the reduction of iron oxide from EAF slag. Fe, Cr, Mn, and a portion of Si were produced from waste EAF slag. Hence, two major streams of industrial solid wastes were combined, and the output was a metallic ferrous alloy contained Cr, Si, and Mn. The major environmental hazards of the e-waste material were also investigated.

2. Experimental

2.1. Feed Materials Preparation and Experiments

Fe source: The source of Fe was from EAF slag as waste material. The delivered slag was crushed into a fine powder (100–200 µm) using a ring mill prior to mixing with a carbon source (i.e., the residue of waste PCBs) for reduction.

C source: The flowchart of the separation process of waste PCBs is presented in Figure 1. Waste PCBs from scrap motherboards were collected from a reverse e-waste company. Polymer slots, capacitors, connectors, the heatsink, backup batteries, the CPU, and the CPU socket of each board were dismantled manually or using a heat gun. The resulting board was sent to be crushed into particles less than 1 mm using ring mill and knife mill in a 3-step process [5]. The crushed powder was then classified into metal-rich and ceramic-rich powders using a sieve shaker (Fritsch, Pulverisette 15) [5]. For the recovery of metal content, the metal-rich powder was treated with acid and all of its metal contents (i.e., Cu, Al, Fe, Sn, Ni, Pb, and Zn) were digested [19]. The resulting solution and the leaching residue were separated by filtration [19]. Lastly, the residue, containing ceramics and polymers, was washed with deionized (DI) water and ethanol using a table-top centrifuge (Hermle Z206 A) for the removal of remaining ions, followed by drying in an oven at 60 °C overnight.

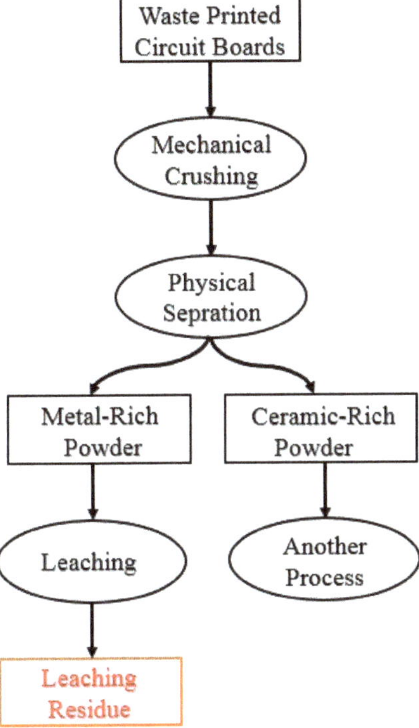

Figure 1. The flowchart of the chemical and physical separation of the waste printed circuit boards (PCBs). The waste PCB leaching residue (WPCBR) was the e-waste used as the source of carbon in this research.

Reduction experiments: To optimize the ratio between the waste slag (WS) and the waste PCB leaching residue (WPCBR), 0.5 g WS was mixed with 0.25, 0.375, 0.5, 0.625, and 0.75 g WPCBR, to provide a $\frac{WPCBR}{WS}$ value between 0.5 and 1.5. Each ratio was calculated twice and the average was reported. The mixture was then placed in a zirconia-based crucible located in a graphite rod and gently inserted into a vertical tube furnace under 1 L/min Ar atmosphere at 1550 °C, which was selected based on the kinetics and thermodynamics of steelmaking slag reduction [20]. After 30 min, the sample was discharged from the furnace for manual separation and the weight of the metal droplets was measured. The weight of recycled metal was measured as an indicator of recovery efficiency.

2.2. Characterization

(a) X-ray diffraction analysis (XRD, PANalytical X'Pert Pro, Malvern, UK) with Cu Kα wavelength and 2θ between 10–100° was applied for phase identification of waste materials and the resulting alloy.

(b) To increase the precision of phase identification, X-ray fluorescence analysis (XRF, PANalytical PW2400 Sequential Wavelength Dispersive, Malvern, UK) was utilized to quantify the approximate percentage of oxide phases.

(c) Scanning electron microscopy (SEM, Hitachi S3400, Tokyo, Japan) equipped with energy dispersive spectroscopy (EDS, Bruker) was used for the morphological observation and distribution analysis of elements in the WPCBR.

(d) LECO analysis was employed to measure the sulfur (S), nitrogen (N) (LECO TruSpec Analyser, Michigan, USA), and carbon contents (LECO CS 444 Analyser, Michigan, USA) of the WPCBR.

(e) The structural and elemental concentrations of the recycled metal were measured by electron probe microanalysis (EPMA) or wavelength dispersive spectroscopy (WDS, JEOL JXA-8500F) for accurate quantitative mapping and point analysis. The sample was mirror polished and all measured elements were calibrated using standard samples.

(f) Moreover, the composition of the recycled metal was analyzed using laser-induced breakdown spectroscopy (LIBS, Z-200 under Ar atmosphere), which is known as a very reliable method even for carbon content analysis. The surface of the sample was mirror polished and the device was calibrated before analysis.

(g) The thermal behavior and proximate analysis of the WPCBR were carried out using thermogravimetric analysis (TGA, PerkinElmer STA 8000 and TGA 8000, Massachusetts, USA) and differential thermogravimetric (DTG) in an alumina crucible under a controlled atmosphere (N_2 or O_2) at a heating rate of 20 °C/min.

(h) The exhaust of the TGA was connected to Fourier transform infrared spectroscopy (FT-IR, PerkinElmer, Frontier, Massachusetts, USA) with continuous wavelength analysis between 4000 and 500 cm^{-1} to identify the functional groups of any evaporated organic species and to measure the volume of the evolved gas.

(i) The exhaust of the FT-IR was joined to a gas chromatograph-mass spectrometer (GC/MS, PerkinElmer, Clatus 500 and Clarus SQ 8 s, Massachusetts, USA) for more accurate recognition of organic species based on separation time. TurboMass V 6.1 equipped with the National Institute of Standards and Technology (NIST) library was used for the identification of the compounds.

3. Result and Discussion

3.1. Characterization of Waste Materials

The composition and phase identification of both waste materials are represented in Tables 1 and 2, and Figure 2. Almost all of the ceramic compounds in both wastes were in oxide form. The ceramic portion of the WPCBR acted like a typical slag for the molten metal, since the main oxides present in the WPCBR, i.e., SiO_2 (silica), CaO (lime), and Al_2O_3 (alumina), were similar to the main oxide in steelmaking slags [21]. However, due to the high fraction of SiO_2 and Al_2O_3, the acidity of the final molten slag was higher, leading to lower viscosity. This issue could be balanced with the CaO content of the WS.

Table 1. The composition, measured by XRF analysis, of the two waste materials used in the reduction process.

Phase (wt%) Waste Material	Fe_2O_3	SiO_2	MnO	CaO	Al_2O_3	MgO	Na_2O	P_2O_5	Cr_2O_3
Ash content of WPCBR	0.9	61.3	0.1	15.5	11.8	0.7	0.7	0.1	0.1
WS	38.6	12.0	7.2	30.0	4.3	7.0	0.3	0.4	2.0

Table 2. The portion of different materials in the WPCBR driven by LECO [1] and proximate [2] analyses.

Phase or Element	N [1]	S [1]	Total Carbon [1]	Total Volatile [2]	Fixed Carbon [2]	Ash [2]
Weight Percentage (wt%)	2.2	0.08–0.1	29.3	54	12	34

The XRD patterns (Figure 2) showed that the WS mainly contained calcium silicate (Ca_2SiO_4), Magnetite (Fe_3O_4), and wustite (FeO), which was in accordance with other studies [4]. However, Al_2O_3, MnO, and Cr_2O_3 were not detected, either due to the amorphous structure or due to them being below the detection limit. Meanwhile, the main phases of the WPCBR were SiO_2, CaO, and calcium aluminum silicate ($Al_3Ca_{0.5}Si_3O_{11}$).

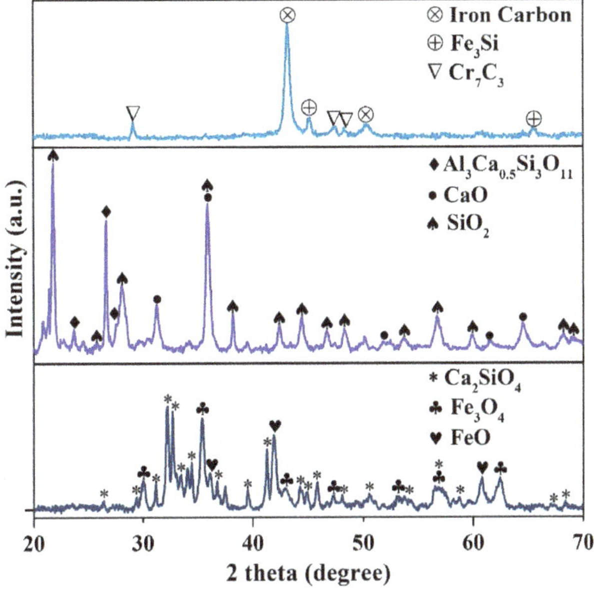

Figure 2. The analyzed XRD patterns matched with the reference patterns of the waste slag (WS) (bottom), ash content WPCBR after burning (middle), and the recycled metal (top). The reference patterns were iron carbon (00-052-0512), Fe_3Si (iron silicon, 00-035-0519), Cr_7C_3 (chromium carbide, 00-036-1482), $Al_3Ca_{0.5}Si_3O_{11}$ (calcium aluminium silicate, 00-046-0744), CaO (lime, 96-900-6731), SiO_2 (cristobalite, 96-900-8228), Fe_3O_4 (magnetite, 01-089-0688), FeO (wustite, 96-900-2670), and Ca_2SiO_4 (calcium silicate, 00-003-0761 and 00-029-0371).

The microscopic images and elemental distribution of both waste materials are provided in Figure 3. As shown, the different elements were distributed almost homogeneously through the WS. In the WPCBR, one of the main ingredients was the fiber, called fiberglass, which was mainly made of Si and Ca [22].

Table 2 provides the percentages of nitrogen (N) and sulfur (S) present in the WPCBR; since these two elements are totally detrimental in steel structures and must be considered during the production process for removal. N and S were transferred in the volatile portion at low temperatures and were less likely to affect the quality of the iron (discussed in the next section). The total volatile of the WPCBR was 54 wt% and the residual materials encompassed 12 wt% fixed carbon and 34 wt% ash; this composition is provided in Table 1. The important value of the WPCBR in this research was related to the carbon content. Around 29.3% of the WPCBR was carbon, of which 12 wt% was fixed carbon. Since the operating temperature of our furnace was high enough (i.e., 1550 °C), all volatile forms of carbon decomposed into simple carbonous components, such as CO, CO_2, or CH_4 [23]. Hence, in addition to the fixed carbon for the reduction, there was a likelihood of the reduction by this reducing gaseous carbon.

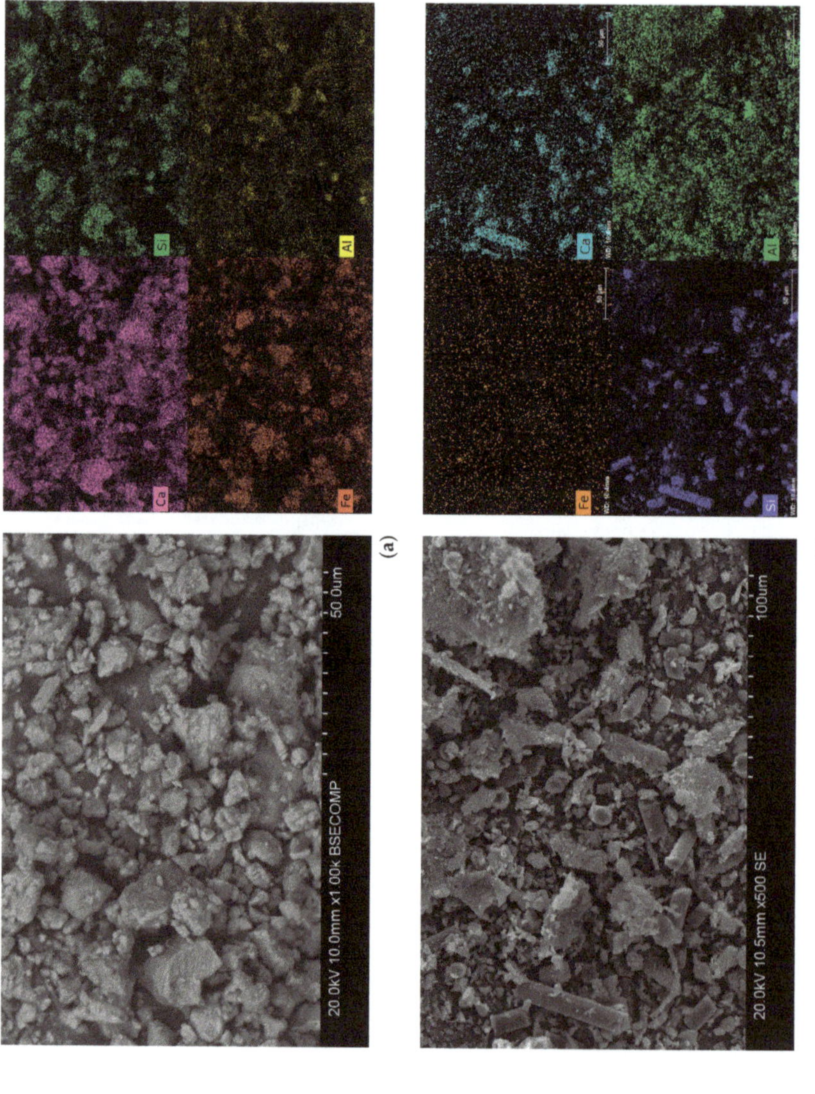

Figure 3. The SEM image and EDS elemental analysis of (**a**) the WS (red: F; violet: Ca; green: Si; yellow: Al) and (**b**) the WPCBR (orange: Fe; turquoise: Ca; blue: Si; green: Al).

The TGA and DTG analyses of the WPCBR are presented in Figure 4a. The weight loss from room temperature to 100 °C was due to the release of adsorbed moisture and physical water [24]. The organic degradation of the WPCBR was gradually triggered from 100 °C and accelerated to 300 °C, which was in good agreement with other research [25]. Between 100 and 150 °C, the degradation rate was lower and started to increase after 150 °C. The weight loss peaked at 360 °C, at which point a significant volume of gas evolved (Figure 4b). With further heating, the degradation rate decreased and almost leveled out before 900 °C, where O_2 was purged for the calculation of the fixed carbon. Due to very high activity, this carbon was considered to be a favorable carbon in the reduction process [18].

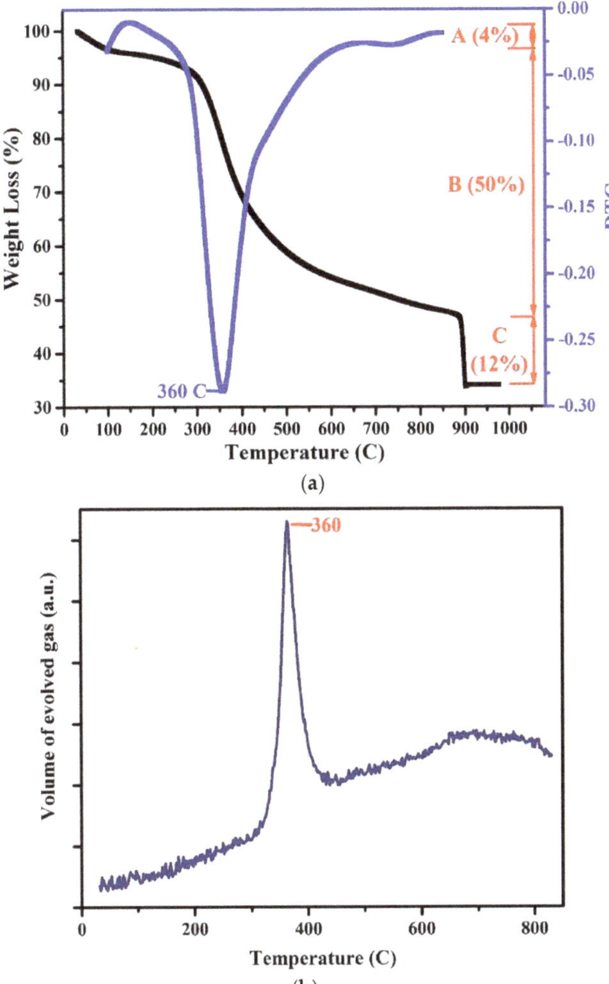

Figure 4. (a) Thermogravimetric analysis (TGA) and differential thermogravimetric (DTG) curves; (b) the volume of gas that evolved during the pyrolysis of the WPCBR under a heating rate of 20 °C/min and N_2 atmosphere.

While the WPCBR at a high temperature (i.e., 1550 °C) was supposed to completely decompose to simple gases, including CO, CO_2, and CH_4, these waste materials contained Br and some other hazardous elements; thus, the evaporated organic compounds were analyzed using FT-IR and GC/MS.

The polymer content of the WPCB had a complex matrix made of various resins [5]. FT-IR analysis was carried out on the exhaust gas from the TGA analysis to identify the functional groups of the organic compounds, particularly toxins. Figure 5 represents the FT-IR spectrum of the evolved gas from the WPCBR when it was at the maximum degradation rate (i.e., 360 °C), accompanied by a detailed investigation of the functional groups, as listed in Table 3. The sharp band at 502 cm^{-1} was assigned to the C-Br stretching vibration, which corresponded to organobromide compounds as identified by MS results. The bands at 664 and 1603 cm^{-1} were assigned to the bending and stretching vibrations of the C=C in alkene and benzene ring compounds, respectively. Two bands at 748 and 829 cm^{-1} were ascribed to C–H vibrations, while two bands at 1086 and 1177 cm^{-1} were assigned to the C-O-C stretching vibrations in aromatic compounds [26]. The vibration in the region of 1540–1174 cm^{-1} was ascribed to the O-H, C-O, C-H, and C=C stretching and bending of the phenolic materials in the sample [27,28]. The vibration in the region of 2900–3700 cm^{-1} was due to O-H starching and bending of carboxylic acids and phenolic groups [29,30].

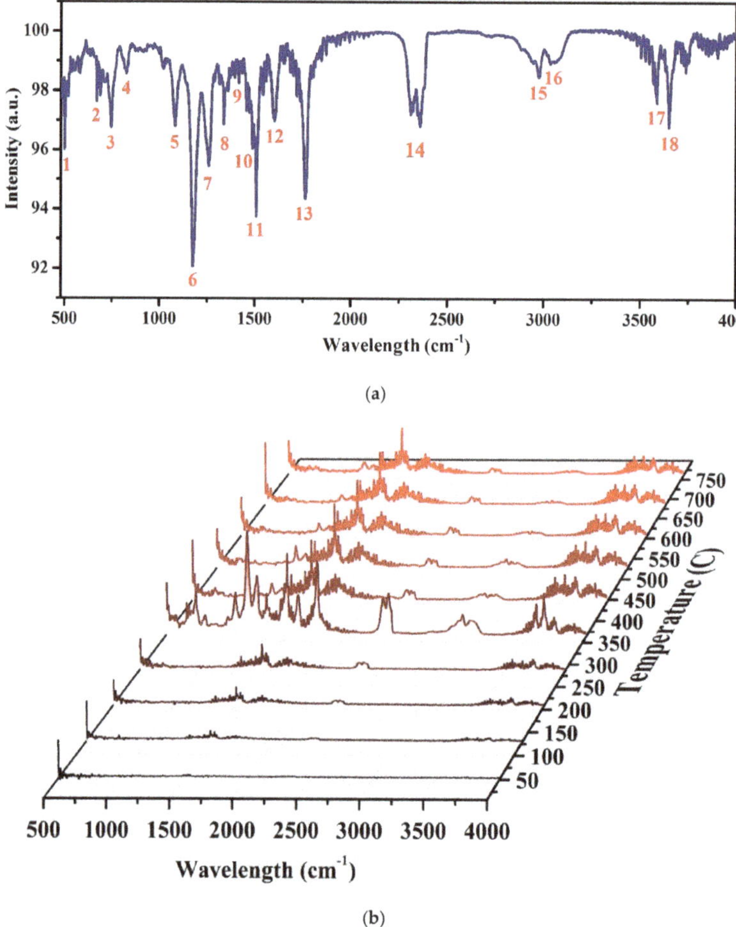

Figure 5. The FT-IR spectra of the WPCBR under a heating rate of 20 °C/min and N$_2$ atmosphere. (**a**) The spectrum at 360 °C (Ads); the corresponding details of the peaks are available in Table 3. (**b**) The sequential plot of spectra at different temperatures (T%).

Table 3. The corresponding details of the FT-IR spectrum represented in Figure 5a, which is the FT-IR spectrum of the WPCBR at 360 °C and N_2 atmosphere.

No.	Functional Group	Absorption Location (cm^{-1})	Type of Vibration
1	C-Br	502	Stretching
2	C=C alkene	664	Bending
3	C-H	748	Bending
4	C-H	829	Bending
5	C-O-C	1086	Stretching
6	C-O-C	1177	Stretching
7	C-O phenolic	1258	Stretching
8	C-O phenolic	1339	Bending
9	C-H aliphatic	1417	Bending
10	C=C	1487	Stretching
11	C=C	1504	Stretching
12	C=C benzene	1603	Stretching
13	C=O carboxylic acid	1756	Stretching
14	O=C=O carbon dioxide	2356, 2310	Stretching
15	O-H phenol (H-bond)	2974	Stretching
16	O-H carboxylic acid (H-bond)	3032	Stretching
17	O-H	3588	Stretching
18	O-H phenol	3649	Stretching

The time-shifting in the FT-IR spectrum of the WPCBR exhausted gas is illustrated in Figure 5b. The sharpest spectrum was taken at 360 °C, which agreed with the highest volume of evolved gas (Figure 4b). Obviously, the intensity of almost all of the peaks declined after 360 °C, and the spectrum was similar after 450 °C, indicating that the evolved gas remained almost unchanged. The only peak that correlated with Br compounds was the first peak, which signified the stability of C-Br. At temperatures higher than 450 °C, it seemed that the rest of the peaks belonged to CO, CO_2, H_2O, and the leftover gases at lower temperatures. Thus, since this exhaust gas contains Br compounds, it should be captured before being released into the atmosphere. It was concluded that at high temperatures, the solid fixed carbon that was produced was stable.

The mass spectrum of the gas evolved from the thermally decomposed WPCBR at 360 °C and the corresponding organic molecular structures are respectively provided in Figure 6 and Table S1. As a matter of fact, there were two types of peaks in the mass spectrum, including a molecular ion peak and a base peak with the highest intensity. According to Figure 6, an ion peak assigned to styrene appeared at a retention time of around 176 s (number 2). The base peak with the highest intensity, which was attributed to phenol, appeared at a retention time of 207 s (number 3). Since the absolute height of all of the ion peaks was dependent on the concentration of the relevant compound, in the WPCBR sample, phenol had the highest concentration and was considered to be the base peak. All the other peaks were referenced as a percentage of the base peak. After phenol, there were four other organic compounds that showed high concentrations in terms of the height of their peaks, including 2-methyl-3,3,5-trimethyl cyclohexyl-2 propenoic acid, 2,3-fihydro-2-methyl-benzofuran, 4-(1-methyl ethyl)-phenol, and p-isopropenylphenol (numbers 4, 7, 10, and 11). Based on the molecular structure of peaks 7, 10, and 11, it was assumed that the major part of the WPCBR sample was made of phenol derivatives. Accordingly, the FT-IR and MS results were in good agreement.

From thermodynamic and kinetic viewpoints, the reduction of iron oxide into Fe using solid C at a low temperature (<900 °C) is impossible. Since our charging process, which was similar to EAF in steelmaking, had a temperature of molten metal higher than 1500 °C, it is correct to assume that all the organic compounds in Table S1 inevitably broke into the simplest gases, i.e., CO, CO_2, and CH_4.

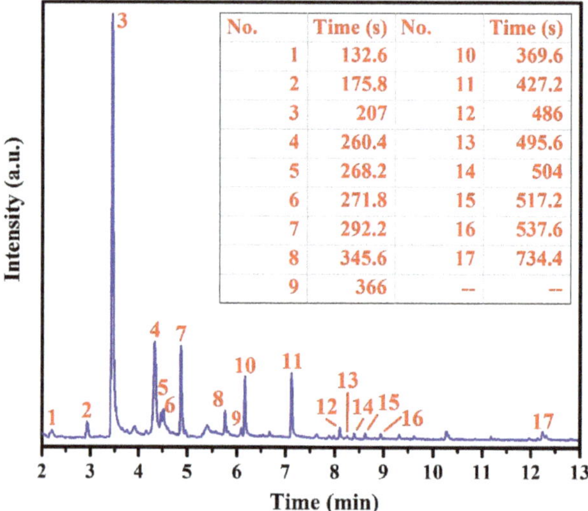

No.	Time (s)	No.	Time (s)
1	132.6	10	369.6
2	175.8	11	427.2
3	207	12	486
4	260.4	13	495.6
5	268.2	14	504
6	271.8	15	517.2
7	292.2	16	537.6
8	345.6	17	734.4
9	366	--	--

Figure 6. The chromatogram of the WPCBR volatile gas at 360 °C (heating rate of 20 °C/min) as a function of separation time. The corresponding compounds of peaks are listed in Table S1.

Considering the FT-IR and GC/MS results and the fact that all of the identified organic compounds could be decomposed into simple greenhouse gases, we concluded that the outlet gas, except CO and the Br-containing gas, were not poisonous at our working temperature (1550 °C). Hence, the exiting gas must have been depleted from Br, which was derived from 1-bromo-butane (number 1), 2,4-dibromo-phenol (number 13), and 2-bromo-p-cymene (number 14), before being released into the atmosphere. The effects of Br and other impurities must be considered for the purity level of the recycled metal; this is investigated in the next section.

3.2. Reduction and Recycling Process

After the carbothermal reduction experiments for different ratios of $\frac{WPCBR}{WS}$ (i.e., 0.5, 0.75, 1, 1.25, and 1.5), the optimum ratio of 1 was chosen based on the weight of the resulting metallic alloy. Figure 7a illustrates the recovery efficiency and the slag residue after reduction at 1550 °C for 30 min and under 1 L/min Ar flow. At lower ratios (0.5–1), the Fe recovery was not complete. By contrast, at higher ratios (1–1.5), the C was more than sufficient and the separation process was more challenging due to the content of SiO_2 and CaO in the WPCBR.

The IR gas analysis was used to monitor the CO, CO_2, and CH_4 gases evolved from the sample under the optimum ratio ($\frac{WPCBR}{WS} = 1$) during reduction; these results are shown in Figure 7b. The concentrations of CO_2 and CH_4 compared to CO were negligible. The reduction behavior was similar to the reduction of iron oxide with rubber-derived carbon or coke reported in previous work [31]; the reaction was complete after less than 10 min, revealing the high activity of the produced carbon. After about 100 s, the concentration of the CO gas surged due to the rapid decomposition of iron oxides (Fe_3O_4 and FeO) and other oxides to metal, gas, and slag phases. In 205 and 300 s reaction times, there were two more peaks, which were possibly attributed to the carbothermal reduction [31] of other oxides, such as SiO_2, MnO, and Cr_2O_3. The detailed mechanism of carbothermal reduction is discussed in reference [31].

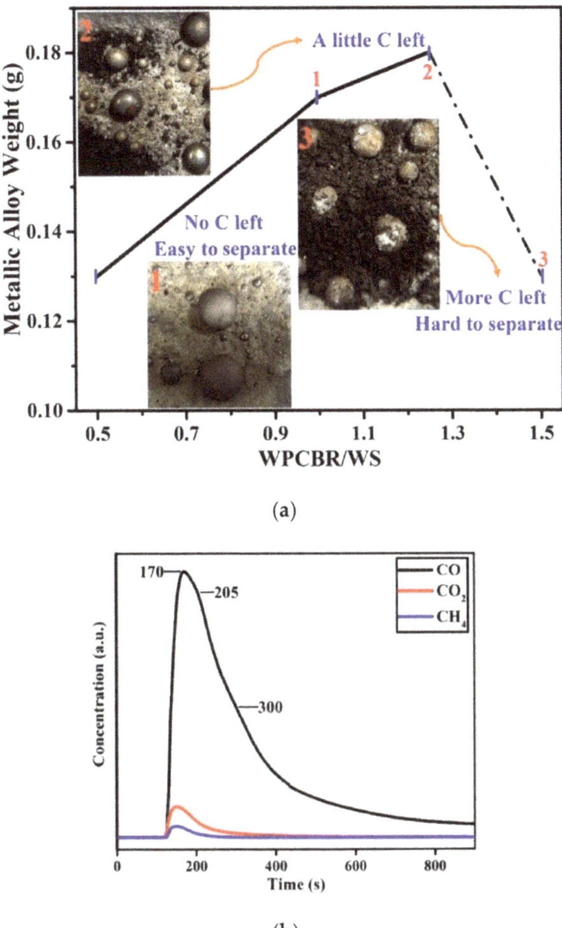

(a)

(b)

Figure 7. (**a**) The weight of metallic alloy recycled from the mixture of the WS and WPCBR as a function of WPCBR/WS at 1550 °C for 30 min under a flow of 1 L/min Ar. (**b**) The gas infrared (IR) analysis of the exhaust gas for the optimum condition (i.e., WPCBR/WS = 1).

From a thermodynamic viewpoint, the reduction of the iron oxide and other oxides involved in our recycling process at 1550°C with a source of solid carbon seemed possible, as per the following reactions [32–35]:

$$Fe_2O_3 + 3/2C = 2Fe + 3/2CO_2(g) \qquad \Delta G^{\circ}_{1550\,°C} = -237 \text{ kJ}, \tag{1}$$

$$2FeO + C = 2Fe + CO_2(g) \qquad \Delta G^{\circ}_{1550\,°C} = -98 \text{ kJ}, \tag{2}$$

$$Cr_2O_3 + C = 2Cr + 3CO(g) \qquad \Delta G^{\circ}_{1550\,°C} = -142 \text{ kJ}, \tag{3}$$

$$MnO + C = Mn + CO(g) \qquad \Delta G^{\circ}_{1550\,°C} = -25 \text{ kJ}, \tag{4}$$

$$SiO_2 + 2C = Si + 2CO(g) \qquad \Delta G^{\circ}_{1550\,°C} = 43 \text{ kJ}. \tag{5}$$

Thermodynamically, Fe, Cr, Mn, and a portion of Si oxides were expected to reduce into their metallic forms. The ΔG of the reactions depended on the activity of the dissolved metal and partial pressures of CO and CO_2 [36]; hence, despite a positive $\Delta G°$ for Reaction (5) at 1550 °C, the very low activity of Si (<0.02, almost equal to wt% of Si dissolved in molten metal) and low partial pressure of CO (ppm level) led to a negative ΔG value. The mean composition of the recycled metal is presented in Table 4. It was concluded that the recovered metallic alloy was an 89Fe-4C-4Cr-2Si-1Mn alloy, which is called high carbon steel. This alloy can be applied for abrasion resistance applications directly or as a master alloy for the production of another type of abrasion resistant alloy.

Table 4. The chemical composition of the recycled metal, as analyzed by LIBS. These results are the averages of 5 points (2 runs).

Element	Fe	Si	Cr	Mn	C
Weight Percent (%)	88.9 ± 0.8	1.7 ± 0.1	3.9 + 0.1	1.3 ± 0.1	3.8 ± 0.4

Figure 8 and Table 5 indicate the distribution and composition of the recovered elements and different phases in the resulting metallic alloy. The WDS results indicate the presence of chromium carbide in the carbide phase and iron silicide in the iron matrix. Figure 2 displays the XRD pattern of the resulting metallic alloy, showing the presence of chromium carbide (Cr_7C_3) and iron silicide (Fe_3Si), which was in accordance with the WDS results. These intermetallics increased the mechanical properties of the alloy.

Table 5. WDS point analysis for C, Mn, Si, Cr, Br, P, and Fe corresponding to Figure 8. Iron phase, carbide phase, and white dot compositions are the averages of 5, 5, and 2 points, respectively.

Element Phase	C	Mn	Si	Cr	Br	P	Fe
Atomic Percent (at%)							
Iron Matrix	1.16 ± 0.04	1.79 ± 0.06	11.82 ± 0.23	0.92 ± 0.04	<0.002	0.84 ± 0.06	83.46 ± 0.19
Carbide Phase	26.62 ± 0.17	3.35 ± 0.23	0.05 ± 0.02	6.34 ± 0.71	<0.002	0.00 ± 0.00	63.65 ± 0.60
White Dots	3.65 ± 0.45	2.15 ± 0.50	9.32 ± 0.19	1.16 ± 0.35	<0.008	1.99 ± 0.67	81.73 ± 2.15
Weight Percent (wt%)							
Iron Matrix	0.27 ± 0.01	1.90 ± 0.06	6.41 ± 0.14	0.92 ± 0.05	<0.003	0.50 ± 0.03	89.94 ± 0.27
Carbide Phase	7.29 ± 0.08	4.20 ± 0.29	0.03 ± 0.01	7.51 ± 0.84	<0.003	0.00 ± 0.00	81.08 ± 0.76
White Dots	0.84 ± 0.10	2.26 ± 0.51	5.02 ± 0.06	1.15 ± 0.34	<0.013	1.18 ± 0.39	87.53 ± 3.00

However, due to the presence of P, which was mainly in the WS, the recycled metal presented a small amount of P as an impurity. It seemed that the P impurity was distributed throughout the iron matrix, mainly in the white dots. The removal of this P is well-established in the steelmaking industry using post-treatment [37,38]. Neither the WDS map nor the point analyses confirmed traces of Br and S in the recycled metal, indicating that all Br and S were expelled in the gaseous form, which was in accordance with the GC/MS and FTIR analyses.

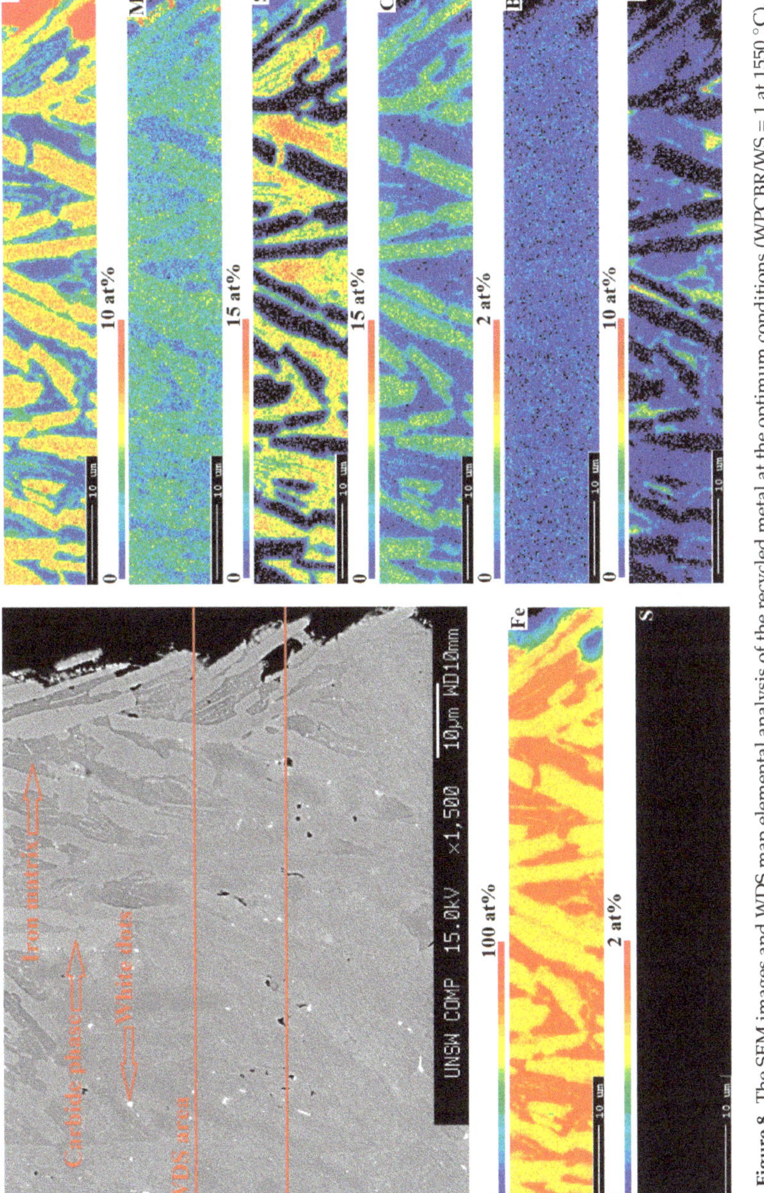

Figure 8. The SEM images and WDS map elemental analysis of the recycled metal at the optimum conditions (WPCBR/WS = 1 at 1550 °C).

4. Conclusions

In this research, two types of solid wastes, i.e., waste EAF slag and the polymeric residue of waste PCBs, were successfully applied for the recovery of the waste EAF slag. The waste slag contained Fe_2O_3 and FeO as sources of Fe, accompanied by Cr_2O_3, MnO, and SiO_2.

During the characterizations of the waste PCBs, the main concern was the intake of hazardous elements, such as Br and S, in the recovered metal, since the volatile gas evolved from the WPCBR contained Br, S, N, Cl, and other hazardous gaseous material. Using FT-IR, GC/MS, and IR analyses, all organic species in the outlet gas were characterized, revealing that, at a high temperature (steelmaking temperature), i.e., 1550 °C, all of the organic species turned into simple gases, such as CO, CO_2, and CH_4. The fixed carbon, volatile materials, and the ash content in the WPCBR were 12 wt%, 54 wt%, and 34 wt%, respectively. The composition of ash in the WPCBR contained mainly SiO_2, CaO, and Al_2O_3, which are common compounds in steelmaking slags.

In the thermal isolation/reduction tests, the ratio between the two wastes was optimized and an optimum ratio of 1 was selected based on the amount of fixed carbon and the easy separation of metal from slag. Regarding the elemental analyses, the recycled metal was an 89Fe-4C-4Cr-2Si-1Mn alloy containing a small amount of P. Moreover, Br, and S, which were the main concerns, were transferred to the exhaust gas and left an insignificant trace in the recycled metals. This research successfully represents the potential of thermal isolation of a clean ferrous alloy from EAF slag using e-waste residuals containing carbon.

Supplementary Materials: The following are available online at http://www.mdpi.com/2227-9717/8/1/53/s1.

Author Contributions: Conceptualization, R.K.N., S.M., and V.S.; experimental design, R.K.N.; GC/MS analysis, M.A.; XRD, TGA, FT-IR, SEM, and TEM analysis, R.K.N., elemental analysis, R.K.N. and F.P., writing (original draft preparation), R.K.N., S.M., and M.A., writing (review and editing), S.M., F.P., and V.S.; funding acquisition, V.S. All authors have read and agreed to the published version of the manuscript.

Funding: This research was funded by the Australian Research Council's Industrial Transformation Research Hub under the project IH130200025.

Acknowledgments: We also gratefully acknowledge the technical support of the Mark Wainwright Analytical Centre (MWAC), in particular, the Solid State & Elemental Analysis Unit and Electron Microscopy Unit (EMU) at the University of New South Wales, Sydney, Australia.

Conflicts of Interest: The authors declare no conflict of interest.

References

1. Yildirim, I.Z.; Mrezzi, M. Experimental evaluation of EAF ladle steel slag as a geo-fill material: Mineralogical, physical & mechanical properties. *Constr. Build. Mater.* **2017**, *154*, 23–33. [CrossRef]

2. Nekouei, R.K.; Pahlevani, F.; Rajarao, R.; Golmohammadzadeh, R.; Sahajwalla, V. Direct transformation of waste printed circuit boards to nano-structured powders through mechanical alloying. *Mater. Des.* **2018**, *141*, 26–36. [CrossRef]

3. Wang, Z.; Zhang, B.; Guan, D. Take responsibility for electronic-waste disposal. *Nature* **2016**, *536*, 23–25. [CrossRef] [PubMed]

4. Yildirim, I.Z.; Prezzi, M. Chemical, mineralogical, and morphological properties of steel slag. *Adv. Civ. Eng.* **2011**, 1–13. [CrossRef]

5. Nekouei, R.K.; Pahlevani, F.; Rajarao, R.; Golmohammadzadeh, R.; Sahajwalla, V. Two-step pre-processing enrichment of waste printed circuit boards: Mechanical milling and physical separation. *J. Clean. Prod.* **2018**, *184*, 1113–1124. [CrossRef]

6. Kumar, A.; Holuszko, M.; Espinosa, D.C.R. E-waste: An overview on generation, collection, legislation and recycling practices. *Resour. Conserv. Recycl.* **2017**, *122*, 32–42. [CrossRef]

7. Song, Q.; Li, J. Environmental effects of heavy metals derived from the e-waste recycling activities in China: A systematic review. *Waste Manag.* **2014**, *34*, 2587–2594. [CrossRef]

8. Singh, N.; Li, J.; Zeng, X. Global responses for recycling waste CRTs in e-waste. *Waste Manag.* **2016**, *57*, 187–197. [CrossRef]

9. Sahajwalla, V.; Gaikwad, V. The present and future of e-waste plastics recycling. *Curr. Opin. Green Sustain. Chem.* **2018**, *13*, 102–107. [CrossRef]

10. Senophiyah-Mary, J.; Loganath, R.; Shameer, P.M. Deterioration of cross linked polymers of thermoset plastics of e-waste as a side part of bioleaching process. *J. Environ. Chem. Eng.* **2018**, *6*, 3185–3191. [CrossRef]

11. Verma, R.; Vinoda, K.S.; Papireddy, M.; Gowda, A.N.S. Toxic Pollutants from Plastic Waste—A Review. *Procedia Environ. Sci.* **2016**, *35*, 701–708. [CrossRef]

12. Maroufi, S.; Mayyas, M.; Nekouei, R.K.; Assefi, M.; Sahajwalla, V. Thermal Nanowiring of E-Waste: A Sustainable Route for Synthesizing Green Si_3N_4 Nanowires. *ACS Sustain. Chem. Eng.* **2018**, *6*, 3765–3772. [CrossRef]

13. Skaf, M.; Manso, J.M.; Aragón, Á.; Fuente-Alonso, J.A.; Ortega-López, V. EAF slag in asphalt mixes: A brief review of its possible re-use. *Resour. Conserv. Recycl.* **2017**, *120*, 176–185. [CrossRef]

14. Ortega-López, V.; Fuente-Alonso, J.A.; Santamaría, A.; San-José, J.T.; Aragón, Á. Durability studies on fiber-reinforced EAF slag concrete for pavements. *Constr. Build. Mater.* **2018**, *163*, 471–481. [CrossRef]

15. Zuo, M.; Renman, G.; Gustafsson, J.P.; Klysubun, W. Phosphorus removal by slag depends on its mineralogical composition: A comparative study of AOD and EAF slags. *J. Water Process Eng.* **2018**, *25*, 105–112. [CrossRef]

16. Xu, R.; Dai, B.; Wang, W.; Schenk, J.; Xue, Z. Effect of iron ore type on the thermal behaviour and kinetics of coal-iron ore briquettes during coking. *Fuel Process. Technol.* **2018**, *173*, 11–20. [CrossRef]

17. Liou, T.-H.; Jheng, J.-Y. Synthesis of High-Quality Ordered Mesoporous Carbons Using a Sustainable Way from Recycling of E-waste as a Silica Template Source. *ACS Sustain. Chem. Eng.* **2018**, *6*, 6507–6517. [CrossRef]

18. Dong, H.; Geng, Y.; Yu, X.; Li, J. Uncovering energy saving and carbon reduction potential from recycling wastes: A case of Shanghai in China. *J. Clean. Prod.* **2018**, *205*, 27–35. [CrossRef]

19. Nekouei, R.K.; Pahlevani, F.; Golmohammadzadeh, R.; Assefi, M.; Rajarao, R.; Chen, Y.-H.; Sahajwalla, V. Recovery of heavy metals from waste printed circuit boards: Statistical optimization of leaching and residue characterization. *Environ. Sci. Pollut. Res.* **2019**, 1–13. [CrossRef]

20. Maroufi, S.; Nekouei, R.K.; Hassan, K.; Sahajwalla, V. Thermal Transformation of Mixed E-Waste Materials into Clean SiMn/FeMn Alloys. *ACS Sustain. Chem. Eng.* **2018**, *6*, 13296–13301. [CrossRef]

21. Ukwattage, N.L.; Ranjith, P.G.; Li, X. Steel-making slag for mineral sequestration of carbon dioxide by accelerated carbonation. *Meas. J. Int. Meas. Confed.* **2017**, *97*, 15–22. [CrossRef]

22. Wang, H.; Zhang, G.; Hao, J.; He, Y.; Zhang, T.; Yang, X. Morphology, mineralogy and separation characteristics of nonmetallic fractions from waste printed circuit boards. *J. Clean. Prod.* **2018**, *170*, 1501–1507. [CrossRef]

23. Shokri, A.; Pahlevani, F.; Levick, K.; Cole, I.; Sahajwalla, V. Synthesis of copper-tin nanoparticles from old computer printed circuit boards. *J. Clean. Prod.* **2017**, *142*, 2586–2592. [CrossRef]

24. Maroufi, S.; Nekouei, S.K.; Assefi, M.; Sahajwalla, V. Waste-cleaning waste: Synthesis of ZnO porous nano-sheets from batteries for dye degradation. *Environ. Sci. Pollut. Res.* **2018**, *25*, 28594–28600. [CrossRef] [PubMed]

25. Ye, Z.; Yang, F.; Lin, W.; Li, S.; Sun, S. Improvement of pyrolysis oil obtained from co-pyrolysis of WPCBs and compound additive during two stage pyrolysis. *J. Anal. Appl. Pyrolysis* **2018**, *135*, 415–421. [CrossRef]

26. Yang, H.; Yan, R.; Chen, H.; Lee, D.H.; Zheng, C. Characteristics of hemicellulose, cellulose and lignin pyrolysis. *Fuel* **2007**, *86*, 1781–1788. [CrossRef]

27. Preserova, J.; Ranc, V.; Milde, D.; Kubistova, V.; Stavek, J. Study of phenolic profile and antioxidant activity in selected Moravian wines during winemaking process by FT-IR spectroscopy. *J. Food Sci. Technol.* **2015**, *52*, 6405–6414. [CrossRef]

28. Biber, M.V.; Stumm, W. An In-Situ ATR-FTIR Study: The Surface Coordination of Salicylic Acid on Aluminum and Iron(III) Oxides. *Environ. Sci. Technol.* **1994**, *28*, 763–768. [CrossRef]

29. Tahir, H.E.; Xiaobo, Z.; Zhihua, L.; Jiyong, S.; Zhai, X.; Wang, S.; Mariod, A.A. Rapid prediction of phenolic compounds and antioxidant activity of Sudanese honey using Raman and Fourier transform infrared (FT-IR) spectroscopy. *Food Chem.* **2017**, *226*, 202–211. [CrossRef]

30. Movasaghi, Z.; Rehman, S.; Rehman, I. Fourier Transform Infrared (FTIR) Spectroscopy of Biological Tissues. *Appl. Spectrosc. Rev.* **2008**, *43*, 134–179. [CrossRef]

31. Maroufi, S.; Mayyas, M.; Mansuri, I.; O'Kane, P.; Skidmore, C.; Jin, Z.; Fontana, A.; Sahajwalla, V. Study of Reaction Between Slag and Carbonaceous Materials. *Metall. Mater. Trans. B* **2017**, *48*, 2316–2323. [CrossRef]

32. Help of HSC V.9 Software Website. Available online: https://www.outotec.com/products/digital-solutions/hsc-chemistry/ (accessed on 1 June 2019).
33. Fruehan, R.J. Rate of Reduction of Cr_2O_3 by Carbon and Carbon Dissolved in Liquid Iron Alloys. *Metall. Trans. B* **1977**, *8*, 429–433. [CrossRef]
34. Şeşen, F.E. Practical reduction of manganese oxide. *J. Chem. Tech. App.* **2017**, *1*, 1–2.
35. Farzana, R.; Rajarao, R.; Sahajwalla, V. Reaction Mechanism of Ferrosilicon Synthesis Using Waste Plastic as a Reductant. *ISIJ Int.* **2017**, *57*, 1780–1787. [CrossRef]
36. Maroufi, S.; Nekouei, R.K.; Hossain, R.; Assefi, M.; Sahajwalla, V. Recovery of Rare Earth (i.e., La, Ce, Nd, and Pr) Oxides from End-of-Life Ni-MH Battery via Thermal Isolation. *ACS Sustain. Chem. Eng.* **2018**, *6*, 11811–11818. [CrossRef]
37. Schrama, F.N.H.; Beunder, E.M.; Van den Berg, B.; Yang, Y.; Boom, R. Sulphur removal in ironmaking and oxygen steelmaking. *Ironmak. Steelmak.* **2017**, *44*, 333–343. [CrossRef]
38. Penn, C.; Chagas, I.; Klimeski, A.; Lyngsie, G. A review of phosphorus removal structures: How to assess and compare their performance. *Water (Switzerland)* **2017**, *9*, 583. [CrossRef]

Article

Risk Assessment of Potentially Toxic Elements Pollution from Mineral Processing Steps at Xikuangshan Antimony Plant, Hunan, China

Saijun Zhou [1,*], Renjian Deng [1] and Andrew Hursthouse [2,3]

[1] College of Civil Engineering, Hunan University of Science and Technology, Xiangtan 411201, China; 800912deng@sina.com

[2] Hunan Provincial Key Laboratory of Shale Gas Resource Utilization, Hunan University of Science and Technology, Xiangtan 411201, China; Andrew.Hursthouse@uws.ac.uk

[3] School Computing, Engineering & Physical Science, University of the West of Scotland, Paisley PA1 2BE, UK

* Correspondence: zsjaw0745@sina.com

Received: 11 November 2019; Accepted: 13 December 2019; Published: 25 December 2019

Abstract: We evaluated the direct release to the environment of a number of potentially toxic elements (PTEs) from various processing nodes at Xikuangshan Antimony Mine in Hunan Province, China. Sampling wastewater, processing dust, and solid waste and characterizing PTE content (major elements Sb, As, Zn, and associated Hg, Pb, and Cd) from processing activities, we extrapolated findings to assess wider environmental significance using the pollution index and the potential ecological risk index. The Sb, As, and Zn in wastewater from the antimony benefication industry and a wider group of PTEs in the fine ore bin were significantly higher than their reference values. The content of Sb, As, and Zn in tailings were relatively high, with the average value being 2674, 1040, and 590 mg·kg^{-1}, respectively. The content of PTEs in the surface soils surrounding the tailings was similar to that in tailings, and much higher than the background values. The results of the pollution index evaluation of the degree of pollution by PTEs showed that while dominated by Sb, some variation in order of significance was seen namely for: (1) The ore processing wastewater Sb > Pb > As > Zn > Hg > Cd, (2) in dust Sb > As > Cd > Pb > Hg > Zn, and (3) surface soil (near tailings) Sb > Hg > Cd > As > Zn > Pb. From the assessment of the potential ecological risk index, the levels were most significant at the three dust generation nodes and in the soil surrounding the tailings reservoir.

Keywords: antimony; mineral processing; potentially toxic elements; pollution characteristics

---

1. Introduction

Pollution caused by nonferrous metal mining has attracted increasing attention from the public in China. With the issue of *The 12th Five-Year Plan for Comprehensive Prevention and Control of Heavy Metal Pollution*, the pollution caused by nonferrous metal mining, processing, and metallurgy has been again highlighted. As the country hosting the world's largest reserves of Sb resources and production, the mining and processing industry has generated significant pollution loads on the local environment. Antimony ores are of various types, with many associated minerals and element associations [1,2]. Because of this, Sb ore processing is complex, during which PTEs associated with ore minerals can enter the environment, dispersing in air, soil, and groundwater and leading to migration in the food chain, accumulating in animals, plants, or the human body, and increasing risk to human health [3–6].

The characteristics of potentially toxic elements (PTEs) pollution in Xikuangshan (XKS) Antimony Mine in Lengshuijiang, Hunan Province, has recently been the focus of independent researches. It has been observed that the water, vegetable fields, farmland, and soil in the tailings zone around XKS Antimony Mine are all seriously polluted [7–13]. PTEs released to water and soil have directly polluted crops, with PTEs in edible parts of foliage vegetables, fruits, and rhizomes grown in the affected areas of XKS mining area found to be several times higher than those in non-affected areas, among which, Sb and As are the two PTEs dominating this contamination [11,14–16]. When an excessive amount of PTEs enter the human body, it will cause tumors and diseases to the liver, skin, and respiratory and cardiovascular systems, and even cancer [17–21]. Sb, As, Hg, and their compounds are listed as pollutants of priority interest by the United States Environmental Protection Agency [22] and the European Union [23]. However, while spatial assessment across the mine-affected region has identified indicative source–pathway–receptor links [24] a more detailed assessment of release from the antimony beneficiation process is missing.

We report here on an assessment of the release of PTEs in wastewater, dust, and solid waste derived from the beneficiation process. The results provide input to process review for environmental protection and occupational exposure routes for longer-term management of site operations.

2. Material and Methods

2.1. Description of Antimony Processing

The antimony processing plant is located in the north of XKS Antimony Mine, Hunan Province. It covers an area of about 18,000 m^2 and has a beneficiation scale of 1500 t/d (ton/day). The beneficiation ore is of single antimony sulfide ore, a kind of low temperature hydrothermal filling deposit. It mainly consists of metal minerals (mainly stibnite (Sb_2S_3) and a small amount of antimony oxide (Sb_2O_3), pyrite (FeS_2), pyrrhotite ($Fe_{(1-x)}S$), etc.) and gangue minerals (mainly quartz (SiO_2)), followed by calcite ($CaCO_3$), barite ($BaSO_4$), kaolin ($Al_2O_3 \cdot SiO_2 \cdot 2H_2O$, gypsum ($CaSO_4 \cdot 2H_2O$), etc.).

The antimony ore beneficiation process was carried out with the combined methods of manual separation, heavy medium separation, and flotation separation. In manual separation, the waste rock accounts for 40%~45%, and the recovery rate of antimony ores after manual separation is 92%~96%; in a heavy-medium separation, with ferrosilicon used as the weighting agent, the ores at the density of 2.62~2.64 t/m^3 are picked out, with the discard and the recovery rate at this point being 40%~43% and 93%~95%, respectively. After the two separation steps, the enriched antimony ores are ground to the size of less than 0.074 mm for separation by flotation. During the flotation process, each ton of ore consumes 3.5 m^3 water and produces 0.2 kg dust and 920 kg tailings.

Management of wastes and longer-term operation has been outlined in our previous studies [25]. The 120 years of operation at the site have resulted in widespread and sporadic deposition to fill voids generated during mining, covering an area of over 70 km^2. Currently tailings are stored behind a reservoir after transport from separation and flotation processes by wastewater. The wastewater is discharged to the local river after minimal treatment.

2.2. Sampling Point Layout and Sample Collection

Sampling carried out for this study identified locations where the predominant discharge/generation of PTEs emissions occurs. Namely: (1) Wastewater generated at ore concentration, tailings accumulation, ore filtration, and tailings reservoir; (2) the production of dust is at the crushing and screening workshop, the fine ore bin, and from the concentrate transportation; and (3) the solid wastes are dominated by the tailings reservoir. Consequently, 24 samples were collected: 12 Wastewater samples were collected from the ore concentrate tank, ore concentrate filter tank, tailings concentration tank, and tailings reservoir. Three separate samples were collected at random from each of the 4 locations (W1~W3, W4~W6, W7~W9, and W10~W12, respectively). Six dust samples with duplicates were collected from the 3 locations (crushing and screening workshop, fine ore bin, and concentrate transportation, numbered D1–D2, D3–D4, and D5–D6, respectively). Three samples were collected from the surface of the tailings reservoir (S1–S3) and three samples of surface soil (0–20 cm in depth) were collected from around the tailings reservoir (S4–S6). Each solid sample was a composite from 3 subsamples mixed together in the field to make total bulk of 1 kg.

The water samples were immediately packed into rinsed pollution-free polyethylene plastic bottles, acidified with HNO_3, sealed, and stored in a portable thermostated cooler at <4 °C. The dust, tailings, and soil samples collected were sealed in pollution-free polyethylene plastic bags and transported to the laboratory.

2.3. Sample Preparation and Testing

The solid samples (dust, tailings, and soil) were naturally air dried in the laboratory and ground to particle size less than 0.074 mm for later study. The PTEs were determined by hydride generation-atomic fluorescence (AFS-9700, Beijing Haiguang, China) after the digestion of solid samples (dust, tailings, and soil), and conditioning of water samples. Solid materials (0.10 g) were digested in triplicate as follows: 5 mL of concentrated HNO_3 and 0.5 mL of HF in Teflon beakers for 12 h at 170 °C followed by cooling and addition of 1 mL of 30% H_2O_2 and 30 min later 10 mL of 5% *v/v* HNO_3 before sample filtration with a 0.2 μm polyethylene film injection filter. Finally, ultrapure water was added to the samples to a volume of 50 mL and stored at 4 °C before analysis. The acidified water samples were filtered through a 0.2 μm polyethylene film injection filter before analysis.

2.4. Quality Control

In order to ensure data accuracy in the analysis process and stability of test equipment, the standard reference soil (GBW07406) from China's National Institute of Metrology was treated in the same way as the dust (tailings, soil) samples, and the recovery rates of Sb, As, Hg, Pb, Cd, and Zn in standard reference materials were 95%–106%, 94%–107%, 94%–104%, 97%–105%, and 91%–105%, respectively. Calibration standards were produced after dilution of stock multi-element solution (reference) At the same time, reagent blanks were included in each batch of analysis samples, and 20% of the samples were remeasured, with RSD (relative standard deviation) remeasurement less than 10%.

2.5. Evaluation Method of PTEs Pollution

The pollution level of PTEs in the antimony beneficiation process was comprehensively evaluated by the pollution index method and potential ecological risk index method, respectively.

2.5.1. Pollution Index (PI)

To evaluate PTEs levels, the pollution index was calculated [26].

$$\text{Pollution index } (PI) = \frac{C_i}{C_b} \tag{1}$$

in which C_i is element i concentration in the samples, and C_b is its environmental background value [27–29].

The PI of PTE was classified as non-pollution ($PI \leq 1$), which indicated that the level of metals was below the threshold concentration but did not necessarily mean there was no pollution from anthropogenic sources or other enrichment over the background, low level of pollution ($1 < PI \leq 2$), moderate level of pollution ($2 < PI \leq 3$), and high level of pollution ($PI > 3$) [30].

2.5.2. Potential Ecological Risk Index (RI)

Potential ecological risk index method links the PTEs' content, toxicity, and ecological and environmental effect, and evaluates the potential ecological risks of PTEs. It has become the main method to evaluate the ecological hazards of PTEs in the environment and has previously been used to spatially screen soil contamination in the wider area of the mining site [24,25].

The potential ecological risk index was performed using the following equations [31].

$$RI = \sum_{i=1}^{n} E_r^i \tag{2}$$

$$E_r^i = T_r^i \cdot C_r^i \tag{3}$$

$$C_r^i = \frac{C_m^i}{C_n^i} \tag{4}$$

where C_r^i is the single factor pollution index of element i, C_m^i is the ith metal concentration in the sample, C_n^i is the background value of the target element, and T_r^i is the toxic response factor for each of the six metals: Sb (19), As (10), Hg (40), Pb (5), Cd (30) and Zn (1) [32,33]. E_r^i is the single factor potential risk factor. According to E_r^i, the sediments or soils were categorized into five levels: (1) $E_r^i \leq 40$, low risk; (2) $40 < E_r^i \leq 80$, moderate risk; (3) $80 < E_r^i \leq 160$, moderate to high risk; (4) $160 < E_r^i \leq 320$, high risk; (5) $E_r^i > 320$, very high risk. RI is Multi-factor and comprehensive potential ecological hazard index. According to RI, the sediments or soil were categorized into four levels: (1) $RI \leq 150$, low risk; (2) $150 < RI \leq 300$, moderate risk; (3) $300 < RI \leq 600$, high risk; (4) $RI > 600$, very high risk.

3. Results and Discussions

3.1. Characteristics of PTE Pollution in Wastewater

In light of Technical Guidelines for Environmental Impact Assessment Surface Water Environment (HJ/T 2.3-1993) and related literature [34,35] the significance of pollution by As, Hg, Pb, Cd, and Zn was referenced against Quantitative IV water quality standard in Surface Water Environmental Quality Standards (GB3838-2002) as the reference value for ecological risk assessment [27], and for Sb, a value of 0.005 mg·L^{-1} was used as the reference value [28].

The results summarized in Table 1 show that the concentration of Sb, As, and Zn in the wastewater from the whole antimony ore beneficiation process exceeded the relevant standards, with the filtration tank the most contaminated by Sb, As, and Pb. The average concentrations of Sb, As, and Pb in wastewater were 4.415 mg·L^{-1}, 1.006 mg·L^{-1}, and 1.536 mg·L^{-1}, respectively. The maximum concentration of Sb, As, and Pb in wastewater were found at sampling sites W4, W5, and W6, respectively, and the values for Sb, As, and Pb were 950, 10, and 30 times higher than the reference value, respectively. The pollution load for As and Pb in the tailings reservoir, tailings tank, and ore concentrate tank was also serious, their average values being 3.414 mg·L^{-1}, 0.772 mg·L^{-1}, and 0.255 mg·L^{-1}, respectively.

Table 1. The concentration of target potentially toxic elements (PTEs) in water samples collected from discharge points at the Xikuangshan (XKS) antimony mineral processing plant.

Node	Sample No.	PTEs Concentration/(mg·L^{-1})					
		Sb	As	Hg	Pb	Cd	Zn
Ore concentrate tank	W$_1$	2.351	0.792	1.96×10^{-5}	0.254	1.16×10^{-5}	7.986
	W$_2$	1.987	0.695	1.59×10^{-5}	0.263	1.35×10^{-5}	7.521
	W$_3$	2.166	0.692	1.15×10^{-5}	0.249	1.08×10^{-5}	6.924
Average		2.168	0.726	1.57×10^{-5}	0.255	1.20×10^{-5}	7.387
filter tank	W$_4$	4.375	1.018	3.12×10^{-5}	1.472	1.79×10^{-5}	9.145
	W$_5$	4.762	1.006	2.87×10^{-5}	1.554	1.58×10^{-5}	8.927
	W$_6$	4.109	0.994	3.57×10^{-5}	1.583	2.04×10^{-5}	9.005
Average		4.415	1.006	3.19×10^{-5}	1.536	1.80×10^{-5}	9.026
Tailings tank	W$_7$	2.774	0.852	1.02×10^{-5}	0.024	9.57×10^{-6}	6.785
	W$_8$	2.584	0.735	9.54×10^{-6}	0.037	9.75×10^{-6}	6.548
	W$_9$	2.981	0.729	9.96×10^{-6}	0.035	8.76×10^{-6}	7.006
Average		2.780	0.772	9.90×10^{-6}	0.032	9.36×10^{-6}	6.780

Table 1. *Cont.*

Node	Sample No.	PTEs Concentration/(mg·L^{-1})					
		Sb	**As**	**Hg**	**Pb**	**Cd**	**Zn**
Tailing reservoir	W$_{10}$	3.157	0.747	3.27×10^{-5}	0.065	1.07×10^{-5}	10.968
	W$_{11}$	3.336	0.692	3.54×10^{-5}	0.135	1.12×10^{-5}	9.997
	W$_{12}$	3.749	0.701	2.99×10^{-5}	0.104	1.19×10^{-5}	11.065
Average		3.414	0.713	3.27×10^{-5}	0.101	1.13×10^{-5}	10.677
Standard value of surface water class IV		0.005	0.1	0.001	0.05	0.005	2.0

The wastewater from the tailings reservoir was most seriously polluted by zinc, with its average concentration of 10.677 mg·L^{-1} and a maximum of 11.065 mg·L^{-1} at sampling point W12, which is 8.3 times higher than the standard. In addition, the wastewater from concentrate filtration equipment was also seriously polluted by zinc, whose average concentration reached 9.026 mg·L^{-1}. The concentration of Hg and Cd in wastewater from antimony ore beneficiation process conformed to the standard.

PTEs pollution index (*PI*) and potential ecological risk index (E^i_r, *RI*) of wastewater produced at each stage are shown in Figures 1 and 2. The results of Figure 1 show that Sb pollution was the most serious in the wastewater from the Sb beneficiation process overall, with a *PI* of ~400 to >980 and for As and Zn, ~7–>10 and >3–>5 respectively. In the case of Pb, pollution impact was more variable across the process, i.e., the tailings concentration tank (pollution-free), tailings reservoir (low-medium pollution), concentrate concentration tank, and concentrate filter tank (serious pollution), while for Hg and Cd, little pollution impact existed. The order of pollution significance in wastewater was: Sb > Pb > As > Zn > Hg > Cd.

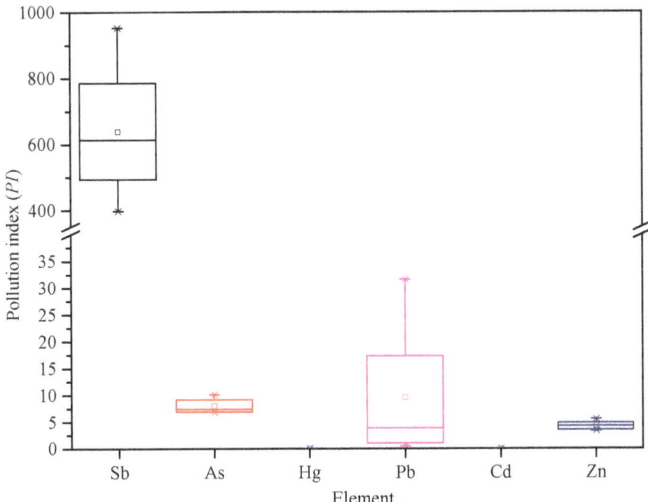

Figure 1. Box-plot showing *PI* for PTEs in all wastewater samples.

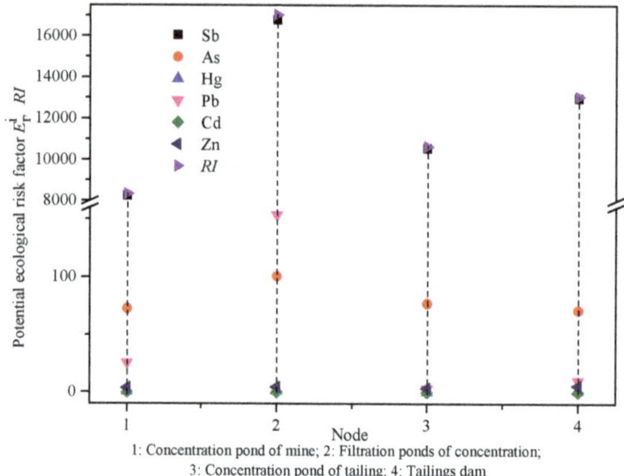

Figure 2. Evaluating results by individual E^i_r and compound *RI*.

Figure 2 shows the individual element (E^i_r) and composite (*RI*) values for ecological risk of the PTES in wastewater for this study. Unsurprisingly, Sb had the greatest contribution to the composite ecological risk at all four wastewater sampling points. The *RI* was dominated by Sb, contributing 98–99% of the compound potential ecological risk (*RI*) at each sampling location, with As and Pb showing moderate risk. Overall the *RI* ranged between 8000–17,000, highlighting incredibly high environmental risk.

3.2. Pollution Characteristics of PTEs in Dust

At present, there is no unified evaluation standard for PTEs pollution in dust at home or abroad, and the reference values are quite different [36]. Some scholars select typical reference points to evaluate PTEs' content according to the evaluation objects [37], but most scholars use local soil background values as reference [38–40]. Hunan is a calcareous lithosol region, and the background value of Sb, Pb, Zn, and other elements in soil and the ore itself is relatively high. Therefore, the background content of soil in Hunan Province was selected as the standard reference.

The PTEs' content of samples from the dust-producing processes is shown in Table 2. It can be seen from the table that the pollution of Sb, As, Hg, Pb, Cd, and Zn in the dust was serious, for all PTEs. Among them, Hg and Cd reached maximum values in the crushing and screening workshop, while for Sb, As, Pb, and Zn the dust of fine ore bin presented the highest concentrations.

The pollution index for dust is shown in Figure 3 and, as expected, Sb was the most enriched in the dusts with an average pollution index >680, with order of magnitude lower values for As, Cd, Pb, Hg, and Zn (69.6, 57.2, 11.1, 11.0, and 6.7), all indicating significant if not extreme pollution.

Table 2. PTEs contents in dust samples from nodes in antimony processing plant.

Node	Sample No.	PTEs Content/(mg·kg⁻¹)					
		Sb	As	Hg	Pb	Cd	Zn
Crushing and screening plant	D1	1845.274	784.026	1.631	238.961	5.781	645.327
	D2	1650.716	655.772	0.954	207.542	6.174	598.182
Average		1747.995	719.899	1.293	223.252	5.978	621.755
Fine ore bin	D3	2478.571	1425.322	1.138	117.264	2.544	758.074
	D4	2275.693	1109.845	1.204	561.029	3.017	812.036
Average		2377.132	1267.584	1.171	339.147	2.781	785.055
Concentrate transportation route	D5	1966.467	958.457	0.413	375.025	4.559	514.637
	D6	2014.278	915.028	0.652	300.416	5.013	498.753
Average		1990.373	936.743	0.533	337.721	4.786	506.695
Hunan Province soil background values [29]		2.98	14	0.09	27	0.079	95

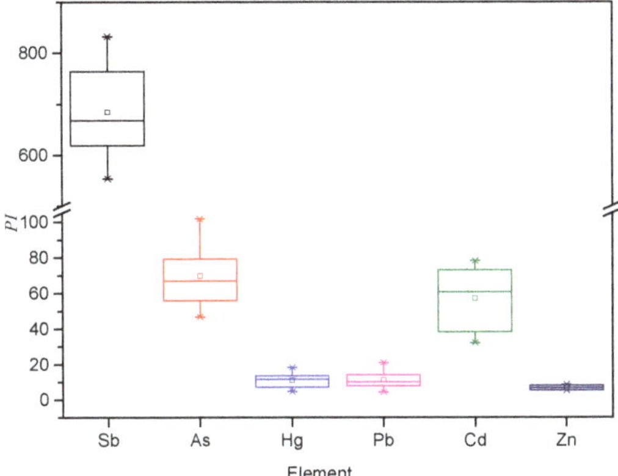

Figure 3. Box-plot of evaluating results by *PI* for dust.

The $E^i{}_r$ of each PTE (Sb, As, Hg, Pb, Cd, Zn) in the dust and the potential ecological risk index (RI) of PTEs (Figure 4) were calculated. The results indicated the following: In the dust, Sb contributed the majority of the ecological risk in the three nodes (crushing and screening plant, fine ore bin, and concentrate transportation route); the $E^i{}_r$ values of Sb, As, and Cd were higher than 320 in the three nodes, suggesting very high risk; for Hg, its $E^i{}_r$ value was higher than 320 in the crushing and screening plant and fine ore bin, suggesting very high risk, and its $E^i{}_r$ value was lower than 320 ($E^i{}_r = 236$) in the concentrate transportation route, indicating high risk; for Pb, its $E^i{}_r$ value was higher than 320 ($E^i{}_r = 415$) in the crushing and screening plant, indicating very high risk, and its $E^i{}_r$ value was lower than 80 in the fine ore bin and transport routes of mine, suggesting moderate risk. The $E^i{}_r$ value for Zn was lower than 40 in the three nodes, which suggested low risk. As can be seen from Figure 4, the pollution made by the three dust-producing nodes was serious, dominated by Sb but depending on location, and different PTEs varied in their individual contribution to the *RI*. The potential ecological risk index of the crushing and screening workshop, fine ore bin, and concentrate transportation route reached >14,000, >17,000 and >15,000, respectively, which means that they are all seriously polluted. The pollution intensity of PTEs in different nodes was: Fine ore bin > concentrate transportation route > crushing and screening workshop.

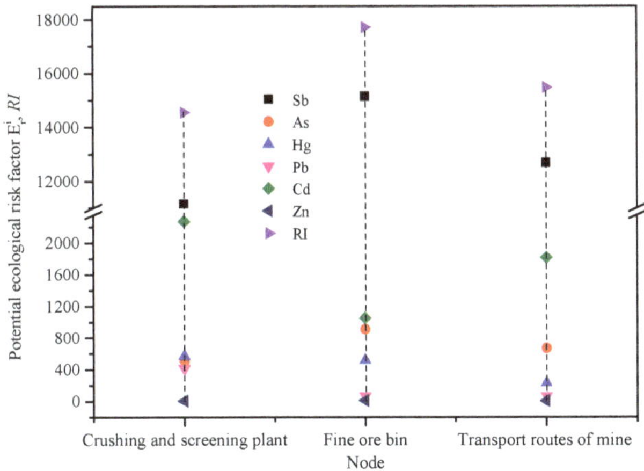

Figure 4. Evaluating results by $E^i{}_r$ and *RI*.

3.3. Pollution Characteristics of PTEs in Tailings

The analysis results of PTEs' content in tailings are shown in Table 3. The results show that the contents of Sb, As, and Zn in tailings were relatively high, with their average contents being 2674.790, 1040.288, and 590.472 mg·kg^{-1}, respectively. The reason may be that the mineral components of the ore were mainly stibnite, accompanied by toxic sand, pyrite, and sphalerite, in which Sb, As, and Zn were most abundant, with significant contribution from Hg, Pb, and Cd.

Table 3. PTEs contents in antimony tailings.

Node	Sample No.	PTEs Content/(mg·kg^{-1})					
		Sb	As	Hg	Pb	Cd	Zn
Tailing dam	S1	2865.591	1082.167	3.273	59.628	2.151	590.492
	S2	2517.722	1141.047	3.017	55.467	2.008	643.057
	S3	2641.058	897.651	2.669	54.235	1.946	537.867
Average		2674.790	1040.288	2.986	56.433	2.035	590.472

The PTEs associated with the tailings can migrate to the surrounding soil through dissolution, leakage, and weathering, resulting in pollution over much wider areas [4,24,41,42]. To understand and more fully evaluate the pollution characteristics of tailings in this area more thoroughly, the PTEs' content in the surface soil around the tailings reservoir was also analyzed in this study. Based on the soil background value of Hunan Province, the pollution index method and potential ecological risk assessment index method were used to evaluate the pollution characteristics (see Tables 4 and 5). Table 4 shows significant enrichment in Sb, As, Hg, Pb, Cd, and Zn in the soil samples. The average contents of Sb, As, Hg, Pb, Cd, and Zn were 1654.462, 105.399, 2.242, 39.311, 1.185, and 523.104 mg·kg^{-1}, respectively. The order of contents in descending order was Sb, Zn, As, Pb, Hg, and Cd, which was similar to the PTEs in the tailings, highlighting an obvious source link.

Table 4. PTEs' contents in soil samples surrounding antimony tailings.

Node	Sample No.	PTEs Content/(mg·kg^{-1})					
		Sb	As	Hg	Pb	Cd	Zn
Soils surrounding the Tailings	S4	1652.147	106.205	2.481	40.028	1.127	499.671
	S5	1834.156	112.055	2.338	38.863	1.326	548.173
	S6	1477.084	97.937	1.907	39.042	1.103	521.467
Average		1654.462	105.399	2.242	39.311	1.185	523.104
Hunan Province soil background values [29]		2.98	14	0.09	27	0.079	95

Table 5. Evaluating results by *PI*, E^i_r, and *RI* for the soils surrounding antimony tailings.

Node	PI						E^i_r						RI
	Sb	As	Hg	Pb	Cd	Zn	Sb	As	Hg	Pb	Cd	Zn	
Soils surrounding the Tailings	555.18	7.52	24.91	1.46	15.00	5.51	10548.42	75.20	996.4	7.30	1500.00	5.51	13132.83

As can be seen from Table 5, the *PI* of Sb had a maximum value of 555.18, higher than the standard [29], and the *PI* of As, Hg, Cd, and Zn also exceeded 5, indicating high level of pollution, while the *PI* of Pb was 1.46, indicating low level of pollution. The E^i_r values of Sb, Hg, and Cd all exceeded 320, indicating very high risk. The E^i_r value of As was 75.20, indicating moderate risk. The E^i_r values of Pb and Zn were less than 40, indicating low risk. The *RI* value of the soil was >13,000, indicating that the soil was at very high risk. Therefore, antimony tailings exerted great influence on the surrounding soils, and should be paid special attention by relevant departments.

3.4. Comparison of the Total Amount of PTEs

The release of PTEs in different stages in the Sb processing plant is shown in Figure 5. The pollution was dominated by Sb, As, and Zn with tailings contributing the most to risk. At the same time, the production process presented variation in contribution from individual PTEs.

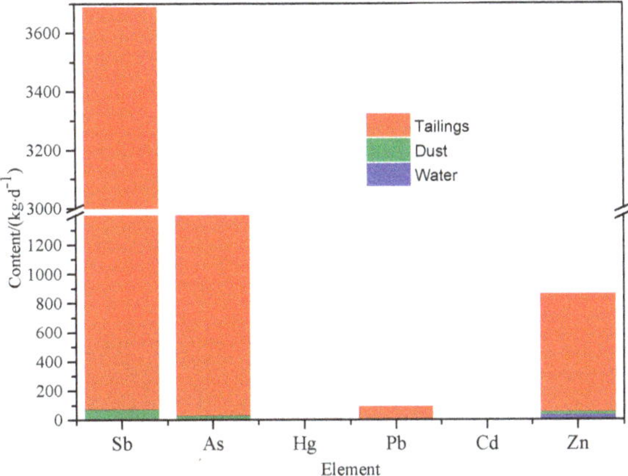

Figure 5. Individual PTEs contents in different waste streams.

This contribution is emphasized in Figure 6, where tailings accounted for 97.00% of the total amount of PTEs in the wastes. The results show that PTEs in tailings played a key role in the pollution degree of PTEs in the whole antimony beneficiation industry. With the increasing awareness of environmental protection and the development and application of integrated recovery and utilization of technology, antimony ore tailings are one of the key links of clean production and nonwaste (or less waste) mining, which is the most effective method to reduce the harm of antimony ore tailings.

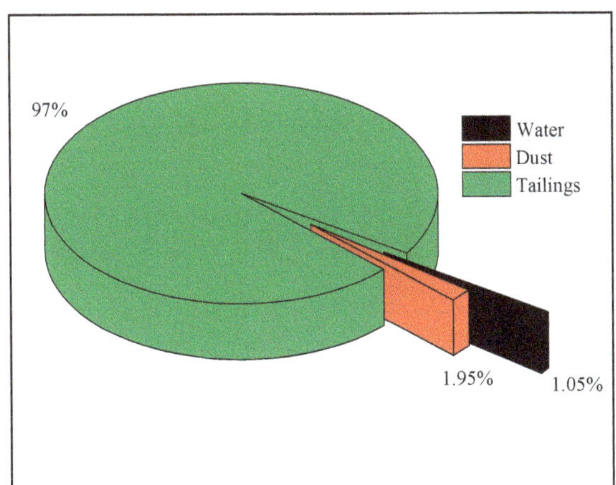

Figure 6. Distribution of PTEs between major site processing wastes.

4. Conclusions

The evaluation of processing wastes in the Sb benefication steps at the XKS site confirmed the significance of the tailings-hosting residues from incomplete ore processing. These materials contribute more widely to environmental contamination as seen in localized soil assessment and agree with findings of other studies on widespread environmental contamination in the wider region. In addition, the significance of dust and wastewater as pollution pathways was confirmed and the magnitude, while lower than for tailings, was still of potential ecological significance. The observation that the partition of PTEs other than Sb between these pathways varies from element to element is of value in considering treatment approach. Control and management of surface tailings with dust suppression and more effective wastewater treatment should be combined in a comprehensive management process to reduce the threat to the wider environment and also to occupationally exposed individuals and inhabitants of the region.

Author Contributions: Conceptualization, S.Z. and A.H.; methodology, S.Z.; software, S.Z. and R.D.; validation, A.H.; formal analysis, S.Z.; investigation, S.Z.; resources, S.Z.; data curation, S.Z. and A.H.; writing—original draft preparation, S.Z.; writing—review and editing, A.H.; visualization, S.Z.; supervision, S.Z.; project administration, S.Z.; funding acquisition, S.Z. and R.D. All authors have read and agreed to the published version of the manuscript.

Funding: This study was financially supported by the National Natural Science Foundation of China (No. 41672350), Chinese Postdoctoral Science Foundation (No. 2018M632961), Doctoral Fund of Hunan University of Science and Technology (No. E57109) and the scientific research project of the Hunan Provincial Education Department (No. 18A184).

Conflicts of Interest: The authors declare that there is no conflict of interest regarding the publication of this paper.

References

1. He, M.C.; Wang, X.Q.; Wu, F.C.; Fu, Z.Y. Antimony pollution in China. *Sci. Total Environ.* **2012**, *41*, 421–422. [CrossRef]

2. Hu, X.Y.; He, M.C.; LI, S.S.; Guo, X.J. The leaching characteristics and changes in the leached layer of antimony-bearing ores from China. *J. Geochem. Explor.* **2017**, *176*, 76–84. [CrossRef]

3. Quina, A.S.; Durão, A.F.; Muñoz, F.; Ventura, J.; Mathias, M.D.L. Population effects of heavy metal pollution in wild Algerian mice (Mus spretus). *Ecotoxicol. Environ. Saf.* **2019**, *171*, 414–424. [CrossRef]

4. Sun, Z.H.; Xie, X.D.; Wang, P.; Hu, Y.N.; Cheng, H.F. Heavy metal pollution caused by small-scale metal ore mining activities: A case study from a polymetallic mine in South China. *Sci. Total Environ.* **2018**, *639*, 217–227. [CrossRef]

5. Zhang, X.; Yang, H.H.; Cui, Z.J. Evaluation and analysis of soil migration and distribution characteristics of heavy metals in iron tailings. *J. Clean. Prod.* **2018**, *172*, 475–480. [CrossRef]

6. Deng, R.J.; Jin, C.S.; Ren, B.Z.; Hou, B.L.; Hursthous, A. The potential for treatment of antimony-containing wastewater by iron-base adsorbents. *Water* **2017**, *9*, 794. [CrossRef]

7. Zhou, J.W.; Nyirenda, M.T.; Xie, L.N.; Li, Y.; Zhou, B.L.; Zhu, Y.; Liu, H.L. Mine waste acidic potential and distribution of antimony and arsenic in waters of the Xikuangshan mine, China. *Appl. Geochem.* **2017**, *77*, 52–61. [CrossRef]

8. Qi, C.C.; Wu, F.C.; Deng, Q.J.; Liu, G.J.; Mo, C.L.; Liu, B.J.; Zhu, J. Distribution and accumulation of antimony in plants in the super-large Sb deposit areas, China. *Microchem. J.* **2011**, *97*, 44–51. [CrossRef]

9. Wei, C.Y.; Ge, Z.F.; Chu, W.S.; Feng, R.W. Speciation of antimony and arsenic in the soils and plants in an old antimony mine. *Environ. Exp. Bot.* **2015**, *109*, 31–39. [CrossRef]

10. Tan, D.; Long, J.M.; Li, B.Y.; Ding, D.; Du, H.H.; Lei, M. Fraction and mobility of antimony and arsenic in three polluted soils: A comparison of single extraction and sequential extraction. *Chemosphere* **2018**, *213*, 533–540. [CrossRef]

11. Okkenhaug, G.; Zhu, Y.G.; Luo, L.; Lei, M.; Li, X.; Mulder, J. Distribution, speciation and availability of antimony (Sb) in soils and terrestrial plants from an active Sb mining area. *Environ. Pollut.* **2011**, *159*, 2427–2434. [CrossRef] [PubMed]

12. Li, J.N.; Wei, Y.; Zhao, L.; Zhang, J.; Shangguan, Y.X.; Li, F.S.; Hou, H. Bioaccessibility of antimony and arsenic in highly polluted soils of the mine area and health risk assessment associated with oral ingestion exposure. *Ecotoxicol. Environ. Saf.* **2014**, *110*, 308–315. [CrossRef]

13. Guo, W.J.; Fu, Z.Y.; Wang, H.; Song, F.H.; Wu, F.C.; Gisey, J.P. Environmental geochemical and spatial/temporal behavior of total and speciation of antimony in typical contaminated aquatic environment from Xikuangshan, China. *Microchem. J.* **2018**, *137*, 181–189. [CrossRef]

14. Wang, X.Q.; Li, F.B.; Yuan, C.L.; Li, B.; Liu, T.X.; Liu, C.S.; Du, Y.H.; Liu, C.P. The translocation of antimony in soil-rice system with comparisons to arsenic: Alleviation of their accumulation in rice by simultaneous use of Fe(II) and NO_3^-. *Sci. Total Environ.* **2019**, *650*, 633–641. [CrossRef]

15. Basnet, P.; Amarasiriwardena, D.; Wu, F.C.; Fu, Z.Y.; Zhang, T. Elemental bioimaging of tissue level trace metal distributions in rice seeds (*Oryza sativa* L.) from a mining area in China. *Environ. Pollut.* **2014**, *195*, 148–156. [CrossRef]

16. Zeng, D.F.; Zhou, S.J.; Ren, B.Z.; Chen, T.S. Bioaccumulation of antimony and arsenic in vegetables and health risk assessment in the superlarge antimony-mining area, China. *J. Anal. Methods Chem.* **2015**, *2015*, 909724. [CrossRef]

17. Kuroda, K.; Endo, G.; Okamoto, A.; Yoo, Y.; Horiguchi, S. Genotoxicity of beryllium, gallium and antimony in short-term assays. *Mutat. Res.* **1991**, *264*, 163–170. [CrossRef]

18. Schnorr, T.M.; Steenland, K.; Thun, M.J.; Rinsky, R.A. Mortality in a cohort of antimony smelter workers. *Am. J. Ind. Med.* **1995**, *27*, 759–770. [CrossRef]

19. Nigra, A.E.; Ruiz-Hernandez, A.; Redon, J.; Navas-Acien, A.; Tellez-Plaza, M. Environmental metals and cardiovascular disease in adults: A systematic review beyond lead and cadmium. *Curr. Environ. Health Rep.* **2016**, *3*, 416–433. [CrossRef]

20. Parvez, F.; Chen, Y.; Brandt-Rauf, P.W.; Bernard, A.; Dumont, X.; Slavkovich, V.; Argos, M.; D'Armiento, J.; Foronjy, R.; Hasan, M.R.; et al. Nonmalignant respiratory effects of chronic arsenic exposure from drinking water among never-smokers in Bangladesh. *Environ. Health Perspect.* **2008**, *116*, 190–195. [CrossRef]

21. Nordberg, G.; Fowler, B.; Nordberg, M. *Handbook on the Toxicology of Metals*, 4th ed.; Academic Press: San Diego, CA, USA, 2015.

22. USEPA. *Water Related Fate of the 129 Priority Pollutants*; USEPA: Washington, DC, USA, 1979.

23. European Council. Directive 76/464/EEC—Water Pollution by Discharges of Certain Dangerous Substance discharged into the aquatic environment of the community. *OJ L* **1976**, *129*, 23–29.

24. Tang, Z.E.; Deng, R.J.; Zhang, J.; Ren, B.Z.; Hursthouse, A.S. Regional distribution characteristics and ecological risk assessment of heavy metal pollution of different land use in an antimony mining area-Xikuangshan, China. *Hum. Ecol. Risk Assess.* **2019**. [CrossRef]

25. Zhou, S.J.; Hursthouse, A.S.; Chen, T.S. Pollution characteristics of Sb, As, Hg, Pb, Cd, and Zn in soils from different zones of Xikuangshan antimony mine. *J. Anal. Methods Chem.* **2019**, *2019*, 2754385. [CrossRef] [PubMed]

26. Islam, M.S.; Ahmed, M.K.; Raknuzzaman, M.; Habibullah-al-mamun, M.; Kundu, G.K. Heavy metals in the industrial sludge and their ecological risk: A case study for a developing country. *J. Geochem. Explor.* **2017**, *172*, 41–49. [CrossRef]

27. China State Environmental Protection Administration. *Environmental Quality Standards for Surface Water*; CSEPA: Beijing, China, 2002. (In Chinese)

28. Ungureanu, G.; Santos, S.; Boaventura, R.; Botelho, C.M.S. Arsenic and antimony in water and wastewater: Overview of removal techniques with special reference to latest advances in adsorption. *J. Environ. Manag.* **2015**, *151*, 326–342. [CrossRef]

29. Pan, Y.; Yang, G. *Background Value of Soils in Hunan and Their Investigation Methods*; China Environmental Science Press: Beijing, China, 1988. (In Chinese)

30. He, J.Y.; Yang, Y.; Christakos, G.; Liu, Y.J.; Yang, X. Assessment of soil heavy metal pollution using stochastic site indicators. *Geoderma* **2019**, *337*, 359–367. [CrossRef]

31. Häkanson, L. An ecological risk index for aquatic pollution control. A sedimentological approach. *Water Res.* **1980**, *14*, 975–986. [CrossRef]

32. Huang, X.F.; Hu, J.W.; Li, C.X.; Deng, J.J.; Long, J.; Qin, F.X. Heavy-metal pollution and potential ecological risk assessment of sediments from Baihua Lake, Guizhou, P.R. China. *Int. J. Environ. Health Res.* **2009**, *19*, 405–419. [CrossRef]

33. Wang, N.N.; Wang, A.H.; Kong, L.H.; He, M.C. Calculation and application of Sb toxicity coefficient for potential ecological risk assessment. *Sci. Total Environ.* **2018**, *610–611*, 167–174. [CrossRef]

34. Sun, J.X.; Luo, T.; Zhou, F.; Zou, W.M.; Yan, J.L. Pollution characteristics and assessment of heavy metals in surface water of Yancheng coastal area in Jiangsu. *Environ. Pollut. Control* **2018**, *40*, 1294–1299. (In Chinese)

35. Sun, Q.Z.; Zang, S.Y. Pollution evalution and forecast of heavy metal in lake of Zhalong Wetland, China. *J. Agro-Environ. Sci.* **2012**, *31*, 2242–2248. (In Chinese)

36. Entwistle, J.A.; Hursthouse, A.S.; Marinho Reis, P.A.; Steward, A.G. Matalliferous mine dust: Human health impacts and potential determinants of disease in mining communities. *Curr. Pollut. Rep.* **2019**, *5*, 67–83. [CrossRef]

37. Zhang, J.Q.; Bai, S.Q.; Du, B.Q. Heavy metal pollution of dust topsoil and roadside tree near by main city roadways. *J. Southwest Jiaotong Univ.* **2006**, *41*, 68–73. (In Chinese)

38. Lu, X.W.; Pan, H.Y.; Wang, Y.W. Pollution evaluation and source analysis of heavy metal in roadway dust from a resource-typed industrial city in Northwest China. *Atmos. Pollut. Res.* **2017**, *8*, 587–595. [CrossRef]

39. Shi, D.Q.; Lu, X.W. Accumulation degree and source apportionment of trace metals in smaller than 63 μm road dust from the areas with different land uses: A case study of Xi'an, China. *Sci. Total Environ.* **2018**, *636*, 1211–1218. [CrossRef] [PubMed]

40. Lanzerstorfer, C. Heavy metals in the finest size fractions of road-deposited sediments. *Environ. Pollut.* **2018**, *239*, 522–531. [CrossRef]

41. Nikolaidis, C.; Zafiriadis, I.; Mathioudakis, V.; Constantinidis, T. Heavy Metal Pollution Associated with an Abandoned Lead-zinc Mine in the Kirki Region, NE Greece. *Bull. Environ. Contam. Toxicol.* **2010**, *85*, 307–312. [CrossRef]

42. Garcí-carmona, M.; Garcí-robles, H.; Turpí torrano, C.; Fernández ondoño, E.; Lorite moreno, J.; Sierra aragón, M.; Martín peinado, F.J. Residual pollution and vegetation distribution in amended soils 20 years after a pyrite mine tailings spill (Aznalcólar, Spain). *Sci. Total Environ.* **2019**, *650*, 933–940. [CrossRef]

Article

Porous Aromatic Melamine Schiff Bases as Highly Efficient Media for Carbon Dioxide Storage

Raghad M. Omer [1], Emaad T. B. Al-Tikrity [1], Gamal A. El-Hiti [2,*], Mohammed F. Alotibi [3,*], Dina S. Ahmed [4] and Emad Yousif [5,*]

[1] Department of Chemistry, College of Science, Tikrit University, Tikrit 34001, Iraq; rm90618@gmail.com (R.M.O.); emaad1954@tu.edu.iq (E.T.B.A.-T.)
[2] Cornea Research Chair, Department of Optometry, College of Applied Medical Sciences, King Saud University, P.O. Box 10219, Riyadh 11433, Saudi Arabia
[3] National Center for Petrochemicals Technology, King Abdulaziz City for Science and Technology, P.O. Box 6086, Riyadh 11442, Saudi Arabia
[4] Department of Medical Instrumentation Engineering, Al-Mansour University College, Baghdad 64021, Iraq; dinasaadi86@gmail.com
[5] Department of Chemistry, College of Science, Al-Nahrain University, Baghdad 64021, Iraq
* Correspondence: gelhiti@ksu.edu.sa (G.A.E.-H.); mfalotaibi@kacst.edu.sa (M.F.A.); emad_yousif@hotmail.com (E.Y.); Tel.: +966-11469-3778 (G.A.E.-H.); Fax: +966-11469-3536 (G.A.E.-H.)

Received: 28 November 2019; Accepted: 19 December 2019; Published: 20 December 2019

Abstract: High energy demand has led to excessive fuel consumption and high-concentration CO_2 production. CO_2 release causes serious environmental problems such as the rise in the Earth's temperature, leading to global warming. Thus, chemical industries are under severe pressure to provide a solution to the problems associated with fuel consumption and to reduce CO_2 emission at the source. To this effect, herein, four highly porous aromatic Schiff bases derived from melamine were investigated as potential media for CO_2 capture. Since these Schiff bases are highly aromatic, porous, and have a high content of heteroatoms (nitrogen and oxygen), they can serve as CO_2 storage media. The surface morphology of the Schiff bases was investigated through field emission scanning electron microscopy, and their physical properties were determined by gas adsorption experiments. The Schiff bases had a pore volume of 0.005–0.036 cm^3/g, an average pore diameter of 1.69–3.363 nm, and a small Brunauer–Emmett–Teller surface area (5.2–11.6 m^2/g). The Schiff bases showed remarkable CO_2 uptake (up to 2.33 mmol/g; 10.0 wt%) at 323 K and 40 bars. The Schiff base containing the 4-nitrophenyl substituent was the most efficient medium for CO_2 adsorption and, therefore, can be used as a gas sorbent.

Keywords: porosity properties; adsorption capacity; carbon dioxide storage; melamine Schiff bases; surface area; energy

---

1. Introduction

Fuel consumption has been increasing over the years to meet the high demand for energy required for various human activities. Therefore, CO_2 concentration has increased drastically to unprecedented levels in the atmosphere [1]. Fossil fuel combustion is the main contributor (60%) to the increased CO_2 concentration level in the environment [2]. Chemical, agro, power, and pharmaceutical industries contribute to approximately 70% greenhouse gas emission, which primarily causes climate changes and global warming [3,4]. The rise of sea and ocean levels, increased acidity of water, and drastic global weather changes are the main environmental problems associated with the increased CO_2 emission, which will consequently lead to economic collapse [5,6]. Additionally, the CO_2 levels in the environment cannot be lowered rapidly due to the large-scale and high consumption of fuels, which are difficult to be reduced. Novel strategies must be developed to not only resolve the environmental problems arising from global warming but also reduce carbon emission at the source. In addition, it is important to devise new technologies and design novel materials that can be used as a media to capture CO2 effectively [7–10].

Various technologies have been developed to capture and store CO_2 that can efficiently reduce its atmospheric level [11–16]. Recently, researchers from both academia and industry directed their attention toward the capture and storage of CO_2 [17–19]. Various chemical absorbents have been used as media for CO_2 capture, in which amines (e.g., ethanolamine) are the most common ones [20]. The use of amines involves a simple process; however, it is limited because of high operational cost, energy requirement, and the use of very volatile chemicals [21]. Therefore, other techniques that involve the use of adsorbents were developed. Such materials reportedly exhibit adsorption capacity of >4.4% by weight, long life duration, recyclability, and reusability [22–24].

Chemical adsorption of CO_2 is a simple as well as cost and energy effective process. Metal-based adsorbents such as metal oxides are known as common capture media for CO_2 because of their basic and ionic nature [25]. For example, calcium and magnesium oxides can adsorb CO_2 stoichiometrically to produce the corresponding metal carbonate through an exothermic reaction [26]. However, the adsorption capacity of materials varies on the basis of kinetic factors [25]. The adsorption capacity of calcium oxide is limited but sufficiently high to facilitate its use as an effective medium for CO_2 capture. Several other materials such as ionic liquids in a solid matrix [27], zeolites [28], silica [29], and those containing activated carbons [30–32] have been evaluated as CO_2 sorbents. Some of these materials possess unique thermal properties, high chemical stability, high surface area, tunable chemical structures, recyclability, and reusability. However, zeolites are not suitable for CO_2 capture from flue gases because of their excellent hydrophilic properties [33]. In addition, materials containing activated carbon exhibit poor selectivity [34].

Activated carbon has been prepared from different materials such as polymers, resins, and biomass and can be used as an efficient adsorbent for CO_2 [30]. Various chemical and physical processes have been conducted to activate and modify the surface area and pore volume of such adsorbents to increase their capacity for CO_2 capture. The chemical process of activation requires the use of a base, while the physical one requires an appropriate carbonization gas [35,36]. The adsorption capacity of activated carbon depends on the distribution of the chemical activator within the matrix. Polyacrylonitrile in the presence of a base (e.g., potassium hydroxide; KOH) was used as an effective medium to capture CO_2 and exhibited good CO_2 uptake at 25 °C and under 1 bar [31]. The CO_2 uptake was even higher for the resorcinol–formaldehyde resin at the same temperature and pressure in the presence of potassium carbonate as an activator [37].

Metal–organic frameworks (MOFs), synthesized from different molecular building units, have been investigated as adsorbents for CO_2 because of their extended surface area [38–40]. The interaction between MOFs and CO_2 is strong because it occurs through hydrogen bonding and requires a low heat of adsorption, similar to that observed for zeolites [25]. The CO_2 storage capacity of MOFs can be enhanced through the addition of polar residues within their surfaces [41]. Porous-organic polymers (POPs) are highly stable chemically and thermally as well as have low density, tunable structure with a desirable surface area, and different functional groups; therefore, they act as good adsorbents for CO_2 [33]. The presence of heteroatoms (e.g., nitrogen, oxygen, sulfur, phosphorus) within the skeleton of POPs enhances CO_2 capture capacity [33]. The surface polarity of POPs can be increased by the addition of organic moieties containing polar groups or inorganic ions, which facilitates the strong interaction between CO_2 and adsorbent materials [33]. Various POPs showed good CO_2 capture capacity; however, the use of metals in the synthesis of POPs produces toxic pollutants. More research is still needed to optimize the synthetic procedures for POP production by employing simple and effective processes [42].

Nitrogen-rich heterocycles such as triazines have potential use in supramolecular applications because they interact with many chemicals through π–π interactions, hydrogen bond formation, and chelation [43]. Melamine has a high nitrogen content (66% by weight) and has been used in various applications such as the production of raw materials with high nitrogen content, plastic, medicinal products, metal-free catalysts, and CO_2 adsorbents [44–46]. Melamine Schiff bases can be easily synthesized through the reaction of melamine and aromatic carbonyl compounds in the presence of a catalyst. Recently, we have synthesized various Schiff bases and investigated their use as additives to stabilize polymeric films against irradiation [47–53]. Melamine Schiff bases have all the qualities needed for their use as efficient adsorbents for CO_2. In this study, we report the use of melamine Schiff bases, which are highly aromatic and porous, as an efficient media for the capture of CO_2 at 40 bars and 323 K.

2. Materials and Methods

2.1. Materials

Chemicals, reagents, and solvents were purchased from Merck (Schnelldorf, Germany) and were used as received.

2.2. Physiochemical Measurements

The surface morphology of Schiff bases was observed through field emission scanning electron microscopy (FESEM, TESCAN MIRA3, Kohoutovice, Czech Republic) at an accelerating voltage of 10 kV. The N_2 adsorption–desorption isotherms of the Schiff bases were recorded on a Quantchrome chemisorption analyzer (Quantachrome Instruments, Boynton Beach, FL, USA) at 77 k. The Schiff bases **1–4** were degassed in a vacuum oven for a long period (6 h) at a high temperature (100 °C) under a flow of N_2 gas (Cascade TEK, Cornelius, OR, USA) to ensure the removal of any residues or small molecules such as water from the pores of materials. The surface area of the Schiff bases was calculated using the Brunauer–Emmett–Teller (BET) equation at a relative pressure ($P/P°$) of 0.98. The pore size of the Schiff bases was verified by the Barrett–Joyner–Halenda (BJH) method. The CO_2 uptake was measured at 40 bars and 323 K using the H-sorb 2600 high-pressure volumetric adsorption analyzer (Gold APP Instruments Corp., Beijing, China), which has two degassing and analyzing ports that can be simultaneously operated. The experiment of CO_2 storage was repeated for at least 10 times for pressure optimization. A known quantity of gas was injected into a measurement tube that contained the Schiff base sample until an equilibrium between the adsorbed gas and the Schiff base sample was established. The final equilibrium pressure was recorded automatically using a software program and the adsorbed quantity of gas was calculated from the obtained data.

2.3. Synthesis of Schiff Bases **1–4**

Schiff bases **1–4** were synthesized using a reported procedure by the condensation of melamine and 3 molar equivalents of aromatic aldehydes; 4-nitrobenzaldehyde, 2-hydroxybenzaldehyde, 3-hydroxybenzaldehyde, and 4-hydroxybenzaldehyde, in boiling dimethylformamide containing acetic acid as a catalyst under reflux for 6 h [47].

3. Results

3.1. Synthesis of Schiff Bases **1–4**

The spectroscopic data from the ^{1}H-NMR and FT-IR spectra, elemental analysis results, and physical properties (e.g., melting points and colors) of the synthesized Schiff bases **1–4** were identical to those of the previously reported bases [47]. Figure 1 represents the chemical structures of the synthesized Schiff bases **1–4**. Schiff base 1 contained a nitro group, while Schiff bases 2–4 contained a hydroxyl group with different arrangements (*ortho*, *meta*, and *para*).

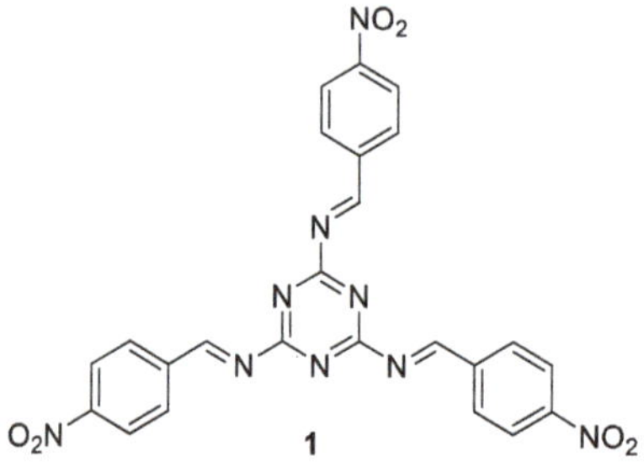

Figure 1. Schiff bases 1–4 [47].

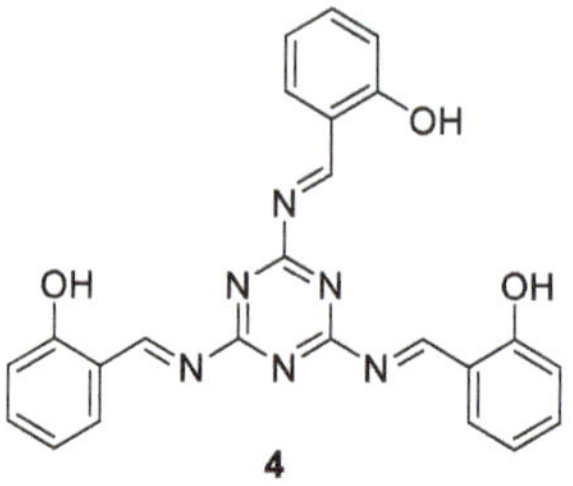

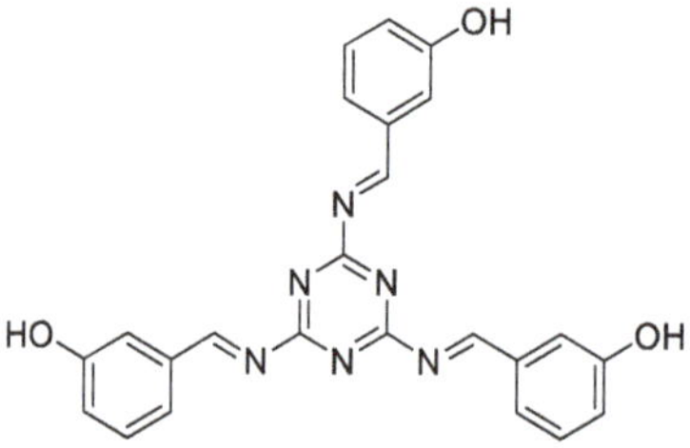

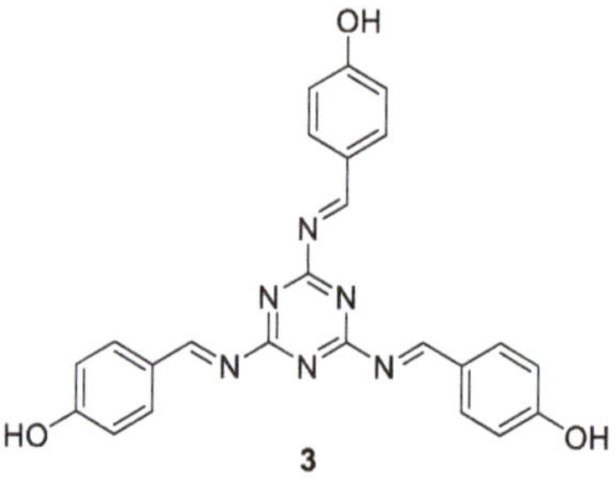

3.2. FESEM of Schiff Bases 1–4

The morphologies of Schiff bases **1–4** were investigated through FESEM. Figures 2–5 show that Schiff bases **1–4** had a relatively uniform and amorphous surface with micro-size particles. The pore dimensions of the Schiff base samples varied and were found to be 20–392 nm. It was clear that the particle size of Schiff base **1** (Figure 2) was smaller than those of Schiff bases **2–4**. Schiff bases **2** and **4** (Figures 3–5) have the largest pore dimensions. Schiff base **1** had a different morphology compared to the other Schiff bases because it contains a nitro group, which causes more noticeable irregularity in particle size and shape and, thus, a highly porous structure. The presence of the functional group that had a high content of nitrogen (nitro group) could improve not only the porosity but also the surface area and efficiency for CO_2 uptake [33].

Figure 2. Field emission scanning electron microscopy (FESEM) image of Schiff base **1**.

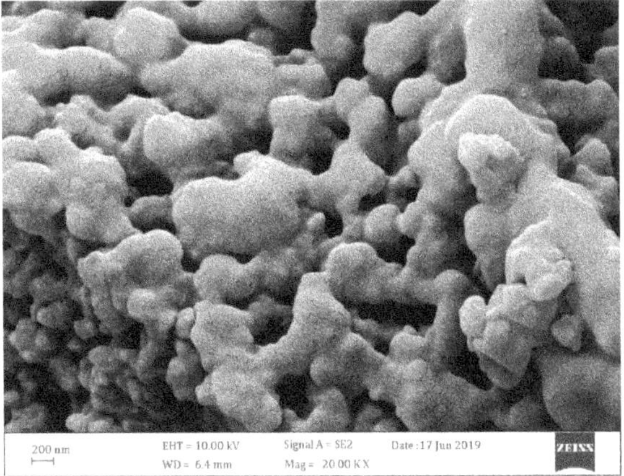

Figure 3. FESEM image of Schiff base **2**.

Figure 4. FESEM image of Schiff base **3**.

Figure 5. FESEM image of Schiff base **4**.

The pore dimensions of Schiff bases **1–4** were smaller than those reported for some POPs and larger than those for telmisartan tin complexes [54–56]. For example, POPs containing polyphosphates derived from 1,4-diaminobenzene showed irregular and porous structures with pore dimensions of 49–981 nm [54]. In addition, polyphosphates derived from benzidine showed porous structures with pore dimensions of 28–806 nm [55]. In contrast, the pore dimensions of telmisartan tin complexes ranged from 20 to 51 nm.

3.3. N_2 Adsorption–Desorption of Schiff Bases 1–4

The N_2 adsorption–desorption measurements for the Schiff bases **1–4** were conducted at 77 K. The N_2 isotherms and pore sizes and volumes of Schiff bases **1–4** are represented in Figures 6–9. The shape of the N_2 isotherm for **1** was similar to the type IV isotherm. Schiff bases **2–4** showed N_2 sorption isotherms that are almost identical to the type III isotherm, in which monolayer formation was not identified.

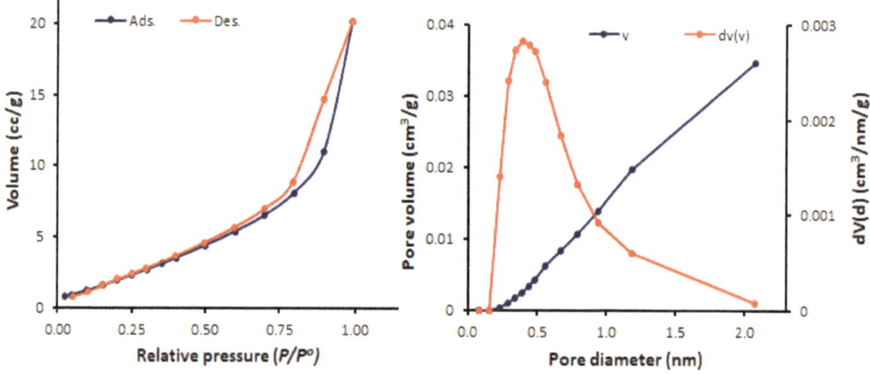

Figure 6. N_2 isotherms and pore size and volume for Schiff base **1**.

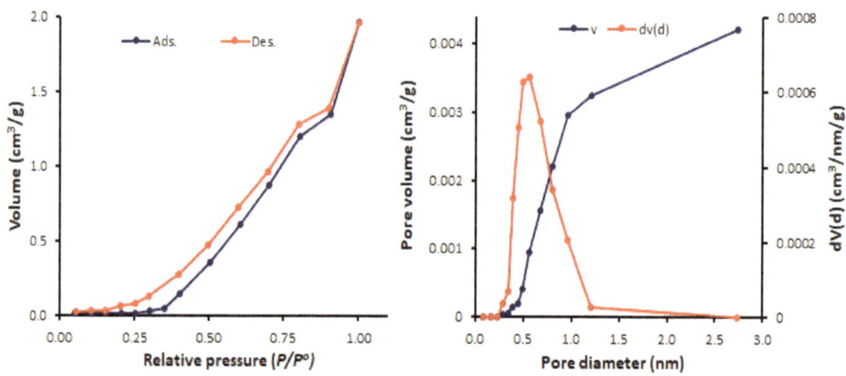

Figure 7. N_2 isotherms and pore size and volume for Schiff base **2**.

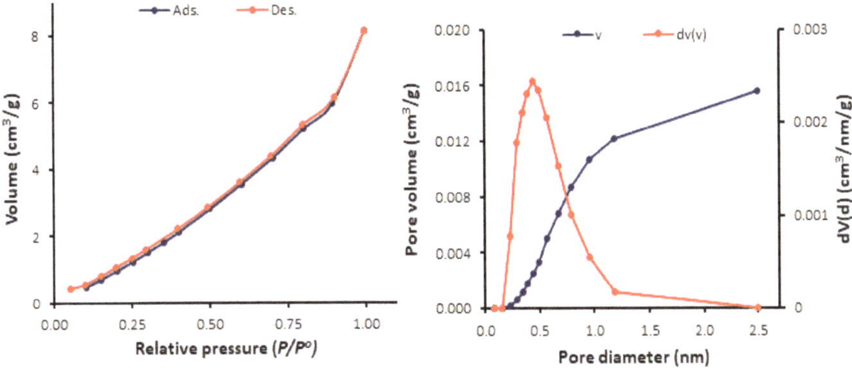

Figure 8. N_2 isotherms and pore size and volume for Schiff base **3**.

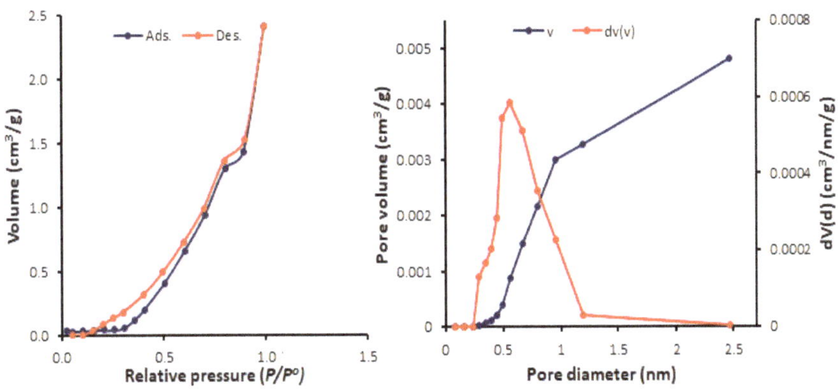

Figure 9. N_2 isotherms and pore size and volume for Schiff base **4**.

The BET surface area (S_{BET}), pore volumes, and average pore diameters of Schiff bases **1–4** were calculated (Table 1). Among the synthesized Schiff bases, **1** (containing a nitro group) exhibited the highest surface area (S_{BET} = 11.6 m^2/g) and total pore volume (0.036 cm^3/gm), but the lowest pore diameter (1.69 nm). Schiff base **1** had a mesoporous structure, while, **2–4** (containing a hydroxy group at *ortho-*, *meta-* and *para-*position of the aryl ring) had microporous structures (pore diameter = 2.44–3.63 nm). Some POPs and tin complexes showed porous structures with similar pore diameters. For example, porous polyphosphates derived from either 1,4-diaminobenzene or benzidine exhibited a pore diameter of 1.96–2.43 nm [54] or 2.43–2.86 nm [55], respectively, compared to that of 2.43 nm for telmisartan tin complexes [56].

Table 1. Porosity properties of **1–4**.

Schiff Base	S_{BET} (m^2/g)	Pore Volume (cm^3/g)	Average Pore Diameter (nm)
1	11.6	0.036	1.69
2	5.2	0.004	3.63
3	10.2	0.016	2.44
4	8.5	0.005	3.62

A gravimetric technique was used to detect the gas uptake quantity and, therefore, determine the gas adsorption isotherm [57]. In addition, the gas quantity that has been removed from the gas phase was used to estimate the physisorption isotherms of the gas. The desorption or adsorption branch of the isotherm can be used to calculate the pore size distribution. The CO_2 sorption isotherms for Schiff bases **1–4** are shown in Figures 10–13 and their CO_2 uptake are reported in Table 2.

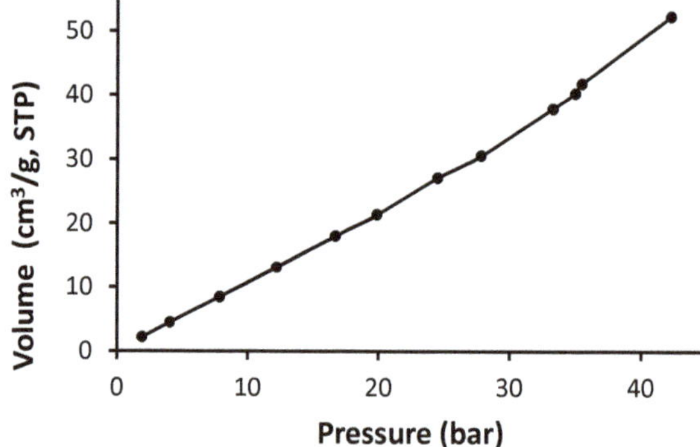

Figure 10. Adsorption isotherm of CO_2 for Schiff base **1**.

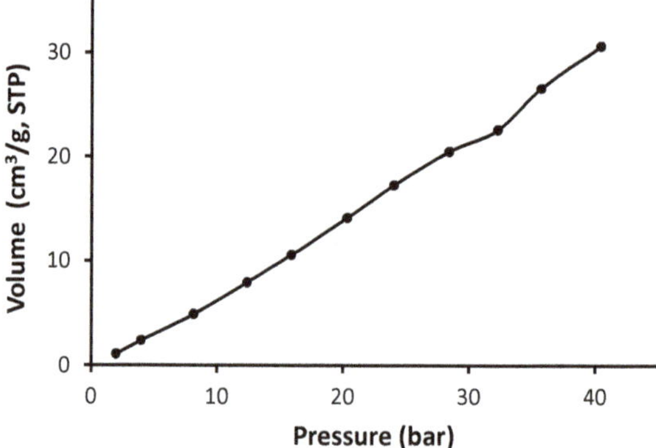

Figure 11. Adsorption isotherm of CO_2 for Schiff base **2**.

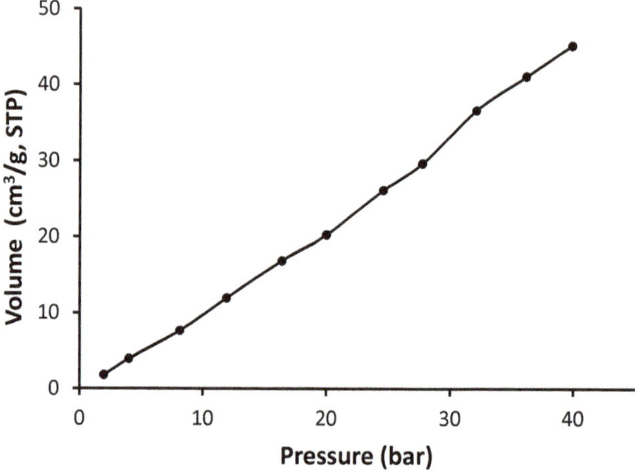

Figure 12. Adsorption isotherm of CO_2 for Schiff base **3**.

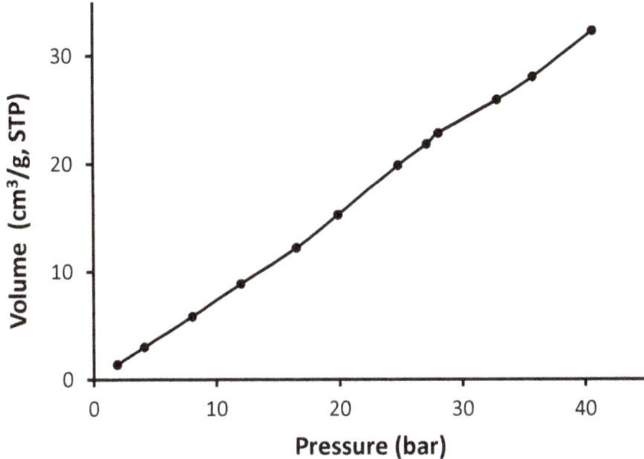

Figure 13. Adsorption isotherm of CO_2 for Schiff base **4**.

Table 2. CO_2 adsorption capacity of Schiff bases **1–4** at 323 K and 40 bars.

Schiff Base	CO_2 Uptake (cm^3/g)	CO_2 Uptake (mmol/g)	CO_2 Uptake (wt%)
1	52.29	2.33	10.0
2	30.64	1.36	6.1
3	45.15	2.01	9.0
4	32.27	1.44	6.4

As seen in Figures 10–13, Schiff bases **1–4** do not have an apparent adsorption–desorption hysteresis, which indicates the reversible adsorption of CO_2 within the Schiff base pores at the temperature and pressure used (323 K and 40 bars). The CO_2 uptake for Schiff bases **1–4** was high (6.1–10.0 wt%), possibly because of the excellent pore diameter and the strong van der Waals interactions and hydrogen bonding between the Schiff bases and CO_2. In addition, Schiff bases **1–4** contain strong Lewis base sites that aid the capture of CO_2. Indeed, porous materials containing heteroatoms such as oxygen, nitrogen, and phosphorous can selectively capture CO_2 over methane and nitrogen gases [54–56].

The surface area for the Schiff bases was relatively low (5.2–11.6 m^2/g); however, they showed remarkable CO_2 uptake (1.36–2.33 mmol/g; 6.1–10.0 wt%). Similar observations have been previously reported at similar temperature and pressure. For example, porous polyphosphates containing benzidine showed low surface area (27.5–30.0 m^2/g) and high CO_2 uptake (up to 14.0 wt%) [55]. On the other hand, polyphosphates containing 1,4-diaminobenzene exhibited high surface area (82.7–213.5 m^2/g), but the CO_2 uptake was limited to 0.6 wt% [54]. Telmisartan tin complexes showed surface area of 32.4–130.4 m^2/g and up to 7.1 wt% CO_2 uptake [56]. Materials with the highest surface area showed the most effective CO_2 uptake. Polyacrylonitrile carbon fibers in the presence of a base provided a CO_2 uptake of 2.74 mmol/g at room temperature and normal pressure [31]. In contrast, porous nanocarbons with a high surface area (1114 m^2/g) in the presence of potassium oxalate and ethylenediamine provided a CO_2 uptake as 4.60 mmol/g at a similar temperature and pressure [30]. Porous nanocarbons with a small surface area (439 m^2/g) provided a low CO_2 uptake (1.94 mmol/g [30]. Ionic liquids in a silica matrix led to materials having a very small surface area (1–9 m^2/g) and relatively poor sorption capacity towards CO_2 as 0.35 g of CO_2 per g of adsorbent [27].

4. Conclusions

Four melamine Schiff bases have been investigated as potential media for CO_2 storage at 323 K and 40 bars. These Schiff bases have a relatively low surface area (S_{BET} = 5.2–11.6 m^2/g) and varied porous structures, showing pore volumes of 0.004–0.036 cm^3/g and diameters of 1.69–2.63 nm. The Schiff bases showed remarkable CO_2 uptake (6.1–10.0 wt%), possibly because of their high aromaticity and heteroatom contents. The Schiff base containing a nitro group showed the most effective CO_2 uptake (10.0 wt%) owing to the high content of nitrogen (heteroatom) within the porous material. The Schiff bases containing a hydroxy group have a lower surface area and pore volume, but higher pore diameter compared to the one containing a nitro group. Such Schiff base is inexpensive and easily producible in high yield and, therefore, can be used at an industrial scale.

Author Contributions: Conceptualization and experimental design: G.A.E.-H., M.F.A., D.S.A., and E.Y.; Experimental work and data analysis: R.M.O. and E.T.B.A.-T.; writing: G.A.E.-H., D.S.A. and E.Y. All authors discussed the results and have approved the final version of the paper. All authors have read and agreed to the published version of the manuscript.

Funding: The authors are grateful to the Deanship of Scientific Research, King Saud University for funding through Vice Deanship of Scientific Research Chairs.

Acknowledgments: We thank Al-Nahrain and Al-Mansour Universities for the technical support.

Conflicts of Interest: The authors declare that they have no conflict of interest.

References

1. Mardani, A.; Streimikiene, D.; Cavallaro, F.; Loganathan, N.; Khoshnoudi, M. Carbon dioxide (CO_2) emissions and economic growth: A systematic review of two decades of research from 1995 to 2017. *Sci. Total Environ.* **2019**, *649*, 31–49. [CrossRef] [PubMed]

2. Yaumi, A.L.; Bakar, M.Z.A.; Hameed, B.H. Recent advances in functionalized composite solid materials for carbon dioxide capture. *Energy* **2017**, *124*, 461–480. [CrossRef]

3. Boamah, K.B.; Du, J.; Bediako, I.A.; Boamah, A.J.; Abdul-Rasheed, A.A.; Owusu, S.M. Carbon dioxide emission and economic growth of China—The role of international trade. *Environ. Sci. Pollut. Res.* **2017**, *24*, 13049. [CrossRef] [PubMed]

4. Sun, H.; Xin, Q.; Ma, Z.; Lan, S. Effects of plant diversity on carbon dioxide emissions and carbon removal in laboratory-scale constructed wetland. *Environ. Sci. Pollut. Res.* **2019**, *26*, 5076. [CrossRef] [PubMed]

5. Sanz-Perez, E.S.; Murdock, C.R.; Didas, S.A.; Jones, C.W. Direct capture of CO_2 from ambient air. *Chem. Rev.* **2016**, *116*, 11840–11876. [CrossRef]

6. Intergovernmental Panel on Climate Change. Climate Change. Climate Change 2007: Synthesis Report. In *Contribution of Working Groups I, II and III to the Fourth Assessment Report of the Intergovernmental Panel on Climate Change*; Core Writing Team, Pachauri, R.K., Reisinger, A., Eds.; Intergovernmental Panel on Climate Change: Geneva, Switzerland, 2008; p. 104.

7. Okesola, A.A.; Oyedeji, A.A.; Abdulhamid, A.F.; Olowo, J.; Ayodele, B.E.; Alabi, T.W. Direct air capture: A review of carbon dioxide capture from the air. *Mater. Sci. Eng.* **2018**, *413*, 12077. [CrossRef]

8. Shukla, S.K.; Khokarale, S.G.; Bui, T.Q.; Mikkola, J.-P.T. Ionic liquids: Potential materials for carbon dioxide capture and utilization. *Front. Mater.* **2019**, *6*, 42. [CrossRef]

9. Mukherjee, A.; Okolie, J.A.; Abdelrasoul, A.; Niu, C.; Dalai, A.K. Review of post-combustion carbon dioxide capture technologies using activated carbon. *J. Environ. Sci.* **2019**, *83*, 46–63. [CrossRef]

10. Goh, K.; Karahan, H.E.; Yang, E.; Bae, T.-H. Graphene-based membranes for CO_2/CH_4 separation: Key challenges and perspectives. *Appl. Sci.* **2019**, *9*, 2784. [CrossRef]

11. Variny, M.; Jediná, D.; Kizek, J.; Illés, P.; Lukáč, L.; Janošovský, J.; Lesný, M. An investigation of the techno-economic and environmental aspects of process heat source change in a refinery. *Processes* **2019**, *7*, 776. [CrossRef]

12. Shukrullah, S.; Naz, M.Y.; Mohamed, N.M.; Ibrahim, K.I.; Abd El-Salam, N.M.; Ghaffar, A. CVD synthesis, functionalization and CO_2 adsorption attributes of multiwalled carbon nanotubes. *Processes* **2019**, *7*, 634. [CrossRef]

13. Osman, A.; Eltayeb, M.; Rajab, F. Utility paths combination in HEN for energy saving and CO_2 emission reduction. *Processes* **2019**, *7*, 425. [CrossRef]

14. Kelektsoglou, K. Carbon capture and storage: A review of mineral storage of CO_2 in Greece. *Sustainability* **2018**, *10*, 4400. [CrossRef]

15. Aminua, M.D.; Nabavia, S.A.; Rochelleb, C.A.; Manovica, V. A review of developments in carbon dioxide storage. *Appl. Energy* **2017**, *208*, 1389–1419. [CrossRef]

16. Leung, D.Y.C.; Caramanna, G.; Maroto-Valer, M.M. An overview of current status of carbon dioxide capture and storage technologies. *Renew. Sustain. Energy Rev.* **2014**, *39*, 426–443. [CrossRef]

17. Thomas, D.M.; Mechery, J.; Paulose, S.V. Carbon dioxide capture strategies from flue gas using microalgae: A review. *Environ. Sci. Pollut. Res.* **2016**, *23*, 16926. [CrossRef]

18. Sabouni, R.; Kazemian, H.; Rohani, S. Carbon dioxide capturing technologies: A review focusing on metal organic framework materials (MOFs). *Environ. Sci. Pollut. Res.* **2014**, *21*, 5427. [CrossRef]

19. Samanta, A.; Zhao, A.; Shimizu, G.K.H.; Sarkar, P.; Gupta, R. Post-combustion CO_2 capture using solid sorbents: A review. *Ind. Eng. Chem. Res.* **2012**, *51*, 1438–1463. [CrossRef]

20. Luis, P. Use of monoethanolamine (MEA) for CO_2 capture in a global scenario: Consequences and alternatives. *Desalination* **2016**, *380*, 93–99. [CrossRef]

21. Figueroa, J.D.; Fout, T.; Plasynski, S.; McIlvried, H.; Srivastava, R.D. Advances in CO_2 capture technology—The U.S. department of energy's carbon sequestration program: A review. *Int. J. Greenh. Gas Control* **2008**, *2*, 9–20. [CrossRef]

22. Asadi-Sangachini, Z.; Galangash, M.M.; Younesi, H.; Nowrouzi, M. The feasibility of cost-effective manufacturing activated carbon derived from walnut shells for large-scale CO_2 capture. *Environ. Sci. Pollut. Res.* **2019**, *26*, 26542–26552. [CrossRef] [PubMed]

23. Gibson, J.A.A.; Mangano, E.; Shiko, E.; Greenaway, A.G.; Gromov, A.V.; Lozinska, M.M.; Friedrich, D.; Campbell, E.E.B.; Wright, P.A.; Brandani, S. Adsorption materials and processes for carbon capture from gas-fired power plants: AMPgas. *Ind. Eng. Chem. Res.* **2016**, *551*, 33840–33851. [CrossRef]

24. Lee, S.-Y.; Park, S.-J. A review on solid adsorbents for carbon dioxide capture. *J. Ind. Eng. Chem.* **2015**, *23*, 1–11. [CrossRef]

25. Choi, S.; Drese, J.H.; Jones, C.W. Adsorbent materials for carbon dioxide capture from large anthropogenic point sources. *ChemSusChem* **2009**, *2*, 796–854. [CrossRef] [PubMed]

26. Lee, S.C.; Chae, H.J.; Lee, S.J.; Choi, B.Y.; Yi, C.K.; Lee, J.B.; Ryu, C.K.; Kim, J.C. Development of regenerable MgO-based sorbent promoted with K_2CO_3 for CO_2 capture at low temperatures. *Environ. Sci. Technol.* **2008**, *42*, 2736–2741. [CrossRef]

27. Aquino, A.S.; Vieira, M.O.; Ferreira, A.S.D.; Cabrita, E.J.; Einloft, S.; de Souza, M.O. Hybrid ionic liquid–silica xerogels applied in CO_2 capture. *Appl. Sci.* **2019**, *9*, 2614. [CrossRef]

28. Hauchhum, L.; Mahanta, P. Carbon dioxide adsorption on zeolites and activated carbon by pressure swing adsorption in a fixed bed. *Int. J. Energy Environ. Eng.* **2014**, *5*, 349–356. [CrossRef]

29. Lu, C.; Bai, H.; Su, F.; Chen, W.; Hwang, J.F.; Lee, H.-H. Adsorption of carbon dioxide from gas streams via mesoporous spherical-silica particles. *J. Air Waste Manag. Assoc.* **2010**, *60*, 489–496. [CrossRef]

30. Staciwa, P.; Narkiewicz, U.; Moszyński, D.; Wróbel, R.J.; Cormia, R.D. Carbon spheres as CO_2 sorbents. *Appl. Sci.* **2019**, *9*, 3349. [CrossRef]

31. Chiang, Y.-C.; Yeh, C.Y.; Weng, C.H. Carbon dioxide adsorption on porous and functionalized activated carbon fibers. *Appl. Sci.* **2019**, *9*, 1977. [CrossRef]

32. Al-Ghurabi, E.H.; Ajbar, A.; Asif, M. Enhancement of CO_2 removal efficacy of fluidized bed using particle mixing. *Appl. Sci.* **2018**, *8*, 1467. [CrossRef]

33. Wang, W.; Zhou, M.; Yuan, D. Carbon dioxide capture in amorphous porous organic polymers. *J. Mater. Chem. A* **2017**, *5*, 1334–1347. [CrossRef]

34. Wang, R.; Lang, J.; Yan, X. Effect of surface area and heteroatom of porous carbon materials on electrochemical capacitance in aqueous and organic electrolytes. *Sci. China Chem.* **2014**, *57*, 1570–1578. [CrossRef]

35. Wickramaratne, N.P.; Jaroniec, M. Activated carbon spheres for CO_2 adsorption. *ACS Appl. Mater. Interfaces* **2013**, *5*, 1849–1855. [CrossRef] [PubMed]

36. Pari, G.; Darmawan, S.; Prihandoko, B. Porous Carbon spheres from hydrothermal carbonization and KOH activation on cassava and tapioca flour raw material. *Procedia Environ. Sci.* **2014**, *20*, 342–351. [CrossRef]

37. Choma, J.; Kloske, M.; Dziura, A.; Stachurska, K.; Jaroniec, M. Preparation and studies of adsorption properties of microporous carbon spheres. *Eng. Prot. Environ.* **2016**, *19*, 169–182. [CrossRef]

38. Dawson, R.; Cooper, A.I.; Adams, D.J. Nanoporous organic polymer networks. *Prog. Polym. Sci.* **2012**, *37*, 530–563. [CrossRef]

39. Férey, G. Hybrid porous solids: Past, present, future. *Chem. Soc. Rev.* **2008**, *37*, 191–214. [CrossRef]

40. Millward, A.R.; Yaghi, O.M. Metal-organic frameworks with exceptionally high capacity for storage of carbon dioxide at room temperature. *J. Am. Chem. Soc.* **2005**, *127*, 17998–17999. [CrossRef]

41. Lu, W.; Yuan, D.; Sculley, J.; Zhao, D.; Krishna, R.; Zhou, H.-C. Sulfonate-grafted porous polymer networks for preferential CO_2 adsorption at low pressure. *J. Am. Chem. Soc.* **2011**, *133*, 18126–18129. [CrossRef]

42. Ahmed, D.S.; El-Hiti, G.A.; Yousif, E.; Ali, A.A.; Hameed, A.S. Design and synthesis of porous polymeric materials and their applications in gas capture and storage: A review. *J. Polym. Res.* **2018**, *25*, 75. [CrossRef]

43. Mooibroek, T.J.; Gamez, P. The s-triazine ring, a remarkable unit to generate supramolecular interactions. *Inorg. Chim. Acta* **2007**, *360*, 381–404. [CrossRef]

44. Jurgens, B.; Irran, E.; Senker, J.; Kroll, P.; Muller, H.; Schnick, W. Melem (2,5,8-triamino-tri-s-triazine), an important intermediate during condensation of melamine rings to graphitic carbon nitride: Synthesis, structure determination by X-ray powder diffractometry, solid-state NMR, and theoretical studies. *J. Am. Chem. Soc.* **2003**, *125*, 10288–10300. [CrossRef] [PubMed]

45. Wang, X.; Maeda, K.; Thomas, A.; Takanabe, K.; Xin, G.; Carlsson, J.M.; Domen, K.; Antonietti, M. A metal-free polymeric photocatalyst for hydrogen production from water under visible light. *Nat. Mater.* **2009**, *8*, 76–80. [CrossRef]

46. Pevida, C.; Drage, T.C.; Snape, C.E. Silica-templated melamine–formaldehyde resin derived adsorbents for CO_2 capture. *Carbon* **2008**, *46*, 1464–1474. [CrossRef]

47. El-Hiti, G.A.; Alotaibi, M.H.; Ahmed, A.A.; Hamad, B.A.; Ahmed, D.S.; Ahmed, A.; Hashim, H.; Yousif, E. The morphology and performance of poly(vinyl chloride) containing melamine Schiff bases against ultraviolet light. *Molecules* **2019**, *24*, 803. [CrossRef]

48. Yousif, E.; Ahmed, D.S.; El-Hiti, G.A.; Alotaibi, M.H.; Hashim, H.; Hameed, A.S.; Ahmed, A. Fabrication of novel ball-like polystyrene films containing Schiff bases microspheres as photostabilizers. *Polymers* **2018**, *10*, 1185. [CrossRef]

49. Hashim, H.; El-Hiti, G.A.; Alotaibi, M.H.; Ahmed, D.S.; Yousif, E. Fabrication of ordered honeycomb porous poly(vinyl chloride) thin film doped with a Schiff base and nickel(II) chloride. *Heliyon* **2018**, *4*, e00743. [CrossRef]

50. Shaalan, N.; Laftah, N.; El-Hiti, G.A.; Alotaibi, M.H.; Muslih, R.; Ahmed, D.S.; Yousif, E. Poly(vinyl chloride) photostabilization in the presence of Schiff bases containing a thiadiazole moiety. *Molecules* **2018**, *23*, 913. [CrossRef]

51. Ahmed, D.S.; El-Hiti, G.A.; Hameed, A.S.; Yousif, E.; Ahmed, A. New tetra-Schiff bases as efficient photostabilizers for poly(vinyl chloride). *Molecules* **2017**, *22*, 1506. [CrossRef]

52. Ali, G.Q.; El-Hiti, G.A.; Tomi, I.H.R.; Haddad, R.; Al-Qaisi, A.J.; Yousif, E. Photostability and performance of polystyrene films containing 1,2,4-triazole-3-thiol ring system Schiff bases. *Molecules* **2016**, *21*, 1699. [CrossRef] [PubMed]

53. Yousif, E.; El-Hiti, G.A.; Hussain, Z.; Altaie, A. Viscoelastic, spectroscopic and microscopic study of the photo irradiation effect on the stability of PVC in the presence of sulfamethoxazole Schiff's bases. *Polymers* **2015**, *7*, 2190–2204. [CrossRef]

54. Satar, H.A.; Ahmed, A.A.; Yousif, E.; Ahmed, D.S.; Alotibi, M.F.; El-Hiti, G.A. Synthesis of novel heteroatom-doped porous-organic polymers as environmentally efficient media for carbon dioxide storage. *Appl. Sci.* **2019**, *9*, 4314. [CrossRef]

55. Ahmed, D.S.; El-Hiti, G.A.; Yousif, E.; Hameed, A.S.; Abdalla, M. New eco-friendly phosphorus organic polymers as gas storage media. *Polymers* **2017**, *9*, 336. [CrossRef] [PubMed]

56. Hadi, A.G.; Jawad, K.; Yousif, E.; El-Hiti, G.A.; Alotaibi, M.H.; Ahmed, D.S. Synthesis of telmisartan organotin(IV) complexes and their use as carbon dioxide capture media. *Molecules* **2019**, *24*, 1631. [CrossRef] [PubMed]

57. Thommes, M.; Kaneko, K.; Neimark, A.V.; Olivier, J.P.; Rodriguez-Reinoso, F.; Rouquerol, J.; Sing, K.S.W. Physisorption of gases, with special reference to the evaluation of surface area and pore size distribution (IUPAC Technical Report). *Pure Appl. Chem.* **2015**, *87*, 1051–1069. [CrossRef]

 *processes*

Article

Vapor Liquid Equilibrium Measurements of Two Promising Tertiary Amines for CO_2 Capture

Diego D.D. Pinto [1,2], Znar Zahraee [1], Vanja Buvik [1], Ardi Hartono [1] and Hanna K. Knuutila [1,*

[1]   Department of Chemical Engineering, Norwegian University of Science and Technology, 7491 Trondheim, Norway; diego.pinto@hovyu.com (D.D.D.P.); znarzahraee@gmail.com (Z.Z.); vanja.buvik@ntnu.no (V.B.); ardi.hartono@ntnu.no (A.H.)

[2]   Hovyu B.V., Watermunt, 72, 2408 LS Alphen aan den Rijn, The Netherlands

*   Correspondence: hanna.knuutila@ntnu.no; Tel.: +47-7359-4119

Received: 4 November 2019; Accepted: 2 December 2019; Published: 12 December 2019

Abstract: Post combustion CO_2 capture is still a rather energy intense, and therefore expensive, process. Much of the current research for reducing the process energy requirements is focused on the regeneration section. A good description of the vapor liquid equilibrium of the solvent is necessary for the accurate representation of the process. 3-(Diethylamino)-1,2-propanediol (DEA-12-PD) and 1-(2-hydroxyethyl)piperidine (12-HEPP) have been proposed as potential components in solvent blends for the membrane contactor. However, there are few available experimental data for these two tertiary amines making difficult to accurate simulate such process. In this work, we provide experimental data on the pure component saturation pressure (383 to 443 K) and on VLE of aqueous solutions of these amines (313 to 373 K) in order to fill part of the data gap. The data were used to estimate model parameters used to represent the data. The saturation pressure was modeled using the Antoine equation and the deviation is calculated lower than 2%. The NRTL model was used in this work to calculate the activity coefficients in the aqueous systems. The deviations in pressure for the aqueous systems were lower than 5% in both systems.

Keywords: VLE; CO_2 capture; amine; DEA-12-PD; 12-HEPP

---

1. Introduction

Chemical absorption is widely used and the most mature technology to remove CO_2 from gas streams. However, it is known that the energy consumption to regenerate the solvent is one of the biggest concerns of this type of process. Other major concerns include solvent emissions, stability, and equipment size. Several studies on solvent development (e.g., phase change solvent [1], new blends [2,3] and water lean solvents [4]) and on process modification [5,6] aiming to address these issues been done since Bottoms patented in 1930 a process to remove acid gases from natural gas.

To design and simulate effectively an absorption process, information is required on both the solvent properties (e.g., physical properties, vapor liquid equilibrium, and absorption kinetics) and the equipment (e.g., absorber type and packing material). Traditionally, in the absorption process, the absorption and desorption are performed in columns: the absorber and the stripper columns, respectively. However, some new designs have been recently proposed. For example, Lin et al. (2016) [7] studied the desorption step using the advanced flash stripper configuration for a 5 m and a 8 m piperazine (PZ) solution. They showed that the process was more energy efficient than the benchmark solvent monoethanolamine (MEA) and previous process configuration, namely the two-staged flash. A reboiler duty of 2.1–2.5 GJ/ton CO_2 was achieved. Recently, membrane contactors have been studied to substitute the absorber tower [8]. Using a membrane contactor could potentially reduce significantly the equipment size as calculated in Hoff and Svendsen (2013) [9].

Traditional solvents such as monoethanolamine (MEA), methyldiethanolamine (MDEA), and piperazine (PZ) have been widely studied and their behavior is well represented by several commercial softwares. However, new solvents/solvent blends are being developed to improve the efficiency and safety of the absorption process. In the solvent development the vapor-liquid equilibria, heat of absorption, corrosion tendencies, solvent degradation, absorption kinetics, and solvent volatility are among the properties that are experimentally measured. In recent years, solvent volatility has gained a lot of attention. Volatility causes solvent losses requiring water wash sections to control the emissions. Volatility in combination with mist formation can significantly magnify the solvent losses. There are three strategies to overcome the volatility issue. One is to develop solvents with very low volatility, like aqueous amino acid salt solutions [10]. The second is to develop systems/operations that minimize the formation of mist, like the anti-mist design developed by Aker Solutions [11]. In recent years, a third strategy has been proposed: the use of a non-porous thin composite membrane [8]. This type of membrane can potentially reduce the amine evaporation towards the gas phase.

Independently of the equipment used for the absorption process, a good description of the vapor liquid equilibrium, together with other properties, is essential. 3-(Diethylamino)-1,2-propanediol (DEA-12-PD) and 1-(2-hydroxyethyl)piperidine (12-HEPP) have been proposed as potential components in solvent blends for membrane contactors [12]. Very few experimental data are found for these two tertiary amines. DEA-12-PD had been identified by Chowdhury et al. (2013) [13] as a potential tertiary amine since the absorption rate and capacity were good. Li et al. (2015) [14] studied the reaction kinetics of aqueous solutions of DEA-12-PD with CO_2 and it was observed that the reaction was faster than MDEA. They also performed pKa measurements at different temperatures.

Later Hartono et al. (2017) [15] performed a series of screening tests with different solvents that could intensify the formation of bicarbonate. DEA-12-PD and 12-HEPP were among the tested solvent candidates. Knuutila et al. (2019) [3] tested DEA-12-PD and 12-HEPP promoted with primary amines and showed that by using a short cut method [16], the tested solvent blends could be regenerated with reboiler duties 2.5–2.6 MJ/kg CO_2. These values are similar to those values measured with several novel solvent blends, 2.5–3.0 MJ/kg CO_2 [17–19].

No data was found for the vapor liquid equilibrium of aqueous DEA-12-PD and 12-HEPP systems. Understanding the volatility of solvent components is an important parameter as discussed above. The volatility will be influenced by the CO_2 loading and the degree of solvent degradation, both depending on the actual industrial application. However, a good estimation of the potential challenges related to volatility can be gained by measuring the vapor–liquid equilibria of non-degraded binary systems.

In this work, we provide experimental data on the pure component saturation pressure and on VLE of aqueous solutions of these amines. The data were then used to estimate model parameters used to represent the data. The saturation pressure was modeled using the Antoine equation and the deviation was calculated lower than 2%. The NRTL model was used in this work to calculate the activity coefficients in the aqueous systems. The deviations in pressure for the aqueous systems were lower than 5% in both systems.

2. Materials and Methods

2.1. Chemicals

DEA-12-PD and 12-HEPP are tertiary amines and their molecular structures are given in Figure 1. In order to validate the apparatus, measurements with pure water and aqueous solutions of MEA were performed and compared to in-house data and literature. DI-water produced in the lab was used to measure the saturation pressure of pure water and to dilute the chemicals to the desired solution concentration. The suppliers and purities (mass basis) of the purchased chemicals are given in Table 1. The chemicals were used in the experiments without any further purification.

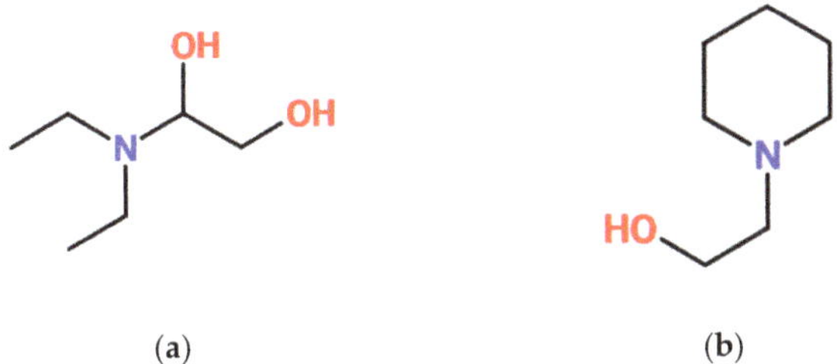

(a)   (b)

Figure 1. Molecular structure of: (**a**) 3-(diethylamino)-1,2-propanediol (DEA-12-PD) and (**b**) 1-(2-hydroxyethyl)piperidine (12-HEPP).

Table 1. Chemicals used.

Component	CAS	MW (g/mol)	Purity (%)	Supplier
MEA	141-43-5	61.08	≥99	Sigma–Aldrich
DEA-12-PD	621-56-7	147.22	≥98	TCI
12-HEPP	3040-44-6	129.20	≥99.5	Sigma–Aldrich

2.2. Apparatus

The modified Swietoslawski ebulliometer previously described in Kim et al. (2008) [20] was used to measure the vapor liquid equilibrium of pure and aqueous solutions. The apparatus was used several times for this type of measurement [21,22] and proven to be accurate and reliable. For further description of the apparatus, the reader is referred to the cited references.

About 80 mL of liquid was added to the ebulliometer for a measurement. When pure component was used, the first solution added was left boiling for some time until it reached the equilibrium at a given temperature. The equilibrium temperature and pressure were recorded as a reference and the solution was later discarded. A new fresh pure liquid solvent was inserted in the apparatus and the procedure was repeated until the equilibrium temperature and pressure from the new round and the old round was the same. This procedure was to ensure that all water trapped in the apparatus was removed and the apparatus was filled with only pure solvent. Usually after the third charge, the measurements of the pure solvent could be started. This procedure was, however, not necessary for aqueous solutions as a small dilution of the initial concentration would not affect the measurements.

Aqueous solutions of approximately 0.2 to 0.8 mol fraction concentration were prepared gravimetrically using a Mettler PM1200 scale ($u(m) = \pm0.005$ g) for DEA-12-PD and 12-HEPP. A VLE experiment in this concentration range was performed by inserting the prepared solution into the ebulliometer and setting a desired equilibrium temperature. Once equilibrium was reached, the temperature and pressure were recorded; samples from the gas (approximately 1 mL) and liquid (approximately 5 mL) phases were taken for further alkalinity analysis and a new desired temperature was set. The equilibrium was considered achieved once the temperature and pressure were constant for more than 10 min. After the measurements at all desired temperatures were performed, the solution was removed from the apparatus and a new solution with a different concentration was inserted for a new measurement cycle.

For concentrations lower than 0.2 (mol fraction), the solution was not discarded, but diluted by removing a certain amount of solution from the sampling point and replacing it with DI water at the same sampling point with the help of a syringe. The measurements, then, followed the same procedure as for the more concentrated solutions.

2.3. Analysis

The samples from the solution preparation, gas phase, and liquid phase from the measurements were analyzed using an acid–base titration. This method is described elsewhere (e.g., [23]) and it is an inexpensive, fast, and reliable method to quantify the amine concentration in aqueous solutions. To understand the accuracy of the titration method in case of very dilute samples, 0.05 M and 0.1 M solutions of MEA were prepared by weighing. Then 1 mL of the solutions was analyzed four times. The deviations in the parallels for the 0.1 M and 0.05 M solutions were 1.9% and 9.2%, respectively. There were in total 4 samples where the amine concentration during analyses was below 0.05 M, leading into analytical uncertainty in the gas phase higher than 10%. These were with DEA-12-PD at 40 and 60 °C. In the tables presenting the data, uncertainty of the data is given.

3. Modeling

The Antoine equation, Equation (1), was used to correlate the saturation pressure of both pure DEA-12-PD and 12-HEPP. This correlation is frequently used to represent the saturation pressure of pure components and, in this work, it was chosen to represent the produced data.

$$log_{10}(p_{sat}) = A + \frac{B}{T + C}$$ (1)

In this work, a non-electrolyte non-reactive system was assumed for the representation of the VLE of aqueous DEA-12-PD and 12-HEPP. The phase equilibrium is solved iteratively for the pressure and vapor phase composition by solving the system of equations described in Equation (2). In that equation the subscript i stands for H_2O and the amine, γ is the activity coefficient, ϕ is the fugacity coefficient, x and y are the liquid and gas phase mol fractions, respectively, and the exponential term is the so-called Poynting correction factor where the liquid volume $\left(\vartheta_i^L\right)$ was fixed as the respective component molar volume. As expected, since the experiments were carried out under low pressures, the Poynting correction factor for the conditions studied in this work was negligible.

The Peng–Robinson equation of state (EoS) [24] is used to correct the gas phase non-idealities while the non-random two-liquids (NRTL) model [25] is used to account for the liquid phase non-ideal behavior. The Van der Waals mixing rule with all binary parameters set to zeros was used in the Peng–Robinson EoS. As a result, all adjustable parameters are from the NRTL model.

$$p y_i \phi_i = p_i^{sat} x_i \gamma_i \phi_i^{sat} exp\left(\int_{p_i^{sat}}^{p} \frac{\vartheta_i^L dp}{RT}\right)$$ (2)

The activity coefficient calculated through the NRTL model is given in Equation (3). The binary energy parameters are assumed to have a temperature dependency as shown in Equation (5) where a_{ij} and b_{ij} are adjustable parameters. The non-randomness parameter $\left(\alpha_{ij}\right)$ can also be used as an adjustable parameter, but in this work it was fixed at a given value.

$$\ln(\gamma_i) = \frac{\sum_{j=1}^{N} \tau_{ji} G_{ji} x_j}{\sum_{k=1}^{N} G_{ki} x_k} + \sum_{j=1}^{N} \frac{x_j G_{ji}}{\sum_{k=1}^{N} G_{kj} x_k}\left(\tau_{ij} - \frac{\sum_{l=1}^{N} \tau_{lj} G_{lj} x_l}{\sum_{k=1}^{N} G_{kj} x_k}\right)$$ (3)

$$G_{ij} = exp\left(-\alpha_{ij}\tau_{ij}\right)$$ (4)

$$\tau_{ij} = a_{ij} + \frac{b_{ij}}{T}$$ (5)

Besides the rigorous framework, the choice of the set of models used in this work, among others, was based on the ease of exporting the parameters to process simulation software. Most process simulators have the models used in this work already implemented.

3.1. Optimization Routine

The adjustable parameters from the NRTL model were fitted to the experimental data using the particle swarm optimization (PSO) routine with the local best topology. The method is widely used for parameter estimation [23,26–28] and information about it can be found elsewhere (e.g., [23,29]). In this work, we fixed the non-randomness parameters at 0.1, 0.2, and 0.3 using four different objective functions. The best results are given in the results section while the results from all optimizations can be found in Appendix B.

Equations (6) and (7) show the general form of the objective functions used in this work. In those equations, the parameter q was set to zero if the vapor phase composition was not included in the objective function. In the case that the vapor phase composition should be considered in the optimization, the parameter q was set to one. A total of four objective functions were used per non-randomness parameter, giving a total of 12 optimizations and a set of parameters for each binary system.

The results were evaluated by means of the average absolute relative deviation (AARD) function (Equation (8)) where φ_i is the variable from which the deviation is calculated.

$$F_{obj.}^{I} = \frac{100}{N}\left(\sum_{i=1}^{N}\frac{\left|p_i^{exp} - p_i^{calc}\right|}{p_i^{exp}} + q\sum_{i=1}^{N}\frac{\left|y_i^{exp} - y_i^{calc}\right|}{y_i^{exp}}\right) \tag{6}$$

$$F_{obj.}^{II} = \sum_{i=1}^{N}\frac{\left(p_i^{exp} - p_i^{calc}\right)^2}{p_i^{exp} \cdot p_i^{calc}} + q\sum_{i=1}^{N}\frac{\left(y_i^{exp} - y_i^{calc}\right)^2}{y_i^{exp} \cdot y_i^{calc}} \tag{7}$$

$$AARD\ (\%) = \frac{100}{N}\left(\sum_{i=1}^{N}\frac{\left|\varphi_i^{exp} - \varphi_i^{calc}\right|}{\varphi_i^{exp}}\right) \tag{8}$$

3.2. Critical Properties

Since the Peng–Robinson EoS was used, the critical properties of the components were required. These properties were not found for the tertiary amines studied in this work. Therefore, the Joback group contribution method [30] was used for this purpose. For 12-HEPP, the tertiary amine contribution was considered as a "non-ring" tertiary amine contribution since the method has no "ring" tertiary amine contribution. Nevertheless, the estimations given here should be treated with care as these results must be experimentally confirmed. The normal boiling point for the amines was calculated with the Antoine equation fitted with the respective experimental points generated in this work.

As a comparison, the critical properties for MEA were estimated using the Joback group contribution method. Using a normal boiling temperature of 443.97 K, the critical temperature, pressure, and volume were estimated to be 637.1 K, 62.89 bar, and 1.96 m^3/mol, respectively. The reported values for the critical temperature, pressure, and volume for MEA [31] are, respectively, 678.2 K, 71.24 bar, and 2.25 m^3/mol. This shows that using the Joback group contribution method gives a good initial estimation for the critical point of a substance.

Once the critical properties are known, the acentric factor can be calculated using Equation (9). The results are summarized in Table 2.

$$\omega = -log_{10}\left(p_r^{sat}(T = 0.7 \cdot T_c)\right) - 1 \tag{9}$$

Table 2. Thermodynamic and critical properties of compounds.

Property	H_2O [a]	DEA-12-PD	12-HEPP [d]
T_b (K)	-	496.42 [b]	475.11 [b]
T_c (K)	647.10	638.14 [c]	655.71 [c]
p_c (bar)	220.64	34.60 [c]	39.71 [c]
v_c (cm³/mol)	55.95	477.50 [c]	406.50 [c]
ω	0.3449	1.2470 [c]	0.7562 [c]

[a] DIPPR [32], [b] Estimated with Antoine equation (this work), [c] Joback group contribution method, [d] Critical properties estimated using non-ring tertiary amine.

4. Results

All experimental data are given in Appendix A.

4.1. Validation

Before the measurements with the unknown tertiary amines, the apparatus was validated by measuring the equilibrium with pure water and aqueous MEA solutions. Equation (10) shows the form of the Riedel equation to calculate the saturation pressure of pure water at a given temperature. Along with previous literature data, the model is used for comparison with the measured data.

$$p_{sat,H_2O} = exp\left(73.649 - \frac{7258.2}{T} - 7.3037\ln(T) + 4.1653E - 06T^2\right) \tag{10}$$

The results from pure water show that the measurements were in line with what was previously reported in the literature. Figure 2 shows the comparison of the different measured data and the Riedel equation for water. The calculated deviation (Equation (8)) from the experimental points generated in this work and the correlation was 0.7%.

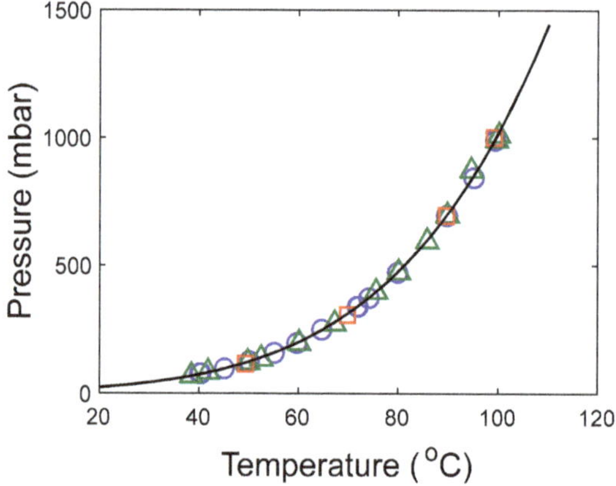

Figure 2. Saturation pressure of pure water. Measurements of: (o) This work, (Δ) [20], and (□) [33]. (-) Model calculated with Equation (10).

A last validation experiment with an aqueous solution of MEA was performed. The VLE at 80 °C was measured. In this case, samples of the gas and liquid phases were taken and analyzed for amine concentration. The results are shown in Figure 3. It is possible to see that the results from MEA in this work are in excellent agreement with previous results reported in the literature. Given all the validation results, it was possible to conclude that the apparatus was accurate and reliable to measure VLE of pure and aqueous solutions of amines.

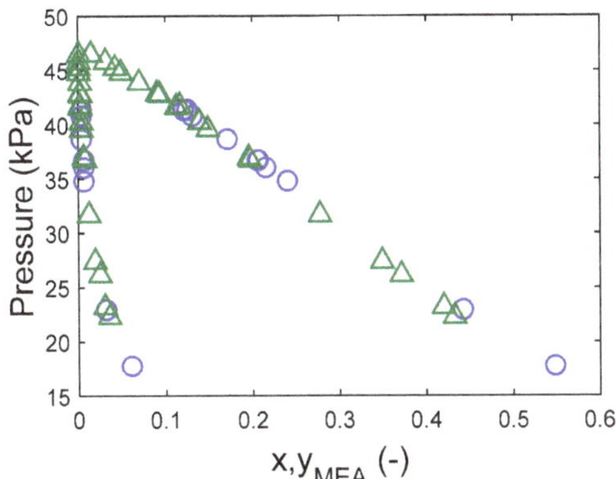

Figure 3. Vapor liquid equilibrium of aqueous monoethanolamine (MEA) solutions at 80 °C. Experimental points from: (o) this work and (Δ) [20].

4.2. Saturation Pressure of Pure Component

The saturation pressure of the pure tertiary amines was modeled using the Antoine equation (Equation (1)). The optimized parameters together with the deviations are given in Table 3 where the pressure is given in Pa and the temperature in K. Figures 4 and 5 show, respectively, the excellent agreement between the experimental data and the model for DEA-12-PD and 12-HEPP. The calculated deviations are below 2%. Two experiment runs were performed for measuring the pure component volatility. In the first run, the measurements were aimed for a 10 °C temperature interval. This was done in order to identify the lower and upper temperature limits of the apparatus for the specific solvent. For the second run, performed with a fresh solvent, the temperature interval between the measurements was reduced to 5 °C.

Table 3. Optimized parameters for the Antoine equation (Equation (1))—T in K and p in Pa.

Parameter	DEA-12-PD	12-HEPP
A	12.3979 ± 0.4979	8.4381 ± 0.1639
B	-4121.3892 ± 476.9160	-1219.5810 ± 95.6280
C	61.1149 ± 27.6857	-119.7961 ± 11.3916
AARD (%)	1.097	1.7831

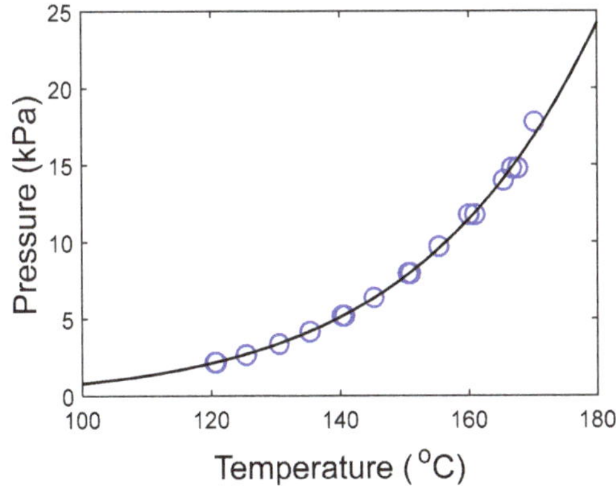

Figure 4. Saturation pressure of DEA-12-PD. (o) Experimental points from this work and (-) model (Equation (1)).

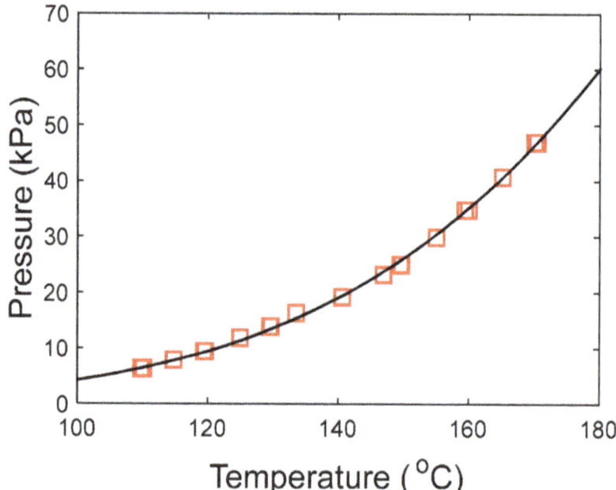

Figure 5. Saturation pressure of 12-HEPP. (□) Experimental points from this work and (-) model (Equation (1)).

It is possible to see from the figures that both runs agree very nicely. Moreover, it can be observed that pure 12-HEPP is more volatile than DEA-12-PD. However, both tertiary amines are less volatile than the benchmark MEA.

4.3. Aqueous Amine Solution

As previously mentioned, 12 optimizations were performed for each system where the objective function and the non-randomness parameter were varied. The optimized parameters are given in Appendix B for DEA-12-PD and 12-HEPP, respectively.

For DEA-12-PD, the calculated deviations in pressure and vapor phase composition are given in Table A7. It is possible to see that using the objective function I and II with $q = 0$ (only pressure) and objective function II with $q = 1$ produces similar results where the pressure is the variable that is prioritized. When using objective function I with $q = 1$, the optimization tries to balance the deviation in the pressure and the vapor phase composition. This is valid for all non-randomness parameters tested and this same behavior is observed for 12-HEPP. Since a good representation of the vapor phase is also important, we have chosen here to use the best results from the objective function I with $q = 1$. These results are highlighted in bold fonts in the respective tables in Appendix B. For DEA-12-PD the best results were found with a non-randomness parameter value of 0.1.

In Figure 6, it is seen that the model is able to represent the experimental data with good accuracy. It is also seen that the model deviates from the experimental data at high amine concentrations. This is caused due to the objective function used and the way we solve the phase equilibrium (for y and p). The mol fraction of DEA-12-PD in the vapor phase is very small. Therefore, it is expected that the deviations calculated for the vapor phase composition is higher compared to the deviation in the pressure and a trade-off must be made. In this case, the deviation in pressure is reasonably good and calculated to be 4.6% while the deviation in the amine vapor composition is around 20%. It is possible to obtain deviations below 2% at the expense of higher deviations in the vapor phase composition (>40%).

As done for DEA-12-PD, we chose the best results from objective function I with $q = 1$ for showing the results of 12-HEPP. In this case, the best result was found using 0.3 as the non-randomness parameter value. Figure 7 shows the good agreement between the model and experimental data for the VLE of 12-HEPP. In aqueous solution, 12-HEPP seems to be more volatile than DEA-12-PD as the vapor phase concentrations are higher. It is also seen that the model does not capture the vapor phase compositions as nicely as for DEA-12-PD.

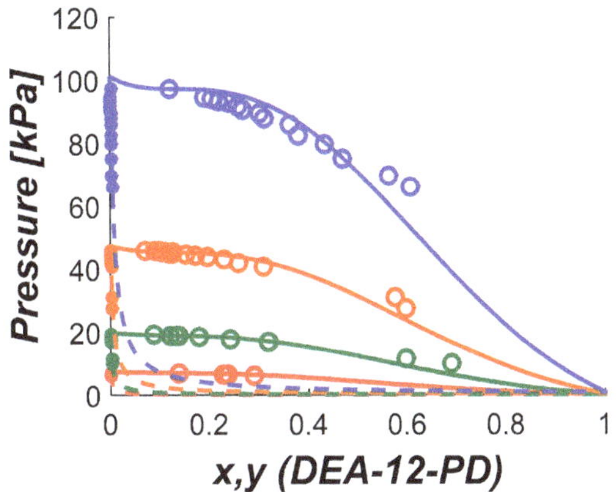

Figure 6. VLE of DEA-12-PD at 40 (red), 60 (green), 80 (orange), and 100 °C (blue). (o) Experimental data from this work and (-) model.

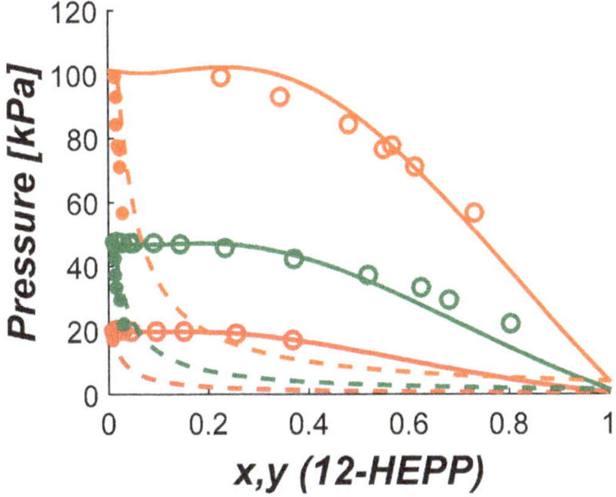

Figure 7. VLE of 12-HEPP at 60 (red), 80 (green), and 100 (orange). (o) Experimental data from this work and (-) model.

5. Conclusions

The VLE of two promising tertiary amines for CO_2 capture were measured in the ebulliometer apparatus. The apparatus was previously used for this type of measurements with great accuracy. Prior to the measurements, the apparatus was validated by measuring VLE from known components (e.g., water and MEA). The results from the validation confirmed the accuracy and reliability of the ebulliometer. Pure component saturation pressures and VLE of aqueous solutions of DEA-12-PD and 12-HEPP were measured up to atmospheric pressure and temperatures ranging from 40 to 170 °C. The Antoine equation was used to correlate the vapour pressure of the pure component while the NRTL was used to calculate the activity coefficient of the components in aqueous solutions. Both models were able to correlate the experimental data with reasonably accuracy. The vapour pressure was correlated within 2% while the aqueous solutions deviations were lower than 5% deviation with respect to pressure.

Author Contributions: Conceptualization, D.D.D.P. and H.K.K.; methodology, D.D.D.P, H.K.K. and A.H.; supervision, D.D.D.P. and H.K.K.; experimental work, Z.Z. and V.B.; writing—original draft preparation, D.D.D.P. writing—review and editing, D.D.D.P, H.K.K., A.H. and V.B.

Funding: This work is supported by the Research Council of Norway through the CLIMIT program (Project No. 239789, Project: 3rd generation membrane contactor).

Acknowledgments: This work is supported by the Research Council of Norway through the CLIMIT program (Project No. 239789, Project: 3rd generation membrane contactor).

Conflicts of Interest: The authors declare no conflict of interest.

Appendix A. Experimental Data

Table A1. VLE measurements of pure water.

Temperature (°C)	Pressure (mbar)
Run 1	
40.47	76.9
50.05	124.9
59.72	197.9
71.93	339.9
79.89	472.9
89.81	697.8
Run 2	
40.20	76.9
45.07	97.0
50.05	124.9
55.02	158.0
59.67	198.0
59.71	197.9
64.74	248.2
71.70	340.0
74.23	373.1
79.88	473.0
89.71	698.0
95.05	843.8
99.50	993.3

Uncertainties u:
$u(T) = \pm 0.1$ K; $u(P) = \pm 0.3$ kPa;

Table A2. VLE measurements of aqueous MEA solution.

Temperature (°C)	Pressure (mbar)	x_{MEA}	y_{MEA}
80.09	347.9	0.2400	0.0065
80.08	366.8	0.2072	0.0065
80.02	386.8	0.1706	0.0041
79.91	407.8	0.1301	0.0031
80.04	413.2	0.1210	0.0036
80.04	413.2	0.1254	0.0037
80.08	177.8	0.5485	0.0606
80.07	229.8	0.4417	0.0305
80.00	360.9	0.2156	0.0066
79.97	367.9	0.2050	0.0059

Uncertainties u:
$u(T) = \pm 0.1$ K; $u(P) = \pm 0.3$ kPa
$u(c_{amine}) = 10^{-3}$ mol amine/kg solution

Table A3. VLE measurements of pure DEA-12-PD.

Temperature (°C)	Pressure (mbar)
Run 1	
120.63	21.9
130.56	33.9
140.62	51.9
150.90	79.6
160.03	118.0
167.49	147.90
Run 2	
120.76	21.9
125.37	26.9
130.57	33.9
135.37	41.9
140.37	51.9
145.27	63.9
150.58	79.8
155.45	96.9
160.87	117.9
160.01	118.1
165.33	139.8
166.62	147.8
170.17	178.0

Uncertainties u:
$u(T) = \pm 0.1$ K; $u(P) = \pm 0.3$ kPa;

Table A4. VLE measurements of pure 12-HEPP.

Temperature (°C)	Pressure (mbar)
Run 1	
110.18	63.9
119.63	93.9
129.67	139.9
140.60	191.8
149.73	249.9
159.99	346.9
170.48	469.9
Run 2	
109.92	63.9
114.89	77.9
119.50	93.9
125.02	117.9
129.85	139.9
133.70	163.0
140.60	191.8
146.91	232.9
149.51	249.9
155.09	297.8
159.53	346.9
165.08	407.8
170.08	469.9

Uncertainties u:
$u(T) = \pm 0.1$ K; $u(P) = \pm 0.3$ kPa;

Table A5. VLE measurements of aqueous DEA-12-PD solutions.

Temperature (°C)	Pressure (mbar)	$x_{DEA-12-PD}$	$y_{DEA-12-PD} \times 10^3$
39.9	67.8	0.1387	0.5108
40.0	64.7	0.2273	0.5389
40.0	63.9	0.2373	1.4471
40.0	61.8	0.2901	0.6525
60.0	191.3	0.0894	1.0125
60.1	189.8	0.1217	0.9385
60.0	187.8	0.1346	0.7198
60.1	187.8	0.1397	2.0186
60.0	183.8	0.1804	0.9504
60.0	177.8	0.2425	1.1351
60.0	169.8	0.3208	0.9970
60.0	115.9	0.5984	4.5560
60.0	98.8	0.6896	2.9977
80.1	459	0.0709	1.4838
80.1	456.9	0.0911	1.3258
80.1	455.9	0.1011	1.2130
80.1	455.1	0.1052	1.2267
80.0	452	0.1129	1.2505
79.9	450.8	0.1189	1.2185
80.1	452.8	0.1257	1.5474
80.0	451.6	0.1265	1.5299
80.1	447.6	0.1545	1.3028
80.1	443.8	0.1729	1.3533
80.1	438.8	0.1968	1.4011
80.0	430.8	0.2295	1.6782
80.0	416.9	0.2604	1.7641
79.9	408.9	0.3116	2.4360
80.0	309.9	0.5749	3.1407
80.0	275.7	0.5989	4.5964
100.0	970.3	0.1220	4.6368
100.0	945.7	0.1928	2.1977
100.0	941.7	0.2059	2.1810
100.1	934.7	0.2199	2.2293
100.1	927.8	0.2396	2.3456
100.0	917.7	0.2599	2.3428
100.0	907.7	0.2687	2.4043
100.0	892.8	0.3024	2.6490
100.0	877.8	0.3121	2.6125
100.0	857.8	0.3630	2.8391
100.0	823.8	0.3806	3.0852
99.9	797.8	0.4343	3.4116
100.0	751.8	0.4701	3.9239
100.0	693.8	0.5650	4.6368
100.0	661.8	0.6085	5.4423

Uncertainties u:
$u(T) = \pm 0.1$ K; $u(P) = \pm 0.3$ kPa;
$u(c_{amine}) = 10^{-3}$ mol amine/kg solution;

Table A6. VLE measurements of aqueous 12-HEPP solutions.

Temperature (°C)	Pressure (mbar)	x_{12_HEPP}	$y_{12_HEPP} \times 10^2$
60.1	199.9	0.0153	0.4533
60.0	198.9	0.0270	0.5184
60.0	197.9	0.0439	0.5600
60.0	196.9	0.0945	0.6209
60.0	194.9	0.1517	0.6483
60.1	189.9	0.2539	0.7201
60.0	167.9	0.3678	0.7671
80.1	475.9	0.0151	0.7230
80.1	474.9	0.0267	0.8355
80.1	473.9	0.0427	0.8787
80.1	473.2	0.0497	0.8782
80.1	471.9	0.0902	0.9223
80.1	468.9	0.1448	0.9106
80.0	458.9	0.2341	1.0020
80.1	420.9	0.3718	1.2120
80.0	373.9	0.5188	1.3502
80.1	334.9	0.6246	1.6388
80.0	294.9	0.6789	2.2342
80.0	218.9	0.8040	2.9447
99.9	989.8	0.2257	1.3699
100.0	928.8	0.3449	1.5584
100.0	841.8	0.4786	1.6826
100.0	764.8	0.5498	2.2406
100.0	776.8	0.5655	1.8139
100.0	711.8	0.6132	2.3135
100.0	566.9	0.7304	2.6767

Uncertainties u:
$u(T) = \pm 0.1$ K; $u(P) = \pm 0.3$ kPa;
$u(c_{amine}) = 10^{-3}$ mol amine/kg solution

Appendix B. Optimization Results

Table A7. NRTL optimized parameters for different non-randomness parameters and objective functions (where subscript 1 = H_2O and 2 = DEA-12-PD).

Objective Function	$\alpha_{12} = \alpha_{21}$	a_{12}	a_{21}	b_{12}	b_{21}
Fobj I and $q = 0$		12.9646	−5.7143	−2356.0227	884.9709
Fobj II and $q = 0$		7.8135	−2.9628	−734.9812	28.0716
Fobj I and $q = 1$	0.1	**5.8630**	**−1.0737**	**784.5002**	**−1043.7131**
Fobj II and $q = 1$		8.4954	−3.0922	−805.0346	−13.5152
Fobj I and $q = 0$		7.8175	−2.3517	−1333.2329	285.8961
Fobj II and $q = 0$		3.002	1.1663	−64.1834	−691.2653
Fobj I and $q = 1$	0.2	6.5318	0.1695	−401.8003	−791.6571
Fobj II and $q = 1$		5.5323	−0.9082	−526.1700	−211.7622
Fobj I and $q = 0$		−5.9015	4.7508	7991.0547	−1230.7370
Fobj II and $q = 0$		5.5323	−0.9082	−526.1700	−211.7622
Fobj I and $q = 1$	0.3	5.9285	1.0922	−541.2799	−847.2236
Fobj II and $q = 1$		4.1419	−0.6188	−329.4525	−89.4873

Table A8. Calculated deviations for the optimized parameters for aqueous DEA-12-PD systems.

Objective Function	$\alpha_{12} = \alpha_{21}$	AARD (%)		
		Pressure	y_{H_2O}	$y_{DEA-12-PD}$
Fobj I and $q = 0$		1.9859	0.0798	42.6090
Fobj II and $q = 0$	0.1	2.0665	0.1002	54.0863
Fobj I and $q = 1$		**4.6043**	**0.0534**	**19.5826**
Fobj II and $q = 1$		2.0626	0.0874	46.0847
Fobj I and $q = 0$		1.8253	0.0859	45.1207
Fobj II and $q = 0$	0.2	3.6501	0.1459	80.3952
Fobj I and $q = 1$		4.7029	0.0566	20.7956
Fobj II and $q = 1$		1.9447	0.0904	46.9822
Fobj I and $q = 0$		2.3678	0.1489	66.6363
Fobj II and $q = 0$	0.3	1.9447	0.0904	46.9822
Fobj I and $q = 1$		4.9901	0.0594	21.5117
Fobj II and $q = 1$		1.8613	0.0930	48.8845

Table A9. NRTL optimized parameters for different non-randomness parameters and objective functions (where subscript 1 = H_2O and 2 = 12-HEPP).

Objective Function	$\alpha_{12} = \alpha_{21}$	a_{12}	a_{21}	b_{12}	b_{21}
Fobj I and $q = 0$		−7.2665	3.0658	4466.2080	−1939.0730
Fobj II and $q = 0$	0.1	9.5567	−5.3598	−1639.0083	1119.2290
Fobj I and $q = 1$		8.7617	0.9294	−155.6712	−1646.4983
Fobj II and $q = 1$		7.1458	−3.3269	−651.0712	328.9348
Fobj I and $q = 0$		−5.1771	2.3299	3092.4113	−1182.4124
Fobj II and $q = 0$	0.2	6.4065	−3.0141	−1101.4384	758.2339
Fobj I and $q = 1$		4.718	2.7861	263.5349	−1603.8336
Fobj II and $q = 1$		0.9967	−1.1560	684.1510	145.7886
Fobj I and $q = 0$		3.8504	−2.5384	−211.8262	732.1378
Fobj II and $q = 0$	0.3	2.2028	−2.6463	107.1788	849.8800
Fobj I and $q = 1$		**0.3148**	**4.2127**	**1400.8632**	**−1828.0520**
Fobj II and $q = 1$		4.3337	−0.6368	−502.0548	69.7773

Table A10. Calculated deviations for the optimized parameters for aqueous 12-HEPP systems.

Objective Function	$\alpha_{12} = \alpha_{21}$	AARD (%)		
		Pressure	y_{H_2O}	$y_{12-HEPP}$
Fobj I and $q = 0$		2.7415	0.9927	66.6669
Fobj II and $q = 0$	0.1	2.9052	0.9948	68.8495
Fobj I and $q = 1$		6.6651	0.8018	39.9976
Fobj II and $q = 1$		2.9547	0.9627	65.1831
Fobj I and $q = 0$		2.5836	0.9912	65.4845
Fobj II and $q = 0$	0.2	2.7575	0.9960	68.7378
Fobj I and $q = 1$		5.7172	0.8419	42.1513
Fobj II and $q = 1$		3.0621	1.0516	71.6277
Fobj I and $q = 0$		2.7795	0.9968	71.8631
Fobj II and $q = 0$	0.3	2.8971	1.0550	74.5644
Fobj I and $q = 1$		**4.7528**	**0.8633**	**43.7299**
Fobj II and $q = 1$		2.6826	0.9653	64.6265

References

1. Bernhardsen, I.M.; Krokvik, I.R.T.; Perinu, C.; Pinto, D.D.D.; Jens, K.J.; Knuutila, H.K. Influence of pKa on solvent performance of MAPA promoted tertiary amines. *Int. J. Greenh. Gas Control* **2018**, *68*, 68–76. [CrossRef]
2. Pinto, D.D.D.; Zaidy, S.A.H.; Hartono, A.; Svendsen, H.F. Evaluation of a phase change solvent for CO_2 capture: Absorption and desorption tests. *Int. J. Greenh. Gas Control* **2014**, *28*, 318–327. [CrossRef]
3. Knuutila, H.K.; Rennemo, R.; Ciftja, A.F. New solvent blends for post-combustion CO_2 capture. *Green Energy Environ.* **2019**, *4*, 439–452. [CrossRef]
4. Wanderley, R.R.; Pinto, D.D.D.; Knuutila, H.K. Investigating opportunities for water-lean solvents in CO_2 capture: VLE and heat of absorption in water-lean solvents containing MEA. *Sep. Purif. Technol.* **2020**, *231*, 115883. [CrossRef]
5. Pinto, D.D.D.; Knuutila, H.; Fytianos, G.; Haugen, G.; Mejdell, T.; Svendsen, H.F. CO_2 post combustion capture with a phase change solvent. Pilot plant campaign. *Int. J. Greenh. Gas Control* **2014**, *31*, 153–164. [CrossRef]
6. Madan, T.; Van Wagener, D.H.; Chen, E.; Rochelle, G.T. Modeling pilot plant results for CO_2 stripping using piperazine in two stage flash. *Energy Procedia* **2013**, *37*, 386–399. [CrossRef]
7. Lin, Y.-J.; Chen, E.; Rochelle, G.T. Pilot plant test of the advanced flash stripper for CO_2 capture. *Faraday Discuss.* **2016**, *192*, 37–58. [CrossRef]
8. Ansaloni, L.; Rennemo, R.; Knuutila, H.K.; Deng, L. Development of membrane contactors using volatile amine-based absorbents for CO_2 capture: Amine permeation through the membrane. *J. Memb. Sci.* **2017**, *537*, 272–282. [CrossRef]
9. Hoff, K.A.; Svendsen, H.F. CO_2 absorption with membrane contactors vs. packed absorbers-Challenges and opportunities in post combustion capture and natural gas sweetening. *Energy Procedia* **2013**, *37*, 952–960. [CrossRef]
10. Van Holst, J.; Versteeg, G.F.; Brilman, D.W.F.; Hogendoorn, J.A. Kinetic study of CO_2 with various amino acid salts in aqueous solution. *Chem. Eng. Sci.* **2009**, *64*, 59–68. [CrossRef]
11. Bade, O.M.; Knudsen, J.N.; Gorset, O.; Askestad, I. Controlling Amine Mist Formation in CO_2 Capture from Residual Catalytic Cracker (RCC) Flue Gas. *Energy Procedia* **2014**, *63*, 884–892. [CrossRef]
12. Ansaloni, L.; Hartono, A.; Awais, M.; Knuutila, H.K.; Deng, L. CO_2 capture using highly viscous amine blends in non-porous membrane contactors. *Chem. Eng. J.* **2019**, *359*, 1581–1591. [CrossRef]
13. Chowdhury, F.A.; Yamada, H.; Higashii, T.; Goto, K.; Onoda, M. CO_2 Capture by Tertiary Amine Absorbents: A Performance Comparison Study. *Ind. Eng. Chem. Res.* **2013**, *52*, 8323–8331. [CrossRef]
14. Li, J.; Liu, H.; Liang, Z.; Luo, X.; Liao, H.; Idem, R.; Tontiwachwuthikul, P. Experimental study of the kinetics of the homogenous reaction of CO_2 into a novel aqueous 3-diethylamino-1,2-propanediol solution using the stopped-flow technique. *Chem. Eng. J.* **2015**, *270*, 485–495. [CrossRef]
15. Hartono, A.; Vevelstad, S.J.; Ciftja, A.; Knuutila, H.K. Screening of strong bicarbonate forming solvents for CO_2 capture. *Int. J. Greenh. Gas Control* **2017**, *58*, 201–211. [CrossRef]
16. Kim, H.; Hwang, S.J.; Lee, K.S. Novel Shortcut Estimation Method for Regeneration Energy of Amine Solvents in an Absorption-Based Carbon Capture Process. *Environ. Sci. Technol.* **2015**, *49*, 1478–1485. [CrossRef]
17. Feron, P.; Conway, W.; Puxty, G.; Wardrefrence 32haugh, L.; Green, P.; Maher, D.; Fernandes, D.; Cousins, A.; Shiwang, G.; Lianbo, L.; et al. Amine Based Post-combustion Capture Technology Advancement for Application in Chinese Coal Fired Power Stations. *Energy Procedia* **2014**, *63*, 1399–1406. [CrossRef]
18. Nakamura, S.; Yamanaka, Y.; Matsuyama, T.; Okuno, S.; Sato, H.; Iso, Y.; Huang, J. Effect of Combinations of Novel Amine Solvents, Processes and Packing at IHI's Aioi Pilot Plant. *Energy Procedia* **2014**, *63*, 687–692. [CrossRef]
19. Knudsen, J.N.; Andersen, J.; Jensen, J.N.; Biede, O. Evaluation of process upgrades and novel solvents for the post combustion CO_2 capture process in pilot-scale. *Energy Procedia* **2011**, *4*, 1558–1565. [CrossRef]
20. Kim, I.; Svendsen, H.F.; Børresen, E. Ebulliometric Determination of Vapor−Liquid Equilibria for Pure Water, Monoethanolamine, N-Methyldiethanolamine, 3-(Methylamino)-propylamine, and Their Binary and Ternary Solutions. *J. Chem. Eng. Data* **2008**, *53*, 2521–2531. [CrossRef]
21. Trollebø, A.A.; Saeed, M.; Hartono, A.; Kim, I.; Svendsen, H.F. Vapour-Liquid Equilibrium for Novel Solvents for CO_2 Post Combustion Capture. *Energy Procedia* **2013**, *37*, 2066–2075. [CrossRef]

22. Aronu, U.E.; Hoff, K.A.; Svendsen, H.F. Vapor–liquid equilibrium in aqueous amine amino acid salt solution: 3-(methylamino)propylamine/sarcosine. *Chem. Eng. Sci.* **2011**, *66*, 3859–3867. [CrossRef]

23. Monteiro, J.G.M.-S.; Pinto, D.D.D.; Zaidy, S.A.H.; Hartono, A.; Svendsen, H.F. VLE data and modelling of aqueous N,N-diethylethanolamine (DEEA) solutions. *Int. J. Greenh. Gas Control* **2013**, *19*, 432–440. [CrossRef]

24. Peng, D.-Y.; Robinson, D.B. A New Two-Constant Equation of State. *Ind. Eng. Chem. Fundam.* **1976**, *15*, 59–64. [CrossRef]

25. Renon, H.; Prausnitz, J.M. Local compositions in thermodynamic excess functions for liquid mixtures. *AIChE J.* **1968**, *14*, 135–144. [CrossRef]

26. Pinto, D.D.D.; Johnsen, B.; Awais, M.; Svendsen, H.F.; Knuutila, H.K. Viscosity measurements and modeling of loaded and unloaded aqueous solutions of MDEA, DMEA, DEEA and MAPA. *Chem. Eng. Sci.* **2017**, *171*, 340–350. [CrossRef]

27. Pinto, D.D.D.; Svendsen, H.F. An excess Gibbs free energy based model to calculate viscosity of multicomponent liquid mixtures. *Int. J. Greenh. Gas Control* **2015**, *42*, 494–501. [CrossRef]

28. Pinto, D.D.D.; Monteiro, J.G.M.-S.; Bersås, A.; Haug-Warberg, T.; Svendsen, H.F. eNRTL Parameter Fitting Procedure for Blended Amine Systems: MDEA-PZ Case Study. *Energy Procedia* **2013**, *37*, 1613–1620. [CrossRef]

29. Poli, R.; Kennedy, J.; Blackwell, T. Particle swarm optimization. An Overview. *Swarm Intell.* **2007**, *1*, 33–57. [CrossRef]

30. Joback, K.G. *A Unified Approach to Physical Property Estimation Using Multivariate Statistical Techniques*; Massachusetts Institute of Technology: Cambridge, MA, USA, 1984.

31. Hessen, E.T. *Thermodynamic Models for CO$_2$ Absorption*; NTNU: Trondheim, Norway, 2010.

32. *DIPPR The DIPPR Information and Data Evaluation Manager for the Design Institute for Physical Properties.* 2004. Available online: https://www.aiche.org/dippr (accessed on 16 December 2019).

33. Klepáčová, K.; Huttenhuis, P.J.G.; Derks, P.W.J.; Versteeg, G.F. Vapor Pressures of Several Commercially Used Alkanolamines. *J. Chem. Eng. Data* **2011**, *56*, 2242–2248. [CrossRef]

Article

Copper Adsorption by Magnetized Pine-Needle Biochar

Eleni Nicolaou, Katerina Philippou *, Ioannis Anastopoulos and Ioannis Pashalidis

Department of Chemistry, University of Cyprus, P.O. Box 20537, 1678 Nicosia, Cyprus;
eleninicol@hotmail.com (E.N.); anastopoulos_ioannis@windowslive.com (I.A.); pspasch@ucy.ac.cy (I.P.)
* Correspondence: kphili03@ucy.ac.cy

Received: 17 September 2019; Accepted: 26 November 2019; Published: 2 December 2019

Abstract: The Cu(II) adsorption from aqueous solutions by magnetic biochar obtained from pine needles has been studied by means of batch-type experiments. The biochar fibers have been magnetized prior (pncm: carbonized-magnetized pine needles) and after oxidation (pncom: carbonized-oxidized-magnetized pine needles) and have been used as adsorbents to study the presence of carboxylic moieties on the magnetization and following adsorption process. The effect of pH (2–10), initial metal concentration (10^{-5}–$9 \cdot 10^{-3}$ mol·L^{-1}) and contact time (0–60 min) has been studied by varying the respective parameter, and the adsorbents have been characterized by Fourier transform infrared (FTIR) and X-ray diffraction (XRD) measurements prior and after Cu(II)-adsorption. FTIR measurements were performed to investigate the formation of surface species and XRD measurements to record possible solid phase formation and characterize formed solids, including the evaluation of their average crystal size. The data obtained from the batch-type studies show that the oxidized magnetic biochar (pncom) presents significantly higher adsorption capacity (1.0 mmol g^{-1}) compared to pncm (0.4 mmol g^{-1}), which is ascribed to the synergistic effect of the carboxylic moieties present on the pncom surface, and the adsorption process follows the pseudo-second order kinetics. On the other hand, the FTIR spectra prove the formation of inner-sphere complexes and XRD diffractograms indicate Cu(II) solid phase formation at pH 6 and increased metal ion concentrations.

Keywords: copper adsorption; magnetized pine needle biochar; isotherms; kinetics; FTIR and XRD studies

---

1. Introduction

The presence of toxic metal ions in wastewaters is a major environmental issue due to their toxicity, which may harm the environment and affect human health. Therefore, the treatment of contaminated wastewaters before disposal into environmental compartments such as soils and natural water systems is a vital necessity. There are many techniques used for the decontamination of the wastewater from toxic metals/metalloids, such as precipitation and flocculation using chemical reagents, flotation techniques, ion exchange procedures, and membrane filtration processes [1].

Adsorption-based remediation technologies are of particular interest because of the existing know-how and their relatively simple and effective application in waste water treatment. Recently, there have been several studies related to abundant and low cost biomass by-products [2], including biochars [3]. Activated carbon and biochar materials are used in many technological applications and analytical processes because of their high affinity for metal/metalloid ions and organic pollutants, which is associated with their large surface area and the presence of active surface groups, particularly after appropriate chemical modification [4–9].

The application of activated carbons (AC) to treat large volumes of wastewater produced by domestic water use and industry is restricted due to its relatively high production costs. More specifically, the biochar and AC cost is \$350–\$1200 and \$1100–\$1700 per tonne, respectively [10]. Moreover, AC materials are produced from non-renewable coal and are thermally activated to improve/promote their adsorption characteristics [10]. On the other hand, biochar materials can be produced from biomass by-products and are also applied as soil amendments to enhance soil carbon sequestration and improve soil productivity [11]. In wastewater treatment applications, biochars are usually chemically modified to increase their adsorption capability and selectivity towards specific pollutants [12]. These techniques include physical activation (steam, gas), chemical activation (alkali and acid treatments) and impregnation methods or combined methods [12].

Hence, modified biochar materials are very attractive adsorbents because they could effectively be used for the removal and recovery of precious and industrial metals from process solutions and wastewaters, and they could replace activated carbons widely used in the treatment and purification of waters. Generally, all types of biomass could be converted into biochar. However, lignocellulosic biomasses are much better precursors because of their higher density and hardness. In addition, the adsorption capacity and selectivity of the biochars towards cations can be significantly increased after chemical oxidation of their surface, which results in the formation of oxygen-containing moieties (e.g., hydroxyl and carbonyl groups) [4,13,14]. Moreover, deposition of metal oxides (e.g., MnO_2 and Fe_3O_4) on the biochar surface may further enhance the adsorption capacity and selectivity of the material towards certain metal ions. Moreover, magnetization of the material allows easy separation of the adsorbed contaminants using an external magnetic field [6,7,15].

This study aims at investigating the adsorption and removal of Cu(II) from water solutions using magnetized biochar adsorbents. Aliquots of the biochar carbonized pine needles (pnc), which has been obtained from pine needles, were chemically oxidized (pnco) and both materials pnc and pnco were magnetized resulting in pncm (carbonized-magnetized pine needles) and pncom (carbonized-oxidized-magnetized pine needles), respectively. Cu(II) has been used as adsorbate as it can be easily measured using the Cu(II) ion selective electrode, has low toxicity for humans, and could be a good analogue for Cd(II) and Pb(II). The fibrous structure of the pine needles, which is an abundant and otherwise useless biomass, is expected to remain in the derived biochar resulting in a robust, high surface biochar, which is further modified by precipitating iron-oxides on its surface and forming a magnetic biochar composite, with desired properties such as magnetism and enormous external surface due to the presence of iron-oxide nanoparticles. The study is based on the investigation of physicochemical parameters such as pH, copper concentration and contact time, which affect the Cu(II) removal by means of batch-type experiments. Moreover, Fourier transform infrared (FTIR) and X-ray diffraction (XRD) measurements are performed to understand the adsorption mechanism, which is of fundamental importance for the development of water treatment processes regarding the removal and recovery of precious and toxic metals/metalloids.

2. Experimental

2.1. Materials

The experiments were carried out under ambient conditions. The Cu(II) stock solutions were prepared by dissolution of $CuSO_4·5H_2O$ (Merck, >98%, Darmstadt, Germany). The needles were obtained from a domestic pine tree (*Pinus brutia Pegeia*, Cyprus) and after cleaning and washing, the needles were carbonized and oxidized as previously described [8] to produce the carbonized pine needles (pnc) and following the chemically oxidized pine needle (pnco) biochar. Iron(II) sulfate heptahydrate, $FeSO_4·7H_2O$ (Sigma-Aldrich, 99%, St. Louis, MO, USA) and anhydrous iron(III) chloride, $FeCl_3$ (Merck, 98%, Darmstadt, Germany) were used as received. The corresponding magnetic biochars (pncm and pncom) were produced according to Oliveira et al. [16]. Specifically, certain amounts of $FeSO_4·7H_2O$ (3.89 g) and $FeCl_3$ (4.54 g) were mixed together with biochars (9.72 g) (weight ratio 3:1) and deionized water (400 mL), at 70 °C, under constant stirring. Then, 100 mL of 5 M NaOH (Merck, Darmstadt, Germany) was dropwise added to the mixture in order to precipitate the iron oxides on the biochar surface. Finally, the products were filtered using a glass frit funnel and washed with distilled water until nearly neutral pH and dried overnight in a vacuum-oven at 110 °C. The materials were characterized by Fourier transform infrared spectroscopy (FTIR, spectrometer 8900, Shimadzu Europa GmbH, Duisburg, Germany)) and powder X-ray diffraction (XRD 6000, Shimadzu Europa GmbH, Duisburg, Germany) before and after Cu(II) adsorption.

2.2. Adsorption Experiments

The study of the Cu(II) adsorption by pncom and pncm was performed by investigating the effect of pH (2–10), the initial copper concentration (10^{-5}–$9\cdot10^{-3}$ mol·L^{-1}) and the adsorption kinetics (0–60 min). The batch-type experiments were carried out in 50 mL polyethylene beakers under ambient conditions as described elsewhere [8]. Specifically, the initial volume of the aqueous suspensions of magnetic biochars was 30 mL and the adsorbent mass 0.01 g. The effect of the studied parameters was investigated at constant total copper concentration $5\cdot10^{-4}$ M (except of the experiments related to the initial copper concentration) at pH 3 and pH 6 (except of the pH experiments). The pH was controlled by addition of NaOH and/or HClO$_4$ (0.01 to 1 M, Merck, Darmstadt, Germany). To reach equilibrium, the samples were placed in a thermostatic orbital shaker (100 r/min, Gallenkamp) for three days. The Cu(II) ion concentration was determined potentiometrically using a Cu(II) ion selective electrode as described elsewhere [4,5,9]. Graphing and data analysis were carried out using Kaleidagraph (4.5.2, Synergy Software, Reading, PA, USA, 2014).

The amount of copper adsorbed at time t, q_t (mg/g), was evaluated using the Equation (1):

$$q_t = \frac{(C_i - C_t)V}{m} \tag{1}$$

where C_i (mg/L) is the initial Cu(II) concentration in solution, C_t (mg/L) is the final Cu(II) concentration in solution at time t, V (L) is the solution volume, and m (g) is the dry weight of the modified biochar.

3. Results and Discussion

3.1. Effect of pH

The pH is a crucial factor regarding Cu(II) speciation in solution and the deprotonation of the acidic surface moieties (e.g., carboxylic-, phenolic-, hydroxy- groups) and determines the surface charge of the biochar materials. Generally, at relatively low Cu(II) concentrations ((Cu(II)) < 1 mmol) and up to pH 6, Cu(II) is present in solution mainly as Cu^{2+} cations and hydrolysis becomes predominant in near neutral and alkaline solutions. The surface charge of pncm is basically related to the point of zero charge (pzc) of the magnetite present on the biochar's surface, and because the pzc of magnetite is 3.8 [17] for pH > 3.8 the negative charged moieties are expected to dominate on the pncm surface. On the other hand, the surface charge of the oxidized biochar composite (pncom) depends on both, the deprotonation of the hydroxyl groups, which are present on the magnetite surface, and the carboxylic moieties present on the surface of oxidized composite [8,16].

According to Figure 1 there are significant differences regarding the adsorption behavior of pncm and pncom, basically in the acidic pH range (pH < 6). In contrast to Cu(II) adsorption on pncom, which increases linearly with pH and reaches maximum relative adsorption values (~100%) at pH 5, the Cu(II) adsorption reaches a plateau (~35%) at pH values between 4 and 5 and then increases steeply and reaches the maximum relative adsorption values (~100%) for pH > 6. The difference in the adsorption behavior between pncm and pncom in the acidic pH range is ascribed to the presence of the carboxylic groups on the pncom surface, which present increased chemical affinity for Cu(II) ions even in the acidic pH range [4,5]. On the other hand, the sharp increase of the relative adsorption observed for pncm at near neutral pH values could be attributed to solid phase formation and surface precipitation of copper salts (e.g., Cu(OH)$_2$), which is corroborated by XRD measurements, as discussed in more detail below.

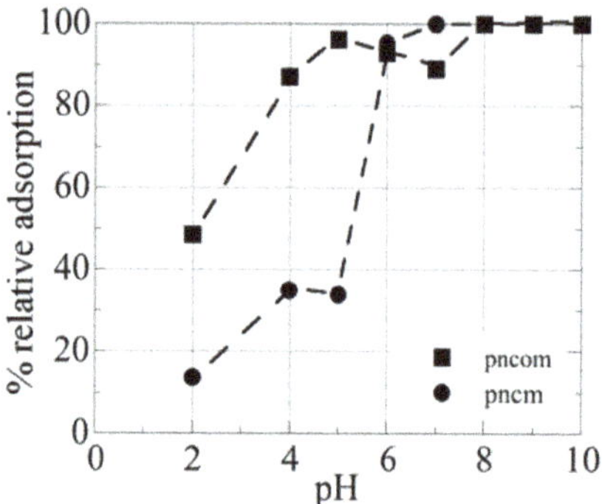

Figure 1. The pH effect on the adsorption efficiency of the magnetic biochar composites pncm (carbonized-magnetized pine needles) and pncom (carbonized-oxidized-magnetized pine needles) for Cu(II). The experiments were performed under ambient conditions and $(Cu(II))_{tot} = 5 \cdot 10^{-4}$ M.

3.2. XRD Measurements

Figure 2 shows the diffractograms of adsorbent aliquots corresponding to Cu(II) adsorption by magnetic pine needle biochar composites pncm and pncom at pH 3 and pH 6 and different Cu(II) concentrations. According to the diffractograms of samples corresponding to experiments performed at pH 3 there is no formation of copper solid phases, because under such acidic conditions the Cu(II) concentrations in solution are below the Cu(II) concentration determined by the formed solid phase. Nevertheless, the diffractograms include a peak at $2\theta = 36°$, which corresponds to magnetite [18]. Interestingly, the magnetite peaks corresponding to the pncm samples are very small compared to the peaks in the diffractograms of pncom, indicating that on the oxidized biochar surface the magnetite formation was more effective and has resulted in the formation of larger particles.

On the other hand, the XRD diffractograms of the samples corresponding to experiments performed at pH 6 include after a certain Cu(II) concentration sharp peaks, which correspond to Cu(II) solid phases [19]. Notably, the Cu(II) concentration above which the solid formation is observed is for pncm 1×10^{-4} M and for pncom 1×10^{-3} M total Cu(II) concentration, clearly indicating the higher Cu(II) adsorption capacity of pncom compared to the adsorption capacity of pncm for the same metal ions. It should be clear that when solid phase formation and surface precipitation occurs the corresponding data are useless regarding the evaluation of the maximum monolayer capacity of the adsorbents.

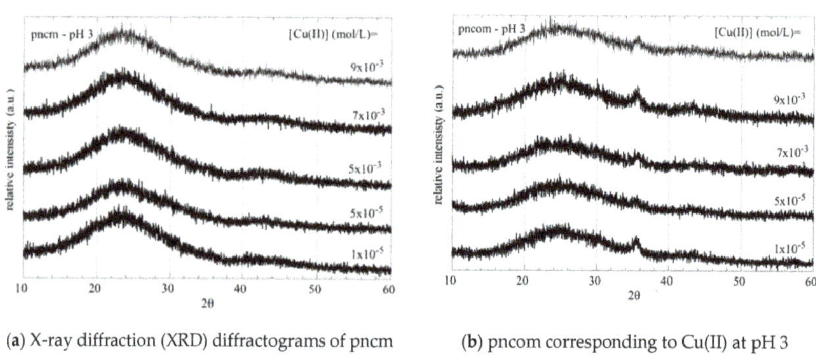

(a) X-ray diffraction (XRD) diffractograms of pncm

(b) pncom corresponding to Cu(II) at pH 3

Figure 2. *Cont.*

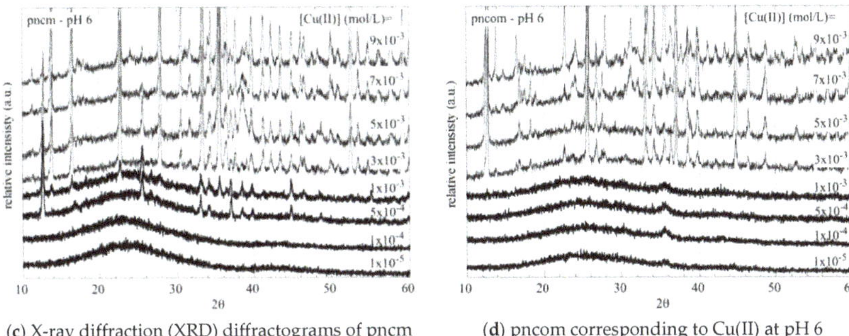

(**c**) X-ray diffraction (XRD) diffractograms of pncm   (**d**) pncom corresponding to Cu(II) at pH 6

Figure 2. X-ray diffraction (XRD) diffractograms of adsorbent aliquots corresponding to Cu(II) adsorption by the magnetic biochar composites pncm and pncom at two different pH values and different Cu(II) concentration.

3.3. Adsorption Data

In order to determine the maximum monolayer adsorption capacity, the experimental data were fitted with the Langmuir [20] adsorption model (Equation (2)).

$$q_e = q_m \frac{K_L C_e}{1 + K_L C_e} \tag{2}$$

where C_e is the concentration (mg/L) of Cu(II) in solution at equilibrium, q_e is the amount of the Cu(II) adsorbed per mass adsorbent (mg/g) at equilibrium, q_m is the maximum monolayer adsorption capacity (mg/g) and K_L is an equilibrium constant associated with the energy of adsorption. The model has been applied only to data obtained from experiments at pH 3 (Figure 3), because at pH 6 surface precipitation of Cu(II) solids occurs and only at very low Cu(II) concentration the Cu(II) adsorption by the magnetic biochars is the dominating surface-related process. The value maximum adsorption capacity evaluated applying the Langmuir adsorption model has been found to be $q_{max} = 0.40$ mmol·g^{-1} (25.4 mg·g^{-1}) and $q_{max} = 1.0$ mmol·g^{-1} (63.5 mg·g^{-1}) for the Cu(II) adsorption by pncm and pncom, respectively. This values are in the range of values determined for Cu(II) adsorption by magnetized and chemically modified biochars [15,21–24].

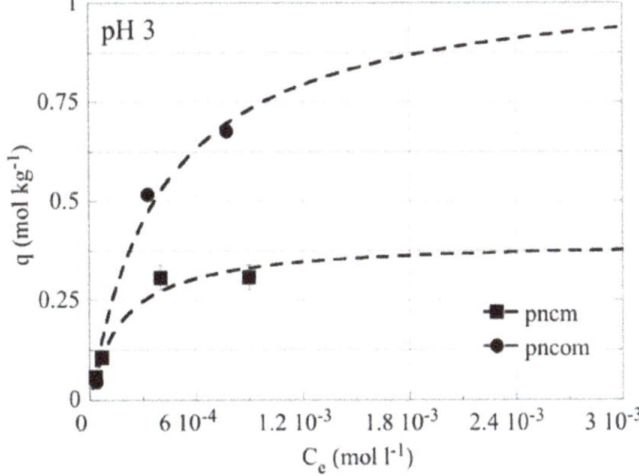

Figure 3. Experimental adsorption data and fitted curves evaluated using the Langmuir isotherm model for the Cu(II) adsorption by the magnetic biochar composites pncm and pncom. The adsorption experiments were performed under ambient conditions and pH 3.

3.4. FTIR Measurements

Figure 4 shows FTIR spectra of solid samples corresponding to experiments performed at pH 3 and pH 6, and different Cu(II) concentrations. The FTIR spectra and particularly those corresponding to pncom show significant changes in the carbonyl range indicating also the participation of the carboxylic groups, which are present on the biochar composite surface and bind Cu(II) [4,5].

On the other hand, the additional peaks around 3500 cm^{-1} and between 1100 cm^{-1} and 500 cm^{-1} observed in the FTIR spectra of the solid samples corresponding to experiments performed at pH 6 are attributed to the formation of the Cu(II) solid phase (e.g., $Cu(OH)_2$) [19]. These results are in agreement with the results obtained from XRD measurements of the corresponding samples and prove the solid phase formation and surface precipitation of $Cu(OH)_2$ at increased pH values, which favor the Cu(II) hydrolysis [25].

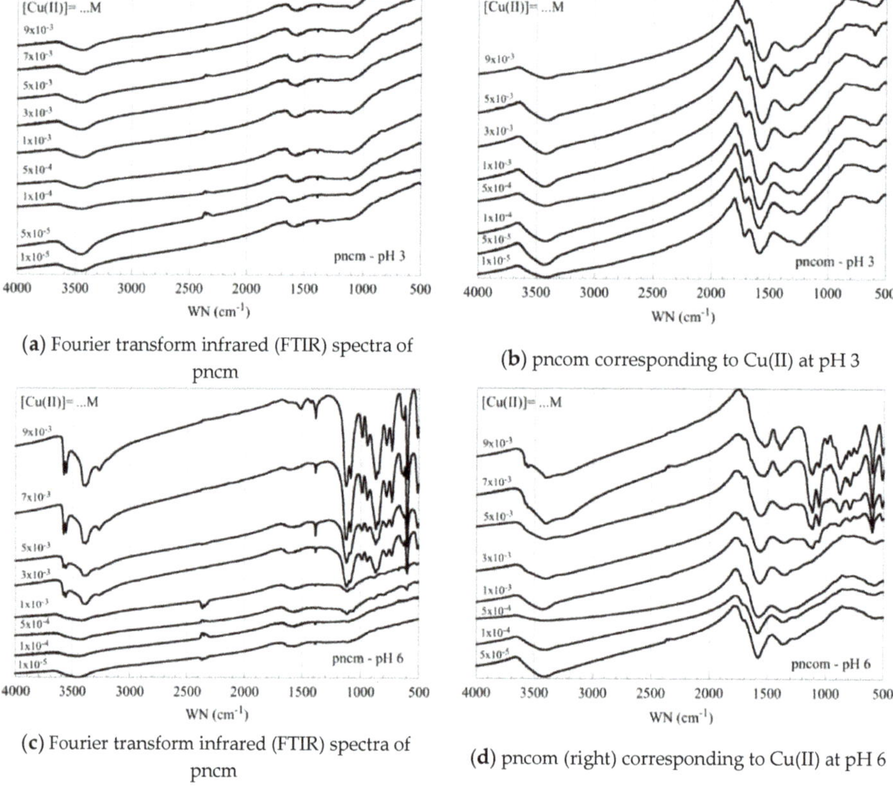

(**a**) Fourier transform infrared (FTIR) spectra of pncm

(**b**) pncom corresponding to Cu(II) at pH 3

(**c**) Fourier transform infrared (FTIR) spectra of pncm

(**d**) pncom (right) corresponding to Cu(II) at pH 6

Figure 4. Fourier transform infrared (FTIR) spectra of adsorbent aliquots corresponding to Cu(II) adsorption by the magnetic biochar composites pncm and pncom at two different pH values and different Cu(II) concentrations. The spectra have been obtained in the wavenumber (WN) range between 4000 and 500 cm^{-1}.

Generally, Cu(II) binding by magnetized biochars can occur through pi-cation interactions and complex formation between the carboxylic moieties [26], which are present on the surface of oxidized biochar composite and the metal cations, as well as surface complexation of Cu(II) cations by the hydroxo groups [26] present on the magnetite particle surface (Figure 5).

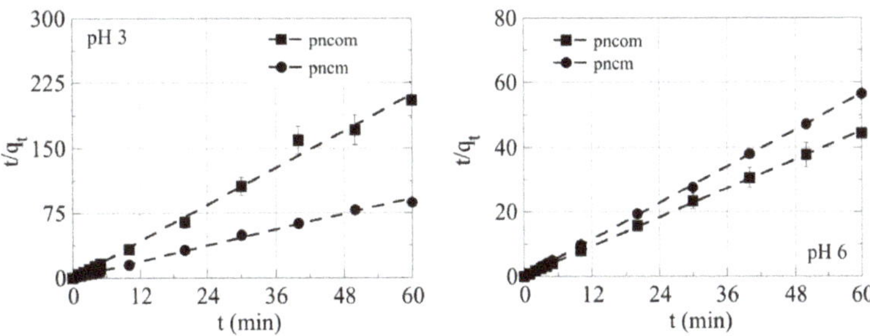

Figure 5. Schematic illustration of the different binding modes of Cu(II) cations by magnetized biochars.

3.5. Kinetic Studies

The effect of contact time was examined in a range between 0 and 60 min at two different pH values. The adsorption was found to be relatively fast and Cu(II) was adsorbed almost quantitatively in the first 10 min. The estimation of copper uptake is an essential parameter for the development of an effective and accurate design for large scale adsorption processes.

For this purpose the experimental kinetic data were fitted with the pseudo-first-order (Equation (3)) [27] and pseudo-second-order kinetic (Equation (4)) models [28,29], in their linearized form:

$$\ln(q_e - q_t) = \ln(q_e) - k_1 t \tag{3}$$

$$\frac{t}{q_t} = \frac{1}{k_2 q_e^2} + \frac{1}{q_e} t \tag{4}$$

where q_e (mg g^{-1}) and q_t are the amounts of adsorbed Cu(II) on the biochar at the equilibrium and at any time t, respectively. For the pseudo-first-order kinetic model the q_e and k_1 can be calculated from the slope and intercept obtained from the plots $\ln(q_e - q_t)$ versus t. For the pseudo-second-order kinetic model the q_e and k_2 can be calculated from the slopes and the intercept of the plot $\frac{t}{q_t}$ versus t.

The constants of the kinetic models are summarized in Table 1. According to Table 1, the experimental data are well fitted with the pseudo-second order kinetics and the associated curves are shown in Figure 6. The pseudo-second order kinetics is characteristic for magnetic biochar adsorbents and the evaluated k_2 values are similar to corresponding literature k_2 values [15,21–24,30].

Table 1. Kinetic parameters evaluated for the Cu(II) adsorption by pncm and pncom at pH 3 and pH 6, and evaluated by fitting the experimental data with the pseudo-first and pseudo-second order kinetic models.

	Pseudo-First-Order				Pseudo-Second-Order			
	pncm		pncom		pncm		pncom	
pH	k_1 (min^{-1})	R	k_1 (min^{-1})	R	k_2 (g·mg^{-1}·min^{-1})	R	k_2 (g·mg^{-1}·min^{-1})	R
3	0.199	0.583	0.002	0.187	1.525	0.998	3.525	0.996
6	0.001	0.126	0.001	0.992	0.939	0.999	0.746	0.999

Figure 6. Kinetic experimental data and fitted curves using the pseudo second kinetic model for the Cu(II) adsorption by the magnetic biochar composites pncm and pncom. The adsorption experiments were performed under ambient conditions and two different pH values.

4. Conclusions

The conclusions that can be drawn from the present study are: (a) Biochar magnetization after oxidation results in significantly higher adsorption capacities due to the synergistic effect of the carboxylic moieties, which favor iron oxide deposition on the biochar surface and complex Cu(II) ions even at low pH (pH < 4); (b) The adsorption process follows the pseudo-second order kinetics; and (c) FTIR measurements denote the formation of inner-sphere complexes and XRD data show Cu(II) surface precipitation at pH 6 and increased Cu(II) concentration. Generally, magnetisation of biochar results in better adsorption capacity and desired properties regarding magnetic-based separation.

Author Contributions: Conceptualization, I.P.; Data curation, E.N., K.P. and I.P.; Formal analysis, E.N., K.P. and I.P.; Investigation, E.N. and K.P.; Project administration, K.P. and I.P.; Supervision, I.P.; Validation, K.P. and I.P.; Visualization, E.N.; Writing—original draft, I.P.; Writing—review & editing, K.P. and I.A.

Funding: This research received no external funding.

Conflicts of Interest: The authors declare no conflict of interest.

References

1. Kurniawan, T.A.; Chan, G.Y.S.; Lo, W.H.; Babel, S. Physico–chemical treatment techniques for wastewater laden with heavy metals. *Chem. Eng. J.* **2006**, *118*, 83–98. [CrossRef]
2. Michalak, I.; Chojnacka, K.; Witek-Krowiak, A. State of the art for the biosorption process—A review. *Appl. Biochem. Biotechnol.* **2013**, *170*, 1389–1416. [CrossRef] [PubMed]
3. Liu, W.J.; Jiang, H.; Yu, H.Q. Development of biochar-based functional materials: Toward a sustainable platform carbon material. *Chem. Rev.* **2015**, *115*, 12251–12285. [CrossRef] [PubMed]
4. Hadjittofi, L.; Prodromou, M.; Pashalidis, I. Activated biochar derived from cactus fibres–preparation, characterization and application on Cu (II) removal from aqueous solutions. *Bioresour. Technol.* **2014**, *159*, 460–464. [CrossRef]
5. Liatsou, I.; Constantinou, P.; Pashalidis, I. Copper binding by activated biochar fibres derived from Luffa Cylindrica. *Water AirSoil Pollut.* **2017**, *228*, 255. [CrossRef]
6. Zhou, Q.; Liao, B.; Lin, L.; Qiu, W.; Song, Z. Adsorption of Cu (II) and Cd (II) from aqueous solutions by ferromanganese binary oxide–biochar composites. *Sci. Total Environ.* **2018**, *615*, 115–122. [CrossRef]
7. Hadjiyiannis, P.; Pashalidis, I. Copper (II) adsorption by Opuntia ficus-indica biochar fiber-MnO_2 composites. *Desalin. Water Treat.* **2019**, *159*, 60–65. [CrossRef]
8. Philippou, K.; Savva, I.; Pashalidis, I. Uranium(VI) binding by pine needles prior and after chemical modification. *J. Radioanal. Nucl. Chem.* **2018**, *318*, 2205–2211. [CrossRef]
9. Liatsou, I.; Pashalidis, I.; Dosche, C. Cu(II) adsorption on 2-thiouracil-modified Luffa Cylindrica biochar fibres from artificial and real samples, and competition reactions with U(VI). *J. Hazard. Mater.* **2019**. [CrossRef]
10. Thompson, K.A.; Shimabuku, K.K.; Kearns, J.P.; Knappe, D.R.; Summers, R.S.; Cook, S.M. Environmental comparison of biochar and activated carbon for tertiary wastewater treatment. *Environ. Sci. Technol.* **2016**, *50*, 11253–11262. [CrossRef]
11. Nair, V.D.; Nair, P.K.R.; Dari, B.; Freitas, A.M.; Chatterjee, N.; Pinheiro, F.M. Biochar in the Agroecosystem-Climate-Change-Sustainability Nexus. *Front. Plant Sci.* **2017**, *8*, 2051. [CrossRef] [PubMed]
12. Rangabhashiyam, S.; Balasubramanian, P. The potential of lignocellulosic biomass precursors for biochar production: Performance, mechanism and wastewater application—A review. *Ind. Crops Prod.* **2019**, *128*, 405–423. [CrossRef]
13. Hadjittofi, L.; Pashalidis, I. Uranium sorption from aqueous solutions by activated biochar fibres investigated by FTIR spectroscopy and batch experiments. *J. Radioanal. Nuclear Chem.* **2015**, *304*, 897–904. [CrossRef]
14. Liatsou, I.; Pashalidis, I.; Oezaslan, M.; Dosche, C. Surface characterization of oxidized biochar fibers derived from Luffa Cylindrica and lanthanide binding. *J. Environ. Chem. Eng.* **2017**, *5*, 4069–4074. [CrossRef]
15. Kołodyńska, D.; Bąk, J. Use of three types of magnetic biochar in the removal of copper (II) ions from wastewaters. *Sep. Sci. Technol.* **2018**, *53*, 1045–1057. [CrossRef]
16. Oliveira, L.C.A.; Rios, R.V.R.A.; Fabris, J.D.; Garg, V.; Sapag, K.; Lago, R.M. Activated carbon/iron oxide magnetic composites for the adsorption of contaminants in water. *Carbon* **2002**, *40*, 2177–2183. [CrossRef]

17. Milonjić, S.; Kopečni, M.; Ilić, Z. The point of zero charge and adsorption properties of natural magnetite. *J. Radioanal. Chem.* **1983**, *78*, 15–24. [CrossRef]

18. Majumder, S.; Sardar, M.; Satpati, B.; Kumar, S.; Banerjee, S. Magnetization Enhancement of Fe_3O_4 by Attaching onto Graphene Oxide: An Interfacial Effect. *J. Phys. Chem. C* **2018**, *122*, 21356–21365. [CrossRef]

19. Devamani, R.H.P.; Alagar, M. Synthesis and characterisation of copper II hydroxide nano particles. *Nano Biomed. Eng.* **2013**, *5*, 116–120. [CrossRef]

20. Langmuir, I. The adsorption of gases on plane surfaces of glass, mica and platinum. *J. Am. Chem. Soc.* **1918**, *40*, 1361–1403. [CrossRef]

21. Nyamunda, B.C.; Chivhanga, T.; Guyo, U.; Chigondo, F. Removal of Zn (II) and Cu (II) Ions from Industrial Wastewaters Using Magnetic Biochar Derived from Water Hyacinth. *J. Eng.* **2019**. [CrossRef]

22. Son, E.B.; Poo, K.M.; Chang, J.S.; Chae, K.J. Heavy metal removal from aqueous solutions using engineered magnetic biochars derived from waste marine macro-algal biomass. *Sci. Total Environ.* **2018**, *615*, 161–168. [CrossRef] [PubMed]

23. Song, Q.; Yang, B.; Wang, H.; Xu, S.; Cao, Y. Effective removal of copper (II) and cadmium (II) by adsorbent prepared from chitosan-modified magnetic biochar. *J. Residuals Sci. Technol* **2016**, *13*, 197–205. [CrossRef]

24. Yin, Z.; Liu, Y.; Liu, S.; Jiang, L.; Tan, X.; Zeng, G.; Li, M.; Liu, S.; Tian, S.; Fang, Y. Activated magnetic biochar by one-step synthesis: Enhanced adsorption and coadsorption for 17β-estradiol and copper. *Sci. Total Environ.* **2018**, *639*, 1530–1542. [CrossRef] [PubMed]

25. Baes, C.F.; Mesmer, R.E. *The Hydrolysis of Cations*; John Wiley & Sons: New York, NY, USA, 1976; Volume 81, pp. 245–246.

26. Philippou, K.; Anastopoulos, I.; Dosche, C.; Pashalidis, I. Synthesis and characterization of a novel Fe_3O_4-loaded oxidized biochar from pine needles and its application for uranium removal. Kinetic, thermodynamic, and mechanistic analysis. *J. Environ. Manag.* **2019**, *252*, 109677. [CrossRef] [PubMed]

27. Lagergren, S. About the theory of so-called adsorption of soluble substances. *Sven. Vetensk. Handingarl* **1898**, *24*, 1–39.

28. Blanchard, G.; Maunaye, M.; Martin, G. Removal of heavy metals from waters by means of natural zeolites. *Water Res.* **1984**, *18*, 1501–1507. [CrossRef]

29. Ho, Y.S.; McKay, G. Pseudo-second order model for sorption processes. *Process Biochem.* **1999**, *34*, 451–465. [CrossRef]

30. Xiao, F.; Cheng, J.; Cao, W.; Yang, C.; Chen, J.; Luo, Z. Removal of heavy metals from aqueous solution using chitosan-combined magnetic biochars. *J. Colloid Interface Sci.* **2019**, *540*, 579–584. [CrossRef]

Article

Filtration Performances of Different Polysaccharides in Microfiltration Process

Shujuan Meng [1], Hongju Liu [1], Qian Zhao [2], Nan Shen [3] and Minmin Zhang [4,*]

[1] School of Space and Environment, Beihang University, No. 37 Xueyuan Road, Beijing 100191, China; mengsj@buaa.edu.cn (S.M.); liuhj@buaa.edu.cn (H.L.)

[2] School of Municipal and Environmental Engineering, Shandong Jianzhu University, 1000 Fengming Road, Jinan 250101, China; zhaoqian@sdjzu.edu.cn

[3] School of Environment, Nanjing Normal University, 122 Ninghai Rd, Nanjing, Jiangsu 210023, China; nanshen@mail.ustc.edu.cn

[4] Suntar Membrane Technology (Xiamen) Co. Ptd, Fujian 361022, China

* Correspondence: mzhang6@e.ntu.edu.sg

Received: 10 November 2019; Accepted: 28 November 2019; Published: 2 December 2019

Abstract: Membrane technology has been widely applied for water treatment, while membrane fouling still remains a big challenge. The polysaccharides in extracellular polymeric substances (EPS) have been known as a significant type of foulant due to their high fouling propensity. However, polysaccharides have many varieties which definitely behave differently in membrane filtration. Therefore, in this study, different polysaccharides alginate sodium and xanthan gum were chosen to study their effects on membrane fouling in a wide concentration range. The results demonstrated that the filtration behaviors of alginate sodium and xanthan gum were completely different, which was due to their different molecular structures. Alginate had a small molecular weight and it was easy for alginate to penetrate membrane pores resulting in pore blocking. A series of concentrations of alginate including 5 mg/L, 10 mg/L, 20 mg/L, 30 mg/L, 40 mg/L, and 50 mg/L were examined and it was found that the permeate flux decline highly depended on the level of alginate in the feed water. While for the filtration of xanthan gum, the same concentration of xanthan gum led to more serious fouling than that observed in alginate, which might be due to its large molecule. In addition, calcium chloride was added in the solutions of both alginate and xanthan gum to examine the influence of a divalent cation on polysaccharide fouling. A "unimodal" peak can be observed in the fouling propensity caused by Ca^{2+} and alginate with increasing the concentration of alginate. Such a phenomenon was not found in the fouling of xanthan gum and Ca^{2+} led to more serious fouling for all concentrations of xanthan gum. In light of this, this study gave new insights into the fouling propensities of different polysaccharides.

Keywords: polysaccharides; membrane fouling; microfiltration process; calcium ion

---

1. Introduction

With the increasing population in the world, there is also an increasing demand for the quantity and quality of clean water and many techniques are explored to deal with this issue [1–5]. Membrane technology has provided a great solution for water treatment [6–8]. In the recent years, membrane technology has become a promising technology due to some advantages over the conventional methods such as coagulation, sedimentation, and clarification processes [9]. The benefits of membrane systems include less complexity, easier operation, less man power needed, less chemical usage, the ability in treating a broad range of contaminants etc. [10–12]. Nowadays, the membrane-based technology is widely used to meet the growing demand of clean water [9,13].

However, one of the main problems of membrane technology is membrane fouling, leading to reduced permeate flux and high operation costs [14–17]. Membrane fouling is caused by the settling of suspended particles or dissolved substances on its surface, pore openings, or within its pores. Membrane fouling is affected by many factors including foulant type, operation modes, composition of feed water, membrane materials and so on, which makes it very complex [12,18]. Extensive effort has been dedicated to the control and mitigation of membrane fouling which comprises chemical/physical cleaning, biologically-based methods, fabrication of anti-fouling membranes, and optimization of operation parameters [19]. The feed water to membrane systems often contains a wide spectrum of foulants including inorganic, organic, colloidal, and microbial substances, which is the key factor affecting fouling. Among these foulants, the organic substances play an important role in both organic fouling and biofouling [20]. Additionally, organic substances are the main contributor to irreversible fouling, which is difficult to be cleaned [21]. However, in all types of fouling, the organic fouling is perhaps the most poorly understood. There are many studies focusing on organic fouling and recently it has been noticed that not only the foulant concentration but also the foulant type matters in organic fouling [22]. More importantly, the interaction between foulant molecules also plays a significant role in fouling development [18,23]. Considering the abundance of organic foulants in natural water, more efforts should be devoted to explore this problem.

Extracellular polymeric substances (EPS) represent the most problematic organic foulants which are believed to cause serious membrane fouling [22]. EPS is synthesized from the excretion of high weight mucous by the microbial cells and the main components of EPS are polysaccharides, proteins humic acid, and other polymeric compounds [24]. The polysaccharides usually cause more serious fouling problems than other substances in EPS due to their long-chain molecules as well as their special gelling properties [21,25]. However, there are various types of polysaccharides in natural water environments and the molecule structures of polysaccharides are different from one to another. It has been found that the molecule structure of polysaccharide is important in fouling membrane and even alginate blocks derived from the same source behaved differently in filtration tests [26,27]. Therefore, an in-depth investigation is required to identify the fouling propensities of different polysaccharides to deepen the understanding of polysaccharides fouling and the relevant fouling mechanisms.

In order to further comprehend the polysaccharide fouling, two different polysaccharides, which are, alginate sodium and xanthan gum are applied as model foulants in this study. Both of them are the typical foulant models that have been widely used in fouling studies. Alginate is a typical model polysaccharide and xanthan gum is another natural polysaccharide employed as the model foulant of EPS [28,29]. Alginate and xanthan gum share some similarities and differences. Both alginate and xanthan gum are long-chain molecules and have some same function groups such as carboxylic group (COO-) and hydroxyl groups (-OH) which are possible binding sites for divalent cations. In addition, alginate and xanthan gum differ from each other in some aspects which in turn would provide a better comparison about the different polysaccharides in membrane fouling. Alginate is a linear chain without branches and has a relatively small molecule weight in the range of 12–80 kDa [30]. Differently, xanthan gum has a trisaccharide side chain and is larger than alginate with molecular weight of ~500 kDa [29]. Additionally, the effect of calcium ions on polysaccharide fouling was also investigated in this study. Alginate has been widely known to interact with Ca^{2+} in an "egg-box" model [31] while the binding pattern between xanthan gum and Ca^{2+} is not clear at this time. In this study, the two chosen polysaccharides were prepared in a series of different levels and the filtration behaviors of them were carefully examined. Moreover, the effect of calcium ion on their fouling propensities were analyzed to determine the influence of Ca^{2+} on different types of polysaccharide fouling.

2. Materials and Methods

2.1. Alginate and Xanthan Gum

In this study, two types of polysaccharides, namely alginate sodium (Sigma, St. Louis, MO, USA) and xanthan gum (Sigma, St. Louis, MO, USA) were employed as model foulants to identify potential differences in the fouling propensities of different types of polysaccharides. The two compounds could work as the model polysaccharides in which alginate was the one representing polysaccharides with small molecule weight and xanthan gum represented polysaccharides with large molecular weight. Molecular weights of alginate and xanthan gum are ~12–80 kDa and ~500 kDa, respectively [23]. As can be seen in Figure 1, alginate is type of polysaccharide with a linear chain. There are two monomers in the alginate chain, (1→4) linked β-D-mannopyranuronic acid (M) and (1→4) linked α-L-gulopyranuronic acid (G), which are randomly arranged into homopolymeric blocks (MM-block, GG-block) and heteropolymeric blocks (MG-block) [32]. Differently, xanthan gum is a high molecule weight polysaccharide with a backbone chain consisting of (1–4) β-D-glucose units linked at the first and the fourth position, which is identical to that of cellulose (Figure 1b) [33]. The C-3 position of the alternate glucose residues is replaced by a trisaccharide side chain containing a D-glucuronic acid unit between two D-mannose units linked at the O-3 position of every other glucose residue in the main chain [29].

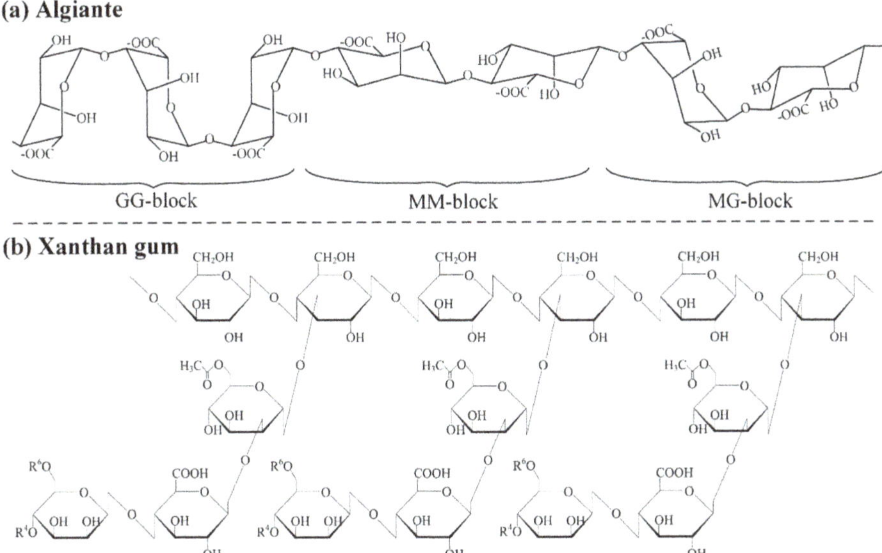

Figure 1. Parts of molecule structures of (**a**) alginate and (**b**) xanthan gum.

2.2. Dead End Filtration Setup

A dead end filtration system was used in this study which comprised a reservoir tank connecting to the filtration cell with the membrane at the bottom and a balance at the end of the setup to weigh the amount of filtrate collected. The entire setup was connected to a computer with software that records the filtration data. Nitrogen gas was supplied at a constant pressure of 1 bar. In this study, the 0.2 μm flat-sheet nylon membrane with a contact angle of 22.6 ± 0.7° at 20 °C was used which provides an effective area of 11.94 cm^2 (FilTrex, Singapore). The whole experiment was conducted for a total of 5 h. For the first 3 h, the membrane used was flushed with deionized water (MilliQ water) to achieve a stable permeate flux and thus to yield reliable results. The polysaccharide solutions prepared at desired concentrations were filtered for at least 2 h until the stable stage achieved. All the data were collected by the data logger. All filtration tests were repeated at least three times and the representative data are shown in this study.

2.3. Preparation of Feed Solutions

An amount of 0.005 g of alginate sodium powder was weighed on a weighing balance, after which it was poured into a beaker of 1 L and using deionized water to fill the beaker up to 1 L. This beaker was then placed on a stirrer to mix the alginate sodium solution for 2 h. At the same time, when alginate sodium solution was being mixed, the reservoir was filled with deionized water. The deionized water was used to flush through the filtration cell and membrane for 3 h to stabilize the membrane. Subsequently, mixed alginate sodium solution was poured into the reservoir and run for 2 h at a pressure of 1 bar. Data was collected from the system using a data logger during the 2 h. This process was again repeated for the other concentrations of alginate sodium: 10 mg/L, 20 mg/L, 30 mg/L, 40 mg/L, and 50 mg/L. As for the preparation of calcium ion feed water, alginate powder and calcium chloride (Sigma, St. Louis, MO, USA) was weighed and poured into the same beaker at a level of 1 mM Ca^{2+}. This beaker was placed on the stirrer to mix the alginate sodium and calcium chloride for 2 h. Feed solutions were then added into the reservoir and run for two hours. The preparation procedure of xanthan gum solutions was actually the same as described above. Different from the preparation of alginate with calcium chloride, the addition of calcium in the xanthan gum solutions must be done after the sufficient dissolution of xanthan gum to prevent the formation of large gels. The same filtration tests were conducted and the permeate data was collected using the data logger. In addition, the viscosities of the xanthan gum solutions were measured with a viscometer (DV2T, Brookfield, Middleboro, MA, USA) since the xanthan gum had been known to increase the viscosity of aqueous solutions. An analysis between the filtration resistances caused by xanthan gum and the viscosities of corresponding solutions would be conducted.

2.4. Field Emission Scanning Electron Microscopy Observation of Fouled Membrane

In order to examine the morphology of the fouled membrane surface, the clean membrane and the fouled membrane were cut into small pieces to be observed via an FESEM. Before observation, all membrane samples were totally freeze-dried and coated with Pt. To obtain the representative graphs, more than 10 pictures were randomly taken from each sample.

3. Results and Discussion

3.1. Effect of Alginate Concentration on Membrane Fouling

In order to examine the effect of alginate concentration on membrane fouling, a set of alginate sodium solutions at 5 mg/L, 10 mg/L, 20 mg/L, 30 mg/L, 40 mg/L, and 50 mg/L were prepared. Before we conduced filtration tests, each membrane filter was contracted for three hours to exclude the interference of membrane and to achieve a stable stage of the membrane. This contraction could provide a relatively stable permeate flux and yield an accurate result. As shown in Figure 2, the filtration behaviors of these alginate samples were evaluated with a filtration time of 2 h. From the result in Figure 2, it demonstrated that with increasing of the alginate sodium concentration, the declines of the permeate flux became more serious. The reason for the slowdown in the permeate flux was possibly due to the fact that the pores of the membrane were blocked by alginate chains, thus the amount of water that can pass through membrane reduced. The more foulant, the more serious fouling was observed. From Figure 2, for the first 15 min, we can see that there was a drastic decrease of flux in the dead-end filtration system. However, from 60 min onwards, the flux started to be more constant. The fouling extent was determined by the filtration performance of the first 15 min. These results suggested that the fouling development in the filtration process was profoundly influenced by the initial stage of filtration. This set of experiment was run for 2 h. It shows that by 60 min, most of the pores on the membrane were relatively blocked and there was no drastic decline of flux anymore. It had come to a plateau by 120 min. Similarly, research done by Susanto et al., suggests that there is a possibility that a small amount of alginate could have entered the pores of the membrane causing the flux to decline drastically especially for 50 mg/L alginate sodium [34]. The flux went to the steady state at a much earlier time compared to the other concentrations of alginate feed solutions.

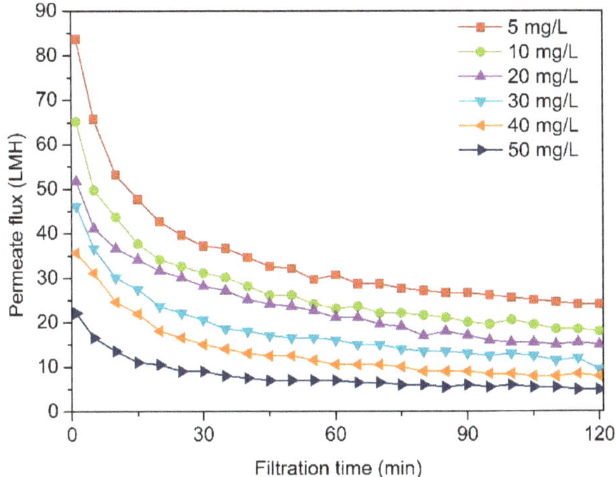

Figure 2. The effect of alginate concentration on membrane fouling.

Figure 3 showed the initial flux decline rate of alginate sodium at constant pressure of 1 bar. The decline rate was calculated using the gradient of the flux in Figure 2 for the first 10 min. This trend demonstrated that the initial flux decline rate increased with increasing concentration of alginate and was directly proportional to the level of alginate in feed with a good fit. The steady state is the point when the flux starts to stay relatively constant without drastic changes. The flux at the steady state was also determined by the concentration of alginate in the feed water and showed an inverse proportional relationship, as shown in Figure 3b. All these results suggested that for the microfiltration of alginate, the fouling problem depends on the level of foulant in feed water to membrane system. The molecular weight of alginate sodium is in the range of 12–80 kDa and the interaction between alginate molecules is weak, which suggests that the main fouling mechanism of alginate is pore blocking for the 0.2 μm membrane [23]. Therefore, at a high level of alginate, many alginate molecules entered the pores of the membrane at the very beginning of the filtration, adhered to the walls of pores, and ultimately caused rapid decline of permeate flux in the dead end filtration mode. At a low concentration of alginate, there were few alginate molecules thus the blocking rate of membrane pores was slow and less fouling happened during the filtration.

3.2. Effects of Calcium Ion on Alginate Fouling

The effect of Ca^{2+} on alginate fouling was examined with 5 mg/L, 10 mg/L, 20 mg/L, 30 mg/L, 40 mg/L, and 50 mg/L alginate solutions. Figure 4 shows the initial flux decline rates of alginate solutions with and without 1 mM Ca^{2+}. First, it can be seen that the addition of calcium ion aggravated the membrane fouling for all alginate samples except for 50 mg/L. In general, it had been widely known that divalent cations promoted the development of membrane fouling in which they acted as bridges and enhanced the interaction between foulant molecules and/or the interaction between foulant and membrane surfaces [31,35,36]. Surprisingly, at 50 mg/L of alginate, more serious fouling was observed in the filtration without calcium. It can be found that in the presence of Ca^{2+}, the fouling rate increased first and then decreased by increasing the concentration of alginate, achieving the maximum value at the concentration of 30 mg/L of alginate. The fouled membrane was examined to provide a direct observation of the fouling situation caused by alginate. As can be seen in Figure 5, compared to the clean membrane, the fouled membrane with 10 mg/L of alginate did not present an obvious fouling layer on the surface, while a gel layer can be observed in the filtration of alginate with Ca^{2+}.

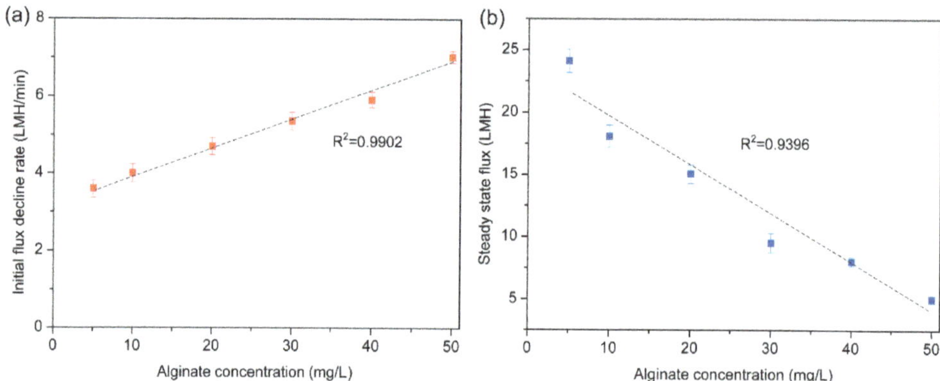

Figure 3. (**a**) The relationship between initial flux decline rate and alginate concentration and (**b**) the relationship between the steady state flux and alginate concentration.

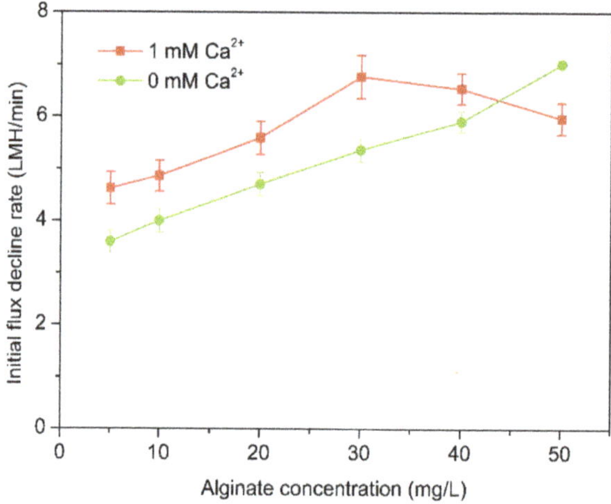

Figure 4. The initial fouling rate of alginate solutions with and without 1 mM Ca^{2+}.

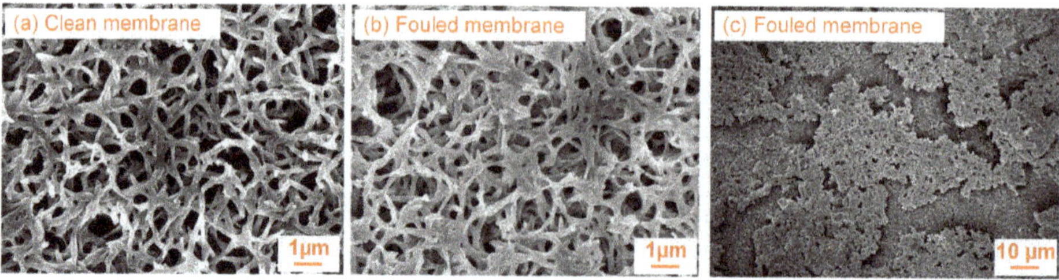

Figure 5. FESEM observations of the (**a**) clean membrane; membrane surfaces fouled by (**b**) alginate and (**c**) alginate with Ca^{2+}.

This unimodal mode of fouling was different from the trend showed in the filtration of alginate only. We deduced that this was due to the formation of gel-like substance from alginate and calcium ion as suggested in Figure 6. At the low concentration of alginate, there were few alginate chains that could crosslink together to form agglomeration via the "bridges" of Ca^{2+}. With further increasing of the level of alginate, more and more alginate molecules were included to form gel-like substances which efficiently blocked the pores of membrane and formed a fouling layer on the membrane surface. Thus, the pathway for permeate flux to pass through the membrane was narrowed. At high a concentration of alginate, the networks formed from alginate and Ca^{2+} were much larger and denser. It would be difficult for them to penetrate membrane pores, and the fouling layer they formed was more porous. As a result, the permeate flux increased again. Consequently, this finding indicated the existence of an "optimal" mixing ratio between alginate sodium and Ca^{2+} that would cause the most serious membrane fouling. Recently, the unimodal mode of fouling in the filtration of fixed concentration of foulant with varying Ca^{2+} level was also reported [37,38]. However, the exact interaction between foulant and Ca^{2+} and the inner structure of these networks should be further explored. Additionally, these results suggested that the gel layer formed on the membrane surface played a significant role in membrane fouling. The structure and permeability of gel layer would relate with the foulant type and operation conditions [39], which should be symmetrically investigated in future work.

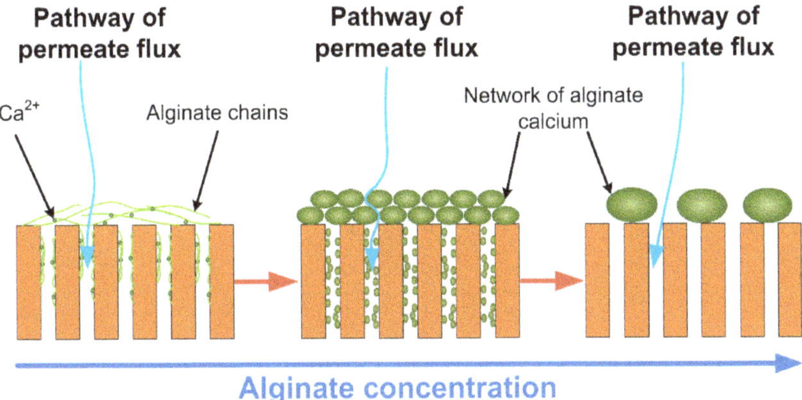

Figure 6. Schematic description of the pathways of permeate flux with increasing the alginate concentrations.

3.3. Effects of Concentrations of Xanthan Gum on Membrane Fouling

Another polysaccharide model xanthan gum was also investigated, and the effect of xanthan gum concentration on membrane fouling is shown in Figure 7. It can be found that the flux decline caused by xanthan gum was much higher than that resulted from alginate at the same concentration of foulant. Different from alginate, xanthan gum is a type of polysaccharide with a much higher molecular weight (~1,000,000 Da) [33]. Xanthan gum has been reported to be a good model polysaccharide to represent the EPS, because xanthan gum showed a similar rheological characteristic to that of activated sludge obtained from an membrane bio-reactor (MBR) plant [28]. The typical properties of xanthan gum are high molecular weight, unique thickening property, and gel formation when at rest. This gel is different from what we discussed in the alginate and calcium mixture. It suggested that xanthan gum would turn viscous when it is untouched like, the texture of chili source. In addition, the viscosities of xanthan gum solutions were measured and the relationship between viscosity and filtration resistance were also analyzed (Figure 8). It can be seen that increasing the concentration of xanthan gum slightly enhanced the viscosity of the solution. Furthermore, the filtration resistance seemed to be related with the viscosity of feed water, which was also reported in the literature [40,41]. Consequently, at the same concentration, xanthan gum resulted in more serious fouling compared to alginate, and the fouling mechanism of xanthan gum may be due to the gel layer formed by this kind of polysaccharide with large molecules on the membrane instead of pore blocking [23].

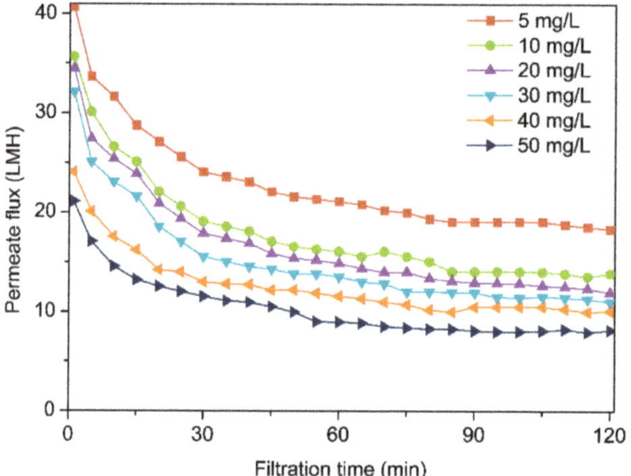

Figure 7. The effect of xanthan gum concentration on membrane fouling.

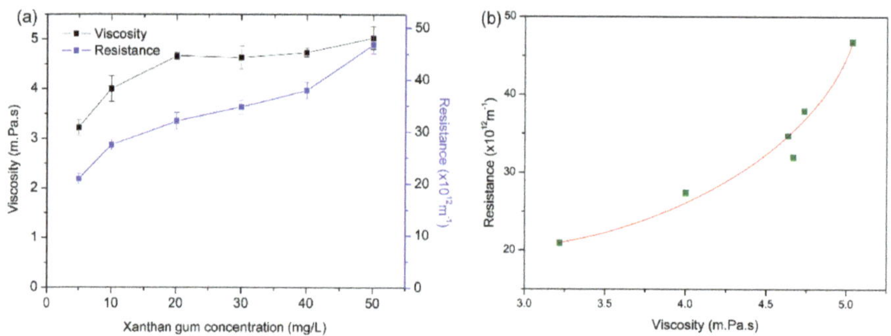

Figure 8. (**a**) The viscosities of xanthan gum solutions and the total filtration resistances of xanthan gum caused after filtration tests; (**b**) The relationship between filtration resistance and viscosities of xanthan gum solutions.

However, increasing the concentration of xanthan gum did not greatly exacerbate the fouling problem as shown in Figure 9. Compared to the alginate, the initial fouling rate of xanthan gum increased at a smaller slope with continuous increasing of the xanthan gum concentration. Additionally, the change of the flux at final steady state was also slower than that of alginate. It may be due to the fact that the permeate flux was reaching the limiting value at this operation conditions used in our study. Xanthan gum had a higher molecular weight and was highly viscous at low concentrations. When it was mixed with water, the viscosity of the solution got thicker therefore affecting filtration flux. With its pseudo-plastic property, xanthan gum makes itself a strong membrane fouling agent [29]. There were few studies in literature using xanthan gum as foulant model and most of them employed alginate. Considering the complexity of the feed water to membrane systems and the difference lying in the fouling propensities of alginate and xanthan gum, more types of polysaccharide should be included in the investigation of membrane fouling.

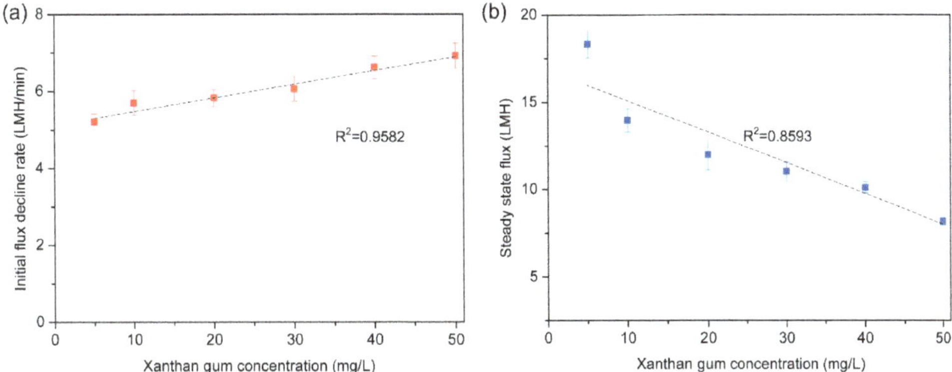

Figure 9. The changes of (**a**) the initial fouling rate and (**b**) final flux at the steady stage with increasing the concentration of xanthan gum.

3.4. Effects of Calcium Ion on Xanthan Gum Fouling

The effect of calcium ions on the fouling of xanthan gum was not obvious, as seen in the filtration of alginate. As shown in Figure 10, the initial flux decline rates of xanthan gum mixed with 1 mM Ca^{2+} were relatively stable. Not like the alginate feed solutions, no big difference was observed among different concentrations of xanthan gum samples. More importantly, the effect of Ca^{2+} on xanthan gum did not show a unimodal mode. As discussed above, the xanthan gum had much higher molecular weights than alginate. Additionally, the molecule of alginate was a linear chain while xanthan gum molecule had branches. It turned out the interaction between Ca^{2+} and xanthan gum was much weaker [42]. Therefore, the influence of Ca^{2+} on the xanthan gum fouling was not obvious as that observed in the filtration of alginate.

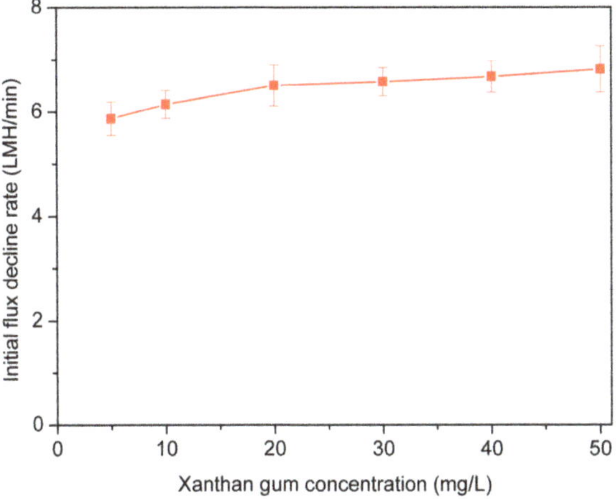

Figure 10. The initial fouling rate of xanthan gum at the presence of 1 mM Ca^{2+}.

4. Conclusions

Polysaccharides are one of the main foulants that cause membrane fouling during the water treatment process employing membrane filtration. The current study investigated the filtration behaviors of different polysaccharides and the influence of calcium ions on them. It was found that alginate and xanthan gum behaved differently in the same microfiltration process and the effects of Ca^{2+} on them were not the same. From the results of alginate, it showed a directly proportional relationship between the flux decline and the polysaccharide concentrations, that is, the more foualnt in the feed water, the more fouling occurred.

The fouling caused by xanthan gum was much severer than that observed in alginate at the same concentration, but increasing the xanthan gum level in the feed water did not promote the fouling as that found in alginate. More importantly, the addition of Ca^{2+} dramatically amplified the difference lying in the filtration of alginate and xanthan gum. A "unimodal" structure was observed in the membrane fouling when increasing the concentration of alginate at a constant level of Ca^{2+}. However, for the filtration of xanthan gum, the addition of Ca^{2+} caused more serious fouling for all concentrations of xanthan gum. This study demonstrates that it is important to combine the structure-function knowledge of polysaccharides with their fouling propensity, especially at the presence of divalent cations. More deep studies are still needed to provide insights into the fouling problems caused by various polysaccharide foulants.

Author Contributions: S.M. and H.L. conceived and designed the experiments; S.M. performed the experiments; S.M. and M.Z. analyzed the data; Q.Z. and N.S. contributed reagents/materials/analysis tools; S.M. wrote the paper.

Funding: This work was financially supported by grants from the National Natural Science Foundation of China (number 51808019) and this research received no external funding.

Conflicts of Interest: The authors declare no conflict of interest.

References

1. Hu, R.; Gwenzi, W.; Sipowo-Tala, V.R.; Noubactep, C. Water Treatment Using Metallic Iron: A Tutorial Review. *Processes* **2019**, *7*, 622. [CrossRef]
2. Yin, H.; Qiu, P.; Qian, Y.; Kong, Z.; Zheng, X.; Tang, Z.; Guo, H. Textile Wastewater Treatment for Water Reuse: A Case Study. *Processes* **2019**, *7*, 34. [CrossRef]
3. Qu, C.; Lu, S.; Liang, D.; Chen, S.; Xiang, Y.; Zhang, S. Simultaneous electro-oxidation and in situ electro-peroxone process for the degradation of refractory organics in wastewater. *J. Hazard. Mater.* **2019**, *364*, 468–474. [CrossRef]
4. Qu, C.; Soomro, G.S.; Ren, N.; Liang, D.W.; Lu, S.-F.; Xiang, Y.; Zhang, S.-J. Enhanced electro-oxidation/peroxone (in situ) process with a Ti-based nickel-antimony doped tin oxide anode for phenol degradation. *J. Hazard. Mater.* **2019**. [CrossRef]
5. Mo, J.; Yang, Q.; Zhang, N.; Zhang, W.; Zheng, Y.; Zhang, Z. A review on agro-industrial waste (AIW) derived adsorbents for water and wastewater treatment. *J. Environ. Manag.* **2018**, *227*, 395–405. [CrossRef]
6. Wagh, P.; Zhang, X.; Blood, R.; Kekenes-Huskey, P.M.; Rajapaksha, P.; Wei, Y.; Escobar, I.C. Increasing Salt Rejection of Polybenzimidazole Nanofiltration Membranes via the Addition of Immobilized and Aligned Aquaporins. *Processes* **2019**, *7*, 76. [CrossRef]
7. Bis, M.; Montusiewicz, A.; Piotrowicz, A.; Łagód, G. Modeling of Wastewater Treatment Processes in Membrane Bioreactors Compared to Conventional Activated Sludge Systems. *Processes* **2019**, *7*, 285. [CrossRef]
8. Tran, T.; Nguyen, T.B.; Ho, H.L.; Le, D.A.; Lam, T.D.; Nguyen, D.C.; Hoang, A.T.; Do, T.S.; Hoang, L.; Nguyen, T.D.; et al. Integration of Membrane Bioreactor and Nanofiltration for the Treatment Process of Real Hospital Wastewater in Ho Chi Minh City, Vietnam. *Processes* **2019**, *7*, 123. [CrossRef]
9. Shannon, M.A.; Bohn, P.W.; Elimelech, M.; Georgiadis, J.G.; Mariñas, B.J.; Mayes, A.M. Science and technology for water purification in the coming decades. *Nature* **2008**, *452*, 301–310. [CrossRef]
10. Daigger, G.T.; Lozier, J.C.; Crawford, G.V. Water reuse applications using membrane technology. *Proc. Water Environ. Federation.* **2006**, *2006*, 2625–2633. [CrossRef]
11. Wang, J.; Cahyadi, A.; Wu, B.; Pee, W.; Fane, A.G.; Chew, J.W. The roles of particles in enhancing membrane filtration: A review. *J. Membr. Sci.* **2019**. [CrossRef]
12. Zhang, W.; Jiang, F. Membrane fouling in aerobic granular sludge (AGS)-membrane bioreactor (MBR): Effect of AGS size. *Water Res.* **2019**, *157*, 445–453. [CrossRef] [PubMed]
13. Tang, C.Y.; Yang, Z.; Guo, H.; Wen, J.J.; Nghiem, L.D.; Cornelissen, E. Potable Water Reuse through Advanced Membrane Technology. *Environ. Sci. Technol.* **2018**, *52*, 10215–10223. [CrossRef] [PubMed]

14. Martin Vincent, N.; Tong, J.; Yu, D.; Zhang, J.; Wei, Y. Membrane Fouling Characteristics of a Side-Stream Tubular Anaerobic Membrane Bioreactor (AnMBR) Treating Domestic Wastewater. *Processes* **2018**, *6*, 50. [CrossRef]

15. Utoro, P.A.R.; Sukoyo, A.; Sandra, S.; Izza, N.; Dewi, S.R.; Wibisono, Y. High-Throughput Microfiltration Membranes with Natural Biofouling Reducer Agent for Food Processing. *Processes* **2018**, *7*, 1. [CrossRef]

16. Gkotsis, P.K.; Banti, D.C.; Peleka, E.N.; Zouboulis, A.I.; Samaras, P.E. Fouling Issues in Membrane Bioreactors (MBRs) for Wastewater Treatment: Major Mechanisms, Prevention and Control Strategies. *Processes* **2014**, *2*, 795–866. [CrossRef]

17. Said, S.A.; Emtir, M.; Mujtaba, I.M. Flexible Design and Operation of Multi-Stage Flash (MSF) Desalination Process Subject to Variable Fouling and Variable Freshwater Demand. *Processes* **2013**, *1*, 279–295. [CrossRef]

18. Meng, S.; Liu, Y. Transparent exopolymer particles (TEP)-associated membrane fouling at different Na+ concentrations. *Water Res.* **2017**, *111*, 52–58. [CrossRef]

19. Bagheri, M.; Mirbagheri, S.A. Critical review of fouling mitigation strategies in membrane bioreactors treating water and wastewater. *Bioresour. Technol.* **2018**, *258*, 318–334. [CrossRef]

20. Al-Amoudi, A.S. Factors affecting natural organic matter (NOM) and scaling fouling in NF membranes: A review. *Desalination* **2010**, *259*, 1–10. [CrossRef]

21. Meng, F.; Zhang, S.; Oh, Y.; Zhou, Z.; Shin, H.-S.; Chae, S.-R. Fouling in membrane bioreactors: An updated review. *Water Res.* **2017**, *114*, 151–180. [CrossRef] [PubMed]

22. Amy, G. Fundamental understanding of organic matter fouling of membranes. *Desalination* **2008**, *231*, 44–51. [CrossRef]

23. Meng, S.; Fan, W.; Li, X.; Liu, Y.; Liang, D.; Liu, X. Intermolecular interactions of polysaccharides in membrane fouling during microfiltration. *Water Res.* **2018**, *143*, 38–46. [CrossRef]

24. Le-Clech, P.; Chen, V.; Fane, T.A. Fouling in membrane bioreactors used in wastewater treatment. *J. Membr. Sci.* **2006**, *284*, 17–53. [CrossRef]

25. Meng, F.; Zhou, Z.; Ni, B.-J.; Zheng, X.; Huang, G.; Jia, X.; Li, S.; Xiong, Y.; Kraume, M. Characterization of the size-fractionated biomacromolecules: Tracking their role and fate in a membrane bioreactor. *Water Res.* **2011**, *45*, 4661–4671. [CrossRef] [PubMed]

26. Meng, S.; Liu, Y. Alginate block fractions and their effects on membrane fouling. *Water Res.* **2013**, *47*, 6618–6627. [CrossRef] [PubMed]

27. Meng, S.; Wang, R.; Zhang, M.; Meng, X.; Liu, H.; Wang, L. Insights into the Fouling Propensities of Natural Derived Alginate Blocks during the Microfiltration Process. *Processes* **2019**, *7*, 858. [CrossRef]

28. Mayer, M.; Braun, R.; Fuchs, W. Comparison of various aeration devices for air sparging in crossflow membrane filtration. *J. Membr. Sci.* **2006**, *277*, 258–269. [CrossRef]

29. Nataraj, S.; Schomäcker, R.; Kraume, M.; Mishra, I.M.; Drews, A. Analyses of polysaccharide fouling mechanisms during crossflow membrane filtration. *J. Membr. Sci.* **2008**, *308*, 152–161. [CrossRef]

30. Wang, Y.-N.; Tang, C.Y. Nanofiltration Membrane Fouling by Oppositely Charged Macromolecules: Investigation on Flux Behavior, Foulant Mass Deposition, and Solute Rejection. *Environ. Sci. Technol.* **2011**, *45*, 8941–8947. [CrossRef]

31. Meng, S.; Winters, H.; Liu, Y. Ultrafiltration behaviors of alginate blocks at various calcium concentrations. *Water Res.* **2015**, *83*, 248–257. [CrossRef] [PubMed]

32. Draget, K.I.; Smidsrød, O.; Skjåk-Bræk, G. Alginates from algae. *Biopolym. Online Biol. Chem. Biotechnol. Appl.* **2005**, *6*.

33. García-Ochoa, F.; Santos, V.E.; Casas, J.A.; Gómez, E. Xanthan gum: Production, recovery, and properties. *Biotechnol. Adv.* **2000**, *18*, 549–579. [CrossRef]

34. Susanto, H.; Arafat, H.; Janssen, E.M.L.; Ulbricht, M. Ultrafiltration of polysaccharide–protein mixtures: Elucidation of fouling mechanisms and fouling control by membrane surface modification. *Sep. Purif. Technol.* **2008**, *63*, 558–565. [CrossRef]

35. Jin, X.; Huang, X.; Hoek, E.M. Role of specific ion interactions in seawater RO membrane fouling by alginic acid. *Environ. Sci. Technol.* **2009**, *43*, 3580–3587. [CrossRef]

36. Wang, R.; Liang, D.; Liu, X.; Fan, W.; Meng, S.; Cai, W. Effect of magnesium ion on polysaccharide fouling. *Chem. Eng. J.* **2020**, *379*, 122351. [CrossRef]

37. Miao, R.; Li, X.; Wu, Y.; Wang, P.; Wang, L.; Wu, G.; Wang, J.; Lv, Y.; Liu, T. A comparison of the roles of Ca2$^+$ and Mg2$^+$ on membrane fouling with humic acid: Are there any differences or similarities? *J. Membr. Sci.* **2018**, *545*, 81–87. [CrossRef]

38. Zhang, M.; Lin, H.; Shen, L.; Liao, B.-Q.; Wu, X.; Li, R. Effect of calcium ions on fouling properties of alginate solution and its mechanisms. *J. Membr. Sci.* **2017**, *525*, 320–329. [CrossRef]

39. Martí-Calatayud, M.C.; Schneider, S.; Wessling, M. On the rejection and reversibility of fouling in ultrafiltration as assessed by hydraulic impedance spectroscopy. *J. Membr. Sci.* **2018**, *564*, 532–542. [CrossRef]

40. Martí-Calatayud, M.C.; Schneider, S.; Yüce, S.; Wessling, M. Interplay between physical cleaning, membrane pore size and fluid rheology during the evolution of fouling in membrane bioreactors. *Water Res.* **2018**, *147*, 393–402. [CrossRef]

41. Pritchard, M.; Howell, J.A.; Field, R.W. The ultrafiltration of viscous fluids. *J. Membr. Sci.* **1995**, *102*, 223–235. [CrossRef]

42. Dário, A.F.; Hortêncio, L.M.A.; Sierakowski, M.R.; Neto, J.C.Q.; Petri, D.F.S. The effect of calcium salts on the viscosity and adsorption behavior of xanthan. *Carbohydr. Polym.* **2011**, *84*, 669–676. [CrossRef]

Article

Numerical Study on Separation Performance of Cyclone Flue Used in Grate Waste Incinerator

Dong-mei Chen [1,2], Jing-yu Ran [1,2,*], Jun-tian Niu [1,2], Zhong-qing Yang [1,2], Ge Pu [1,2] and Lin Yang [3]

[1] Key Laboratory of Low-grade Energy Utilization Technologies and Systems, Chongqing University, Chongqing 400044, China; dongmeichen@cqu.edu.cn (D.-m.C.); juntianniu@cqu.edu.cn (J.-t.N.); zqyang@cqu.edu.cn (Z.-q.Y.); pujiayi@163.com (G.P.)
[2] School of Energy and Power Engineering, Chongqing University, Chongqing 400044, China
[3] Coll Comp Sci & Informat Engn, Chongqing Technol & Business University, Chongqing 400067, China; ctbuyanglin@163.com
* Correspondence: jyran@189.cn

Received: 9 September 2019; Accepted: 8 November 2019; Published: 20 November 2019

Abstract: The traditional treatment of waste incineration flue gas is mostly carried out in low temperatures, but there are some problems such as corrosion of the heating surface at high and low temperatures, re-synthesis of dioxins, and low efficiency. Therefore, it is necessary to remove the pollutants at high temperatures. For the grate waste incinerator, this study proposes an adiabatic cyclone flue arranged at the exit of the first-stage furnace of the grate waste incinerator to pre-remove the fly ash at high temperatures, so as to alleviate the abrasion and corrosion of the tail heating surface. In this paper, computational fluid dynamics (CFD) method is applied to study the performance of a cyclone flue under different structural parameters, and the comprehensive performance of the cyclone flue is evaluated by the technique for order preference by similarity to an ideal solution (TOPSIS) method. The results show that particle separation efficiency increases at first and then decreases with the increase of the vortex finder length, the vortex finder diameter, and the distance between vortex finder and gas outlet tube, while it decreases with the increase of the gas outlet tube diameter. The pressure drop increases with the increase of the vortex finder length, and the vortex finder diameter, while decreases with the increase of the distance between the vortex finder, the gas outlet tube, and the gas outlet tube diameter. In the scope of this study, when $h_1/a = 1.1$, $D_1/A = 0.33$, $h_2/A = 1.5$, and $D_2/A = 0.50$, the comprehensive performance of the cyclone flue is much better.

Keywords: waste incineration; cyclone flue; gas-solid separation; numerical simulation

---

1. Introduction

In recent years, waste incineration technology has been widely used in the world by virtue of reduction (about 90% reduction in volume, 70% reduction in weight), harmlessness, and energy recovery [1,2]. Due to the complex compositions of waste, a variety of harmful gases and toxic substances will be formed during the incineration process. These are inorganic acid gases such as HCl, HF, NO_x, SO_2, and gaseous heavy metals such as mercury, highly toxic organic substances like Polycyclic Aromatic Hydrocarbons (PAHs), Polychlorinated Biphenyls (PCBs), Polychlorinated Dibenzo-Dioxins/Polychlorinated Dibenzofurans (PCDD/PCDFs), fly ash which equivalent to 2–5% of the original waste weight and required for special treatment [3–5]. Therefore, secondary pollution problem is urgent to be solved in the waste incineration process.

The traditional treatment of flue gas is mostly to remove the pollutants at low temperatures (100–350 °C), and the process is very complicated. The corrosion of superheater caused by acid gases and fly ash in incineration flue gas at high temperatures limits the further improvement of steam parameters and power generation efficiency [6]. In addition, the low temperature environment is easy to promote the re-synthesis of dioxins, and the presence of chlorine and heavy metals in fly ash provide favorable conditions for dioxins re-synthesis [7–10]. Therefore, it is necessary to pre-remove acidic gases and fly ash from flue gas at high temperatures.

The grate waste incinerator is the mainstream equipment for waste incineration [11,12]. At present, the grate waste incinerator relies on a three-stage furnace to reduce the flue gas temperature at the entrance of convective heating surface generally. In order to prevent corrosion at low temperatures, the flue gas temperature at the exit of the waste heat boiler is approximately designed to be at temperatures between 190–230 °C, which will limit the improvement of the efficiency of waste heat boiler. According to the structural characteristics of the grate waste incinerator, this research proposes a square adiabatic cyclone flue arranged at the first-stage furnace exit of the grate waste incinerator (as shown in Figure 1) to pre-remove the fly ash in the flue gas. In addition, calcium and ammonia can be injected into the flue to remove acidic gases synergistically, which alleviates corrosion of heating surface at high and low temperatures, as well as inhibits dioxin formation and improves furnace parameters. In order to facilitate the arrangement, the cyclone flue is designed as a square, and downward exhaust is adopted to conform to the flow direction of flue gas in the grate waste incinerator, the original grate waste incinerator can be reformed directly. There is no heating surface inside the flue, the inner layer is refractory, and the outer layer is insulation materials. These can prevent heat loss and ensure that the flue gas exists at a higher temperature, which can crack dioxins effectively. In order to understand the flow field characteristics and gas–solid separation performance of cyclone flue, this paper simulates the internal flow field and gas–solid separation performance of cyclone flue with the CFD method.

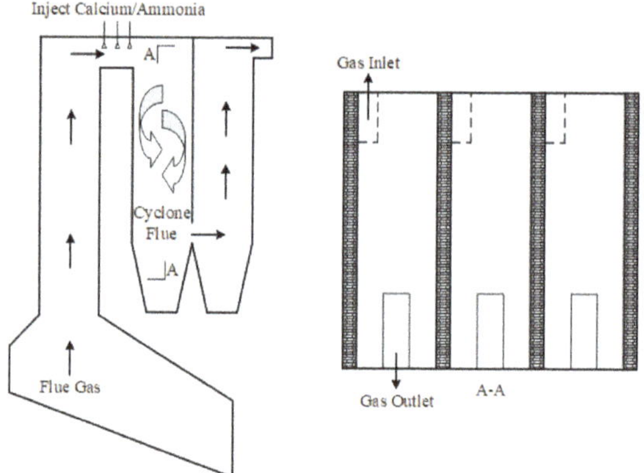

Figure 1. Arrangement diagram of cyclone flue in grate waste incinerator.

The adiabatic cyclone flue proposed in this paper is similar in structure to the cyclone separator, so the simulation method of the cyclone separator can be referred to. The traditional cyclone separator is circular and has an upper exhaust, which has been subjected to a large number of experiments and numerical studies by many researchers, and the research methods are relatively reasonable and reliable. Xiang et al. [13] used the Reynold's stress model (RSM) to simulate the internal flow field of cyclone separators with different heights. The results showed that as the height of the cyclone separator increases, the tangential velocity will decrease, then the centrifugal force on the particles is reduced, which will result in a decrease in dust removal performance. Raoufi et al. [14] studied the effect of the gas outlet tube structure on the separation efficiency of the cyclone separator based on the Euler–Lagrangian method; the results indicated that the separation efficiency of the separator decreases with the increase of the diameter of gas outlet tube. Moreover, Safikhani et al. [15] applied a simple pressure correction algorithm, using the Reynold's stress turbulence model and the random walk model to study the internal flow field and particle motion trajectory of cyclone separators with different number of inlets. In addition, Sakura et al. [16] used the large eddy simulation (LES) model, based on Ansys CFX to study the gas–solid flow characteristics of cyclone separators, and compared the effects of different dust outlet shapes on separation efficiency and pressure drop. Zhang et al. [17] used the transient Reynold's stress model and discrete phase model to study the flow field and particle motion trajectory in the cyclone separator, and proposed a new cyclone separator, which can improve efficiency and reduce pressure drop. Furthermore, Parvaz et al. [18] studied the performance of a cyclone separator which had an inner cone located at the bottom of the cyclone, and simulated the influence of the inner cone with different heights and diameters on the performance of the separator.

Compared to conventional cyclone separator, there is less research on square and downward exhaust cyclones. Safikhani et al. [19] compared the internal flow field, separation efficiency of a square cyclone separator and a circular cyclone separator based on the CFD method, the results showed that the separation efficiency of the square cyclone separator is lower than the circular cyclone separator, but the pressure drop is rather small. Su et al. [20] used the Euler–Lagrange method to simulate the gas–solid flow characteristics of three inlet structures of square cyclones. Raoufi et al. [21] studied the internal flow field of a square cyclone separator by numerical simulation method, and analyzed the similarities and differences of the internal flow field of the upper exhaust type and the downward exhaust type cyclone separator. Moreover, Oh et al. [22] investigated the internal flow field and particle separation efficiency of the downward exhaust cyclone based on the Euler–Lagrange method. Fatahian et al. [23] studied the performance that using the laminarizer in the square and circular cyclones, the results suggested that square cyclone is more effective. Additionally, Mokni et al. [24] used the CFD method to study the effects of cylinder height on flow field, pressure drop and separation performance in a turbulent hydrocyclone.

Here we simulated by way of a realizable k-ε model and random walk model. Firstly, the geometric model and mathematical model are described, then the internal flow field characteristics of the cyclone flue and the influence of various structural parameters on the separation performance and pressure drop are analyzed, finally, the comprehensive performance of the cyclone flue is evaluated by technique for order preference by similarity to an ideal solution (TOPSIS) method.

2. Model Description

2.1. Adiabatic Cyclone Flue Model

As shown in Figure 1, the cyclone flue distributes along the width of the grate waste incinerator evenly, and each adiabatic cyclone flue works independently without any influence on each other, so only one of the flues needs to be studied. Its structure is shown in Figure 2. In this paper, the influences of the vortex finder diameter, vortex finder length, distance between vortex finder and gas outlet tube, gas outlet tube diameter on the performance of cyclone flue were studied. The dimensions of each cyclone flue are given in Table 1. In order to reduce the computational complexity, the numerical studies only were carried out on the parts above the cone. The study of Oh et al. [22] showed that neglecting the cone part has little effect on the results and can be neglected.

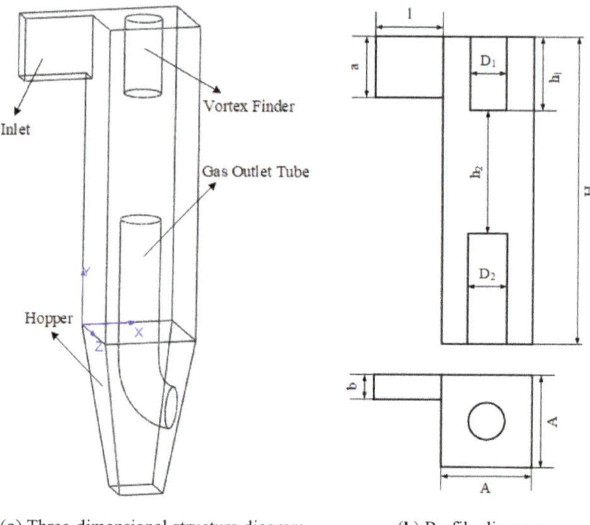

(**a**) Three-dimensional structure diagram   (**b**) Profile diagram

Figure 2. Structure diagram of adiabatic cyclone flue: (**a**) Three-dimensional structure diagram, (**b**) Profile diagram.

Table 1. Dimension of the adiabatic cyclone flue.

Dimension	Selection Principle [25]	Length (m)	Dimension Ration (Dimension/A)
Body edge length, A	Consistent with depth of grate furnace (Usually for 3–4 m)	3.0	1
Inlet height, a	$a = 2.2\text{–}2.5b$	2.0	2.5b
Inlet width, b	$b = (A - D_1)/3 - (A - D_1)/2$	0.8	0.27
Inlet length, l	$l = 0.75\,A$	2.25	0.75
Vortex finder length, h_1	$h_1 = 1\text{–}1.5a$	2.2 2.4 2.6 2.8 3.0	1.1a 1.2a 1.3a 1.4a 1.5a
Vortex finder diameter, D_1	$D_1 = 0\text{–}0.5A$	1.0 1.1 1.2 1.3 1.4	0.33 0.37 0.40 0.43 0.47
Distance between vortex finder and gas outlet tube, h_2	Ensure that the gas flow section is larger than the gas outlet tube section	2.5 3.0 3.5 4.0 4.5	0.83 1.00 1.17 1.33 1.50
Gas outlet tube diameter, D_2	$D_2 = 0.3\text{–}0.5A$	1.0 1.1 1.2 1.3 1.4 1.5	0.33 0.37 0.40 0.43 0.47 0.50
Cyclone flue height, H	—	10.0	3.33

2.2. Governing Equation

The CFD method was used to simulate a three-dimensional flow field in the adiabatic cyclone flue. For steady incompressible flow, the continuity equation and momentum equation are as follows:

$$\frac{\partial \overline{u}_i}{\partial x_i} = 0 \tag{1}$$

$$\overline{u}_j \frac{\partial \overline{u}_i}{\partial x_j} = -\frac{1}{\rho} \frac{\partial \overline{P}}{\partial x_i} + v \frac{\partial^2 \overline{u}_i}{\partial x_j \partial x_j} - \frac{\partial}{\partial x_j} R_{ij} \tag{2}$$

In the above equations, $\overline{u}_i$ is the average velocity of the fluid; x_i is the coordinate position; $\overline{P}$ is the average pressure; ρ is the fluid density; v is the kinematic viscosity of the fluid; $R_{ij} = \overline{u'_i u'_j}$, is the Reynold's stress tensor, which represents the influence of turbulence on the flow field, here, $u'_i = u_i - \overline{u}_i$.

The turbulence models for cyclone separator simulation were the k-ε (standard, RNG, realizable) model and the Reynold's stress model (RSM) commonly. In this paper, the realizable k-ε model was adopted, the transport equations of turbulent kinetic energy (k) and turbulent dissipation rate (ε) were as follows:

$$\frac{\partial}{\partial t}(\rho k) + \frac{\partial}{\partial x_j}(\rho k u_j) = \frac{\partial}{\partial x_j}[(\mu + \frac{\mu_t}{\sigma_k}) \frac{\partial k}{\partial x_j}] + G_k - \rho \varepsilon \tag{3}$$

$$\frac{\partial}{\partial t}(\rho \varepsilon) + \frac{\partial}{\partial x_j}(\rho \varepsilon u_j) = \frac{\partial}{\partial x_j}[(\mu + \frac{\mu_t}{\sigma_\varepsilon}) \frac{\partial \varepsilon}{\partial x_j}] - \rho C_2 \frac{\varepsilon^2}{k + \sqrt{v\varepsilon}} \tag{4}$$

In the above equations, G_k represents the turbulent energy generated by the velocity gradient; μ is the dynamic viscosity of the fluid; C_2 is a constant; σ_k, and σ_ε are turbulent Prandtl numbers in k equation and ε equation respectively.

There are two main methods simulating gas–solid two-phase flow. They are the Euler–Euler method and the Euler–Lagrange method. In order to track the trajectory of particles, the Euler–Lagrange method was used to simulate gas–solid two-phase flow. In this method, gases were regarded as continuous phase and particles as dispersed discrete phase. Because the particle concentration was small, it belonged to the dilute phase flow and the particle size was small too, so the influence of particle on flow field can be neglected, unidirectional coupling mode was adopted in this paper, and the interaction force between particles was ignored. In Lagrangian coordinates, the force balance equations of particles are as follows:

$$\frac{du_p}{dt} = F_D(u_A - u_p)_x + \frac{g_x(\rho_p - \rho_A)}{\rho_p} \tag{5}$$

$$F_D = \frac{18\mu_A}{\rho_p d_p^2} \frac{C_D Re_p}{24} \tag{6}$$

$$C_D = C_1 + \frac{C_2}{Re_p} + \frac{C_3}{Re_p} \tag{7}$$

$$Re_p = \frac{\rho_A d_p |u_p - u_A|}{\mu_A} \tag{8}$$

In the above equations, u_A is the fluid velocity; u_p is the particle velocity; μ_A is the fluid dynamic viscosity; d_p is the particle diameter; ρ_p is the particle density; ρ_A is the fluid density; Re_p is the relative Reynold's number (particle Reynold's number); C_D is the resistance coefficient; C_1–C_3 is a constant, depending on Reynolds number; g_x is the gravitational acceleration.

2.3. Boundary Conditions and Numerical Schemes

Semi-implicit method pressure-linked equations consistent (SIMPLEC) were used as a method for pressure-velocity coupling. The Pressure staggering option (PRESTO) was used for pressure interpolation. The quadratic upstream interpolation for convective kinetics (QUICK) was used for discrete difference format.

In this paper, we assumed that the gas flow enters the flue at a constant velocity, the gas phase is a high temperature flue gas, its temperature is 850 °C, its velocity is 20 m/s, and the particles and fluids have the same inlet velocity. The setting of specific boundary conditions are shown in Table 2, non-slip adiabatic boundary conditions are adopted for all walls.

Table 2. The boundary conditions in numeric simulation.

Name	Boundary Condition Type	Discrete Phase Model (DPM) Boundary Condition Type
Gas–solid inlet	velocity-inlet	wall-jet
gas outlet	outflow	escape
particle outlet	wall	trap
wall	wall	reflect

2.4. Grid Independence Study

In this paper, the structured hexahedral mesh of cyclone flue was generated by ICEM. In order to ensure the accuracy of calculation and shorten the calculation time, five different numbers of structured grids were generated to verify the irrelevance between the calculated results and the number of grids, which are 890,000 cells, 1,190,000 cells, 1,400,000 cells, 1,670,000 cells and 1,960,000 cells. Figure 3 shows the weighted average turbulence intensity (I) and weighted average turbulent kinetic energy (K) at z = 1.5 m cross section area under different mesh numbers, with the increase of the number of grids, the values of K and I decreased gradually, the differences between the 1,670,000 grids and the 1,960,000 grids were 0.17% and 0.06% respectively, indicating that the number of grids had little impact on the calculation results. The model with 1,670,000 grids is selected for numerical simulation.

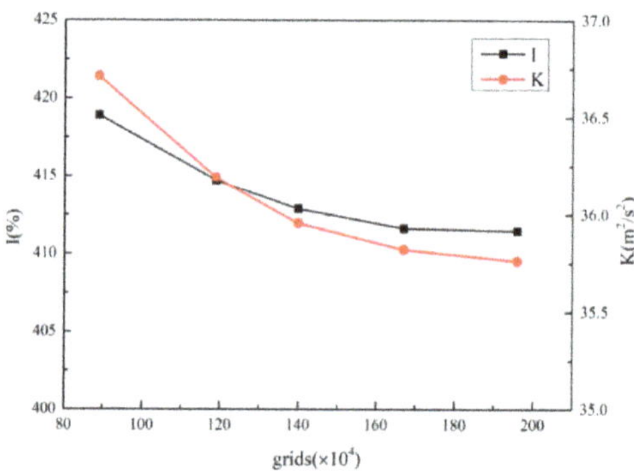

Figure 3. The weighted average turbulent kinetic energy (*K*) and weighted average turbulence intensity (*I*) at z = 1.5 m cross-section under different grids.

2.5. Model Validation

In order to verify the accuracy of the model, the flow characteristics were compared with the experimental data of Su [26]. Su et al. measured the axial velocity distribution of Plane 1 at an inlet velocity of 25.3 m/s using a particle dynamic analyzer (PDA). The geometric model is shown in Figure 4.

Figure 5 compares the computed axial velocity at different positions in Plane 1 with the corresponding experimental data of Su. The two positions were $x = 30$ mm and $x = 90$ mm. It is clearly seen that the model prediction was in close quantitative agreement with the experimental data. Overall, the adopted model can accurately and reliably predict the performance of the cyclone flue.

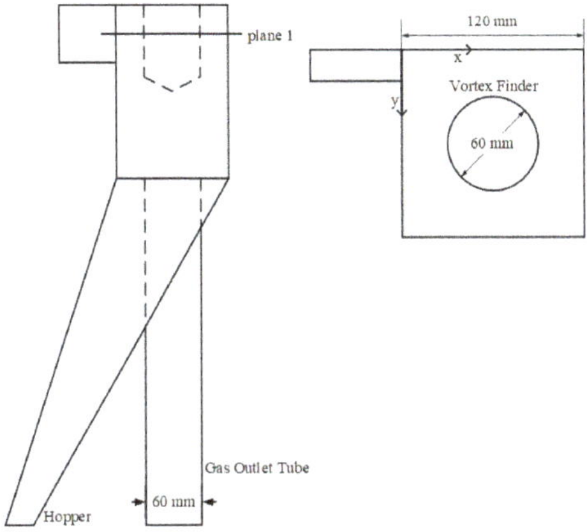

Figure 4. Geometric model of Su et al.

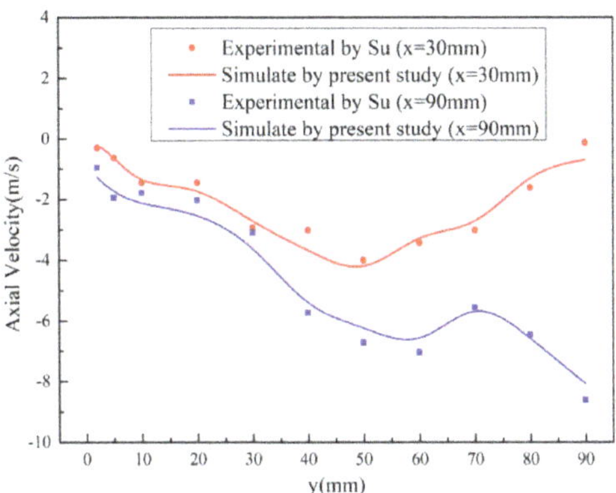

Figure 5. Comparison of simulated velocity with measurements.

3. Results and Discussion

3.1. Characteristics of Flow Field

Figure 6a shows the velocity distribution of the cyclone flue at $z = 1.5$ m cross section. It can be seen from the figure that the maximum velocity appeared at the entrance of the gas outlet tube. The velocity below the vortex finder was lower, because the recirculation zone was formed here. The obvious flow separation was found at the edge of the vortex finder near the inlet, resulting in a lower velocity. Oh et al. [22] also found this phenomenon in the simulation of a circular uniflow cyclone. The gas flow went into the flue and rotated around the vortex finder, flow separation appeared due to the centrifugal and inertial forces, and this phenomenon was weakened gradually in the downstream direction.

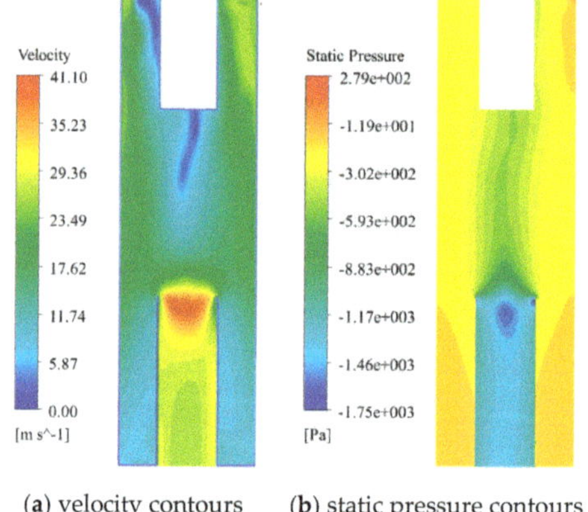

(**a**) velocity contours   (**b**) static pressure contours

Figure 6. Flow field distribution at z = 1.5 m: (**a**) velocity contours, (**b**) static pressure contours.

Figure 6b shows the static pressure distribution of the cyclone flue at z = 1.5 m cross section. It can be seen from the figure that the static pressure distributed symmetrically around the central axis, which is similar to traditional cyclone separators [14,27]. Along the radius, the pressure decreased slowly, thus forming a low pressure zone in the center. The minimum static pressure appeared at the entrance of the gas outlet tube, reaching −1750 Pa. Combined with the velocity distribution, it can be seen that the fluid velocity reached the highest here, and the static pressure was converted into dynamic pressure.

Figure 7 shows the velocity vector diagram of cyclone flue at y = 9 m cross section. It can be seen from the figure that when the gas flow went into the flue from the inlet, the velocity increased firstly, and then decreased gradually in the course of the flow. There was a strong eddy at the corner facing the inlet, which has been confirmed in Su's discovery [20]. As the gas flow goes into the flue, it impinged the corner opposite the inlet directly, and the sharp changed the flow direction at the corner, causing the local vortex. Some particles were collected in the corner and fell down, while some particles were rebounded in the corner and rotated continuously. In the other corners, there was also a weaker local vortex.

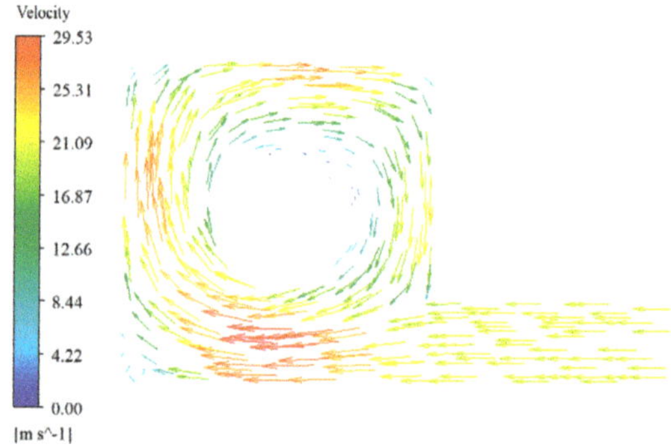

Figure 7. Velocity vector at y = 9 m.

3.2. Particle Separation Efficiency and Pressure Drop

Particle separation efficiency and pressure drop are two important indices to study the performance of the adiabatic cyclone flue. This paper used the control variable method to study the effects of structural parameters such as vortex finder length (h_1), vortex finder diameter (D_1), distance between vortex finder and gas outlet tube (h_2), and gas outlet tube diameter (D_2) on the separation performance and pressure drop of the cyclone flue, which provided guiding significance for the optimal design of the cyclone flue. According to the literature [28–30], the particle size of fly ash in the waste incineration flue gas ranged from 1 to 100 μm, which can be assumed that the particle size follows the Rosin–Rammler distribution.

3.2.1. The Influence of the Vortex Finder Length

Figure 8 shows the effects of the vortex finder length on the separation efficiency and pressure drop of the cyclone flue, and the h_1/a varying from 1.1 to 1.5. It can be seen from the figure that the separation efficiency increased firstly and then decreased with the increase of the vortex finder length. There was an optimum value in the middle. When $h_1/a = 1.2$, the separation efficiency reached its highest point. The pressure drop increased linearly with the increase of the vortex finder length, and the increase ratio was about 1.5%. The experiments conducted by Trieseh et al. also found that the pressure drop increased proportionally to the vortex finder length [31].

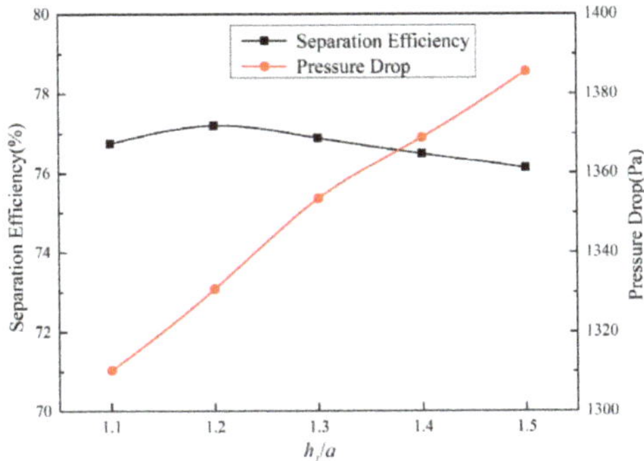

Figure 8. Effect of the vortex finder length (h_1/a) on efficiency and pressure drop.

The function of the vortex finder was mainly embodied in two aspects. On the one hand, the gas flow was rotated under the guidance of the vortex finder. On the other hand, the annular space formed by the vortex finder and the body affected the rotation speed of gas flow. As the increase of the vortex finder length appropriated, enough time was ensured to separate particles to the wall. However, if the vortex finder length was too long, it caused back-flushing of the bottom due to the extension of the swirling gas flow, and resulted in a decrease in separation efficiency. The longer the length of vortex finder was, the larger the area of the wall contacted with the gas flow was. The rotational kinetic energy loss of the gas flow increased due to friction, resulting in an increase in the pressure drop.

3.2.2. The Influence of the Vortex Finder Diameter

Figure 9 shows the effects of the vortex finder diameter on the separation efficiency and pressure drop of the cyclone flue, and the D_1/A varying from 0.33 to 0.47. It can be seen from the figure that the separation efficiency increased firstly and then decreased with the increase of the vortex finder diameter. While $D_1/A = 0.4$, the separation efficiency was the highest, reaching to 77.2%. The pressure drop increased with the increase of vortex finder diameter, but while D_1/A was higher than 0.4, the change of the diameter of vortex finder had little effect on the pressure drop. Compared with the vortex finder length, the vortex finder diameter had a weaker influence on the pressure drop.

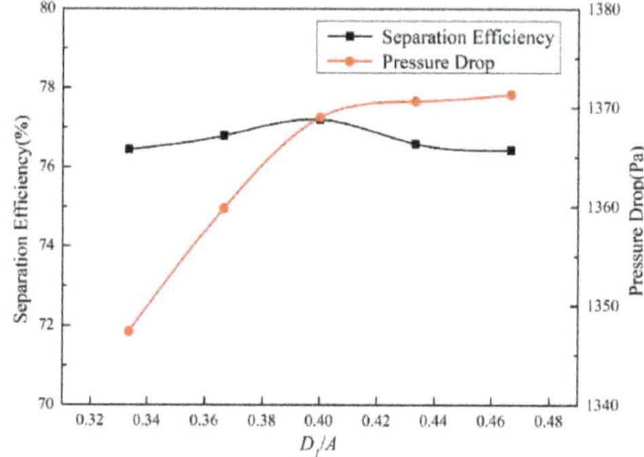

Figure 9. Effect of the vortex finder diameter (D_1/A) on efficiency and pressure drop.

With the increase of the vortex finder diameter, the annular cross-section area of the vortex finder was reduced, and the velocity of the flow increased when entering the flue, in addition, the centrifugal force strengthened, which helped to improve separation efficiency. In addition, the increase in velocity of flow led to an increase in the loss of rotational kinetic energy, resulting in an increase in pressure loss. However, the too large vortex finder diameter led to the impact of part of the gas flow on the vortex finder directly, resulting in the deterioration of the flow field and the rebound of particles, which was not conducive to the separation of particles.

3.2.3. The Influence of the Gas Outlet Tube Diameter

Figure 10 shows the effects of gas outlet tube diameter on separation efficiency and the pressure drop of the cyclone flue, and the D_2/A varying from 0.33 to 0.5. It can be seen from the figure that the separation efficiency decreased with the increase of the gas outlet tube diameter. But in the range of 0.33–0.4 and 0.43–0.5, the variation range of separation efficiency was small, while D_2/A increased from 0.4 to 0.43, the separation efficiency decreased from 77.2% to 75.8%. Compared with the vortex finder, the gas outlet tube diameter had a greater influence on the pressure drop. When D_2/A increased from 0.33 to 0.5, the pressure drop decreased from 2065 Pa to 934 Pa, and the variation decreased gradually.

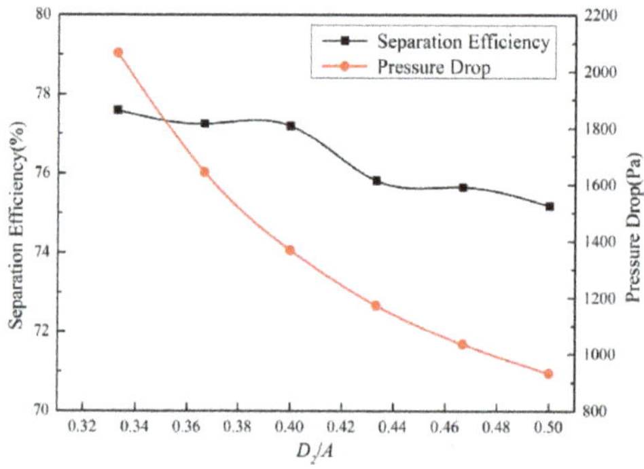

Figure 10. Effect of the gas outlet tube diameter (D_2/A) on efficiency and pressure drop.

The entrance of the gas outlet tube was a high-speed and low-pressure area, as shown in Figure 6, and the gas flow converged here. If the diameter of gas outlet tube was small, the fly ash particles collided with the wall of gas outlet tube easily when the particles moved to the entrance of the gas outlet tube, and then falling into the hopper and being separated, which helped to improve the separation efficiency. With the increase of the diameter of the gas outlet tube, the phenomenon of the flow rate rising sharply when the gas reached the gas outlet tube was alleviated, thus making the flow field more stable.

3.2.4. The Influence of the Distance between Vortex Finder and Gas Outlet Tube

Figure 11 shows the effects of the distance between vortex finder and gas outlet tube on the separation efficiency and pressure drop of the cyclone flue, and the h_2/A varying from 0.83 to 1.5. It can be seen from the figure that the separation efficiency increased at first and then decreased with the increase of the distance between the vortex finder and the gas outlet tube. When $h_2/A = 1.33$, the separation efficiency was the highest, reaching to 76.5%. The pressure drop decreased linearly with the increase of the distance between the vortex finder and the gas outlet tube, and the reduction ratio was about 5%.

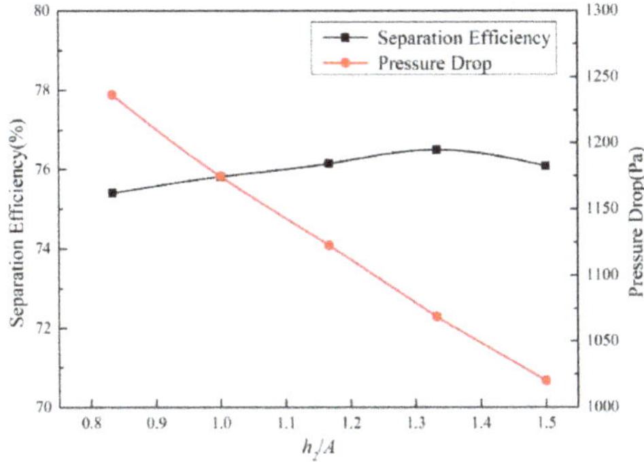

Figure 11. Effect of the distance between vortex finder and gas outlet tube (h_2/A) on efficiency and pressure drop.

Gauthier studied the effect of the length of the separation section (the sum of h_1 and h_2) on the performance of the uniflow cyclone. The results showed that the separation efficiency increased first and then decreased with the increase of the separation section [32].

As the increase in the distance between the vortex finder and gas outlet tube appropriated, it allowed enough time for particles to converge into the wall. However, the excessive distance led the particles that have been separated to re-enter the central gas flow and escape from the gas outlet tube. In addition, the longer the distance between the vortex finder and gas outlet tube was, and the shorter the length of the gas outlet tube was then the weaker the interference to the flow field was, which made the flow field more stable and decreased the pressure loss.

3.3. Comprehensive Performance Evaluation

The highest separation efficiency and the lowest pressure drop cannot always exist at the same time, it is necessary to analyze and evaluate the comprehensive performance of the cyclone flue, and to ensure high efficiency and low energy loss. The TOPSIS evaluation method is a scientific method used in multi-objective decision analysis of limited schemes commonly, and it was developed by Yoon and Hwang [33]. This method has no special requirement on the sample data, it can make full use of the original data, and it can obtain better results [34–36]. Therefore, this method was used to analyze and evaluate the comprehensive performance of the cyclone flue in this study. The steps of the TOPSIS model were as follows:

(1) Establish the initial decision matrix $X = \{x_{ij}\}$. Two evaluation indexes of separation efficiency and pressure drop of cyclone flue were considered, and x_{ij} is the j-th evaluation index of the i-th evaluation object.

(2) Assimilation of indicator attributes. The inverse method was used to convert low-quality indicators into high-quality indicators, that is, the higher the value is, the better the performance will be. The pressure drop was a low-quality indicator, which needed to be converted into a high-quality indicator to obtain a new matrix X'.

(3) Calculating the standard decision matrix Z. The dimensionless processing of each indicator can eliminate the effects of dimension and magnitude between different attribute indicators. The equation was as follows:

$$z_{ij} = x'_{ij} / \sqrt{\sum_{i=1}^{m} x'^2_{ij}} \tag{9}$$

(4) Calculate the set of positive ideal (S^+) and negative ideal (S^-) solutions of the decision matrix. S^+ is the set of maximum values of each index, that is, when the separation efficiency of the cyclone flue is the highest, the pressure drop will be the smallest. S^- is the set of minimum values of each index, that is, when the efficiency is lowest, the pressure drop will be largest.

(5) Calculating the Euclidean distances of each evaluation scheme to the S^+ and S^-. The equations are as follows:

$$d_i^+ = \sqrt{\sum_{j=1}^{n} (s_j^+ - z_{ij})^2} \tag{10}$$

$$d_i^- = \sqrt{\sum_{j=1}^{n} (s_j^- - z_{ij})^2} \tag{11}$$

(6) Calculating the relative proximity (C_i) of each evaluation scheme to the S^+. The equation is as follows:

$$C_i = d_i^- / (d_i^+ + d_i^-) \tag{12}$$

(7) Ranking according to the value of C_i. the larger the value of C_i is, the closer the scheme to S^+ and the farther away from S^-, the better the scheme is.

In this paper, the vortex finder length, the vortex finder diameter, the gas outlet tube diameter, and the distance between the vortex finder and the gas outlet tube were taken as the evaluation objects. The C_i values were calculated under different structural parameters. The calculation results are listed in Table 3.

Table 3. Comprehensive performance evaluation of cyclone flue.

Variable 1		Variable 2		Variable 3		Variable 4	
h_1/a	C_i	D_1/A	C_i	h_2/A	C_i	D_2/A	C_i
1.1	0.91	0.33	0.65	0.83	0	0.33	0.04
1.2	0.73	0.37	0.48	1.00	0.25	0.37	0.21
1.3	0.43	0.40	0.39	1.17	0.48	0.40	0.42
1.4	0.22	0.43	0.11	1.33	0.74	0.43	0.63
1.5	0	0.47	0.01	1.5	0.97	0.47	0.82
/	/	/	/	/	/	0.50	0.96

According to the data in Table 3, the comprehensive performance of the cyclone flue was negatively correlated with the vortex finder diameter and the vortex finder length, while it was positively correlated with the distance between the vortex finder and the gas outlet tube, and gas outlet tube diameter. When designing the cyclone flue, the vortex finder diameter and the vortex finder length should be adopted to smaller values, but the distance between the vortex finder and the gas outlet tube, and the gas outlet tube diameter, should be taken to a larger value. In the scope of this study, when $h_1/a = 1.1$,

$D_1/A = 0.33$, $h_2/A = 1.5$, $D_2/A = 0.50$, the separation efficiency reached to 75% and the pressure drop was about 900 Pa, reaching a better comprehensive performance.

4. Conclusions

In this paper, the internal flow field and performance of the adiabatic cyclone flue were studied by the CFD method. The effects of the vortex finder diameter, vortex finder length, distance between vortex finder and gas outlet tube, and gas outlet tube diameter on the separation efficiency and pressure drop of cyclone flue were also studied. Then the TOPSIS method was applied to evaluate the comprehensive performance of the cyclone flue. The main conclusions are shown as follows:

(1) The inlet of the gas outlet tube of the cyclone flue was a high-speed and low-pressure zone, where the velocity was the largest and the pressure was the smallest. The internal pressure was symmetrically distributed along the central axis, which is consistent with the conventional cyclone. In addition, there was a local vortex at the corner in the flue.

(2) The particle separation efficiency increased at first and then decreased with the increase of the vortex finder length, the vortex finder diameter, and the distance between the vortex finder and the gas outlet tube, while it decreased with the increase of the gas outlet tube diameter. Above all, the gas outlet tube diameter had the most important influence on the separation efficiency.

(3) The pressure drop increased with the increase of the vortex finder length, and the vortex finder diameter, while it decreased with the increase of the distance between vortex finder and gas outlet tube, and the gas outlet tube diameter. In addition, the gas outlet tube diameter had the greatest influence on the pressure drop.

(4) The comprehensive performance of the cyclone flue was negatively correlated with the vortex finder diameter and the vortex finder length, while it was positively correlated with the distance between the vortex finder and the gas outlet tube, and the gas outlet tube diameter.

(5) In the scope of this study, when $h_1/a = 1.1$, $D_1/A = 0.33$, $h_2/A = 1.5$, $D_2/A = 0.50$, the comprehensive performance of the cyclone flue was better, and the separation efficiency reached to 75%, which can remove the fly ash in the flue gas effectively, alleviate erosion wear and ash corrosion of the tail heating surface, and reduce the burden of the bag filter.

Author Contributions: Conceptualization, D.-m.C., J.-y.R. and L.Y.; methodology, D.-m.C., J.-y.R., J.-t.N. and Z.-q.Y.; software, D.-m.C. and J.-t.N.; validation, D.-m.C., J.-y.R. and J.-t.N.; formal analysis, Z.-q.Y. and G.P.; investigation, D.-m.C. and J.-t.N.; resources, J.-y.R., G.P. and L.Y.; data curation, D.-m.C. and Z.-q.Y.; writing—original draft preparation, D.-m.C.; writing—review and editing, J.-t.N. and J.Y.R; visualization, G.P. and Z.-q.Y.; supervision, L.Y.; project administration, J.-y.R.; funding acquisition, J.-y.R.

Funding: This research was funded by KEY INDUSTRIAL GENERIC TECHNOLOGY INNOVATION PROJECT OF CHONGQING, grant number Cstc2016zdcy-ztzx20006.

Conflicts of Interest: The authors declare no conflicts of interest.

Abbreviations

Nomenclature

A	body edge length (m)
a	inlet height (m)
b	inlet width (m)
l	inlet length (m)
h_1	vortex finder length (m)
D_1	vortex finder diameter (m)
h_2	distance between vortex finder and gas outlet tube (m)
D_2	gas outlet tube diameter (m)
H	cyclone flue height (m)

u	velocity (m/s)
P	pressure (Pa)
ρ	density (kg/m^3)
ν	kinematic viscosity (m^2/s)
μ	dynamic viscosity (Pa·s)
k	turbulent kinetic energy (m^2/s^2)
ε	turbulent dissipation rate (m^2/s^3)
σ	turbulent Prandtl number
Re_p	relative Reynold's number
g_x	gravitational acceleration (m/s^2)
I	turbulence intensity (%)
C_i	relative proximity coefficient

References

1. Li, M.; Xiang, J.; Hu, S.; Sun, L.S.; Su, S.; Li, P.S.; Sun, X.X. Characterization of solid residues from municipal solid waste incinerator. *Fuel* **2004**, *83*, 1397–1405. [CrossRef]
2. He, J.; Lin, B. Assessment of waste incineration power with considerations of subsidies and emissions in China. *Energy Policy* **2019**, *126*, 190–199. [CrossRef]
3. Niu, J.; Liland, S.E.; Yang, J.; Rout, K.R.; Ran, J.; Chen, D. Effect of oxide additives on the hydrotalcite derived Ni catalysts for CO_2 reforming of methane. *Chem. Eng. J.* **2019**, *377*, 119763. [CrossRef]
4. Makarichi, L.; Jutidamrongphan, W.; Techato, K.A. The evolution of waste-to-energy incineration: A review. *Renew. Sustain. Energy Rev.* **2018**, *91*, 812–821. [CrossRef]
5. Wang, P.; Hu, Y.; Cheng, H. Municipal solid waste (MSW) incineration fly ash as an important source of heavy metal pollution in China. *Environ. Pollut.* **2019**, *252*, 461–475. [CrossRef] [PubMed]
6. Valente, T. Fireside corrosion of superheater materials in chlorine containing flue gas. *J. Mater. Eng. Perform.* **2001**, *10*, 608–613. [CrossRef]
7. Stieglitz, L.; Bautz, H.; Roth, W.; Zwick, G. Investigation of precursor reactions in the de-novo synthesis of PCDD/PCDF on fly ash. *Chemosphere* **1997**, *34*, 1083–1090. [CrossRef]
8. Gullett, B.K.; Bruce, K.R.; Beach, L.O.; Drago, A.M. Mechanistic steps in the production of PCDD and PCDF during waste combustion. *Chemosphere* **1992**, *25*, 1387–1392. [CrossRef]
9. Zhang, Z.E.; Yan, Y.F.; Zhang, L.; Ju, S.X. Hollow fiber membrane contactor absorption of CO_2 from the flue gas: Review and perspecrive. *Glob. Nest J.* **2014**, *16*, 354–373.
10. Niu, J.; Du, X.; Ran, J.; Wang, R. Dry (CO_2) reforming of methane over Pt catalysts studied by DFT and kinetic modeling. *Appl. Surf. Sci.* **2016**, *376*, 79–90. [CrossRef]
11. Wissing, F.; Wirtz, S.; Scherer, V. Scherer. Simulating municipal solid waste incineration with a DEM/CFD method-Influences of waste properties, grate and furnace design. *Fuel* **2017**, *206*, 638–656. [CrossRef]
12. Zhang, N.; Pan, Z.; Zhang, Z.; Zhang, W.; Zhang, L.; Baena-Moreno, F.M.; Lichtfouse, E. CO_2 capture from coalbed methane using membranes: A review. *Environ. Chem. Lett.* **2019**, *8*, 1–18. [CrossRef]
13. Xiang, R.B.; Lee, K.W. Numerical study of flow field in cyclones of different height. *Chem. Eng. Process.* **2005**, *44*, 877–883. [CrossRef]
14. Raoufi, A.; Shams, M.; Farzaneh, M.; Ebrahimi, R. Numerical simulation and optimization of fluid flow in cyclone vortex finder. *Chem. Eng. Process. Process. Intensif.* **2008**, *47*, 128–137. [CrossRef]
15. Safikhani, H.; Zamani, J.; Musa, M. Numerical study of flow field in new design cyclone separators with one, two and three tangential inlets. *Adv. Powder Technol.* **2018**, *29*, 611–622. [CrossRef]
16. Bogodage, S.G.; Leung, A.Y. CFD simulation of cyclone separators to reduce air pollution. *Powder Technol.* **2015**, *286*, 488–506. [CrossRef]
17. Zhang, G.; Chen, G.; Yan, X. Evaluation and improvement of particle collection efficiency and pressure drop of cyclones by redistribution of dustbins. *Chem. Eng. Res. Des.* **2018**, *139*, 52–61. [CrossRef]
18. Parvaz, F.; Hosseini, S.H.; Elsayed, K.; Ahmadi, G. Numerical investigation of effects of inner cone on flow field, performance and erosion rate of cyclone separators. *Sep. Purif. Technol.* **2018**, *201*, 223–237. [CrossRef]
19. Safikhani, H.; Shams, M.; Dashti, S. Numerical simulation of square cyclones in small sizes. *Adv. Powder Technol.* **2011**, *22*, 359–365. [CrossRef]

20. Su, Y.; Zheng, A.; Zhao, B. Numerical simulation of effect of inlet configuration on square cyclone separator performance. *Powder Technol.* **2011**, *210*, 293–303. [CrossRef]

21. Raoufi, A.; Shams, M.; Kanani, H. CFD analysis of flow field in square cyclones. *Powder Technol.* **2009**, *191*, 349–357. [CrossRef]

22. Oh, J.; Choi, S.; Kim, J. Numerical simulation of an internal flow field in a uniflow cyclone separator. *Powder Technol.* **2015**, *274*, 135–145. [CrossRef]

23. Fatahian, H.; Fatahian, E.; Nimvari, M.E. Improving efficiency of conventional and square cyclones using different configurations of the laminarizer. *Powder Technol.* **2018**, *339*, 232–243. [CrossRef]

24. Fatahian, H.; Fatahian, E.; Nimvari, M.E. Numerical investigation of the effect of the cylindrical height on separation performances of uniflow hydrocyclone. *Chem. Eng. Sci.* **2015**, *122*, 500–513.

25. Cen, K.F. *Theoretical Design and Operation of CFB Boiler*; China Electric Power Press: Beijing, China, 1998. (In Chinese)

26. Su, Y.; Mao, Y. Experimental study on the gas-solid suspension flow in a square cyclone separator. *Chem. Eng. J.* **2006**, *121*, 51–58. [CrossRef]

27. Zhou, H.; Hu, Z.; Zhang, Q.; Wang, Q.; Lv, X. Numerical study on gas-solid flow characteristics of ultra-light particles in a cyclone separator. *Powder Technol.* **2019**, *344*, 784–796. [CrossRef]

28. Vavva, C.; Voutsas, E.; Magoulas, K. Process development for chemical stabilization of fly ash from municipal solid waste incineration. *Chem. Eng. Res. Des.* **2017**, *125*, 57–71. [CrossRef]

29. Bayuseno, A.P.; Schmahl, W.W. Characterization of MSWI fly ash through mineralogy and water extraction. *Resour. Conserv. Recycl.* **2011**, *55*, 524–534. [CrossRef]

30. Loginova, E.; Proskurnin, M.; Brouwers, H.J.H. Municipal solid waste incineration (MSWI) fly ash composition analysis: A case study of combined chelatant-based washing treatment efficiency. *J. Environ. Manag.* **2019**, *235*, 480–488. [CrossRef]

31. Triesch, O.; Bohnet, M. Measurement and CFD prediction of velocity and concentration profiles in a decelerated gas-solid flow. *Powder Technol.* **2001**, *155*, 101–113. [CrossRef]

32. Gauthier, T.A.; Briens, C.L.; Bergougnou, M.A.; Galtier, P. Uniflow cyclone efficiency study. *Powder Technol.* **1990**, *62*, 217–225. [CrossRef]

33. Hwang, C.L.; Yoon, K. *Multiple Attribute Decision Making: Methods and Applications a State-of-the-Art Survey*; Springer: Berlin/Heidelberg, Germany, 1981.

34. Wang, C.N.; Tsai, T.T.; Huang, Y.F. A Model for Optimizing Location Selection for Biomass Energy Power Plants. *Processes* **2019**, *7*, 353. [CrossRef]

35. Wang, C.N.; Huang, Y.F.; Cheng, I.; Nguyen, V. A Multi-Criteria Decision-Making (MCDM) Approach Using Hybrid SCOR Metrics, AHP, and TOPSIS for Supplier Evaluation and Selection in the Gas and Oil Industry. *Processes* **2018**, *6*, 252. [CrossRef]

36. Jin, L.; Zhang, C.; Fei, X. Realizing Energy Savings in Integrated Process Planning and Scheduling. *Processes* **2019**, *7*, 120. [CrossRef]

Article

Insights into the Fouling Propensities of Natural Derived Alginate Blocks during the Microfiltration Process

Shujuan Meng [1], Rui Wang [1], Minmin Zhang [2], Xianghao Meng [1], Hongju Liu [1] and Liang Wang [3,*]

1   School of Space and Environment, Beihang University, Beijing 100191, China; mengsj@buaa.edu.cn (S.M.); rui_wang@buaa.edu.cn (R.W.); SY1930212@buaa.edu.cn (X.M.); liuhj@buaa.edu.cn (H.L.)
2   Chemical Engineering, Xiamen University, Xiamen 361005, China; mmzhang666@gmail.com
3   State Key Laboratory of Separation Membranes and Membrane Processes, Tiangong University, Tianjin 300387, China
*   Correspondence: mashi7822@163.com; Tel.: +86-186-026-50716

Received: 15 October 2019; Accepted: 14 November 2019; Published: 17 November 2019

Abstract: Membrane technology has been one of the most promising techniques to solve the water problem in future. Unfortunately, it suffers from the fouling problem which is ubiquitous in membrane systems. The origin of the bewilderments of the fouling problem lies in the lack of deep understanding. Recent studies have pointed out that the molecular structure of foulant affects its fouling propensity which has been ignored in the past. In this study, the filtration behaviors of alginate blocks derived from the same source were comprehensively explored. Alginate blocks share the same chemical composition but differ from each other in molecular structure. The alginate was first extracted from natural seaweed using calcium precipitation and ion-exchange methods. Extracted alginate was further fractionized into MG-, MM- and GG-blocks and the characteristics of the three blocks were examined by Fourier transform infrared spectroscopy (FTIR) and field emission scanning electron microscopy (FESEM) observations, and transparent exopolymer particles' (TEPs) measurements. Results showed that MG-, MM- and GG-blocks had the same functional groups, but they showed different intermolecular interactions. TEP formation from MG-, MM- and GG-blocks revealed that the molecule crosslinking of them decreased in the order of MM-blocks > GG-blocks > MG-blocks. It was further found from microfiltration tests that these alginate blocks had completely different fouling propensities which can be explained by the TEP formation. TEPs would accumulate on membrane surfaces and worked as a pre-filter to avoid serious pore blocking of membrane. That all suggested that the membrane fouling was closely related to the molecular structure of foulant. It is expected that this study can provide useful insights into the fouling propensities of different types of polysaccharides during filtration processes.

Keywords: membrane fouling; molecular composition of foulant; transparent exopolymer particles (TEP); fouling propensities

---

1. Introduction

Water scarcity is one of the most serious crises in the world as a result of the uneven distribution of water resources, poor water management and climate change. To cope with this situation, scientists and engineers have been working hard to develop treatment methods of every sort [1–4]. Most of the efforts aim to remove the pollutants in water bodies and to increase water supplies via the reliable reuse of wastewater and efficient desalination of seawater and brackish water [5,6]. In the past few decades, membrane technology has been widely used around the world as a promising technique for water treatment [7–9]. However, membrane fouling still remains one of the obstacles to the successful operation of membrane systems, as it has since the day that membrane filtration was employed [10]. Among all fouling problems, organic fouling has often been reported to be a major type and polysaccharides play a key role in organic fouling. Compared to other organic foulants, such as proteins and humic acid, the chains of polysaccharides are usually much longer and prone to gelling to form gel layer on a membrane surface via interaction of the molecular chain. However, the exact fouling propensities of polysaccharides are not clear, especially the influence of molecular structure on fouling. As a consequence of this, more efforts should be devoted to the investigation of the filtration behaviors of polysaccharides, relating the information of molecular structure with fouling tendencies.

Recently, it has been found that polysaccharides can aggregate together to form three-dimensional networks; i.e., transparent exopolymer particles (TEPs). The role of TEPs in membrane fouling has attracted more and more attention since Berman and Holenberg first pointed out that they may participate in the fouling problems in membrane systems [11]. TEPs are planktonic hydrogels in water environments which are mainly formed by polysaccharides via an abiotic pathway [12]. TEPs possess a high viscosity; thus, they can attach to the membrane surface easily to develop a gel layer [13]. More importantly, it is difficult to remove TEPs from a membrane system. TEPs are deformable; thus, they also can penetrate into membrane pores whose sizes actually are smaller than the dimensions of TEP under pressure. In addition, some of TEPs will break into small pieces at the feed side and then reassemble into larger forms at the filtrate side of membrane at the aid of stream turbulence and divalent cations [14]. Thus, TEPs are dynamic in membrane systems, which poses difficulties in identifying the exact nature of TEP fouling. The effect of TEPs on membrane fouling needs to be further explored.

Alginate is a typical model polysaccharide that is commonly employed in studies of membrane fouling. Alginate is unbranched binary copolymer consisting of b-D-mannopyranuronic acid (M) and a-L-gulopyranuronic acid (G) which combines into MG-, MM- and GG-blocks in varying proportions [15]. These blocks play different roles in alginate chains; i.e., GG-blocks are responsible for the gel-forming capacity, and MG- and MM-blocks provide flexibility to the chains [16]. Their proportions, distributions and lengths determine the chemical and physical properties of alginate molecules. So far, there are more than 200 kinds of alginates derived from different sources [17] which are different in MG-, MM- and GG-blocks contents [18]. It is reasonable to consider that alginates extracted from different algal sources may behave differently during membrane filtration processes. Since alginate has been commonly used as a model organic foulant in numerous studies of membrane fouling, a new challenge on the interpretation and reliability of filtration data would appear if alginate used in filtration experiments was not well characterized. Therefore, in this study, alginate was directly prepared from dry seaweed using alkaline extraction and then block fractionation was performed. Subsequently, three kinds of alginate blocks were systematically characterized with Fourier transform infrared spectroscopy (FTIR) and field emission scanning electron microscopy (FESEM). Dead-end filtration tests were employed to examine the fouling propensities of MG-, MM- and GG-blocks under the same experimental conditions. In addition, TEP measurement was also performed to evaluate the influence of TEP on membrane filtration.

2. Materials and Methods

2.1. Alginate Extraction from Raw Seaweed

Alginate used in this study was extracted from Fuerji dried seaweed according to the following steps. Seaweed was chopped into small pieces and 5 g of chopped dry seaweed was soaked in 300 mL of 1% formaldehyde (Sigma, City of Saint Louis, MO, USA) for 4 h. After that, seaweed pieces were washed by deionized water (Milli-Q, Burlington, MA, USA) to remove excess formaldehyde residue. Subsequently, rinsed seaweed was soaked in 300 mL of 3% Na_2CO_3 (Merck, Kenilworth, NJ, USA) solution for 3 h, and alginate was converted to a soluble form so that it could be separated from the insoluble seaweed residue. After alkaline soaking, the solution became very sticky and a clean cloth was used first to remove the bulk seaweed residue, and that was followed by filtration using filter papers. After careful addition of 50 mL of 10% $CaCl_2$ (Merck, Kenilworth, NJ, USA) into the above alginate solution, wooly cloudlike aggregates formed. The aggregates were quickly transferred to a clean beaker. Ion exchange was employed to convert calcium alginate to sodium alginate. A series of ion exchanges, 20 min each, were carried out to get rid of the Ca^{2+} until no white precipitate formed when Na_2CO_3 was added into the exchanged solution. Finally, an equal volume of ethanol (Merck, Kenilworth, NJ, USA) was added to precipitate sodium alginate.

2.2. Block Fractionation of Alginate

The extracted alginate was fractionated according to the method proposed by Leal et al. as presented in Figure 1 [19]. Firstly, 10 g/L alginate was prepared with Millipore water. The solution was stirred for 2 h for complete dissolution. Then, HCl (Merck, Kenilworth, NJ, USA) was slowly added into the sodium alginate solution with stirring, forming white cloudlike aggregations. The solid-liquid mixture was subsequently heated at 100 °C in an oil bath for 0.5 h. After heating, the cloudlike aggregations were partially dissolved and the mixture was centrifuged at 10,000 rpm for 30 min. The supernatant solution was neutralized with 1 M NaOH (Merck, Kenilworth, NJ, USA) and precipitated with an equal volume of ethanol. The white precipitation was separated by another 30 min of centrifugation at 10,000 rpm and freeze-dried (MG-blocks). The insoluble residue from the first centrifugation was re-dissolved in 1 M NaOH and the pH was readjusted to 2.85 by adding 1 M HCl. At pH 2.85, new precipitation developed and centrifugation was used for separation. The soluble fraction was again neutralized with 1 M NaOH, precipitated by ethanol and freeze-dried (MM-blocks). The solid fraction was re-dissolved in 1 M NaOH, neutralized by 1 M HCl, precipitated by ethanol and freeze-dried (GG-blocks).

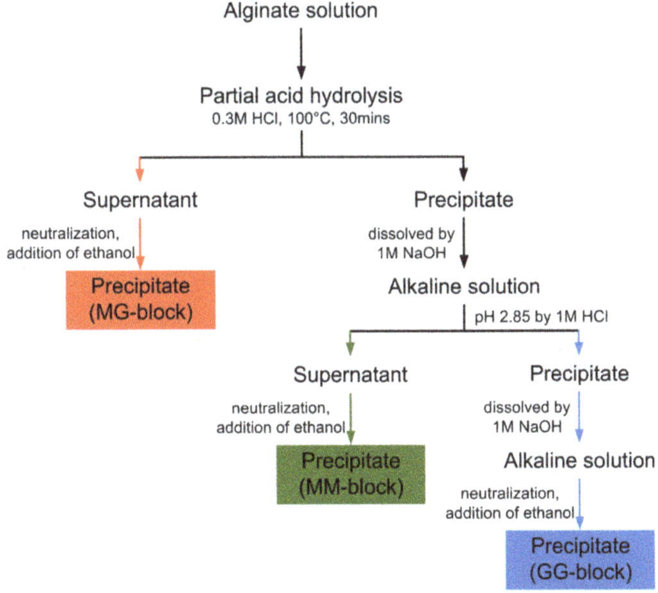

Figure 1. The alginate fractionation procedure employed to obtain MG-, MM- and GG-blocks.

2.3. FTIR Spectroscopy of Alginate Blocks

In order to characterize the alginate blocks, the FTIR spectra of the MG-, MM- and GG-blocks fractionated from the extracted alginate were measured. The alginate blocks obtained from commercial alginate sodium (Wako, Tokyo, Japan) via the same procedure were also measured by FTIR to provide a direct comparison. Sample preparation was done by the FTIR Start Kit and 300 mg of KBr (Merck, Kenilworth, NJ, USA) was used to create the baseline. An amount of 2–4 mg of sample was mixed with 295 mg KBr and ground for a short time in order to mix it thoroughly. Then, the pure KBr disc and alginate-KBr mixture pellets were transferred to the disc holder which was subsequently inserted into the spectrometer. The FTIR spectra were recorded using a PerkinElmer FTIR Spectrum GX 50905 (PerkinElmer, Waltham, MA, USA). A total of 60 scans were taken for both background establishment and sample measurements.

2.4. Field Emission Scanning Electron Microscope (FESEM) Observations

In order to visualize the micro-structures of the alginate, MG-, MM- and GG-blocks we prepared, the powders of them were completely freeze-dried and they were examined by an FESEM (Jeol JSM-7600F, Tokyo, Japan). These powder samples were coated with Pt just before observation to increase electrical conductivity. Besides, the TEPs formed from alginate blocks were also observed by the FESEM. Solutions of MG-, MM- and GG-blocks were gently filtered by the 0.05 μm filters under a pressure below 0.2 bars and then freeze-dried to keep the morphology of TEP. Subsequently, these samples were observed and at least 10 pictures for each sample were recorded. In addition, the fouled membrane pieces were also examined by the FESEM to provide a direct observation of the fouling development of the membrane. Membrane samples were taken from the same location from the fouled membrane and representative images were taken.

2.5. TEP Determination

TEP measurements were carried out for the MG-, MM- and GG-blocks samples. The determination procedure was the same as in our previous study [20,21]. Alcian blue was used to stain the TEP sample and preparation involved dissolving alcian blue 8 GX (0.02%) (Sigma, City of Saint Louis, MO, USA) into 0.06% acetic acid before the tests. To quantify the TEP formed from different alginate block samples, solutions of MG-, MM- and GG-blocks were prepared. The alginate blocks had smaller sizes than a whole molecule chain of alginate; thus, filters with smaller sizes were used to determine the TEP. To look into the detailed size distributions of TEPs formed from alginate blocks, 0.2 μm and 0.05 μm polycarbonate filters (Whatman, Maidstone, UK) were used to conduct the TEP tests. Alginate block solutions were filtered through 0.2 μm and 0.05 μm filters at a constant vacuum of 0.2 bars. The retained TEP on filter was stained by alcian blue solution and subsequently rinsed with 1 mL of Milli-Q water after ~5 s staining. The filter together with stained TEPs were immersed in a beaker with 5 mL of 80% H_2SO_4 solution for 2 h, and we made sure all alcian blue was dissolved into the sulfuric solution. Finally, the absorbance of sulfuric acid solution was measured with a UV–Vis spectrophotometer (Shimadzu, Kyoto, Japan) at 787 nm wavelength and TEP concentration was expressed as mg gum xanthan equivalent per liter of water (mg $X_{eq.}L^{-1}$).

2.6. Dead-End Filtration Tests

Standard dead-end filtration, as shown in Figure 2, using nylon membrane with a pore size of 0.2 μm, was chosen to examine the different fouling potentials of the MG-, MM- and GG-blocks. The effective membrane surface area was 11.94 cm^2, and the filtration process was driven by compressed nitrogen gas at 1 bar. The mass of the filtrate was recorded at the time interval of 10 s by an electronic balance with data-logging system installed in the connected computer. Feed solution was prepared with 0.01 M NaCl and 0.05 g/L alginate samples of respective blocks. A 20 min long initializing period with 0.01 M NaCl was conducted to make sure that stable performance of the membrane was achieved. Subsequently, the samples of alginate blocks were filtered and the mass change of the filtrate was recorded.

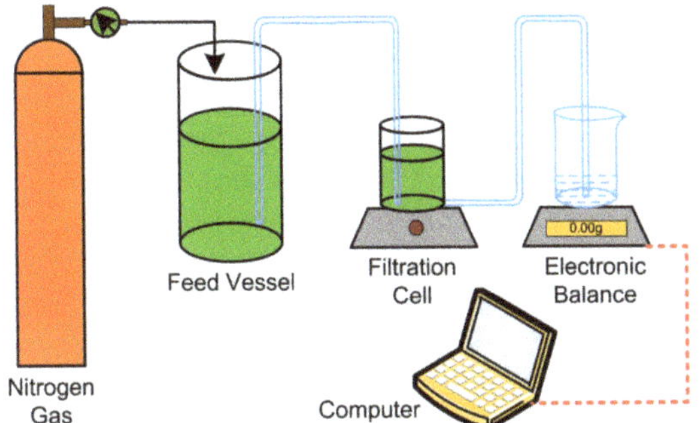

Figure 2. Schematic diagram of the dead-end filtration system used in this study.

3. Results and Discussion

3.1. Block Composition of Alginate

Extracted alginate from the seaweed was fractionated using partial acid hydrolysis. The results showed (Table 1) that 19.2% MG-blocks, 62.4% MM-blocks and 19.3% GG-blocks made up the alginate chains, suggesting that MM-block is the main component of the sodium alginate we extracted. The commercial alginate (Sigma, City of Saint Louis, MO, USA) contained 13.8% MG-blocks, 53.5% MM-blocks and 32.7% GG-blocks, which also indicated that the main component in alginate was MM-blocks. These results were consistent with the literature reports that for most alginates MM-blocks dominated molecular composition [15]. According to the above result, the M/G ratios in the extracted alginate could be calculated to be 2.51, which was consistent with the literature in which M/G ratio was within the range of 0.33–9 [15]. It should be noted that different seaweed species could yield different M/G ratios, even the same species from different seasonal and growth conditions. For example, the M/G values of alginates from *Lessonia vadosa* blades collected at the same locality in winter and spring varied between 0.79 and 0.33 [22]. Therefore, it is of primary importance to pay attention to the chemical composition of the polysaccharide that was subjected to filtration. The alginate used in this study was extracted from the natural seaweed. It provided a different and more natural source for alginate blocks compared to the commercial alginate. In natural water environments, most of the TEPs form from the precursor materials (most of them are polysaccharides) secreted by algae via an abiotic pathway. Therefore, the alginate directly extracted from seaweed would be a good model for the TEP formation and polysaccharide fouling study.

Table 1. Molecular composition of extracted and commercial alginate.

	MG	MM	GG
Extracted alginate from the seaweed	19.2 ± 2.1%	62.4 ± 3.5%	19.3 ± 0.9%
Commercial alginate	13.8 ± 1.9%	53.5 ± 1.3%	32.7 ± 0.6%

3.2. FTIR Spectrums of MG-, MM- and GG-Blocks

Figure 3 shows the respective FTIR spectrums of the MG-, MM- and GG-blocks derived from the seaweed and the commercial alginate. It can be seen that the alginate blocks showed similar absorption peaks to those obtained from commercial alginate. This observation suggested that the alginate we extracted had the same material composition with the commercial one, although all the alginate blocks had more impure peaks than commercial ones. The broad bands centered at wave numbers of 3393 cm^{-1}, 3401 cm^{-1} and 3381 cm^{-1} were assigned to hydrogen bonded O–H stretching vibrations. The weak signals at wavenumbers of 2923 cm^{-1}, 2935 cm^{-1} and 2939 cm^{-1} were due to C–H stretching vibrations and the asymmetric stretching of carboxylate O–C–O vibration contributed to the strong absorptions at 1618 cm^{-1}, 1610 cm^{-1} and 1615 cm^{-1}. According to literature [19,23–25], the band at 1420/1412/1419 cm^{-1} was assigned to C–OH deformation vibration with contribution of O–C–O symmetric stretching vibration of carboxylate group. The medium absorption at 1305 cm^{-1}, 1303 cm^{-1} and 1320 cm^{-1} in the three figures, respectively, may be assigned to O–C stretching vibration of carboxylic acid and derivatives [23]. In addition, the medium to strong IR absorption bands at 1200–970 cm^{-1} are mainly due to C–C and C–O stretching in the pyranoid ring and C–O–C stretching of glycosidic bonds [25]. An intense absorption in this spectral region is common for all polysaccharides [26,27]. The fingerprint or anomeric region at 950–750 cm^{-1} showed three characteristic absorption bands in all polysaccharide standards and alginate polysaccharides. The band at 957/938/949 cm^{-1} was assigned to the C–O stretching vibration of uronic acid residues. The one at 886/889/904 cm^{-1} was assigned to the C1–H deformation vibration of β-mannuronic acid residues. Finally, the band at 814/820/812 cm cm^{-1} is characteristic of mannuronic acid residues [22]. Consequently, alginate was the main polysaccharide extracted from seaweed and the three blocks had the same functional groups as the commercial ones.

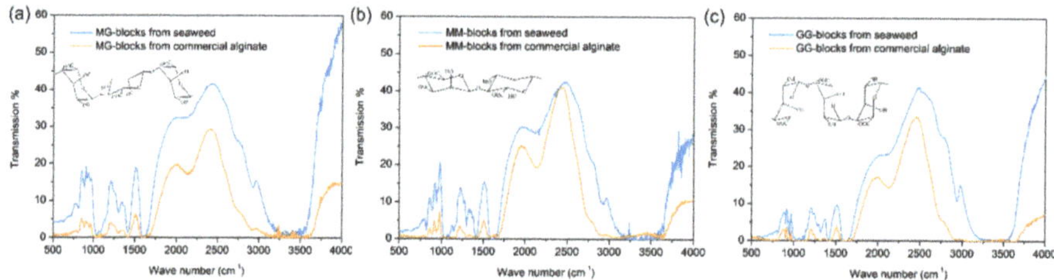

Figure 3. FTIR spectrum of alginate blocks derived from seaweed, and the commercial ones. (**a**) MG-blocks; (**b**) MM-blocks; (**c**) GG-blocks

3.3. The Micro-Structures of the Alginate, MG-, MM- and GG-Blocks

The dry powders of the alginate extracted from seaweed and the MG-, MM- and GG-blocks fractionated from the alginate were observed under an FESEM. As can be seen in Figure 4, the micro-structures of alginate and alginate blocks shared some similarities and discrepancies. All of the samples showed particle like structures (Figure 4a–d) but the detailed structures were different from each other (Figure 4e–h). These observations indicated that MG-, MM- and GG-blocks would have different chemical and physical properties which should have significant effects on their filtration behaviors.

Figure 4. The micro-structures of the alginate, MG-, MM- and GG-blocks derived from the natural seaweed. (**a**) Alginate; (**b**) MG-blocks; (**c**) MM-blocks; (**d**) GG-blocks; (**e**)–(**h**) were the amplificatory observations of (**a**)–(**d**).

3.4. The TEP Formation from MG-, MM- and GG-Blocks

The TEP level formed from the 50 mg/L alginate blocks was measured to quantitatively assess their aggregation ability. To look into the size distribution of the TEP formed from MG-, MM- and GG-blocks, two sizes of filters were employed to conduct the measurements. As observed in Figure 5a, the highest TEP level was recorded in the MM-blocks determined with a 0.05 μm filter, achieving 13.6 mg X_{eq}/L from the 50 mg/L alginate blocks. The MG- and GG-blocks produced 1.1 and 7.8 mg X_{eq}/L TEP respectively. This finding revealed that in the water environments, the polysaccharide chains would aggregate through molecule crosslinking instead of existing as single chains even at a low concentration. The TEP they formed via molecule crosslinking was much larger than their molecule size. From the TEP results, it could be found that the extent of molecule crosslinking decreased in the order of MM-blocks > GG-blocks > MG-blocks, which is consistent with the literature [28]. As shown in Figure 5, the highest TEP concentration was observed in the solution of MM-blocks, which was also the main component of alginate, as determined in the alginate fractionation. It revealed that although fractionated from the same alginate source, the MG-, MM- and GG-blocks possessed different abilities in molecular crosslinking due to their discrepancies in chain characteristics [28]. In addition, the TEP detected with a size larger than 0.2 μm was remarkably smaller than the TEP formed with size bigger than 0.05 μm, as shown in Figure 5b. That suggested that the TEPs formed from all alginate blocks were small in size and most of the TEPs had sizes ranging from 0.05 μm to 0.2 μm.

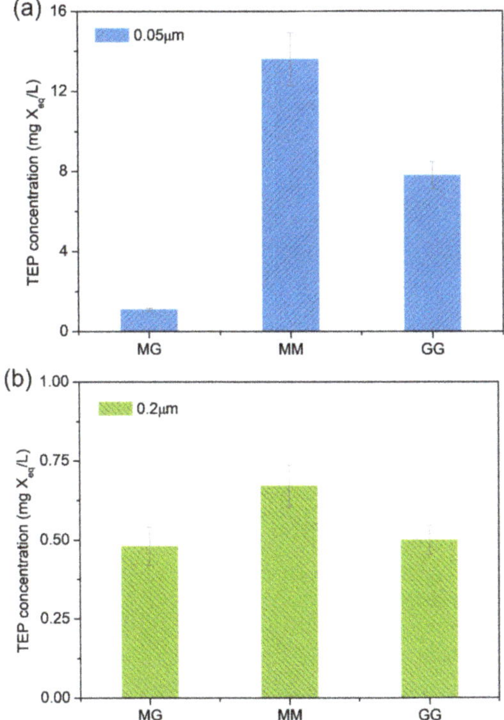

Figure 5. The TEP formation from MG-, MM- and GG-blocks. (**a**) TEP measured using 0.05 μm filter and (**b**) TEP measured with 0.2 μm filter.

Besides, the TEPs formed from alginate blocks were gently retained on the 0.05 μm filters to provide a direct observation. As shown in Figure 6, the TEPs seemed like the result of crosslinked alginate blocks and had dimensions larger than a single molecule. TEPs actually were the aggregations of alginate blocks. The formation of TEPs from these alginate blocks would definitely affect their filtration behaviors in membrane filtration process. The intermolecular interactions of foulants was essential in understanding fouling mechanism [29,30] and more efforts should be devoted into exploring the complex crosslinking of various foulants in water environments.

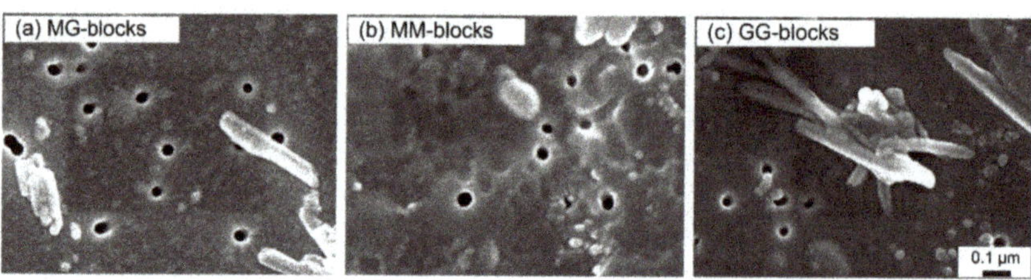

Figure 6. The observation of TEPs formed from (**a**) MG-, (**b**) MM- and (**c**) GG-blocks with FESEM (0.05 µm filters).

3.5. Filtration Behaviors of MG-, MM- and GG-Blocks

Although the monomers M and G are isomers, MG-, MM- and GG-blocks displayed completely different filtration behaviors. As shown in Figure 7, the severest fouling happened in filtration of MG-blocks, and the least flux decline was observed during the filtration of MM-blocks. GG-blocks showed a fouling tendency between those of MG-blocks and MM-blocks. All these results suggest that differences lying in the molecular structures of these blocks lead to their diverse filtration properties. Combined with the TEP determination results, it can be deduced that the more TEPs formed, the less fouling that occurred during membrane filtration. TEPs possessed large sizes and easily attached on a membrane surface to form a fouling layer, which in turn prevented the serious pore clogging of membrane [28]. Thus, the least membrane fouling was recorded in the filtration of MM-blocks which showed the largest TEPs. On the contrary, the MG-blocks, which possessed the smallest size, would penetrate into the membrane pores and be absorbed onto the membrane pore walls resulting with the reduced radius and number of the effective membrane pores. The fouling propensities of alginate blocks derived from the commercial alginate decreased at the same order: MG-blocks > GG-blocks > MM-blocks, as can be seen in Figure 7b,c. These results were further evident from the FESEM observations of the fouled membrane surfaces as can be seen in Figure 8. After filtration with MM-blocks, some of a fouling layer can be observed on the membrane surface while there was very little fouling layer development on membrane surfaces after they were filtered by MG- and GG-blocks. Instead, the membrane pores seemed to be filled by foulants (Figure 8b,d). As a consequence of these results, it was found that the more TEP formed, the less serious of a fouling problem was observed in the filtration process in this study. In this study, it was found that the TEP may work as a pre-filter to reduce the amount of alginate blocks that penetrated into membrane pores; thus, mitigating the membrane fouling. The TEP themselves certainly did not lead to serious membrane fouling. It depended on the sizes of membrane pores. Figure 6 showed the TEPs gently retained by 0.05 µm filters and then freeze-dried to keep the morphology of TEPs. It could be seen that TEPs formed from the aggregation of alginate blocks and had bigger sizes than single molecules of alginate blocks. The TEP formation was the result of the intermolecular interactions between these blocks. Accordingly, in this study we deduced that TEPs possessed large sizes and easily attached on a membrane surface to form a fouling layer which in turn prevented the serious pore clogging of the membrane. In addition, the TEP level indicated the extent of intermolecular interactions of polysaccharides, which was an important factor influencing membrane fouling.

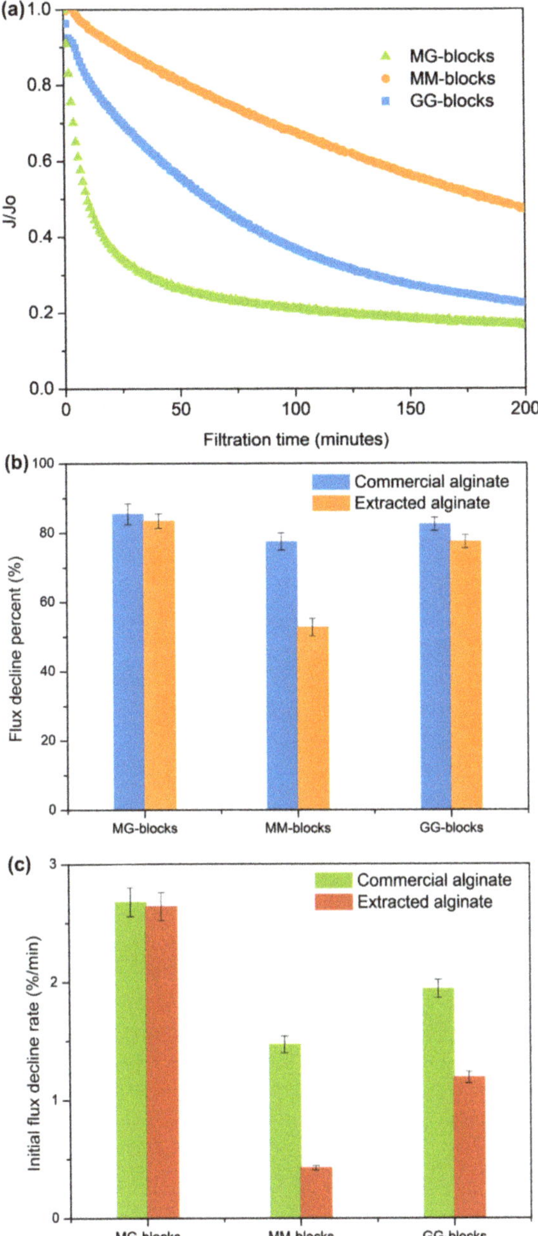

Figure 7. (**a**) Flux profiles of MG-, MM- and GG-blocks at 50 mg/L; (**b**) the final flux decline percentage and the (**c**) initial flux decline rates of alginate blocks derived from commercial and extracted alginates.

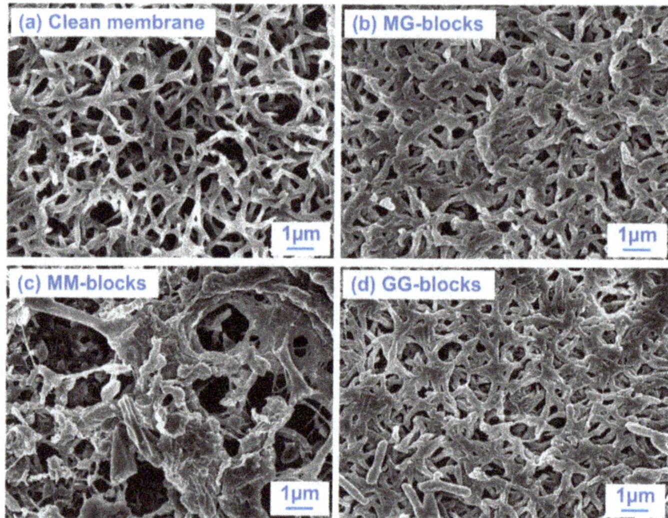

Figure 8. FESEM observations of the (**a**) clean membrane and membrane surfaces fouled by (**b**) MG-, (**c**) MM- and (**d**) GG-blocks.

It appears that this conclusion is different from the general understanding of TEPs—that they accelerate the membrane fouling [31–33]. However, it should be noted that the role of TEPs in membrane fouling is not a black and white issue. TEPs are easy to attach onto a membrane surface and promote fouling layer formation on a membrane. For the non-porous membrane and the membrane with small pores, TEPs may promote fouling layer development during membrane filtration which contributes to the filtration resistance. Differently, it turned out that the fouling layer acted as pre-filter to prevent the serious pore blocking of membranes in this study, which on the contrary, did not lead to serious fouling problems. Consequently, the TEP fouling is a complicated issue which also depends on the characteristics of membrane and the operation conditions and so on. Nevertheless, with high viscosity and big size, TEPs are prone to attach onto membranes compared to other foulants. More importantly, the TEP formation indicates the potential of polysaccharide crosslinking, which is an important issue in membrane fouling. These findings reveal that the molecular chains of polysaccharides do not exist singly, by themselves in a free state, but they usually combine together to form big three dimensional networks. The crosslinking abilities of polysaccharides vary from one to another dramatically, which in turn affects their fouling propensities. Combined with the results obtained in this study, it can be concluded that polysaccharides with different molecular compositions would have different characteristics, including their effects on membrane fouling. Therefore, future studies should pay attention to the molecule structures of foulant and the crosslinking between foulant molecules.

4. Conclusions

Alginate is abundant in natural water environments and consists of three different blocks; i.e., MG-, MM- and GG-blocks. In this study, an alginate sample was first extracted from the raw seaweed and then well characterized with fractionation, FTIR and FESEM. Results confirmed the existence of alginate in the seaweed and showed that the main component in this alginate was the MM-blocks. FTIR spectra of extracted MG-, MM- and GG-blocks illustrated a similar chemical composition with commercial alginate blocks. What is more important, is that the TEP measurement showed that MG-, MM- and GG-blocks possessed different abilities in forming TEP via molecular crosslinking. Furthermore, the filtration tests of the three kinds of alginate bocks demonstrated totally different flux profiles, indicating that they do have diverse fouling propensities. So far, many studies have employed alginate as a model foulant. However, our results reveal that alginates with different molecular compositions behave differently in membrane filtration experiments. It is a reasonable consideration that future work should include analysis of the alginate in question's molecular composition in order to obtain more reliable results.

Author Contributions: S.M. and L.W. conceived and designed the experiments; R.W. and S.M. performed the experiments; R.W. and X.M. analyzed the data; H.L. and M.Z. contributed reagents/materials/analysis tools; S.M. and L.W. wrote the paper.

Acknowledgments: This work was financially supported by grants from the National Natural Science Foundation of China (number 51808019) and this research received no external funding.

Conflicts of Interest: The authors declare no conflict of interest.

References

1. Qu, C.; Lu, S.; Liang, D.; Chen, S.; Xiang, Y.; Zhang, S. Simultaneous electro-oxidation and in situ electro-peroxone process for the degradation of refractory organics in wastewater. *J. Hazard. Mater.* **2019**, *364*, 468–474. [CrossRef] [PubMed]

2. Liu, S.; Cui, M.; Li, X.; Thuyet, D.Q.; Fan, W. Effects of hydrophobicity of titanium dioxide nanoparticles and exposure scenarios on copper uptake and toxicity in Daphnia magna. *Water Res.* **2019**, *154*, 162–170. [CrossRef] [PubMed]

3. Hu, R.; Gwenzi, W.; Sipowo-Tala, V.R.; Noubactep, C. Water Treatment Using Metallic Iron: A Tutorial Review. *Processes* **2019**, *7*, 622. [CrossRef]

4. Qu, C.; Soomro, G.S.; Ren, N.; Liang, D.-W.; Lu, S.-F.; Xiang, Y.; Zhang, S.-J. Enhanced electro-oxidation/peroxone (in situ) process with a Ti-based nickel-antimony doped tin oxide anode for phenol degradation. *J. Hazard. Mater.* **2019**, 121398. [CrossRef] [PubMed]

5. Elimelech, M.; Phillip, W.A. The Future of Seawater Desalination: Energy, Technology, and the Environment. *Science* **2011**, *333*, 712–717. [CrossRef]

6. Zamora, S.; Sandoval, L.; Marín-Muñíz, J.L.; Fernández-Lambert, G.; Hernández-Orduña, M.G. Impact of Ornamental Vegetation Type and Different Substrate Layers on Pollutant Removal in Constructed Wetland Mesocosms Treating Rural Community Wastewater. *Processes* **2019**, *7*, 531. [CrossRef]

7. Bis, M.; Montusiewicz, A.; Piotrowicz, A.; Łagód, G. Modeling of Wastewater Treatment Processes in Membrane Bioreactors Compared to Conventional Activated Sludge Systems. *Processes* **2019**, *7*, 285. [CrossRef]

8. Song, Y.; Motuzas, J.; Wang, D.K.; Birkett, G.; Smart, S.; Diniz da Costa, J.C. Substrate Effect on Carbon/Ceramic Mixed Matrix Membrane Prepared by a Vacuum-Assisted Method for Desalination. *Processes* **2018**, *6*, 47. [CrossRef]

9. Wagh, P.; Zhang, X.; Blood, R.; Kekenes-Huskey, P.M.; Rajapaksha, P.; Wei, Y.; Escobar, I.C. Increasing Salt Rejection of Polybenzimidazole Nanofiltration Membranes via the Addition of Immobilized and Aligned Aquaporins. *Processes* **2019**, *7*, 76. [CrossRef]

10. Martin Vincent, N.; Tong, J.; Yu, D.; Zhang, J.; Wei, Y. Membrane Fouling Characteristics of a Side-Stream Tubular Anaerobic Membrane Bioreactor (AnMBR) Treating Domestic Wastewater. *Processes* **2018**, *6*, 50. [CrossRef]

11. Berman, T.; Holenberg, M. Don't fall foul of biofilm through high TEP levels. *Filtr. Sep.* **2005**, *42*, 30–32. [CrossRef]

12. Passow, U. Transparent exopolymer particles (TEP) in aquatic environments. *Prog. Oceanogr.* **2002**, *55*, 287–333. [CrossRef]

13. Wang, R.; Liang, D.; Liu, X.; Fan, W.; Meng, S.; Cai, W. Effect of magnesium ion on polysaccharide fouling. *Chem. Eng. J.* **2020**, *379*. [CrossRef]

14. Bar-Zeev, E.; Passow, U.; Castrillon, S.R.; Elimelech, M. Transparent exopolymer particles: From aquatic environments and engineered systems to membrane biofouling. *Environ. Sci. Technol.* **2015**, *49*, 691–707. [CrossRef] [PubMed]

15. Draget, K.I.; Smidsrød, O.; Skjåk-Bræk, G. Alginates from Algae. In *Biopolymers Online*; Wiley-VCH Verlag GmbH & Co. KGaA: Weinheim, Germany, 2005. [CrossRef]

16. Melvik, J.; Dornish, M. Alginate as a Carrier for Cell Immobilisation. In *Fundamentals of Cell Immobilisation Biotechnology*; Nedović, V., Willaert, R., Eds.; Springer: Amsterdam, The Netherlands, 2004; Volume 8A, pp. 33–51.

17. Tønnesen, H.H.; Karlsen, J. Alginate in Drug Delivery Systems. *Drug Dev. Ind. Pharm.* **2002**, *28*, 621–630. [CrossRef] [PubMed]

18. Lee, K.Y.; Mooney, D.J. Alginate: Properties and biomedical applications. *Prog. Polym. Sci.* **2012**, *37*, 106–126. [CrossRef] [PubMed]

19. Leal, D.; Matsuhiro, B.; Rossi, M.; Caruso, F. FT-IR spectra of alginic acid block fractions in three species of brown seaweeds. *Carbohydr. Res.* **2008**, *343*, 308–316. [CrossRef]

20. Meng, S.; Liu, Y. New insights into transparent exopolymer particles (TEP) formation from precursor materials at various Na+/Ca2+ ratios. *Sci. Rep.* **2016**, *6*, 19747. [CrossRef]

21. Meng, S.; Winters, H.; Liu, Y. Ultrafiltration behaviors of alginate blocks at various calcium concentrations. *Water Res.* **2015**, *83*, 248–257. [CrossRef]

22. Chandía, N.P.; Matsuhiro, B.; Mejías, E.; Moenne, A. Alginic acids in Lessonia vadosa: Partial hydrolysis and elicitor properties of the polymannuronic acid fraction. *J. Appl. Phycol.* **2004**, *16*, 127–133. [CrossRef]

23. Steyermark, A. *Spectrometric Identification of Organic Compounds*, 3rd ed.; Silverstein, R.M., Bassler, G.C., Morrill, T.C., Eds.; Wiley: New York, NY, USA, 1974; p. 340. [CrossRef]

24. Mathlouthi, M.; Koenig, J.L. Vibrational Spectra of Carbohydrates. In *Advances in Carbohydrate Chemistry and Biochemistry*; Tipson, R.S., Derek, H., Eds.; Academic Press: Cambridge, MA, USA, 1987; Volume 44, pp. 7–89.

25. Gómez-Ordóñez, E.; Rupérez, P. FTIR-ATR spectroscopy as a tool for polysaccharide identification in edible brown and red seaweeds. *Food Hydrocoll.* **2011**, *25*, 1514–1520. [CrossRef]

26. Coimbra, M.A.; Barros, A.; Barros, M.; Rutledge, D.N.; Delgadillo, I. Multivariate analysis of uronic acid and neutral sugars in whole pectic samples by FT-IR spectroscopy. *Carbohydr. Polym.* **1998**, *37*, 241–248. [CrossRef]

27. Synytsya, A.; Kim, W.-J.; Kim, S.-M.; Pohl, R.; Synytsya, A.; Kvasnička, F.; Čopíková, J.; Il Park, Y. Structure and antitumour activity of fucoidan isolated from sporophyll of Korean brown seaweed Undaria pinnatifida. *Carbohydr. Polym.* **2010**, *81*, 41–48. [CrossRef]

28. Meng, S.; Liu, Y. Alginate block fractions and their effects on membrane fouling. *Water Res.* **2013**, *47*, 6618–6627. [CrossRef] [PubMed]

29. Meng, S.; Fan, W.; Li, X.; Liu, Y.; Liang, D.; Liu, X. Intermolecular interactions of polysaccharides in membrane fouling during microfiltration. *Water Res.* **2018**, *143*, 38–46. [CrossRef]

30. Wang, X.; Fan, W.; Dong, Z.; Liang, D.; Zhou, T. Interactions of natural organic matter on the surface of PVP-capped silver nanoparticle under different aqueous environment. *Water Res.* **2018**, *138*, 224–233. [CrossRef]

31. Berman, T.; Mizrahi, R.; Dosoretz, C.G. Transparent exopolymer particles (TEP): A critical factor in aquatic biofilm initiation and fouling on filtration membranes. *Desalination* **2011**, *276*, 184–190. [CrossRef]

32. Villacorte, L.O.; Kennedy, M.D.; Amy, G.L.; Schippers, J.C. Measuring transparent exopolymer particles (TEP) as indicator of the (bio)fouling potential of RO feed water. *Desalin. Water Treat.* **2009**, *5*, 207–212. [CrossRef]

33. Bar-Zeev, E.; Berman-Frank, I.; Liberman, B.; Rahav, E.; Passow, U.; Berman, T. Transparent exopolymer particles: Potential agents for organic fouling and biofilm formation in desalination and water treatment plants. *Desalin. Water Treat.* **2009**, *3*, 136–142. [CrossRef]

Article

Modifying Nanoporous Carbon through Hydrogen Peroxide Oxidation for Removal of Metronidazole Antibiotics from Simulated Wastewater

Teguh Ariyanto [1,2,*], Rut Aprillia Galuh Sarwendah [1], Yove Maulana Novirdaus Amimmal [1], William Teja Laksmana [1] and Imam Prasetyo [1,2]

[1] Department of Chemical Engineering, Faculty of Engineering, Universitas Gadjah Mada, Jl Grafika No 2, Yogyakarta 55281, Indonesia; rutruut23@gmail.com (R.A.G.S.); yove.maulana.n@mail.ugm.ac.id (Y.M.N.A.); william.teja.l13@gmail.com (W.T.L.); Imampras@ugm.ac.id (I.P.)

[2] Carbon Material Research Group, Department of Chemical Engineering, Faculty of Engineering, Universitas Gadjah Mada, Jl Grafika No 2, Yogyakarta 55281, Indonesia

*   Correspondence: teguh.ariyanto@ugm.ac.id; Tel.: +62-274-649-2171

Received: 6 October 2019; Accepted: 5 November 2019; Published: 8 November 2019

Abstract: This study examined change in pore structure and microstructure of nanoporous carbon after surface oxidation and how it affects the adsorption performance of metronidazole antibiotics. The surface oxidation was performed by hydrogen peroxide at 60 °C. The properties of porous carbon were investigated by N_2-sorption analysis (pore structure), scanning electron microscopy (surface morphology), the Boehm titration method (quantification of surface functional group), and Fourier transform infrared spectroscopy (type of surface functional group). The results showed that the oxidation of porous carbon by hydrogen peroxide has a minor defect in the carbon pore structure. Only a slight decrease in specific surface area (8%) from its original value (973 m^2g^{-1}) was seen but more mesoporosity was introduced. The oxidation of porous carbon with hydrogen peroxide modified the amount of oxide groups i.e., phenol, carboxylic acid and lactone. Moreover, in the application the oxidized carbon exhibited a higher the metronidazole uptake capacity of up to three-times manifold with respect to the pristine carbon.

Keywords: adsorption; metronidazole; porous carbon; surface modification; wastewater treatment

---

1. Introduction

The occurrence of pharmaceutical compounds in the aquatic environment has been a high concern. When water containing pharmaceutical compounds is released into river, it can influence the quality of water and harm the ecosystem balance. Studies showed that pharmaceutical compounds in an aquatic system could have serious effects such as antibiotic-resistant microbes [1], the change of fish reproduction [2], and inhibition of photosynthesis in aquatic organisms [3]. When a waterbody containing antibiotics flows to agriculture land, this could affect the maturity of plants [4]. One of the main sources of pharmaceutical effluents is hospitals which, in Europe, typically discharges pharmaceutical components in the range concentrations of 0.2–11 µg L^{-1} [5]. This could be even worse in other countries when using improper wastewater treatment, resulting in a higher concentration of pharmaceutical substances released to the environment.

Metronidazole is one of the most used pharmaceutical compounds. This is an antibiotic which hinders the growth of parasitic microbes like bacteria and protozoa [6]. Metronidazole is highly soluble in water and persist even after biodegradation [7]. The molecular structure of metronidazole is given in Figure 1.

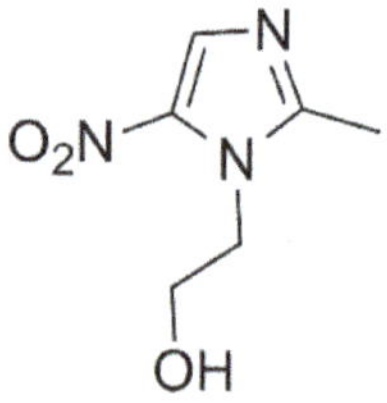

Figure 1. Molecular structure of metronidazole.

Adsorption is a popular technique for removal of contaminants from wastewater. The method has the advantages of a simple design and ease of operation. Among adsorbents, porous carbon has been regarded as the most promising materials, taking advantages of its excellent physical and chemical stability and high specific surface area (500–2000 m^2g^{-1}) [8–10]. Studies showed porous carbon could work to remove metronidazole from aqueous systems [11–13]. The capacity of adsorption of organic compounds is influenced e.g. pore textural properties and surface functional groups of porous carbon [10,14]. Therefore, there is possibility to further increase adsorption capacity by modification of pore- and surface chemistry of nanoporous carbon. A high uptake capacity is beneficial providing longer adsorption time, hence more efficient. However, no detail study yet is presented so far to modify pore- and surface chemistry of porous carbon for adsorption of antibiotic substances.

A presence of functional groups like acidic sites could increases wettability of pore surfaces. To introduce acidic groups onto nanoporous carbon, surface oxidation using sulphuric acid, nitric acid and hydrogen peroxide can be performed [15,16]. Oxidation using strong acids is effective to increase functional groups of oxygen, but this relatively harsh oxidation induces a destruction of pore structures hence remarkably decreasing specific surface area [17,18]. Oxidation of nanoporous carbon using hydrogen peroxide can give a good trade-off between introduction of oxygen-functional groups and change in pore structures. Furthermore, this oxidation method does not require a further step of washing of nanoporous carbon.

This work presents a study of ease modification of nanoporous carbon and its efficacy on metronidazole adsorption. Oxidation was performed by hydrogen peroxide. Effects of this mild oxidation on the pore structure, surface morphology and functional groups and their influence on the performance of metronidazole are described.

2. Materials and Methods

2.1. Material

Nanoporous carbon from coconut shell (CSC) was obtained from PT Home System, Indonesia. CSC was then crushed and sieved to obtain a particle size in the range of 355–710 μm. Hydrogen peroxide solution of 50% was obtained from Merck, Germany. Hydrogen peroxide solution of 30% was prepared by dilution of highly concentrated hydrogen peroxide of 50%. Metronidazole 99% purity was a product of Sigma Aldrich, Singapore.

2.2. Oxidation of Nanoporous Carbon

Fifty milliliters of 30% hydrogen peroxide solution was poured in a three-neck flask. After this, the solution was heated to the desired oxidation temperature of 60 °C. After oxidation was finished (2 h), carbon was separated from the solution and dried overnight in an oven (60 °C).

2.3. Characterization of Nanoporous Carbon

Morphologies of material CSC, both before and after the oxidation process, were probed by scanning electron microscope (SEM) of JSM-6510LA (JEOL, Tokyo, Japan). Pore analysis of nanoporous carbon was performed by using Quantachrome Nova 2000 (Quantachrome Instruments, Boynton

Beach, FL, USA) through the nitrogen sorption method. The multipoint Brunauer–Emmett–Teller (BET) method was employed to evaluate the specific surface area (SSA) from the isotherm data. Micropore volume and micropore SSA were evaluated using the t-plot method, which is a correlation between the adsorbed volume of nitrogen and statistical thickness. BET and statistical thickness of the Hasley model are shown in Equations (1) and (2) [19]. Calculation of pore textural parameters were carried out using a software provided by Quantachrome NovaWin version 11.03 (Quantachrome Instruments, Boynton Beach, FL, USA).

$$\frac{P}{V(P_0 - P)} = \frac{1}{V_m C} + \frac{C-1}{V_m C}\left(\frac{P}{P_0}\right) \tag{1}$$

$$t(A) = 3.54\left[\frac{5}{2.303\log(P_o/P)}\right]^{1/3} \tag{2}$$

where P is the pressure, V is the adsorbed volume of nitrogen, Po is the saturated pressure of nitrogen at 77 K, Vm is the monolayer adsorbed volume of nitrogen, t is the statistical thickness, and C is constant.

The qualitative analysis of functional groups in porous carbon was evaluated through Fourier-Transform Infrared Spectroscopy (FTIR) using Nicolet iS10 (Thermo Fisher Scientific, MA, USA), while the quantitative analysis was performed through the Boehm titration method [20]. Fifty milligrams of carbons were soaked in 50 mL of 0.001 N Na_2CO_3 (prepared from solid Na_2CO_3, Merck, Darmstadt, Germany), 0.001 N $NaHCO_3$ (prepared from solid $NaHCO_3$, Merck, Darmstadt, Germany), 0.001 N NaOH (prepared from solid NaOH, Merck, Darmstadt, Germany), for 24 h. Next, the carbons were separated and the solutions of Na_2CO_3, $NaHCO_3$, and NaOH of the separation's result were titrated with 0.001 N HCl (prepared from concentrated HCl, Merck, Darmstadt, Germany). The number of functional groups present were determined through a balance of acid in the solution.

2.4. Adsorption of Antibiotic Substances

The metronidazole adsorption was performed through a preparation of metronidazole solution with various concentrations, from 20 to 200 ppm. This initial concentration range was employed to determine the equilibrium curve between the amount of adsorbate in the solid phase and the amount of adsorbate in the liquid phase, until the maximum adsorption capacity was reached. Fifty milliliters of metronidazole solution were put into Erlenmeyer of 125 mL and 50 mg of carbon samples was added for each of the solution samples. The Erlenmeyer was shaken for 24 h in the water bath shaker, (Memmert, Schwabach, Germany) at room temperature. After adsorption, the metronidazole solution was then analyzed by the method described in literature [21]. The solution was mixed with reagents of 1 N HCl, zinc dust, 1% $NaNO_2$ (prepared from solid $NaNO_2$, Merck, Darmstadt, Germany) solution, 2% sulfamic acid solution, 1% N-(1-naphthyl)ethylenediamine dihydrochloride or NEDA (prepared from ~100% NEDA, Merck, Darmstadt, Germany) solution, and 10% NaOH solution until it was colored. The metronidazole concentration in the solution after the adsorption was measured with UV–Vis spectrophotometer Shimadzu UV Mini 1240 (Shimadzu, Kyoto, Japan) with a wavelength of 510 nm. The experimental set-up and adsorption conditions were based on the reports in [22,23].

2.5. Adsorption Model of Antibiotic Substances

Adsorption of metronidazole was modeled using Langmuir and Freundlich isotherm models, as shown in Equations (3) and (4) [19].

$$Q_e = \frac{Q_m K_L C_e}{1 + K_L C_e} \tag{3}$$

$$Q_e = K_F C_e^{1/n} \tag{4}$$

where K_L is the Langmuir equilibrium constant (L/mol), Q_m is the maximum adsorbate adsorption capacity of the adsorbent (mg adsorbate/g adsorbent), K_F is the Freundlich equilibrium constant (mg adsorbate/(g adsorbent $(mol/L)^n$)), n is the correction factor, Q_e is the adsorbate adsorption capacity (mg adsorbate/g adsorbent) in the solid state in an equilibrium condition with a concentration of adsorbates in the liquid, and *Ce* is the adsorbate concentration in the solution (mol/L).

The mass balance of metronidazole in the solution and adsorbent were derived to get the adsorption variable. The variable was used to evaluate the adsorption parameter of Equations (3) and (4). The mass balance derivative of metronidazole for the batch system was written as Equation (5).

$$(Co - Ct)V = (Qt - Qo)w \tag{5}$$

In the initial condition, before metronidazole was adsorbed by carbon, $Qo = 0$, so Equation (5) became Equation (6).

$$Qe = \frac{(Co - Ce)V}{w} \tag{6}$$

where *Co* is the initial concentration of metronidazole in the solution (mg adsorbate/L), *V* is the solution volume (L), and *w* is the adsorbent mass (g).

3. Results and Discussion

3.1. Pore Structure of Carbon Material

The pore structure of CSC, before and after oxidation with hydrogen peroxide was characterized by low-temperature nitrogen physisorption at 77 K. The sorption isotherms are displayed in Figure 2A. The CSC and CSC-OX feature an identical isotherm of Type I, according to the IUPAC classification [24], indicating that the materials possess substantial pores in a micropore regime. Of note is that the amount of nitrogen adsorbed on the CSC-OX was low. This indicated a decrease in the specific surface area (SSA). The pore-sized distributions (PSDs) determined through the QSDFT model are shown in Figure 2B. PSD of CSC and CSC-OX are dominantly in micropore regime around (<2 nm). When comparing the PSDs of CSC and CSC-OX in the mesopore range (2–50 nm), it could be seen that a higher fraction of mesopores were present in the CSC-OX. These pores were likely created when performing the oxidation process using hydrogen peroxide.

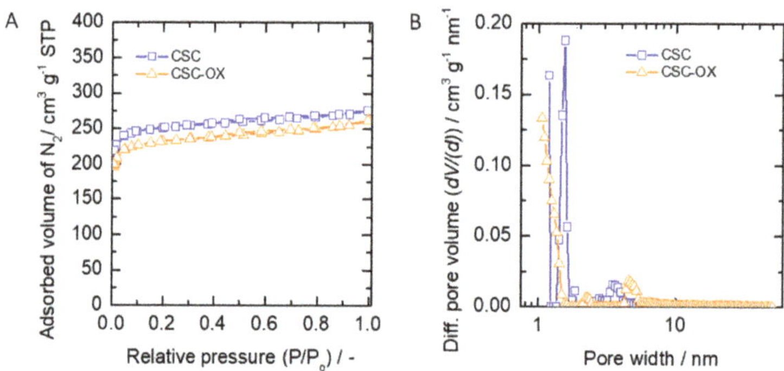

Figure 2. (**A**) N$_2$-sorption isotherm sorption. (**B**) The pore size distribution of carbon evaluated by the N$_2$-QSDFT method. In Figure 2A, *P* is the pressure and *Po* is the saturated pressure of nitrogen at 77 K.

Table 1 shows the pore parameters of the carbon both before and after the oxidation process, which were calculated based on the N$_2$-sorption isotherm data. From the analysis result, the SSA, total pore volume, and percentage of micropore fraction after the oxidation process were decreased when compared to the pristine one. The decreases occurred likely due to the oxidation process. Oxidation causes a removal of the carbon element and the formation of functional groups, which then produce bigger pores [25]. When voids larger than 300 nm are created by the oxidation of pores, they cannot be detected by nitrogen sorption, resulting in a small decrease in the total pore volume from 0.43 to 0.41 cm^3 g^{-1}.

Table 1. Analysis result of surface area with the nitrogen sorption method.

No.	Carbon Type	SSA [a] (m²/g)	Total Pore Volume [b] (cm³/g)	Percentage of Micropore SSA [c] (%)	Percentage of Micropore Volume [c] (%)
1.	CSC	973	0.43	95	84
2.	CSC-OX	895	0.41	94	80

[a] Specific surface area; [b] Pore volume at 0.995 P/Po; [c] Determined by the t-plot method.

3.2. Surface Morphology

The morphology of CSC before and after oxidation was investigated using scanning electron microscopy. From Figure 3, it can be seen that the microstructures in the CSC's surface before the oxidation process were different. In the 100 times magnification (Figure 3A), the CSC's surface was quite rough while in the 1000 times magnification (Figure 3B), the granules looked smoother and no void appears.

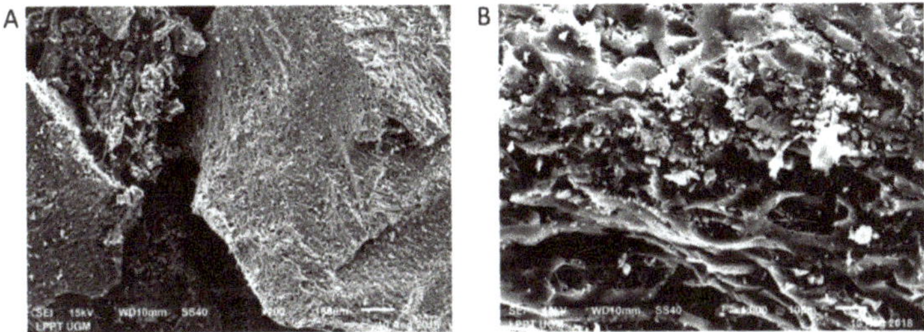

Figure 3. SEM micrograph of CSC at 100× magnification (**A**) and 1000× magnification (**B**).

In Figure 4, the surface of CSC-OX in the 100 times magnification (Figure 4A) appears rougher and looks even more damaged than before the oxidation process. In the 1000 times magnification (Figure 4B), the granules still looked smooth but some voids started to appear. Compared to CSC, CSC-OX looked rougher and had a bigger void than CSC. The pore structures also looked more crumbled than before. The results indicated that performing the oxidation process could alter the surface morphology.

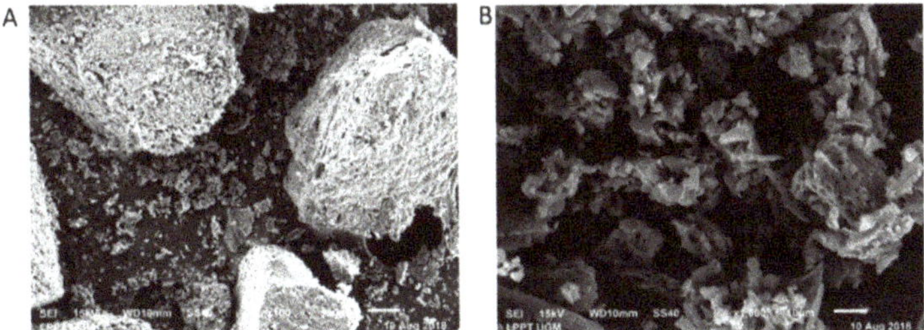

Figure 4. SEM micrograph of CSC-OX at 100× magnification (**A**) and 1000× magnification (**B**).

3.3. Oxygen Functional Groups

Functional groups in the porous carbon of CSC and CSC-OX were analyzed qualitatively and quantitatively. The qualitative analysis used Fourier-Transform Infrared spectroscopy (FTIR) while the quantitative analysis used the Boehm Titration.

From Figure 5, in the CSC before oxidation, there are a group of [C–O] showed by the wave number around 1200 cm^{-1}, a group of [C=C] showed by the wave number around 1600 cm^{-1}, and a group of [C–H] showed by the wave number around 3,000 cm^{-1}. These indicate the presence of a phenol group (at ca. 1200 cm^{-1}) and an aromatic ring of carbon structures (at ca. 1600 cm^{-1}). After the oxidation process, there is a new peak in the wave number around 1700 cm^{-1} which shows that after the oxidation process, the group of [C=O] appeared, which is a carbonyl group. When comparing the CSC and the CSC-OX spectra, it could be noticed that before the oxidation, the carbon already had groups of oxygen functional groups. It is understandable because CSC is formed from biomass of lignocellulose material and contains oxygen elements.

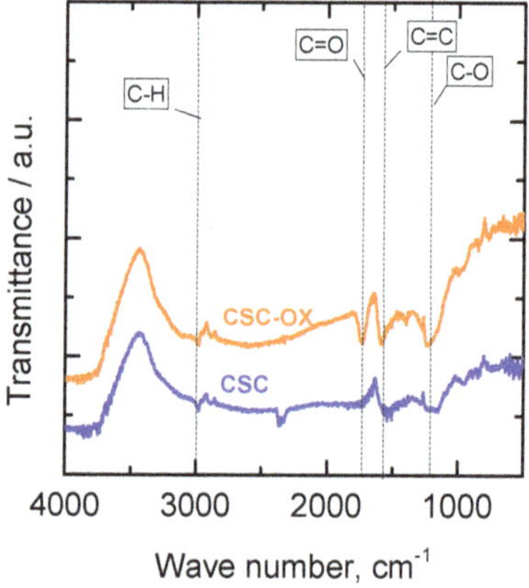

Figure 5. CSC and CSC-OX's FTIR analysis

Figure 6 displays the number of oxygen functional groups in CSC and CSC-OX. There are changes in the group of lactones, carboxyl, and phenols due to the oxidation process of carbons. When the oxidation occurs, carbon lose its electron and makes it more positively charged. The occurrence of oxygen and hydrogen molecules in the peroxide solution causes the formation of oxygen groups in the carbon's surface. It was noticed that carboxylic and lactone groups are more presence in CSC-OX and phenol group is decreasing as the oxidation process takes place.

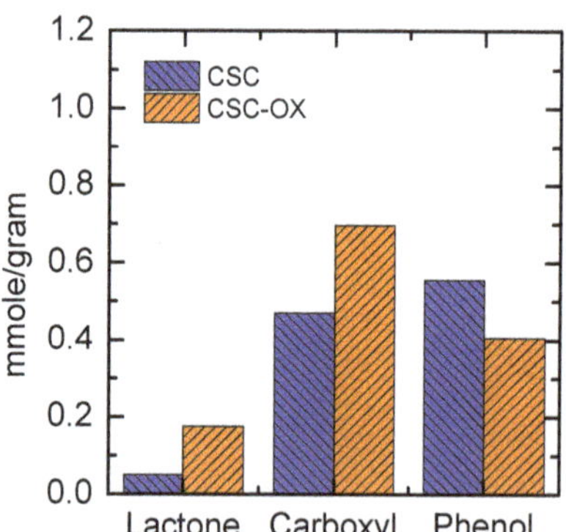

Figure 6. Analysis result of the acid group before and after the oxidation process with the Boehm titration method of CSC and CSC-OX.

3.4. Metronidazole Adsorption Performance

The oxidation process tended to increase the mesoporosity of porous carbon (shown in Table 1) and also increased the number of oxygen functional groups (Figures 5 and 6). CSC and CSC-OX were then employed for metronidazole adsorption. It can be seen from Figure 7 that the adsorption performance of oxidized carbon is better than the pristine carbon in the term of uptake capacity. At equilibrium of 45 mg/L concentration, adsorbed metronidazole in CSC-OX is ca. 310 mg/g adsorbent while in CSC is only ca. 100 mg/g adsorbent. This shows that modifications of porous carbon through hydrogen peroxide improve the adsorption performance. The high adsorption capacity of nanoporous carbon obtained by the simple oxidation process could compete with other activated carbons, having a range of uptake capacity of 150–300 mg/g adsorbent [11,12].

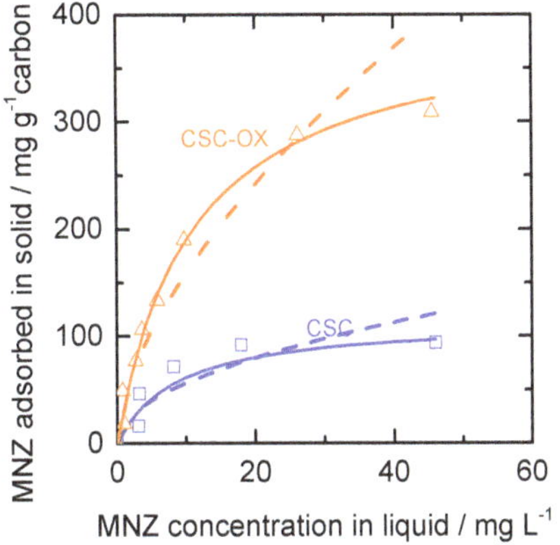

Figure 7. Adsorption isotherm of metronidazole on CSC and CSC-OX (data: square and triangle; Langmuir-fitted curve—full line; Freundlich-fitted curve—dash line).

The distribution of molecular adsorption between the liquid phase and the solid phase could be described with an adsorption isotherm when it is in the equilibrium condition. The parameter that was obtained through the isotherm model is important information in the sorption mechanism; its surface characteristics and also showed adsorbent affinity [26]. The constants values of the metronidazole adsorption models of Langmuir and Freundlich are shown in Table 2. The Langmuir models is more appropriate for metronidazole adsorption models because its R^2 is closer to 1. For Langmuir model, the value of the maximum adsorption capacity (Q_m) for CSC-OX showcases the 400 mg/g adsorbent which is more than 3-times than that of CSC (ca. 115 mg/g adsorbent). However, the value of Langmuir constant of K_L for oxidized carbon is 25% lower with respect to pristine carbon. For this phenomenon, it is likely that an increase in uptake capacity of metronidazole seems to be more related to the change in pore textural property rather than surface chemistry. An increase in the amount of oxygen functional groups in CSC-OX tends to decrease affinity of metronidazole onto surface of pores as can be seen from the decrease of value of Langmuir constant of K_L (from 0.12 to 0.09 L/mol). This negative effects of oxygen functional groups to adsorption of organic compounds are also observed from literatures [14,27]. But alternation of porosity in oxidized carbon (see Table 1 and SEM image) likely induces more occupation of pore surfaces as indicated by the increase of the Q_m value. Freundlich fitted parameters also agree with the findings. Freundlich constant of K_F increases for oxidized carbon but the value of correction factor of n decreases from 2.00 to 1.64.

Figure 7 displays the Langmuir and Freundlich fitted curves of CSC and CSC-OX suggesting that Langmuir model is more fitted to the equilibrium data. Because the Langmuir model is compatible (indicated by value of R^2), this shows that the adsorption process happened in a specific homogenous area to the adsorbent that is in the interface layer between adsorbate and adsorbent or called monolayer adsorption [19].

Table 2. Constant value of various fitting model of metronidazole adsorption.

No.	Material	Langmuir Model			Freundlich Model		
		K_L, L/mol	Qm, mg Adsorbate/g Adsorbent	R^2	K_F mg Adsorbate/(g Adsorbent $(mol/L)^n$)	n	R^2
1.	CSC	0.12	114.94	0.90	17.99	2.00	0.64
2.	CSC-OX	0.09	400.00	0.95	39.03	1.64	0.85

4. Conclusions

Modification of nanoporous carbon was conducted using mild oxidation of hydrogen peroxide. The results showed that the oxidation process alters the porosity of nanoporous carbon and amount of oxygen functional group. After oxidation, a higher amount of oxygen functional groups of porous carbon is seen and only a little decrease in the specific surface area, but more porosity is exhibited. In the application, oxidized carbon showed remarkably higher metronidazole uptake capacity. The increase in the adsorption capacity seems to be more related to the change in pore textural property rather than amount of oxygen functional group.

Author Contributions: Conceptualization, T.A. and I.P.; methodology, T.A. and I.P.; formal analysis, R.A.G.S. and Y.M.N.A.; investigation, R.A.G.S. and Y.M.N.A.; resources, T.A. and I.P.; writing—original draft preparation, T.A.; writing—review and editing, T.A. and I.P.; visualization, W.T.L.; supervision, T.A. and I.P; funding acquisition, T.A.

Funding: This research and APC were funded by program of Rekognisi Tugas Akhir (RTA), Universitas Gadjah Mada, grant number 2129/UN1/DITLIT/DIT-LIT/LT/2019.

Acknowledgments: The authors thank the PT Home System Indonesia for providing the porous carbon synthesized from coconut shell as a gift.

Conflicts of Interest: The authors declare no conflict of interest.

References

1. Gadipelly, C.; Pérez-González, A.; Yadav, G.D.; Ortiz, I.; Ibáñez, R.; Rathod, V.K.; Marathe, K.V. Pharmaceutical industry wastewater: Review of the technologies for water treatment and reuse. *Ind. Eng. Chem. Res.* **2014**, *53*, 11571–11592. [CrossRef]

2. Kidd, K.A.; Blanchfield, P.J.; Mills, K.H.; Palace, V.P.; Evans, R.E.; Lazorchak, J.M.; Flick, R.W. Collapse of a fish population after exposure to a synthetic estrogen. *Proc. Natl. Acad. Sci. USA* **2007**, *104*, 8897–8901. [CrossRef] [PubMed]

3. Escher, B.I.; Baumgartner, R.; Koller, M.; Treyer, K.; Lienert, J.; McArdell, C.S. Environmental toxicology and risk assessment of pharmaceuticals from hospital wastewater. *Water Res.* **2011**, *45*, 75–92. [CrossRef] [PubMed]

4. Yakubu, O.H. Pharmaceutical wastewater effluent-source of contaminants of emerging concern: Phytotoxicity of metronidazole to soybean (Glycine max). *Toxics* **2017**, *5*, 10. [CrossRef] [PubMed]

5. Munoz, M.; Garcia-Muñoz, P.; Pliego, G.; Pedro, Z.M.D.; Zazo, J.A.; Casas, J.A.; Rodriguez, J.J. Application of intensified Fenton oxidation to the treatment of hospital wastewater: Kinetics, ecotoxicity and disinfection. *J. Environ. Chem. Eng.* **2016**, *4*, 4107–4112. [CrossRef]

6. Goolsby, T.A.; Jakeman, B.; Gaynes, R.P. Clinical relevance of metronidazole and peripheral neuropathy: A systematic review of the literature. *Int. J. Antimicrob. Agents* **2018**, *51*, 319–325. [CrossRef] [PubMed]

7. Lanzky, P.F.; Halting-Sørensen, B. The toxic effect of the antibiotic metronidazole on aquatic organisms. *Chemosphere* **1997**, *35*, 2553–2561. [CrossRef]

8. Prasetyo, I.; Rochmadi, R.; Wahyono, E.; Ariyanto, T. Controlling synthesis of polymer-derived carbon molecular sieve and its performance for CO_2/CH_4 separation. *Eng. J.* **2017**, *21*, 83–94. [CrossRef]

9. Amelia, S.; Sediawan, W.B.; Prasetyo, I.; Munoz, M.; Ariyanto, T. Role of the pore structure of Fe/C catalysts on heterogeneous Fenton oxidation. *J. Environ. Chem. Eng.* **2019**, 102921. [CrossRef]

10. Ariyanto, T.; Kurniasari, M.; Laksmana, W.T.; Prasetyo, I. Pore size control of polymer-derived carbon adsorbent and its application for dye removal. *Int. J. Environ. Sci. Technol.* **2019**, *16*, 4631–4636. [CrossRef]
11. Carrales-Alvarado, D.H.; Ocampo-Pérez, R.; Leyva-Ramos, R.; Rivera-Utrilla, J. Removal of the antibiotic metronidazole by adsorption on various carbon materials from aqueous phase. *J. Colloid Interface Sci.* **2014**, *436*, 276–285. [CrossRef] [PubMed]
12. Çalişkan, E.; Göktürk, S. Adsorption characteristics of sulfamethoxazole and metronidazole on activated carbon. *Sep. Sci. Technol.* **2010**, *45*, 244–255. [CrossRef]
13. Ilomuanya, M.; Nashiru, B.; Ifudu, N.; Igwilo, C. Effect of pore size and morphology of activated charcoal prepared from midribs of Elaeis guineensis on adsorption of poisons using metronidazole and Escherichia coli O157:H7 as a case study. *J. Microsc. Ultrastruct.* **2017**, *5*, 32. [CrossRef] [PubMed]
14. Liu, G.; Li, X.; Campos, L.C. Role of the functional groups in the adsorption of bisphenol A onto activated carbon: Thermal modification and mechanism. *J. Water Supply Res. Technol. AQUA* **2017**, *66*, 105–115. [CrossRef]
15. Peng, Y.; Liu, H. Effects of oxidation by hydrogen peroxide on the structures of multiwalled carbon nanotubes. *Ind. Eng. Chem. Res.* **2006**, *45*, 6483–6488. [CrossRef]
16. Rehman, A.; Park, M.; Park, S.-J. Current progress on the surface chemical modification of carbonaceous materials. *Coatings* **2019**, *9*, 103. [CrossRef]
17. Moreno-Castilla, C.; Ferro-Garcia, M.A.; Joly, J.P.; Bautista-Toledo, I.; Carrasco-Marin, F.; Rivera-Utrilla, J. Activated carbon surface modifications by nitric acid, hydrogen peroxide, and ammonium peroxydisulfate treatments. *Langmuir* **1995**, *11*, 4386–4392. [CrossRef]
18. Ros, T.G.; Van Dillen, A.J.; Geus, J.W.; Koningsberger, D.C. Surface oxidation of carbon nanofibres. *Chem. A Eur. J.* **2002**, *8*, 1151–1162. [CrossRef]
19. Do, D.D. *Adsorption Analysis: Equilibria and Kinetics*, 2nd ed.; Imperial College Press: London, UK, 1998; ISBN 1860941303.
20. Allwar, A. Characteristics of pore structures and surface chemistry of activated carbons by physisorption, FTIR And Boehm methods. *IOSR J. Appl. Chem.* **2012**, *2*, 9–15. [CrossRef]
21. Saffaj, T.; Charrouf, M.; Abourrche, A.; Aboud, Y.; Bennamara, A.; Berrada, M. Spectrophotometric determination of metronidazole and secnidazole in pharmaceutical preparations based on the formation of dyes. *Dyes Pigments* **2006**, *70*, 259–262. [CrossRef]
22. Sarwendah, R.A.G. Oksidasi Karbon Berpori dari dari Tempurung Kelapa dengan Hidrogen Peroksida serta Pengaruhnya untuk Adsorpsi Metronidazol. Bachelor's Thesis, Universitas Gadjah Mada, Yogyakarta, Indonesia, 2018.
23. Amimmal, Y.M.N. Oksidasi Karbon Berpori dari Kayu Mlanding dengan Hidrogen Peroksida serta Pengaruhnya untuk Adsorpsi Metronidazol. Bachelor's Thesis, Universitas Gadjah Mada, Yogyakarta, Indonesia, 2018.
24. Thommes, M.; Kaneko, K.; Neimark, A.V.; Olivier, J.P.; Rodriguez-Reinoso, F.; Rouquerol, J.; Sing, K.S.W. Physisorption of gases, with special reference to the evaluation of surface area and pore size distribution (IUPAC Technical Report). *Pure Appl. Chem.* **2015**, *87*, 1051–1069. [CrossRef]
25. Hasse, B.; Glasel, J.; Kern, A.M.; Murzin, D.Y.; Etzold, B.J.M. Preparation of carbide-derived carbon supported platinum catalysts. *Catal. Today* **2015**, *249*, 30–37. [CrossRef]
26. Li, Z.; Jia, Z.; Ni, T.; Li, S. Adsorption of methylene blue on natural cotton based flexible carbon fiber aerogels activated by novel air-limited carbonization method. *J. Mol. Liq.* **2017**, *242*, 747–756. [CrossRef]
27. Ania, C.O.; Parra, J.B.; Pis, J.J. Influence of oxygen-containing functional groups on active carbon adsorption of selected organic compounds. *Fuel Process. Technol.* **2002**, *79*, 265–271. [CrossRef]

 *processes*

Article

Knowledge Mapping of Carbon Footprint Research in a LCA Perspective: A Visual Analysis Using CiteSpace

Shihu Zhong [1,*,†], Rong Chen [1,*], Fei Song [2,*,†] and Yanmin Xu [3]

[1] School of Public Economics and Administration, Shanghai University of Finance and Economics, Shanghai 200433, China
[2] School of Information Engineering, Jiangsu Open University, Nanjing 210017, China
[3] School of Humanities, Shanghai University of Finance and Economics, Shanghai 200433, China; 2018310003@live.sufe.edu.cn
* Correspondence: zhongshihu@163.sufe.edu.cn (S.Z.); 2017311013@live.sufe.edu.cn (R.C.); songf@jsou.cn (F.S.)
† These two authors contributed equally to this work.

Received: 26 August 2019; Accepted: 29 October 2019; Published: 5 November 2019

Abstract: Carbon emissions are inevitably linked to lifestyle and consumption behaviours, and the concept of "carbon footprinting" is now well-recognised beyond academia. Life cycle assessment (LCA) is one of the primary tools for assessing carbon footprints. The aim of this paper is to present a systematic review of literatures focusing on carbon footprint calculated with life cycle assessment. We used CiteSpace software to draw the knowledge map of related research to identify and trace the knowledge base and frontier terminology. It was found that the LCA application in respects of carbon footprint studies was completed mainly for the following aspect: beef production and dairy industry, seafood and fishery, nutrition, urban structure and energy use. The CiteSpace analysis showed the development path of the above aspects, for example, beef production and dairy industry has been a long-term topic in this kind of research, while the topic of nutrition appeared in recent years. There was also a cluster of literature discussing footprint evaluation tools, such as comparing LCA with input–output analysis. The CiteSpace analysis indicated that earlier methodological literature still plays an important role in recent research. Moreover, through the analysis of burst keywords, it was found that agriculture productions (dairy, meat, fish, crop) as well as global climate issues (greenhouse gases emission, global warming potential) have always been the areas of concern, which matches the result of co-citation analysis. Building materials (low-carbon building, natural buildings, sustainable buildings) and soil issues (soil carbon sequestration, soil organic carbon) are the topics of recent concern, which could arouse the attention of follower-up researchers.

Keywords: carbon footprint; life cycle assessment; CiteSpace; a visual analysis

---

1. Introduction

Greenhouse gas emissions have been an environmental issue of great concern in the past two decades. According to the Intergovernmental Panel on Climate Change (IPCC) report, without action, the global temperature will rise by more than 1.5 °C in the future, while in the past 10,000 years, the climate change was only 1 °C [1]. The emergence of the carbon footprint concept facilitates the measurement of greenhouse gas emissions. The term "carbon footprint" evolved from the ecological footprint proposed by Wackernagel in 1996 [2]. It is a concept formed by the integration of ecological footprint and carbon emissions. With the deepening of the global division of labour and the complexity of the commodity supply chain, the assessment of the environmental impact of modern economic activities has become more complicated. The unique advantages of life cycle assessment (LCA) as a "cradle-to-grave" measurement method are taken seriously by researchers. The concept of the carbon footprint involves the carbon emissions throughout all the stages of a product's life, which is similar to the core idea of the LCA approach. The combination of the two can be an effective tool for measuring the environmental impact of economic activities. When discussing the relationship between carbon footprint and LCA, carbon footprint can be regarded as either a carbon emission measurement method or a result calculated by this method. When it is regarded as a carbon emission measurement method, the relationship between carbon footprint and LCA can be regarded as a subordinate relationship between measurement methods. It is the application of LCA with a single impact category (climate change). When it is considered as a result of calculation, it can be measured by carbon inventory methods (this method uses the corresponding emission factors to calculate the emissions of various greenhouse gases based on the guidelines for national greenhouse gas inventories compiled by IPCC) [3], input–output method (IOA is a "from top to bottom" analysis method, which reflects the relationship between initial input, intermediate input, total input and intermediate output, final output and total output. It transforms the economic relationship between production sectors or regions into the physical relationship of greenhouse gas emissions, which clearly reflects the exchange process of emissions and distributes them to various sectors or regions, thus making the direct and indirect carbon emission relationship clear) [4] and the LCA method. The relationship between carbon footprint and LCA is the relationship between measurement results and methods. In this paper, the concept of "carbon footprint" is regarded as the measurement result, and the related literature of "carbon footprint" micro-measurement is concerned, that is, carbon footprint research from the perspective of LCA.

Some progress has been made by researchers in this field. The related literature has reviewed this topic from various perspectives such as the expansion of the carbon footprint concept and the application of LCA in earlier literature, city carbon emission [5], diet carbon footprint [6] and the re-examination of LCA [7] in recent literature. Few studies have adopted a knowledge map to carry out a visual analysis of the intellectual structure of the related fields, but a knowledge map is beneficial for showing where knowledge can be found within a group or organization, and how to find those with the most expertise. Specifically, a knowledge map is an emerging bibliometric tool, which can provide a visual knowledge graph of the existing literature and a serialized knowledge spectrum [8]. CiteSpace, a relatively widely used bibliometric tool based on network theory, is a Java application which provides a variety of analysis methods such as co-citation analysis, keyword co-occurrence analysis and collaboration network analysis. Therefore, we utilized this software to visualize the existing literature on carbon footprint research in a LCA perspective, and explored the intellectual structure of this field and its possible development trend by focusing on the scope, core literature, collaboration network and research front.

The rest of this paper is organized in four sections. Section 2 describes the data collection process and illustrates the result of basic analysis. Section 3 establishes the disciplines and terms network, co-citation network and collaboration network to show intellectual structure of the focus topic. Section 4 gives a document co-citation network and keyword co-occurrence network based on the results of Section 3 to analyse the emerging trends of focus topic. In the last section, the key findings are summarized and discussed to implicate further study on carbon footprint in a LCA perspective.

2. Data Collection and Basic Analysis

The literature analysed in this paper is from the core collection of Web of Science, including SCI-EXPANDED and SSCI. In the process of retrieval, we searched the "Article" category published in English by retrieving the keyword "carbon footprint", "carbon footprints" or "carbon footprinting", and a total of 6180 papers were identified with their published time ranging from 2003 to 2019. Then we narrowed the results by retrieving the above keywords separately together with "life cycle assessment" or "LCA", and 1563 papers were identified with their published time ranging from 2006 to 2019 (retrieval on 9 June 2019). The year-by-year distribution of these papers is shown in Figure 1.

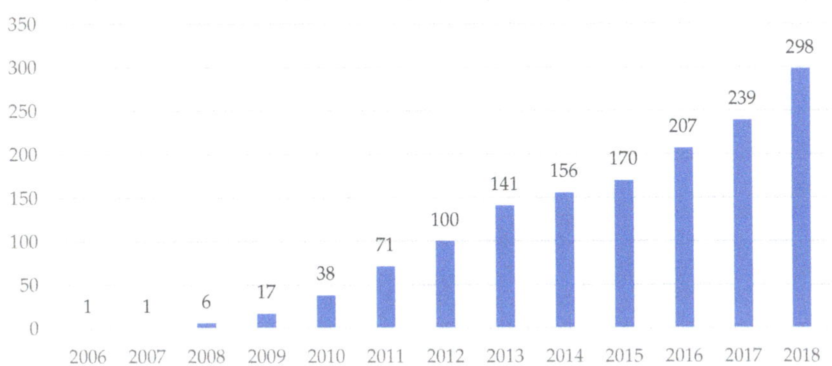

Figure 1. The number of published papers on carbon footprint and life cycle assessment (2006–2018).

Figure 1 shows the number of papers published in corresponding years. The sample literature had an explosive growth in 2008, after which it entered a stage of rapid development. With regard to the relative number, the increase of papers in 2008 compared to a past year reached 500%, and the growth rate (compared to a past year) exceeded 100% (183.33% and 123.53% respectively) in both 2009 and 2010 as well, after which the growth slowed down. In terms of the absolute number, papers had a double-digit growth per year from 2009; the top five countries/regions were the USA (340), People's Republic of China (PRC) (177), Italy (159), Spain (148) and England (137) (counting first and co-authors); the top five journals were *Journal of Cleaner Production* (375), *International Journal of Life Cycle Assessment* (125), *Science of the Total Environment* (66), *Sustainability* (64) and *Journal of Industrial Ecology* (52); the top five authors were VAZQUEZ-ROWE I (18), KUCUKVAR M (18), TATARI O (13), FEIJOO G (13) and HERTWICH EG (13).

3. The Intellectual Structure of the Focus Topic (Carbon Footprint Research in a LCA Perspective)

This paper focuses on the analysis of TYPE II papers through the bibliometric method. The scientometric software, CiteSpace, was utilized to describe the intelligent structure of the focus topic. Firstly, we constructed a subject categories co-occurrence network to identify the fields and research contents involved in the focus topic. Secondly, a co-citation network was established to analyse the features of clusters. Finally, we built a collaboration network to analyse the partnerships between the source regions, institutions and authors.

3.1. The Disciplines and Terms Network Based on the Focus Topic

3.1.1. Noun Term Co-Occurrence Network

Noun term co-occurrence analysis refers to the extraction of noun terms from the titles, keywords and abstracts of the sample papers and the construction of a noun term network for literature analysis in accordance with the criterion that two noun terms are adopted in the same document. When two noun terms appear in the same paper, it is considered that there is a non-negligible relationship between the two terms [9]. The noun term network can display hot topics and topic distribution of a discipline.

We used Citespace to construct the noun term network. Citespace uses three technology means to simplify the process of calculation during the process of network generation. Firstly, Citespace uses the idea of divide-and-conquer algorithms to separate the whole co-citation network into individual networks which are called time slices at the first stage, and integrate them at the second stage after some treatments. In the construction of noun term co-occurrence network, the whole network was divided into 14 individual networks from 2006 to 2019. Secondly, Citespace provides two main criteria to filter data in individual network—citation threshold and the rank of being cited. In the construction of the noun term co-occurrence network, this paper used citation threshold. It contains three specific criteria—citation quantity, together with co-citation frequency and co-citation coefficient (c, cc and ccv) and the selected articles were above the threshold on these indicators. The threshold was set as (2,3,15), (3,3,20), (3,3,20) at three levels in chronological order to form the noun term co-occurrence network, and the rest was determined by linear interpolation. This paper used the criterion of the rank of being cited in later sections as well. Thirdly, Citespace provides minimum spanning tree mode and pathfinder mode to prune the network to highlight the key points. In noun term co-occurrence network, this paper used the minimum spanning tree mode [10].

Next, we extracted noun terms from the sample papers and merged those with the same meaning in the network (singular and plural forms, abbreviation and full name, etc.) to construct a noun term network as shown in the upper right corner of Figure 2. The connections in the network illustrate which words often appear together in the same paper, such as the connections between LCA and terms like climate change and global warm as shown in Figure 2. Table 1 reveals the 12 words with the highest frequency in the network, demonstrating that environmental impact was the most concerned topic in the field, which may be explained by considering that LCA was used to evaluate the environmental impact of carbon footprint. These high-frequency noun terms first appeared between 2008 and 2011 during which the field experienced an explosive development, which is consistent with the previous analysis. For more detailed analysis, we set the thresholding (c, cc, ccv) parameter as (10,10,15), (10,10,20) and (10,10,20) to extract the noun terms in the relatively more important co-cited literature and drew a network diagram, as shown in Figure 3. The font size in Figure 3 represents the frequency of occurrence of noun terms and the node size the centrality of noun terms, which means that when the font size is bigger, the noun term appears more frequently and the square of the yellow box is lager, the centrality of noun term is higher. The centrality measures the number of links between focus noun term and other noun terms, which shows the power of the focus noun term in network, and nodes with higher centrality are called hub nodes. For example, environmental impact not only appeared frequently, but also had high centrality in the network, which implied that environmental impact was a term related to many issues of the focus topic. It can also be understood that the focus topic was about the environmental impact. In addition, greenhouse gas emissions, climate change, global warming and supply chain were also important noun terms.

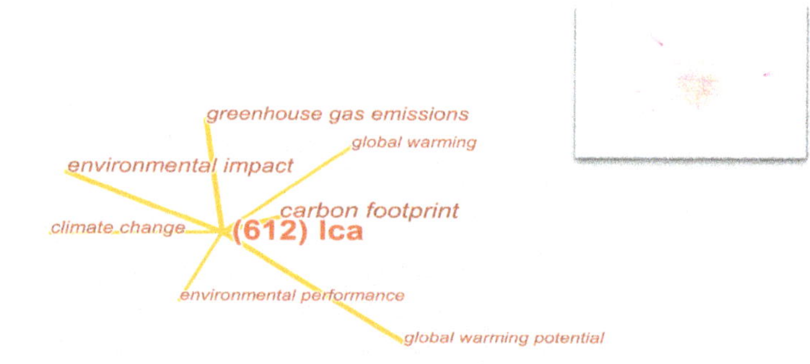

Figure 2. Life cycle assessment (LCA) in the whole noun term co-occurrence network.

Table 1. Top 12 highest frequency noun terms.

Frequency	Year	Noun Phrases
701	2009	carbon footprint
612	2009	LCA
466	2008	environmental impact
394	2009	greenhouse gas emissions
267	2009	greenhouse gas
158	2009	life cycle
155	2008	climate change
131	2010	global warming potential
108	2010	environmental performance
99	2010	global warming
92	2008	carbon emissions
87	2011	energy consumption

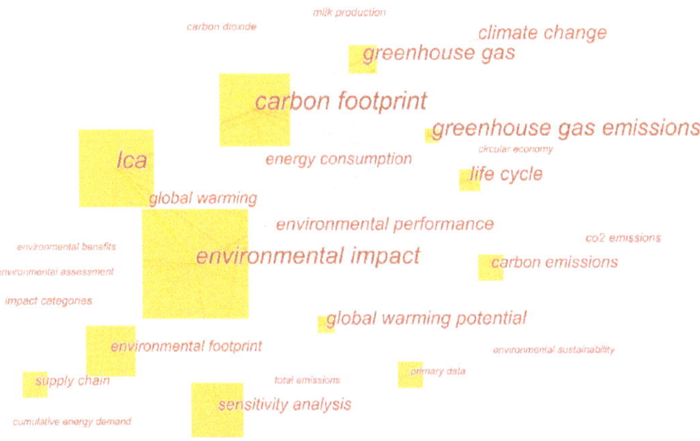

Figure 3. Streamlined noun term co-occurrence network.

3.1.2. Subject Categories Co-Occurrence Network

The text file provided by Web of Science contains a Subject Categories (SC) field which represents the name of the discipline to which the included documents belong, and each paper is assigned to one or more such names. This information can be used to analyse which subjects are covered by the literature on carbon footprint assessment in a LCA perspective.

Under the same parameter setting as that of the noun term co-occurrence network, a subject categories network as shown in Figure 4 was constructed. The size of node represents the occurrence number, which means that when the node radius is larger the subject category contains more sample papers, and the several biggest nodes are called landmark nodes. A node with a purple ring is called a pivot node which owns higher value of betweenness centrality. Betweenness centrality is an index to measure the sum of probability of the focus to be on an arbitrary shortest path in a network, which shows the importance of the focus node as a bridge role. As we can see from Figure 4, Environmental Science and Ecology is the most common subject category to which the sample literature belongs, followed by Engineering, Green and Sustainable Science and Technology and Energy and Fuels. Food Science and Technology is a subject category with high centrality. That is to say, the sample papers often contain interdisciplinary contents of these disciplines and other disciplines. The thickness and colour of lines which link two nodes make sense as well. The thicker the line, the deeper the relationship between the two categories, and the darker the line, the earlier the relationship between the two categories is established. Figure 4 illustrates that Environmental Sciences and Ecology has a deep and long-time relationship between Business and Economics.

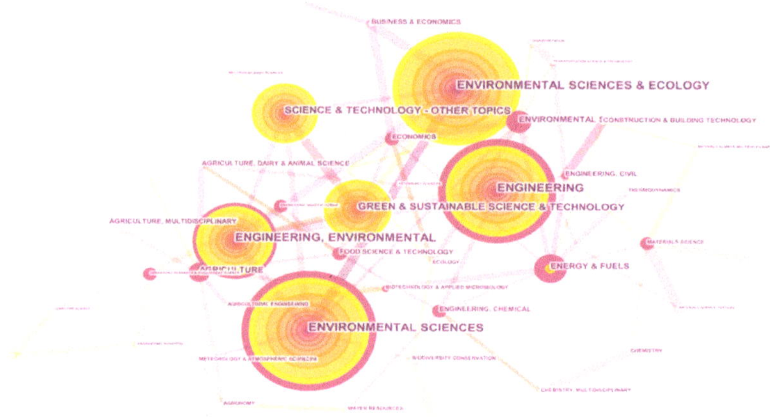

Figure 4. Disciplines shown as a minimum spanning tree network of subject categories.

3.2. The Co-Citation Network based on the Focus Topic

3.2.1. Document Co-Citation Network

Co-citation refers to the relationship between two documents cited by the same later paper. The co-citation relationships between a series of documents form a document co-citation network. Such a network is a useful tool for document analysis, which can provide clues for understanding literature development and is more reliable than simple analysis of citation relationships [11]. We put two years per slice, selected the top 50 levels of the most cited references from each slice and used pathfinder to prune (set as base settings), and then got document co-citation network with clusters shown in Figure 5. As shown in Figure 5, Finnveden [12], Vries [13], Weidema et al. [14], Finkbeiner [15] and Roy [16] have higher centrality. Finnveden [12], Vries [13] and Roy [16] discussed the LCA approach. Finkbeiner [15] deepened the field's understanding of "carbon footprint". Weidema et al. [14] analysed the relationship between "carbon footprint" and "LCA". These documents have been cited together with many different documents and have become the basic literature in the field.

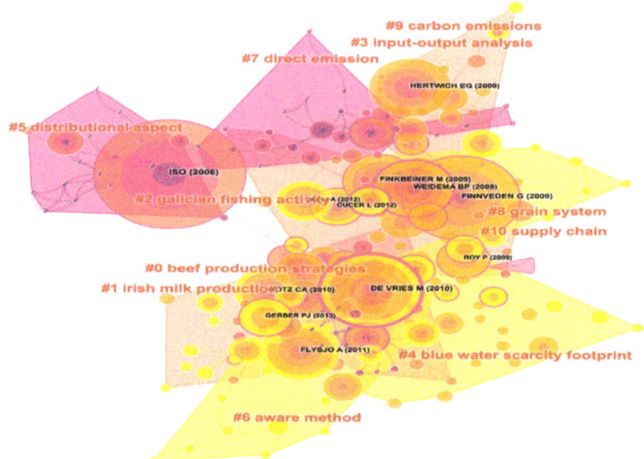

Figure 5. Document co-citation network with clusters.

The top 10 most frequently cited documents related to the focus topic are shown in Table 2. The most frequently cited one is International Organization for Standardization (ISO) [17], which provides the international standard for the LCA approach. Since then, LCA has been widely used in the measurement of carbon footprint and all later sample documents have cited ISO [17]. The far-reaching impact of ISO [17] on the subsequent literature can be seen more clearly in Figure 6. The sample documents from 2008 to 2011 were cited relatively frequently, and eight of the top 10 most frequently cited were published during this period. This shows that the literature had experienced an explosive growth in quantity during this period, providing a solid foundation for future research in this field.

Table 2. Top 10 references based on cited frequency.

References	Citation Counts	Source
ISO [17]	115	14044 ISO
Finnveden et al. [12]	89	*Journal of Environment Management*
Vries et al. [13]	75	*Livestock Science*
Hertwich et al. [18]	62	*Environment Science and Technology*
Flysjo et al. [19]	60	*Agriculture System*
Weidema et al. [14]	59	*Journal of Industrial Ecology*
Finkbeiner [15]	55	*The International Journal of Life Cycle Assessment*
Galli et al. [20]	42	*Ecological Indicators*
Rotz et al. [21]	52	*Journal of Dairy Science*
Roy et al. [16]	39	*Journal of Food Engineering*

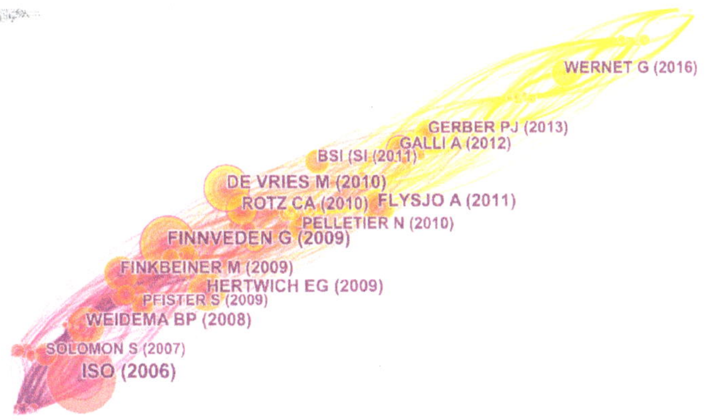

Figure 6. Time-zones of document co-citation.

References with top cited frequency after ISO [17] can be simply divided into three categories: the research focusing on carbon footprint, LCA and their relationship. The first category includes Hertwich et al. [18], Flysjo et al. [19], Finkbeiner [15] and Galli [20] et al. Hertwich et al. [18] discussed the quantification of greenhouse gas emissions associated with the final consumption of goods and services in 73 countries of 14 regions from a global trade perspective. The analysis indicated that food accounts for 20% of greenhouse gas emissions, the operation and maintenance of residential areas 19% and transportation 17%. Food and services are more important in developing countries, while transportation and food products are growing rapidly with increasing income and are dominant in developed countries. Flysjo et al. [19] adopted the LCA approach to analyse the carbon footprint of dairy production by two different farm systems in New Zealand and Sweden. Finkbeiner [15] explored the significance of discussing the concept of carbon footprint, pointing out that it is not without limitations (even pollution treatment will produce carbon emissions), and then summarized the opportunities and challenges in the relevant research. Galli et al. [20] extended the concept of carbon footprint to ecological, carbon and water footprint, and analysed the similarities and differences between the three, thus defining the concept of "footprint family" and developing an integrated Footprint approach to assess the environmental impact of human behaviours; the second category includes Finnveden et al. [12], Vries et al. [13], Roy et al. [16], etc. Finnveden et al. [12] and Vries et al. [13] are highly cited documents after ISO [17]. Both review the LCA approach and Finnveden et al. [12] is more comprehensive, while Vries et al. [13] only reviews the literature on livestock products. Roy et al. [16] also reviewed the LCA approach, focusing on the topic of food products; the third category includes Weidema et al. [14], Rotz et al. [21], etc. Weidema et al. [14] discussed whether the emergence of the carbon footprint concept would facilitate the development of the LCA approach. He believed that the concept of carbon footprint made the measurement easier and simpler to master than the traditional LCA approach. Therefore, it is very suitable for governments to show consumers the environmental impact of their consumption, which explains that the concept of carbon footprint promotes the application of the LCA approach. Rotz et al. [21] introduced the Dairy Greenhouse Gas model and discussed the specific application of the partial life cycle assessment in the carbon footprint of dairy production systems.

As discussed above, the basic documents in the field has contributed mainly to literature review and definition. Next, we performed cluster analysis on the sample references and the major clusters are shown in Table 3. The second column is the cluster size and the third column the Silhouette index, which is used to measure the cluster quality and ranges from 0 to 1. The closer the index is to 1, the better the cluster quality. The Silhouette scores of all the clusters in Table 3 are above 0.7, indicating that the cluster quality was good. The fourth to sixth columns are cluster labels extracted from abstracts via the LSI, LLR, and MI methods respectively [22]. It can be seen that the most concerned topics in the sample literature were the carbon footprint assessment related to beef production, dairy industry and fishery (cluster#0, cluster#1, cluster#2), as well as urban systems and grain system (cluster#7, cluster#8), based on which other factors in the environment (cluster#4, cluster#5) were discussed in the references. In addition, the literature related to the carbon footprint measurement (cluster#3) and the carbon footprint warning mechanism (cluster#6) also formed clusters.

Table 3. Major clusters of co-citation references.

Cluster ID	Size	Silhouette Score	Label (LSI)	Label (LLR)	Label (MI)
0	32	0.836	butter	beef production strategies	pigmeat
1	23	0.851	production systems	irish milk production	smallholder dairying
2	22	0.737	land use	Galician fishing activity	green competitiveness
3	21	0.872	measurement	input-output analysis	footprint analysis
4	18	0.796	eutrophication	blue water scarcity footprint	water use and water management
5	18	0.895	nutrient recovery	distributional aspect	nutrient recycling
6	17	0.882	indicators	aware method	decision-making units
7	17	0.872	building	direct emission	urban systems
8	11	0.882	crop planting	grain system	saline agriculture

3.2.2. Author Co-Citation Network

The document co-citation lays a foundation for the analysis of the author co-citation and the journal co-citation relationships. We put one year per slice, selected the top 50 levels of the most cited references from each slice with no pruning, and established the author co-citation network. The network contained 269 authors, among who there were 1116 co-citation relationships. As can be seen from Figure A1 of Appendix A and Table 4, group authors also provided important cited literature for the field. As analysed above, the document published by ISO that provided the LCA standard is a reference a researcher much read if they study this field, and ISO has become the most important group author. In addition, IPCC has also contributed to the research in this field. For instance, it published "The IPCC Guidelines for National Greenhouse Gas Inventories". Food and Agriculture Organization of the United Nations (FAO), European Commission (EC), British Standards Institution (BSI) are other important group authors. Ecoinvent provides important data resources for the research in this field.

Table 4. Top 6 group authors based on cited frequency.

Author	Frequency	Centrality
ISO (International Organization for Standardization)	875	0.2
IPCC (Intergovernmental Panel on Climate Change)	488	0.03
FAO (food and agriculture organization of the united nations)	169	0.06
EC (European Commission)	150	0.02
BSI (British Standards Institution)	57	0
Ecoinvent	57	0

The cited frequency of some group authors was high and that of personal authors on average was not low. The citation count of each of the top 10 cited individuals was more than 100. As shown in Table 5, only C. Cederberg, E.G. Hertwich and I. Vazquez-rowe were among the top 20 personal authors in terms of the number of published papers and citations. The rest of the highly cited authors were widely recognized in the field for one or two high-quality papers. For example, B.P. Weidema, R. Frischknecht and S. Suh became highly cited authors because of the co-authored paper, "Life Cycle Assessment: Part 1: Framework, Goal and Scope Definition, Inventory Analysis, and Applications", which was published in *Environment International*. In addition, the key papers of the top five highly cited authors focus on methods, including those by B.P. Weidema and R. Frischknecht, and T. Wiedmann's "A Review of Recent Multi-Region Input–Output Models Used for Consumption-Based Emission and Resource Accounting", J.B. Guinee's "Handbook on Life Cycle Assessment Operational Guide to the ISO Standards", and M. Lenzen's "System Boundary Selection in Life-Cycle Inventories Using Hybrid Approaches".

Table 5. Top 20 personal authors based on cited frequency.

Author	Frequency	Centrality	Author	Frequency	Centrality
B.P. Weidema	197	0.15	E.G. Hertwich	123	0
R. Frischknecht	196	0.02	N. Pelletier	110	0.06
T. Wiedmann	189	0.02	C.L. Weber	99	0.12
J.B. Guinee	155	0.03	S. Solomon	99	0.01
M. Lenzen	152	0.06	M. Finkbeiner	99	0.03
M. Goedkoop	149	0.09	T. Nemecek	92	0.03
C. Cederberg	136	0.1	M.A. Thomassen	87	0.09
G. Finnveden	128	0.03	M.D. Vries	79	0.01
S. Suh	124	0.2	I. Vazquez-Rowe	77	0.07
A. Flysjo	123	0.04	A. Tukker	70	0.01

3.2.3. Journal Co-Citation Network

We set the same parameters as those of the author co-citation network to build a journal co-citation network, in which there were 169 journals and 635 connections. As shown in Figure A2 of Appendix A and Table 6, the most cited journal was *Journal of Cleaner Production* (1135 citations), which also had a high degree of centrality (0.14). *Environmental Science and Technology* and *Science of the Total Environment* were also with high citations and centrality. The second most cited was the *International Journal of Life Cycle Assessment* with 1037 citations, but its centrality in the network was not high (0.09). That is to say, its co-citations with other journals was relatively low. From the basic analysis, *Journal of Cleaner Production*, *International Journal of Life Cycle Assessment*, *Science of the Total Environment* and *Journal of*

Industrial Ecology were not only the journals with the most publications, but also the most cited ones. In addition, they were the major journals in which papers on the focused topic were published. Table 6 also illustrates the impact factor of each journal in 2018. According to Table 6, journals with higher impact factor were usually those with high citations and centrality in the focus topic. Only *Renewable and Sustainable Energy Reviews* in Table 6 did not comply with this rule, which may be due to the deviation caused by the time factor.

Table 6. Top 10 journals based on cited frequency.

Journal	Frequency	Centrality	Impact Factor
Journal of Cleaner Production	1135	0.14	6.395
International Journal of Life Cycle Assessment	1037	0.09	4.868
Environmental Science and Technology	795	0.15	7.147
Journal of Industrial Ecology	551	0.1	4.826
Journal of Environmental Management	444	0.04	4.865
Energy Policy	439	0.05	4.880
Ecological Economics	398	0.04	4.281
Energy	394	0.06	5.537
Science of the Total Environment	393	0.14	5.589
Renewable and Sustainable Energy Reviews	390	0.05	10.556

3.3. The Collaboration Network Based on the Focus Topic

3.3.1. Country Collaboration Network

The collaboration network was analysed at the macroscopic, midscopic and microscopic level, corresponding to the collaborations between countries/regions, institutions and authors respectively. The collaboration between countries was first analysed. We put one year per slice, selected the top 100 levels of the most cited references from each slice with no pruning, and obtained the following country/region collaboration network, where there were 51 countries/regions and 263 connections. We used authors' affiliation to determine countries associated with the analysed papers. For example, a 2014 article, entitled "Emerging approaches, challenges and opportunities in life cycle assessment", is an international multi-authors publication, and has co-authors from Switzerland and France respectively, a link has been established between the two countries in country collaboration network and the geographical scope of analysed paper is determined as both Switzerland and France. If two authors of an analysis paper come from the same country, the occurred frequency of the country is recorded twice in Table 7, which means that the unit of measurement of Table 7 is person–time. USA was exceptionally outstanding in the network in that it not only had the largest number of publications (340), but also had actively participated in the collaborations with various countries/regions, thus becoming the node with the highest centrality in the network (the centrality was 0.44). PRC was the only Asian country among the top 10 with the largest number of publications. Although PRC ranked second in terms of the number of publications, its absolute number was only about half of the USA, and its centrality was only 0.1 without so many partners. Referring to the year–colour correspondence bar in Figure A6 of Appendix A, the circle/figure refers to the colour of circle in Figure A3 of Appendix A and represents the corresponding year in which sample literature appeared. It can be seen in Figure A3 that USA was among the first counties/regions to study the carbon footprint in a LCA perspective, and the lighter colour of PRC's circle indicates that it entered the field later. Australia and Canada ranked sixth and tenth respectively in term of publication number, and were also countries with a large number of publications but low centrality. The rest of the top 10 countries were in Europe, including England, Italy, Spain, Netherlands, Germany, Sweden. This may be because Europe has developed animal husbandry, which makes scholars pay more attention to carbon emissions. Most European countries were similar to the USA in that they had a large number of publications and collaborated with different nodes in the network and their centrality was higher. For example, England published 137 papers and its centrality was 0.31.

Table 7. Top 10 countries based on occurred frequency.

Country	Frequency	Centrality	Year
USA	340	0.44	2008
PEOPLES R CHINA	177	0.1	2011
ITALY	159	0.12	2010
SPAIN	148	0.18	2010
ENGLAND	137	0.31	2009
AUSTRALIA	111	0.06	2010
GERMANY	87	0.1	2009
NETHERLANDS	82	0.12	2011
SWEDEN	80	0.09	2010
CANADA	59	0.04	2010

3.3.2. Institution Collaboration Network

We set the same parameters as those of the country/region collaboration network to build an institution collaboration network, which contained 274 institutions and 372 connections. Taking into consideration the number of publications and centrality, we identified four representative institutions, namely Chinese Academy of Sciences (29), University of Santiago de Compostela (28), Aarhus University (27) and Wageningen University (19). However, they did not have direct cooperation with each other. Instead, four sub-networks of institution collaboration were formed with each of them as the centre. Tech Univ Denmark had direct cooperation with Chinese Academy of Sciences and Aarhus University, thus becoming a more important part of the institution collaboration network. University of Central Florida, despite being among the top five institutions with the largest number of publications, had only cooperated with six institutions. Therefore, it was not deeply integrated with the institution collaboration network. In general, the network was relatively loose and the research community less mature.

3.3.3. Author Collaboration Network

We set the same parameters as those of the country/region network to build an author collaboration network, where there were 380 authors and 507 connections. Among them, VAZQUEZ-ROWE I and KUCUKVAR M were the two authors who published the largest number of papers. From 2006 to the present, the average number of publications per year on the focus topic was more than one. The two authors were from University of Santiago de Compostela and University of Central Florida, both of which were among the top five research institutions. Similar to the institution collaboration network, the structure of the author cooperation network was also loose, as indicated by the centrality in Table 8 and shown in Figure A4 of Appendix A. As shown in Table 9 and Figure A5 of Appendix A, there were only three large sub-networks, namely the sub-network centred on VAZQUEZ-ROWE I, FEIJOO G and MOREIRA MT, the one centred on KUCUKVAR M, TATARI O and EGILMEZ G and the one with CEDERBERG C and DE BOER IJM as the important nodes. Some authors published more papers, although they cooperated less with others, such as KENDALL A, HEIJUNGS R, etc.

Table 8. Top 20 institution based on occurred frequency.

Institution	Frequency	Centrality	Institution	Frequency	Centrality
Chinese Acad Sci	29	0.28	Univ Perugia	15	0.07
Univ Santiago de Compostela	28	0.15	Tech Univ Denmark	14	0.28
Aarhus Univ	27	0.13	KTH Royal Inst Technol	14	0.04
Univ Cent Florida	20	0.01	Univ Sydney	14	0.19
Wageningen Univ	19	0.23	Univ Manchester	13	0.01
Univ Calif Davis	19	0.01	Texas A and M Univ	12	0.01
China Agr Univ	17	0.06	Michigan Technol Univ	12	0
Swedish Univ Agr Sci	16	0.01	Norwegian Univ Sci and Technol	11	0.04
Leiden Univ	15	0.2	Hong Kong Polytech Univ	11	0.02
Aalto Univ	15	0	Univ Ghent	11	0.11

Table 9. Top 20 authors based on occurred frequency.

Author	Frequency	Centrality	Author	Frequency	Centrality
VAZQUEZ-ROWE I	18	0.01	DE BOER IJM	9	0.01
KUCUKVAR M	18	0	JUNNILA S	8	0
TATARI O	13	0	HALL CR	8	0
FEIJOO G	13	0	ONAT NC	8	0
HERTWICH EG	13	0	HEINONEN J	8	0
CEDERBERG C	12	0.01	YAN MJ	7	0
MOREIRA MT	11	0	GABARRELL X	7	0
EGILMEZ G	11	0	HOLDEN NM	7	0
KENDALL A	10	0	MOGENSEN L	7	0
INGRAM DL	10	0	HOSPIDO A	7	0

4. The Emerging Trends of Carbon Footprint Research in a LCA Perspective

This paper analysed the emerging trends of the focus topic from the document co-citation and keyword co-occurrence perspectives. First, we reorganized the document co-citation network in Section 3 from the perspective of timeline to analyse the clusters and bursts of documents. Second, we established a keyword co-occurrence network based on the disciplines and terms network in Section 3 from the perspective of time-zone to analyse the bursts and development trend of keywords.

4.1. References Analysis with Citation Bursts

Based on the above efforts, we obtained a document co-citation network from the timeline perspective as shown in Figure 7. As can be seen from the distribution of major clusters along the time axis, beef production and dairy industry, as well as assessment method, are long-term topics in this kind of research. Topics of water footprint, aware method and grain system appeared in recent years.

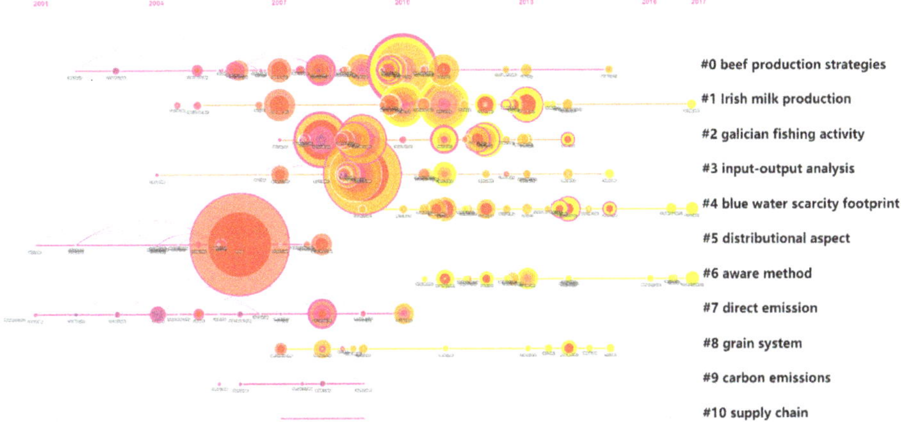

Figure 7. Timelines of co-citation clusters.

From the reference bursts as shown in Table 10 [23–32], the documents widely cited in recent years are S. Hellweg [25] and D. Tilman [29]. S. Hellweg [25] reviewed and summarized the latest LCA methods, arguing that LCA methods could locate environmentally relevant hotspots in complex value chains, but at the expense of simplification and uncertainty. In the future, the LCA approach needs to be more practical and accurate, and its application in the assessment of economic and social aspects should be expanded. D. Tilman [29] discussed the impact of changes in dietary trends on carbon emissions. More and more foods in the diet structure of modern people are refined sugar and fat. If not controlled, such diet structure will lead to a substantial increase in agricultural greenhouse gas emissions. Meanwhile, it will have a negative impact on health. Thus, the paper advocated a healthy alternative diet to solve the diet–environment–health trilemma.

Table 10. Top 10 references based on citation burst strength.

References	Year	Strength	Begin	End	2006–2019
S. Suh [23]	2004	8.5618	2008	2011	
H. Steinfeld [24]	2006	7.6963	2008	2013	
S. Hellweg [25]	2014	6.9636	2016	2019	
T. Kristensen [26]	2011	6.8461	2014	2015	
A.G. Williams [27]	2006	6.4231	2010	2014	
T.O. Wiedmann [28]	2011	6.3867	2014	2015	
D. Tilman [29]	2014	6.33	2017	2019	
A. Laurent [30]	2012	6.2713	2015	2017	
M. Goedkoop [31]	2009	6.2359	2013	2015	
M.A. Thomassen [32]	2008	6.2121	2010	2014	

4.2. Keyword Analysis with Citation Bursts

We set the time slicing in CiteSpace as one year per slice and extracted the top 50 cited keywords from each slice before pruning the sliced network in the pathfinder mode to obtain the keyword co-occurrence network; the network in the time-zones is as shown in Figure 8. Most of the high-frequency keywords appeared in the early stages of research, such as the two keywords of this study, namely "LCA" and "carbon footprint". In line with the analysis of noun terms, "environment impact" was the earliest high-frequency word and the origin and essence of the research in this field, i.e., how to adopt a proper LCA method in accordance with the carbon footprint concept to evaluate the environmental impact of an activity. In recent years, there were fewer new high-frequency keywords, some of which were "water footprint", "anaerobic digestion" and "plant". Their emergence indicates that the research in the field has become more extensive and in-depth. "Water footprint" is a horizontal extension of "carbon footprint", which makes the measurement system more complete. "Plant" and "anaerobic digestion" is an upward and downward extension of the research respectively, with the former focusing on the important sources of carbon emissions and the later ways to improve the environment.

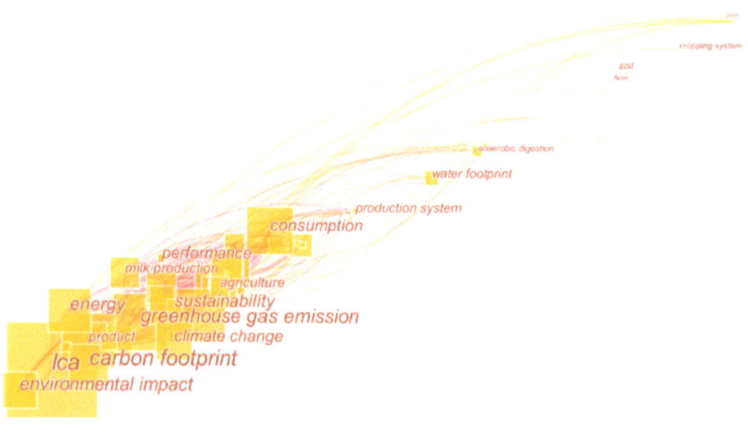

Figure 8. Time-zones of keywords.

In addition, the keyword bursts can be analysed based on the citation frequency. As shown in Table 11, "inventory", "industrial ecology" and "sustainable development" became keywords during the period of document bursts in this field and have been keywords for a long time. In recent years, "crop", "soil" and "China" have become the bursting keywords. "China" has emerged as a keyword mainly because of the rapid development of the Asian research collaboration network in recent years, which has shifted the research focus from Europe and America to Asia. "Crop" became a focus due to the concerns for feed production in farm carbon footprint measurement and for the carbon footprint of crop cultivation [33,34]. "Soil" attracted much attention due to the fact that, like atmosphere, soil is another final destination for carbon. For example, if the soil is of potential for organic carbon sequestration, it is possible that the net greenhouse gas emissions from pasture systems decrease drastically [35].

Table 11. Top 10 keywords based on citation burst strength.

Keywords	Strength	Begin	End	2006–2019
inventory	7.4049	2010	2015	
crop	6.2555	2016	2017	
soil	5.8014	2017	2019	
china	5.6248	2017	2019	
industrial ecology	5.4509	2010	2015	
land use change	5.3122	2014	2015	
perspective	5.0877	2013	2014	
food product	5.0877	2013	2014	
global warming potential	5.0534	2010	2014	
sustainable development	4.7419	2011	2012	

As indicated by literature and the trend analysis above, the latest hot topics related to the focus topic involve several aspects—the combination of LCA and carbon footprint and its application, carbon footprint measurement of different products, and environmental impact and carbon footprint.

The combination of LCA and carbon footprint and its application. At present, there are still some limitations in the application of quantitative LCA methods to study the environmental impacts caused by resource utilization and waste discharge during product production. For example, social and economic factors are evaluated only via qualitative analysis, and time and space factors are not given sufficient attention [36]. These limitations have not been resolved by the combination of LCA and carbon footprint. At present, some scholars try to do so through the dynamic LCA method [37]. LCA is an approach that has been applied to the measurement of a carbon footprint since the proposal of this concept. In addition to the above limitations, this approach is more suitable for the measurement of specific products at the microscopic level, because it is still difficult to define the boundary at the macro level (such as the carbon footprint of organizations). For instance, it is hard to decide whether the carbon emissions of aircraft manufacturing should be included when calculating the flight carbon footprint. In contrast, the input–output method can offset the calculation errors caused by the complex division of labour, so some researchers have combined the two to develop the hybrid LCA method, which makes the boundary clearer to some extent [38]. However, this method requires high-quality data and technology, so how to better apply LCA in the carbon footprint field will still be a hot topic related to the focus topic in the future.

Carbon footprint measurement for different products. Since LCA is more suitable for calculating carbon footprint at the product level, there is a large body of literature discussing different types of specific products, such as beef and milk which are classic measured products. Both of them are important sources of protein, and other sources of protein, such as fish and shrimp, are also the subject of carbon footprint measurement. Due to the differences in production systems, the environmental friendliness of different protein sources differs remarkably [39]. Crops are another subject of carbon footprint measurement, such as wheat [40], tomatoes [41] and organic farming crops. All of the products mentioned above are related to diet and nutrition, so the adjustment of diet structure has an impact on carbon emissions, and the relationship between food carbon footprint, demographic characteristics and environment will gain more attention. As stated in the section of disciplinary analysis, this is an interdisciplinary field of Environmental Science and Ecology and Food Science and Technology, and a potential point of growth for those disciplines. In recent years, cities as an important source of carbon emissions have also received more and more attention from researchers. Specific research topics include the impact of urban green space projects on carbon emissions [42], and the relationship between citizen behaviours and urban carbon emissions [43]. The carbon footprint of citizen behaviours is primarily related to private driving, housing-related activities, and consumer behaviours. Private driving, which involves the interdisciplinary field of Environmental Science and Ecology and Energy and Fuels, has attracted much attention. With the emergence of electric vehicles [44], it is likely to become one of the hot research directions. In terms of housing-related activities, a concerned aspect of the focus topic is how to use LCA to intervene in the early stages of building design to order to reduce carbon emissions of building construction [45]; the comparison of carbon footprints between environmentally friendly homes and ordinary ones is also one of the hot topics [46]. In addition, the measurement of the carbon footprint of consumer goods is also a concerned aspect of the focus topic, and changes in consumer demand can have an impact on carbon emissions [47]. However, the application of the LCA approach has high requirements on data. The literature on the focus topic studies urban issues only at the micro level. With the application of big data and the advancement of LCA, the use of LCA to measure urban carbon footprint may be one of the future research directions.

Environmental impact and carbon footprint. Carbon footprint was proposed to measure the environmental impact of economic activities, so environmental impact has always been a hot word of the focus topic. Carbon footprint is an indicator or a concept. As a measurement indicator of greenhouse gas emissions, carbon footprint has already adopted the concept of carbon equivalent and taken carbon-free greenhouse gases into consideration. Nevertheless, it still cannot be the sole indicator of environmental impact. However, the concept of carbon footprint has contributed to alleviating the environmental burden. On the one hand, compared with obscure terms like Multi-Region Input–Output Model (MRIO) and LCA, carbon footprint is simple, easy to understand and interesting. The wide spreading and extensive application of this concept may have influenced people's behaviours of environmental impact [48]. On the one hand, the concept of carbon footprint provides new ideas for assessing other environmental impact factors and has inspired the proposal of water footprint, ecology footprint, energy footprint and even the family of footprints. A growing body of literature is attempting to use a more comprehensive evaluation system to measure the environmental impact of an economic activity based on the combination of LCA and carbon footprint.

5. Conclusions and Discussion

We have conducted a bibliometric analysis of the relevant literature on the focus topic by utilizing CiteSpace, and come to the following conclusions. First, from the time dimension, the burst period of related literature was between 2008 and 2011, after which it was in a stage of stable development. The literature published during the period was not only in a large number, but also highly cited, which has laid a solid foundation for the later research. Second, as for the source of literature, the United States, China, Italy, Spain, the United Kingdom and Australia were the major countries where the research was conducted. These countries/regions have had the most productive authors and institutions, and a considerable number of academic collaborations have been carried out between them. However, as far as institutions and authors are concerned, only a few collaboration sub-networks have been established in the field, and the research community needs further development. As indicated by the co-citation analysis, productive institutions or authors may not be highly cited. For example, the Chinese Academy of Sciences ranks second in terms of the number of publications, but it does not have any highly cited papers. This may be due to the fact that the field has developed for only 16 years and is still in its early stages, and some institutions and authors as newcomers have not received much attention. In addition, group authors (such as ISO and IPCC) have contributed to the development of this field. Finally, regarding the research content, most of the research belongs to the field of Environmental Science and Ecology, and some studies have involved interdisciplinary fields, such as Energy and Fuels. Studies in the related fields had environment impact as their keyword and mainly focused on the discussion of methods or their application. Although a certain research base in this field has been laid, there are still a lot of gaps to be filled.

As the citation bursts analysis showed in Section 4, the application situation of focus topic ranged from farm to city, from Europe and America to Asia, while there were fewer new high frequency keywords in recent years' literatures. Combined with the more detailed literature interpretation in keywords analysis part in Section 4, we could see that innovation of method may widen the depth of research in this field and breadth of the application situation. As a result, researchers in the field need to pay more attention to the latest development of methods and the expansion of their application. In the literature focusing on the discussion of methods, the technical routes from static to dynamic measurement and from micro to macro measurement were developed based on the clarification of the definition of carbon footprint and the discussion of the LCA methods. Nowadays, LCA has become the mainstream method of carbon footprint measurement. In future research, the effort should be devoted to complete the system of the LCA method on carbon footprint measurement. For example, in the macro-scenario carbon footprint measurement at the urban level, some scholars have combined LCA with the input–output method to form the hybrid LCA method [38]. Hybrid LCA requires great amounts of data and technology. Therefore, the use of computer technology for hybrid LCA analysis is one of the future directions of this field. Moreover, in the literature focusing on the method application, the carbon emission measurement for multi-category products of crops, urban houses and other industries has developed, based on animal husbandry, the classic source of carbon emission. In fact, the design of life cycle inventory was critical in the method application. For example, Europe built The

European Platform on Life Cycle Assessment and released the International Life Cycle Data System (ILCD) to be a guidance. How to design different product life cycle lists in different socio-economic environments and how to improve the universality of LCA accounting standard are also the future research contents.

This study still has some space for improvement: (1) Analysing more papers related to carbon footprint, and comparing their results with this study to gain a deeper understanding of the application of LCA in carbon footprint measurement; or analysing more documents related to the footprint family to better understand the characteristics of the carbon footprint indicators through a comparative analysis. (2) Improving the bibliometric technology by, for instance, taking into account the self-citation of literature; solving the problem of considering only the first author in the collaboration analysis; and the combined application of various visualization technologies.

Author Contributions: S.Z. and R.C. drafted the manuscript. S.Z., R.C. and F.S. contributed to the construction of the framework and designed the research methodology. F.S. and Y.X. polished the language. S.Z. and R.C. assisted in data collection and data analysis, S.Z., F.S. and Y.X. revised the manuscript. All authors read and approved the final version of the manuscript.

Funding: This research was funded by the General Project of Philosophy and Social Science Research in Colleges and Universities in Jiangsu Province, grant number No. 2019SJA0678 and the Fundamental Research Funds for the Central Universities, grant number CXJJ-2018-361.

Conflicts of Interest: The authors declare no conflict of interest.

Appendix A

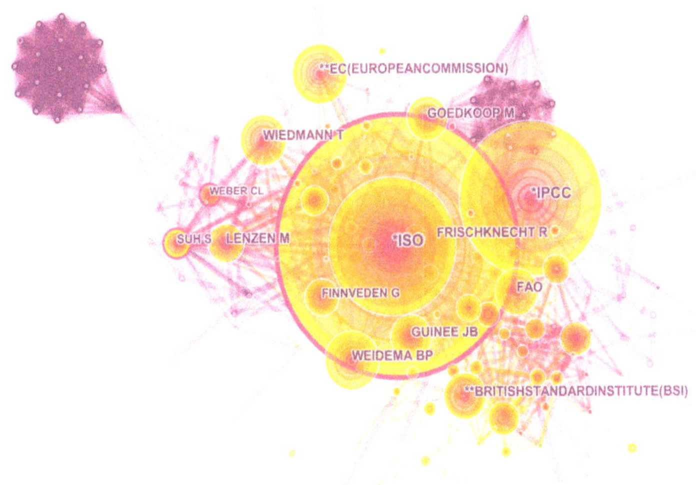

Figure A1. Author co-citation network.

Figure A2. Journal co-citation network.

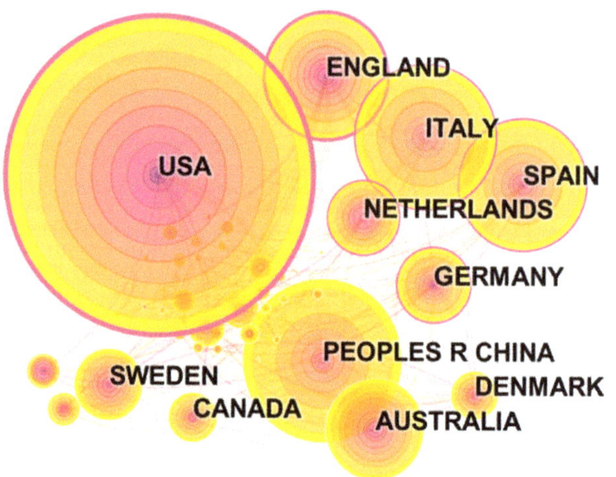

Figure A3. Country collaboration network.

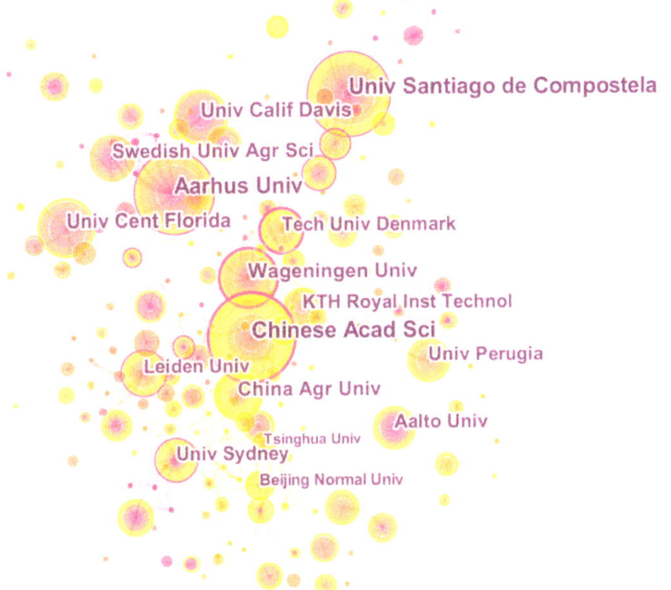

Figure A4. Institution collaboration network.

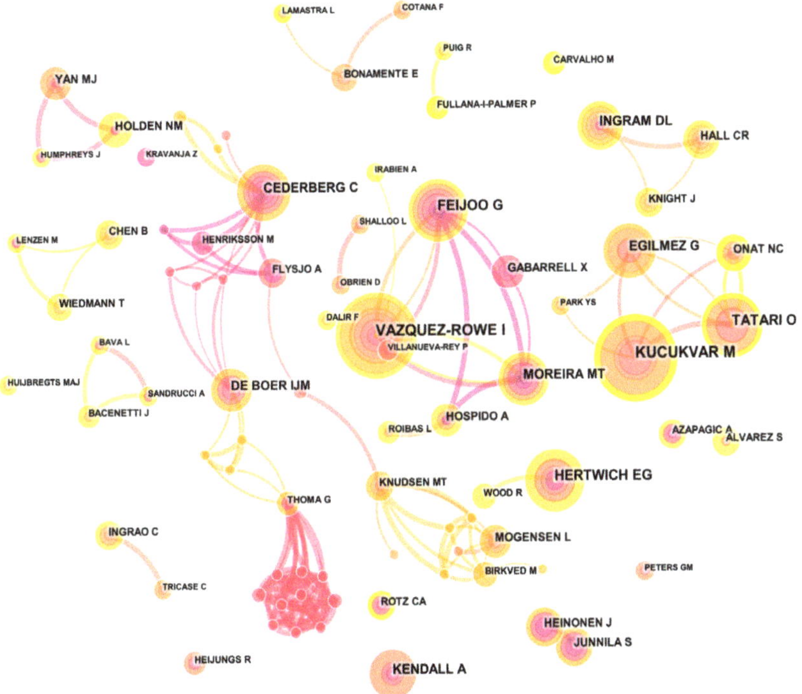

Figure A5. Author collaboration network.

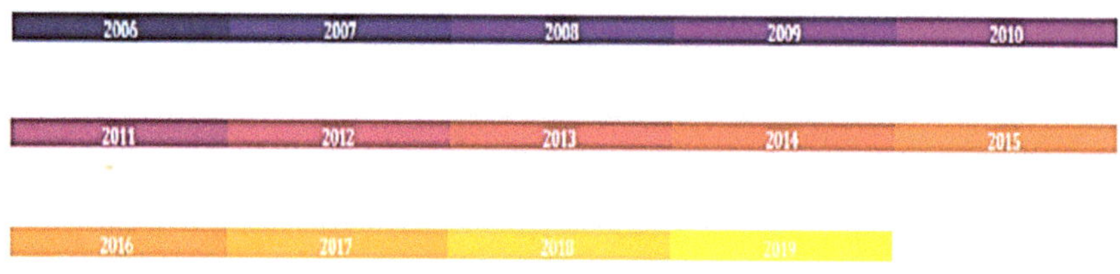

Figure A6. Year–colour correspondence bar.

References

1. IPCC. Glob. Warm. 1.5 °C. 2018. Available online: https://www.ipcc.ch/sr15/ (accessed on 23 June 2019).
2. Pandey, D.; Agrawal, M.; Pandey, J.S. Carbon footprint: Current methods of estimation. *Environ. Monit. Assess.* **2011**, *178*, 135–160. [CrossRef]
3. IPCC. 2019 Refinement to the 2006 IPCC Guidelines for National Greenhouse Gas Inventories. 2019. Available online: https://www.ipcc.ch/report/2019-refinement-to-the-2006-ipcc-guidelines-for-national-greenhouse-gas-inventories/ (accessed on 23 June 2019).
4. Bourque, P.J. Embodied energy trade balances among regions. *Int. Reg. Sci. Rev.* **1981**, *6*, 121–136. [CrossRef]
5. Chen, G.; Shan, Y.; Hu, Y.; Tong, K.; Wiedmann, T.; Ramaswami, A.; Guan, D.; Shi, L.; Wang, Y. Review on city-level carbon accounting. *Environ. Sci. Technol.* **2019**, *53*, 5545–5558. [CrossRef] [PubMed]
6. Chai, B.C.; van der Voort, J.R.; Grofelnik, K.; Eliasdottir, H.G.; Klöss, I.; Perez-Cueto, F.J.A. Which Diet Has the Least Environmental Impact on Our Planet? A Systematic Review of Vegan, Vegetarian and Omnivorous Diets. *Sustainability* **2019**, *11*, 4110. [CrossRef]
7. Pohl, J.; Hilty, L.M.; Finkbeiner, M. How LCA contributes to the environmental assessment of higher order effects of ICT application: A review of different approaches. *J. Clean. Prod.* **2019**, *219*, 698–712. [CrossRef]
8. Chen, C. Searching for intellectual turning points: Progressive Knowledge Domain Visualization. *Proc. Nat. Acad. Sci. USA* **2004**, *101*, 5303–5310. [CrossRef]
9. Callon, M.; Law, J.; Rip, A. *Mapping the Dynamics of Science and Technology: Sociology of Science in the Real World*; Macmillan Press: London, UK, 1986.
10. Chen, C. Predictive effects of structural variation on citation counts. *J. Am. Soc. Inf. Sci. Technol.* **2012**, *63*, 431–449. [CrossRef]
11. Chen, C.; Hu, Z.; Liu, S.; Tseng, H. Emerging trends in regenerative medicine: A scientometric analysis in CiteSpace. *Expert Opin. Biol. Ther.* **2012**, *12*, 593–608. [CrossRef]
12. Finnveden, G.; Hauschild, M.Z.; Ekvall, T.; Guinée, J.; Heijungs, R.; Hellweg, S.; Koehler, A.; Pennington, D.; Suh, S. Recent developments in Life Cycle Assessment. *J. Environ. Manag.* **2009**, *91*, 1–21. [CrossRef]
13. Vries, M.D.; Boer, I.J.M.D. Comparing environmental impacts for livestock products: A review of life cycle assessments. *Livest. Sci.* **2010**, *128*, 1–11. [CrossRef]
14. Weidema, B.P.; Thrane, M.; Christensen, P.; Schmidt, J.; Løkke, S. Carbon footprint: A catalyst for life cycle assessment? *J. Ind. Ecol.* **2008**, *12*, 3–6. [CrossRef]
15. Finkbeiner, M. Carbon footprinting-opportunities and threats. *Int. J. Life Cycle Assess.* **2009**, *14*, 91–94. [CrossRef]
16. Roy, P.; Nei, D.; Orikasa, T.; Xu, Q.; Okadome, H.; Nakamura, N.; Shiina, T. A review of life cycle assessment (LCA) on some food products. *J. Food Eng.* **2009**, *90*, 1–10. [CrossRef]
17. ISO. Environmental Management-Life Cycle Assessment-Requirements and Guidelines. 2006. Available online: https://www.iso.org/standard/38498.html (accessed on 23 June 2019).
18. Hertwich, E.G.; Peters, G.P. Carbon Footprint of Nations: A Global, Trade-Linked Analysis. *Environ. Sci. Technol.* **2009**, *43*, 6414–6420. [CrossRef] [PubMed]
19. Flysjö, A.; Henriksson, M.; Cederberg, C.; Ledgard, S.; Englund, J. The impact of various parameters on the carbon footprint of milk production in New Zealand and Sweden. *Agric. Syst.* **2011**, *104*, 459–469. [CrossRef]
20. Galli, A.; Wiedmann, T.; Ercin, E.; Knoblauch, D.; Ewing, B.; Giljum, S. Integrating Ecological, Carbon and Water footprint into a "Footprint Family" of indicators: Definition and role in tracking human pressure on the planet. *Ecol. Indic.* **2012**, *16*, 100–112. [CrossRef]

21. Rotz, C.A.; Montes, F.; Chianese, D.S. The carbon footprint of dairy production systems through partial life cycle assessment. *J. Dairy Sci.* **2010**, *93*, 1266–1282. [CrossRef]
22. Chen, C.; Ibekwe-SanJuan, F.; Hou, J. The structure and dynamics of co-citation clusters: A multiple-perspective co-citation analysis. *J. Am. Soc. Inf. Sci. Technol.* **2010**, *61*, 1386–1409. [CrossRef]
23. Suh, S.; Lenzen, M.; Treloar, G.J.; Hondo, H.; Horvath, A.; Huppes, G.; Jolliet, O.; Klann, U.; Krewitt, W.; Moriguchi, Y.; et al. System boundary selection in life-cycle inventories using hybrid approaches. *Environ. Sci. Technol.* **2004**, *38*, 657–664. [CrossRef]
24. Steinfeld, H.; Wassenaar, T.; Jutzi, S. Livestock production systems in developing countries: Status, drivers, trends. *Rev. Sci. Tech.* **2006**, *25*, 505–516. [CrossRef]
25. Hellweg, S.; Mila, C.L. Emerging approaches, challenges and opportunities in life cycle assessment. *Science* **2014**, *344*, 1109–1113. [CrossRef] [PubMed]
26. Kristensen, T.; Mogensen, L.; Knudsen, M.T.; Hermansen, J.E. Effect of production system and farming strategy on greenhouse gas emissions from commercial dairy farms in a life cycle approach. *Livest. Sci.* **2011**, *140*, 136–148. [CrossRef]
27. Williams, A.G. *Determining the Environmental Burdens and Resource Use in the Production of Agricultural and Horticultural Commodities*; Cranfield University and Defra: London, UK, 2006.
28. Wiedmann, T.O.; Suh, S.; Feng, K.; Lenzen, M.; Acquaye, A.; Scott, K.; Barrett, J.R. Application of hybrid life cycle approaches to emerging energy technologies–the case of wind power in the UK. *Environ. Sci. Technol.* **2011**, *45*, 5900–5907. [CrossRef]
29. Tilman, D.; Clark, M. Global diets link environmental sustainability and human health. *Nature* **2014**, *515*, 518–522. [CrossRef] [PubMed]
30. Laurent, A.; Olsen, S.I.; Hauschild, M.Z. Limitations of carbon footprint as indicator of environmental sustainability. *Environ. Sci. Technol.* **2012**, *46*, 4100–4108. [CrossRef] [PubMed]
31. Goedkoop, M.; Heijungs, R.; Huijbregts, M.; De Schryver, A.; Struijs, J.V.Z.R.; Van Zelm, R. A Life Cycle Impact Assessment Method Which Comprises Harmonised Category Indicators at the Midpoint and the Endpoint Level. 2009. Available online: http://www.lcia-recipe.net (accessed on 23 June 2019).
32. Thomassen, M.A.; Van Calker, K.J.; Smits, M.C.; Iepema, G.L.; de Boer, I.J. Life cycle assessment of conventional and organic milk production in the Netherlands. *Agric. Syst.* **2008**, *96*, 95–107. [CrossRef]
33. Thoma, G.; Popp, J.; Nutter, D.; Shonnard, D.; Ulrich, R.; Matlock, M.; SooKim, D.; Neiderman, Z.; Kemper, N.; East, C.; et al. Greenhouse gas emissions from milk production and consumption in the United States: A cradle-to-grave life cycle assessment circa 2008. *Int. Dairy J.* **2013**, *31*, S3–S14. [CrossRef]
34. Knudsen, M.T.; Meyer-Aurich, A.; Olesen, J.E.; Chirinda, N.; Hermansen, J.E. Carbon footprints of crops from organic and conventional arable crop rotations–using a life cycle assessment approach. *J. Clean. Prod.* **2014**, *64*, 609–618. [CrossRef]
35. Pelletier, N.; Pirog, R.; Rasmussen, R. Comparative life cycle environmental impacts of three beef production strategies in the Upper Midwestern United States. *Agric. Syst.* **2010**, *103*, 380–389. [CrossRef]
36. Beloin, S.P.; Heijungs, D.R.; Blanc, I. The ESPA (Enhanced Structural Path Analysis) method: A solution to an implementation challenge for dynamic life cycle assessment studies. *Int. J. Life Cycle Assess.* **2014**, *19*, 861–871. [CrossRef]
37. Mery, Y.; Tiruta-Barna, L. An integrated "process modelling-life cycle assessment" tool for the assessment and design of water treatment processes. *Int. J. Life Cycle Assess.* **2013**, *18*, 1062–1070. [CrossRef]
38. Fry, J.; Lenzen, M.; Jin, Y.; Wakiyama, T.; Baynes, T.; Wiedmann, T.; Malik, A.; Chen, G.; Wang, Y.; Geschke, A. Assessing carbon footprints of cities under limited information. *J. Clean. Prod.* **2017**, *176*, 1254–1270. [CrossRef]
39. Nijdam, D.; Rood, T.; Westhoek, H. The price of protein: Review of land use and carbon footprints from life cycle assessments of animal food products and their substitutes. *Food Policy* **2012**, *37*, 760–770. [CrossRef]
40. Casolani, N.; Pattara, C.; Liberatore, L. water and carbon footprint perspective in Italian durum wheat production. *Land Use Policy* **2016**, *58*, 394–420. [CrossRef]
41. Ronga, D.; Gallingani, T.; Zaccardelli, M.; Perrone, D.; Francia, E.; Milc, J.; Pecchioni, N. Carbon footprint and energetic analysis of tomato production in the organic vs the conventional cropping systems in Southern Italy. *J. Clean. Prod.* **2019**, *220*, 836–845. [CrossRef]
42. Strohbach, M.W.; Arnold, E.; Haase, D. The carbon footprint of urban green space—A life cycle approach. *Landsc. Urban. Plan.* **2012**, *104*, 220–229. [CrossRef]

43. Heinonen, J.; Junnila, S. A life cycle assessment of carbon mitigation possibilities in metropolitan areas. In Proceedings of the Conference on Sustainable Building, Helsinki, Finland, 18 December 2010.

44. Onat, N.C.; Kucukvar, M.; Tatari, O. Conventional, hybrid, plug-in hybrid or electric vehicles? State-based comparative carbon and energy footprint analysis in the United States. *Appl. Energy* **2015**, *150*, 36–49. [CrossRef]

45. Basbagill, J.; Flager, F.; Lepech, M.; Fischer, M. Application of life-cycle assessment to early stage building design for reduced embodied environmental impacts. *Build. Environ.* **2013**, *60*, 81–92. [CrossRef]

46. Blengini, G.A.; Carlo, T.D. The changing role of life cycle phases, subsystems and materials in the LCA of low energy buildings. *Energy Build.* **2010**, *42*, 869–880. [CrossRef]

47. Deng, M.; Zhong, S.; Xiang, G. Carbon emission reduction effect of China's final demand structure change from 2013 to 2020: A scenario-based analysis. *Carbon Manag.* **2019**, *10*, 387–404. [CrossRef]

48. Zhong, S.; Chen, J. How environmental beliefs affect consumer willingness to pay for the greenness premium of low-carbon agricultural products in china: Theoretical model and survey-based evidence. *Sustainability* **2019**, *11*, 592. [CrossRef]

Article

Mechanism and Kinetics of Ammonium Sulfate Roasting of Boron-Bearing Iron Tailings for Enhanced Metal Extraction

Xiaoshu Lv [1,2], Fuhui Cui [3,*], Zhiqiang Ning [1], Michael L. Free [4] and Yuchun Zhai [1,5]

[1] School of Metallurgy, Northeastern University, Shenyang 110819, China; lvsiaoshu@163.com (X.L.); hainuo@tom.com (Z.N.); zhaiyc@smm.neu.edu.cn (Y.Z.)

[2] School of Biomedical and Chemical Engineering, Liaoning Institute of Science and Technology, Benxi 117004, China

[3] School of Metallurgy and Environment, Central South University, Changsha 410083, China

[4] Department of metallurgical engineering, University of Utah, Salt Lake City, UT 84112, USA; michael.free@utah.edu

[5] College of Resources and Materials, Northeastern University at Qinhuangdao, Qinhuangdao 066004, China

* Correspondence: fuhuicui@csu.edu.cn or cfhty0501@126.com

Received: 4 October 2019; Accepted: 24 October 2019; Published: 4 November 2019

Abstract: The large amount of boron-bearing iron tailings in China is a resource for metals that needs to be more completely and efficiently utilized. In this evaluation, the ammonium sulfate roasting process was used to make a controllable phase transformation to facilitate the subsequent extraction of valuable metals from boron-bearing iron tailings. The effects of roasting temperature, roasting time, the molar ratio of ammonium sulfate to tailings, and the particle size on the extraction of elements were investigated. The orthogonal experimental design of experiments was used to determine the optimal processing conditions. XRD (X-Ray Diffractomer), scanning electron microscope (SEM), and simultaneous DSC–TG analyzer were used to assist in elucidating the mechanism of ammonium sulfate roasting. The experimental results showed that nearly all Fe, Al, and Mg were extracted under the following conditions: (1) the molar ratio of ammonium sulfate to iron tailings was 3:1; (2) the roasting temperature was 450 °C; (3) the roasting time was 120 min.; and, (4) the particle size was less than 80 μm. The kinetics analysis indicated that the sulfation of metals was controlled by internal diffusion, with the apparent activation energies of 17.10 kJ·mol^{-1}, 17.85 kJ·mol^{-1}, 19.79 kJ·mol^{-1}, and 29.71 kJ·mol^{-1} for Fe, Al, Mg, and B, respectively.

Keywords: iron tailings; ammonium sulfate roasting process; reaction mechanism; kinetics

1. Introduction

Boron resources are abundant, but they are unevenly distributed around the world [1]. For example, Turkey, the United States, Russia, and China were the countries with the largest proved boron reserves, and collectively these countries account for about 90% of the world's total Boron reserves. China's boron reserves are the fourth largest in the world [2–4].

There are large amounts of reserves of boron resources in China, which account for 18% of the world's boron resources [4–6]. The known boron reserves are mainly distributed in northeast Liaoning, Jilin, Qinghai, and Tibet [6]. Among them, 56% of the national boron reserves are in the Liaoning province. Although the boron reserves in our country are extensive, the boron grade of the ore is very low, and the ore is mineralogically complex with a significant occurrence of economically significant metals. The complex nature of the ore requires extensive mineral processing to effectively recover the boron and the desirable metals [7,8].

There are two types of boron mineral resources in Liaoning, namely, ascharite and paigeite, which are collectively known as ludwigite. Ascharite accounts for 20% of the total reserves and it has higher boron content. After years of resource utilization, the reserve of ascharite has been depleted, and the major remaining mineral is paigeite. After magnetic separation, boron concentrates and boron-bearing iron tailings are obtained [9–11].

The boron-bearing iron tailings often contain magnesium, iron, silicon, boron, and aluminum. Currently, the boron-bearing iron tailings are processed by hydrometallurgical and pyrometallurgical methods [12]. The common hydrometallurgical methods include sulfuric acid leaching, soda leaching, ammonium sulfate leaching, and alkaline medium leaching [13–19]. The sulfuric acid leaching makes the majority of magnesium, iron, boron, and aluminum dissolve into water, and then followed by a chemical separation process. Soda and ammonium sulfate leaching have a selectivity on boron leaching, but the recovery of boron is low, and the generated slag is difficult to utilize. Alkaline leaching uses the alkali medium to dissolve boron in the style of sodium borate, the leaching efficiency of this method is high, but the serious corrosion of alkali on equipment and the huge but unrecoverable consumption of alkali restricts its developing. These methods generate waste slag and metal-bearing waste solutions, which is harmful to the environment. For the pyrometallurgical method [7,20–25], ferro-boron alloys are prepared by electric furnace smelting from boron concentrates. Smelting in a blast furnace produces ferro-boron alloys and boron-rich slags. However, these processes only recover boron from virgin ore resources. The challenges with these methods include a large amount of waste slag that is generated in the smelting process and the loss of iron and aluminum to the tailings [26].

Herein, we proposed a new process to more fully utilize boron-bearing slag. Ammonium sulfate roasting was used to make a controllable phase transformation for iron, aluminum, and boron. In the following leaching process, the silicon components in the tailings are not soluble in water. The separation of iron, magnesium, and aluminum are precipitated at different pH levels [27]. Boron is recycled based on the solubility differences between boric acid and ammonium sulfate [28,29]. The kinetics and reaction mechanisms of the roasting process were also studied, and these mechanisms have theoretical significance and reference value in the development and utilization of boron-bearing iron tailings.

2. Experimental Procedures and Materials

2.1. Materials

The boron-bearing iron tailings in this roasting evaluation were taken from Dandong, Liaoning province, China. The chemical composition of the tailings was analyzed by X-Ray Fluorescence (XRF, Bruker, S6 JAGUAR, German), and Table 1 shows the results. As seen in Table 1, the contents of boron and magnesium are high in the tailings. Ammonium sulfate that was used in this work was of analytical grade. The ammonium sulfate used in this study was obtained from Sinopharm Chemical Regent Co., Ltd. (purity > 99.0%, AR). The other regent used in the titration method was sourced from Sinopharm Chemical Regent Co., Ltd. (purity > 99.0%, AR).

Table 1. Chemical composition of the boron-bearing tailings (mass fraction, %).

MgO	CO_2	SiO_2	Fe_2O_3	B_2O_3	Al_2O_3	CaO	Na_2O	else
36.40	5.38	32.30	9.89	8.92	2.56	1.77	0.11	2.67

Figure 1 shows the X-ray diffraction patterns of the boron-bearing tailings. As presented in Figure 1, the main phases in the tailings were lizardite ($Mg_3Si_2O_5(OH)_4$), magnesium borate ($Mg_2(B_2O_5)(H_2O)$), hematite (Fe_2O_3), and cordierite ($Mg_2Si_5Al_4O_{18}$).

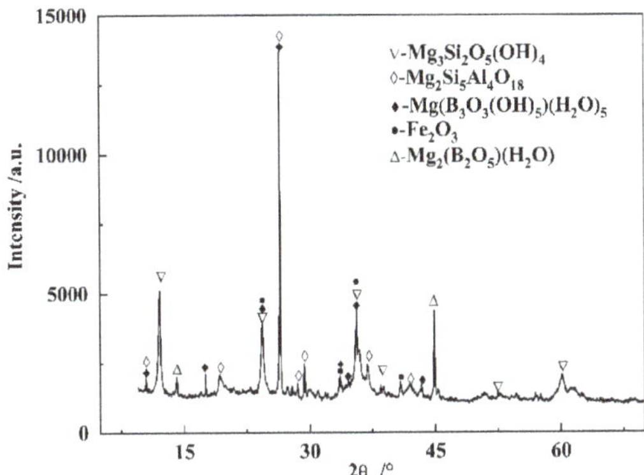

Figure 1. XRD patterns of the boron-bearing iron tailings.

Figure 2a,b show the scanning electron microscope (SEM) images of the boron-bearing iron tailings. It can be seen that the ore particles have rough surfaces, loose structures, and irregular shapes.

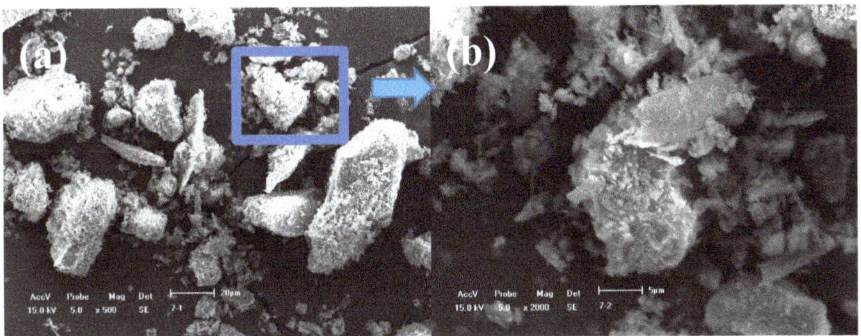

Figure 2. Scanning electron microscope (SEM) pattern of the boron-bearing iron tailings: (**a**)-particle distribution; (**b**)-partial region enlarged.

2.2. Product Analysis

In the final leach liquor, the concentration of Fe was measured by an ultraviolet-visible photometer (Beijing PUXI, SF2-7520, Beijing China). The concentration of Mg was determined while using an EDTA complexation titration method (Mg and Ca complexation method, the error was less than 0.2%). The concentration of Mg was determined while using an EDTA titration method (the error was less than 0.2%). The roasting products and residues were identified using a D/max RB X-ray diffraction instrument (Rigaku Corporation, D/max-RB, Tokyo, Japan) with Cu-Kα radiation ranging from 5~80° and a scanning speed of 10°/min. The analysis of the morphology of the roasted products was conducted while using a Scanning Electron Microscope (Shimadzu, SSX-550, Tokyo, Japan).

Thermogravimetric-differential scanning calorimeter (TG-DSC) analysis of the boron-bearing iron tailings was determined using a differential scanning calorimetry- thermal Gravity (NETZSCH STA449F3A-1001-M, Selb, German) instrument operated in a flowing Ar (purity > 99.999%) atmosphere, with a heating rate of 10 °C/min. from room temperature to 800 °C.

2.3. Methods and Procedures

The boron-bearing iron tailings were grounded into fine powder in ball mill equipment (BS2308, TENCAN POWDER, China). Subsequently, the ore powder and a certain amount of ammonium sulfate were blended and put it into a ceramic crucible. The crucible was placed in a resistance furnace (DB2, Tianjin Zhonghuan Furnace Corp., Tianjin, China) with an automated temperature control unit and a tail gas instrument. After roasting at the specified temperature for a certain amount of time, the roasted products were cooled to room temperature. Afterwards, the roasted products were ground and leached with deionized water in a constant temperature water bath (DK-524-type, Changzhou Jintan Scientific Instrument Technology, Changzhou, China) at 90 °C for 30 min. The resultant liquor was leached, filtered using a vacuum filtration system, and the remaining solids were then washed twice with deionized water. The filtrate was analyzed to determine the concentration of iron, aluminum, and magnesium leached. Figure 3 shows the schematic diagram of the roasting reaction system.

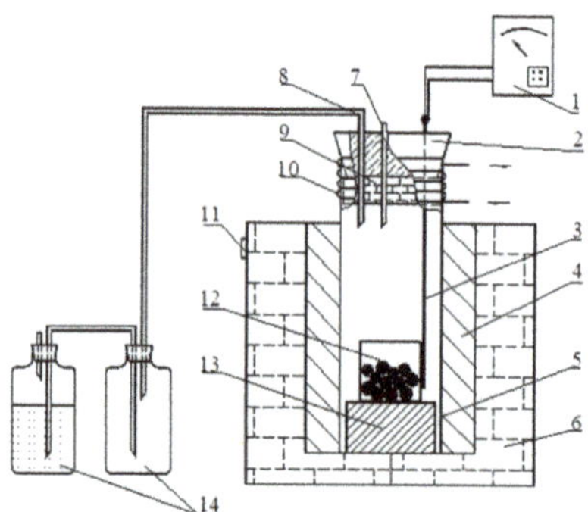

Figure 3. Schematic diagram of the roasting reaction system. 1. intelligent programmable regulator; 2. rubber stopper; 3. thermocouple; 4. Brusque; 5. reaction still; 7. resistance wire heater; 7. gas inlet; 8. gas outlet; 9. refractory bricks; 10. Copper cooling system; 11. binding post; 12. corundum crucible; 13. refractory bricks; 14. gas collecting bottle.

3. Results and Discussion

3.1. Roasting Process

The roasting products were leached at 90 °C for 30 min.

3.1.1. Effect of the Molar Ratio of Ammonium Sulfate to Ore

The effect of the molar ratio of ammonium sulfate to ore on the extractions of magnesium, aluminum, iron, and boron were investigated at the roasting temperature of 450 °C, the roasting time of 120 min., and the ore particles size of less than 80 μm. Figure 4 presents the results.

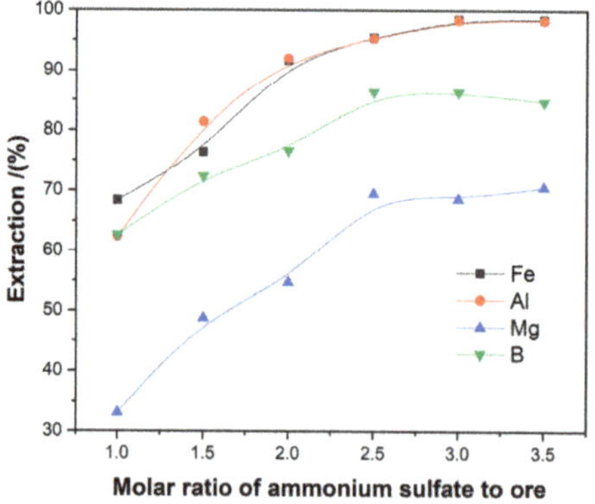

Figure 4. The effect of the molar ratio ammonium sulfate to ore on the extraction of elements.

It is shown in Figure 4 that the extractions rate of iron, aluminum, magnesium, and boron increase with the ammonium-to-ore molar ratio, and reach their maximum at 3:1.

3.1.2. Effect of Roasting Temperature

The influence of the roasting temperature from 250 °C to 500 °C on the extractions of iron, aluminum, magnesium, and boron was studied under the condition of the molar ratio of ammonium sulfate to tailings of 3:1, roasting time of 120 min., and the ore particles size of less than 80 μm.

As seen in Figure 5, there are significant improvements in the extraction of iron, aluminum, and magnesium when the roasting temperature is increased from 250 °C to 450 °C, and the extraction reaches the maximum values for each of the metals at 450 °C. The extraction of boron firstly increases and then decreases with the enhanced roasting temperature, and reaches the maximum at 400 °C. The increase of the extractions of iron, aluminum, and magnesium results from the increasing reaction activity of ammonium sulfate with increased temperature. However, the decrease in the extraction of boron can be attributed to the volatilization of boron oxide at high temperature. Thus, the roasting temperature of 450 °C was applied in the following experiments. The roasting mechanism will be discussed in detail in the following section.

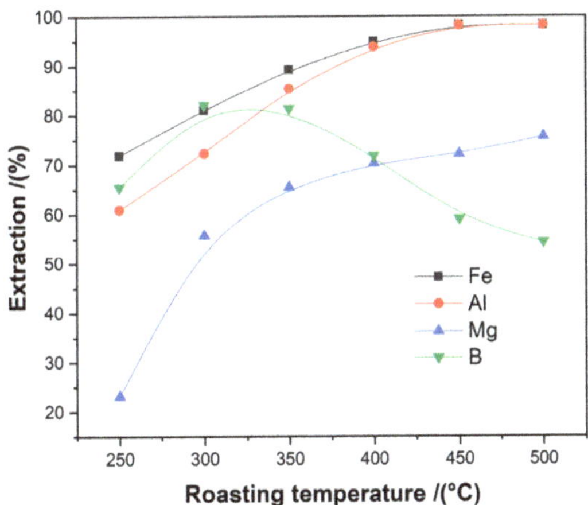

Figure 5. The effect of roasting temperature on the extraction of metals.

3.1.3. Effect of Roasting Time

The effect of roasting time from 20 min. to 160 min. on the extraction of iron, aluminum, and magnesium was studied under the conditions of the molar ratio of ammonium sulfate to tailings of 3:1, the roasting temperature of 450 °C, and the ore particles size of less than 80 μm.

As can be seen from Figure 6, the extraction of iron, aluminum, and magnesium rapidly increased from 20 min. to 120 min. The maximum extraction was reached at 120 min., indicating that the sulfating process of iron, aluminum, and magnesium are close to completion at 120 min. While the extraction of boron firstly increases but then decreases with the prolonging of roasting time, resulting from the increasing of the volatilization of boron oxide. Therefore, the roasting time of 120 min. was applied in subsequent experiments.

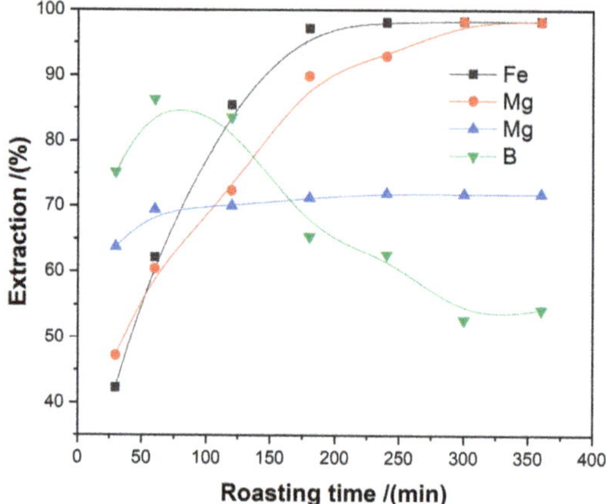

Figure 6. The effect of roasting time on the extraction of metals.

3.1.4. Effect of the Particle Size the Tailings

The effect of the particle size from 40 μm to 120 μm on the extractions of iron, aluminum, magnesium, and boron was studied under the condition of ammonium-to-ore mole ratio of 3:1, roasting temperature of 450 °C, and roasting time of 120 min.

The results (as seen in Figure 7) show that all metal extractions are increasing with declining particle size. All elements extractions reach the maximum at the particle size of less than 80 μm. The extractions were not improved with the reduced size.

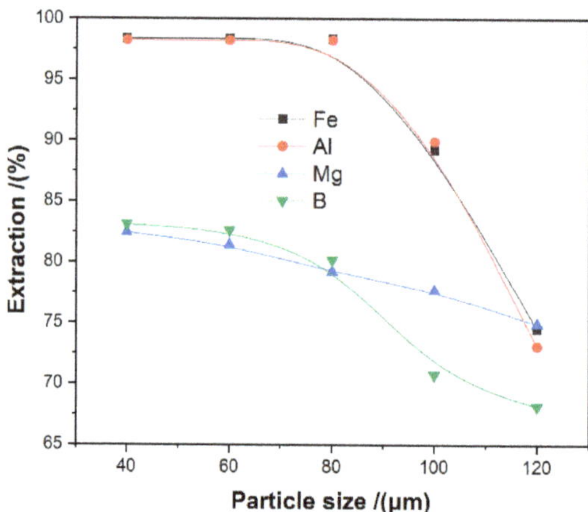

Figure 7. The effect of the particle size on the extraction of metals.

3.1.5. Orthogonal Design of Experiments

An orthogonal experiment of $L_9(3^4)$ was designed based on the single factor experiment in order to obtain the optimal reaction conditions of the roasting process. The factors of roasting temperature (A), the molar ratio of the ammonium sulfate to boron-bearing iron tailings (B), the roasting time (C), and particle size of tailings (D), the selected factors and levels are listed in Table 2. Table 3 shows the orthogonal experimental results.

Table 2. Experimental factors and levels.

Level No.	Factor			
	A/°C	B/h	C/min	D/μm
1	350	2	80	120
2	400	2.5	100	100
3	450	3	120	80

Table 3. Results of $L_9(3^4)$ orthogonal experiments.

No.	Factor							
	A/°C	B/h	C/min	D/μm	Fe	Al	Mg	B
1	350	2	80	120	71.40	72.91	57.72	76.58
2	350	2.5	100	100	74.37	76.31	69.14	80.39
3	350	3	120	80	83.89	85.22	79.89	81.07
4	400	2	100	80	75.27	76.87	71.72	82.47
5	400	2.5	120	120	78.34	78.63	82.47	84.71
6	400	3	80	100	80.02	82.41	81.71	83.01
7	450	2	120	100	80.13	70.73	78.41	70.14
8	450	2.5	80	80	78.01	79.44	84.98	69.58
9	450	3	100	120	82.09	82.68	85.31	68.31
Fe Average1	76.55	75.60	76.81	77.28				
Fe Average2	78.21	76.91	77.24	78.51				
Fe Average3	80.08	82.30	80.79	79.06				
R of Fe extraction	3.53	6.70	3.98	1.78	$R_B>R_C>R_A>R_D$			
Al Average1	78.14	76.83	78.25	78.07				
Al Average2	79.30	78.12	78.62	79.81				
Al Average3	80.95	83.43	81.52	80.51				
R of Al extraction	2.803	6.600	3.274	2.437	$R_B>R_C>R_A>R_D$			
Mg Average1	76.243	75.013	76.287	76.540				
Mg Average2	77.843	76.727	76.513	77.720				
Mg Average3	78.787	81.133	80.073	78.613				
R of Mg extraction	2.544	6.120	3.786	2.073	$R_B>R_C>R_A>R_D$			
B Average1	79.347	76.397	76.390	76.533				
B Average2	83.397	78.277	77.057	77.847				
B Average3	69.343	77.643	78.640	77.707				
R of B extraction	17.054	1.830	2.250	1.314	$R_A>R_C>R_B>R_D$			

As shown in Table 3, the orthogonal experimental results show the effect of the four factors on the extraction of metals, as follows: the molar ratio of ammonium sulfate to tailings, the roasting time, the roasting temperature, and the particle size of the tailings. Figure 8 shows the trend of the range. It can be seen that the optimum conditions of the roasting process are the roasting temperature of 450 °C, the roasting time of 120 min., the molar ratio of ammonium sulfate to tailings of 3:1, and the particle size of the tailings of less than 80 μm. According to the optimized process conditions, the experimental verification shows that the extractions of Fe and Al all exceed 98%, while the extraction of Mg and B all exceed 80%.

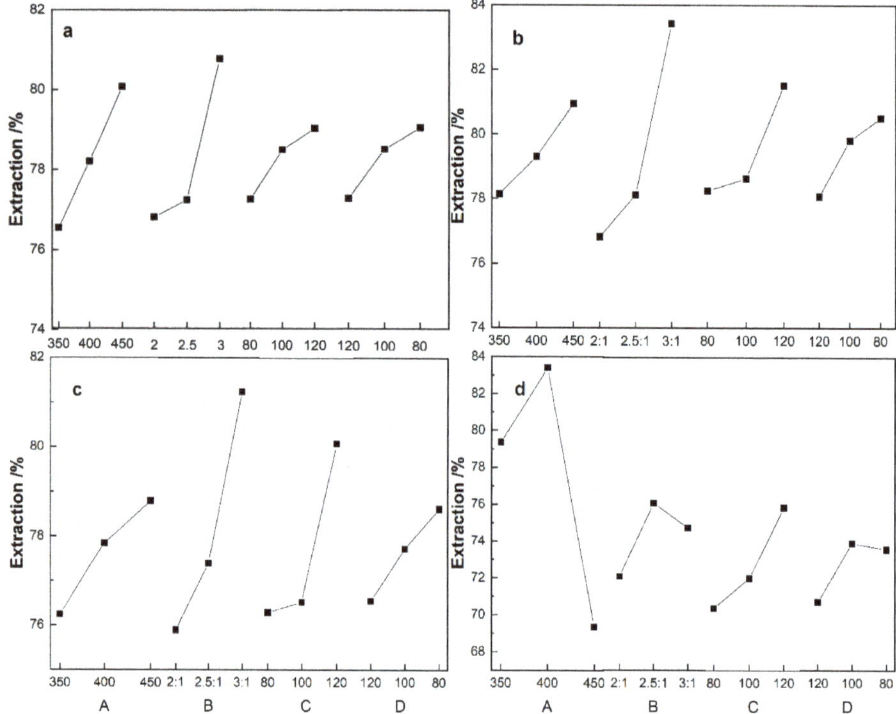

Figure 8. Trend chart of range: a-Fe; b-Al; c-Mg; d-B for the factors of roasting temperature (**a**), the molar ratio of the ammonium sulfate to boron-bearing iron tailings (**b**), the roasting time (**c**) and particle size of tailings (**d**).

3.2. Roasting Mechanism and Kinetics

3.2.1. The Roasting Mechanism

Figure 9 shows the DSC-TG curves of the ammonium sulfate (Figure 9a) and the mixture of ammonium sulfate and tailings with a molar ratio of 1:3 (Figure 9b) from room temperature to 800 °C at a heating rate of 10 °C/min. As seen in Figure 8a, there are two obvious endothermic peaks at approximately 289 °C and 393 °C, which corresponded to the decomposition of $(NH_4)_2SO_4$ (Equation (1)) and the decomposition of NH_4HSO_4 (Equation (2)), respectively, while there are two stages of mass loss corresponding to the two stages decomposition [30]. It can be seen from Figure 9b that there are two endothermic peaks at approximately 318 °C and 439 °C on the DSC curve, corresponding two large weight losses that were observed from the TG curve at two different temperature ranges of 240–359 °C and 359–448 °C. The first endothermic peak can be attributed to the sulfation reactions between $(NH_4)_2SO_4$ and the tailings, while the second endothermic peak can be attributed to the sulfation reactions between NH_4HSO_4 and the tailings, the sulfation reactions are shown in Equations (3)–(8). The followed mass loss on the TG curve can be attributed to the decomposition of ammonium metal sulfates (as shown in Equations (9)–(11)).

Figure 10 shows the XRD patterns of samples that were obtained at different roasting temperatures. It can be seen that the main phases at the roasting temperature of 300 °C are lizardite ($Mg_3Si_2O_5(OH)_4$) phase, ferric sulfate ($Fe_2(SO_4)_3$), ammonium ferric sulfate (($NH_4)_3Fe(SO_4)_3$), ammonium magnesium sulfate (($NH_4)_2Mg(SO_4)_2$), and ammonium hydrogen sulfate (NH_4HSO_4), indicating that the decomposition of ammonium sulfate is not complete and that part of the mineral phases have reacted and generated corresponding metal sulfates. The diffraction peaks of szaibelyite ($MgBO_2(OH)$), magnetite phase (Fe_3O_4), and halloysite phase ($Al_2Si_2O_5(OH)_4(2H_2O)$) appear on the samples that were produced at the roasting temperature of 350 °C. When the roasting temperature is 400 °C, the diffraction peaks of ammonium sulfate disappear and the diffraction peaks of metal sulfate intensify, which indicates that the decomposition of ammonium sulfate is complete and the generation of metal sulfates is increased. Additionally, when the roasting temperature is 450 °C, the diffraction peaks of ammonium sulfate become weak, but the diffraction peaks of ferric sulfate significantly intensify. These results show that parts of the ammonium metal sulfates are decomposed at this temperature. The sulfation roasting process with ammonium sulfate demonstrates that the metal-bearing mineral phases react with ammonium sulfate and ammonium bisulfate to form ammonium metal sulfates. As the roasting temperature increases, ammonium metal sulfates begin to decompose into corresponding metal sulfate. However, the diffraction peaks of boron-bearing phases are not detected.

The total chemical reactions between the minerals of the tailings and ammonium sulfate can be given, as follows.

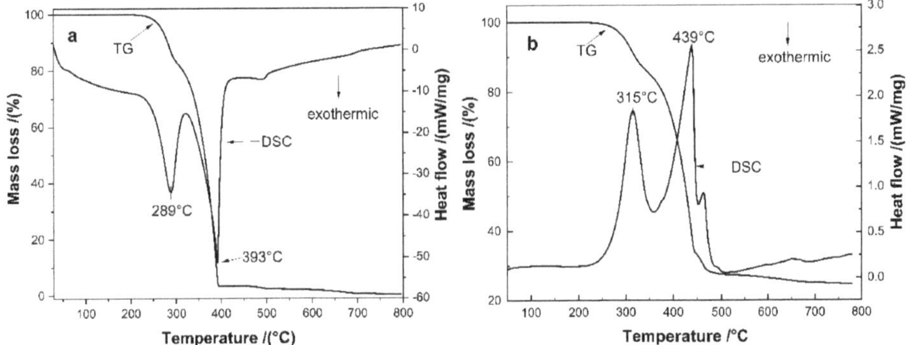

Figure 9. Differential thermal analysis curve of ammonium sulfate and the mixture of and raw tailings and ammonium sulfate: (**a**) ammonium sulfate; and (**b**) the mixture of ammonium sulfate and raw tailings.

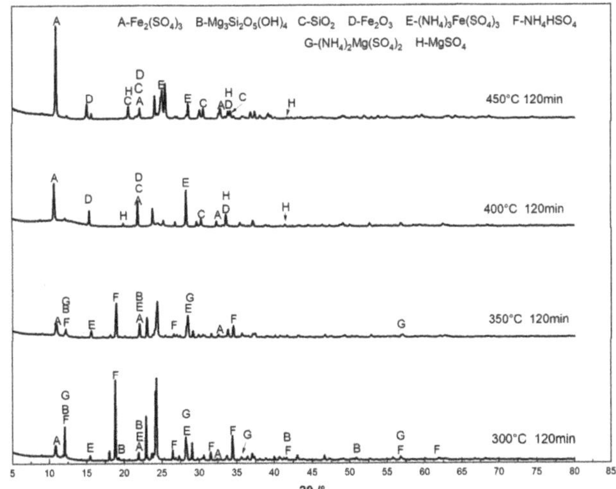

Figure 10. XRD of residue obtained at different roasting temperatures.

$$(NH_4)_2SO_4 \rightarrow NH_4HSO_4 + NH_3(g) \tag{1}$$

$$NH_4HSO_4 \rightarrow NH_3(g) + SO_3(g) + H_2O(g) \tag{2}$$

$$Fe_2O_3 + 4(NH_4)_2HSO_4 \rightarrow 2NH_4Fe(SO_4)_2 + 3H_2O(g) + 6NH_3(g) \tag{3}$$

$$Fe_2O_3 + 4NH_4HSO_4 \rightarrow 2NH_4Fe(SO_4)_2 + 3H_2O(g) + 2NH_3(g) \tag{4}$$

$$Mg_3Si_2O_5(OH)_4 + 6(NH_4)_2SO_4 \rightarrow 3(NH_4)_2Mg(SO_4)_2 + 2SiO_2 + 5H_2O(g) + 6NH_3(g) \tag{5}$$

$$Mg_3Si_2O_5(OH)_4 + 6NH_4HSO_4 \rightarrow 3(NH_4)_2Mg(SO_4)_2 + 2SiO_2 + 5H_2O(g) \tag{6}$$

$$Mg_2(B_2O_5)(H_2O) + (NH_4)_2SO_4 \rightarrow 2(NH_4)_2Mg(SO_4)_2 + B_2O_3 + 4NH_3(g) + 3H_2O(g) \tag{7}$$

$$Mg_2Si_5Al_4O_{18} + 12(NH_4)_2SO_4 \rightarrow 4NH_4Al(SO_4)_2 + 2(NH_4)_2Mg(SO_4)_2 + 5SiO_2 + 4NH_3(g) + 8H_2O(g) \tag{8}$$

$$2NH_4Fe(SO_4)_2 \rightarrow Fe_2(SO_4)_3 + H_2O(g) + 2NH_3(g) + SO_3(g) \tag{9}$$

$$(NH_4)_2Mg(SO_4)_2 \rightarrow MgSO_4 + H_2O(g) + 2NH_3(g) + SO_3(g) \tag{10}$$

$$2NH_4Al(SO_4)_2 \rightarrow Al_2(SO_4)_3 + H_2O(g) + 2NH_3(g) + SO_3(g) \tag{11}$$

Figure 11 gives the SEM images of the samples that were roasted at different temperature under the conditions that the roasting time of 120 min., the molar ratio of ammonium sulfate to tailings of 3:1, and the particle size of the tailings of less than 80 μm; as seen in Figure 10a, the roasted product at 300 °C presents a thick, cubic plate structure, and smooth surface. With roasting temperature continually increasing, the plate is cracked, the size of roasted product shrinks (400 °C, Figure 10c), and the surface of the particle changes from smooth to rough from 300 to 450 °C (Figure 10a–d).

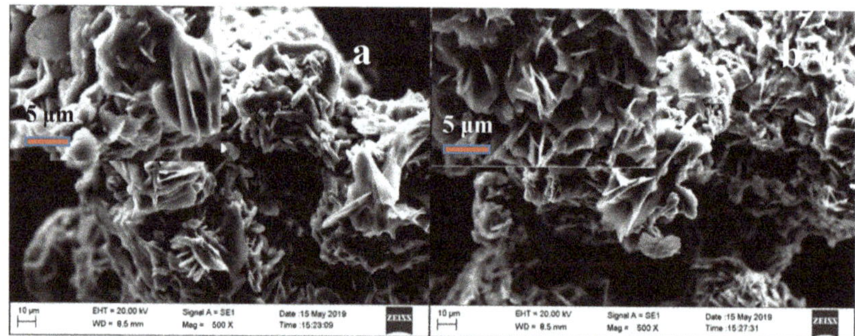
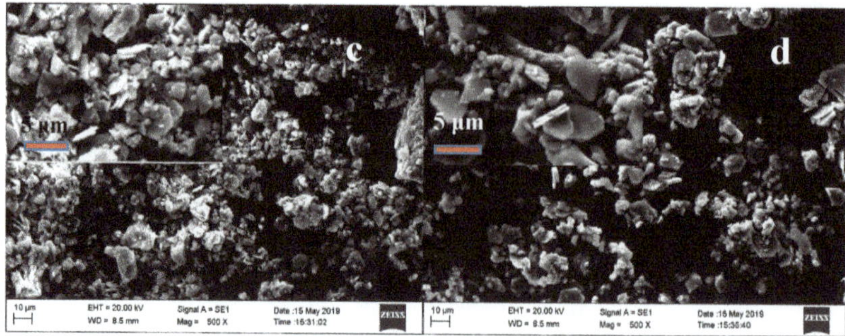

Figure 11. SEM images of residues after roasting at specified temperatures: (**a**). 300 °C; (**b**). 350 °C; (**c**). 400 °C; and, (**d**). 450 °C.

3.2.2. Kinetic Analysis

Figure 12 shows the extractions of Fe, Al, Mg, and B at 250–450 °C varying the different time (molar ratio of ammonium sulfate to ore of 3:1, particle size of less than 80 μm).

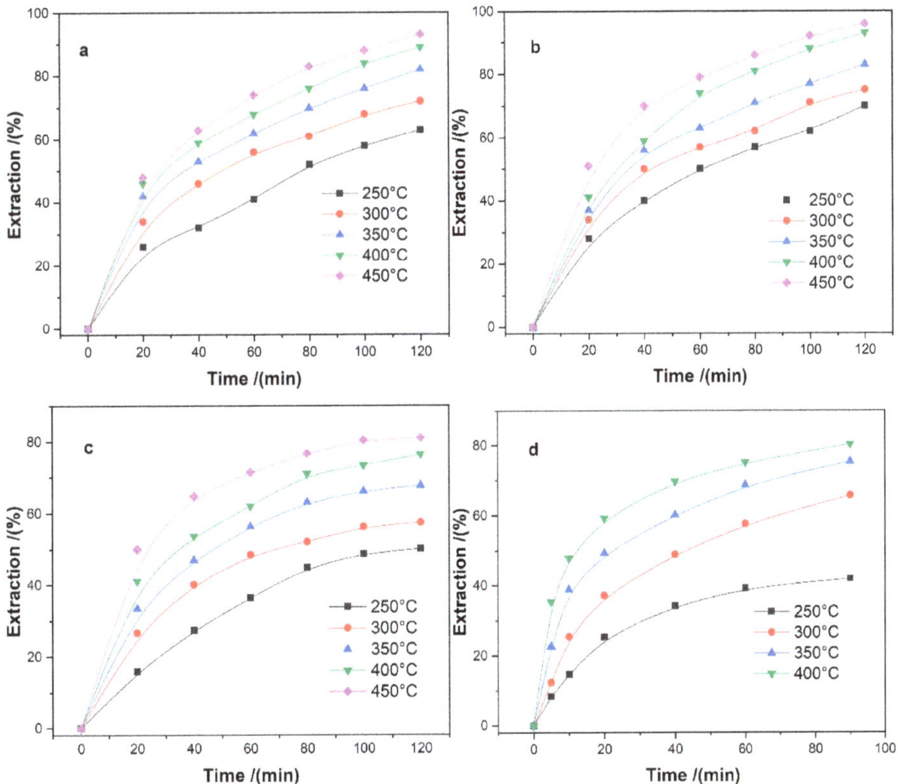

Figure 12. Extraction of metals at different roasting temperatures as a function of roasting time: (**a**)-Fe; (**b**)-Al; (**c**)-Mg; and, (**d**)-B.

The system where sulfation roasting to boron-bearing iron tailings with ammonium sulfate is a liquid-solid reaction, which can be analyzed by the shrinking core model. If the sulfation reactions are controlled by the phase boundary reaction (R_3), the following expression of the model can be used to describe the kinetics of the process:

$$1 - (1-\alpha)^{1/3} = k_c t \tag{12}$$

However, if the sulfation reactions are controlled by the three-dimensional diffusion (internal diffusion), the G-B equation (D_4) [31] can be described as:

$$1 - 2/3\alpha - (1-\alpha)^{2/3} = k_p t \tag{13}$$

where α is the extraction ratio of Fe, Al, and Mg, k_c and k_p is the reaction rate constant (min.$^{-1}$), t is reaction time (min.).

The experimental data were used to fit the suitable mechanism function in the sulfation roasting process. The results showed that the fitting results by the G-B equation have more positive correlation coefficients than the phase boundary reaction equation; hence, the G-B equation was applied in the fitting of the experimental data and the results are shown in Figure 13. As seen, $1 - 2/3\alpha - (1-\alpha)^{2/3}$

has a good linear relationship against the reaction time for Fe, Al, and Mg, which indicates that the sulfation process to boron-bearing iron tailings is controlled by internal diffusion. According to Arrhenius equation $k = Ae^{-E_a/RT}$, $\ln k = \ln A - E_a/RT$, the apparent activation energy can be obtained from the slope of the plot of $\ln k$ vs. $1/T$, and the pre-exponential factor can be obtained by intercept, the fitting results of $\ln k$ vs. $1/T$ are shown in Figure 14. The apparent activation energy of each metal is calculated as 17.10 kJ·mol^{-1}, 17.85 kJ·mol^{-1}, 19.79 kJ·mol^{-1}, and 29.71 kJ·mol^{-1} for Fe, Al, Mg, and B, respectively. Additionally, the pre-exponential factor is calculated as 0.0338, 0.0392, 0.0339, and 0.3265 for Fe, Al, Mg, and B, respectively. Therefore, the equation between the reaction rate (α) and reaction time (t) can be written as Equations (13)–(15) for Fe, Al, and Mg, respectively.

$$1 - 2/3\alpha - (1-\alpha)^{2/3} = 0.0338\exp(-17100/RT)t \tag{14}$$

$$1 - 2/3\alpha - (1-\alpha)^{2/3} = 0.0392\exp(-17850/RT)t \tag{15}$$

$$1 - 2/3\alpha - (1-\alpha)^{2/3} = 0.0339\exp(-19790/RT)t \tag{16}$$

$$1 - 2/3\alpha - (1-\alpha)^{2/3} = 0.3265\exp(-29710/RT)t \tag{17}$$

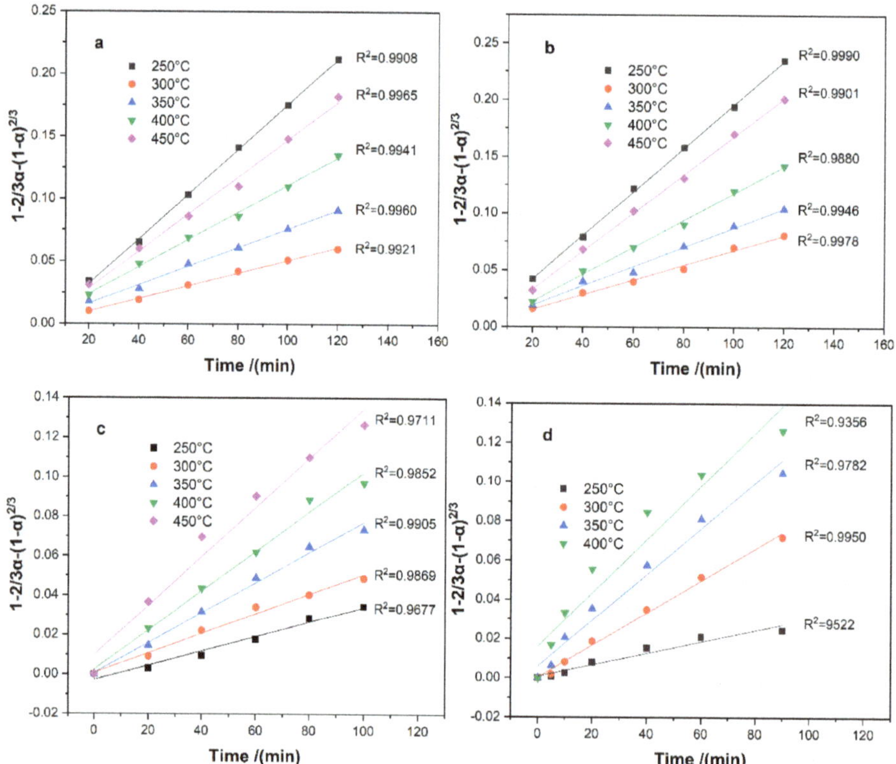

Figure 13. Plots of 1-2/3α-(1-α)$^{2/3}$ vs roasting time at different roasting temperatures: (**a**)-Fe; (**b**)-Al; (**c**)-Mg; and, (**d**)-B.

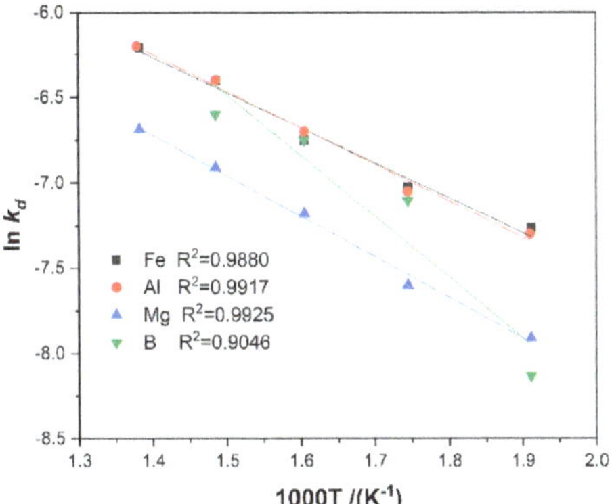

Figure 14. Arrhenius plot for roasting boron-bearing iron tailings with ammonium sulfate.

3.3. Leaching Residue Analysis

The leaching residue was obtained under the conditions that the roasting temperature of 450 °C, the roasting time of 120 min., the molar ratio of ammonium sulfate to tailings of 3:1 and the particle size of the tailings of less than 80 μm, the XRD patterns and the SEM image are shown in Figure 15. As seen, the main phases in the residue are ferric oxide serpentine ($Mg_3SiO_5(OH)_4$) talc ($Mg_3Si_4O_{10}(OH)_2$) and quartz (SiO_2). The leaching residue presents fluffy bulbous that reunite together.

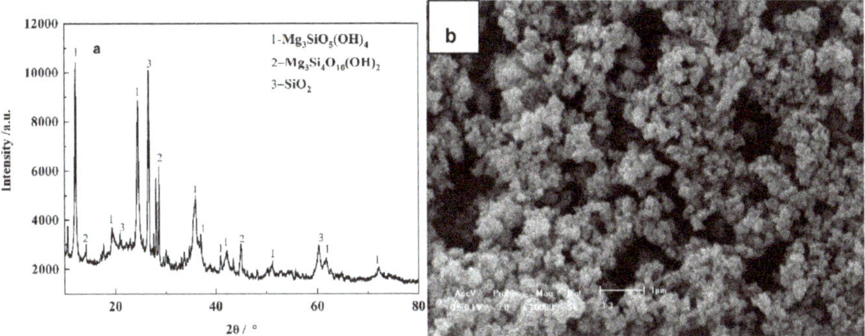

Figure 15. XRD patterns and SEM image of the leaching residue: a-XRD patterns; b-SEM image.

4. Conclusions

In this study, a clean process was proposed to process a large amount of boron bearing iron tailings in China. The results that were obtained in this study are as follows:

(1) The optimal sulfation roasting conditions using ammonium sulfate for boron-bearing iron tailings are that the roasting temperature is 450 °C, the roasting time is 120 min., the molar ratio of ammonium sulfate to tailings is 3:1, and the particle size is less than 80 μm. These conditions yield more than 98% extraction of Fe and Al, and more than 80% extraction of Mg and B. These findings were obtained from both single factor and orthogonal experiments.

(2) The sulfation reactions between ammonium sulfate and the tailings can be divided into two steps, the reactions between ammonium sulfate and the tailings at 240–359 °C, and the reactions between ammonium bisulfate at 359–448 °C. The mineral phase transformation in the roasting process can be described by the sequence of mineral phases→ammonium metal sulfates→metal sulfates.

(3)   The kinetics analysis indicates that internal diffusion controls the sulfation reactions of metals. The apparent activation energies of the reactions are 17.10 kJ·mol^{-1}, 17.85 kJ·mol^{-1}, 19.79 kJ·mol^{-1}, and 29.71 kJ·mol^{-1} for Fe, Al, Mg, and B, respectively.

The proposed process can comprehensively utilize the elements in the born-bearing iron tailings at low temperatures and with no additional waste slag being generated, which could be given a guide to process similar waste solid slags.

Author Contributions: X.L. performed the experiment, analysis of the data and contributed to writing the draft paper; F.C. contributed to the aspects related to the processing figures, kinetic analysis, and writing the paper; Z.N. performed the generating the results and microstructural work; M.L.F. performed the language and the quality of the paper; Y.Z. provided the fund and the experimental design.

Funding: This research received no external funding.

Conflicts of Interest: The authors declare no conflict of interest.

References

1.   Jiang, S.Y. Boron isotope geochemistry of hydrothermal ore deposits in China: A preliminary study. *Phys. Chem. Earth Part A Sol. Earth Geod.* **2001**, *26*, 851–858. [CrossRef]
2.   Kistler, R.B.; Helvaci, C. Boron and borates. *Ind. Miner. Rocks.* **1994**, *6*, 171–186.
3.   Kar, Y.; Şen, N.; Demirbaş, A. Boron minerals in Turkey, their application areas and importance for the country's economy. *Miner. Energy Raw Mater. Rep.* **2006**, *20*, 2–10. [CrossRef]
4.   Qiao, X.; Li, W.; Zhang, L.; White, N.C.; Zhang, F.; Yao, Z. Chemical and boron isotope compositions of tourmaline in the Hadamiao porphyry gold deposit, Inner Mongolia, China. *Chem. Geol.* **2019**, *519*, 39–55. [CrossRef]
5.   Li, G.; Liang, B.; Rao, M.; Zhang, Y.; Jiang, T. An innovative process for extracting boron and simultaneous recovering metallic iron from ludwigite ore. *Miner. Eng.* **2014**, *56*, 57–60. [CrossRef]
6.   Ding, Y.G.; Wang, J.S.; Wang, G.; Ma, S.; Xue, Q.G. Comprehensive utilization of paigeite ore using iron nugget making process. *J. Iron Steel Res. Int.* **2012**, *19*, 9–13. [CrossRef]
7.   Wang, G.; Ding, Y.G.; Wang, J.S.; She, X.F.; Xue, Q.G. Effect of carbon species on the reduction and melting behavior of boron-bearing iron concentrate/carbon composite pellets. *Int. J. Miner. Metallurg. Mater.* **2013**, *20*, 522–528. [CrossRef]
8.   Jie, L.; Fan, Z.G.; Liu, Y.L.; Liu, S.L.; Jiang, T.; Xi, Z.P. Preparation of boric acid from low-grade ascharite and recovery of magnesium sulfate. *Tran. Nonferrous Met. Soc. China* **2010**, *20*, 1161–1165.
9.   Sivrikaya, O.; Arol, A.I. Use of boron compounds as binders in iron ore pelletization. *Open Min. Process. J.* **2010**, *3*, 25–35. [CrossRef]
10.   Ucbeyiay, H.; Ozkan, A. Two-stage shear flocculation for enrichment of fine boron ore containing colemanite. *Sep. Purif. Technol.* **2014**, *132*, 302–308. [CrossRef]
11.   Celik, M.; Yasar, E. Effect of temperature and impurities on electrostatic separation of boron minerals. *Miner. Eng.* **1995**, *8*, 829–833. [CrossRef]
12.   Boncukcuoğlu, R.; Kocakerim, M.M.; Kocadağistan, E.; Yilmaz, M.T. Recovery of boron of the sieve reject in the production of borax. *Resour. Conserv. Recycl.* **2003**, *137*, 147–157. [CrossRef]
13.   Guliyev, R.; Kuşlu, S.; Çalban, T.; Çolak, S. Leaching kinetics of colemanite in potassium hydrogen sulphate solutions. *J. Ind. Eng. Chem.* **2012**, *18*, 38–44. [CrossRef]
14.   Liang, B.; Li, G.; Rao, M.; Peng, Z.; Zhang, Y.; Jiang, T. Water leaching of boron from soda-ash-activated ludwigite ore. *Hydrometallurgy* **2017**, *167*, 101–106. [CrossRef]
15.   Erdoğan, Y.; Aksu, M.; Demirbaş, A.; Abalı, Y. Analyses of boronic ores and sludges and solubilities of boron minerals in CO2-saturated water. *Resour. Conserve Recycle* **1998**, *24*, 275–283. [CrossRef]
16.   Kavcı, E.; Calban, T.; Colak, S.; Kuşlu, S. Leaching kinetics of ulexite in sodium hydrogen sulphate solutions. *J. Ind. Eng. Chem.* **2014**, *20*, 2625–2631. [CrossRef]
17.   Qin, S.; Yin, B.; Zhang, Y.; Zhang, Y. Leaching kinetics of szaibelyite ore in NaOH solution. *Hydrometallurgy* **2015**, *157*, 333–339. [CrossRef]
18.   Xu, Y.; Jiang, T.; Zhou, M.; Wen, J.; Chen, W.; Xue, X. Effects of mechanical activation on physicochemical properties and alkaline leaching of boron concentrate. *Hydrometallurgy* **2017**, *173*, 32–42. [CrossRef]

19. Xu, Y.; Jiang, T.; Wen, J.; Gao, H.; Wang, J.; Xue, X. Leaching kinetics of mechanically activated boron concentrate in a NaOH solution. *Hydrometallurgy* **2018**, *179*, 60–72. [CrossRef]
20. Liu, S.; Cui, C.; Zhang, X. Pyrometallurgical separation of boron from iron in ludwigite ore. *ISIJ Int.* **1998**, *38*, 1077–1079. [CrossRef]
21. Wang, G.; Xue, Q.; She, X.; Wang, J. Carbothermal reduction of boron-bearing iron concentrate and melting separation of the reduced pellet. *ISIJ Int.* **2015**, *55*, 751–757. [CrossRef]
22. Zhang, X.; Li, G.; You, J.; Wang, J.; Luo, J.; Duan, J.; Zhang, T.; Peng, Z.; Rao, M.; Jiang, T. Extraction of Boron from Ludwigite Ore: Mechanism of Soda—Ash Roasting of Lizardite and Szaibelyite. *Minerals* **2019**, *9*, 533. [CrossRef]
23. Gao, P.; Li, G.; Gu, X.; Han, Y. Reduction Kinetics and Microscopic Properties Transformation of Boron-Bearing Iron Concentrate—Carbon-Mixed Pellets. *Miner. Process. Extr. Metallurg. Rev.* **2019**, *40*, 1–9. [CrossRef]
24. Liu, Y.; Jiang, T.; Liu, C.; Huang, W.; Wang, J.; Xue, X. Effect of microwave pre-treatment on the magnetic properties of Ludwigite and its implications on magnetic separation. *Metallurg. Res. Technol.* **2019**, *116*, 107. [CrossRef]
25. Xu, Y.; Jiang, T.; Zhou, M.; Gao, H.; Liu, Y.; Xue, X. Surface properties changes during a two-stage mechanical activation and its influences on B2O3 activity of boron concentrate. *Miner. Eng.* **2019**, *131*, 1–7. [CrossRef]
26. Su, Q.; Zheng, S.; Li, H.; Hou, H. Boron resource and prospects of comprehensive utilization of boron mud as a resource in China. *Earth Sci. Front.* **2014**, *21*, 325–330.
27. Kasemann, S.; Erzinger, J.; Franz, G. Boron recycling in the continental crust of the central Andes from the Palaeozoic to Mesozoic, NW Argentina. *Contrib. Mineralog. Petrol.* **2000**, *140*, 328–343. [CrossRef]
28. Lyman, J.W.; Palmer, G.R. *Recycling of Neodymium Iron Boron Magnet Scrap*; US Department of the Interior, Bureau of Mines: Salt Lake City, UT, USA, 1993.
29. Allen, R.P.; Morgan, C.A. Boric Acid Process. U.S. Patent 3,953,580, 27 April 1976.
30. Cui, F.; Mu, W.; Wang, S.; Xin, H.; Xu, Q.; Zhai, Y.; Luo, S. Sodium sulfate activation mechanism on co-sulfating roasting to nickel-copper sulfide concentrate in metal extractions, microtopography and kinetics. *Miner. Eng.* **2018**, *123*, 104–116. [CrossRef]
31. Ginstling, A.M.; Brounshtein, B.I. Concerning the diffusion kinetics of reactions in spherical particles. *J. Appl. Chem. USSR* **1950**, *23*, 1327–1338.

Article

Determination of the Least Impactful Municipal Solid Waste Management Option in Harare, Zimbabwe

Trust Nhubu [1,*] **and Edison Muzenda** [1,2]

[1] Department of Chemical Engineering Technology, University of Johannesburgy, Johannesburg 2001, South Africa; emuzenda@uj.ac.za or muzendae@biust.ac.bw

[2] Department of Chemical, Materials and Metallurgical Engineering, Botswana International University of Science and Technology, Private Mail Bag 16, Palapye, Botswana

* Correspondence: 217078674@student.uj.ac.za or nhubutrust@gmail.com; Tel.: +26-37-7734-2295

Received: 12 September 2019; Accepted: 12 October 2019; Published: 1 November 2019

Abstract: Six municipal solid waste management (MSWM) options (A1–A6) in Harare were developed and analyzed for their global warming, acidification, eutrophication and human health impact potentials using life cycle assessment methodology to determine the least impactful option in Harare. Study findings will aid the development of future MSWM systems in Harare. A1 and A2 considered the landfilling and incineration, respectively, of indiscriminately collected MSW with energy recovery and byproduct treatment. Source-separated biodegradables were anaerobically treated with the remaining non-biodegradable fraction being incinerated in A3 and landfilled in A4. A5 and A6 had the same processes as in A3 and A4, respectively, except the inclusion of the recovery of 20% of the recoverable materials. The life cycle stages considered were collection and transportation, materials recovery, anaerobic digestion, landfilling and incineration. A5 emerged as the best option. Materials recovery contributed to impact potential reductions across the four impact categories. Sensitivity analysis revealed that doubling materials recovery and increasing it to 28% under A5 resulted in zero eutrophication and acidification, respectively. Increasing material recovery to 24% and 26% under A6 leads to zero acidification and eutrophication, respectively. Zero global warming and human health impacts under A6 are realised at 6% and 9% materials recovery levels, respectively.

Keywords: municipal solid waste management; life cycle assessment; life cycle impacts; life cycle stages; eutrophication; global warming; human health; acidification; Harare; Zimbabwe

---

1. Introduction

The annual global municipal solid waste (MSW) generation rate is projected to reach 2.2 billion metric tons per annum by 2025 from 1.3 billion metric tons per annum in 2012 [1]. Member countries of the Organization for Economic Co-operation and Development (OECD) however, are reporting a reduction in MSW generation [2]. Dramatic population increase in urban areas within Africa and Asia was singled out by the United Nations [3] as a typical phenomenon that leads to the astronomical increase in MSW generation. Standards of living, rapid urbanization, ever increasing population and obtaining economic environments in a given locality were cited as some of the factors that influence MSW generation [4–7]. Dongquing et al. [8] also cited the type and abundance of a region's natural resources apart from the above mentioned factors as a factor that influences MSW generation.

The best way to identify and manage solid waste streams is the fundamental environmental issue globally, both in industrialised and developing nations [9]. Global initiatives are supporting the prioritization of solid waste management (SWM) because it is viewed an important facet for the sustainable development of any country [10]. Sustainable development is the reduction of ecological footprints while improving quality of life for current and future generations within the earth's capacity limit [11]. UNDESA [12] Agenda 21 of the Rio Declaration on Environment and Development affirmed the need for environmentally friendly waste management since it is an environmental issue of major concern in maintaining the quality of the earth's environment.

1.1. Solid Waste Management Dynamics in Developed Nations

Solid waste (SW) mass production characterised human life since the formation of non-nomadic communities around 10,000 BC [13]. Seadon [14] argued that small communities could bury the SW they generated in environments surrounding their settlements or dispose it in rivers, which could not prevent the wide spread of diseases or foul odours from accumulated SW and filth emanating from increased population densities that characterised the formation of non-nomadic communities. Exceptional cases on waste management existed Worrell and Vesilind [13] reported that by 200 BC, organised (SWM) systems had been under implementation in Mohenjo–Daro, an ancient Indus Valley metropolis, and the Chinese had established disposal police to enforce waste disposal laws. Melosi [15] also reported that by 500 BC, the Greeks had issued a decree that banned the disposal of waste in streets and organised first accepted MSW dumps in the Western world.

Middle Ages' city streets were characterised by odorous mud with stagnant water, soil, household waste and excreta from both humans and animals creating favorable conditions for disease vectors [16]. Therefore, the disposal of biodegradable or organic waste in streets is argued to have partly contributed to the Black Death of the 1300s that occurred in Europe [13,16,17]. Developments in SWM in developed nations were and are initiated to address environmental, land use, natural resources depletion, human health, climate change, waste value, aesthetic, economic, public information and participation issues associated with improper waste disposal [13,16,18–20]. SWM has evolved in developed nations driven by historical forces and mechanisms which can possibly inform the development of SWM strategies in developing nations [20]. Marshall and Farahbakhsh [21] noted five drivers for integrated SWM paradigm in developed nations, namely the environment, climate change, resource scarcity, public health and public awareness and participation.

Public health concerns remain a driver of SWM transformation in the developed world characterised with continued review of public health legislation. The need to reduce land, air and water contamination [20,22] was a primary driver of policy changes in SWM development in the 1970s and beyond [20]. Waste control characterised the SWM policy framework between the 1970s and mid-1980s focusing on daily landfill compacting and covering and incinerator retrofitting for dust control. The SWM policies enacted from the 1980s to date focus on increasing technical standards, starting with control of landfill leachate and gas, reduction of incinerator flue gas and dioxin and the current span covering control of odour at composting and anaerobic digestion (AD) facilities [20]. The last decade of the 20th century saw the increased focus and attention towards the adoption of integrative policy due to the inadequacies of advocating for continued increase in environmental protection only from both the technical (engineering and scientific) and environmental perspective without considering the political, economic, social, cultural and institutional dimensions of SWM [20,23,24]. The waste hierarchy upon which the European Union (EU) current policy on waste is based reignited materials recycling and reuse of the 19th century in the 1970s [20,25] in light of the increasing scarcity of resources. The EU's Second Environment Action Programme of 1977 introduced the waste hierarchy model for SWM priorities derived from the "Ladder of Lansik" [26].

Climate change has also driven the development of SWM from the early 1990s to address greenhouse gas (GHG) emissions from biodegradable waste landfilling, a major contributor of methane gas emissions, complimented with a strong focus on the recovery of energy from SW [20,22]. The concerns by the public on poor SWM practices with their increased awareness have also contributed in driving the developments in SWM [20]. The public became concerned with the location of SWM facilities in the vicinity of their households, 'not in my backyard' (NIMBY), though they appreciate the need of SWM facilities. Therefore, effective communication, wide public knowledge of SWM needs, the active engagement of all stakeholders during the entire SWM cycle have been successful in overcoming

NIMBY public behavior and opposition to numerous developmental projects [27] thereby acting as drivers for developments in SWM [21].

1.2. Solid Waste Management Dynamics in Developing Nations

Despite the increase in waste generation, global call and acceptance that waste management must take an integrated approach to derive economic benefits while reducing environmental burdens, Africa is still lagging in this regard. This lag is also being witnessed despite the reported increased globalization as poor SWM challenges and their associated public health impacts are affecting urban environments in many developing nations [22,28,29] one and a half centuries after the sanitary revolution in the EU [30]. Unlike developed nations that are concerned with diseases associated with affluence (cancer, cardiovascular disease, alcohol and drug abuse), poor SWM derived public health impacts in developing nations are evidently manifesting in the form of communicable diseases giving the double headache of dealing with both communicable diseases and emerging diseases of affluence [30]. Public health mostly drives SWM development in developing nations, though other factors as in developed nations are considered because the key priority is waste collection and removal from population centres as it was in European and American cities before the 1960s [20,31–33]. Wilson [20] noted that environmental protection remained relatively low on the SWM priorities despite the presence of legislation prohibiting unregulated waste disposal with minor changes towards its prioritization taking place. The value of waste as a resource is also another vital driver within developing nations currently providing livelihoods to the urban poor through informal recycling [20,22]. Climate change is a significant driver globally with a number of nations having incorporated the municipal solid waste (MSW) sector amongst the sectors considered for low-emission development strategies (LEDs) on the national emission reduction commitments or targets within the nationally determined contributions (NDCs) framework of the Paris agreement under the United Nations Framework on Climate Change Convention (UNFCCC).

A number of similarities do exist between the current conditions characterizing many cities in developing nations and those experienced in European and American cities during the 19th century with regards to increased urbanisation levels, degraded sanitary environment emanating from lack of adequate sanitation and environmental services, inequalities and social exclusions in SWM systems, unprecedented mortality and morbidity levels due to inadequate sanitation, potable water supply and waste disposal services [30]. Thus, developing nations are likely to go through almost similar SWM development pathways as those developed nations went through. However, Marshall and Farahbakhsh [21] argued that despite these similarities, complex local-level-specific technical, political, social, economic and environmental challenges in developing nations have been created from rapid urbanization, increasing population, the fight for economic growth, institutional, governance and authority issues, international influences, along with their interaction with diverse economic, cultural, political and social dynamics which are bringing associated SWM complexities in developing nations.

In developing countries therefore, SWM is complicated by levels of urbanization, economic growth and inequality as well as socio-economic dimensions, governance, policy and institutional issues coupled with international interferences [21] which limit the application of SWM approaches that succeeded in SWM development pathways for developed nations. The understanding of the origins and critical drivers in the past developments in SWM in developed nations provides contextual knowledge on the current changes occurring in developing nations. Simelane and Mohee [34] identified African social norms with their associated concerns including economic and environmental issues, national and regional legislative deficiencies, technological and human resources developments and historical influences among other factors necessitating this lag. Iriruaga [35], on another note, cited low private investment in infrastructure, industry linkages and academic research as the drivers of Africa's inability to effectively derive benefits from the waste it generates. Muzenda et al. [36] identified the increased demand for SWM provision, MSW minimization, and recovery of materials for reuse and recycle, constraining factors including physical, land use and environmental constraints, as well as demographic and socio-economic factors as the core drivers for the need of integrated waste management (IWM) techniques.

MSW generation and its disposal are causing enormous environmental and human health challenges in urban environments of developing countries [37–39]. It is considered hazardous and to have toxic impacts on the biological environment, thereby affecting lifestyles and economic activities [40]. This, therefore, calls for the need to sustainably manage waste to reduce its impact in the ecosystem and human health [41]. The need to design and develop integrated waste management (IWM) options that seeks to meet the economic, technical, environmental and social constraints of products or production processes has become paramount and urgent. McDougall et al. [42] defined IWM as a combination of technically sound, economically feasible, environmentally sustainable and socially acceptable collection and treatment processes that handle materials constituting MSW.

1.3. Municipal Solid Waste Management in Zimbabwe

Like many developing countries facing enormous MSW generation and disposal associated environmental and human health challenges in urban environments, the Government of Zimbabwe acknowledged that its urban local authorities (city municipalities, town councils, district councils and local boards) are experiencing major challenges in managing MSW due to rapid population growth. Most of Zimbabwe's local authorities fail to cope with the ever increasing volumes of waste being generated by the public [43]. Several studies have also affirmed that municipal solid waste management (MSWM) is one of the greatest challenges facing urban environments in Zimbabwe [41,44–53]. In Zimbabwe, about 60% of the MSW generated in urban environments is disposed at official dumpsites with the remaining waste being dumped illegally in undesignated areas namely storm water drains, open spaces, alleys and road verges [45]. The dumping of waste in open and illegal dumpsites is not only an eyesore but creates an environment where disease causing vectors can thrive, contribute to air, soil and water pollution and emit greenhouse gases that cause global warming [43].

MSW problems in Harare specifically are evidently manifesting in the form of both surface and groundwater pollution due to the dumping of MSW in waterways and untreated leachate from dumpsites. The storage capacity of the sole official MSW dumpsite in Harare is expected to reach its limit in the next five years [54]. This calls for the need to redefine future MSWM options as well as redefining the models of operating the MSWM facilities considering biogas recovery for electricity generation as well as the production of saleable products from MSW. To date no or few studies have been carried out focusing on determining the most probable integrated MSWM option with the least environmental impacts for Harare. Such study results could possibly inform future decisions and policies on MSWM considering the increasing population, changing lifestyles, global pressure for the need for sustainable cities, the impacts the current MSWM practices have on both the environment and human health as well as the imminent closure of the existing dumpsite whose service life is anticipated to come to an end in 2020. This study, therefore, is a life cycle-based comparative assessment of the various probable MSWM scenarios to be implemented in Harare. The study seeks to identify the scenario with the least burden with regards to human health, acidification, eutrophication and global warming impact categories.

1.4. Life Cycle Assessment

Life cycle assessment (LCA) is a tool that could be used in the design and development of IWM options. LCA holistically quantifies the environmental burdens and impacts for entire products' or processes' life cycles [55]. Winkler and Bilitewski [56] described LCA as a science-based impact assessment methodology for the impacts of a product or system on the environment, which is not purely a scientific tool. LCA application in sustainable MSWM started over two decades ago, as argued by Güereca et al. [57] that it has been applied for MSWM since 1995. The use of LCA for decision making and strategy development in MSWM systems has expanded rapidly over the recent past years as a tool with the capacity to capture and address complexities and interdependencies characterizing modern IWM systems [58]. Mendes et al. [59] noted the appropriateness of LCA application as a tool for decision making and strategy development in MSWM because of the associated wide differences in spatial locations, waste composition and characteristics, sources of energy, waste disposal options available as well as available nature and size of products from various waste treatment methods. Therefore LCA has emerged as an appropriate holistic method increasingly being applied in MSWM decision making and strategy development processes [60].

LCA has been previously applied to assess the associated impacts of MSWM systems thereby assisting in comparing alternative MSWM systems and/or identifying areas of major concerns that need potential improvements [61]. It has been applied to identify and probe likely negative impacts of various MSWM practices [62] because it is capable of calculating and comparing impacts of different MSWM scenarios [63]. It incorporates environmental impact weighing or valuation to estimate the performance of a specific MSWM scenario [62]. The intensification of MSWM policies in Europe and global call for the implementation of LCA methodology ISO 14044: 2006 standards have resulted in a positive trend towards the adoption of life cycle studies on MSWM [64]. To date, numerous studies have been undertaken worldwide applying LCA to the different MSW life cycle stages that cover the entire life cycle of MSW [60,62–66]. Khandelwal et al. [64] reviewed 153 studies that applied LCA on MSWM, undertaken globally and published between 2013 and 2018. The distribution of the selected LCA studies reviewed by continents showed that 72 were in Asia, 53 in Europe, 10 in North America, 9 in South America, 3 in America, 2 in Africa, 2 addressed generic cities assuming MSW generation, characteristics and associated environmental emissions together with other remaining studies that focused on at least one country. Very few life cycle studies on MSWM were found in Africa and poor LCA methodology penetration in Africa was cited as the cause of the limited LCA studies on MSW. The only two LCA studies found for Africa were done in Nigeria.

2. Materials and Methods

2.1. Description of the Study Area

The study area comprises of Harare (the capital city of Zimbabwe), Chitungwiza, Norton, Ruwa and Epworth local boards with an estimated total population of 2,133,802 people, as shown in Table 1. Harare urban, Chitungwiza and Epworth local boards are located within Harare metropolitan province while Ruwa and Norton local boards are located in Mashonaland East and West respectively, as illustrated in Figure 1. An estimated 60% of the MSW generated in the study area is indiscriminately disposed at official dumpsites, except for Norton, whose MSW is disposed in an engineered sanitary landfill. The remaining 40% of the MSW is illegally dumped in undesignated areas, namely storm water drains, open spaces, alleys and road verges [45]. The capacity of the sole official dumpsite for Harare city, Pomona dumpsite, which covers an estimated area of 100 hectares having been operational since 1985 is expected to be exhausted by 2020 [54].

Table 1. Population figures for the study area [67].

Town/Local Board	Estimated Population (Male)	Estimated Population (Female)	Estimated Total Population
Harare	716,595	768,636	1,485,231
Chitungwiza	168,600	188,240	356,840
Norton	32,382	35,209	67,591
Ruwa	26,745	29,933	56,678
Epworth	83,983	83,479	167,462
Total	**1,028,305**	**1,105,497**	**2,133,802**

One unique feature of the study area is that it sits on the water catchment that drains into water reservoirs (Lake Chivero and Manyame) that supply the study area with potable water as shown in Figure 1. MSW problems in the study area are evidently manifesting in the form of both surface and groundwater pollution. Lake Chivero has been reported to have reached super eutrophic levels partly due to the deposition of MSW which constitutes in excess of biodegradable waste laden in runoff. The underground water in the study area has also been reported to have been compromised from untreated leachate from dumpsites [52].

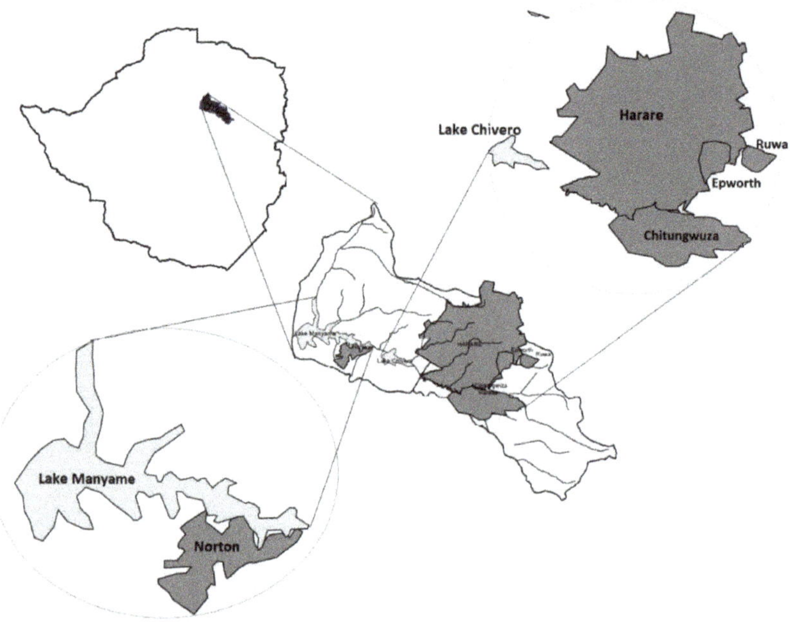

Figure 1. Location of the study area.

2.2. Definition of MSW

Definitions of MSW vary within countries and between countries and regions making it difficult and confusing to estimate MSW generation in various countries [68]. The variations in definitions bring along challenges and difficulties in LCA studies. Therefore for the purpose of this study, MSW is regarded as the waste that is managed by or on behalf of municipalities as a public service [69] comprising waste generated at households, offices, supermarkets and restaurants. Consequently, in Zimbabwe, local authorities are mandated to manage such MSW [70].

2.3. Quantity of MSW Generated in Harare and Its Dormitory Towns

The MSW annual generation for a given locality, communities, cities or countries, is a core indicator of the pressure exerted by MSW on the environment. It is useful for LCA when the annual generation of MSW is considered the functional unit. Obtaining reliable data on estimates and characteristics of MSW generated in developing countries is a challenge due to incomplete data, lack of equipment like weighbridges, rural to urban migration and low efficiency rates of MSW collection [71]. The development of initiatives that derive benefits from the promotion of sustainable use and management of MSW is hindered by the low availability and quality of data regarding MSW generation and management [34]. In Harare, Zimbabwe, quality and reliable MSW data on waste generation, characteristics and composition necessary for LCA that could inform effective planning for sustainable MSWM are unavailable. In addition the unreliable data available are only from official records of MSW collected and delivered at the official dump site. This MSW data does not capture much of the MSW managed outside the dumpsite management process that would have been generated at various sources [72]. Afon [72] further observed enormous variations of MSW generation on temporal scales (weekday, week of month and month of year) across localities highlighting the need for longitudinal collection of MSW generation data measurements over a year at sampled households according to their life styles and levels of income if resources and time permit to acquire reliable MSW data. In this study literature data was considered for the estimation of the annual MSW generation for Harare using literature reported per capita MSW generation figures.

Therefore the average per capita MSW generation rate of 0.6 kg/capita/day (0.5 kg/capita/day [52,73,74], 0.65 kg/capita/day [1,75] and 0.7 kg/capita/day [76]) was considered for the study. The average figure of 0.6 kg/capita per day, though slightly on the higher side of observed figures of 0.42 ± 0.15 kg/capita/day MSW generation in Zimbabwe by Muchandiona et al. [52], it is a reasonable estimate when considering other reported figures from literature. Miezah et al. [77] reported Ghana's daily MSW generation of 12,710 tons considering a daily per capita waste generation rate of 0.47 kg and a population of 27,043,093. Harare and its dormitory towns have a population of 2,133,802 translating to daily and annual MSW generation of 1280 and 467,303 tons, respectively, as shown in Table 2. Due to uncertainties on population data serviced with MSW collection, MSW data normalisation was assumed to have been enabled in the calculations of the per capita waste generation rate datasets that were used to calculate the daily average per capita (0.6 kg) MSW generation for this study to factor in the effects of population changes as proposed by the European Environment Agency [78].

Table 2. Estimates of daily and annual municipal solid waste (MSW) generation.

Town/Local Board	Estimated Population (People) [67]	Estimated Daily MSW Generation (tons)	Estimated Annual MSW Generation (tons)
Harare	1,485,231	891	325,266
Chitungwiza	356,840	214	78,148
Norton	67,591	41	14,802
Ruwa	56,678	34	12,412
Epworth	167,462	100	36,674
Total	**2,133,802**	**1280**	**467,303**

2.4. Composition of MSW Generated in Harare and Its Dormitory Towns

The composition of MSW is a vital aspect in MSW management as it is necessary for examining sustainable options for MSW reduction, recovery (reuse and recycle) as well for identifying the most appropriate and sustainable treatment and disposal method [79]. Hoornweg and Bhada-Tata [1] observed that organic waste fraction of MSW in developing countries constitutes a much larger fraction as compared to developed countries. However, like MSW generation data, reliable MSW composition data are absent in the study area. Estimates of averages from the Environmental Management Agency and notable literature studies conducted in Harare, Bulawayo and Chinhoyi were considered for the study as illustrated in Table 3.

Table 3. Estimates of percentage composition of MSW generated in Harare.

MSW Fraction	Percentage MSW Composition from Literature Studies Reviewed			Average % Composition
	Harare [80–84]	Bulawayo [85]	Chinhoyi [49]	
Organic	40	40	45	42
Plastic	26	50	24	33
Metals	7	3	13	8
Paper	15	7	14	14
Glass	2	0	3	3
Other	10	0	1	0

2.5. Integrated MSW Management Options and Treatment Processes/Life Cycle Stages

The integrated MSWM options and their associated processes or life cycle stages are described in the sections below and summarised in Table 4. The transportation system considered is the municipal waste collection service by municipal waste collection trucks [86]. Transport figures for waste collection were derived from the product of annual MSW generation for Harare and the estimated average distance the waste will be transported to the MSW management facility giving 21,028,500 t·km (product of distance to be travelled by the MSW to the treatment facility and the weight of the MSW transported) as waste collection trucks were estimated to travel an average distance of 90 km to and from the MSW management facility. The return trip was modelled only for an empty waste collection truck for the 22,252 trips of 45 km distance each carried out annually.

Table 4. Description of considered MSW management options.

MSW Management Scenario	Description of the Life Cycle Stages Considered for the MSW Management Options
A1	This option involves the disposal of 467,303 metric tons of MSW that would have been indiscriminately collected (both organic and nonorganic municipal solid waste) without any prior treatment in a landfill, recovering biogas energy and treating landfill leachate.
A2	The 467,303 metric tons of indiscriminately collected MSW undergoes incineration with recovery of energy and the treatment of the gaseous emissions and leachate produced during bottom ash recovery.
A3	Organic fraction of MSW generated amounting to 196,167 metric tons is anaerobically treated to produce biogas. The remaining 271,036 metric tons of mixed bag MSW (154,210 metric tons of plastics, 37,384 metric tons of metals, 65,422 metric tons of paper and 14,019 metric tons of glass) undergo incineration as in A2.
A4	Organic fraction of MSW generated amounting to 196,167 metric tons is anaerobically treated to produce biogas. The remaining 271,036 metric tons of mixed bag MSW (154,210 metric tons of plastics, 37,384 metric tons of metals, 65,422 metric tons of paper and 14,019 metric tons of glass) is landfilled as in A1.
A5	Organic fraction of MSW generated amounting to 196,167 metric tons is anaerobically treated to produce biogas. The 20% of the nonorganic waste amounting to 54,207 metric tons of MSW (30,842 metric tons of plastics, 7477 metric tons of metals, 13,084 metric tons of paper and 2804 metric tons of glass) are recovered for reuse and recycle in the mixed bag sorting plant. 216,829 metric tons of mixed bag MSW (123,368 metric tons of plastics, 17,346 metric tons of metals, 30,356 metric tons of paper and 15,178 metric tons of glass) which is not recovered in the mixed bag sorting plant undergoes incineration as in A2.
A6	Organic fraction of MSW generated amounting to 196,167 metric tons is anaerobically treated to produce biogas. The 20% of the nonorganic waste amounting to 54,207 metric tons of MSW (30,842 metric tons of plastics, 7477 metric tons of metals, 13,084 metric tons of paper and 2804 metric tons of glass) are recovered for reuse and recycle in the mixed bag sorting plant. 216,829 metric tons of mixed bag MSW (123,368 metric tons of plastics, 17,346 metric tons of metals, 30,356 metric tons of paper and 15,178 metric tons of glass) which is not recovered in the mixed bag sorting plant undergoes landfilling as in A1.

The recovery of the recoverable materials considered a mixed bag sorting plant equipped with relevant filters to treat waste gases produced during the recovery of the recoverable materials. The materials considered for recovery are metals, paper, plastics and glass at a recovery rate of 20% of their annual estimated generation. The anaerobic digestion plant considered the anaerobic digestion of the estimated biodegradable fraction of MSW amounting to 196,166 metric tons that is generated annually in Harare and its dormitory towns to produce biogas at an estimated average production rate of 115 m^3/metric ton [87–90]. The biogas produced will be burnt to produce electrical and heat energy. The digestate or solid residue from the anaerobic digestion process will undergo a compositing process to obtain quality compost for sale as a biological fertilizer or soil enhancer. Gases from the anaerobic digestion process will undergo bio-filtration before being scrubbed or washed with sulphiric acid to produce a generally acceptable leachate that is assumed or considered decontaminated. The mixed bag fraction that reaches the incineration plant will be combusted in a furnace. Combustion engines transform the flue gases from the furnace into electrical energy. Combustion furnace bottom ash will be used in road construction as aggregates prior to its treatment with physical chemical treatment methods applied to treat the leachate produced during bottom ash recovery. Gaseous emissions from the combustion furnace are treated using appropriate methods such as lime based dry adsorption, bag house filtration, activated carbon-based adsorption and selective noncatalytic reduction. Mixed bag MSW is landfilled with energy recovery. The landfill leachate undergoes nitrification–denitrification process under pressure. Ultrafiltration is used to separate the sludge from the leachate. The treated leachate is sent to a wastewater treatment plant. The transportation of treated leachate from the landfill to the wastewater treatment plant is considered negligible.

2.6. Life Cycle Assessment

LCA was used to estimate and compare the potential acidification, eutrophication, global warming and human health impacts of the various six MSW management scenarios. ISO 14040 standards [91] were the basis for the LCA study. Several studies have been carried out using LCA to assess different MSW management scenarios in a number of countries, namely Spain [57,92,93], Italy [58,94,95], China [96,97], Brazil [59], Australia [98], Indonesia [99], Canada [100], United States of America [101], Lithuania [102] and Nigeria [103,104] to mention just but a few. LCA was therefore applied to assess the human health, acidification, eutrophication and global warming potential of the various MSW management scenarios in Harare and its dormitory towns of Chitungwiza, Epworth, Norton and Ruwa.

2.6.1. Goal and Scope

LCA was performed to assess the acidification, eutrophication, global warming and human health impact potentials of the proposed six MSWM scenarios that could be implemented in Harare and its dormitory towns. The LCA results could possibly inform decisions for future MSWM in Harare and its dormitory towns considering the increasing population, lifestyles, global pressure for the need for sustainable cities, the impacts the current MSWM option has on both the environment and human health as well as the imminent closure of Pomona dumpsite whose capacity will be exhausted by 2020 [54].

2.6.2. The LCA System Boundaries

The processes that fall under the scope of the study are within the MSWM system boundary as denoted by the dotted line on Figure 2. The entire management processes of all MSW which is not managed by or on behalf of municipalities fall outside the system boundary and study scope. Associated impacts from emissions emanating from the construction of MSWM facilities were assumed negligible compared to those produced from the actual operation of the facilities, hence they were not considered under the study as noted by Mendes et al. [59].

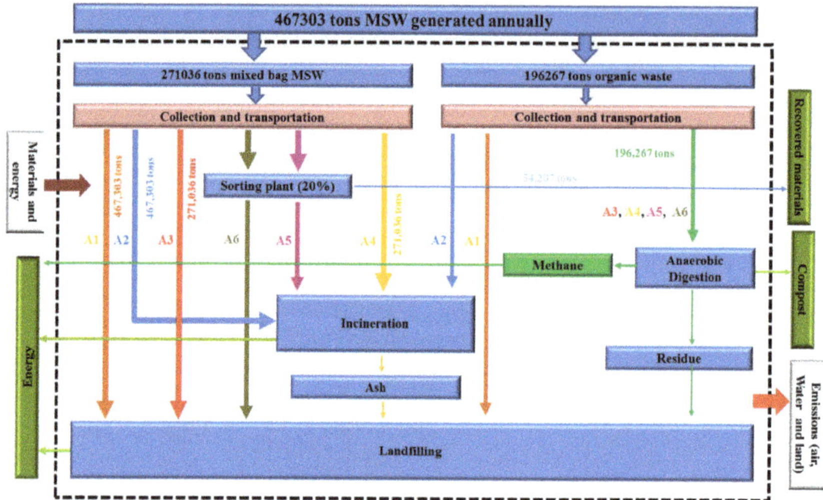

Figure 2. Life cycle assessment (LCA) system boundaries.

2.6.3. LCA Functional Unit and Software Model

The annual MSW generation for Harare and its dormitory towns of 467,303 tons was considered the functional unit for LCA. Quite a number of studies applied the annual MSW generation as the functional unit [92,95,105]. SimaPro software Version 8.5.2 analyst and its associated database update852 produced by Pre'Sustainability consultants B.V in Amersfoort, Netherlands were used to undertake the LCA. The impacts loads associated with the materials and processes were gathered from the Ecoinvent 3 database (2018) [106]. The detailed input–output pathways for the LCA are as shown in Figure 3.

The anaerobic digestion project database modelled for the rest of the world found on the processing, waste, biowaste and transformation pathway was utilised for the LCA with 2.26×10^7 m^3 of biogas produced annually being the inputs for MSWM options A3, A4, A5 and A6 where AD was incorporated. Alternatively the AD project database modelled for the rest of the world on processes, waste treatment, waste, transformation and finally biowaste pathway can also be used if the amount of biowaste to be digested is used as input. For waste incineration, the respective individual waste types i.e., metals, glass, paper, biodegradable and plastics that constituted MSW were modelled using their corresponding project databases modelled for the rest of the world on the product selection pathway processes, waste treatment, waste, transformation, incineration then finally municipal incineration with the specific MSW fraction quantities provided in Table 4 under MSWM options A2, A3 and A5 being the inputs. The reason being that Ecoinvent MSW incineration database modelled for the rest of the world is only recommended to be used for MSW with an average of 92.8% burnable fraction which is not a characteristic of the MSW generated in Harare; MSW generated in Harare has a combustible fraction of just over 75%, as reported by Makarichi et al. [81]. The MSW fraction-specific Ecoinvent database modelled for the rest of the world on the processes, waste treatment, waste, transformation, landfilling and then finally sanitary landfilling pathway was used for landfilling with the waste-specific quantities provided in Table 4 for the scenarios that incorporated landfilling being used as model inputs. Waste collection and transportation average distance of 45 km was considered giving a total of 2.10×10^7 t·km input on the Ecoinvent transport model for the rest of the world on the processes, transport, road and transformation pathway for all the MSWM options.

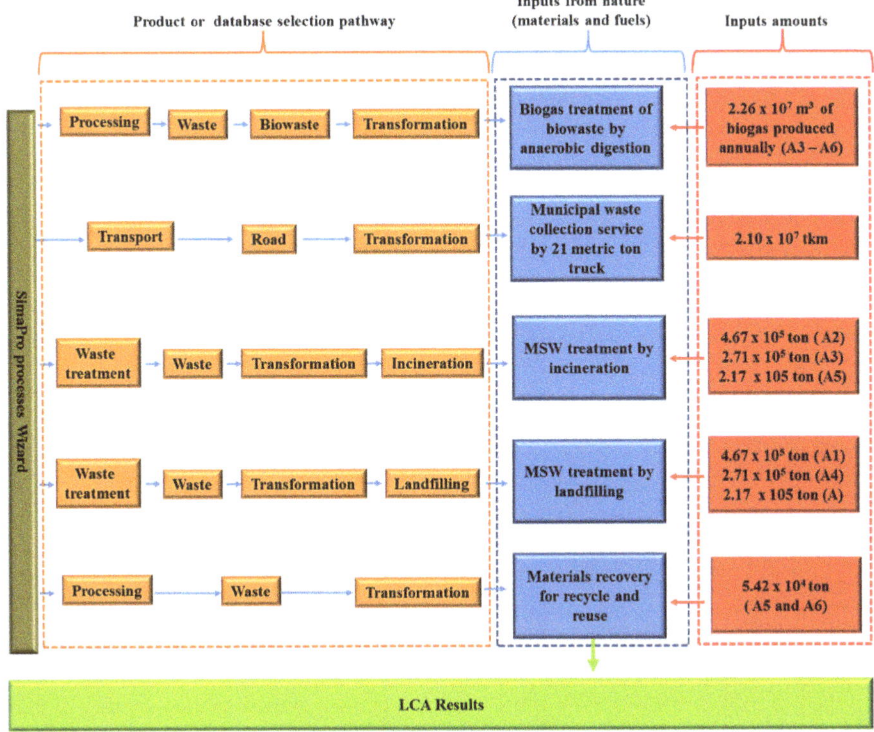

Figure 3. LCA methodological framework.

2.6.4. Life Cycle Impact Assessment (LCIA) Method

The LCIA for all the processes under the MSWM scenarios was undertaken using the ReCiPe 2016 v1.02 endpoint method, Hierarchist version, which is the default ReCiPe endpoint method. ReCiPe 2016 method is a new version of ReCiPe 2008 that was created by RIVM, Radboud University, Norwegian University of Science and Technology and PRé Consultants [107,108]. The method has 22 defined endpoint impact categories which are grouped into three damage categories, namely human health, ecosystems and resources. ReCiPe2016 has characterization factors that are globally representative rather than being representative only for Europe while at the same time providing the possible implementation of characterisation factors at national or continental scale for a handful of impact categories. The choices of values used in deriving characterisation factors and the midpoint characterization factors are provided by Huijbregts et al. [107] with the endpoint characterisation factors directly derived from the midpoint characterisation factors according to Equation (1). Therefore, constant global midpoint to endpoint characterisation factors were determined for all the impact categories save for fossil resource scarcity due to limited cause–effect pathway knowledge. The derivation of individual impact category midpoint to endpoint characterisation factors is provided [107,108].

$$CFe_{x,c,a} = CFm_{x,c} \times F_{M\to,E,c,a'} \tag{1}$$

where; CFe and CFm are the end and midpoint characterisation factors respectively, c is the cultural perspective; a is the area of protection, namely human health, freshwater ecosystems, marine ecosystems, terrestrial ecosystems or resource scarcity; x is the stressor of concern; and $F_{M\to,E,c,a}$ is the conversion factor from midpoint to endpoint impact for c and a.

3. Results

Figure 3 shows the LCIA results for the acidification, eutrophication, global warming and human health impact potentials for the six MSW management options under consideration. All the MSW management options under consideration lead to a reduction in global warming and human health endpoint impact categories. Detailed results for the endpoint impact categories for acidification, eutrophication, global warming and human health are presented below.

3.1. Acidification

Figures 4–7 show that MSW management options A1, A5 and A6 lead to reduction in acidification while A2, A3 and A4 contribute to increased acidification. The acidification impact potential is measured using the species extinction rates (species-years). A6 contributes the highest reduction in acidification potential of -3.9×10^{-2} species-years, followed by A5 with an acidification potential reduction of -2.97×10^{-2} species-years. Results show that A1 contributes the least acidification potential of -8.94×10^{-3} species-years, which is consistent with findings by Mendes el al. that landfilling with gas recovery and leachate treatment leads to reduced acidification impacts. The recovery of metals plays a crucial role in reducing the eutrophication impacts under A5 and A6 as observed by Beigl and Salhofer [105]. A2 leads to the greatest acidification potential of 4.13×10^{-2} species-years, with A3 giving an acidification increase of 2.48×10^{-2} species-years. A4 leads to the least increase in acidification of 8.57×10^{-3} species-years. Sensitivity analysis results from Table 5 show that increasing materials recovery levels for A5 and A6 to 28% and 24%, respectively, will result in zero acidification impact potentials.

Table 5. Sensitivity analysis.

Impact Category	Percentage Materials Recovery Levels for Zero Impact Potential to be Realised	
	A5	A6
Acidification	28	24
Eutrophication	40	26
Global warming	0	6
Human health	0	9

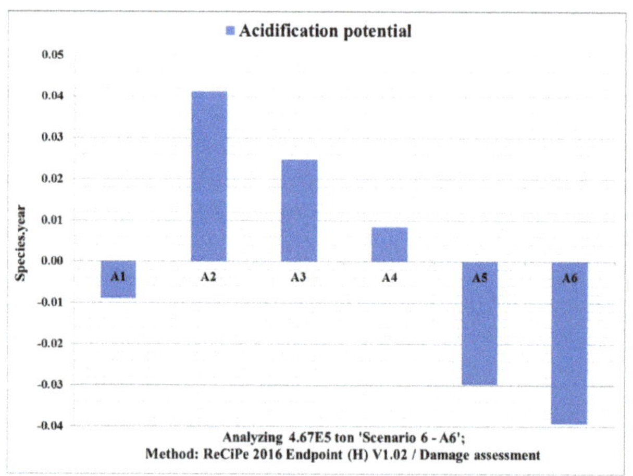

Figure 4. Acidification impact potentials.

3.2. Eutrophication

Figure 5 shows that MSW management options A1, A4, A5 and A6 bring about a reduction in eutrophication, with A2 and A3 leading to increased eutrophication. The eutrophication impact potential is measured using the species extinction rate (species-years). A1 has the highest eutrophication reduction potential of -2.16×10^{-2} species-years followed by A6 with eutrophication potential reduction of -6.12×10^{-3} species-years. A4 and A5 have eutrophication reduction potentials of -3.77×10^{-3} and -2.81×10^{-3} species-years, respectively. A2 and A3 result in eutrophication potential increases of 2.55×10^{-4} and 1.60×10^{-3} species/year, respectively, indicating that the incineration of MSW leads to increased eutrophication, which was also noted by Hong et al. [109]. This confirms that materials recovery contributes to reduced eutrophication potential as it contributes to the reduced eutrophication potential characterizing A5 consisting of incineration, materials recovery and the AD of the biodegradable fraction of MSW. Sensitivity analysis results from Table 5 show that doubling the materials recovery levels under A5 and increasing it to 26% under A6 will result in zero eutrophication impact potentials.

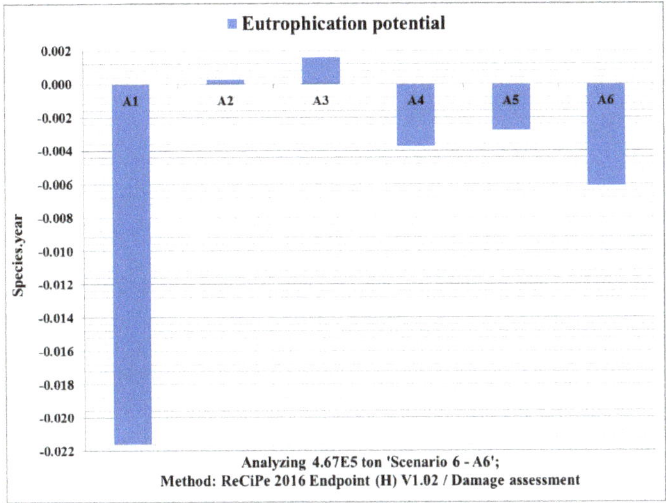

Figure 5. Eutrophication impact potentials.

3.3. Global Warming

As shown in Figure 6, all six scenarios lead to reductions in global warming, with A5 having the highest global warming reduction potential estimated at -9.05×10^{-1} species-years followed by A3 that has a reduction potential in global warming of -8.28×10^{-1} species-years. A2 brings about a global warming reduction potential of -7.68×10^{-1} species-years and A1 has a -5.04×10^{-1} species-years reduction potential. A6 has the second from least global warming reduction potential of -2.03×10^{-1} species-years with A4 having the least reduction potential of -1.46×10^{-1} species-years. It is therefore evident that the scenarios that combine other MSW treatment technologies with incineration perform better compared to those combined with landfilling, which is consistent with findings by Wittmaier et al. [110]. The materials recovery also contributed to reduced global warming potential as indicated by the increase in the reductions in global warming potential from A3 to A5 and A4 to A6. Results from Table 5 sensitivity analysis show that no materials recovery effort is necessary under A5 as reduction in global warming impact potential will be realised in its absence. However, under A6, sensitivity analysis indicates that a 6% materials recovery is sufficient to attain zero global warming impact potential.

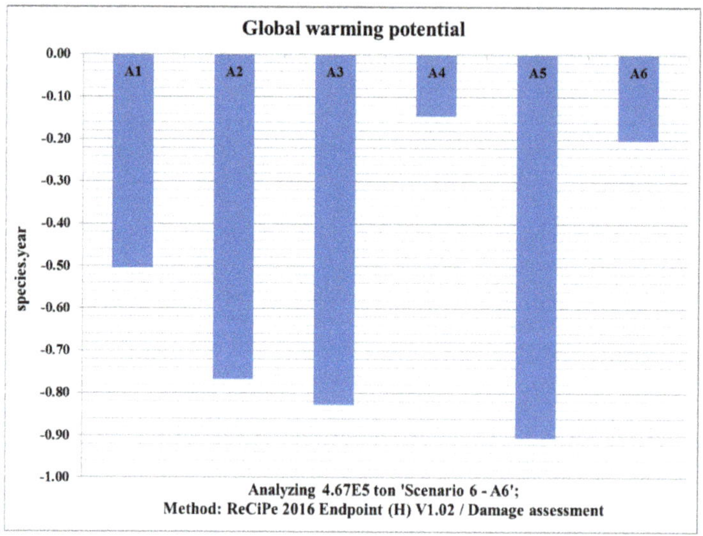

Figure 6. Global warming impact potentials.

3.4. Human Health

Figure 7 shows that all the MSW management options have negative human health impact potentials with option A5 having the highest reduction of −268 DALYs (an overall disease burden measure quantitatively expressed as the total number of years lost due to ill-health, disability or premature or early death) followed by A3 and A2 with human health reduction potentials of −247 and −216 DALYs, respectively. A1 and A6 have human health reduction potentials of −174 and −119 DALYs, respectively. A4 leads to the least reduction in human health of −36 DALYs. Results from Table 5 sensitivity analysis show that no materials recovery effort is necessary under A5 as reduction in human impact potential will be realised even without materials recovery. However, under A6, sensitivity analysis indicates a 9% materials recovery is sufficient to attain zero human health impact potential.

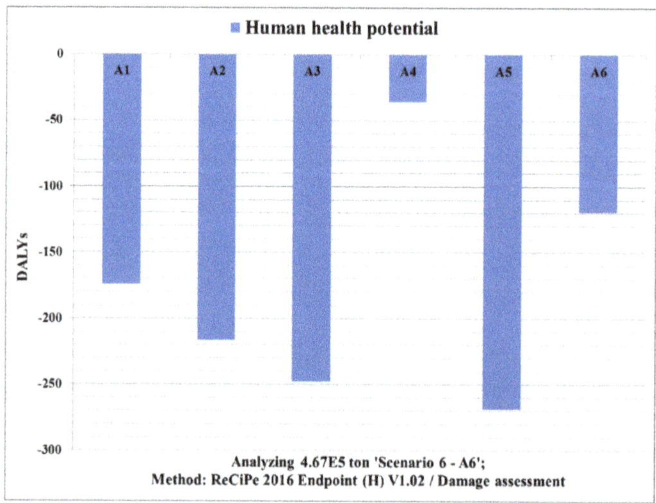

Figure 7. Human health impact potentials.

4. Discussion

LCIA results show that scenario A6 is the best option with regards to acidification while scenario A2 is the worst option. MSWM option A1 is the best scenario considering eutrophication potential and A3 is the worst. In terms of global warming and human health impact potential, A5 is the best option and A4 is the worst MSWM option. Overall, MSWM option A5 emerges as the best option for managing MSW in Harare as shown in Table 6. This is confirmed from findings by Sharma and Chandel [111] that MSWM systems that combines incineration, anaerobic digestion, composting and materials recovery have the least environmental impacts.

Table 6. Ranking of MSW management options.

Impact Category	Scenario Rank Number					
	A1	A2	A3	A4	A5	A6
Acidification	3	6	5	4	2	1
Eutrophication	1	5	6	4	2	3
Global Warming	4	3	2	6	1	5
Human health	4	3	2	6	1	5
Average Rank	3.0	4.3	3.8	5.0	1.5	3.5
Average rank	2	5	4	6	1	3

The recovery of landfill gas for combined heat and power (CHP) generation under the current study is attributed to the reduction of impact potentials across all the impact categories under consideration, except for A4 under acidification, in the MSWM options that incorporated landfilling because energy recovery from waste bring significant environmental benefits [95,110,112–116]. Khandelwal et al. [64], in their review of 153 LCA based MSWM studies published between 2013 and 2018, had 9 studies concluding the appropriateness of AD compared to biodegradable waste landfilling. The same review noted the conclusions from 11 studies regarding the appropriateness of landfilling with landfill gas recovery for CHP generation. This was also noted by Yadav and Samadder [62] in their review analysis of 91 LCA studies on MSWM undertaken from 2006 to 2017 in Asian countries with 5% of the reviewed studies reporting the relative environmental friendliness and sustainability of landfilling with landfill gas recovery—an observation that was also observed by Menikpura et al. [117].

Yadav and Samadder [62] further observed that incineration was reported as a better option than landfilling by 9% of the reviewed studies largely due to the reduced methane emissions associated with incineration. This observation is in agreement with this study's conclusions with regards to human health and global warming impact categories since MSWM options A2, A3 and A5 that incorporated incineration bring more global warming and human health impact potential reductions than A1, A4 and A6 which incorporated landfilling. Cleary [65], like Yadav and Samadder [62], also noted the better performance of thermal treatment with regards to global warming, which is consistent with this study's findings. Thermal treatment was also reported to perform better than landfilling in a critical review of 222 published LCA studies on SWM systems in general, accessed from 216 peer reviewed articles and 15 public reports undertaken by Laurent et al. [63] and Laurent et al. [66].

Overall review results by Yadav and Samadder [62] show that 71% of the reviewed LCAs found landfilling to be the worst or least preferred MSWM treatment option with 8% of the studies concluding incineration to be the worst or least preferred MSW treatment option among other treatment options due to its associated harmful emissions in the form of dioxins and furans as well as human toxicity. Cleary [65], in their review of 20 LCA-based MSWM assessments undertaken and published in peer-reviewed journals between 2002 and 2008, observed that 19 studies confirmed the low environmental performance of landfilling. A review by Abeliotis [60] of 21 LCA studies further observed that landfilling was reported as the worst option for managing and treating MSW, as was observed by Mendes et al. [118], Hong et al. [109], Wanichpongpan and Gheewala [116], Cherubini et al. [95] and Miliūtė and Staniškis [102]. However, despite these reported low environmental performances of landfilling, it performed better than incineration with regards to acidification and eutrophication impact potentials under this study. This is also contrary to observations made by Cleary [65], who noted the better performance of thermal treatment compared to landfilling with regards to eutrophication and acidification impact categories.

The better environmental performance of recycling and thermal treatment of plastics and paper compared to landfilling, as shown by the best performance of A5 which combined incineration and recovery of materials together with AD, was observed by Laurent et al. [63] and Laurent et al. [66] in their reviews consistent with findings by Michaud et al. [119], Lazarevic et al. [120] and Tyskeng and Finnveden [121]. Materials recovery and recycling are environmentally appropriate and sustainable as they lead to reduced environmental impacts potentials [60,62–66,102]. This is confirmed by the better performances of A5 compared to A3 and of A6 compared A4 under this study; sensitivity analysis results that reveal an inverse relation between materials recovery levels and the magnitudes of environmental impact potentials.

Differences in results from LCA studies were observed by Laurent et al. [66] who noted little agreements with regards to the conclusions and no definite agreement except for landfilling with regards to which amongst thermal treatment, anaerobic digestion and recycling is most preferable for managing or treating plastic, paper, organics and metals. De Feo and Malvano [122] observed that the best IMSWM option is subject to the examined impact categories, hence the differences amongst impact categories considered render other MSWM or treatment methods environmentally sustainable while simultaneously rendering others as unsustainable. Khandelwal et al. [64] singled out the heterogeneous nature of MSW as a factor that makes no single MSW treatment method capable to be applied to all the MSW fractions, inevitably resulting in different LCA results from region to region due to differences in MSW generation and composition, MSWM structures, system boundaries, MSWM practices and the choice of impact categories.

5. Conclusions

LCIA results show that scenario A6 is the best option with regards to acidification while scenario A2 is the worst option. MSW management option A1 is the best scenario considering eutrophication potential and A3 is the worst. In terms of global warming and human health impact potential, A5 is the best option and A4 is the worst MSW management option. Overall, MSW management option A5 emerges the best option for managing MSW in Harare as shown in Table 6. This is confirmed from findings by Sharma and Chandel [111] that MSW management that combines incineration, anaerobic digestion, composting and materials recovery has the least environmental impacts. Therefore, the LCA results from this study will inform the design and development of future integrated MSWM systems with reduced environmental and human health impacts. Furthermore, the study will provide a baseline for design and development of further studies to assess economic affordability, social acceptability, renewable energy and job creation potential of the LCA-identified integrated MSWM system with least environmental impact potential. The compositing option for the organic fraction of the MSW instead of anaerobic digestion should also be incorporated in future LCA studies. The study had its own limitations due to the unavailability of quality and reliable data on waste generation and transportation. Therefore, studies to quantify the waste generation and composition in Harare must be undertaken to give reliable data that could be used for further LCIA of MSWM options for Harare.

Author Contributions: The work is part of T.N.'s study towards a PhD under the supervision of E.M. at the University of Johannesburg. T.N. conceptualized the research proposal with E.M. giving guidance regarding the methodological framework. T.N. applied for software from the Life Cycle Initiative in partnership with the UN Environment and Pre-sustainability under the 2017 Life Cycle Awards and his proposal was awarded the Life Cycle award that provided the Simapro 8.5.2 and associated databases. T.N. also applied for research funds from the National Geographic Society under their Early Careers scientist which was awarded to facilitate fieldwork and data collection for the study. E.M. gave guidance in scenario development with T.N. responsible for setting up the input output systems on the Simapro software interface for simulation. T.N. actively analysed the results and undertook sensitivity analysis under the supervision of E.M. T.N. prepared the manuscript with E.M. doing the editing, supervision and reviewing.

Funding: This research was funded by Life Cycle Initiative in partnership with the United Nations Environment Programme (UNEP), PRé Sustainability and Federation of Indian Chambers of Commerce and Industry (FICCI) through the Life Cycle Assessment award. The National Geographic Society funded fieldwork, grant number HJ-170ER-17. The University of Johannesburg funded Trust Nhubu's PhD studies.

Acknowledgments: The authors acknowledge the Life Cycle Initiative for awarding the project the 2017 Life Cycle award in the form of Simapro software that was used to carry out the LCIA. The authors are also grateful to the National Geographic Society for awarding the project the early career grant. The Zimbabwe Environmental Management Agency, Zimbabwe National Statistical Agency, Harare City Council are greatly appreciated for assistance with data for the study. The University of Johannesburg is greatly appreciated for funding these studies.

Conflicts of Interest: No conflict of interest declared.

References

1. Hoornweg, D.; Bhada-Tata, P. *What a Waste: A Global Review of Solid Waste Management*; Urban Development Series; Knowledge Papers; World Bank: Washington, DC, USA, 2012; Available online: https://openknowledge.worldbank.org/handle/10986/17388License:CCBY3.0IGO (accessed on 22 March 2018).
2. OECD. Municipal Waste. In *Environment at a Glance 2013: OECD Indicators*; OECD Publishing: Paris, France, 2013.
3. UN. Urban and Rural Areas. 2009. Available online: http://www.un.org/en/development/desa/population/publications/urbanization/urban-rural.shtml (accessed on 14 July 2018).
4. UNEP; IETC. *Urban Waste Management Strategy*; United Nation Environment Programme (UNEP)-International Environment Technology Centre (IETC): Nairobi, Kenya, 2003.
5. Dickerson, G.W. Solid Waste: Trash to Treasury in an Urban Environment. *N. M. J. Sci.* **1999**, *8*, 166–199.
6. Ndum, E.A.; Busch, G.; Voigt, H. Bottom-Up Approach to Sustainable Solid Waste Management in African Countries. Ph.D. Thesis, Brandenburg University of Technology, Cottbus, Germany, 2013.
7. Afolayan, O.; Ogundele, F.; Odewumi, S. Spatial variation in landfills leachate solution in urbanized area of Lagos State, Nigeria. *Am. Int. J. Contemp. Res.* **2012**, *2*, 178–184.
8. Dongquing, Z.; Tan, S.; Gersberg, R. A comparison of municipal solid waste management in Berlin and Singapore. *Waste Manag.* **2010**, *30*, 921–933.
9. Twardowska, I. Solid Waste: Assessment, Monitoring and Remediation. *Waste Manag. Ser.* **2004**, *4*, 3–32.
10. UNEP. Thematic Focus: Ecosystem Management, Environmental Governance, Harmful Substances and Hazardous Waste. Municipal Solid Waste: Is It Garbage or Gold? 2013. Available online: http://na.unep.net/geas/getUNEPPageWithArticleIDScript.php?article_id=105 (accessed on 13 February 2018).
11. Lundström. Sustainable Waste Management: International Perspectives. In Proceedings of the International Conference on Sustainable Solid Waste Management, Chennai, India, 5–7 September 2007; pp. 1–8.
12. UNDESA. Agenda 21- Chapter 21 Environmentally Sound Management of Solid Wastes and Sewage-related Issues. Division for Sustainable Development, United Nations Department of Economic and Social Affairs. 2005. Available online: http://www.un.org/esa/sustdev/documents/agenda21/index.htm (accessed on 15 February 2018).
13. Worrell, W.A.; Vesilind, P.A. *Solid Waste Engineering*, 2nd ed.; Cengage Learning; Stamford: Dhaka, Bangladesh, 2011.
14. Seadon, J.K. Integrated waste management—Looking beyond the solid waste horizon. *Waste Manag.* **2006**, *26*, 1327–1336. [CrossRef]
15. Melosi, M.V. *Garbage in the Cities: Refuse, Reform, and the Environment, 1880–1980*, 1st ed.; Texas A&M University Press: College Station, TX, USA, 1981.
16. Louis, G.E. A historical context of municipal solid waste management in the United States. *Waste Manag. Res.* **2004**, *22*, 306–322. [CrossRef]
17. Tchobanoglous, G.; Theisen, H.; Eliassen, R. *Solid Wastes: Engineering Principles and Management Issues*; Mcgraw-Hill: New York, NY, USA, 1977.
18. Henry, R.K.; Yongsheng, Z.; Jun, D. Municipal solid waste management challenges in developing countries—Kenyan case study. *Waste Manag.* **2006**, *26*, 92–100. [CrossRef]
19. Nemerow, N.L. *Environmental Engineering: Environmental Health and Safety for Municipal Infrastructure, Land Use and Planning, and Industry*, 6th ed.; Wiley: Hoboken, NJ, USA, 2009.
20. Wilson, D.C. Development drivers for waste management. *Waste Manag. Res.* **2007**, *25*, 198–207. [CrossRef]

21. Marshall, R.E.; Farahbakhsh, K. Systems approaches to integrated solid waste management in developing countries. *Waste Manag.* **2013**, *33*, 988–1003. [CrossRef]

22. Rodic, L.; Wilson, D.C.; Scheinberg, A. *Solid Waste Management in the World's Cities*; UN-HABITAT: Nairobi, Kenya, 2010.

23. McDougall, F.; White, P.R.; Franke, M.; Hindle, P. *Integrated Solid Waste Management: A Lifecycle Inventory*; Blackwell Science Publishing: Oxford, UK, 2001.

24. van de Klundert, A.; Anschutz, J. *Integrated Sustainable Waste Management—The Concept: Tools for Decision-Makers. Experiences from the Urban Waste Expertise Programme (1995–2001)*; Scheinberg, A., Ed.; WASTE: Gouda, The Netherlands, 2001.

25. Wolsink, M. Contested environmental policy infrastructure: Socio-political acceptance of renewable energy, water, and waste facilities. *Environ. Impact Assess. Rev.* **2010**, *30*, 302–311. [CrossRef]

26. CEC. *Second EC Environment Action Programme*; Commission of the European Communities: Brussels, Belgium, 1977.

27. Noto, F. *Overcoming NIMBY Opposition*; Public Sector Digest: London, ON, Canada, 2010; pp. 1–5.

28. Tsiko, R.G.; Togarepi, S. A situational analysis of waste management in Harare, Zimbabwe. *J. Am. Sci.* **2012**, *8*, 692–706.

29. Manzungu, E.; Chioreso, R. Internalising a Crisis? Household Level Response to Water Scarcity in the City of Harare, Zimbabwe. *J. Soc. Dev. Afr.* **2012**, *27*, 111.

30. Konteh, F.H. Urban sanitation and health in the developing world: Reminiscing the nineteenth century industrial nations. *Health Place* **2009**, *15*, 69–78. [CrossRef] [PubMed]

31. Rodic, L.; Scheinberg, A.; Wilson, D.C. Comparing Solid Waste Management in the World's Cities. In Proceedings of the ISWA World Congress 2010, Urban Development and Sustainability—A Major Challenge for Waste Management in the 21st Century, Hamburg, Germany, 15–18 November 2010.

32. Coffey, M.; Coad, A. *Collection of Municipal Solid Waste in Developing Countries*; United Nations Human Settlement Programme—UN-HABITAT: Nairobi, Kenya, 2010.

33. Memon, M.A. Integrated solid waste management based on the 3R approach. *J. Mater. Cycles Waste Manag.* **2010**, *12*, 30–40. [CrossRef]

34. Simelane, T.; Mohee, R. *Future Directions of Municipal Solid Waste Management in Africa*; Africa Institute of South Africa (AISA): Pretoria, South Africa, 2012; pp. 1–6.

35. Iriruaga, E.T. Solid Waste Management in Nigeria. D-WASTE. 2012. Available online: www.d-waste.com (accessed on 17 August 2019).

36. Muzenda, E.; Ntuli, F.; Pilusa, T.J. Waste Management, Strategies and Situation in South Africa: An Overview, World Academy of Science, Engineering and Technology. *Int. J. Environ. Ecol. Eng.* **2012**, *6*, 552–555.

37. Ogwueleka, T. *Municipal Solid Waste Characteristics and Management in Nigeria, Abuja*; University of Abuja: Abuja, Nigeria, 2009.

38. Senkoro, H. Solid Waste Management in Africa. A WHO/AFRO Perspective. In Proceedings of the CWR Workshop on Solid Waste Collection that Benefits the Urban Poor, St. Gallen, Switzerland, March 2003.

39. Achankeng, E. Globalisation, Urbanisation and Municipality Solid Waste management in Africa. In Proceedings of the African Studies Association of Australasia and the Pacific, Africa on a Global Scale, Adelaide, Australia, January 2003.

40. Makwara, E.C.; Snodia, S. Confronting the reckless gambling with people's health and lives: Urban solid waste mangement in Zambabwe. *Eur. J. Sustain. Dev.* **2013**, *2*, 67–98.

41. Mapira, J. Challenges of solid Waste Disposal and Management in the city of Masvingo. *J. Soc. Dev. Afr.* **2012**, *26*, 67–91.

42. McDougall, F.R.; White, P.R.; Franke, M.; Hindle, P. *Integrated Solid Waste Management: A Life Cycle Inventory*; John Wiley & Sons: Hoboken, NJ, USA, 2008.

43. GoZ. *Zimbabwe Environment Outlook: Our Environment, Everybody's Responsibility*; Feresu, S.B., Ed.; Government of Zimbabwe (GoZ) Ministry of Environment & Natural Resources Management: Harare, Zimbabwe, 2010.

44. Tevera, D.S. Solid waste disposal in Harare and its effects on the environment: Some preliminary observations. *Zimb. Sci. News* **1991**, *25*, 9–13.

45. Masocha, M.; Tevera, D.S. Open waste dumps in Victoria Falls Town: Spatial patterns, environmental threats and public health implications. *Geogr. J. Zimb.* **2003**, *33/34*, 9–19.

46. Mapira, J. Pollution of the Sakubva River, Mutare(Zimbabwe): Culprits, Penalties and Consequences. *Zimb. J. Geogr. Res.* **2007**, *1*, 9–19.

47. Makwara, E.C. *Work Related Environmental Health Risks:The Case of Garbage Handlers in the City of Masvingo*; Lambert Academic Publishing: Saarbrucken, Germany, 2011.

48. Mafume, P.N.; Zendera, W.; Mutetwa, M.; Musimbo, N. Challenges of solid waste management in Zimbabwe: A case study of Sakubva high density suburb. *J. Environ. Waste Manag.* **2016**, *3*, 142–155.

49. Musademba, D.; Musiyandaka, S.; Muzinda, A.; Nhemachena, B.; Jambwa, D. Municipality Solid Waste (MSW) Management Challenges of Chinhoyi Town in Zimbabwe: Opportunities of Waste Reduction and Recycling. *J. Sustain. Dev. Afr.* **2011**, *13*, 168–180.

50. Mangizvo, R.V. Challenges of Solid Waste Management in the Central Business District of the City of Gweru in Zimbabwe. *J. Sustain. Dev. Afr.* **2007**, *9*, 134–145.

51. Mangizvo, R.V. Illegal Dumping of Solid Waste in the Alleys in the Central Business District of Gweru, Zimbabwe. *J. Sustain. Dev. Afr.* **2010**, *12*, 110–123.

52. Muchandiona, A.; Nhapi, I.; Misi, S.; Gumindoga, W. *Challenges and Opportunities in Solid Waste Management in Zimbabwe's Urban Councils*; University of Zimbabwe: Harare, Zimbabwe, 2013; p. 87.

53. Tanyanyiwa, V.I. Not In My Backyard (NIMBY)?: The Accumulation of Solid Waste in the Avenues Area, Harare, Zimbabwe. *Int. J. Innov. Res. Dev.* **2015**, *4*, 122–128.

54. Chijarira, S.R. *The Impact of Dumpsite Leachate on Ground and Surface Water: A Case Study of Pomona Waste Dumpsite*; Department of Geography, Bindura University of Science and Technology: Bindura, Zimbabwe, 2013; p. 63.

55. Rebitzer, G.; Ekvall, T.; Frischknecht, R.; Hunkeler, D.; Norris, G.; Rydberg, T.; Schmidt, W.P.; Suh, S.; Weidema, B.P.; Pennington, D.W. Life cycle assessment: Part 1: Framework, goal and scope definition, inventory analysis, and applications. *Environ. Int.* **2004**, *30*, 701–720. [CrossRef]

56. Winkler, J.; Bilitewski, B. Comparative evaluation of life cycle assessment models for solid waste management. *Waste Manag.* **2007**, *27*, 1021–1031. [CrossRef]

57. Güereca, L.P.; Gassó, S.; Baldasano, J.M.; Jiménez-Guerrero, P. Life cycle assessment of two biowaste management systems for Barcelona, Spain. *Resour. Conserv. Recycl.* **2006**, *49*, 32–48. [CrossRef]

58. Blengini, G.A.; Fantoni, M.; Busto, M.; Genon, G.; Zanetti, C. Participatory approach, acceptability and transparency of waste Management LCAs: Case studies of Torino and Cuneo. *Waste Manag.* **2012**, *32*, 1712–1721. [CrossRef]

59. Mendes, M.R.; Aramaki, T.; Hanaki, K. Comparison of the environmental impact of incineration and landfilling in Sao Paulo City as determined by LCA. *Resour. Conserv. Recycl.* **2004**, *41*, 47–63. [CrossRef]

60. Abeliotis, K. Life cycle assessment in municipal solid waste management. In *Integrated Waste Management*; Kumar, S., Ed.; INTECH Open Access Publisher: London, UK, 2011.

61. Koci, V.; Trecakova, T. Mixed municipal waste management in the Czech Republic from the point of view of the LCA method. *Int. J. Life Cycle Assess.* **2011**, *16*, 113–124. [CrossRef]

62. Yadav, P.; Samadder, S.R. A critical review of the life cycle assessment studies on solid waste management in Asian countries. *J. Clean. Prod.* **2018**, *185*, 492–515. [CrossRef]

63. Laurent, A.; Clavreul, J.; Bernstad, A.; Bakas, I.; Niero, M.; Gentil, E.; Christensen, T.H.; Hauschild, M.Z. Review of LCA studies of solid waste management systems–Part II: Methodological guidance for a better practice. *Waste Manag.* **2014**, *34*, 589–606. [CrossRef] [PubMed]

64. Khandelwal, H.; Dhar, H.; Thalla, A.K.; Kumar, S. Application of life cycle assessment in municipal solid waste management: A worldwide critical review. *J. Clean. Prod.* **2019**, *209*, 630–654. [CrossRef]

65. Cleary, J. Life cycle assessments of municipal solid waste management systems: A comparative analysis of selected peer-reviewed literature. *Environ. Int.* **2009**, *35*, 1256–1266. [CrossRef] [PubMed]

66. Laurent, A.; Bakas, I.; Clavreul, J.; Bernstad, A.; Niero, M.; Gentil, E.; Hauschild, M.Z.; Christensen, T.H. Review of LCA studies of solid waste management systems–Part I: Lessons learned and perspectives. *Waste Manag.* **2014**, *34*, 573–588. [CrossRef]

67. Zimstat. 2012 Zimbabwe Census National Report, Zimbabwe National Statistics Agency. 2013. Available online: www.zimstat.co.zw/sites/default/files/img/publications/Population/National_Report.pdf (accessed on 23 August 2017).

68. UNEP; CalRecovery. *Solid Waste Management*; UNEP: Nairobi, Kenya, 2005.

69. Hester, R.E.; Harrison, R.M. *Environmental and Health Impact of Solid Waste Management Practices*; Royal Society of Chemistry: Cambridge, UK, 2002.

70. EMA. Environmental Management Agency Statutory Instrument 10 of 2007: Environmental Management (Hazardous waste management) Regulations. 2007. Available online: https://www.ema.co.zw/about-us/law/environmental-regulations/item/17-hazardous-waste-management-regulations-si-10-2007 (accessed on 1 September 2019).

71. Kawai, K.; Tasaki, T. Revisiting estimates of municipal solid waste generation per capita and their reliability. *J. Mater. Cycles Waste Manag.* **2015**, *2016*, 1–13. [CrossRef]

72. Afon, A. An analysis of solid waste generation in a traditional African city: The example of Ogbomoso, Nigeria. *Environ. Urban.* **2007**, *19*, 527–537.

73. Mshandete, A.M.; Parawira, W. Biogas Technology Research in Selected Sub-Saharan African Countries—A Review. *Afr. J. Biotechnol.* **2009**, *8*, 116–125.

74. Pawandiwa, C.C. Municipal Solid Waste Disposal Site Selection the Case of Harare. Master's Thesis, University of the Free State, Bloemfontein, South Africa, 2013.

75. Mbiba, B. Urban solid waste characteristics and household appetite for separation at source in Eastern and Southern Africa. *Habitat Int.* **2014**, *43*, 152–162. [CrossRef]

76. Emenike, C.U.; Iriruaga, E.T.; Agamuthu, P.; Fauziah, S.H. Waste Management in Africa: An Invitation to Wealth Generation. In Proceedings of the International Conference on Waste Management and Environment, ICWME, Kuala Lumpur, Malaysia, 26–27 August 2013; pp. 1–6.

77. Miezah, K.; Obiri-Danso, K.; Kádár, Z.; Fei-Baffoe, B.; Mensah, M.Y. Municipal solid waste characterization and quantification as a measure towards effective waste management in Ghana. *Waste Manag.* **2015**, *46*, 15–27. [CrossRef] [PubMed]

78. EEA. *Managing Municipal Solid Waste—A Review of Achievements in 32 European Countries*; European Environment Agency: Copenhagen, Denmark, 2013; Available online: http://www.eea.europa.eu/publications/managing-municipal-solid-waste (accessed on 11 November 2018).

79. Al-Khatib, I.A.; Monou, M.; Abu Zahra, A.F.; Shaheen, H.Q.; Kassinos, D. Solid waste characterization, quantification and management practices in developing countries. A case study of Nablus district, Palestine. *J. Environ. Manag.* **2010**, *91*, 1131–1138. [CrossRef] [PubMed]

80. Nyanzou, P.; Steven, J. Solid waste management practices in high density suburbs of Zimbabwe: A focus on Budiriro 3, Harare. *Dyke* **2014**, *8*, 17–54.

81. Makarichi, L.; Kan, R.; Jutidamrongphan, W.; Techato, K.A. Suitability of municipal solid waste in African cities for thermochemical waste-to-energy conversion: The case of Harare Metropolitan City, Zimbabwe. *Waste Manag. Res.* **2019**, *37*, 83–94. [CrossRef] [PubMed]

82. EMA. *Waste Generation and Management in Harare, Zimbabwe: Residential Areas, Commercial Areas and Schools*; Environmental Management Agency: Harare, Zimbabwe, 2016; in press.

83. GoZ. *Zimbabwe's Integrated Solid Waste Management Plan*; Government of Zimbabwe, Environmental Management Agency, Institute of Environmental Studies, University of Zimbabwe: Harare, Zimbabwe, 2014; pp. 1–98.

84. Tirivanhu, D.; Feresu, S. *A Situational Analysis of Solid Waste Management in Zimbabwe's Urban Centres*; Institute of Environment Studies, University of Zimbabwe: Harare, Zimbabwe, 2013; pp. 1–153.

85. Mudzengerere, F.H.; Chigwenya, A. Waste management in Bulawayo City Council in Zimbabwe: In search of sustainable waste management in the city. *J. Sustain. Dev. Afr.* **2012**, *14*, 228–244.

86. Moreno Ruiz, E.; Lévová, T.; Reinhard, J.; Valsasina, L.; Bourgault, G.; Wernet, G. *Documentation of Changes Implemented in Ecoinvent Database v3.3*; Ecoinvent: Zürich, Switzerland, 2016.

87. Stucki, M.; Jungbluth, N.; Leuenberger, M. *Life Cycle Assessment of Biogas Production from Different Substrates*; Final report; Federal Department of Environment, Transport, Energy and Communications, Federal Office of Energy: Bern, Switzerland, 2011.

88. Achinas, S.; Achinas, V.; Euverink, G.J.W. A technological overview of biogas production from biowaste. *Engineering* **2017**, *3*, 299–307. [CrossRef]

89. GMI. *Summary of Findings Anaerobic Digestion for MSW*; Global Methane Initiative: Florianopolis, Brazil, 2014.

90. Kigozi, R.; Aboyade, A.; Muzenda, E. Biogas production using the organic fraction of municipal solid waste as feedstock. *Int'l J. Res. Chem. Metall. Civ. Eng.* **2013**, *1*, 107–114.

91. ISO. (ISO 14044:2006). Ginebra, Suiza. 2006. Available online: www.iso.org/iso/home.htm (accessed on 11 July 2019).

92. Fernández-Nava, Y.; Del Rio, J.; Rodríguez-Iglesias, J.; Castrillón, L.; Marañón, E. Life Cycle Assessment (LCA) of different municipal solid waste management options: A case study of Asturias (Spain). *J. Clean. Prod.* **2014**, *81*, 178–189. [CrossRef]

93. Montejo, C.; Tonini, D.; Márquez, M.C.; Astrup, T.F. Mechanical biological treatment: Performance and potentials. An LCA of 8 MBT plants including waste characterization. *J. Environ. Manag.* **2013**, *128*, 661–673. [CrossRef]

94. Arena, U.; Mastellone, M.L.; Perugini, F. The environmental performance of alternative solid waste management options: A life cycle assessment study. *Chem. Eng. J.* **2003**, *96*, 207–222. [CrossRef]

95. Cherubini, F.; Bargigli, S.; Ulgiati, S. Life cycle assessment (LCA) of waste management strategies: Landfilling, sorting plant and incineration. *Energy* **2009**, *34*, 2116–2123. [CrossRef]

96. Han, H.; Long, J.; Li, S.; Qian, G. Comparison of green-house gas emission reductions and landfill gas utilization between a landfill system and an incineration system. *Waste Manag. Res.* **2010**, *28*, 315–321.

97. Song, Q.; Wang, Z.; Li, J. Environmental performance of municipal solid waste strategies based on LCA method: A case study of Macau. *J. Clean. Prod.* **2013**, *57*, 92–100. [CrossRef]

98. Lundie, S.; Peters, G.M. Life cycle assessment of food waste management options. *J. Clean. Prod.* **2005**, *13*, 275–286. [CrossRef]

99. Gunamantha, M.; Sarto. Life cycle assessment of municipal solid waste treatment to energy options: Case study of Kartamantul region, Yogyakarta. *Renew. Energy* **2012**, *41*, 277–284. [CrossRef]

100. Assamoi, B.; Lawryshyn, Y. The environmental comparison of landfilling vs. incineration of MSW accounting for waste diversion. *Waste Manag.* **2012**, *32*, 1019–1030. [CrossRef]

101. Vergara, S.E.; Damgaard, A.; Horvath, A. Boundaries matter: Greenhouse gas emission reductions from alternative waste treatment strategies for California's municipal solid waste. *Resour. Conserv. Recycl.* **2011**, *57*, 87–97. [CrossRef]

102. Miliute, J.; Kazimieras Staniškis, J. Application of life-cycle assessment in optimisation of municipal waste management systems: The case of Lithuania. *Waste Manag. Res.* **2010**, *28*, 298–308. [CrossRef]

103. Ogundipe, F.O.; Jimoh, O.D. Life Cycle Assessment of Municipal Solid Waste Management in Minna, Niger State, Nigeria. *Int. J. Environ. Res.* **2015**, *9*, 1305–1314.

104. Ayodele, T.R.; Ogunjuyigbe, A.S.O.O.; Alao, M.A. Life cycle assessment of waste-to-energy (WtE) technologies for electricity generation using municipal solid waste in Nigeria. *Appl. Energy* **2017**, *201*, 200–218. [CrossRef]

105. Beigl, P.; Salhofer, S. Comparison of ecological effects and costs of communal waste management systems. *Resour. Conserv. Recycl.* **2004**, *41*, 83–102. [CrossRef]

106. Moreno Ruiz, E.; Valsasina, L.; Brunner, F.; Symeonidis, A.; FitzGerald, D.; Treyer, K.; Bourgault, G.; Wernet, G. *Documentation of Changes Implemented in Ecoinvent Database v3.5*; Ecoinvent: Zürich, Switzerland, 2018.

107. Huijbregts, M.A.; Steinmann, Z.J.; Elshout, P.M.; Stam, G.; Verones, F.; Vieira, M.; Zijp, M.; Hollander, A.; van Zelm, R. ReCiPe2016: A harmonised life cycle impact assessment method at midpoint and endpoint level. *Int. J. Life Cycle Assess.* **2017**, *22*, 138–147. [CrossRef]

108. Huijbregts, M.A.J.; Steinmann, Z.J.N.; Elshout, P.M.F.; Stam, G.; Verones, F.; Vieira, M.D.M.; Hollander, A.; Zijp, M.; Van Zelm, R. ReCiPe 2016: A Harmonized Life Cycle Impact Assessment Method at Midpoint and Endpoint Level Report I: Characterization. RIVM Report 2016-0104. 2016. Available online: http://www.rivm.nl/en/Topics/L/Life_Cycle_Assessment_LCA/Downloads/Documents_ReCiPe2017/Report_ReCiPe_Update_2017 (accessed on 27 July2017).

109. Hong, R.J.; Wang, G.F.; Guo, R.Z.; Cheng, X.; Liu, Q.; Zhang, P.J.; Qian, G.R. Life cycle assessment of BMT-based integrated municipal solid waste management: Case study in Pudong, China. *Resour. Conserv. Recycl.* **2006**, *49*, 129–146. [CrossRef]

110. Wittmaier, W.; Langer, S.; Sawilla, B. Possibilities and limitations of life cycle assessment (LCA) in the development of waste utilization systems—Applied examples for a region in Northern Germany. *Waste Manag.* **2009**, *29*, 1732–1738. [CrossRef]

111. Sharma, B.K.; Chandel, M.K. Life cycle assessment of potential municipal solid waste management strategies for Mumbai, India. *Waste Manag. Res.* **2017**, *35*, 79–91. [CrossRef]

112. Buttol, P.; Masoni, P.; Bonoli, A.; Goldoni, S.; Belladonna, V.; Cavazzuti, C. LCA of integrated MSW management systems: Case study of the Bologna District. *Waste Manag.* **2007**, *27*, 1059–1070. [CrossRef]

113. Fruergaard, T.; Astrup, T. Optimal utilization of waste-to-energy in an LCA perspective. *Waste Manag.* **2011**, *31*, 572–582. [CrossRef]

114. Khoo, H.H. Life cycle impact assessment of various conversion technologies. *Waste Manag.* **2009**, *29*, 1892–1900. [CrossRef]

115. Liamsanguan, C.; Gheewala, S.H. LCA: A decision support tool for environmental assessment of MSW management systems. *J. Environ. Manag.* **2008**, *87*, 132–138. [CrossRef]

116. Wanichpongpan, W.; Gheewala, S.H. Life cycle assessment as a decision support tool for landfill gas-to energy projects. *J. Clean. Prod.* **2007**, *15*, 1819–1826. [CrossRef]

117. Menikpura, S.N.M.; Sang-Arun, J.; Bengtsson, M. Assessment of environmental and economic performance of Waste-to-Energy facilities in Thai cities. *Renew. Energy* **2016**, *86*, 576–584. [CrossRef]

118. Mendes, M.R.; Aramakib, T.; Hanakic, K. Assessment of the environmental impact of management measures for the biodegradable fraction of municipal solid waste in São Paulo City. *Waste Manag.* **2003**, *23*, 403–409. [CrossRef]

119. Michaud, J.; Farrant, L.; Jan, O.; Kjær, B.; Bakas, I.Z. *Environmental Benefits of Recycling—2010 Update*; Waste and Resources Action Programme (WRAP): Banbury, UK, 2010.

120. Lazarevic, D.; Aoustin, E.; Buclet, N.; Brandt, N. Plastic waste management in the context of a European recycling society: Comparing results and uncertainties in a life cycle perspective. *Resour. Conserv. Recycl.* **2010**, *55*, 246–259. [CrossRef]

121. Tyskeng, S.; Finnveden, G. Comparing energy use and environmental impacts of recycling and waste incineration. *J. Environ. Eng.* **2010**, *136*, 744–748. [CrossRef]

122. De Feo, G.; Malvano, C. The use of LCA in selecting the best management system. *Waste Manag.* **2009**, *29*, 1901–1915. [CrossRef]

Article

Chromium VI and Fluoride Competitive Adsorption on Different Soils and By-Products

Ana Quintáns-Fondo [1], Gustavo Ferreira-Coelho [1], Manuel Arias-Estévez [2],
Juan Carlos Nóvoa-Muñoz [2], David Fernández-Calviño [2], Esperanza Álvarez-Rodríguez [1],
María J. Fernández-Sanjurjo [1] and Avelino Núñez-Delgado [1,*]

[1] Department of Soil Science and Agricultural Chemistry, Engineering Polytechnic School, Universidade de Santiago de Compostela, 27002 Lugo, Spain; anaquintansfondo@hotmail.com (A.Q.-F.); gf_coelho@yahoo.com.br (G.F.-C.); esperanza.alvarez@usc.es (E.Á.-R.); mf.sanjurjo@usc.es (M.J.F.-S.)

[2] Department of Plant Biology and Soil Science, Faculty of Sciences, Campus Ourense, Universidade de Vigo, 32004 Ourense, Spain; mastevez@uvigo.es (M.A.-E.); edjuanca@uvigo.es (J.C.N.-M.); davidfc@uvigo.es (D.F.-C.)

* Correspondence: avelino.nunez@usc.es; Tel.: +34-982-823-140

Received: 24 September 2019; Accepted: 12 October 2019; Published: 15 October 2019

Abstract: Chromium (as Cr(VI)) and fluoride (F$^-$) are frequently found in effluents from different industrial activities. In cases where these effluents reach soil, it can play an important role in retaining those pollutants. Similarly, different byproducts could act as bio-adsorbents to directly treat polluted waters or to enhance the purging potential of soil. In this work, we used batch-type experiments to study competitive Cr(VI) and F$^-$ adsorption in two different soils and several kinds of byproducts. Both soils, as well as mussel shell, oak ash, and hemp waste showed higher adsorption for F$^-$, while pyritic material, pine bark, and sawdust had a higher affinity for Cr(VI). Considering the binary competitive system, a clear competition between both elements in anionic form is shown, with decreases in adsorption of up to 90% for Cr(VI), and of up to 30% for F$^-$. Adsorption results showed better fitting to Freundlich's than to Langmuir's model. None of the individual soils or byproducts were able to adsorbing high percentages of both pollutants simultaneously, but it could be highly improved by adding pine bark to increase Cr(VI) adsorption in soils, thus drastically reducing the risks of pollution and deleterious effects on the environment and on public health.

Keywords: adsorption; chromium; competition; fluoride; soil and water pollution

1. Introduction

In recent years there has been growing concern regarding F$^-$ and Cr(VI) pollution, due to both substances being transported into effluents from industries related to the extraction of minerals, foundries, dyes and pigments, semiconductors, and glass manufacturing [1,2]. These effluents can reach surface- and ground-waters by direct discharge or after passing through soils. Authors such as Rafique et al. [3] or Kumar et al. [4] indicate that there is a global hazard as regards fluoride and chromium pollution, taking into account that their permissible limits in drinking water (1.5 and 0.05 mg L^{-1}, respectively, as per the World Health Organization) are widely exceeded in occasions, some of them referenced for countries such as India, China, USA, Mexico, or Argentina.

A F$^-$ concentration between 0.5–1.0 mg L^{-1} in drinking water can be considered beneficial for bones and teeth, but it can cause fluorosis and even neurological damage when it is higher than 1.5 mg L^{-1} [5]. In the case of Cr, although Cr(III) is indispensable in low quantities, Cr(VI) is considered highly toxic due to its mutagenic, carcinogenic, and teratogenic potential [6]. Given that it is extremely improbable that a ban will be implemented in the short term in order to remove F$^-$ and Cr(VI) from industrial use (mainly in aluminum, textile, or leather tanning factories), it is of main importance to determine the capacity of soils to retain both anions, aiding to prevent their entry into waterbodies and plant uptake, as well as to develop low-cost methodologies to increase soil retention capacity and to remove these toxics when they reach waters [4].

Although different methodologies have been developed to remove F^- and Cr(VI) from waters, such as precipitation, electrocoagulation, ion exchange, or electro-dialysis, the use of adsorbent materials has been considered as the most economical and sustainable alternative [2]. Previous works have dealt with individual adsorption of F^- and Cr(VI), separately, both in soils and in different waste materials [3,7–13]. In addition, some studies focused on simultaneous retention of F^- and Cr(VI), using adsorbents such as a chitosan-alginate aluminum complex (CSAlg-Al) [4], or synthetic mesoporous alumina [2]. However, there is not enough information on simultaneous retention of these two anions in different soils, as well as on the adsorbent capacity of different byproducts derived from industrial activities.

Galicia (NW Spain) is one of the geographical areas affected by activities which cause environmental pollution by F^- and Cr(VI) (mining, aluminum factories, glass, dyes, leather tanning), and where it is also easy to obtain locally different by-products that could be used as low-cost sorbents for both pollutants [10–13]. Taking into account this previous background, the objectives of this work are: (1) to determine the capacity of forest and vineyard soils to adsorb F^- and Cr(VI) simultaneously; (2) to determine the adsorption capacity for both anions of byproducts from forestry (oak ash, pine bark, and pine sawdust), from agriculture (hemp waste), from mining (pyritic material), and from the food industry (mussel shell). The results of this study could aid to solve environmental issues due to both pollutants in waters and soils, and at the same time promote the productive recycling of by-products.

2. Materials and Methods

2.1. Materials

In Galicia (NW Spain), some environmental problems related to F^- and Cr(VI) pollution have been previously pointed out [10–13]. In fact, the soils and sorbent materials used in this work to study F^- and Cr(VI) competitive adsorption were the same previously described when performing individual adsorption tests for these two anions [13], in addition to pine sawdust, also previously described [12]. Specifically, in the current work we used samples of forest and vineyard soils, pyritic material, fine mussel shell, pine bark, oak ash, hemp waste, and pine sawdust. Taking into account that detailed descriptions for all these materials were previously published in the referred works, data on it are included in Supplementary Material. In fact, specific references regarding Supplementary Material are also included [14–37].

2.2. Methods

2.2.1. Characterization of the Soil Samples and Sorbent Materials

Details on all methods used to characterize soils and by-products, as well as the results of those procedures, are shown in Supplementary Material. Specifically, total C and N contents, pH in distilled water, pH of the point of zero charge (pH_{pzc}), exchangeable Na, K, Ca, Mg, and Al, effective cation exchange capacity (eCEC), total P, total concentrations of Na, K, Ca, Mg, Al, Fe, Mn, As, Cd, Cr, Cu, Ni, Pb, and Zn, non-crystalline Al and Fe (Al_o, Fe_o), and particle-size distribution for forest and vineyard soils. Also, infrared spectroscopy was used to determine the main functional groups present in each soil and byproduct.

2.2.2. Competitive Adsorption Experiments for F^- and Cr(VI)

A methodology similar to that previously described by Romar-Gasalla et al. [13] was used for these experiments. Specifically, each of the individual samples of soils and by-products were added with F^- and Cr(VI) simultaneously. To do that, 3 g of each sample were stirred with 30 mL of a 0.01 M $NaNO_3$ solution in which F^- and Cr(VI) were incorporated at the same concentration (0.5, 1.5, 3, and 6 mmol L^{-1} of each anion), using analytical grade KF (Panreac, Spain) and $K_2Cr_2O_7$ (Panreac, Spain). All these suspensions were stirred for 24 h, centrifuged for 15 min (6167× g), and filtered by acid washed paper (Whatman, Spain). ICP Mass (Varian 820-NS, USA) was used to determine Cr(VI) concentration in the filtrated liquid, and F^- was quantified by means of an ion-selective electrode and TISAB IV (Orion Research, Cambridge, MA, USA). Triplicate determinations were carried out in all cases.

The amounts of F^- and Cr(VI) adsorbed were calculated as the difference between those added and those remaining in solution at equilibrium.

2.2.3. Modeling Adsorption

Adsorption data were firstly adjusted to the Langmuir and Freundlich models. The Langmuir model corresponds to the following equation:

$$Q_{eq} = \frac{Q_{max} \times K_L \times C_e}{(1 + K_L \times C_e)} \tag{1}$$

where Q_{eq} is the amount of F^- or Cr(VI) adsorbed (mmol kg^{-1}), K_L is the Langmuir's constant related to the adsorption energy ((L $mmol^{-1}$), C_e is the concentration of F^- or Cr(VI) in the equilibrium solution (mmol L^{-1}), and Q_{max} is the maximum adsorption capacity (mmol kg^{-1}).

The Freundlich model (Equation (2)) corresponds to the equation:

$$Q_{eq} = K_F \times C_e^{1/n} \tag{2}$$

where K_F is the Freundlich's constant related to the adsorption energy (L^n kg^{-1} $mmol^{(1-n)}$), and n is the constant related to the adsorption intensity (dimensionless).

In a second step, adsorption data was adjusted to the Tempkin model (Equation (3)):

$$q_a = \beta \ln K_T + \beta \ln C_e \tag{3}$$

where $\beta = RT/b_t$ and b_t is the Temkin isotherm constant; K_t is the Tempkin isotherm equilibrium binding constant (L g^{-1}); T is Temperature (25 °C) (K = 298°), and R is the universal gas constant (8314 Pa m^3/mol K).

3. Results and Discussion

3.1. F^- and Cr(VI) Competitive Adsorption in a Binary System

Figure 1 shows F^- and Cr(VI) adsorption in a competitive (binary) system, as well as data corresponding to an individual (simple) system (from Romar-Gasalla et al. [13]).

In the binary competitive system, two kinds of behavior can be observed. On the one hand, both soils, mussel shell, oak ash, and hemp waste present higher adsorption for F^- (between 15 and 42 mmol kg^{-1}) than for Cr(VI) (between 8 and 19 mmol kg^{-1}) for the highest concentrations added (Figure 1), which becomes more evident when results are expressed as percentages (25–70% for F^-, and 1–30% for Cr (VI)). On the other hand, pine bark, sawdust and, in general, pyritic material, have a higher affinity for Cr(VI) (39–60 mmol kg^{-1}, representing between 22–99% of the highest concentration added) than for F^- (1.5–11 mmol kg^{-1}, representing 1–67% of the highest concentration added) (Figure 1). In the previous study, carried out in individual (simple) systems for Cr(VI) and F^- separately (Romar-Gasalla et al. [13]), the highest F^- adsorption corresponded to forest soil, pyritic material, pine bark, and oak ash, with values between 30 and 40 mmol kg^{-1}, which represented between 60–72% of the highest concentration of F^- added, while pine bark, pine sawdust, and pyritic material showed the highest adsorption of Cr(VI) for the highest concentration of Cr(VI) added, with 69, 40, and 32 mmol kg^{-1}, representing 98, 68, and 55%, respectively. Li et al. [2] found higher affinity for F$^-$ than for Cr(VI) in synthetic alumina gels, while Mohapatra et al. [38] obtained the opposite result in alumina nanofibers.

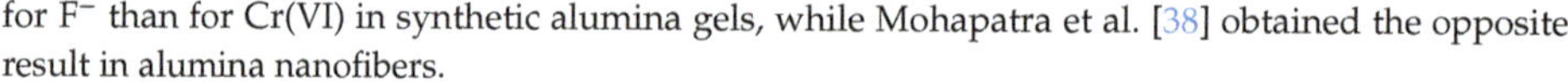

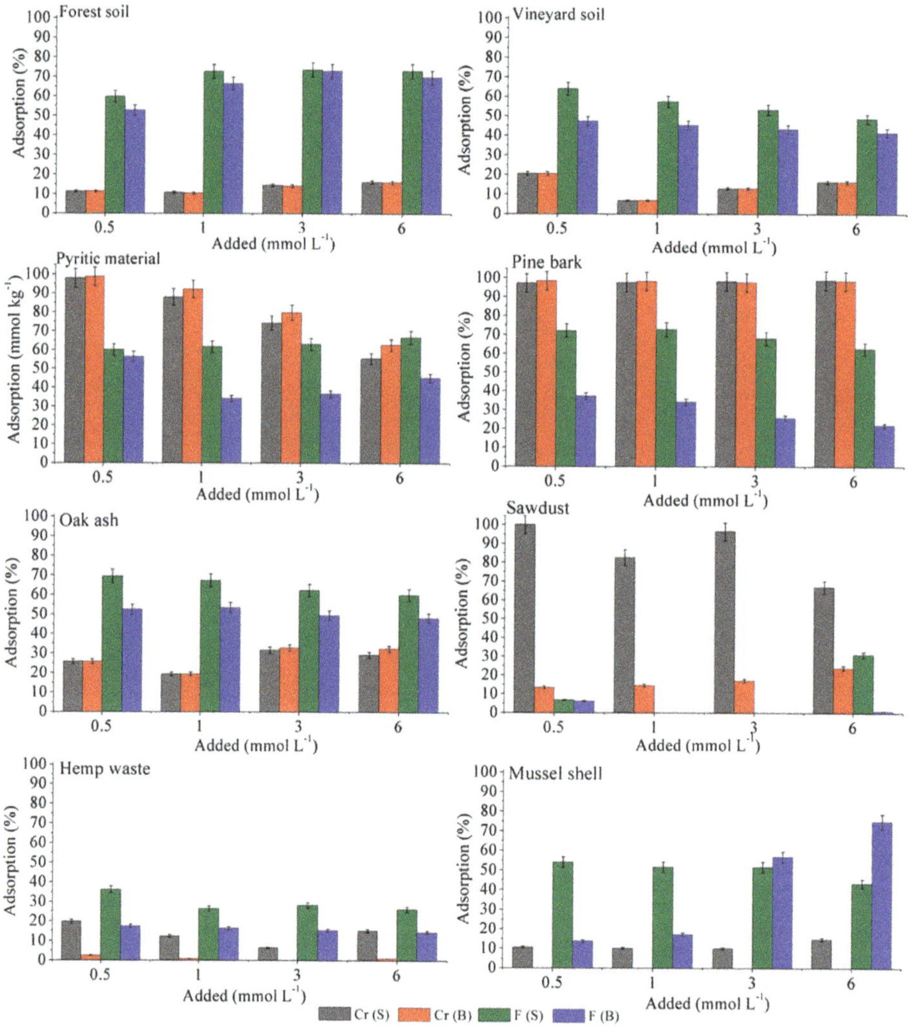

Figure 1. Percentage adsorption for Cr(VI) and F$^-$ in the various soils and sorbents tested, in simple (S) and binary (B) systems. Average values for three replicates, with error bars showing that coefficients of variation were <5%.

The behavior observed in the present study can be related to the different composition of the sorbent materials used, and to their pH values (Table S1, Supplementary Material). Specifically, those sorbents having more acidic pH (pine bark, sawdust, and pyritic material) show higher adsorption for Cr(VI) than for F$^-$, because Cr(VI) adsorption is higher in clearly acid media, decreasing in alkaline and slightly acidic conditions [39,40], as the HCrO$_4^-$ species, present in acid media, is more intensively adsorbed than CrO$_4^{2-}$, which predominates in alkaline conditions [41].

In the case of pine bark and sawdust, acidic conditions favor the protonation of the phenolic groups present in organic compounds [42,43], electrostatically attracting negatively charged Cr(VI) species. In the case of the pyritic material, the positive charges of non-crystalline Fe minerals favor Cr(VI) adsorption, as observed in previous studies [44,45]. However, when the materials have a high concentration of non-crystalline Al minerals, such as in forest soil, vineyard soil, and oak ash (Table S1, Supplementary Material), the affinity for F^- is very high, as previously pointed out [10,11,46]. In the case of mussel shell, high F^- adsorption may be related to binding to the surface of carbonates through inner-sphere complexes with octahedral Ca [11,47].

Comparing percentage adsorption in binary and simple systems (Figure 1), it is clear that in the binary system F^- interferes Cr(VI) retention in sawdust, hemp waste, and mussel shell, decreasing Cr(VI) adsorption up to 90, 20, and 12%, respectively. Likewise, Cr(VI) decreases F^- adsorption in the binary system up to 30% in the pyritic material and pine bark, up to 12% in hemp waste and vineyard soil, and less than 10% in forest soil.

Li et al. [2] found for synthetic alumina gels that the coexistence of F^- and Cr(VI) in solution decreased Cr(VI) adsorption, but enhanced F^- adsorption, due to the fact that alumina has higher affinity for F^- than for Cr(VI), and Cr can form $\equiv CrOH_2^+$ groups, which could be new adsorption sites for F^-.

Deng et al. [48], studying competitive adsorption in fibers impregnated with cerium, found that F^- adsorption increased in a binary system in relation to a simple system, whereas Cr(VI) adsorption decreased (i.e., Cr(VI) favored F^- adsorption, but F^- hindered Cr(VI) adsorption). In addition, Wu et al. [49] reported that Cr(VI) enhanced the adsorption of another anion (anionic As) on activated carbon in a binary system.

In the present study, Cr(VI) hinders F^- adsorption (except in mussel shell for the highest concentrations of both pollutants), and F^- decreases Cr(VI) adsorption in sawdust, hemp waste, and mussel shell (Figure 1), and overall results indicate that Cr(VI) and F^- compete for adsorption sites. The decrease in the adsorption of each anion in the binary system (in relation to the simple system) can be related to the fact that F^- and some of the Cr(VI) species ($HCrO_4^-$, $Cr_2O_7^{2-}$, CrO_4^{2-}) have similar mechanisms of adsorption, and compete with each other [2]. Liu et al. [50], studying competitive adsorption between F^- and As(V) in Fe, Al, and Fe and Al oxyhydroxides, found that Fe oxyhydroxides have a high capacity to adsorb another element in anionic form (As(V)), but very little to adsorb F^-, while Al oxyhydroxides (AlO_xH_y) have a high capacity to adsorb both anions, but the efficiency depends on pH value, and there is an important competition between both anions for adsorption sites; and, finally, the mixed oxyhydroxides $FeAlO_xH_y$ have a high capacity to adsorb F^- and As(V) over a wide range of pH, and no competition was observed.

Taking into account that in the forest soil there is a lower decrease of Cr(VI) and F^- adsorption in the binary system as compared to the simple system, it would indicate absence of competition between F^- and Cr(VI) for adsorption sites in that soil, which could be due to its high concentrations of Fe and Al oxyhydroxides (Table S1, Supplementary Material), with a high capacity to adsorb F^- and Cr(VI) without competition at the concentrations tested (up to 6 mmol L^{-1} for each one). However, it must be noted that none of the materials here studied show a high capacity to adsorb simultaneously both pollutants.

Overall, in cases of contamination where both pollutants are present simultaneously, the forest soil would be able to retain up to 70% of F^- (Figure 1), but the incorporation of pine bark would be needed to also adsorb high percentages of Cr(VI). Regarding the vineyard soil, the incorporation of pine bark would increase Cr(VI) and F^- adsorption up to 94% and 86%, respectively, while oak ash could increase F^- adsorption up to 74%, as shown in a previous study [13].

3.2. Fitting to Adsorption Models

Table 1 shows fitting of adsorption data to the Langmuir model for F^- and Cr(VI) in binary and simple systems, while Table 2 presents fitting to the Freundlich model.

As shown in this Table, the existence of high errors prevents valid adjustments to the Langmuir model in most sorbents. In the case of Cr(VI), R^2 values corresponding to the Langmuir model are higher than 0.9 for the two sorbents, which are adjustable. Table 2 shows that in the Freundlich model R^2 values are higher than 0.93, except for hemp waste. Therefore, Cr(VI) adsorption is better adjusted to the Freundlich's equation, coinciding with what was pointed out by other authors [13,44,45,51,52]. In the case of F^-, R^2 values are generally high for both models (in those sorbents which are adjustable), as found in previous studies [10,13,46], but in the current work there are several sorbents for which the errors associated to parameters in the Langmuir equation are also too high, preventing from valid adjustments being made in these cases.

Table 1 also shows that in the materials for which the adjustment of F^- adsorption to Langmuir is good, Q_{max} (maximum adsorption capacity) is higher in the simple than in the binary system, coinciding with that reported by Li et al. [2], which would indicate a competition between F^- and Cr(VI) for the adsorption sites in these materials. The K_L parameter is related to the energy of adsorption [53], and in the present study it is generally higher in the binary than in the simple system (Table 1), indicating that this anion occupies the places of higher binding energy, even though F^- adsorption is smaller in the binary than in the simple system. In addition, in the binary system, K_L values for F^- correlated significantly with non-crystalline Al (r = 0.63, $p < 0.01$), and also with cation exchange capacity (r = 0.75, $p < 0.05$).

Table 1. Parameters of the Langmuir model for Cr(VI) and F^- in individual simple (S) and competitive binary (B) systems.

Adsorbent	Adsorbate	Langmuir Parameters				
		Q_{max} (mmol kg^{-1})	Error	K_L (L mmol^{-1})	Error	R^2
Forest soil	Cr (S)	-	-	-	-	-
	Cr (B)	-	-	-	-	-
	F (S)	-	-	-	-	-
	F (B)	-	-	-	-	-
Vineyard soil	Cr (S)	-	-	-	-	-
	Cr (B)	-	-	-	-	-
	F (S)	82.050	9.960	0.188	0.030	0.999
	F (B)	43.418	10.653	0.407	0.188	0.959
Pyritic material	Cr (S)	36.594	4.049	2.636	1.012	0.970
	Cr (B)	40.779	5.260	3.320	1.540	0.947
	F (S)	-	-	-	-	-
	F (B)	-	-	-	-	-
Pine bark	Cr (S)	-	-	-	-	-
	Cr (B)	-	-	-	-	-
	F (S)	93.130	12.790	0.297	0.061	0.997
	F (B)	24.991	4.055	0.225	0.063	0.989
Hemp waste	Cr (S)	-	-	-	-	-
	Cr (B)	-	-	-	-	-
	F (S)	107.829	57.080	0.038	0.023	0.997
	F (B)	-	-	-	-	-
Pine sawdust	Cr (S)	-	-	-	-	-
	Cr (B)	-	-	-	-	-
	F (S)	31.90557	4.480	0.345	0.151	0.932
	F (B)	-	-	-	-	-
Mussel shell	Cr (S)	-	-	-	-	-
	Cr (B)	-	-	-	-	-
	F (S)	37.900	3.170	0.420	0.020	0.997
	F (B)	-	-	-	-	-
Oak ash	Cr (S)	-	-	-	-	-
	Cr (B)	26.756	2.752	0.287	0.055	0.994
	F (S)	139.780	37.080	0.140	0.050	0.997
	F (B)	5.808	0.551	0.648	0.074	0.980

Q_{max}: maximum adsorption capacity; K_L: parameter related to the strength of interaction adsorbent/adsorbate; R^2: coefficient of determination; -: error values too high for fitting.

Regarding the Freundlich model, K_F is a main parameter for which higher values are indicative of higher adsorption capacity of the adsorbents [54,55]. Overall, and contrary to that reported by Li et al. [2] for synthetic alumina gels, in the present study K_F values were lower for both anions in the binary than in the simple system (Table 2), which indicates a reduction in adsorption capacity in the binary in relation to the simple system. However, both soils, pine bark, and oak ash did not evidence variation in Cr(VI) adsorption in the presence of F^- (Figure 1). In addition, in the binary system, K_F values (and therefore adsorption capacities) were higher for F^- than for Cr(VI) in both soils, hemp waste, pine sawdust, mussel shell, and oak ash, whereas K_F was much higher for Cr(VI) in the pyritic material and in pine bark.

Table 2. Parameters of the Freundlich model for Cr(VI) and F^- in individual simple (S) and competitive binary (B) systems.

Adsorbent	Adsorbate	Freundlich Parameters				
		K_F (L^n kg^{-1} $mmol^{(1-n)}$)	Error	n	Error	R^2
Forest soil	Cr (S)	1.380	0.260	1.200	0.130	0.990
	Cr (B)	1.303	0.246	1.224	0.125	0.986
	F (S)	26.370	1.260	1.075	0.009	0.999
	F (B)	22.988	1.642	1.026	0.123	0.977
Vineyard soil	Cr (S)	0.890	0.249	1.482	0.182	0.990
	Cr (B)	0.815	0.394	1.349	0.309	0.936
	F (S)	12.180	0.250	0.778	0.002	0.999
	F (B)	11.660	1.737	0.639	0.138	0.926
Pyritic material	Cr (S)	23.343	1.171	0.381	0.045	0.980
	Cr (B)	28.504	0.075	0.344	0.002	0.999
	F (S)	17.240	0.3929	1.208	0.003	0.998
	F (B)	4.584	0.974	1.480	0.191	0.980
Pine bark	Cr (S)	8436.359	3298.200	1.976	0.153	1.000
	Cr (B)	1357.763	1269.334	1.385	0.387	0.941
	F (S)	20.238	0.837	0.771	0.005	0.994
	F (B)	4.675	0.233	0.662	0.038	0.995
Hemp waste	Cr (S)	1.515	0.263	1.094	0.109	0.993
	Cr (B)	0.059	0.053	2.037	0.889	0.876
	F (S)	4.073	0.260	0.910	0.005	0.997
	F (B)	2.875	0.612	0.687	0.145	0.968
Pine sawdust	Cr (S)	10.010	0.000	-	-	1.000
	Cr (B)	1.257	0.149	1.599	0.082	0.997
	F (S)	4.073	0.253	0.908	0.047	0.996
	F (B)	2.875	0.612	0.687	0.145	0.968
Mussel shell	Cr (S)	0.2817	0.1362	2.117	0.301	0.990
	Cr (B)	-	-	9.341	0.892	0.999
	F (S)	10.34	1.260	0.562	0.083	0.971
	F (B)	1.215	0.922	8.843	1.896	0.947
Oak ash	Cr (S)	3.536	0.766	1.112	0.163	0.980
	Cr (B)	5.808	0.551	0.648	0.074	0.980
	F (S)	17.11	0.480	0.842	0.035	0.999
	F (B)	10.585	0.216	10.585	0.020	0.999

K_F: parameter related to adsorption capacity; n: parameter related to the heterogeneity of the sorbent; R^2: coefficient of determination; -: error values too high for fitting.

The Freundlich's n parameter is related to the affinity between the adsorbent and the adsorbate (lower n values indicate higher affinity) [55], informing on the reactivity and heterogeneity of the active sites of the adsorbent. If $n = 1$, the adsorption is linear, while if $n > 1$, the adsorption is chemical, and values of $n < 1$ are indicative of high-energy heterogeneous sites, with strong interactions between molecules of adsorbate, where physical adsorption will be the most favorable, and high-energy sites are the first to be occupied [53,56,57]. Table 2 shows that n is >1 for Cr(VI) in most cases (except for oak ash in the binary system, and for the pyritic material in both binary and simple systems), which would indicate that Cr(VI) adsorption is mainly chemical in both soils and most of the other sorbents studied, while it is mainly physical in oak ash and the pyritic material. For Cr(VI), the n parameter presents very similar values for binary and simple systems in both soils and in the pyritic material, whereas n decreases in the binary system for pine bark and oak ash, and increases for pine sawdust, mussel shell, and hemp waste. Therefore, the presence of F^- generally modifies Cr(VI) affinity for the sorbents (except for both soils and the pyritic material), increasing affinity for pine bark and oak ash, and decreasing it for mussel shell, hemp waste, and pine sawdust.

This was confirmed in practice by the adsorption experiments (Figure 1), with the exception of pine bark and mussel shell, which had a high affinity for Cr(VI) not further affected by the presence of F^-. In the case of F^-, n was <1 in both the binary and the simple systems for most sorbents (except forest soil, which is around 1, and the pyritic material and mussel shell, which is >1) (Table 2). The presence of Cr(VI) does not modify the value of n in forest soil, but caused a decrease in vineyard soil, pine bark, hemp waste, and pine sawdust, and caused an increase in pyritic material, mussel shell, and oak ash. These results would indicate that for most of the sorbents, F^- is adsorbed on high energy heterogeneous sites in both binary and simple systems, and the presence of Cr(VI) affects the interaction with sorbate, favoring it in some cases, and hampering it in other, which coincides with experimental data for forest soil, pyritic material, oak ash, and mussel shell, but not for vineyard soil, pine bark, hemp waste, and pine sawdust.

Li et al. [2], and Deng et al. [48], reported increased F^- adsorption in binary as compared to simple systems, while Cr(VI) adsorption decreased. In addition, in the present study, n values in the binary system were lower for F^- than for Cr(VI) in most sorbents, with the exception of pyritic material and oak ash, which would indicate that the highest energy sites are occupied by F^- in most materials.

In addition, Table 3 shows the fitting of adsorption data to the Temkin model. This model assumes that adsorption is characterized by uniform binding energies distribution up to the maximum level [58], and that adsorption energy decreases linearly with surface occupation. As shown in Table 3, in some cases error values were too high to allow adjustment. Taking into account that this model is considered appropriate for chemical adsorption based on strong electrostatic interactions, those cases where the model fits well can be considered indicative of the relevance of chemisorption, as previously reported by Gao et al. [59] and by Rajapaksha et al. [60].

Table 3. Parameters of the Temkin model for Cr(VI) and F^- in individual simple (S) and competitive binary (B) systems.

Adsorbent	Adsorbate	Temkin Parameter				
		b_t	Error	K_t (L g^{-1})	Error	R^2
Forest soil	Cr (S)	3,176,374.39	0.485	-	-	0.680
	Cr (B)	813,385.42	1.022	2.361	1.303	0.897
	F (S)	126,626.39	2.385	4.083	0.610	0.983
	F (B)	128,112.72	2.604	4.905	0.853	0.980
Vineyard soil	Cr (S)	780,829.49	1.226	2.075	1.242	0.866
	Cr (B)	1,035,341.41	0.911	2.316	1.458	0.874
	F (S)	280,522.19	1.661	5.973	2.167	0.965
	F (B)	272,680.16	0.924	4.480	0.789	0.989
Pyritic material	Cr (S)	463,270.75	0.727	112.489	60.409	0.983
	Cr (B)	-	-	235.022	170.959	0.974
	F (S)	-	-	4.574	1.657	0.939
	F (B)	325,525.16	3.093	4.189	3.104	0.862

Table 3. *Cont.*

Adsorbent	Adsorbate	Temkin Parameter				
		b_t	Error	K_t (L g^{-1})	Error	R^2
Pine bark	Cr (S)	189,257.65	6.035	222.501	214.420	0.781
	Cr (B)	134,453.35	5.739	120.715	66.304	0.912
	F (S)	207,675.77	1.763	7.664	2.080	0.977
	F (B)	-	-	4.268	1.207	0.982
Hemp waste	Cr (S)	-	-	2.755	2.094	0.824
	Cr (B)	-	-	-	-	-
	F (S)	-	-	-	-	-
	F (B)	-	-	2.430	1.127	0.943
Pine sawdust	Cr (S)	-	-	25.597	29.425	0.464
	Cr (B)	-	-	-	-	-
	F (S)	-	-	-	-	-
	F (B)	-	-	-	-	-
Mussel shell	Cr (S)	-	-	-	-	-
	Cr (B)	-	-	1.144	0.763	0.717
	F (S)	-	-	-	-	-
	F (B)	-	-	2.052	1.302	0.693
Oak ash	Cr (S)	-	-	2.221	0.818	0.935
	Cr (B)	-	-	3.808	0.417	0.996
	F (S)	-	-	15.540	9.444	0.910
	F (B)	-	-	4.866	1.858	0.956

b_t: Temkin isotherm constant; K_t: Tempkin isotherm equilibrium binding constant; R^2: coefficient of determination; -: adjustment not allowed due to high error values.

4. Conclusions

Competitive adsorption was studied for F$^-$ and Cr(VI) in binary systems where the same concentrations of both pollutants were added simultaneously. The pH values of the adsorbent materials used determine their affinity for both pollutants, causing those with higher pH to have a higher affinity for F$^-$, while lower pH values favor Cr(VI) adsorption. In the soils studied, the simultaneous presence of F$^-$ and Cr(VI) does not modify Cr(VI) adsorption, but decreases that of F$^-$ compared to the simple systems, especially in the vineyard soil. The presence of Cr(VI) also hinders F$^-$ adsorption in all by-products, while F$^-$ decreased Cr(VI) retention in pine sawdust, hemp waste, and mussel shell. When both pollutants are present simultaneously, Cr(VI) occupies the highest energy adsorption sites in both soils, and also in pine bark, while they are occupied by F$^-$ in the other adsorbent materials studied here. In episodes of contamination in which F$^-$ and Cr(VI) are involved simultaneously, the use of only one of the bioadsorbents could be not effective to successfully retain both pollutants. However, the problem could be addressed using mixtures of two bioadsorbents. Specifically, pine bark would adsorb most Cr(VI), and oak ash would perform very effectively in removing F$^-$ from binary systems, which could be of great aid in the case of soils having low adsorption capacity for these two pollutants.

Supplementary Materials: The following are available online at http://www.mdpi.com/2227-9717/7/10/748/s1, Table S1: General characteristics of the sorbent materials (average values for 3 replicates, with coefficients of variation always <5%), Figure S1: Infrared spectrum of forest soil, Figure S2: Infrared spectrum of vineyard soil, Figure S3: Infrared spectrum of pyritic material, Figure S4: Infrared spectrum of fine mussel shell, Figure S5: Infrared spectrum of pine bark, Figure S6: Infrared spectrum of oak ash, Figure S7: Infrared spectrum of hemp waste, Figure S8. Infrared spectrum of pine sawdust, Figure S9: Fitting of Cr(VI) adsorption data to the Freundlich model for simple (S) and binary (B) experiments, Figure S10: Fitting of F$^-$ adsorption data to the Freundlich model for simple (S) and binary (B) experiments.

Author Contributions: Conceptualization, E.Á.-R., M.J.F.-S., and A.N.-D.; methodology, E.Á.-R., M.J.F.-S., and A.N.-D.; software, G.F.-C., M.A.-E., J.C.N.-M., and D.F.-C.; validation, E.Á.-R., M.J.F.-S., A.N.-D., M.A.-E., J.C.N.-M., and D.F.-C.; formal analysis, A.Q.-F.; investigation, A.Q.-F. and G.F.-C.; resources, E.Á.-R. and M.A.-E.; data curation, E.Á.-R., M.J.F.-S., and A.N.-D.; writing—original draft preparation, A.Q.-F., E.Á.-R., and M.J.F.-S.;

writing—review and editing, A.N.-D.; visualization, All authors; supervision, E.Á.-R. and M.A.-E.; project administration, E.Á.-R. and M.A.-E.; funding acquisition, E.Á.-R. and M.A.-E.

Funding: This research was funded by the SPANISH MINISTRY OF ECONOMY AND COMPETITIVENESS by means of the research projects CGL2012-36805-C02-01 and CGL2012-36805-C02-02. It was also partially financed by the European Regional Development Fund (FEDER in Spain). The APC was not funded but waived by MDPI.

Conflicts of Interest: The authors declare no conflict of interest.

References

1. Aoudj, S.; Khelifa, A.; Drouiche, N.; Belkada, R.; Miroud, D. Simultaneous removal of chromium(VI) and fluoride by electrocoagulation–electroflotation: Application of a hybrid Fe-Al anode. *Chem. Eng. J.* **2015**, *267*, 153–162. [CrossRef]
2. Li, T.; Xie, D.; He, C.; Xu, X.; Huang, B.; Nie, R.; Liu, S.; Duan, Z.; Liu, W. Simultaneous adsorption of fluoride and hexavalent chromium by synthetic mesoporous alumina: Performance and interaction mechanism. *RSC Adv.* **2016**, *6*, 48610–48619. [CrossRef]
3. Rafique, T.; Naseem, S.; Bhanger, M.I.; Usmani, T.H. Fluoride ion contamination in the groundwater of Mithi sub-district, the Thar Desert, Pakistan. *Environ. Geol.* **2008**, *56*, 317–326. [CrossRef]
4. Kumar, A.; Parimal, P.; Nataraj, S.K. Bionanomaterial scaffolds for effective removal of fluoride, chromium, and dye. *ACS Sustain. Chem. Eng.* **2016**, *5*, 895–903. [CrossRef]
5. Wang, Y.; Reardon, E.J. Activation and regeneration of a soil sorbent for defluoridation of drinking water. *Appl. Geochem.* **2001**, *16*, 531–539. [CrossRef]
6. Li, L.; Li, Y.X.; Cao, L.X.; Yang, C.F. Enhanced chromium (VI) adsorption using nanosized chitosan fibers tailored by electrospinning. *Carbohydr. Polym.* **2015**, *125*, 206–213. [CrossRef] [PubMed]
7. Alvarez-Ayuso, E.; Garcia-Sanchez, A.; Querol, X. Adsorption of Cr(VI) from synthetic solutions and electroplating wastewaters on amorphous aluminium oxide. *J. Hazard. Mater.* **2007**, *142*, 191–198. [CrossRef]
8. Teng, S.X.; Wang, S.G.; Gong, W.X.; Liu, X.W.; Gao, B.Y. Removal of fluoride by hydrous manganese oxide-coated alumina: Performance and mechanism. *J. Hazard. Mater.* **2009**, *168*, 1004–1011. [CrossRef]
9. Mohapatra, M.; Anand, S.; Mishra, B.K.; Giles, D.E.; Singh, P. Review of fluoride removal from drinking water. *J. Environ. Manag.* **2009**, *91*, 67–77. [CrossRef]
10. Gago, C.; Romar, A.; Fernández-Marcos, M.L.; Álvarez, E. Fluorine sorption by soils developed from various parent materials in Galicia (NW Spain). *J. Colloid Interface Sci.* **2012**, *374*, 232–236. [CrossRef]
11. Quintáns-Fondo, A.; Ferreira-Coelho, G.; Paradelo-Núñez, R.; Nóvoa-Muñoz, J.C.; Arias-Estévez, M.; Fernández-Sanjurjo, M.J.; Álvarez-Rodríguez, E.; Núñez-Delgado, A. F sorption/desorption on two soils and on different by-products and waste materials. *Environ. Sci. Pollut. Res.* **2016**, *23*, 14676–14685. [CrossRef] [PubMed]
12. Quintáns-Fondo, A.; Santás-Miguel, V.; Nóvoa-Muñoz, J.C.; Arias-Estévez, M.; Fernández-Sanjurjo, M.J.; Álvarez-Rodríguez, E.; Núñez-Delgado, A. Effects of changing pH, incubation time, and As(V) competition, on F−retention on soils, natural adsorbents, by-products, and waste materials. *Front. Chem.* **2018**, *6*, 51–60. [CrossRef] [PubMed]
13. Romar-Gasalla, A.; Santás-Miguel, V.; Nóvoa-Muñoz, J.C.; Arias-Estévez, M.; Álvarez-Rodríguez, E.; Núñez-Delgado, A.; Fernández-Sanjurjo, M.J. Chromium and fluoride sorption/desorption on un-amended and waste amended forest and vineyard soils and pyritic material. *J. Environ. Manag.* **2018**, *222*, 3–11. [CrossRef] [PubMed]
14. Alejano, L.R.; Perucho, A.; Olalla, C.; Jiménez, R. *Rock Engineering and Rock Mechanics: Structures in and on Rock Masses*; CRC Press: London, UK, 2014; p. 372.
15. Álvarez, E.; Fernández-Sanjurjo, M.J.; Núñez, A.; Seco, N.; Corti, G. Aluminium fractionation and speciation in bulk and rhizosphere of a grass soil amended with mussel shells or lime. *Geoderma* **2012**, *173–174*, 322–329. [CrossRef]
16. Banerjee, S.; Chattopadhyaya, M.C. Adsorption characteristics for the removal of a toxic dye, tartrazine from aqueous solutions by a low cost agricultural by-product. *Arab. J. Chem.* **2017**, *10*, S1629–S1638. [CrossRef]
17. Brás, I.; Teixeira-Lemos, L.; Alves, A.; Pereira, M.F.R. Application of pine bark as a sorbent for organic pollutants in effluents. *Manag. Environ. Qual. Int. J.* **2004**, *15*, 491–501. [CrossRef]

18. Chatterjee, A.; Lal, R.; Wielopolski, L.; Martin, M.Z.; Ebinger, M.H. Evaluation of different soil carbon determination methods. *Crit. Rev. Plant Sci.* **2009**, *28*, 164–178. [CrossRef]

19. Coelho, G.F.; ConÇalves, A.C.; Tarley, C.R.T.; Casarin, J.; Nacke, N.; Francziskowski, M.A. Removal of metal ions Cd (II), Pb (II) and Cr (III) from water by the cashew nut shell *Anarcadium occidentale* L. *Ecol. Eng.* **2014**, *73*, 514–525. [CrossRef]

20. Dlapa, P.; Bodí, M.B.; Mataix-Solera, J.; Cerdà, A.; Doerr, S.H. FT-IR spectroscopy reveals that ash water repellency is highly dependent on ash chemical composition. *Catena* **2013**, *108*, 35–43. [CrossRef]

21. Fackler, K.; Stevanic, J.S.; Ters, T.; Hinterstoisser, B.; Schwanninger, M.; Salmén, L. Localisation and characterisation of incipient brown-rot decay within spruce wood cell walls using FT-IR imaging microscopy. *Enzym. Microb. Technol.* **2010**, *47*, 257–267. [CrossRef]

22. Haberhauer, G.; Gerzabek, M.H. Drift and transmission FT-IR spectroscopy of forest soils: An approach to determine decomposition processes of forest litter. *Vib. Spectrosc.* **1999**, *19*, 413–417. [CrossRef]

23. Kamprath, E.J. Exchangeable aluminium as a criterion for liming leached mineral soils. *Soil Sci. Soc. Am. Proc.* **1970**, *34*, 252–254. [CrossRef]

24. Margenot, A.J.; Calderón, F.J.; Goyne, K.W.; Mukome, F.N.D.; Parikh, S.J. IR spectroscopy, soil analysis applications. In *Encyclopedia of Spectroscopy and Spectrometry*, 3rd ed.; Lindon, J., George, E., Tranter, D.K., Eds.; Academic Press: Cambridge, MA, USA, 2017; Volume 2, pp. 448–454. [CrossRef]

25. McLean, E.O. Soil pH and lime requirement. In *Methods of Soil Analysis, Part 2, Chemical and Microbiological Properties*, 2nd ed.; Page, A.L., Miller, R.H., Keeney, D.R., Eds.; ASA: Madison, WI, USA, 1982; Volume 2, pp. 199–223.

26. Mimura, A.M.S.; Vieira, T.V.A.; Martinelli, P.B.; Gorgulho, H.F. Utilization of rice husk to remove Cu^{2+}, Al^{3+}, Ni^{2+} and Zn^{2+} from wastewater. *Química Nova* **2010**, *33*, 1279–1284. [CrossRef]

27. Movasaghi, Z.; Rehman, S.; Rehman, I. Fourier transform infrared (FTIR) spectroscopy of biological tissues. *Appl. Spectrosc. Rev.* **2008**, *43*, 134–179. [CrossRef]

28. Nóbrega, J.A.; Pirola, C.; Fialho, L.L.; Rota, G.; de Campos, C.E.; Pollo, F. Microwave-assisted digestion of organic samples: How simple can it become? *Talanta* **2012**, *98*, 272–276. [CrossRef]

29. Pavia, D.L.; Lampman, G.M.; Kriz, G.S.; Vyvyan, J.R. *Introdução à Espectroscopia*, 4th ed.; Cengage Learning: São Paulo, Bresil, 2010; p. 700.

30. Rubio, F.; GonÇalves, A.C., Jr.; Meneghel, A.P.; Tarley, C.R.T.; Schwantes, D.; Coelho, G.F. Removal of cadmium from water using by-product *Crambe abyssinica* Hochst seeds as biosorbent material. *Water Sci. Technol.* **2013**, *68*, 227–233. [CrossRef]

31. Saikia, B.J.; Parthasarathy, G. Fourier transform infrared spectroscopic characterization of kaolinite from Assam and Meghalaya, Northeastern India. *J. Mod. Phys.* **2010**, *1*, 206–210. [CrossRef]

32. Sila, A.M.; Shepherd, K.D.; Pokhariyal, G.P. Evaluating the utility of mid-infrared spectral subspaces for predicting soil properties. *Chemom. Intell. Lab. Syst.* **2016**, *153*, 92–105. [CrossRef]

33. Smidt, W.; Meissl, K. The applicability of Fourier transform infrared (FT-IR) spectroscopy in waste management. *Waste Manag.* **2007**, *27*, 268–276. [CrossRef]

34. Sumner, M.E.; Miller, W.P. Cation exchange capacity and exchange coefficients. In *Methods of Soil Analysis, Part 3, Chemical Methods*; Bartels, J.M., Bigham, J.M., Eds.; ASA: Madison, WI, USA, 1996; Volume 3, pp. 437–474.

35. Tan, K.H. *Soil Sampling, Preparation, and Analysis*; Marcel Dekker: New York, NY, USA, 1996; p. 408.

36. Tarley, C.R.T.; Arruda, M.A.Z. Biosorption of heavy metals using rice milling by-products. Characterisation and application for removal of metals from aqueous effluents. *Chemosphere* **2004**, *54*, 987–995. [CrossRef]

37. Tinti, A.; Tugnoli, V.; Bonora, S.; Francioso, O. Recent applications of vibrational mid-Infrared (IR) spectroscopy for studying soil components: A review. *J. Cent. Eur. Agric.* **2015**, *16*. [CrossRef]

38. Mahapatra, A.; Mishra, B.G.; Hota, G. Studies on Electrospun Alumina Nanofibers for the Removal of Chromium(VI) and Fluoride Toxic Ions from an Aqueous System. *Ind. Eng. Chem. Res.* **2013**, *52*, 1554–1561. [CrossRef]

39. Wang, X.S.; Li, Z.Z.; Tao, S.R. Removal of chromium (VI) from aqueous solution using walnut hull. *J. Environ. Manag.* **2009**, *90*, 721–729. [CrossRef] [PubMed]

40. Choppala, G.; Bolan, N.; Lamb, D.; Kunhikrishnan, A. Comparative sorption and mobility of Cr(III) and Cr(VI) species in a range of soils: Implications to bioavailability. *Water Air Soil Pollut.* **2013**, *224*, 1699–1711. [CrossRef]

41. Griffin, R.; Au, A.K.; Frost, R. Effect of pH on adsorption of chromium from landfill-leachate by clay minerals. *J. Environ. Sci. Health A* **1977**, *12*, 431–449. [CrossRef]

42. Cutillas-Barreiro, L.; Ansias-Manso, L.; Calviño, D.F.; Arias-Estévez, M.; Nóvoa-Muñoz, J.C.; Fernández-Sanjurjo, M.J.; Álvarez-Rodríguez, E.; Núñez-Delgado, A. Pine bark as bio-adsorbent for Cd, Cu, Ni, Pb and Zn: Batch-type and stirred flow chamber experiments. *J. Environ. Manag.* **2014**, *144*, 258–264. [CrossRef]

43. Paradelo, R.; Cutillas-Barreiro, L.; Soto-Gomez, D.; Novoa-Muñoz, J.C.; Arias-Estevez, M.; Fernández-Sanjurjo, M.J.; Álvarez-Rodríguez, E.; Núñez-Delgado, A. Study of metal transport through pine bark for reutilization as a biosorbent. *Chemosphere* **2016**, *149*, 146–153. [CrossRef]

44. Fernández-Pazos, M.T.; Garrido-Rodriguez, B.; Nóvoa-Muñoz, J.C.; Arias-Estévez, M.; Fernández-Sanjurjo, M.J.; Núñez-Delgado, A.; Álvarez, E. Cr(VI) adsorption and desorption on soils and biosorbents. *Water Air Soil Pollut.* **2013**, *224*, 1366. [CrossRef]

45. Otero, M.; Cutillas-Barreiro, L.; Nóvoa-Muñoz, J.C.; Arias-Estévez, M.; Álvarez-Rodríguez, E.; Fernández-Sanjurjo, M.J.; Núñez-Delgado, A. Cr(VI) sorption/desorption on untreated and mussel-shell-treated soil materials: Fractionation and effects of pH and chromium concentration. *Solid Earth* **2015**, *6*, 373–382. [CrossRef]

46. Gago, C.; Romar, A.; Fernández-Marcos, M.L.; Álvarez, E. Fluorine sorption and desorption on soils located in the surroundings of an aluminium smelter in Galicia (NW Spain). *Environ. Earth Sci.* **2014**, *72*, 4105–4114. [CrossRef]

47. Alexandratos, V.G.; Elzinga, E.J.; Reeder, R.J. Arsenate uptake by calcite: Macroscopic and spectroscopic characterization of adsorption and incorporation mechanisms. *Geochim. Cosmochim. Acta* **2007**, *71*, 4172–4187. [CrossRef]

48. Deng, H.; Yu, X. Adsorption of fluoride, arsenate and phosphate in aqueous solution by cerium impregnated fibrous protein. *Chem. Eng. J.* **2012**, *184*, 205–212. [CrossRef]

49. Wu, Y.; Ma, X.; Feng, M.; Liu, M. Behavior of chromium and arsenic on activated carbon. *J. Hazard. Mater.* **2008**, *159*, 380–384. [CrossRef] [PubMed]

50. Liu, W.; Wnag, T.; Borthwick, A.; Wang, Y.; Yin, X.; Li, X.; Ni, J. Adsorption of Pb^{2+}, Cd^{2+}, Cu^{2+} and Cr^{3+} onto titanate nanotubes: Competition and effect of inorganic ions. *Sci. Total Environ.* **2013**, *456–457*, 171–180. [CrossRef]

51. Ucun, H.; Bayhan, Y.K.; Kaya, Y.; Cakici, A.; Algur, O.F. Biosorption of chromium (VI) form aqueous solution by cone biomass of *Pinus sylvestris*. *Bioresource Technol.* **2002**, *85*, 155–158. [CrossRef]

52. Núñez-Delgado, A.; Fernández-Sanjurjo, M.J.; Álvarez-Rodríguez, E.; Cutillas-Barreiro, L.; Nóvoa-Muñoz, J.C.; Arias-Estévez, M. Cr(VI) sorption/desorption on pine sawdust and oak wood ash. *Int. J. Environ. Res. Public Health* **2015**, *12*, 8849–8860. [CrossRef]

53. Khezami, L.; Capart, R. Removal of chromium (VI) from aqueous solution by activated carbons: Kinetic and equilibrium studies. *J. Hazard. Mater.* **2005**, *123*, 223–231. [CrossRef]

54. Bhaumik, R.; Mondal, N.K.; Das, B.; Roy, P.K.C. Eggshell powder as an adsorbent for removal of fluoride from aqueous solution: Equilibrium, kinetic and thermodynamic studies. *J. Chem.* **2012**, *9*, 1457–1480. [CrossRef]

55. Vijayaraghavan, K.; Palanivelu, K.; Velan, M. Biosorption of copper(II) and cobalt(II) from aqueous solutions by crab shell particles. *Bioresour. Technol.* **2006**, *97*, 1411–1419. [CrossRef]

56. Behnajady, M.A.; Bimeghdar, S. Synthesis of mesoporous NiO nanoparticles and their application in the adsorption of Cr(VI). *Chem. Eng. J.* **2014**, *239*, 105–113. [CrossRef]

57. Foo, K.Y.; Hameed, B.H. Insights into the modeling of adsorption isotherm systems. *Chem. Eng. J.* **2010**, *156*, 2–10. [CrossRef]

58. Ofomaja, A.E.; Unuabonah, E.I. Kinetics and time-dependent Langmuir modeling of 4-nitrophenoladsorption onto *Mansonia* sawdust. *J. Taiwan Inst. Chem. Eng.* **2013**, *44*, 566–567. [CrossRef]

59. Gao, Y.; Li, Y.; Zhang, L.; Huang, H.; Hu, J.; Shah, S.M.; Su, X. Adsorption and removal of tetracycline antibiotics from aqueous solution by graphene oxide. *J. Colloid Interface Sci.* **2012**, *368*, 540–546. [CrossRef] [PubMed]

60. Rajapaksha, A.U.; Vithanage, M.; Ahmad, M.; Seo, D.C.; Cho, J.S.; Lee, S.E.; Ok, Y.S. Enhanced sulfamethazine removal by steam-activated invasive plant-derived biochar. *J. Hazard. Mater.* **2015**, *290*, 43–50. [CrossRef]

 *processes*

Review

Nitrogen Removal from Agricultural Subsurface Drainage by Surface-Flow Wetlands: Variability

Lipe Renato Dantas Mendes

Independent Researcher, Natal 59064-740, Brazil; liperenato@hotmail.com; Tel.: +55-84-99891-8045

Abstract: Agriculture has long been considered a great source of nitrogen (N) to surface waters and a major cause of eutrophication. Thus, management practices at the farm-scale have since attempted to mitigate the N losses, although often limited in tile-drained agricultural catchments, which speed up the N transport, while minimizing natural removal in the landscape. In this context, surface-flow constructed wetlands (SFWs) have been particularly implemented as an edge-of-field strategy to intercept tile drains and reduce the N loads by re-establishing ecosystems services of previously drained water ponded areas. These systems collect the incoming water volumes in basins sufficiently large to prolong the hydraulic residence time to a degree where biogeochemical processes between the water, soil, sediments, plants, macro and microorganisms can mediate the removal of N. Despite their documented suitability, great intra and inter-variability in N treatment is still observed to date. Therefore, it is essential to thoroughly investigate the driving factors behind performance of SFWs, in order to support their successful implementation according to local catchment characteristics, and ensure compliance with N removal goals. This review contextualizes the aforementioned issue, and critically evaluates the influence of hydrochemistry, hydrology and biogeochemistry in the treatment of N by SFWs.

Keywords: surface-flow constructed wetland; nitrogen load; nitrate; ammonium; organic nitrogen; hydraulic load; hydraulic residence time; temperature; denitrification; biological uptake

Citation: Mendes, L.R.D. Nitrogen Removal from Agricultural Subsurface Drainage by Surface-Flow Wetlands: Variability. *Processes* **2021**, *9*, 156. https://doi.org/10.3390/pr9010156

Received: 7 December 2020
Accepted: 10 January 2021
Published: 15 January 2021

Publisher's Note: MDPI stays neutral with regard to jurisdictional claims in published maps and institutional affiliations.

1. Introduction

1.1. Agriculture as Nitrogen Sources

Nitrogen (N) is an essential element for crop systems, and has been progressively used in agriculture as fertilizers, which made agricultural catchments a major source of N with detrimental effects on the quality of inland and coastal surface waters, including an increased incidence of eutrophication [1–3]. In this context, precipitation and mineralization of the soil organic matter are non-controllable factors which highly regulate the level of N leaching and subsequent loss [1]. Application of fertilizers in agricultural fields, on the other hand, can be regulated at the farm-scale, and anticipates the excess N content in the soil profile prone to leaching [1,4,5]. Therefore, efforts have been made to decrease or optimize the use of fertilizers by balancing N inputs with crop uptake through proper timing and rate of application in order to prevent N accumulation (residual N) and loss [1,6].

Management practices at the farm-scale, often supported by agricultural policies and measures such as the Water Framework Directive, are crucial to reduce N losses and mitigate the impact of agriculture. However, these are often insufficient to reduce N losses to desired levels and protect surface waters [5,7,8]. This normally occurs when the N surplus or storage in the soil profile is particularly high. A common challenge is to find a fine balance between N availability and crop growth through the application of fertilizers so that optimal production and minimal N leaching are achieved [4–6]. Furthermore, the heterogeneous distribution of N across agricultural fields complicates this challenge [6]. Therefore, application of fertilizers above the required level, thus producing N surpluses, are normally the case [4]. In spite of that, crop rotations that include perennial crops have

501

demonstrated great capacity to reduce N leaching [5,9], although limitations can occur owing to climatic variations [6] or difficulties to make the activity profitable [7]. Moreover, controlled drainage has been used to control the loss of N by regulating the water table level at the site and the resulting outflow [10]. However, this practice can be somewhat complex (e.g., in steep terrains) and hinder crop production [6,10], besides increasing the risk of surface runoff and phosphorus loss due to the emergence of reducing conditions in the soil [11]. Cover crops, in turn, have been reported to reduce N leaching through uptake and storage of organic N in some plant species between crop seasons [6]. Short and cold periods between crop harvest and new planting, however, may limit the effectiveness of this practice. The effect of tillage, on the other hand, can be negligible [1,4] or even negative by supporting N mineralization and subsequent leaching [5–7]. Finally, these practices may also be limited in catchments rich in organic matter, where N mineralization is promoted [1], or in the so-called critical source areas, where N losses are markedly high.

Subsurface drainage has been largely utilized in agricultural catchments with low-permeable and fine-textured soils (e.g., loamy and clayey soils) to allow proper water infiltration and prevent waterlogging at the site so that agricultural activities can be carried out [12]. However, the drainage networks also function as a direct conduit of nutrients, speeding up the transport of N to surface waters downstream, while minimizing surface runoff and the transport of sediment particles [1,7,10]. Macro-pores in the soil profile and higher soil permeability can also enhance the leaching of N to tile drains and subsequent transport by promoting preferential flow from the soil surface [13,14]. The transport of N occurs according to precipitation events, which control the water flow, and, together with the N content of the soil, regulate the N concentration in tile drains [1,5]. The resulting amount of N lost then refers to the N load in tile drain (Equation (1)). Moreover, subsurface drainage discharge skips natural removal mechanisms for N in the landscape, often resulting in N loads or concentrations sufficiently high to compromise aquatic ecosystems [4,6].

$$\text{N load in tile drain } = \text{ water flow } \left(\text{m}^3 \, \text{yr}^{-1} \right) \times \text{N concentration } \left(\text{g m}^{-3} \right) \qquad (1)$$

Tile-drained agriculture can be highly diverse in space and time in relation to agricultural practices, geology, soil type, topography, hydrology and climate, which all contribute to determine the level of N loss between catchments, as well as within the same due to seasonality and annual differences [7,15]. This consequently results in variable N loads and fractions of the N forms transported in tile drains, as these depend on the local catchment characteristics. Nitrogen is transported in the dissolved forms of nitrate (NO_3^-) and ammonium (NH_4^+), i.e., the bioavailable N forms for crop systems [6], as well as particle-bound or organic N, which is usually transported in low amounts. Ammonium, however, is prone to bind to negatively charged soil particles and become less mobile than NO_3^-, or be nitrified, thus converting into NO_3^- [7,9]. Therefore, N transport consists mainly of NO_3^-, as it is highly mobile and may be generated in situ.

1.2. Need to Recover Ecosystem Services at the Edge-of-Field

Artificial drainage of large water ponded areas has been intensified in the last century in order to allow the expansion and development of agriculture [4,7,12]. This occurred because natural hydrology was often insufficient to lower the water table and promote optimal conditions for crop growth. This process commonly resulted in the conversion of wetlands and peatlands into agricultural fields, which not only disrupted the hydrological regime, but also gradually restricted the natural capacity of the landscape to reduce N loads from highlands to surface waters downstream [4,9,12]. Moreover, mineralization of the pre-existing organic N pool has been favored in response to both water drainage and agricultural practices (e.g., tillage and seasonal vegetation), consequently increasing the potential N content prone to leaching [6]. Thereby, N losses have been increasingly recurrent, especially in soils rich in organic matter.

The reduced capacity of the landscape to decrease N loads, in addition to the complex spatiotemporal dynamics of N losses from tile-drained agriculture and limitations of management practices at the farm-scale described in Section 1.1, have demanded the implementation of strategies at the catchment-scale capable of effectively lowering N loads in tile drains to acceptable levels. Accordingly, construction or restoration of surface-flow systems has been critical in order to re-establish ecosystem services so that significant reduction of N loads from highlands can occur again [4,12]. Therefore, these systems are normally designed to reproduce the N removal mechanisms of natural wetlands. In addition to that, flood control and enhancements in biodiversity are promoted. Surface-flow systems are located at the edge of agricultural catchments and play a fundamental role on increasing the hydraulic residence time (HRT) of the water flow at the outlet of tile drains so that removal of dissolved and any remaining fraction of particle-bound or organic N in water can occur through mechanisms deemed low-cost. The process is primarily promoted by enlarging the area or volume of the surface-flow medium, thus decelerating the water flow, which subsequently favors sedimentation of particles as well as biogeochemical cycling of N between the water, soil, sediments, plants, macro and microorganisms. At this stage, two N removal mechanisms are especially promoted, i.e., assimilation and storage of N into organic forms by the locally existing biota; and denitrification, which depends on carbon availability and anaerobic conditions. Surface-flow systems normally contain hydrophyte plants and hydric soils, which contribute to these mechanisms.

As a result of the above, edge-of-field measures deemed cost-effective are often necessary to intercept the drainage networks and mitigate the effects of N discharge into surface waters [4,6–8]. These measures are, therefore, commonly recommended in critical source areas [9]. It is known that HRT is a key factor regulating N removal [10,12], thus the water flow in tile drains must primarily slow down. Systems that utilize this approach include surface-flow constructed wetlands (SFWs), restored wetlands and drainage ditches, which allow the subsurface drainage discharge to be collected into a basin so that biogeochemical processes ultimately resulting in the removal of N can occur. These systems have been widely used for decades, thus with proven records to suitably reduce the N loads from tile drains.

1.3. Aim of the Review

Among the surface-flow systems mentioned in Section 1.2, SFWs have become the prevalent practice, accounting for a growing body of research in different aspects, as described in Section 2. However, despite the progress, great variability in N treatment within and between SFWs is still observed to date, which often leads to suboptimal performance and uncertain collective effect in watersheds, ultimately complicating estimates of cost-efficiency for planned systems. Therefore, a thorough understanding of this issue is fundamental to ensure that local N removal goals are achieved in the short and long-term. The successful application of SFWs may subsequently minimize interventions in agricultural activities and production. In line with the above, this review aims to (i) describe the SFWs located at the edge of tile-drained agricultural catchments and intended for N removal, (ii) discuss the driving factors behind performance, the causes of variability and related processes, (iii) highlight their strengths and limitations in relation to N treatment, and (iv) suggest plausible outcomes for specific conditions.

This review has no focus on measures that mitigate N losses at the farm-scale, nor on approaches that intensify denitrification at the edge of agricultural catchments (e.g., woodchip bioreactors and riparian buffers), but rather on the use of SFWs as a strategy to increase the HRT of tile drain discharge and promote N removal. Finally, this review avoided citing studies dealing with effluents other than agricultural subsurface or tile drainage, and bases its structure on that found in Mendes [16].

2. Variability in the Performance of Surface-Flow Constructed Wetlands

The past decades have demonstrated that implementation of SFWs at the edge of tile-drained agricultural catchments is a feasible strategy to reduce N loads, especially in critical source areas. This has led to the widespread use of this practice in large national plans—so that a cumulative effect could be achieved—intending to reduce N pollution to coastal and inland surface waters [9,17,18]. Therefore, SFWs are strategically located systems, built in areas where no natural wetland previously existed, and designed to target individual catchments, hence with great potential to cope with large watersheds when used collectively.

The successful implementation of SFWs for this purpose is currently described by numerous studies reporting varying levels of N removal [19–45]. It is generally observed that the performance of these systems largely depends on the N load in the tile drain, climate and SFW design, which all play a role on regulating the system HRT. The varying characteristics between catchments, seasonality and annual differences affect the aforementioned factors, and contribute to the large intra and inter-variability in the performance of SFWs. This consequently complicates estimations for N removal and results in wide variations in area-based N removal rate (mass area^{-2} time^{-1}) and efficiency (%) within and between SFWs (Figure 1a,b). Thereby, in order to understand the performance of SFWs, it is fundamental to critically evaluate the influence of hydrochemistry, hydrology and biogeochemistry.

2.1. Nitrogen Load and Forms

2.1.1. Nitrogen Removal Rate

Among a multitude of factors, it is primarily important to observe how SFWs respond to the N loads in tile drains that reach them. At this point, enlargement of the surface-flow medium makes the SFW area relevant when calculating the N load (Equation (2)). Nitrogen removal rate (Equation (3)) correlates positively to N load, which is normally a major explanatory factor [17,30,33,39,42], as N removal depends on the inputs of N. Therefore, N removal rate has a direct relationship to N concentration at the SFW inlet, and decreases as N is removed and its concentration is reduced through the system [46,47]. In line with this, a relevant correlation between N removal rate and load is observed when testing the relationship from a compilation of SFWs receiving agricultural subsurface drainage (Figure 2a). Therefore, increasing loads of N tend to enhance the removal rate, and SFWs receiving higher N loads tend to outperform others in a rate basis.

$$\text{N load} = \frac{\text{water flow} \left(\text{m}^3 \text{ yr}^{-1}\right) \times \text{N concentration} \left(\text{g m}^{-3}\right)}{\text{SFW area} \left(\text{m}^2\right)} \tag{2}$$

$$\text{N removal rate} = \text{N load} \left(\text{g m}^{-2} \text{ yr}^{-1}\right) - \text{N export} \left(\text{g m}^{-2} \text{ yr}^{-1}\right) \tag{3}$$

Despite the strong correlation between N removal rate and load, Tolomio et al. [28] demonstrated through multiple linear regression models that N concentrations (flow-weighted) at the inlet and outlet of a SFW were still strongly correlated (direct relationship) ($R^2 = 0.60$; regression coefficient = 0.67), especially for NO_3^- ($R^2 = 0.63$; regression coefficient = 0.90). Thus, the study highlighted a relatively small contribution of the SFW to the reduction of N concentration. Similarly, Steidl et al. [26] found a strong negative correlation between N concentration at the inlet and the reduction of N concentration through the system (Kendall's τ coefficient with $p < 0.001 = -0.30$). This study indicated therefore an approximation of the N concentration at the outlet to that at the inlet as the latter increased. This study also displayed contrasting results to the above, whereby N removal rate strongly correlated negatively with N concentration (Kendall's τ coefficient with $p < 0.001 = -0.13$). The study found, however, a significant direct relationship between N concentration and water flow ($p < 0.001$; logarithmic function), thus suggesting that lower N removal rates given higher N concentrations occurred due to an associated shortening of HRT, taking into account the effect of contact time for N removal (Section 2.2.2). Thereby, these studies suggest that the effect of N load on N removal, especially on reducing the N concentration, may be weakened if accompanied by significant increments in water flow (Section 2.2.1). The inverse relationship between N load and reduction of N concentration through the system is indeed supported by the first order model presented in Kadlec [46,47]; by Tanner and Sukias [41], who demonstrated relevant positive correlations between NO_3^- load and concentration at the outlet (R^2 = 0.40–0.58; linear functions); and by Tanner et al. [42], in which this correlation for NO_3^- was significant ($p < 0.05$; analysis of covariance). Hence, it can be expected that increasing N loads support higher N removal rates at the expense of lower reductions in N concentration through the system.

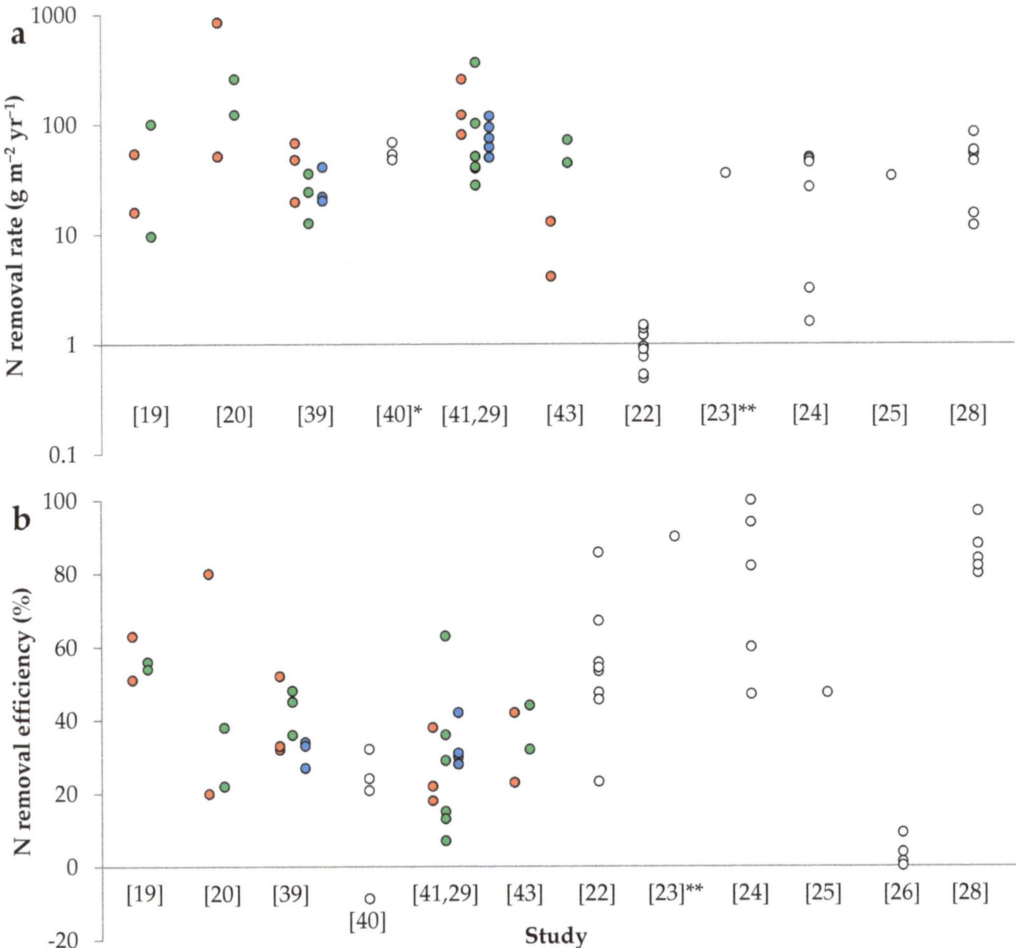

Figure 1. Annual variation in nitrogen (N) removal rate (**a**) and efficiency (**b**) from agricultural subsurface drainage within and between surface-flow constructed wetlands. Reference studies are indicated in square brackets. Wetlands in the same study are distinguished by different colors. * Include a year with net N export (-38.9 g m^{-2}). ** Average.

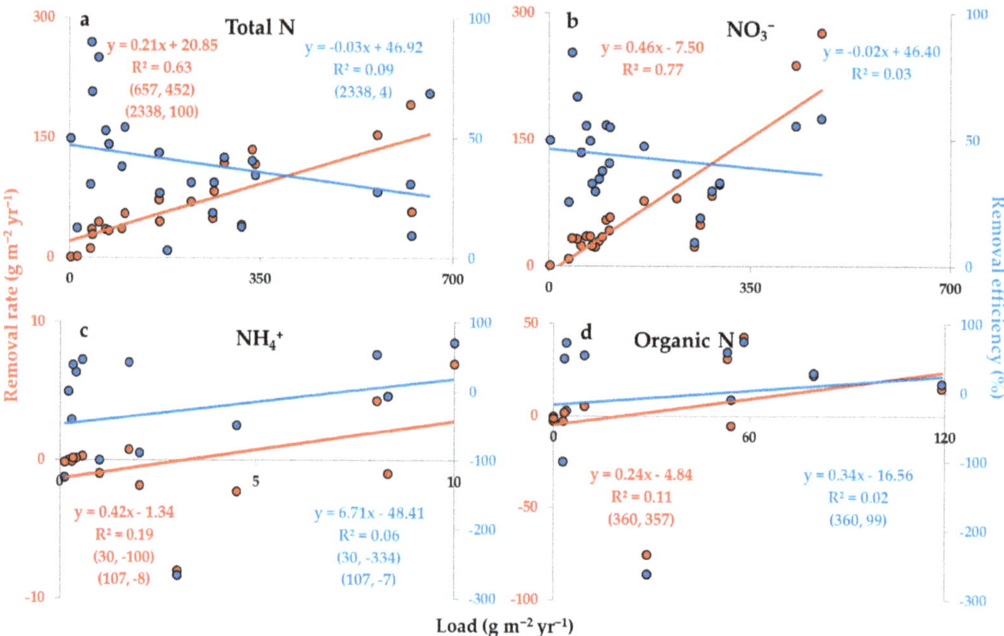

Figure 2. Simple regression between nitrogen (N) load and removal rate (data in red) and efficiency (data in blue) for total N (**a**) (data from Table 1) and the N forms nitrate (NO$_3^-$) (**b**), ammonium (NH$_4^+$) (**c**) and organic N (**d**) (data from Table 2). The numbers in brackets indicate outliers (x, y) removed from the analysis. Note that the scales for total N and NO$_3^-$ are equal.

Nitrogen load varies between SFWs depending on the N load in the tile drain (i.e., the water flow and N concentration) and the SFW area. Nitrogen concentration is sometimes reported to have a direct–yet inconsistent–relationship to water flow up to a certain threshold, above which a dilution effect occurs [22,26,44]—although this effect is not always obvious [42]. When examining a compilation of SFWs receiving agricultural subsurface drainage, it is observed that these parameters vary widely (e.g., 3.4–30.0, median 10.4 mg L^{-1} in N concentration), thus explaining the large differences in N load (2–2338, median 181 g m^{-2} yr^{-1}), which subsequently contribute to the variation in N removal rate (1–452, median 55 g m^{-2} yr^{-1}) (Table 1), according to the strong correlation between N removal rate and load.

Although N load is a strong explanatory factor for N removal rate, it is still common to observe SFWs receiving comparable N loads with large differences in removal rate (Table 1), which probably results in differing correlation strengths between systems. In these cases, ascertaining the fractions of the different N forms at the SFW inlet, i.e., NO$_3^-$, NH$_4^+$ and organic N, may be relevant, as these can affect differently the overall performance [20,48], and contribute to the variability. Specifically, it can be relevant to analyze whether the removal rate of each N form responds differently to its load. When testing the correlation strength between load and removal rate for the different N forms from a compilation of SFWs receiving agricultural subsurface drainage, it is observed that NO$_3^-$ removal rate clearly responds more promptly to the variation of its load (regression coefficient = 0.46) than the other N forms (regression coefficients = 0.24–0.42), whose simple linear regression models are rather weak (R^2 < 0.20) (Figure 2b–d). Other studies also support and clearly state the major role of NO$_3^-$ for overall performance [20,48], thus indicating that higher fractions of NO$_3^-$ from total N at the SFW inlet are expected to enhance the N removal rate. Fortunately, this is generally the case for SFWs receiving agricultural subsurface drainage, i.e., NO$_3^-$ fractions normally higher than 70% (Table 1), which supports the use of SFWs in this context.

Table 1. Characterization of surface-flow constructed wetlands (SFWs) in relation to the treatment of nitrogen (N) from tile-drained agricultural catchments (ACs). Included are the annual averages of hydraulic load, N load, nominal hydraulic residence time and N removal rate (g m^{-2} yr^{-1}); as well as the average N concentration, input fractions of nitrate (NO$_3^-$), ammonium (NH$_4^+$) and organic N from total N, and N removal efficiency (%) for the total monitoring time.

Country	Name	SFW Area	Ratio SFW:AC Area	Monitoring Time	Hydraulic Load	N Concentration	N Load	NO$_3^-$/NH$_4^+$/Org. N Fraction	Hydraulic Residence Time	N Removal		Study
		m^2	%	yr	m yr^{-1}	mg L^{-1}	g m^{-2} yr^{-1}	%	d	g m^{-2} yr^{-1}	%	
USA	Wetland A	6000	4	2	5.8	11.4	66	>96/-/-	56.8	35	53	[19]
USA	Wetland B	3000	3	2	6.4	15.7	101	>96/-/-	22.7	55	55	[19]
New Zealand	Waikato	260	1	2	25.1	12.7 [a]	657	45/0/55	4.4 [b]	452	69	[20,42]
New Zealand	Northland	898	1.6	2	47.4	11.1 [a]	623	76/5/19	-	192	31	[20]
USA	-	1012	1	3	8.4	-	-	-	-	-	68 [c]	[31]
USA	Wetland A	6000	4	3	8.0	14.0 [d]	336	93/7/0	41.1	136	40	[39]
USA	Wetland B	3000	6	3	5.3	10.4 [d]	166	99/1/0	27.5	73	44	[39]
USA	Wetland D	8000	3.2	3	6.3	8.7 [d]	266	98/2/0	37.9	83	31	[39]
Canada	Walbridge	1215	0.004 [e]	4	58.3	3.4 [c]	263	100/0/0	-	49	19	[40]
New Zealand	Titoki	898	1.6	3	53.1	10.7	564	76/19/5	-	154	27	[41]
New Zealand	Toenepi	293	1.1	5	25.3	13.4	342	83/1/16	-	117	34	[41]
New Zealand	Bog Burn	112.5	0.66	4	40.9	5.8	224	73/2/25	-	70	31	[41]
USA	Wetland 1	1600	0.07	2	4.0	9.8 [d]	39	90/0/10	38.1	12	30	[43]
USA	Wetland 2	4000	0.03	2	6.5	13.2 [d]	96	96/0/4	24.8	37	38	[43]
Switzerland	Boden	720	0.86	2.5	30.4	5.5	167	47.5/5/47.5	11.7	45	27	[45]
France	-	4165	1.2	8	0.1	14.1 [c]	2 [c]	-	7240.1	1 [c]	50 [c]	[22]
Italy	-	3200	5.3	5	5.7	7.0	40	87/-/-	-	36	90	[23]
Italy	-	3750	3	6	2.5	13.5	42	-	58.8	29	69	[24]
USA	Wetland B	3000	6	1	8.8	8.2	72	100/0/0	19.4	34	47	[25]
Germany	-	4632	0.4	4	20.1	9.0 [a]	181	89/0/11	-	-	3	[26]
Netherlands	-	64	0.26	2	-	30.0	-	96/-/-	-	166	58	[27]
Italy	-	3200	7.1	6	7.8	5.9 [a]	54	74/-/-	-	45	84	[28]
New Zealand	Toenepi	293	1.1	1	30.4	10.4	316	96/1/3	-	41	13	[29]
New Zealand	Bog Burn	112.5	0.66	1	38.2	7.4	284	78/4/18	-	119	42	[29]
Sweden	Bölarp	2800	0.14	2	-	-	2338	-	-	100	4	[30]
Sweden	Edenberga	2200	0.37	2	-	-	625	-	-	58	9	[30]
Sweden	Södra Stene	21,000	2.1	2	-	-	14	-	-	2	12	[30]

[a] Median; [b] assuming a constant water depth of 0.3 m; [c] nitrate; [d] flow-weighted concentration; [e] only 5% of the water flow is diverted into the SFW.

In line with the above, it is clear that the loads, concentrations and removal rates of NO_3^- are rather superior to the other N forms in most of the cases when examining a compilation of SFWs receiving agricultural subsurface drainage (Table 2). Moreover, a wide range of NO_3^- loads (2–474, median 96 g m^{-2} yr^{-1}) and concentrations (1.4–15.4, median 8.6 mg L^{-1}) can be observed between systems, thus largely contributing to explain the variation not only in NO_3^- removal rate (1–277, median 35 g m^{-2} yr^{-1}), but also in the overall performance. The latter statement is clearly supported by the stronger correlation outputs of NO_3^- ($R^2 = 0.77$) compared to total N ($R^2 = 0.63$), by which the regression coefficient of NO_3^- (0.46) more than doubled that of total N (0.21) (Figure 2a,b). Furthermore, that statement is supported by the generally dominant NO_3^- fractions (Table 1).

In relation to NH_4^+ and organic N, the latter generally reveals higher concentrations and loads at the inlet (Table 2), which leads to higher fractions of total N (Table 1). These two N forms, however, generally represent less than a quarter of the total N load (Table 1), and reveal low ranges of concentration (generally less than 0.5 and 2.0 mg L^{-1} for NH_4^+ and organic N, respectively) and load (generally up to 10 g m^{-2} yr^{-1} for NH_4^+) at the inlet, while organic N loads tend otherwise to highly vary between SFWs (0–360 g m^{-2} yr^{-1}) (Table 2). Suggestively, some studies demonstrated that transient pulses of organic N can be associated to highly pulsed water flows when the agricultural catchment soil undergoes significant mineralization [20,42]. Despite the weak correlation between organic N load and removal rate ($R^2 = 0.11$; Figure 2d), its removal rate was comparably variable (−75–357 g m^{-2} yr^{-1}) to the load between systems (Table 2). This indicates a wide variance in the treatment performance of organic N, from little to highly effective SFWs, which is indeed verified when observing the large differences in its removal efficiency (Table 2). Removal rates of NH_4^+, on the other hand, are rather mild (less than 5–8 g m^{-2} yr^{-1}), nearly zero or slightly negative in most of the cases (Table 2), thus normally negligible compared to those of NO_3^- and organic N. The low NH_4^+ loads and removal rates, as well as the weak correlation between these parameters ($R^2 = 0.19$; Figure 2c), indicate therefore that NH_4^+ plays a smaller role in the overall performance than the other N forms.

According to the discussed above, variation in the load of NH_4^+ and organic N appears to have a minor effect in the overall performance from the perspective of a mass balance analysis. Therefore, dominant fractions of these N forms at the SFW inlet may be undesirable when aiming to achieve relatively high and consistent N removal rates. This is probably due to deficient removal mechanisms for these N forms compared to those for NO_3^-, or by the product of N transformation processes, which may generate NH_4^+ and organic N in situ (Section 2.3.1). The latter statement is indeed supported by the many cases, in which net removals of NH_4^+ and organic N are negative, commonly up to −8 g m^{-2} yr^{-1}, although much larger exports may occur (Table 2). Given the above observations, the variation of N fractions at the inlet of SFWs receiving agricultural subsurface drainage (Table 1) consequently contributes to the variability in overall performance and hinders predictability. Thereby, it is important to acknowledge the incoming N fractions in order to better estimate the removal potential.

Table 2. Treatment of the nitrogen (N) forms nitrate (NO_3^-), ammonium (NH_4^+) and organic N in surface-flow constructed wetlands receiving agricultural subsurface drainage. Included are the annual averages of load and removal rate ($g\ m^{-2}\ yr^{-1}$) of the N forms, and nominal hydraulic residence time; as well as the average concentration and removal efficiency (%) of the N forms for the total monitoring time.

Name	NO_3^- Concentration mg L^{-1}	Load g m^{-2} yr^{-1}	Removal g m^{-2} yr^{-1}	Removal %	NH_4^+ Concentration mg L^{-1}	Load g m^{-2} yr^{-1}	Removal g m^{-2} yr^{-1}	Removal %	Organic N Concentration mg L^{-1}	Load g m^{-2} yr^{-1}	Removal g m^{-2} yr^{-1}	Removal %	Hydraulic Residence Time d	Study
Wetland A	11.0	64	36	56	-	-	-	-	-	-	-	-	56.8	[19]
Wetland B	15.4	99	55	56	-	-	-	-	-	-	-	-	22.7	[19]
Waikato	10.9 [a]	295	97	33	0.0 [a]	2	−2	−90	0.7 [a]	360	357	99	4.4 [b]	[20,42]
Northland	8.4 [a]	474	277	58	0.2 [a]	30	−100	−334	1.6 [a]	119	15	13	-	[20]
-	-	-	-	68	-	-	-	-	-	-	-	-	-	[31]
Wetland A	13 [c]	104	42	41	1	8	4	54	0	0	−1	-	41.1	[39]
Wetland B	10.3 [c]	55	25	45	0.1	1	0	45	0	0	−1	-	27.5	[39]
Wetland D	8.5 [c]	87	30	34	0.2	2	1	42	0	0	−3	-	37.9	[39]
Walbridge	3.4	263	49	19	0.0	0.41	0.11	27	-	-	-	-	-	[40]
Titoki	8.1	429	239	56	2.0	107	−8	−7	0.6	29	−75	−263	-	[41]
Toenepi	11.1	282	84	30	0.1	3	−8	−267	2.2	58	43	74	-	[41]
Bog Burn	4.2	164	78	47	0.1	5	−2	−50	1.4	54	−6	−10	-	[41]
Wetland 1	8.6	34	9	25	0.1	0	0	0	1.1	4	3	73	38.1	[43]
Wetland 2	12.4	92	35	37	0.1	0	0	−42	0.8	3	2	50	24.8	[43]
Boden	2.6	79	23	29	0.3	8	−1	−7	2.6	79	22	28	11.7	[45]
-	14.1	2	1	50	0.1	-	-	-	-	-	-	-	7240.1	[22]
-	1.4 [a]	35	-	-	0.0 [a]	-	-	-	1.0 [a]	-	-	80	-	[23]
-	12.8	48	32	67	0.1	0.3	0.1	38	2.2	9	5	54	58.8	[24]
Wetland B	8.1	71	35	50	0.0	0.1	−0.2	−125	0.0	0	−1	-	19.4	[25]
-	8.0 [a]	161	-	-	0.0 [a]	0.2	-	-	1.0 [a]	20	-	-	-	[26]
-	4.3 [a]	40	33	84	-	-	-	-	-	-	-	-	-	[28]
Toenepi	10.0	253	24	9	0.1	1	−1	−100	0.3	3	−3	−100	-	[29]
Bog Burn	5.8	221	81	37	0.3	10	7	70	1.4	53	31	58	-	[29]
Wetland A	11.2 [c]	106	58	55	-	-	-	-	-	-	-	-	35.8	[35]
Wetland D	7.1 [c]	74	24	33	-	-	-	-	-	-	-	-	23.3	[35]

[a] Median; [b] assuming a constant water depth of 0.3 m; [c] flow-weighted concentration.

2.1.2. Nitrogen Removal Efficiency

Nitrogen removal efficiency is useful to compare the treatment performance of SFWs, as it determines the fraction of N load that is removed in the system (Equation (4)). In the same way as N removal rate, the model of N removal efficiency implies that N load must be higher than N export to achieve positive net removal, and that higher N loads and exports increase and decrease, respectively, the removal efficiency. Moreover, the model implies that N removal efficiency varies according to the fraction of the N load leaving the system, i.e., the proportion of N export in relation to N load. As N export tends to increase under higher N loads (discussion below), N removal efficiency can only increase when the increment in N export is not sufficiently high to raise or stabilize the fraction of the N load leaving the system, thus in this case reducing that fraction. Therefore, N exports promptly responding to N loads can markedly suppress the removal efficiency. As a result, N removal efficiency depends on how N export responds to N load, i.e., the degree of change of the former in relation to the variation of the latter.

$$N \text{ removal efficiency} = \left(1 - \frac{N \text{ export } \left(g\ m^{-2}\ yr^{-1}\right)}{N \text{ load } \left(g\ m^{-2}\ yr^{-1}\right)}\right) \times 100 \qquad (4)$$

Although N load accounts for the N inputs into the system, the relationship between N load and export normally weakens the effect of N load on removal efficiency—differently from that for N removal rate (Section 2.1.1). Therefore, it is common to observe SFWs, in which N load plays a minor role in explaining the variation of N removal efficiency [33,41,49]. This is indeed observed when testing the relationship between N load and removal efficiency from a compilation of SFWs receiving agricultural subsurface drainage for both total N and the different N forms ($R^2 < 0.10$; Figure 2a–d). The weak correlations reflected in well distributed values of N removal efficiency on a scale of 0–100% with little influence of N load, as clearly observed for total N and NO_3^- (3–90, median 38% and 9–84, median 45%, respectively; Tables 1 and 2) (Figure 2a,b). Ammonium and organic N, in turn, presented wider variations in removal efficiency, commonly including negative values (-334–70% and -263–99%, respectively; Table 2) (Figure 2c,d). Despite the above, N load was a major explanatory factor for N removal efficiency in a few studies, as observed in Tanner and Sukias [41] ($R^2 = 0.66$; linear function) and Strand and Weisner [30] ($R^2 = 0.83$; logarithmic function). Taking into account the variation in correlation strength between systems (including the regression coefficient), the relationship between N load and removal efficiency may be closely associated to the efficiency of N removal mechanisms, such as denitrification and biological uptake (Section 2.3.1).

Among the correlation tests between N load and removal efficiency described in this review, that for total N was the strongest and negative ($R^2 = 0.09$; Figure 2a), suggesting that N removal efficiency may tend to have an inverse relationship to N load. This observation is indeed supported by many studies [30,33,37,41,49], indicating that increasing N loads tend to suppress the removal efficiency. According to the model for N removal efficiency (Equation (4)), the inverse relationship implies that higher N loads tend to raise N export above a certain threshold, which increases the fraction of the N load leaving the system and consequently makes the SFW less effective. As N removal depends on the inputs of N, the increasing fraction of the N load leaving the system as a function of higher N loads probably relates to the effect of hydrology, i.e., the load and movement of water through the system, which can promote passage of N without treatment. Therefore, in order to better understand the intra and inter-variability in N removal efficiency, factors unrelated to N inputs should be investigated.

2.2. Effect of Hydrology

2.2.1. Hydraulic Load

In the context of this review, hydraulic load quantifies the amount of agricultural subsurface drainage that is discharged into SFWs, thus accounting for the water flow from the tile drain and the area of the system (Equation (5)). Although hydraulic load does not account for the inputs of N, this parameter is critical to further understand the intra and inter-variability of N treatment by SFWs, as it regulates the water flow dynamics and subsequent level of N mixing throughout the system, thus the active hydrological area or volume of the SFW. Finally, hydraulic load regulates the contact time for N removal mechanisms (Section 2.3.1).

$$\text{Hydraulic load} \; = \; \left(\frac{\text{water flow } (\text{m}^3 \text{ yr}^{-1})}{\text{SFW area } (\text{m}^2)} \right) \tag{5}$$

When integrating Equation (5) into Equation (2), it can be seen that N load is the product of hydraulic load and N concentration. As discussed in Section 2.1.1, N load plays a key role in regulating N removal rate. Therefore, it can be expected that the constituent parameters of N load, i.e., hydraulic load and N concentration, contribute to the regulation of N removal rate. Indeed, the combined effect of hydraulic load and N concentration is clearly verified when modelling the performance data of a compilation of SFWs receiving agricultural subsurface drainage (Figure 3a). The model shows that N removal rate increases when there are increments in both hydraulic load and N concentration, which can be expected due to a cumulative effect of N in the system, which subsequently promotes N removal mechanisms. Moreover, the model shows that N removal rate responds more promptly to variations in N concentration under higher hydraulic loads. This observation is indeed implied when relating N removal rate as a product of N load (Equation (6)). Nitrogen removal efficiency, on the other hand, is expected to respond less promptly to variations in N concentration than N removal rate. Thereby, the performance of SFWs subject to intense hydraulic loads may be more variable. The output of the aforementioned model fits relatively well to the observed data (Table 1), explaining 62% of the variation in N removal rate as a function of hydraulic load and N concentration (Figure 3b).

In addition to verifying the combined effect of hydraulic load and N concentration, it is fundamental to determine the individual effect of these parameters, i.e., the degree of contribution of each parameter in explaining the variation in N removal rate. Therefore, this has been tested in this review through multiple regression models by accounting for the operational data from a compilation of SFWs receiving agricultural subsurface drainage (Table 3). The model for total N was nearly significant at 95% confidence interval ($p = 0.06$) and found that both hydraulic load and N concentration explained about a quarter of the variation in total N removal rate ($R^2 = 0.26$). Interestingly, the effect of total N concentration was over four times higher than that of hydraulic load (regression coefficients = 12.33 and 2.76, respectively). The model for NO_3^- showed that both parameters explained more than half of the variation in NO_3^- removal rate ($R^2 = 0.55$). Similarly, this model demonstrated a superior effect of NO_3^- concentration to the hydraulic load (regression coefficients = 8.03 and 3.38, respectively). Thus, the models for total N and NO_3^- clearly demonstrate that N inputs play a major regulatory role in determining N removal rate due to a cumulative effect of N in the system, despite the importance of hydraulic load for N treatment, as emphasized above. Moreover, the models indicate that hydraulic load may only produce observable variations in N removal rate if accompanied by changes in N concentration. These observations highlight the importance of ascertaining the concentration of N in tile drains prior to constructing a wetland if the goal is to achieve marked N removal rates. Finally, the models for NH_4^+ an organic N showed a low explanatory power ($R^2 < 0.15$) and were insignificant at 95% confidence interval ($p > 0.05$). This demonstrates that the concentration of these N forms and hydraulic load had a minor effect on explaining the variation in their removal rates. The low concentration ranges of NH_4^+ an organic N, especially NH_4^+, highly variable organic N removal rates (Table 2), and the apparent susceptibility of these N forms to be generated in the system, as discussed in Section 2.1.1, probably explain the lack of significant effect of their concentrations. Finally, it can be expected that the weak models for NH_4^+ and organic N suppressed the explanatory power of the model for total N, which is lower than that for NO_3^-.

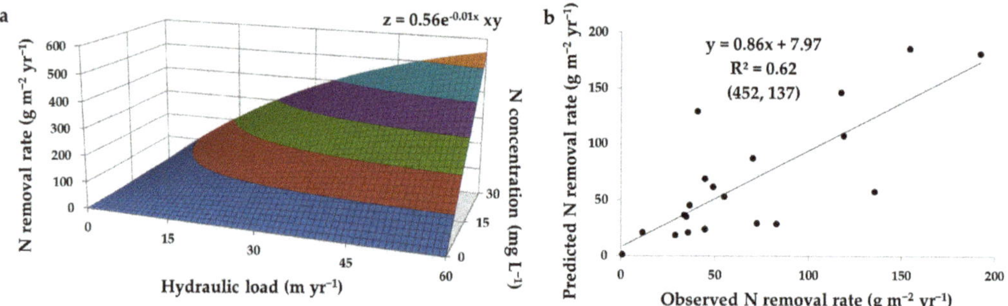

Figure 3. Modelled total nitrogen (N) removal rate as a function of hydraulic load and total N concentration (**a**) according to the relationship between total N removal efficiency and hydraulic load (Figure 4a), and the integration of this relationship into the model of N removal rate as a function of N removal efficiency and N load (Equation (6)); and simple regression between the predicted and observed (data from Table 1) N removal rates (**b**). The numbers in brackets indicate an outlier (x, y) removed from the analysis.

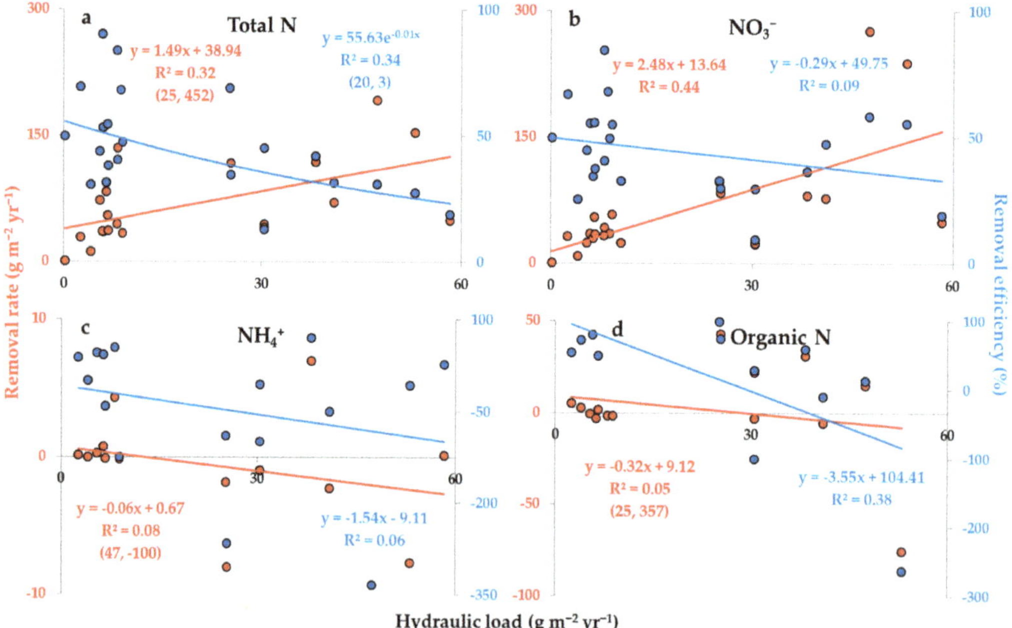

Figure 4. Simple regression between hydraulic load and removal rate (data in red) and efficiency (data in blue) for total nitrogen (N) (**a**) (data from Table 1) and the N forms nitrate (NO_3^-) (**b**), ammonium (NH_4^+) (**c**) and organic N (**d**) (data from Table 2). The numbers in brackets indicate outliers (x, y) removed from the analysis. Note that the scales for total N and NO_3^- are equal.

Table 3. Multiple regression between hydraulic load (m yr^{-1}) and concentration (mg L^{-1}) of the nitrogen (N) forms total N, nitrate (NO$_3^-$), ammonium (NH$_4^+$) and organic N as independent variables, and removal rate (g m^{-2} yr^{-1}) and efficiency (%) of these N forms as dependent variables (data from Tables 1 and 2).

N form	Removal	Number of Observations	R^2	*p*-Value	*p*-Value		Coefficient	
					Hydraulic Load	Concentration	Hydraulic Load	Concentration
Total N	Rate	21	0.26	0.06	0.03 *	0.07	2.76	12.33
	Efficiency	22	0.25	0.06	0.02 *	0.49	−0.65	−1.00
NO$_3^-$	Rate	22	0.55	0.00 *	0.00 *	0.04 *	3.38	8.03
	Efficiency	22	0.07	0.48	0.31	0.95	−0.26	−0.09
NH$_4^+$	Rate	16	0.14	0.38	0.17	0.68	−0.51	5.51
	Efficiency	16	0.14	0.36	0.23	0.29	−2.07	65.38
Organic N	Rate	15	0.00	0.98	0.93	0.86	−0.15	5.79
	Efficiency	12	0.51	0.04 *	0.03 *	0.16	−3.37	51.07

* Significant at 95% confidence interval.

Following the approach described above, this review also tested the individual effect of hydraulic load and N concentration in the variation of N removal efficiency through multiple regression models (Table 3). Similar to the results above for total N, the model for total N removal efficiency was nearly significant at 95% confidence interval ($p = 0.06$), and explained a quarter of the variation in total N removal efficiency ($R^2 = 0.25$). In this case, however, the effect of total N concentration was not even close to being significant at 95% confidence interval ($p = 0.49$), thus being negligible. The effect of hydraulic load, on the other hand, was significant ($p < 0.05$) and slightly negative (regression coefficient = −0.65), thus demonstrating potential to decrease the efficiency of SFWs under higher hydraulic loads. These observations support the assumption made at the end of Section 2.1.2, which relates lower N removal efficiencies to increasing N loads due to the effect of hydrology.

$$\text{N removal rate} = \left(\frac{\text{N removal efficiency (\%)} \times \text{N load} \left(\text{g m}^{-2} \text{ yr}^{-1} \right)}{100} \right) \quad (6)$$

Surprisingly, the models not only for NH$_4^+$ but also for NO$_3^-$ removal efficiency were rather weak ($R^2 < 0.15$) and insignificant at 95% confidence interval ($p > 0.05$). Thereby, hydraulic load and the concentration of these N forms did not show an observable effect in the variation of their removal efficiencies (Table 3). For NH$_4^+$, the same assumptions made above for the model of its removal rate may explain the lack of significant effect here. The variation in NO$_3^-$ removal efficiency, in turn, seems to be unaffected by the inputs of NO$_3^-$ and hydrology, as observed here and by the negligible correlation between NO$_3^-$ removal efficiency and load ($R^2 = 0.03$) (Figure 2b). Hence, these observations strengthen the assumption that NO$_3^-$ removal efficiency highly depends on factors that regulate the functioning of its removal mechanisms. Considering that NO$_3^-$ generally constitutes the largest fraction of total N at the inlet (>70%) (Table 1), the lack of significant effect of its concentration helps explain the lack of significant effect of total N concentration for the model of total N removal efficiency.

Finally, the model for organic N removal efficiency interestingly explained about half of the variation in its removal efficiency ($R^2 = 0.51$) (Table 3). Moreover, only hydraulic load showed a significant effect at 95% confidence interval ($p < 0.05$). Similar to the model for total N, the effect of hydraulic load was negative, although much stronger (regression coefficient = −3.37). These results suggest that removal efficiency of organic N is the most susceptible to vary as a function of hydraulic load among the N forms. Moreover, the results indicate that the negative effect of hydraulic load on total N removal efficiency is caused by significant increases in organic N export under higher hydraulic loads, according to the implications of the model for N removal efficiency (Section 2.1.2). Fortunately, the normally low fractions of organic N at the inlet ($\leq 25\%$) (Table 1) are expected to minimize this undesirable effect. This can indeed be indicated by the much weaker regression coefficient of hydraulic load for the model of total N removal efficiency (-0.65) compared to that for the model of organic N removal efficiency (-3.37). The prompt response of organic N export to variations in hydraulic load may be associated to the apparent susceptibility of this N form to be generated in the system, as discussed in Section 2.1.1, and the loss of suspended organic particles at the SFW outlet, such as plankton and duckweed [45].

As discussed above, hydraulic load can affect the removal rate and efficiency of N and its forms in varying ways. Simple regression models testing the operational data from a compilation of SFWs receiving agricultural subsurface drainage indeed supported the aforementioned findings (Figure 4a–d). Hydraulic load played a relevant role in the removal rate models of total N and especially NO_3^- ($R^2 = 0.32$ and 0.44, respectively), both of which with direct relationships (Figure 4a,b). The removal rate models of NH_4^+ and organic N, on the other hand, indicated a negligible role of hydraulic load ($R^2 < 0.10$) (Figure 4c,d). Removal efficiency, in turn, was only affected by hydraulic load to a relevant degree in the models of total and organic N ($R^2 = 0.34$ and 0.38, respectively), both of which with inverse relationships (Figure 4a,d), while the models of NO_3^- and NH_4^+ showed a low explanatory power for this relationship ($R^2 < 0.10$) (Figure 4b,c).

Although hydraulic load has shown a relevant effect on the removal rate and/or efficiency of total N and its forms, this effect was small, explaining less than 45% of the variation in N removal rate and efficiency (Figure 4a–d). Regarding N removal rate, N load and concentration are still attributed as key factors in this review, indicating again that the cumulative effect of N in SFWs is crucial to regulate N treatment in a rate basis, despite the importance of water flow dynamics and contact time. In contrast, Groh et al. [19] reported hydraulic load as a major factor controlling NO_3^- removal rate ($R^2 = 0.82$; linear function with direct relationship). Khan [50], in turn, supported the relevance of hydraulic load to NO_3^- removal rate through significant positive relationships ($p < 0.05$; t-test and Mann-Whitney U test), although no significant effect was found for total N. Despite the results above, the lack of significant effect of hydraulic load on NO_3^- removal rate ($p > 0.05$; analysis of variance) [51], or even inverse relationships between hydraulic load and NO_3^- and total N removal rates (Kendall's τ coefficients with $p < 0.001 = -0.27$ and -0.10, respectively) [26], have been reported. Amid these different effects of hydraulic load, the aforementioned strong correlation reported in Groh et al. [19] could be explained by a concomitant variation between N load and water flow. Thereby, N loads greatly regulated by hydraulic load—and less by N concentration—may increase the explanatory power of hydraulic load in the variation of N removal rate. This relationship is probably stronger under high and steady N concentrations at the inlet (Figure 3a). This observation may help explaining to some degree the different effects of hydraulic load on N removal rate, and may indicate the conditions by which hydraulic load becomes a strong explanatory factor.

The effect of hydraulic load on N removal efficiency was stronger than that of N load (Figures 2a–d and 4a–d) and concentration (Table 3). Moreover, the significant effect of hydraulic load on N removal efficiency indicated that organic N export particularly varies according to hydraulic load. Thereby, higher loads or concentrations of organic N at the inlet–possibly increasing its fraction—or its generation in the system may enhance the effect of hydraulic load on the efficiency of the system. Taking into account the wide variation in hydraulic load between SFWs receiving agricultural subsurface drainage (0.1–58.3, median 8.4 m yr^{-1}) (Table 1), an increased availability of organic nitrogen in the system could cause mild to severe impacts on the efficiency.

Other studies also described a clear inverse relationship between hydraulic load and N removal efficiency [26,28,37,50], albeit not only for total N, as verified in this review, but also for NO_3^-. Nevertheless, those studies found that total N removal efficiency responded slightly more promptly to variations in hydraulic load than NO_3^- removal efficiency (Kendall's τ coefficients with $p < 0.001 = -0.38$ and -0.34, respectively [26]; and regression coefficients $= -8.22$ ($R^2 = 0.47$) and -7.80 ($R^2 = 0.46$), respectively, in a multiple linear regression model [28]). According to the results of this review, these studies may indicate that the availability of organic N contributed to further decrease the efficiency of the SFWs given higher hydraulic loads, although to a lesser extent than reported here. Moreover, these studies indicate that NO_3^- export is also prone to vary according to hydraulic load, thus with potential effects on the efficiency of the system. Taking into account that NO_3^- fractions at the inlet are normally dominant (Table 1), hydraulic load may play a major role in controlling N removal efficiency in certain cases, as observed in Tolomio et al. [28], although the analyses of this review did not report that. It is important to note, however, that the analyses performed herein accounted for the inter-variability in NO_3^- removal efficiency as a function of hydraulic load, i.e., between SFWs, whereas those studies investigated the intra-variability, i.e., within SFWs. Thus, the latter approach probably explained better the relationship between hydraulic load and NO_3^- removal efficiency, as it was tested within a set of common experimental conditions.

In addition to hydraulic load, the performance of SFWs is also commonly tested by the direct relationship between N concentration at the outlet and water flow, thus implicitly indicating that outlet N concentrations tend to rise under higher water flows. However, the different N forms may respond differently to the variation in water flow. Tanner et al. [42] and Steidl et al. [26], for example, demonstrated that outlet concentrations of total N and NO_3^- strongly correlated to water flow ($p < 0.005$; analysis of covariance [42]; and Kendall's τ coefficient with $p < 0.001 = 0.6$ [26]), whose relationships were best fit by logarithmic functions. Tanner et al. [42], however, did not find a significant correlation between outlet NH_4^+ and organic N concentrations and water flow, whereas Steidl et al. [26] reported an inverse relationship for NH_4^+ (Kendall's τ coefficient with $p < 0.05 = -0.2$). Other studies also observed increasing outlet total N and NO_3^- concentrations given higher water flows [22,44]–yet sometimes inconsistent [44]–including Tanner et al. [42], who reported the relationship through a logarithmic function. Thereby, these results not only indicate that NO_3^- export is prone to vary according to variations in hydraulic load, as discussed above, but also show that outlet NO_3^- concentration tends to respond more promptly to variations in water flow in its lower range. The latter statement likely relates to limitations of the denitrification process under high water flows, which result in short HRTs, potentially insufficient to allow adequate processing rates (Section 2.3.2). Finally, the results above indicate that N concentrations at the outlet tend to approximate those at the inlet given higher water flows, i.e., more untreated N crossing the system. In contrast, Tanner et al. [44] occasionally found very high outlet N concentrations under low water flows, and suggested that water stagnation in the system may cause the release of N to the water column, e.g., by ammonification (Section 2.3.5), thus increasing N concentration at the outlet. Khan [50] reported higher standard deviations for outlet total N and NO_3^- concentrations in a lower range of water flow, which suggests that N release under this condition may indeed occur.

2.2.2. Hydraulic Residence Time

As mentioned in Section 2.2.1, hydraulic load regulates the active hydrological area or volume of the SFW and the contact time for N treatment. Therefore, it has an intrinsic relationship to the actual HRT, which accounts not only for the water flow from the tile drain to a SFW, but also the volume of the SFW with active water flow subject to N treatment (Equation (7)). Measurement of the latter parameter, however, demands complicated methods related to the analysis of water flow dynamics in the system. Therefore, HRT is generally determined as nominal HRT–referred in this review solely as HRT–which simply replaces the SFW volume with active water flow by the SFW volume (Equation (8)). Because of that, nominal HRT does not determine the actual time that a parcel of water takes to cross the system from the inlet to the outlet, considering that plug-flow is not achieved in practice [52]. Moreover, nominal HRT reflects in an inverse and simpler relationship to hydraulic load compared to actual HRT (Equations (5), (7) and (8)).

$$\text{actual HRT} = \left(\frac{\text{SFW volume with active water flow } (m^3)}{\text{water flow } (m^3 \, yr^{-1})} \right) \tag{7}$$

$$\text{nominal HRT} = \left(\frac{\text{SFW volume } (m^3)}{\text{water flow } (m^3 \, yr^{-1})} \right) \tag{8}$$

Hydraulic residence time is a critical parameter in the treatment of N in SFWs, as the N removal mechanisms depend on a contact time to process N. Therefore, hydraulic loads exceeding a certain threshold can compromise the performance of SFWs by severely shortening the HRT. It is worth mentioning, however, that long HRTs as a product of large SFW areas are likely to contribute more to the efficiency of the system than as a product of deep waters. This is due to the need for water to be in contact with the soil, sediments, plants, macro and microorganisms, which implicitly require SFW area (e.g., for hyporheic exchange), so that N biogeochemical processes and removal can occur (Section 2.3.1). In this context, Song et al. [53] and Guo et al. [54,55] investigated the effect of water depth on N removal in experimental SFWs. The former two studies and the latter study found higher N removal in deeper and shallower systems, respectively. The latter study, however, reported a steady decline in N removal efficiency as the water flow increased. These observations indicate that deeper SFWs may particularly benefit N treatment when the SFW has a small area and is subject to high water flows in order to prolong the HRT.

As observed in Section 2.2.1, the effect of hydraulic load is greater on N removal efficiency than on N removal rate. Thus, it can be expected that variations in HRT would affect the SFW efficiency to a greater degree than the removal of N in a rate basis. Lavrnić et al. [24], for example, found that SFW efficiency generally increased due to longer HRTs rather than increasing N loads. Moreover, simple regression models testing the operational data from a compilation of SFWs receiving agricultural subsurface drainage demonstrated that the effect of HRT on total N and NO_3^- removal rates was negligible (graphical representations not shown; data from Tables 1 and 2). The effect of HRT, however, explained 29% of the variation in NO_3^- removal efficiency (linear function with direct relationship), thus clearly stronger than the relationship with hydraulic load (Figure 4b). This finding supports not only the assumption made in Section 2.2.1 that NO_3^- removal efficiency highly depends on factors that regulate the functioning of its removal mechanisms, but also that longer HRTs promote these mechanisms. Interestingly, longer HRTs also showed a relevant positive effect in the removal efficiency of NH_4^+ ($R^2 = 0.46$; linear function), as well as in its removal rate ($R^2 = 0.39$; logarithmic function), despite the low concentrations and loads of NH_4^+ at the inlet (Table 2). This finding indicates that NH_4^+ removal mechanisms are also affected by the water contact time in the SFW. In relation to organic N, the number of observations was too small (5–7) to attempt any interpretation.

Although the observations above are in line with what would be expected from the effect of HRT, it is worth mentioning that analyses accounting the annual averages of HRT may contain a high degree of uncertainty due to marked variations of HRT throughout the year (Section 2.2.3). In this context, Kadlec [47] proposed a first order model (Equation (9), in which N is a hydraulic efficiency parameter) to account for seasonal variations in hydrology, as well as in NO_3^- concentration at the SFW inlet. Therefore, the model allows comparative evaluations for NO_3^- removal within and between SFWs through a first order uptake rate constant (k).

$$\frac{NO_3^- \text{ outlet concentration } (g \, m^{-3})}{NO_3^- \text{ inlet concentration } (g \, m^{-3})} = \left(1 + \frac{k \, (m \, yr^{-1})}{N \, (-) \times \text{hydraulic load } (m \, yr^{-1})} \right)^{-N} \tag{9}$$

The first order model described above implies not only that HRT relates directly to NO_3^- removal efficiency, but also that NO_3^- removal rate and HRT have an inverse relationship, i.e., lower hydraulic loads decrease the NO_3^- removal rate (Section 2.2.1). Khan [50] supports the implications of the model by showing that longer HRTs significantly decreased outlet total N and NO_3^- concentrations in experimental SFWs ($p < 0.05$; t-test and Mann–Whitney U test). Drake et al. [37], in turn, reported lower outlet NO_3^- concentrations during periods of longer HRT, which resulted in an increase in SFW efficiency. Finally, Steidl et al. [26] found a significant positive effect of HRT on the reductions of total N and NO_3^- concentrations in a SFW (Kendall's τ coefficients with $p < 0.001 = 0.38$ and 0.34, respectively). This study, however, surprisingly reported that longer HRTs also contributed to increase the total N and NO_3^- removal rates (Kendall's τ coefficients with $p < 0.001 = 0.10$ and 0.27, respectively), although to a lesser degree compared to the reduction of their concentrations. Thereby, these results highlight the potential of longer HRTs to improve the overall performance of SFWs. Steidl et al. [26] further demonstrated the positive effect of HRT in reducing total N and NO_3^- concentrations (outlet minus inlet) through highly significant ($p < 0.001$) negative logarithmic functions ($R^2 = 0.19$ and 0.11, respectively). Thus, these results indicated that the reduction in total N and NO_3^- concentrations were minimal above a HRT threshold (20 days), whereas large variations in the concentration reductions occurred in a lower range of HRT (in which the relationships were weaker). These findings indicate that HRT is not a limiting factor for the SFW efficiency if there is sufficient contact time for the N removal mechanisms. Under this condition, it can be expected that other factors are more important in controlling N removal, probably related to N biogeochemical processes (Section 2.3.1).

It can be deduced from the above that the appropriate HRT of a SFW to achieve a certain N removal goal varies between systems. Some review studies described, however, that a minimum of two days of HRT is necessary to achieve any substantial NO_3^- removal [32,33,48].

2.2.3. Seasonality

Surface-flow constructed wetlands receiving agricultural subsurface drainage are largely located in the temperate zone, thus subject to the temperate climate. As a result, water flow from tile drains to SFWs can greatly vary throughout the year, especially in non-irrigated agricultural catchments [20,41,44]. This occurs due to large differences in precipitation patterns (sometimes coupled with snowmelt) over the seasons. Thereby, peak or pulse flow events and steady base or no water flow are normally observed during winter and summer, respectively [22,24,26,35,39,42,43,56], except when the water flow stops due to freezing in winter [31]. Therefore, the water depth of SFWs varies seasonally, in which SFWs occasionally dry out during summer owing to the lack of water flow and intense evapotranspiration, sometimes combined with seepage [19,34,35,43]. This seasonal variation in water flow regulates the hydraulic load and HRT of SFWs (Equations (5), (7) and (8)). This results in hydrological regimes generally characterized by peaks of hydraulic load and short HRTs during winter and the opposite during summer, as well as intermediate values during spring and autumn [20,22,23,25,26,32,38,40–42,45]. The variations in hydrology affect the performance of SFWs by influencing the N load over the seasons, and the subsequent removal of N [23,25,26,35,37–39,41–43,45,56]. Thus, N loads typically peak during winter and are mild during summer.

As hydraulic load has a direct relationship to N load—sometimes clearly reported [22]—N removal rate normally increases during periods of intense discharge (e.g., during winter), and declines when hydraulic load is lower (e.g., during summer), according to the relationship between N load and removal rate (Section 2.1.1). Periods of intense discharge, on the other hand, shorten the HRT, thus normally decreasing the efficiency of the system due to significant increments in N export. Periods of low discharge, on the other hand, contribute to raise the SFW efficiency by prolonging the HRT. Thereby, an ideal period would have hydraulic loads sufficiently high to achieve a significant N removal rate without shortening the HRT to a degree that would make the SFW less effective. In reality, however, seasonal variations in N load and removal can greatly differ between years and SFWs, thus contributing to the intra and inter-variability in performance, especially in relation to N removal efficiency. This occurs not only because of the variations in hydrological regime, but also due to temperature fluctuations and the associated effects on N removal mechanisms, such as denitrification and biological uptake (Section 2.3.6). Therefore, poor overall performance during winter can also occur, as clearly demonstrated by Steidl et al. [26], who reported lower reductions in N concentration and N removal rates in that season as a response to high hydraulic loads and cold temperatures.

Tanner and Kadlec [57] stressed the above through a simple first-order dynamic model whereby hydrological regime is a key factor regulating NO_3^- removal. The model clearly suggests that NO_3^- removal rate and efficiency decrease under more variable water flows. Therefore, SFWs with steadier hydrological regimes, i.e., receiving more stable and consistent water flows from tile drains, seem to improve the treatment of N. This could be verified in Drake et al. [37], who reported in a three-year study that the highest NO_3^- removal rate and efficiency occurred in the year with moderate HRT and the least variation in hydraulic load.

Despite the low efficiency of SFWs normally observed during periods of intense discharge, it is important to note that N loads in these periods are often the highest on an annual basis, which results in the highest cumulative removal of N [26,28,35,37–39,41]. Therefore, effective removal of N during periods with long or appropriate HRTs does not necessarily result in an effective SFW on an annual basis. These observations indicate that periods of intense discharge, such as winter, which commonly supplies SFWs with the highest hydraulic and N loads, are critical for the performance of these systems. Periods of low to moderate discharge, on the other hand, have less significance for annual N removal, as these periods supply lower N loads to SFWs, thus resulting in low cumulative removal of N, despite the systems show high N removal efficiencies.

2.2.4. Area Ratio of Surface-Flow Constructed Wetland for the Agricultural Catchment

In addition to seasonality (Section 2.2.3), the water flow in tile drains is also affected by the area of the agricultural catchment, in which larger areas drain more water and result in higher water flows. Therefore, it is critical to ascertain that the SFW area or volume is large enough to accommodate the cumulative volume of the incoming water flows, and therefore ensure sufficient HRT for N removal mechanisms. As observed in Sections 2.2.1 and 2.2.2, SFW sizing affects the hydraulic load and HRT (Equations (5), (7) and (8)). It is verifiable that the area-based N removal rate referred in this review (Equation (3)) would decline if the SFW area increased. Thus, smaller SFWs are more likely to achieve higher N removal rates due to an increase in the cumulative effect of N in the system. However, smaller SFWs may decrease their efficiency as a result of shorter HRTs. Oversized SFWs, on the other hand, may enhance the seasonal variations in hydrological regime, thus compromising the treatment of N and becoming more prone to drying out during periods of low discharge (Section 2.2.3). Thereby, it is fundamental to dimension SFWs according to their agricultural catchments, thus determining the area ratio of the former in relation to the latter, in order to ensure effective N removal. This is because the area ratio of SFW for its catchment influences the hydrology and thus the treatment of N, especially in relation to N removal efficiency, which responds more clearly to variations in HRT (Section 2.2.2). Specifically, it is crucial to determine the area ratio according to periods of intense discharge in order to ensure that HRT is not significantly shortened when the N loads are the highest (Section 2.2.3).

Some studies indeed found that SFWs with larger areas in relation to their agricultural catchments achieved higher N removal efficiencies [39], especially Baker et al. [7], who clearly tested the relationship. This study found that SFWs with smaller areas in relation to their agricultural catchments received higher NO_3^- loads, which resulted in more NO_3^- passing through the system without treatment, i.e., higher NO_3^- export. When analyzing the data from a compilation of SFWs receiving agricultural subsurface drainage, it is also observed that higher area ratios of SFW for the agricultural catchment do contribute to increase the SFW efficiency (Figure 5a,b). The logarithmic functions produced in this analysis indicate, however, that only minor increments in N removal efficiency are achieved above an area ratio of 1%. The models indicate that 40–45% N removal may be achieved with an area ratio of 1%, while 4% area ratio may result in the removal of half the N load (50%).

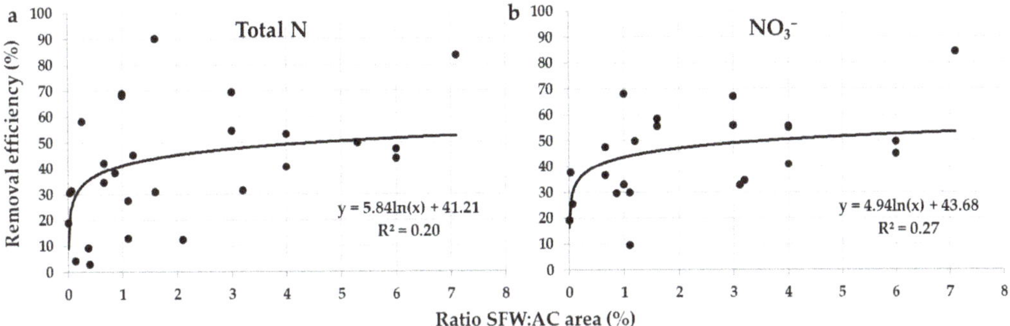

Figure 5. Simple regression between area ratio of surface-flow constructed wetland (SFW) for the agricultural catchment (AC) and removal efficiency of total nitrogen (N) (**a**) (data from Table 1) and nitrate (NO_3^-) (**b**) (data from Tables 1 and 2). Note that the scales for total N and NO_3^- are equal.

Vymazal [33] found a similar relationship between area ratio of SFW for the agricultural catchment and N removal efficiency to that described herein (R^2 = 0.12; logarithmic function). The model in this study indicated that approximately 40% total N removal may be achieved with an area ratio of 1%. Tanner et al. [58], however, reported that a larger area ratio (2.5%) is required to achieve 40% NO_3^- removal. Moreover, this study showed that only 22% NO_3^- removal may be achieved with an area ratio of 1%. The studies above also found that no substantial increase in N removal efficiency is achieved above the aforementioned area ratio thresholds. Tournebize et al. [32], in turn, reported a curve of comparable shape to the relationship described herein (Figure 5a,b) when using the simple first-order dynamic model described in Tanner and Kadlec [57], and recommended a smaller area ratio of 1% to achieve 50% NO_3^- removal, provided the conditions for denitrification are optimal. Finally, Tanner and Kadlec [57] explored the outputs of the simple first-order dynamic model, and these have been found here to relate to the observations above, including a marked decline in NO_3^- removal rate in SFWs with larger areas in relation to their catchments. Moreover, the model predicted that up to 55% NO_3^- removal on average may be achieved with an area ratio of 5%, which is close to the prediction found herein (Figure 5b). Thereby, as discussed in Section 2.2.2, other factors seem to become more important in controlling N removal mechanisms–and thus the SFW efficiency–in case the SFW has an appropriate area for the catchment. In this case, it can be expected that the SFW has sufficient HRT, and therefore factors related to N biogeochemical processes probably regulate the efficiency of the system (Section 2.3.1).

2.2.5. Water Flow Dynamics

So far, attention has been given solely to the effect of incoming water volumes from tile drains in the treatment of N by SFWs. It is important to note, however, that the degree of water distribution through the SFW, i.e., the proximity of water flow dynamics to plug-flow, is also critical for N treatment. This is due to the common incidence of areas or volumes in the SFW with preferential and moderate flows, and stagnant water, which regulate the level of N mixing or homogenization between water parcels, and the contact time of N with the SFW components involved in N removal, i.e., the soil, sediments, plants, macro and microorganisms. As described at the beginning of Section 2.2.1, hydraulic load regulates the active hydrological area or volume of the SFW. Increasing hydraulic loads favor the incidence of preferential flow and stagnant water, both of which restrict N removal; the former by shortening the contact time of N with the aforementioned SFW components, and the latter by reducing the treatment space of the SFW [46]. Taking into account that plug-flow is only theoretical [52], an ideal hydraulic load would allow water to be distributed throughout the SFW with relatively similar contact times between water parcels. This would increase the active hydrological area or volume of the SFW, and consequently improve N treatment.

In the context described above, hydraulic efficiency is an important parameter that estimates the degree of water distribution through the SFW by determining the proximity of the time that a tracer takes to cross a SFW from inlet to outlet in relation to the SFW nominal HRT (Equation (10)) [59]. Similarly, the volumetric efficiency determines the proximity of the actual HRT to the nominal HRT, thus estimating the degree of space utilization of the SFW by the water flow (Equation (11)) [60]. Therefore, these hydraulic parameters are dimensionless and can vary from zero to one, in which values closer to zero indicate more preferential flows and stagnant waters, whereas values approximating one indicate that water flow dynamics is nearing plug-flow. In other words, these hydraulic parameters estimate how much of the SFW space contains active water flow, and is thus being used for N treatment. Thereby, SFWs with high hydraulic parameters tend to better distribute N through the SFW components and equalize the contact time of water parcels, ultimately contributing to improve N treatment. Systems with low hydraulic parameters, on the other hand, normally have preferential flows, stagnant waters, and/or areas or volumes with limited water mobility, where the incoming N loads have little access. As a result, this condition compromises the overall performance of SFWs.

$$\text{hydraulic efficiency} = \left(\frac{\text{time of peak outlet concentration of a tracer (yr)}}{\text{nominal HRT (yr)}} \right) \quad (10)$$

$$\text{volumetric efficiency} = \left(\frac{\text{actual HRT (yr)}}{\text{nominal HRT (yr)}} \right) \quad (11)$$

Despite the importance of the issue described above, studies on water flow dynamics in SFWs receiving agricultural subsurface drainage are scarce. Nevertheless, Lavrnić et al. [61] found a clear difference between the actual and nominal HRT of a 17 year old SFW (6.7 and 8.1 days, respectively). This reflected in hydraulic and volumetric efficiencies of 0.79 and 0.71, respectively, thus indicating that about a quarter of the SFW space was not contributing to N treatment. Further studies, however, are necessary to estimate the potential for increased use of the treatment space of SFWs.

2.2.6. Design Aspects

In addition to hydraulic load (Section 2.2.1), SFW design is critical in controlling hydraulic parameters (Section 2.2.5), and therefore N treatment. Primary design aspects of SFWs that affect hydrology include shape, dimensioning or aspect ratio length to width, position of the inlet and outlet, presence of obstructions to the water flow, bathymetry, and vegetation type and distribution [47,48]. Su et al. [62], for example, reported that aspect ratio length to width can highly affect hydraulic parameters, whereby an aspect ratio higher than five can approximate water flow dynamics to plug-flow (hydraulic parameters ≥ 0.9). This study recommended a minimum aspect ratio of 1.88, in case of site limitations, so that hydraulic parameters are greater than 0.7. Moreover, the study found that a series of parallel inlets can even out the water flow towards a common outlet, and thus optimize hydraulic parameters (reported values = 0.88–0.89). Finally, this study suggested the use of a few slender obstructions in order to improve hydraulic parameters when necessary. Pugliese et al. [63], in turn, reported the relevance of shallow zones between deep areas in a SFW, and the effect of winds opposite the direction of the water flow to enhance water mixing, vertical circulation and hydraulic efficiency. The effect of vegetation type, in turn, was tested by Bodin et al. [64] in experimental SFWs. This study found that emergent vegetation occupied a larger volume in the SFWs, thus significantly reducing the volumetric efficiency compared to SFWs with submerged and free-growing vegetation ($p < 0.05$ and 0.001; method of moments and a modified Gauss model, respectively). Lastly, Guo et al. [54,55] tested the effect of design aspects in experimental SFWs, and support some of the findings described above. These include a higher volumetric efficiency when a number of inlets are evenly distributed and the water flows towards a common outlet, and when the SFW has specific water depths. Moreover, those studies found that volumetric efficiency was significantly affected by aspect ratio length to width ($p < 0.05$; analysis of variance) [55], and that this can increase in SFWs with less scattered vegetation [54].

Despite the relevance of design aspects, few studies investigated the relationship of these to the treatment of N. Guo et al. [54,55] reported that water depth can affect N removal, as discussed in Section 2.2.2. Moreover, N removal rate was found to increase when the inlet and outlet are positioned on opposite sides and in the center of experimental SFWs with rectangular shape [55]. Nitrogen removal efficiency, in turn, was found to increase in SFWs with less scattered vegetation, in which water flow slows down [54], and denitrification and biological uptake are intensified (Sections 2.3.2 and 2.3.3). Regarding the effect of water flow dynamics, none of these studies found a significant correlation between volumetric efficiency and N removal ($p > 0.01$; bivariate analysis). More studies, however, are needed to elucidate these relationships.

2.2.7. Final Remarks

It has been observed that N treatment, especially N removal efficiency, is probably controlled by factors that regulate N removal mechanisms, such as HRT. Therefore, hydrology is still limited in explaining the variability in performance within and between SFWs, as it does not account for N biogeochemical processes, which ultimately lead to N removal. Thus, investigation of the factors that regulate N transformations in SFWs, by which mass balance analyses are unable to cover, is necessary to further understand the variability in N removal. Specifically, by exploring N removal mechanisms and associated factors, differences in the treatment of the various N forms and the effects of seasonality can be further clarified.

2.3. Nitrogen Removal Mechanisms and Biogeochemical Factors
2.3.1. Overview

Nitrogen from agricultural subsurface drainage, once discharged into a SFW, can either pass through the system unprocessed, as promoted by shorter HRTs (Section 2.2.2), or enter the N cycle (Figure 6). At this stage, different N removal mechanisms, i.e., the processes that reduce the concentration of N forms in water, take place. Some of these mechanisms solely transform a N form into another with no direct contribution to the overall N removal, such as ammonification, nitrification and dissimilatory NO_3^- reduction to NH_4^+ (DNRA) (stoichiometric Equations (12)–(14), respectively). This is because the end products generated, i.e., NH_4^+ and NO_3^-, are still mobile and prone to be present in the water column and be exported. Other mechanisms, however, promote short or long-term storage of N in the system, and include settling of particle-bound and organic N onto the topsoil, and biological uptake or assimilation of NH_4^+ and NO_3^- by the locally existing plants and microorganisms, thus turning these into organic N forms. Finally, volatilization and denitrification are exit mechanisms for N, whereby NH_4^+ and NO_3^- are converted into ammonia and dinitrogen gases (stoichiometric Equations (15) and (16), respectively), which eventually cross the water-air interface into the atmosphere. Given the above observations, the N storage and exit mechanisms are particularly important for the overall N removal, as these result in short, long-term or permanent removal of N from water, thus directly contributing to decrease N export and subsequently increase the system efficiency.

$$CH_4N_2O \text{ (urea as an example)} + H_2O \rightarrow 2NH_3 + CO_2 \therefore NH_3 + H_2O \rightarrow NH_4^+ + OH^- \tag{12}$$

$$2NH_4^+ + 3O_2 \rightarrow 2NO_2^- + 2H_2O + 4H^+ \therefore 2NO_2^- + O_2 \rightarrow 2NO_3^- \tag{13}$$

$$C_6H_{12}O_6 + 3NO_3^- + 3H^+ \rightarrow 3NH_3 + 6CO_2 + 3H_2O \therefore NH_3 + H_2O \rightarrow NH_4^+ + OH^- \tag{14}$$

$$NH_4^+ + OH^- \rightarrow NH_3 + H_2O \tag{15}$$

$$C_6H_{12}O_6 + 4NO_3^- \rightarrow 6CO_2 + 6H_2O + 2N_2 \tag{16}$$

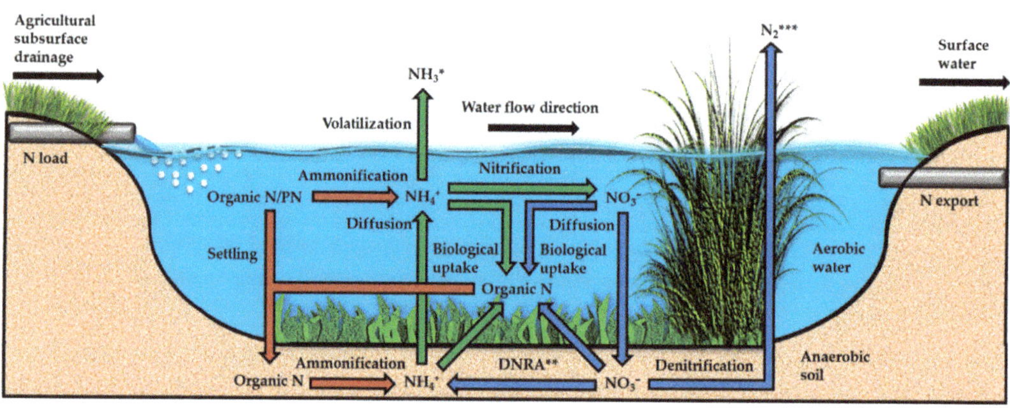

Figure 6. Schematic of nitrogen (N) cycle or removal mechanisms for the N forms nitrate (NO_3^-), ammonium (NH_4^+) and organic N/particle-bound N (PN), represented by blue, green and red arrows, respectively, in a surface-flow constructed wetland receiving agricultural subsurface drainage. * Ammonia. ** Dissimilatory NO_3^- reduction to NH_4^+. *** Dinitrogen.

Among the aforementioned N removal mechanisms, some may process faster than others, thus directly or indirectly contributing to the overall N removal in different degrees. In fact, denitrification is commonly referred to as the main N removal mechanism, therefore responsible for most of the N removal [9,33,48,49,65]. A comparative evaluation of the processing rates of the different N removal mechanisms from a compilation of SFWs receiving agricultural subsurface drainage, however, cannot be properly made to date due to lack of data and/or large differences in study methods (Table 4). Nevertheless, an attempt to make a comparison with the available data indeed suggests that denitrification generally contributes more to the removal of N (0.3–11.8, median/average 3.6/4.7 mg m^{-2} h^{-1}), followed by settling (0.0–4.3, median/average 2.0/2.1 mg m^{-2} h^{-1}) and then biological uptake (0.0–2.5, median/average 0.6/1.0 mg m^{-2} h^{-1}). Moreover, a few studies quantified the recovery (%) of the initial NO_3^- concentration in water by denitrification and biological uptake, and found that the former processed about four to eleven times more available NO_3^- than the latter (Table 4).

Table 4. Average processing rates of nitrogen removal mechanisms in surface-flow constructed wetlands receiving agricultural subsurface drainage, recovery (%) of the initial nitrate (NO_3^-) concentration in water (in brackets), and experimental conditions. Note that the remaining fraction of the total recovery refers to NO_3^- not removed. * Dissimilatory NO_3^- reduction to ammonium.

Name (Scale)	Denitrification	Biological Uptake	Settling	Ammonification	Nitrification	DNRA *	NO_3^- Con-centration	Temperature	Carbon in Sediments	Period	Study
	mg m^{-2} h^{-1} (%)						mg L^{-1}	°C	g m^{-2}	d	
Gully (Laboratory) [a]	38 [b]	-	-	-	-	-	2 [10 [c]]	25	-	20	[66]
Gully (Laboratory) [a]	103 [b]	-	-	-	-	-	2 [40 [c]]	25	-	20	[66]
Gully (Laboratory) [a]	24 [b]	-	-	-	-	-	20 [10 [c]]	25	-	20	[66]
Gully (Laboratory) [a]	221 [b]	-	-	-	-	-	20 [40 [c]]	25	-	20	[66]
Moga, Durbin and Gully (Laboratory) [a]	64 [b]	-	-	-	-	-	10 [20 [c]]	25	-	20	[66]
Moga, Durbin and Gully (Laboratory) [a]	78 [b]	-	-	-	-	-	10 [40 [c]]	25	-	20	[66]
- (Mesocosm 1)	≤9.3 (58)	0.3 (5.5) [d+e]	-	-	-	-	15	-	-	33	[67]
- (Mesocosm 2)		0.5 (9.9) [d+e]	-	-	-	-					[67]
Wetland 1 (Mesocosm)	2.1	-	-	-	-	-	9–20	24 [f]	-	0.7	[67]
Wetland 1 (Mesocosm)	6.2	-	-	-	-	-	9–20	18 [f]	9.1 [e]	0.1	[67]
Wetland 1 (Mesocosm)	11.8	-	-	-	-	-	9–20	25 [f]	9.5 [e]	0.1	[67]
Wetland 2 (Mesocosm)	2.7	-	-	-	-	-	9–20	25 [f]	-	0.7	[67]
Wetland 2 (Mesocosm)	2.0	-	-	-	-	-	9–20	4 [f]	2.1 [e]	0.1	[67]
Wetland 2 (Mesocosm)	3.9	-	-	-	-	-	9–20	17 [f]	5.0 [e]	0.1	[67]
Wetland 2 (Mesocosm)	9.0	-	-	-	-	-	9–20	20 [f]	9.3 [e]	0.1	[67]
Toenepi (Mesocosm)	7.2 (39.3)	1.9 (10.2) [d+e+g+h+i]	-	-	-	0.2 (1.0) [e]	10.4	14 [j]	-	6	[68]
Toenepi (Field)	6.2 [b]	-	-	-	-	-	10.8	-	61 [k]	365	[42]
Boden (Field)	2.9	0.7 [i]	4.3	1.1	3.8	-	1.9–3.0	10	-	210	[45]
Boden (Laboratory) [a]	≤28.8	-	-	-	-	-	1.9–15.6	9	-	4.8	[45]
- (Laboratory) [a]	3.6	-	-	-	-	-	50	20	9.7 [k]	<5	[22]
- (Field)	0.3 (6.3 [l])	2.5 (53.9 [l]) [d+h]	1.3 (27.7 [l]) [m]	-	-	-	1.4	−1.5–27.2	-	1825	[23]
- (Field)	-	0.0 [b, d+h]	0.01 [b, n]	-	-	-	11.8 [l]	2.7–24.3	-	1825	[24]
Wetland B (Field)	-	0.005 [d+h]	0.0	-	-	-	7.5	-	-	153	[25]
- (Field)	-	-	2.7	-	-	-	7.5	9	144.2	577	[26]

[a] Media kept in the dark and anaerobic; [b] value in mg kg^{-1} h^{-1}; [c] carbon concentration; [d] above-ground plant; [e] topsoil (5 cm); [f] sediment-water interface; [g] duckweed; [h] below-ground plant; [i] overlying water; [j] air; [k] value in g kg^{-1}; [l] total nitrogen; [m] topsoil (30 cm); [n] topsoil (15 cm).

The potential of the different N removal mechanisms, however, is better evaluated in individual studies due to common experimental conditions. Reinhardt et al. [45], for example, attested the superiority of denitrification (94% of the N removed) over biological uptake and settling (6% of the N removed). Similarly, Matheson and Sukias [68] reported that denitrification accounted for 77% of the transformed NO_3^-, while biological uptake accounted for a smaller fraction (20%). Besides that, this study found that DNRA contributed with only 2% of the transformed NO_3^-. Xue et al. [67], in turn, implied that 85–91% of the NO_3^- removed had been denitrified, while biological uptake accounted for only 9–15%. Finally, Tournebize et al. [22] described N removal by algae uptake as minimal (0.7–1.5%). Nevertheless, denitrification can sometimes contribute minimally to the overall N removal, as observed by Borin and Tocchetto [23], whose process contributed with around 7% of the N removed, while plant uptake was the main N removal mechanism (61%) followed by settling (31%). This study, however, suggested that the contribution of these processes to the overall N removal will likely change as the SFW matures, which may be a general trend in newly established SFWs when carbon availability for denitrification is still low and vegetation is developing (Sections 2.3.2 and 2.3.3).

Despite the lack of data and comparable study methods for general conclusions (Table 4), it is clear that denitrification has been the most studied process and often appointed as the main N removal mechanism according to individual studies, followed by biological uptake and settling. The processes that transform N into the mobile forms of NH_4^+ and NO_3^-, on the other hand, have been little investigated, as well as the N exit mechanism by volatilization. The available data, however, suggest that DNRA rate and recovery of the initial NO_3^- concentration in water may be normally low (Table 4), probably due to competition with denitrification and biological uptake for the available NO_3^- in the normally anaerobic soils of SFWs (Figure 6). Moreover, the requirement of DNRA for an acidic medium may limit the process (stoichiometric Equation (14)). Nitrification rate, on the other hand, has even been reported to exceed denitrification rate (Table 4), and is promoted by aerobic conditions (stoichiometric Equation (13)). Finally, ammonification rate may be relatively low in certain cases (Table 4), but can vary according to redox conditions and temperature (Section 2.3.5). The lack of studies on volatilization, in turn, presumably suggests that it contributes little to the overall N removal and may therefore play a negligible role [47], probably due to its requirement for basic water (stoichiometric Equation (15)) and the low fractions, concentrations and loads of NH_4^+ at the inlet (Tables 1 and 2). Steidl et al. [26], for example, reported that ammonia was only 3% on average in relation to NH_4^+ at the SFW outlet, indicating that the system pH was not sufficiently high to promote volatilization.

It is recognized that studies on N biogeochemistry in SFWs involve complicated methods in relation to the mass balance approach, by which the relationship between system performance and the incoming water volumes and N inputs can be assessed. However, it is essential to investigate the effect of biogeochemical processes in the context described herein, considering that agricultural subsurface drainage is particularly characterized by high fractions, concentrations and loads of NO_3^- (Tables 1 and 2), while providing little carbon to the system [46,66,69]—with some exceptions [38]—whose availability is required to prompt denitrification (stoichiometric Equation (16)). Moreover, SFWs typically take many years to fully develop their vegetation—i.e., long maturation periods [23,48,70,71]—which is crucial to ensure not only a sufficient source of carbon for effective denitrification, but also a well-established biota for effective biological uptake. Finally, systems located in the temperate zone experience seasonal variations in temperature, which directly affect denitrification and biological uptake rates (Section 2.3.6). Therefore, the performance of SFWs is commonly limited to some degree by suboptimal conditions at the biogeochemical level, which inhibit N removal mechanisms. This observation is supported by the often observed N removal efficiencies below 50% by SFWs receiving agricultural subsurface drainage, despite the varying N inputs and hydrology (Table 1). Thereby, studies accounting for the relationship between biogeochemical factors and the processing rates and/or recoveries of the available N by the different N removal mechanisms are highly encouraged in order to better understand the intra and inter-variability in performance of SFWs. It is especially recommended that these studies take place under field conditions and over several years, as these are rather limited to date (Table 4). The more often conducted laboratory and mesocosm studies (Table 4), on the other hand, are important to investigate the response of N removal mechanisms to certain conditions, but are not able to account for the spatiotemporal complexity at the field-scale, thus producing results that cannot be used to predict the functioning of N removal mechanisms in SFWs [47].

2.3.2. Denitrification

As indicated in Section 2.3.1, denitrification is normally regarded as the main N removal mechanism. Thereby, successful application of SFWs largely depends on the effectiveness of this process. This mechanism is primarily mediated by heterotrophic facultative bacteria that use NO_3^- as the terminal electron acceptor in the absence of oxygen to oxidize organic matter for energy. Therefore, the process requires the presence of NO_3^-, anaerobic conditions, suitable temperatures and a source of organic carbon as electron donor–all conditions normally found in the topsoil of SFWs–so that NO_3^- can be reduced (Figure 6). This consumption of NO_3^- in the anaerobic topsoil creates a concentration gradient [22], by which NO_3^- from the aerobic water diffuses downwards, and typically generates gaseous nitrous oxide (in minor proportions [19,22,49,67]) and dinitrogen as end products, which subsequently leave the system to the atmosphere (Figure 6). Hence, the process ensures permanent removal of N. In line with the above, denitrification rates benefit the higher the NO_3^- concentration in the water column—as attested by Reinhardt et al. [45] ($R^2 = 0.96$), Grebliunas and Perry [66] ($p < 0.05$; analysis of variance) and Xue et al. [67], who reported a sudden increase in denitrification rate after a pulse input of NO_3^-—which largely explains the strong and direct relationship of NO_3^- load and concentration to its removal rate (Sections 2.1.1 and 2.2.1). As a result, denitrification studies in SFWs commonly measure NO_3^- concentration, temperature and carbon availability as part of the methods to estimate the processing rate and recovery of the initial NO_3^- concentration (Table 4).

Among the aforementioned factors, the effect of temperature has been particularly investigated. Overall, temperature has a direct relationship to denitrification rate, as the process is mediated by enzymatic activity [48,49]. Tolomio et al. [28], for example, reported that warmer air temperatures contributed to decrease the concentration of total N at the outlet of a SFW through a multiple linear regression model ($R^2 = 0.60$; regression coefficient = −0.11). Likewise, Drake et al. [37] found that months with higher water temperatures in a SFW contributed to increase the NO_3^- removal efficiency. Finally, Steidl et al. [26] found greater N concentration reductions and removal rates under warmer water temperatures ($p < 0.001$; Kendall's τ coefficients = 0.40 and 0.24 for total N, and 0.40 and 0.26 for NO_3^-, respectively). Moreover, this study described the effect of higher temperatures on increasing the concentration reductions (outlet minus inlet) of total N and NO_3^- through highly significant ($p < 0.001$) negative linear functions ($R^2 = 0.20$ and 0.15, respectively). As noted in this study, the effect of temperature was more pronounced on removal efficiency than on removal rate of N. Although the studies above did not investigate the underlying biogeochemical processes, it can be assumed that denitrification probably intensified under warmer temperatures and contributed to greater N removal. Xue et al. [67], on the other hand, attested the positive effect of temperature specifically on denitrification rate ($R^2 = 0.71$).

Given the varying temperatures in a seasonal basis within and between SFWs (Section 2.3.6), Kadlec [47] proposed a modified Arrhenius temperature relation (Equation (17), in which θ is a temperature factor) in connection with the also proposed first order model (Equation (9)) to calibrate the temperature effect by standardizing the k value at a common temperature of 20 °C (k_{20}), and therefore allow comparative evaluations for NO_3^- removal presumably through denitrification.

$$k = k_{20}\left(\text{m yr}^{-1}\right) \times \theta\,(-)^{(\text{water temperature (°C)}-20)} \tag{17}$$

In addition to the aforementioned factors, HRT, whose role was described as fundamental in the efficiency of SFWs, has been suggested as a regulator of N removal mechanisms, such as denitrification (Section 2.2.2), and determines in this case the contact time of NO_3^- with denitrifying bacteria. Grebliunas and Perry [66] indeed reported a significant direct relationship between denitrification rate and contact time ($p < 0.05$; analysis of variance). Finally, as denitrification takes place in the anaerobic topsoil, the process benefits the larger the bottom area of the SFW for hyporheic exchange.

Vegetation density plays a key role in controlling denitrification by supplying organic matter to the topsoil through plant litter and root exudates, and serving as a substrate for denitrifying bacteria, which stimulate or maintain denitrifying activity [32,48]. The addition of organic matter increases the biochemical oxygen demand, thus contributing to anaerobic conditions, which also promote denitrification. Thereby, SFWs with dense vegetation stands tend to enhance denitrification in relation to others with sparse or without vegetation [46,47]. The studies of Guo et al. [54,55] in experimental SFWs and David et al. [69] support this statement by reporting higher total N and NO_3^- removal efficiencies in SFWs or treatments with the densest vegetation, although the underlying biogeochemical processes were not investigated. The supply of organic matter by vegetation is particularly important in SFWs receiving agricultural subsurface drainage due to the normally low carbon inputs [46,66,69]. Moreover, N and carbon inputs at the SFW inlet may not be significantly correlated in these cases ($p > 0.01$) [38]. These observations thus highlight the need for SFWs to produce their own carbon sources, as insufficient carbon availability can limit denitrification, especially in systems subject to high inputs of NO_3^- [46,66]. Grebliunas and Perry [66], and Xue et al. [67] indeed demonstrated a direct relationship between carbon availability and denitrification rate ($p < 0.05$; analysis of variance and $R^2 = 0.50$, respectively). Furthermore, David et al. [69] reported higher NO_3^- removal after adding carbon to treatments containing plants. As a result, newly constructed wetlands normally reveal low denitrification rates compared to mature systems owing to limited supply of organic matter by plants for denitrification [47]. Consequently, SFWs can take many years to accumulate carbon to a level that supports the maximum denitrification potential [47]. In this context, Nilsson et al. [71] reported a clear increase in N removal during the maturation (time span around 12 years) of experimental SFWs without initial planting, thus allowing free growth of vegetation, and in experimental SFWs initially planted with submerged vegetation. Interestingly, however, the maturation process did not influence the N removal in experimental SFWs initially planted with emergent vegetation. Likewise, Sukias and Tanner [29] could not observe an increase in total N removal over 8–9 years after construction of the investigated SFWs, whereas Tanner et al. [42] reported a significant increase ($p < 0.01$; *t*-test) of about 1.5 times in denitrification rate from 6 to 18 months since the construction of the wetland. Finally, water depth affects the development of vegetation by controlling the penetration of light through the water [48]. Therefore, SFWs with wide deep areas can restrict denitrification by hindering the expansion of vegetation.

Vegetation type, whether submerged or emergent, is expected to vary across SFWs, but may otherwise have little effect on carbon availability, and thus on denitrification [46,47]. In line with this, Guo et al. [54,55] found that different species of emergent plants played a negligible role in the removal efficiency of total N in experimental SFWs. Denitrification rates, however, can increase in the presence of plant litter that readily decomposes, i.e., labile carbon, or decrease in case woody plants are the dominant carbon source [48]. Bastviken et al. [51] tested the effect of vegetation type on NO_3^- removal in experimental SFWs, and found significantly higher NO_3^- removal rates, k and k_{20} values (Equations (9) and (17), respectively) in SFWs with emergent vegetation compared to those with submerged and free-growing vegetation ($p < 0.01$; analysis of variance). Thus, the study indicated that SFWs dominated by emergent vegetation promoted NO_3^- removal mechanisms such as denitrification. Moreover, this study reported an increase in NO_3^- removal as the SFWs matured and the vegetation became denser, therefore in line with the discussion above. Nilsson et al. [71] confirmed these results in the same experimental SFWs by demonstrating that those SFWs initially planted with emergent vegetation indeed enhanced N removal in relation to the other treatments. Surprisingly, however, the effect of vegetation type disappeared once the systems reached a more mature state around eight years after construction, i.e., N removal was no longer significantly different between treatments ($p > 0.10$; analysis of variance). This finding indicated that it is not the vegetation type, but rather the maturation process itself that affects N removal—and likely the denitrification rate—in mature SFWs. Therefore, vegetation type seems to affect denitrification—and possibly biological uptake as well—only at the beginning of the system lifetime.

In addition to carbon, phosphorus may be transported in limited amounts in relation to N in tile drains, although it is required for the growth and maintenance of the population of denitrifying bacteria [38,66]. Song et al. [53] attested the relevance of phosphorus for N removal in experimental SFWs by clearly demonstrating that higher phosphorus availability significantly increased the removal rates and k_{20} values (Equation (17)) of NO_3^- and total N ($p \leq 0.001$; analysis of variance). Hence, the study concluded that low phosphorus inputs may limit the performance of SFWs. Grebliunas and Perry [66], on the other hand, reported no significant effect of phosphorus addition on denitrification rate ($p > 0.05$; analysis of variance), irrespective of the NO_3^- and carbon concentrations in the medium. Therefore, it is possible that phosphorus availability may only affect denitrification when NO_3^- and carbon concentrations exceed a threshold, by which phosphorus becomes a limiting factor.

Given the susceptibility of denitrification to vary according to the many aforementioned controlling factors, it can be expected that this process varies widely within and between SFWs. Indeed, denitrification rates can be about two to three times more variable (standard deviation 3.5 mg m^{-2} h^{-1}; data from Table 4) than biological uptake and settling rates (standard deviations 1.0 and 1.9 mg m^{-2} h^{-1}, respectively; data from Table 4). This highlights the variability or lack of consistency of SFWs in the treatment of NO_3^--rich effluents, such as agricultural subsurface drainage, and emphasizes the importance of biogeochemical factors, especially those related to denitrification, for the performance of these systems.

2.3.3. Biological Uptake

As described in Section 2.3.1, biological uptake and settling are also important N removal mechanisms, as these directly contribute to the overall N removal. These mechanisms are especially promoted in large SFWs whereby significant amounts of organic N can be stored, and when N is stored for long periods [48,49]. Biological uptake mainly occurs during the vegetation growing season in spring-summer through assimilation of NH_4^+ (the preferred N form [46]) and NO_3^- by the root system of plants and by microorganisms, which refer to plant uptake and immobilization, respectively. These N forms are then converted to organic N and become part of the biomass (Figure 6). Thus, studies investigating biological uptake normally measure the N concentration in the biomass of above and below-ground plant, and topsoil (Table 4). The results of Xue et al. [67] suggest that immobilization rapidly removes NO_3^- compared to plant uptake. In line with this, Matheson and Sukias [68] reported immobilization of the transformed NO_3^- (11%) as slightly higher than plant uptake (9%). However, Xue et al. [67] found that plant uptake was able to store much larger amounts of N (in above-ground plant biomass) than immobilization. Besides that, immobilization is expected to store N in the short-term. Thereby, plant biomass is normally regarded as the main N storage location due to greater storage for longer periods compared to microbial biomass.

Given the above observations, vegetation composition and density are major controlling factors for biological uptake rate, as these determine the capacity of the SFW to store N in plant biomass. Moreover, vegetation serves as a substrate for microorganisms, therefore promoting immobilization as well. The maturation process and appropriate water depth, as discussed in Section 2.3.2, are essential for the development of vegetation so as to allow the SFW to reach its maximum biological uptake potential. The superior NO_3^- and total N removals in mature SFWs or in those containing dense emergent vegetation (e.g., *Typha latifolia* and *Phragmites australis*), as described in Bastviken et al. [51], Guo et al. [54,55], David et al. [69] and Nilsson et al. [71] (Section 2.3.2), may support these statements, although the underlying biogeochemical processes were not investigated. Other plant types can also greatly support biological uptake, as attested by Matheson and Sukias [68], who found a markedly higher NO_3^- uptake rate for duckweed (221 mg kg^{-1} d^{-1}) compared to the emergent plant *Typha orientalis* (10 mg kg^{-1} d^{-1}). Assimilation in the latter plant, however, accounted for a greater fraction of the transformed NO_3^- (6%) in relation to duckweed (3%), and is expected to ensure N storage for longer periods, e.g., in below-ground plant biomass.

Other factors may otherwise inhibit biological uptake. Denitrification, for example, competes with biological uptake for the available NO_3^- in the anaerobic topsoil (Figure 6), and can therefore markedly suppress the biological uptake rates there [46]. Moreover, significant biological uptake rates are restricted to the vegetation growing season—prior to the senescence period [25,26]—and can thus only contribute to the overall N removal in a seasonal basis (Section 2.3.6). This seasonal effect results in the variation of N concentration in plant tissues and subsequent storage of N in plant biomass throughout the year as well as between years [24,25]. Finally, SFWs with N-rich soils may largely support plant growth, thereby decreasing the uptake of N from water by plants and the contribution of this process to the overall N removal [25].

2.3.4. Settling

Besides biological uptake (Section 2.3.3), settling not only stores, but also accumulates N in the SFW in the long run [23,24,26,45], which occurs in the topsoil through sedimentation of N bound to organic and inorganic particles (Figure 6). Therefore, the process is partly controlled by the inputs of particle-bound N and benefits the longer the HRT. Under adequate HRTs, sedimentation of particle-bound N occurs mainly near the SFW inlet due to deceleration of the incoming water. The dominant NO_3^- fractions at the SFW inlet (Table 1), however, limit settling rates. In spite of that, plant senescence and plankton death especially contribute to promote settling by adding plant litter, organic particles and associated organic N onto the topsoil after the vegetation growing season [24,45]. In this case, however, the process only transfers the organic N stored in biota through biological uptake to the topsoil with no further contribution to the overall N removal (Figure 6). Nevertheless, these events support denitrification by supplying organic matter and substrate for denitrifying bacteria [23,24,26], and contribute to reaching the maximum denitrification potential of the SFW, as it matures (Section 2.3.2).

2.3.5. Nitrogen Transformation Processes

The storage of N by settling discussed in Section 2.3.4 can be offset by resuspension, in case the hydraulic load is sufficiently high, or by ammonification, which converts organic N to NH_4^+ by ammonifying bacteria (Figure 6). Ammonification is primarily controlled by the availability of organic N and redox conditions, whereby anaerobic conditions slow it down. Thus, ammonification occurs more quickly in the presence of oxygen. This process tends to accumulate NH_4^+ in the anaerobic soil, which may be biologically assimilated or slowly diffuse upwards into the aerobic water or sediment-water interface due to a concentration gradient. At this point, nitrification or the oxidation of NH_4^+ to NO_3^- by nitrifying bacteria can occur (Figure 6). Therefore, ammonification releases N previously assimilated as NH_4^+ or NO_3^- back into the system, thus offsetting biological uptake to some degree, and mainly occurs during the senescence period. A part of the settled organic N, however, is recalcitrant and can be stored in the SFW in the long-term.

As described above, nitrification may follow ammonification and is regulated by the concentration of NH_4^+ and oxygen, which functions as the terminal electron acceptor of the reaction (stoichiometric Equation (13)). Plants can therefore assist nitrification by transporting oxygen to the anaerobic soil through the aerenchyma, albeit restricted to the rhizosphere [48]. The NO_3^- generated by nitrification may diffuse into the anaerobic soil and be used as the terminal electron acceptor by DNRA bacteria when oxidizing organic matter to obtain energy through DNRA. This process, in turn, generates NH_4^+ as the end product and occurs when the pH of the medium is acidic (stoichiometric Equation (14)) and the available NO_3^- is not denitrified or biologically assimilated (Figure 6).

As observed above, ammonification and nitrification are relevant not only to reduce the concentrations of organic N and NH_4^+, but also to generate NO_3^- as the end product, which can be permanently removed from the system by denitrification (Figure 6). In general, the N transformation processes are limited by the diffusion of NH_4^+ in the anaerobic soil, which is considerably slower than that of NO_3^- [4,48]. Reinhardt et al. [45] reported that redox conditions of the water overlying anaerobic soils control nitrification rates and subsequent denitrification. The study found that aerobic water—supported by photosynthesis—promoted nitrification at the sediment-water interface, whereas anaerobic water caused the release of NH_4^+ from the topsoil, which stimulated assimilation by duckweed. This event consequently inhibited generation of NO_3^- by nitrification and subsequent denitrification. Finally, as the N transformation processes are all mediated by enzymatic activity, temperature is also a controlling factor.

2.3.6. Seasonality

As discussed in Section 2.2.3, SFWs under the temperate climate are subject to large variations in water flow, and thus in HRT. In addition to that, these SFWs experience seasonal variations in temperature, which also affect the overall N removal, especially in relation to N removal efficiency [26–28,37,38,45,51,69]. This is because important N removal mechanisms, such as denitrification and biological uptake, are mediated by enzymatic activity. As SFWs receiving agricultural subsurface drainage largely depend on denitrification and partly on biological uptake to reduce N loads, seasonality benefits N removal in the so-called vegetation growing season, characterized by warmer temperatures, and suppresses it during winter [26,38,45,48,49,51,67,71]. Xue et al. [67], for example, estimated the denitrification capacities of three SFWs (based on results from mesocosms) around six times higher in summer (25 °C) than in winter (4 °C). The seasonal variations of temperature and HRT produce even stronger effects on the removal of N throughout the year, by which denitrification and biological uptake rates typically reach their peaks during summer, as a result of warmer temperatures and longer HRTs, whereas cold and short HRTs during winter markedly suppress these processes [22,45,48]. This combined effect could be observed in Ulén et al. [38], who demonstrated a significant direct relationship between the product of soil temperature and HRT, and NO_3^- removal efficiency in a SFW ($p < 0.01$; linear function). This study found that the NO_3^- removal efficiency values in the linear function corresponding to summer were indeed markedly higher than those corresponding to the other seasons.

According to the above, Kadlec [47] proposed that a fixed NO_3^- removal efficiency could be achieved if the temperature variations were compensated by the HRT, i.e., cold and warm temperatures could produce similar NO_3^- removal efficiencies provided the HRT is longer and shorter, respectively. Steidl et al. [26], for example, reported effective N removal in a SFW when the HRT and water temperature were over 20 days and 8 °C, which occurred only from spring to autumn. However, SFWs commonly receive the highest N loads during winter, as discussed in Section 2.2.3, which complicates management plans to optimize performance. Therefore, a proper SFW sizing is not only necessary to ensure sufficient HRT for the N removal mechanisms, but also to compensate for seasonal temperature variations so that the system can achieve an annual N removal goal. Surface-flow constructed wetlands that do not take this into account may risk becoming sources of N during cold periods with short HRTs [26].

2.3.7. Final Remarks

At this point, only intra and inter-variability in N treatment by SFWs receiving agricultural subsurface drainage, and the associated processes have been covered. However, successful application of SFWs for the treatment of N must also take into account other aspects, such as (i) the system lifetime, including maturation and post-maturation effects, and the effectiveness of different maintenance operations; (ii) the collective effect of SFWs in watersheds, including siting and design; (iii) the intrinsic limitations of SFWs, including their common need for large areas, great dependence on the climate for optimal performance, and emission of nitrous oxide (a potent greenhouse gas); and finally (iv) the cost-efficiency of the system. Thereby, further analyses and evaluations are necessary to ensure that SFWs are a viable solution to reduce excess N loads under different agricultural catchment conditions. Lastly, these efforts are essential to help preserve the quality of surface waters with minimal impact on agricultural activities and production.

3. Summary, Conclusions and Final Remarks

Surface-flow constructed wetlands are a prevalent practice in reducing excess N loads at the edge of tile-drained agricultural catchments to surface waters. This is achieved through a combination of increased HRT with N biogeochemical processes. Despite the increasing use of SFWs and growing body of research attesting their capacity, this review clearly observed a large intra and inter-variability in performance, which are the result of system design and local catchment characteristics, including N load and climate.

3.1. Nitrogen Inputs and Incoming Loads

Nitrogen load largely varied between SFWs and proved to be normally a major controlling factor for N removal rate due to a cumulative effect of N in the system, therefore greatly explaining the inter-variability in N removal rate. Thus, increasing hydraulic loads and particularly N concentrations tend to markedly raise the removal rate, and SFWs receiving higher N loads tend to outperform others in a rate basis. Reductions in N concentration at the outlet, however, proved to decrease given higher inputs of N, which suggested more untreated N crossing the system. The review found that this effect is associated with an increase in water flow, which often correlates positively to N concentration and shortens the HRT for effective removal. Therefore, significant reductions in N concentration or lower concentrations at the outlet may only be observed if water flow does not increase concomitantly to N inputs. As a result, increasing N loads are expected to benefit the performance of SFWs in a rate basis, while decreasing the reduction of N concentration at the outlet or efficiency of the system.

Among the different N forms, the review found that NO_3^- clearly contributes most to the relationship between N load and removal rate—as denitrification rate promptly responds to NO_3^- inputs—thus indicating that higher NO_3^- fractions from total N at the inlet enhance the removal rate. Fortunately, the review found that this is often the case, i.e., dominant NO_3^- fractions at the inlet, which reflected in comparable variations of NO_3^- and total N loads. Thereby, NO_3^- largely contributed to the variability in performance. Ammonium and organic N loads, on the other hand, were found to contribute little to the relationship between N load and removal rate. Besides that, NH_4^+ loads are normally low, while organic N loads can be highly variable, albeit weakly correlated to their removal rates, which caused great variance in the removal efficiency of organic N between SFWs. Thus, these N forms contribute less to the overall performance, besides being more prone to be generated in situ and exported. As a result, the performance of SFWs benefits from N loads consisting mainly of NO_3^-.

In relation to N removal efficiency, the effect of N load tends to be negative and minor, thus generally explaining little of its variability. In this case, not even N concentration showed a significant effect. The review found, however, that system efficiency is closely related to hydrological and biogeochemical factors that regulate the functioning of N removal mechanisms, such as denitrification, biological uptake and settling. Increasing hydraulic loads—probably leading to higher N loads—contribute, therefore, to decrease the efficiency of the system or reduction of N concentration at the outlet by significantly raising N export as a result of insufficient HRT to allow proper N processing rates. Thereby, unstable efficiency can be expected in SFWs subject to varying hydrological regimes. Specifically, the review found that exports of NO_3^- and organic N respond to variations in hydraulic load, where the effect on organic N was more prominent. Thus, increasing fractions of organic N from total N at the inlet or the generation of organic N in situ may accentuate the negative effect of hydraulic or N load in the efficiency of SFWs. Finally, the review indicated that N removal efficiency tends to respond more promptly to hydraulic load in a lower range, whereby HRT is a limiting factor for N removal mechanisms.

The review found that hydraulic load regulates the HRT and the active hydrological area or volume of the SFW, therefore the contact time of N with the removal components of the system, and the level of N distribution and treatment. Hence, significantly high hydraulic loads tend to restrict N removal efficiency not only by shortening the HRT, but also by promoting preferential flow and stagnant water, which reduce that contact time and underutilize the treatment space of the system. Despite the importance, the review recognized this issue as little investigated. Therefore, further studies are recommended, mainly regarding the relationship between water flow dynamics and N treatment.

3.2. Design

Design parameters have been evaluated as fundamental to control HRT and water flow dynamics, and thus ensure effective SFWs. Large SFWs allow long HRTs and wide areas for N biogeochemical processes and removal, while deep waters can be particularly relevant for small systems subject to intense hydraulic loads in order to prolong the HRT. In this context, the area ratio of the SFW for the agricultural catchment has been found as critical, as effective systems must be large enough to accommodate and treat the incoming N loads with sufficient HRT, mainly during periods of intense discharge. The review estimated that 40–50% N removal can be generally achieved with an area ratio of 1–4%, with minor increments in N removal above this threshold, as HRT is no longer a limiting factor. Smaller systems may otherwise be advantageous if high N removal rates are the goal. Furthermore, it has been found that aspect ratio length to width, position and configuration of the inlet and outlet, presence of obstructions to the water flow, bathymetry, and vegetation type and distribution, have all proved to influence water flow dynamics in different degrees, therefore with potential implications for N treatment, as indeed reported to some extent. In general, however, the review observed that design parameters have been little investigated in relation to N treatment, and recommends particular attention in future studies. These efforts can potentially contribute to enhance N removal by improving the hydrology of SFWs.

3.3. Removal Processes and Factors

Denitrification has been described as the main N removal mechanism due to normally higher processing rates and recoveries of the available N compared to biological uptake and settling, besides ensuring permanent removal of N from the system. The latter two mechanisms, in turn, can only store N—mainly as organic N—in biomass or onto the topsoil for a certain period, thus with a limited contribution to the overall N removal. Ammonification, nitrification and DNRA, on the other hand, do not contribute directly to the overall N removal, but regulate the availability of NH_4^+ and NO_3^- prone to biological uptake or denitrification, according to local conditions. Finally, although volatilization also removes N permanently from the system, its contribution is generally considered negligible. Investigation of the potential of this process is otherwise recommended in SFWs experiencing anaerobic and basic water.

The review found that maturation process is crucial to ensure effective N removal in SFWs, and mainly involves the development of vegetation, which supports denitrification by supplying organic matter and substrate for denitrifying bacteria. Moreover, this process promotes biological uptake by plants and microorganisms, and settling. Maturation of SFWs, however, typically takes many years to complete, and so the full potential of the system to remove N. Therefore, suboptimal performance is normally observed during initial operation. In line with this, denser vegetation contributes to increasing the system efficiency, where emergent vegetation, in particular, can promote N removal at an early stage. Thereby, wide deep areas should be avoided in SFWs in order to allow vegetation development, and initial planting with emergent vegetation can potentially enhance performance during initial operation. In addition to the above, the review found that denitrification is promoted by warmer temperatures, whereas low inputs of phosphorus may limit this process given sufficient concentrations of NO_3^- and carbon. Finally, the review highlighted the susceptibility of denitrification to vary due to the many controlling factors, which ultimately contribute to the intra and inter-variability in performance of SFWs. In relation to biological uptake, the review observed that N storage takes place mainly in plant biomass during the vegetation growing season, thus contributing to the overall N removal in a seasonal basis. Lastly, settling has demonstrated particular contribution to the overall N removal by accumulating and storing N for long periods in the topsoil, as promoted mainly by senescence events, although ammonification may offset its contribution to some extent by releasing NH_4^+ into the system.

Despite the findings of this review, N biogeochemical processes in SFWs receiving agricultural subsurface drainage have been little investigated, mainly in relation to long-term tests at the field-scale. Therefore, the review strongly recommends further studies, given the susceptibility of the performance to be affected or suppressed at the biogeochemical level in the context described herein, i.e., NO_3^--rich effluents with low carbon inputs, long maturation periods and seasonal variations in temperature. These efforts are essential to support effective SFWs and reduce the intra and inter-variability in performance.

3.4. Climate

Climate has been identified as a major challenge for the successful application of SFWs due to seasonal and annual variations in hydrological regime and temperature, which highly affect controlling factors for N removal, thus the N processing rates, and subsequently the overall performance. Thereby, climate greatly contributes to the intra and inter-variability in performance. The review found that warm periods with long HRTs (e.g., during summer) provide proper conditions for effective removal, although N loads are typically low to contribute significantly to the removal rate and cumulative removal of N in an annual basis. Cold periods with short HRTs (e.g., during winter), on the other hand, do not support effective removal, although N loads are high to enhance the removal rate, which often results in major cumulative removals of N in a year. Moreover, the review found that more variable hydrological regimes tend to worsen the overall performance. Thus, management operations or technological interventions capable not only to enhance the system efficiency during winter, when N loads are often the highest, but also to stabilize it throughout the year, are greatly desired. Furthermore, these efforts are critical to reduce the intra and inter-variability in performance and ensure more predictable N treatments in SFWs.

Funding: This research received no external funding.

Conflicts of Interest: The author declares no conflict of interest.

References

1. Randall, G.W.; Mulla, D.J. Nitrate nitrogen in surface waters as influenced by climatic conditions and agricultural practices. *J. Environ. Qual.* **2001**, *30*, 337–344. [CrossRef]
2. Sun, B.; Zhang, L.; Yang, L.; Zhang, F.; Norse, D.; Zhu, Z. Agricultural non-point source pollution in China: Causes and mitigation measures. *Ambio* **2012**, *41*, 370–379. [CrossRef]
3. Guo, W.; Fu, Y.; Ruan, B.; Ge, H.; Zhao, N. Agricultural non-point source pollution in the Yongding River Basin. *Ecol. Indic.* **2014**, *36*, 254–261. [CrossRef]
4. Mitsch, W.J.; Day, J.W.; Gilliam, J.W.; Groffman, P.M.; Hey, D.L.; Randall, G.W.; Wang, N. *Reducing Nutrient Loads, Especially Nitrate-Nitrogen, to Surface Water, Ground Water, and the Gulf of Mexico: Topic 5 Report for the Integrated Assessment on Hypoxia in the Gulf of Mexico*; Decision Analysis Series no. 19; NOAA Coastal Ocean Office: Silver Spring, MD, USA, 1999.
5. Baker, J.L. Limitations of improved nitrogen management to reduce nitrate leaching and increase use efficiency. In *Optimizing Nitrogen Management in Food and Energy Production and Environmental Protection: Proceedings of the 2nd International Nitrogen Conference on Science and Policy*; TheScientificWorld, Hindawi: London, UK, 2001; pp. 10–16.
6. Dinnes, D.L.; Karlen, D.L.; Jaynes, D.B.; Kaspar, T.C.; Hatfield, J.L.; Colvin, T.S.; Cambardella, C.A. Nitrogen management strategies to reduce nitrate leaching in tile-drained midwestern soils. *Agron J.* **2002**, *94*, 153–171. [CrossRef]
7. Baker, J.L.; Melvin, S.W.; Lemke, D.W.; Lawlor, P.A.; Crumpton, W.G.; Helmers, M.J. Subsurface drainage in Iowa and the water quality benefits and problem. In *Drainage VIII the Proceedings of the Eighth International Drainage Symposium*; Cooke, R., Ed.; American Society of Agricultural and Biological Engineers: Sacramento, CA, USA, 2004; pp. 39–50.
8. McLellan, E.; Robertson, D.; Schilling, K.; Tomer, M.; Kostel, J.; Smith, D.; King, K. Reducing nitrogen export from the corn belt to the gulf of Mexico: Agricultural strategies for remediating hypoxia. *JAWRA J. Am. Water Resour. Assoc.* **2014**, *51*, 263–289. [CrossRef]
9. Mitsch, W.J.; Day, J.W.; Gilliam, J.W.; Groffman, P.M.; Hey, D.L.; Randall, G.W.; Wang, N. Reducing nitrogen loading to the gulf of Mexico from the Mississippi River Basin: Strategies to counter a persistent ecological problem. *Bioscience* **2001**, *51*, 373–388. [CrossRef]
10. Strock, J.S.; Kleinman, P.J.A.; King, K.W.; Delgado, J.A. Drainage water management for water quality protection. *J. Soil Water Conserv.* **2010**, *65*, 131A–136A. [CrossRef]
11. Valero, C.S.; Madramootoo, C.A.; Stämpfli, N. Water table management impacts on phosphorus loads in tile drainage. *Agric. Water Manag.* **2007**, *89*, 71–80. [CrossRef]
12. Pierce, S.C.; Kröger, R.; Pezeshki, R. Managing artificially drained low-gradient agricultural headwaters for enhanced ecosystem functions. *Biology* **2012**, *1*, 794–856. [CrossRef]
13. Frey, S.K.; Hwang, H.-T.; Park, Y.-J.; Hussain, S.I.; Gottschall, N.; Edwards, M.; Lapen, D.R. Dual permeability modeling of tile drain management influences on hydrologic and nutrient transport characteristics in macroporous soil. *J. Hydrol.* **2016**, *535*, 392–406. [CrossRef]
14. Hussain, S.I.; Frey, S.K.; Blowes, D.W.; Ptacek, C.J.; Wilson, D.; Mayer, K.U.; Su, D.; Gottschall, N.; Edwards, M.; Lapen, D.R. Reactive Transport of Manure-Derived Nitrogen in the Vadose Zone: Consideration of Macropore Connectivity to Subsurface Receptors. *Vadose Zone J.* **2019**, *18*, 1–18. [CrossRef]
15. Stålnacke, P.; Aakerøy, P.A.; Blicher-Mathiesen, G.; Iital, A.; Jansons, V.; Koskiaho, J.; Kyllmar, K.; Lagzdins, A.; Pengerud, A.; Povilaitis, A. Temporal trends in nitrogen concentrations and losses from agricultural catchments in the Nordic and Baltic countries. *Agric. Ecosyst. Environ.* **2014**, *198*, 94–103. [CrossRef]
16. Mendes, L.R.D. Edge-of-Field technologies for phosphorus retention from agricultural drainage discharge. *Appl. Sci.* **2020**, *10*, 634. [CrossRef]
17. Weisner, S.E.B.; Johannesson, K.; Thiere, G.; Svengren, H.; Ehde, P.M.; Tonderski, K.S. National large-scale wetland creation in agricultural areas—Potential versus realized effects on nutrient transports. *Water* **2016**, *8*, 544. [CrossRef]
18. Hoffmann, C.C.; Zak, D.; Kronvang, B.; Kjaergaard, C.; Carstensen, M.V.; Audet, J. An overview of nutrient transport mitigation measures for improvement of water quality in Denmark. *Ecol. Eng.* **2020**, *155*, 105863. [CrossRef]
19. Groh, T.A.; Gentry, L.E.; David, M.B. Nitrogen removal and greenhouse gas emissions from constructed wetlands receiving tile drainage water. *J. Environ. Qual.* **2015**, *44*, 1001–1010. [CrossRef]
20. Tanner, C.C.; Nguyen, M.; Sukias, J. Constructed wetland attenuation of nitrogen exported in subsurface drainage from irrigated and rain-fed dairy pastures. *Water Sci. Technol.* **2005**, *51*, 55–61. [CrossRef]
21. Kang, H.; Freeman, C.; Lee, D.; Mitsch, W.J. Enzyme activities in constructed wetlands: Implication for water quality amelioration. *Hydrobiologia* **1998**, *368*, 231–235. [CrossRef]
22. Tournebize, J.; Chaumont, C.; Fesneau, C.; Guenne, A.; Vincent, B.; Garnier, J.; Mander, Ü. Long-term nitrate removal in a buffering pond-reservoir system receiving water from an agricultural drained catchment. *Ecol. Eng.* **2015**, *80*, 32–45. [CrossRef]
23. Borin, M.; Tocchetto, D. Five year water and nitrogen balance for a constructed surface flow wetland treating agricultural drainage waters. *Sci. Total. Environ.* **2007**, *380*, 38–47. [CrossRef]
24. Lavrnić, S.; Braschi, I.; Anconelli, S.; Blasioli, S.; Solimando, D.; Mannini, P.; Toscano, A. Long-term monitoring of a surface flow constructed wetland treating agricultural drainage water in northern Italy. *Water* **2018**, *10*, 644. [CrossRef]
25. Hoagland, C.R.; Gentry, L.E.; David, M.B.; Kovacic, D.A. Plant nutrient uptake and biomass accumulation in a constructed wetland. *J. Freshw. Ecol.* **2001**, *16*, 527–540. [CrossRef]

26. Steidl, J.; Kalettka, T.; Bauwe, A. Nitrogen retention efficiency of a surface-flow constructed wetland receiving tile drainage water: A case study from north-eastern Germany. *Agric. Ecosyst. Environ.* **2019**, *283*. [CrossRef]

27. de Haan, J.; van der Schoot, J.R.; Verstegen, H.; Clevering, O. Removal of nitrogen leaching from vegetable crops in constructed wetlands. *Acta Hortic.* **2010**, *852*, 139–144. [CrossRef]

28. Tolomio, M.; Ferro, N.D.; Borin, M. Multi-Year N and P removal of a 10-year-old surface flow constructed wetland treating agricultural drainage waters. *Agronomy* **2019**, *9*, 170. [CrossRef]

29. Sukias, J.; Tanner, C.C. Surface flow constructed wetlands as a drainage management tool–long term performance. In *Adding to the Knowledge Base for the Nutrient Manager*; Currie, L.D., Christensen, C.L., Eds.; Fertilizer & Lime Research Centre, Occasional Report No 24. Massey University: Palmerston North, New Zealand, 2011; pp. 1–16.

30. Strand, J.A.; Weisner, S.E.B. Effects of wetland construction on nitrogen transport and species richness in the agricultural landscape—Experiences from Sweden. *Ecol. Eng.* **2013**, *56*, 14–25. [CrossRef]

31. Lenhart, C.; Gordon, B.; Gamble, J.; Current, D.; Ross, N.; Herring, L.; Nieber, J.; Peterson, H. Design and hydrologic performance of a tile drainage treatment wetland in Minnesota, USA. *Water* **2016**, *8*, 549. [CrossRef]

32. Tournebize, J.; Chaumont, C.; Mander, Ü. Implications for constructed wetlands to mitigate nitrate and pesticide pollution in agricultural drained watersheds. *Ecol. Eng.* **2017**, *103*, 415–425. [CrossRef]

33. Vymazal, J. The use of constructed wetlands for nitrogen removal from agricultural drainage: A review. *Sci. Agric. Bohem.* **2017**, *48*, 82–91. [CrossRef]

34. Maxwell, E.; Peterson, E.W.; O'Reilly, C.M. Enhanced nitrate reduction within a constructed wetland system: Nitrate removal within groundwater flow. *Wetlands* **2017**, *37*, 413–422. [CrossRef]

35. Larson, A.C.; Gentry, L.E.; David, M.B.; Cooke, R.A.; Kovacic, D.A. The role of seepage in constructed wetlands receiving agricultural tile drainage. *Ecol. Eng.* **2000**, *15*, 91–104. [CrossRef]

36. Carstensen, M.V.; Hashemi, F.; Hoffmann, C.C.; Zak, D.; Audet, J.; Kronvang, B. Efficiency of mitigation measures targeting nutrient losses from agricultural drainage systems: A review. *Ambio* **2020**, *49*, 1820–1837. [CrossRef]

37. Drake, C.; Jones, C.; Schilling, K.; Amado, A.A.; Weber, L. Estimating nitrate-nitrogen retention in a large constructed wetland using high-frequency, continuous monitoring and hydrologic modeling. *Ecol. Eng.* **2018**, *117*, 69–83. [CrossRef]

38. Ulén, B.; Geranmayeh, P.; Blomberg, M.; Bieroza, M. Seasonal variation in nutrient retention in a free water surface constructed wetland monitored with flow-proportional sampling and optical sensors. *Ecol. Eng.* **2019**, *139*, 105588. [CrossRef]

39. Kovacic, D.A.; David, M.B.; Gentry, L.E.; Starks, K.M.; Cooke, R.A. Effectiveness of constructed wetlands in reducing nitrogen and phosphorus export from agricultural tile drainage. *J. Environ. Qual.* **2000**, *29*, 1262–1274. [CrossRef]

40. Kroeger, A.C.; Madramootoo, C.A.; Enright, P.; Laflamme, C. Efficiency of a small constructed wetland in southern Québec for treatment of agricultural runoff waters. In *IWA Specialist Conference: Wastewater Biosolids Sustainability: Technical, Managerial, and Public Synergy*; ResearchGate: Moncton, NB, Canada, 2007; pp. 1057–1062.

41. Tanner, C.C.; Sukias, J.P.S. Multiyear nutrient removal performance of three constructed wetlands intercepting tile drain flows from grazed pastures. *J. Environ. Qual.* **2011**, *40*, 620–633. [CrossRef]

42. Tanner, C.C.; Nguyen, M.; Sukias, J. Nutrient removal by a constructed wetland treating subsurface drainage from grazed dairy pasture. *Agric. Ecosyst. Environ.* **2005**, *105*, 145–162. [CrossRef]

43. Kovacic, D.A.; Twait, R.M.; Wallace, M.P.; Bowling, J.M. Use of created wetlands to improve water quality in the Midwest—Lake Bloomington case study. *Ecol. Eng.* **2006**, *28*, 258–270. [CrossRef]

44. Tanner, C.C.; Nguyen, M.L.; Sukias, J. Using constructed wetlands to treat subsurface drainage from intensively grazed dairy pastures in New Zealand. *Water Sci. Technol.* **2003**, *48*, 207–213. [CrossRef]

45. Reinhardt, M.; Müller, B.; Gächter, R.; Wehrli, B. Nitrogen removal in a small constructed wetland: An isotope mass balance approach. *Environ. Sci. Technol.* **2006**, *40*, 3313–3319. [CrossRef]

46. Kadlec, R.H. Constructed wetlands to remove nitrate. In *Nutrient Management in Agricultural Watersheds: A Wetlands Solution*; Dunne, E.J., Reddy, K.R., Carton, O.T., Eds.; Wageningen Academic Publishers: Wageningen, The Netherlands, 2005; pp. 132–143.

47. Kadlec, R.H. Nitrogen farming for pollution control. *J. Environ. Sci. Heal. Part A* **2005**, *40*, 1307–1330. [CrossRef] [PubMed]

48. O'Geen, A.; Budd, R.; Gan, J.; Maynard, J.; Parikh, S.; Dahlgren, R. Mitigating nonpoint source pollution in agriculture with constructed and restored wetlands. In *Advances in Agronomy*; Elsevier: Amsterdam, The Netherlands, 2010; pp. 1–76.

49. Crumpton, W.G.; Kovacic, D.A.; Hey, D.L.; Kostel, J.A. Potential of restored and constructed wetlands to reduce nutrient export from agricultural watersheds in the corn belt. In *Final Report: Gulf Hypoxia and Local Water Quality Concerns Workshop*; American Society of Agricultural and Biological Engineers: St. Joseph, MI, USA, 2008; pp. 29–42.

50. Khan, I. *Effects of Hydraulic Load on Nitrate Removal in Surface-Flow Constructed Wetlands*; Halmstad University: Halmstad, Sweden, 2011.

51. Bastviken, S.K.; Weisner, S.E.B.; Thiere, G.; Svensson, J.M.; Ehde, P.M.; Tonderski, K.S. Effects of vegetation and hydraulic load on seasonal nitrate removal in treatment wetlands. *Ecol. Eng.* **2009**, *35*, 946–952. [CrossRef]

52. Kadlec, R.H.; Wallace, S. *Treatment Wetlands*, 2nd ed.; Taylor & Francis Group: Boca Raton, FL, USA, 2009.

53. Song, X.; Ehde, P.M.; Weisner, S.E.B. Effects of water depth and phosphorus availability on nitrogen removal in agricultural wetlands. *Water* **2019**, *11*, 2626. [CrossRef]

54. Guo, C.; Cui, Y.; Dong, B.; Luo, Y.; Liu, F.; Zhao, S.; Wu, H. Test study of the optimal design for hydraulic performance and treatment performance of free water surface flow constructed wetland. *Bioresour. Technol.* **2017**, *238*, 461–471. [CrossRef] [PubMed]

55. Guo, C.; Cui, Y.; Shi, Y.; Luo, Y.; Liu, F.; Wan, D.; Ma, Z. Improved test to determine design parameters for optimization of free surface flow constructed wetlands. *Bioresour. Technol.* **2019**, *280*, 199–212. [CrossRef] [PubMed]

56. Haverstock, M.J. *An Assessment of a Wetland-Reservoir Wastewater Treatment and Reuse System Receiving Agricultural Drainage Water in Nova Scotia*; Dalhousie University: Halifax, NS, Canada, 2010.

57. Tanner, C.C.; Kadlec, R.H. Influence of hydrological regime on wetland attenuation of diffuse agricultural nitrate losses. *Ecol. Eng.* **2013**, *56*, 79–88. [CrossRef]

58. Tanner, C.C.; Sukias, J.P.S.; Yates, C.R. *New Zealand Guidelines: Constructed Wetland Treatment of Tile Drainage*; NIWA Information Series No. 75; National Institute of Water & Atmospheric Research Ltd.: Hamilton, New Zealand, 2010.

59. Persson, J.; Somes, N.L.G.; Wong, T.H.F. Hydraulics efficiency of constructed wetlands and ponds. *Water Sci. Technol.* **1999**, *40*, 291–300. [CrossRef]

60. Thackston, E.L.; Shields, F.D.; Schroeder, P.R. Residence time distributions of shallow basins. *J. Environ. Eng.* **1987**, *113*, 1319–1332. [CrossRef]

61. Lavrnić, S.; Alagna, V.; Iovino, M.; Anconelli, S.; Solimando, D.; Toscano, A. Hydrological and hydraulic behaviour of a surface flow constructed wetland treating agricultural drainage water in northern Italy. *Sci. Total. Environ.* **2020**, *702*, 134795. [CrossRef]

62. Su, T.-M.; Yang, S.-C.; Shih, S.-S.; Lee, H.-Y. Optimal design for hydraulic efficiency performance of free-water-surface constructed wetlands. *Ecol. Eng.* **2009**, *35*, 1200–1207. [CrossRef]

63. Pugliese, L.; Kusk, M.; Iversen, B.V.; Kjaergaard, C. Internal hydraulics and wind effect in a surface flow constructed wetland receiving agricultural drainage water. *Ecol. Eng.* **2020**, *144*, 105661. [CrossRef]

64. Bodin, H.; Mietto, A.; Ehde, P.M.; Persson, J.; Weisner, S.E.B. Tracer behaviour and analysis of hydraulics in experimental free water surface wetlands. *Ecol. Eng.* **2012**, *49*, 201–211. [CrossRef]

65. Kalcic, M.; Crumpton, W.; Liu, X.; D'Ambrosio, J.; Ward, A.; Witter, J. Assessment of beyond-the-field nutrient management practices for agricultural crop systems with subsurface drainage. *J. Soil Water Conserv.* **2018**, *73*, 62–74. [CrossRef]

66. Grebliunas, B.D.; Perry, W.L. The role of C:N:P stoichiometry in affecting denitrification in sediments from agricultural surface and tile-water wetlands. *Springerplus* **2016**, *5*, 359. [CrossRef]

67. Xue, Y.; Kovacic, D.A.; David, M.B.; Gentry, L.E.; Mulvaney, R.L.; Lindau, C.W. In Situ Measurements of Denitrification in Constructed Wetlands. *J. Environ. Qual.* **1999**, *28*, 263–269. [CrossRef]

68. Matheson, F.E.; Sukias, J.P. Nitrate removal processes in a constructed wetland treating drainage from dairy pasture. *Ecol. Eng.* **2010**, *36*, 1260–1265. [CrossRef]

69. David, M.B.; Gentry, L.E.; Smith, K.M.; Kovacic, D.A. Carbon, plant, and temperature control of nitrate removal from wetland mesocosms. *Trans. Ill. State Acad Sci.* **1997**, *90*, 103–112.

70. Rutherford, K.; Wheeler, D. Wetland nitrogen removal modules in OVERSEER®. In *Adding to the Knowledge Base for the Nutrient Manager Occasion Report No 24*; Currie, L.D., Christensen, C.L., Eds.; Fertilizer and Lime Research Centre, Massey University: Palmerston North, New Zealand, 2011; p. 12.

71. Nilsson, J.E.; Liess, A.; Ehde, P.M.; Weisner, S.E. Mature wetland ecosystems remove nitrogen equally well regardless of initial planting. *Sci. Total. Environ.* **2020**, *716*, 137002. [CrossRef] [PubMed]

Review

Effective Heavy Metals Removal from Water Using Nanomaterials: A Review

Mohamed A. Tahoon [1], Saifeldin M. Siddeeg [1,2], Norah Salem Alsaiari [3], Wissem Mnif [4,5] and Faouzi Ben Rebah [1,6,*]

[1] Department of Chemistry, College of Science, King Khalid University, P.O. Box 9004, Abha 61413, Saudi Arabia; tahooon_87@yahoo.com (M.A.T.); saif.siddeeg@gmail.com (S.M.S.)

[2] Chemistry and Nuclear Physics Institute, Atomic Energy Commission, P.O. Box 3001, Khartoum 11111, Sudan

[3] Chemistry Department, College of Science, Princess Nourah Bint Abdulrahman University, Riyadh 11671, Saudi Arabia; nsalsaiari@pnu.edu.sa

[4] Department of Chemistry, Faculty of Sciences and Arts in Balgarn, University of Bisha, P.O. BOX 199, Bisha 61922, Saudi Arabia; w_mnif@yahoo.fr

[5] LR11-ES31 Laboratory of Biotechnology and Valorisation of Bio-Geo Resources, Higher Institute of Biotechnology of Sidi Thabet, BiotechPole of Sidi Thabet, University of Manouba, Biotechpole Sidi Thabet, Ariana 2020, Tunisia

[6] Higher Institute of Biotechnology of Sfax (ISBS), Sfax University, P.O. Box 263, Sfax 3000, Tunisia

* Correspondence: benrebahf@yahoo.fr

Received: 22 April 2020; Accepted: 23 May 2020; Published: 29 May 2020

Abstract: The discharge of toxic heavy metals including zinc (Zn), nickel (Ni), lead (Pb), copper (Cu), chromium (Cr), and cadmium (Cd) in water above the permissible limits causes high threat to the surrounding environment. Because of their toxicity, heavy metals greatly affect the human health and the environment. Recently, better remediation techniques were offered using the nanotechnology and nanomaterials. The attentions were directed toward cost-effective and new fabricated nanomaterials for the application in water/wastewater remediation, such as zeolite, carbonaceous, polymer based, chitosan, ferrite, magnetic, metal oxide, bimetallic, metallic, etc. This review focused on the synthesis and capacity of various nanoadsorbent materials for the elimination of different toxic ions, with discussion of the effect of their functionalization on the adsorption capacity and separation process. Additionally, the effect of various experimental physicochemical factors on heavy metals adsorption, such as ionic strength, initial ion concentration, temperature, contact time, adsorbent dose, and pH was discussed.

Keywords: water treatment; nanomaterials; heavy metals; functionalization

---

1. Introduction

Healthy peoples require clean drinking water as one of the essential life requirements [1–3]. However, due to the increased industrialization and the excessive use of chemicals all over the world, unwanted pollutants released into drinking water have became a major concern. One of the major noxious water pollutants are heavy metal ions that cause serious side effects to human and living organisms. Toxic heavy metals are the comparatively high-density metallic elements that can cause toxicity also at lower doses. In the recent years, heavy metals pollution was increased significantly by several folds, resulting from the continuous development of urban, industrial, and agricultural activities. Generally, the main sources of heavy metals toxicity are sewage sludge, pesticides, metallic ferrous ores, fertilizers, municipal wastes, and fossil fuels burning [4,5]. As advised by the Environmental Protection Agency (EPA), there is a standard quality for drinking water, which depends on the level of the present pollutants and the harmful effects on health which could occur when pollutants levels exceed the permitted levels [6]. Industrial contaminants commonly include zinc (Zn), nickel (Ni), lead (Pb), copper (Cu), chromium (Cr), and cadmium (Cd) over other metals. These toxic metals have many poisoning symptoms, with impacts on human and living organisms through their accumulation inside the food chain [7]. The progress and excessive activities of humans has caused the environmental pollution by toxic metals even though these metals naturally exist in the crust of earth. These activities include wood and dyes preservation, photographic materials, electroplating and steel industries, smelting, printing pigments, pesticides, leather tanning, explosive industrial, coating, mining biosolids, atmospheric accumulation, sewage irrigation, textiles, alloys fabrication, fertilizers manufacturing, and batteries construction [8]. Nevertheless, many physiological and biochemical processes require some heavy metals usage in trace amounts, such as molybdenum (Mo), nickel (Ni), manganese (Mn), selenium (Se), zinc (Zn), magnesium (Mg), iron (Fe), chromium (Cr), copper (Cu), and cobalt (Co). These trace metals are considered essential while others are non-essential. However, high concentrations of these trace metals become harmful for humans and the environment. Many harmful effects such as nervous disorders, anemia, renal failure, and cancer are associated with high lead levels in drinking water. Many metals such as mercury [9], lead [10], chromium [11], cadmium [12], and arsenic [13] are classified as having the greatest importance for public health, due to their high toxicity, as exposure to these metals even at small levels can cause several organs damage. According to the International Agency for Research on Cancer (IARC) and the United States Environmental Protection Agency (U.S. EPA), several metals such as lead, arsenic, cadmium, chromium, and mercury are known to be carcinogenic depending on their experimental and epidemiological tests. Therefore, the elevated toxic metals contamination has motivated scientists to develop several remediation methods, such as flocculation and coagulation [14], photocatalysis [15], ion exchange [16], adsorption [17], membrane filtration [18], and chemical precipitation [19]. Because of diverse factors related to economic and technological considerations, a limited number of wastewater treatment processes are universally used by various industries to treat their effluents. In spite of their availability to remove different pollutants from wastewaters, these processes represent some limitations (high cost, process complexity, low efficiency, etc.). Interestingly, the adsorption technique was considered as the most effective one for heavy metals removal from wastewater since is considered as safe, clean, efficient and technical feasible process [20]. Absorbent materials are characterized by their high surface area and porosity, resistance to toxic substances with high effectiveness and simple design and easy operation, making this technology the most worldwide method for heavy metals removal from wastewater [20]. Many adsorbents have been used for toxic metals in wastewater treatment, such as nanocomposites [21], nanofilteration [22], engineered nanomaterials [23], chelating minerals [24], activated carbon [25], and biopolymers [26]. Because the old technologies are conventional and limited, in the past few years nanotechnology science has been applied for heavy metals removal from waters. Nanoparticles (NPs) size-dependent properties provide various opportunities in diverse applications. Interestingly, this science has gained tremendous international interest with the increasing request for nanomaterials in which materials are transferred from the micro- to the nano-scale [27]. Additionally, the existence of the high number of functional groups over their surfaces, availability, and lesser flocculation allowed their wide application for the remediation of wastewater containing toxic metals [28]. Interestingly, the developed nanomaterials provided different geometries with various new characteristics which cannot be observed in bulk materials. These nanomaterials with different geometries have demonstrated high heavy metals removal efficiency. This fact is related to their exceptional properties including high surface area, porosity, specific surface charge, surface functionality, and ions binding capabilities [27,28]. Furthermore, nanomaterials are attractive and cost effective due to their reusability for several cycles. However, these nanomaterials have some limitations hindering their industrial applications, such as their removal from treated water [29]. This problem is overcome through nanomaterials functionalization by the link with biomolecules [30], carbon [31], polymers [32], and inorganic compounds [33], which allow easy separation from treated water as well as enhancing their adsorption power. In this review, we briefly discuss the use of various nanomaterials for the efficient removal of toxic metals from water/wastewater via the adsorption technique. We also discuss the effect of different conditions on the adsorption process, such as the initial metal concentration, temperature, contact time, adsorbent dosage, and pH. Additionally, we summarize the adsorption isotherms used for heavy metals adsorption through different nanomaterials.

2. Nanomaterials for Heavy Metals Removal

In the recent years, heavy metals removal from wastewater has been achieved using different types of nanomaterials such as zeolite, carbonaceous, polymer based, chitosan, ferrite, magnetic, metal oxide, bimetallic, metallic, etc. The heavy metals scavenging from wastewater using nanomaterials is achieved through the adsorption technique over their surfaces. Based on their basic material being used as absorbents and their importance for heavy metal removal, various types of nanomaterials are listed and discussed in detail in the next sections.

2.1. Carbon Based Nanomaterials

Functionalized carbon nanomaterials are considered one of the most promising adsorbents for toxic metals due to their unique chemical and physical features that allow their usage at large scales for wastewater purification. They have attracted much attention in the scientific community and engineering due to their extraordinary physical, chemical, optical, mechanical and thermal properties. In recent years, different types of contaminants were removed using several constructed carbon-based nanomaterials such as graphene, carbon-based nanocomposite, and carbon nanotubes. One of the most important carbon-based nanomaterials used for water treatment is graphene. It is the thinnest two-dimensional material comprised of a one-atom-thick planar sheet of sp^2-bonded carbon atoms, while carbon nanotubes have a cylindrical nanostructure which also consist of sp^2-bonded carbon atoms. Graphene has a high dispersity in aqueous solution, resulting from its high hydrophilic tendency due to the existence of oxygen functional groups on the graphene oxide (GO) surface. The large surface area and existence of functional groups over the GO surface makes it a promising material for water detoxification. Several pollutants are found to be removed with high efficiency from water using graphene-based nanomaterials. The adsorption of toxic metals from water via graphene occurs through the complexation of oxygen functional groups in the graphene and cationic metal. For organic dyes the interaction occurs through the π-electron delocalized arrangement of graphene. Zeng et al. [34] functionalized reduced GO hydrogel by MnO_2 nanotubes to produce 3D nanomaterial with a size of 20 nm used for the high adsorption of zinc, copper, silver, cadmium, and lead ions with the adsorption capacity of 83.9, 121.5, 138.2, 177.4 and 356.37 mg/g, respectively. This high adsorption capacity resulted from the synergetic action between MnO_2 nanotubes and reduced GO of the porous 3D nanostructures. Another research group, prepared the nanosheets of GO for the elimination of cobalt ions with the adsorption capacity of 68.2 mg/g and cadmium ions with an adsorption capacity of 106.3 mg/g [35]. Interestingly, another form of carbon material called carbon nanotubes (CNTs) are the most attractive materials for environmental engineers due to their electrical conductivity, cylindrical hollow structure, small size, and large surface area. Multi-walled carbon nanotubes (MWCNTs) and single-walled carbon nanotubes (SWCNTs) are the two main classes of CNTs. Wastewater treatment from toxic heavy metals via adsorption over CNTs was designated by several researchers. For instance, CNTs allowed the removal of Mn (II), Zn (II), Co (II), Pb (II), and Cu (II) [36]. In another study, the lead ions removal from water with an adsorption capacity equal 70.2 mg/g was reported [37]. Furthermore, MnO_2 was used to coat oxidized multi-walled CNTs, allowing an efficient scavenging of cadmium ions from water with an adsorption capacity equal 41.7 mg/g [38]. Lately, researchers synthesized the nanocomposite of carbon layered silicate that has developed features compared to conventional carbon nanomaterials that attracted the scientists' attention. Interestingly, an environmentally-safe composite, nanoadsorbent montmorillonite/carbon, was synthesized and used for the elimination of lead ions with an adsorption capacity of 247.86 mg/g [39].

2.2. Zeolites Nanoparticles

The high porosity of zeolites allows their use for the removal of toxic metals from wastewater. The negative charge between silicon and aluminum ions can be balanced by the movement of alkaline earth and alkali metals in their framework. However, nanoparticles of zeolites are used for adsorption efficiently due to their active and large surface areas. Recently, the adsorbent NaX nanozeolite/polyvinyl alcohol was synthesized and used efficiently for the removal of cadmium and nickel ions from water with an adsorption capacity for nickel lower than cadmium ions [40]. In another work, zeolite immobilized over alumina nanoparticles was used for the removal of Co (II) and Cr (III). Compared with immobilized nanoparticles, the adsorption power of zeolite and alumina nanoparticles was found to be enhanced by 17.3% for Co (II) and 31.77% for Cr (III) [41]. Additionally, zeolite nanoparticles used to fill ploysulfone membrane to remove lead and nickel ions from water via a combined treatment of filtration and adsorption. The capacity of adsorption of this fabricated membrane reached 122 mg/g for nickel and 682 mg/g for lead [42]. Recently, clinoptilolite zeolite nanoparticles were modified using pentetic acid and characterized then tested for the removal of Cd (II) from water. The modified clinoptilolite zeolite nanoparticles offered an excellent elimination of cadmium pollutants from water showing an adsorption capacity of 138.9 mg/g. The cadmium ion chelation occurred through ion exchange with raw clinoptilolite zeolite nanoparticles and by complexation with functional groups of pentetic acid indicating the important role of functionalization in the enhancement of adsorption capacity. The adsorption process occurred through monolayer and multilayered adsorption [43].

2.3. Polymer Based Nanomaterials

Many factors enhance the application of nanofibrous membranes filled with polymers (such as cellulose, chitosan, etc.) in environmental remediation, such as high surface area, small interfibrous pore size, and high gas penetrability that improve their chelation adsorption capacity. They are an excellent choice for the adsorption process due to their simple degradation behavior, adaptable surface functional groups, and excellent strength skeleton. Nanomembranes filled with polymers have a high selectivity and adsorption capacity resulting from existing functional moieties such as SO_3H, COOH, and NH_2, by which the permeated toxic metals are chelated. Materials used in the polymer-based nanomaterials can be used for their classification.

2.3.1. Cellulose Based Nanomaterials

Scientists tested the removal of toxic heavy metals using biopolymers-based adsorbents such as cellulose. Abou-Zeid et al. [44] synthesized 3 bio adsorbents: tetramethyl piperidine oxide oxidized 2, 3, 6-Tricarboxy cellulose nanofiber (T-CNFs) and TPC-cellulose nanofiber (TPC-CNFs) with and without polyamide-amine-epichlorohydrin crosslinker (PAE) and used for the treatment of water containing calcium, lead, and copper ions. For both lead and copper ions, the adsorption capacities were 82.19, 97.34, and 92.23 mg/g, respectively for Crosslinked TPC-CNFs, TPC-CNFs, and T-CNFs. In another study, phosphoric acid was used to functionalize cellulose nanofibers with a phosphate group to form CNF/P useful for the removal of copper ions from water [45]. Recently, Choi et al. synthesized cellulose nanofibers through electrospun cellulose acetate nanofibers deacetylation process that followed by functionalization of nanofibers with thiol group through esterification reaction. This polymer-based adsorbent used for the removal of Pb (II), Cd (II), and Cu (II) ions with adsorption capacity has been determined via Langmuir isotherm equal to 22.0, 45.9, and 49 mg/g, respectively. The metal ions removal occurred via the complexation reaction between the two surface thiol groups and divalent metals to approve the success of this synthesized functionalized nanomaterial for water remediation process [46]. Additionally, the previous method was used for the synthesis of cellulose nanofibrous mats that are then modified using citric acid. The citric acid modified cellulose nanofibrous were tested for the removal of chromium ions through a batch adsorption experiment. This synthesized nanoadsorbent offered an efficient removal of Cr (VI) from aqueous solution [47].

2.3.2. Dendrimer Based Nanomaterials

The removal of toxic metals has been achieved efficiently using organic polymers with functional groups to greedily chelate these toxics [48]. This system is composed of polymeric support with functional arms. On this basis, different ratios of acrylamide and acrylic acid were used for the synthesis of superadsorbent polymer hydrogel for the elimination of Co (II), Cu (II), Ni (II), and Cd (II) with high adsorption capacity from water. The adsorption efficiency toward cadmium and nickel ions was lower than that of copper and cobalt ions due to the easy penetration of small cations from polymeric support and easily chelated by functional arms [48]. In another study, nanofibers with ion-selectivity behavior including diethylenetriamine, ethyleneglycol, and ethylenediamine supported over a base polymer of polyacrylonitrile showed efficient removals of Zn (II), Pb (II), and Cd (II), with an adsorption capacity of 7.2, 8.8, and 6.1 mmol/g, respectively. Its high adsorption capacity may be attributed to the high surface area of the three nanofibers that allowed large numbers of functional groups for the scavenging of toxic metals [49]. Additionally, two new polymers were prepared by the reaction between 1,3,5-tris(6-isocyanatohexyl)-1,3,5-triazinane-2,4,6-trione with pentaethylenehexamine and diethylenetriamine to form polymeric nanomaterials with ethyleneamine arms to capture metallic ions through complexation reaction. The long amine chains in the nanoadsorbent were prepared using pentaethylenehexamine, which allowed a higher capture of divalent copper, chromium, cobalt, and cadmium ions than that prepared using diethylenetriamine [50]. A co-electro spinning technique was used for the fabrication of nanofibers membranes from metal-organic frameworks (MOF-808) and polyacrylonitrile to produce PAN/MOF-808 nanoadsorbent, offering an efficient removal of heavy metals from water according to the order Hg (II) < Pb (II) < Cd (II) < Zn (II). This removal efficiency order is in line with the metals ionic size, which indicates that the steric effect of metals hampers the access of metals to complex with active sites [51]. In another study, researchers synthesized poly-(ethylene-co-vinyl alcohol) nanofibers that efficiently removed hexavalent chromium pollutants from wastewater with the adsorption capacity of 90.75 mg/g obtained at less than 100 min [52]. In a similar study, nanofibers of polyacrylonitrile were functionalized using amidoxime to enhance its adsorption capacity for metals elimination. This nanoadsorbent efficiently removed lead and copper divalent ions from water [53]. A quantity of 1, 2-diaminoethane or 1, 3-diaminopropane was used to modify the resin of synthesized poly (styrene-alt-maleic anhydride) with 3-aminobenzoic acid. The synthesized nanoadsorbent was used for the elimination of Pb (II), Zn (II), Cu (II), and Fe (II) from water [54]. Sohail et al. [55] synthesized polyamidoamine (PAMAM) dendrimers via a divergent method. The functionalization occurred via two steps: Michael addition reactions of amino groups on methyl acrylate to produce shell of ester then linked to ethylene diamine to form the new developed surface. The characterization indicated that these zero generation dendrimers have 200–400 nm of size. These higher generation dendrimers have more outer functional groups on the external side that allowed the efficient removal of nickel ions from water [55].

2.3.3. Chitosan Based Nanomaterials

Chitosan is a linear polysaccharide composed of randomly distributed β-(1→4)-linked D-glucosamine (deacetylated unit) and N-acetyl-D-glucosamine (acetylated unit). It is made by treating the chitin shells of shrimp and other crustaceans with an alkaline substance, like sodium hydroxide. Chitosan is biocompatible, non-toxic, and hydrophilic polymer that able to interact with different metals via complexation reaction. This complexation reaction easily occurred due to the existence of amino functional groups over chitosan surface. This functionalization of chitosan occurs via connection with different materials to improve its adsorption capacity, selectivity, mechanical properties, and stability at low pH. For example, chitosan was used as a carbon and nitrogen source for the synthesis of nitrogen-doped carbon materials with a 3D hierarchically porous structure, which was tested for the elimination of cadmium and lead ions from water via the electrostatic interactions between surface oxygen and nitrogen functional groups and metallic ions. Therefore, the chelation process is chemisorption over this nanoadsorbent [56]. In another study, chitosan was combined with zinc oxide nanoparticles to produce core shell nanocomposite of ZnO/chitosan that showed to remove effectively Cu (II), Cd (II), and Pb (II) ions from water [26]. Additionally, the same coupling was made using titanium dioxide nanoparticles to fabricate a nanocomposite of chitosan/TiO$_2$ that allowed to adsorb efficiently lead and copper divalent ions [57]. Moreover, the chitosan nanoparticles were functionalized using alginate to form alginate/chitosan nanocomposite that permit the removal of 94.9% of nickel ions from aqueous solution [58]. The nanocomposite of chitosan alginate was fabricated and tested for the removal of divalent mercury from water. Here, alginate is negatively charged while chitosan is positively charged allowing the interaction with different charged pollutants in water as calcium and tripolyphosphate ions. This nanocomposite showed high adsorption capacity of 217.40 mg/g toward mercury ions [59]. Recently, Hussain et al. synthesized chitosan modified with salicylaldehyde in the size of 80 nm for the efficient removal of Pb (II), Cd (II), and Cu (II) from water. The nanocomposite showed high adsorption capacity toward Pb (II), Cd (II), and Cu (II) equal to 63.71, 84.60, and 123.67 mg/g, respectively. Additionally, this nanocomposite was tested for lead removal from tap water of 4.88 ppb [60]. In a separate study, Khalil et al. synthesized two novel chitosan nanocomposites. One was prepared via the functionalization of chitosan using cinnamaldehyde to produce chitosan-cinnamaldehyde (CTS-Cin) and the other nanocomposite was magnetic chitosan cinnamaldehyde (Fe$_3$O$_4$@CTS-Cin). These nanocomposites are promising for the treatment of water containing toxic metals. The two nanocomposites effectively removed hexavalent chromium Cr (VI) from water at 298 K with adsorption capacity of 61.35 mg/g for CTS-Cin and 58.14 mg/g for Fe$_3$O$_4$@CTS-Cin [61].

2.4. Magnetic Nanomaterials

Advanced nanomaterials have several classes in which magnetic nanomaterials are the main class that save the easily magnetic separation as well as the improved removal of toxic metals. This class of nanomaterials introduces excellent recyclability after separation that gives this class ecological benefits and wide applications in environmental remediation. With size decrease, the properties of these nanomaterials change significantly. The efficiency of non-magnetic nanomaterials in water treatment is lower than magnetic nanomaterials due to their hard separation and small surface area. However, the magnetic nanomaterials are easily dispersed, less toxic, with inert chemical behavior, and have a large surface area. These properties of magnetic nanomaterials make them cost effective, efficient, and faithful for water purification. Thus, heavy metals removal is effectively achieved by using these nanomaterials. A vital member of this class is iron oxide nanomaterials that provide ease of reusability, separation, and quick and high adsorption capacity. In this context, Fe$_3$O$_4$ nanoparticles were synthesized and used for the efficient adsorption of lead divalent ions from water with 36 mg/g of adsorption capacity [62]. In addition, Mn (II), Zn (II), Cu (II), and Pb (II) were removed using nanoparticles of Fe$_3$O$_4$ prepared by Giraldo et al. with a minimum adsorption capacity toward Mn (II) and maximum adsorption capacity toward Pb (II), which is related to the different hydrated ionic radius of these metals. This differentiation of metals size affects the electrostatic attraction between adsorption sites and toxic metals [63]. In another study, the nanoparticles of Fe$_3$O$_4$ were synthesized and their adsorption capacity of toxic metals Cu (II) and As (II) was compared with commercial iron oxide. The lower adsorption capacity of commercial iron oxide toward metals adsorption indicated the improved properties of synthesized nanoparticles [64]. Interestingly, nanorods, nanowires, and nanotubes provided great efficiency toward heavy metals removal from water/wastewater due to their higher surface area than other structured nanomaterials. For example, Cu (II), Cd (II), Ni (II), Zn (II), Pb (II), and Fe (II) were efficiently removed from water via adsorption over Fe$_3$O$_4$ nanorods with adsorption capacity of 76, 88, 95, 107, 113, and 127 mg/g, respectively [65]. The real applications of bare magnetic nanoparticles are limited due to their easy aggregation in water, which results from their tendency toward oxidation. Additionally, it is very important to prevent the magnetic nanoparticles from precipitation or aggregation alongside their synthesis development. A functionalization process of bare magnetic nanoparticles was followed to overcome these problems. The selectivity, stability, and adsorption capacity of magnetic nanoparticles increased via surface functionalization. The existence of different functional groups over magnetic surfaces offers different types of interactions between surfaces and metals including Van der Waals, electrostatic, ligand combination, chemical binding, and complex formation. Additionally, surface metal electrostatic interactions can be enhanced via functionalization using charged moieties. Thus, the selectivity of surfaces toward different metals resulted mainly through enhanced electrostatic attraction. Based on this, preventing the aggregation of magnetic nanoparticles can be achieved using capping hydrophilic coatings such as poly vinyl pyrollidine, poly vinyl alcohol, and polyethylene glycol. In addition, magnetic nanoparticles homogeneity and high surface area were enhanced through functionalization. Tailoring the magnetic surfaces caused alteration of their properties for toxic metals selective chelation. The selectivity and the improvement of the adsorption of magnetic nanoparticles toward toxic pollutants depend mainly on their surface are and size that are decided significantly by surface modification. Surface modification of magnetic nanoparticles occurred through many trials, including different modifiers such as carbonaceous, biomolecules, inorganic, organic, and polymers materials, etc.

2.4.1. Nanocomposite Magnetic Nanoparticles

Several metal oxides magnetic nanocomposites have been engineered and applied for heavy metals removal from water. For example, nanocomposite of Fe_3O_4/MnO_2 was fabricated with flower shape for the adsorption of Zn (II), Pb (II), Cu (II), and Cd (II). The adsorption of prepared nanocomposite was compared with that of iron oxide NPs that found to be enhanced by modification process [66]. Researchers coupled ferrite molecules with many metals to produce metal ferrite magnetic nanomaterials to allow its ease magnetic separation for the recyclability. The general formula of metal ferrite nanocomposite is $M(Fe_xO_y)$, where M is the metal atom. Many metal ferrites have been synthesized for heavy metals removal such as $ZnFe_2O_4$, $CuFe_2O_4$, and $Mn_{0.67}Zn_{0.33}Fe_2O_4$ [29,67]. $ZnFe_2O_4$ removed lead divalent ions while $CuFe_2O_4$ removed molybdenum divalent ions from aqueous solution with high adsorption capacity. $Mn_{0.67}Zn_{0.33}Fe_2O_4$ efficiently removed lead, cadmium divalent ions, and arsenic pentavalent ions. In another study, the co-precipitation method used for the synthesis of manganese and cobalt spinel ferrites ($MnFe_2O_4$ and $CoFe_2O_4$) resulted in nanoparticles in the size range of 20–80 nm. The synthesized nanoparticles removed divalent zinc ions with the adsorption capacity of 384.6 and 454.5 mg/g for $CoFe_2O_4$ and $MnFe_2O_4$, respectively [68]. Vamvakidis et al. synthesized the magnetic nanocomposite cobalt ferrite ($CoFe_2O_4$) that was further modified by coating with octadecylamine. The synthesized magnetic nanocomposite removed copper divalent ions with high adsorption capacity of 164.2 mg/g. The nanocomposite showed very easy separation via a magnet and simple regeneration via acidic treatment [69].

2.4.2. Inorganic Functionalized Magnetic Nanoparticles

Magnetic nanoparticles can be functionalized with inorganic materials including silica, metal, nonmetal, metal oxides, etc. These coatings improve the stability of the nanoparticles in solution and provide sites for covalent binding with specific ligands to the nanoparticle surface. For example, amorphous oxide shells of Mn-Co were used for the modification of magnetic nanoparticles surface allowing the exhibition of strong negative charge on their surfaces at various pHs. Interestingly, the obtained absorbent allowed the removal of Pb(II), Cu(II) and Cd(II) with adsorption capacity of 481.2, 386.2 and 345.5, respectively [70]. Additionally, magnetic nanoparticles functionalized with aluminosilicate displayed the higher adsorption capacity of Hg^{2+} [71]. In another study, silica was used to coat magnetic nanoparticles modified by 1, 2, 3-triazole for the removal of toxic divalent metals such as zinc, copper, and lead, allowing a maximum removal for lead ions and a minimum for zinc ions. This order is matching the atoms electronic properties that derive the complexation reaction. Metals with high ionic potentials will tend to acquire ligand electrons and stable alliances will be established. Thus, when the metals have the same charge, the hydrated ionic radius will derive the complexation reaction and the bigger cation will form the less stable complex. Therefore, in this study, the adsorption order followed the metals hydrated ionic radius [72]. Another class of inorganic materials used for the functionalization of magnetic nanoparticles is carbonate. In this context, calcium carbonate is a promising material for the modification of magnetic nanoparticles used for heavy metal removal from water due to its non-toxicity, low price, and high solubility that enhance the precipitation reaction of toxic species. However, their use is limited in water/wastewater remediation due to problematics of separation, sludge generation, and low efficacy. These limitations can be avoided by the combination with magnetic nanomaterials to enhance their adsorption capacity and ease their separation from experimental media. In this direction, Wang et al. synthesized magnetic mesoporous calcium carbonate-based nanocomposites via solvothermal method followed by annealing treatment to produce irregular sphere morphology nanocomposite with a size of 50 nm. This nanoadsorbent caused the removal Cd (II) and Pb (II) from water with maximum adsorption capacity of 821 and 1179 mg/g, respectively [73].

2.4.3. Carbon Materials Magnetic Nanoparticles

Nanoadsorbents efficiency for the adsorption of heavy metals from aqueous solutions can be improved by functionalization with carbon materials, such as activated carbon and graphene oxide. Subsequently, the adsorption capacity of graphene oxide/iron oxide/EDTA (EDTA: ethylenediaminetetraacetic acid) toward lead ion removal was tested. The capacity of adsorption improved with high degree due to the existence of hydroxyl and carboxyl groups over graphene oxide surface that chelate metal ions in addition to the existence of EDTA coordination power. All these functional groups over the surface interact with toxic metals through electrostatic interactions and hence, enhance the modified surface adsorption capacity [31]. In another study, Li et al. [74] synthesized environmentally benign nanocomposite of thiol (SH) functionalized Fe_3O_4 nanoparticles modified with activated carbon (Fe_3O_4@C-SH NPs) via one-step fabrication. The nanocomposite removed divalent copper ions from aqueous solutions with a maximum adsorption capacity of 28.8 mg/g. The nanocomposite showed high adsorption capacity and selectivity to copper ions over co-existing ions [74].

2.4.4. Organic Functionalized Magnetic Nanoparticles

The adsorption capability of magnetic nanomaterials toward heavy metals is believed to be improved via functionalization with functional groups provided from organic molecules. For example, EDTA was used to functionalize magnetic nanoparticles coated with silica for their application for toxic metals removal. In this context, the added dithiocarbamate groups were used for the removal of divalent mercury at trace concentrations, showing high adsorption removal capacity [75]. Likewise, glutathione was used for the modification of magnetic nanoparticles coated with silica to produce the nanoadsorbent of $Fe_3O_4/SiO_2/GSH$ useful for the removal of divalent lead ions from water. Interestingly, the temperature enhanced the adsorption of lead ions over the adsorption sites of synthesized nanoadsorbent. Additionally, the magnetic properties of this adsorbent allowed its ease of separation from the reaction medium via a magnet [76]. Furthermore, iron oxide nanoparticles were modified with amine and metformine and applied for copper ions removal from water. The highest elimination reached 92% of copper ions over Fe_3O_4 coated with silica modified with 0.1 wt % metformine [77]. In another research, Fe_3O_4 nanoparticles modified with 3-aminopropyltriethoxysilane were linked with crotonic acid and acrylic acid for the removal of divalent ions of Cd, Zn, Cu, and Pb. The highest adsorption capacity was reached for lead while the lowest adsorption capacity was found for cadmium due to different complexation affinity of metals with carbonyl group [78]. Recently, Zhang et al. used humic acid to modify nitrogen-doped Fe_3O_4 magnetic porous carbon for the removal of hexavalent chromium ions from water. The existence of functional groups over surfaces enhanced their complexation with Cr (VI) allowing the high adsorption capacity of adsorption equal to 130.5 mg/g [79]. Additionally, magnetic nanoparticles modified with amino groups showed efficient removal of toxic metals. Li et al. synthesized ternary amino-functionalized magnetic nano-sized illite-smectite clay for the effective removal of Pb^{+2} ions from water with the high adsorption capacity of 227.8 mg/g due to the complexation of metals with amino groups on the surfaces [80]. Wang et al. modified the nanocomposites of hollow Fe_3O_4/SiO_2/chitosan with triethylenetetramine for the removal of hexavalent chromium ions from water. The modification of the surface with the ligand triethylenetetramine provided more and stronger sites for the chromium complexation. This modified nanocomposite showed high adsorption capacity of Cr (VI) of 254.6 mg/g within only 15 min [81]. Likewise, triethylenetetramine was used for the modification of mesoporous Fe_3O_4 nanoparticles for the removal of Cu (II) from water with an efficiency of 86% [82]. Tetraethylene pentamine was used to functionalize magnetic polymer surfaces for the adsorption of hexavalent chromium via the reduction of Cr (VI) to non-toxic Cr (III) and via electrostatic interactions as confirmed by experimental data [83]. Similarly, many core-shell magnetic nanoparticles were functionalized with multi amino groups for the chelation of hexavalent chromium and divalent copper ions in water. The adsorption of two metals was found depending on electrostatic interactions and depending on medium pH. The pH value of 2–4 enhanced the adsorption capacity due to the availability of amino binding sites but at lower pH these sites protonated and loss its availability that negatively affect the adsorption process of copper ions. Precipitation of copper ions occurred at pH higher than 4. At the same pH range 2–4, the higher adsorption of chromium occurred via electrostatic interaction while at higher pH the adsorption process retarded due to active sites competitions between hydroxyl ions and chromium ions. In co-metals systems the competition occurred at low pH and high metals concentration while high pH and low metals concentration did not influence the adsorption removal capacity in co-metals systems [84].

2.4.5. Biomolecules Functionalized Magnetic Nanoparticles

Many investigations have been reported for the environmentally-safe use of biomolecules in the functionalization of magnetic nano-scaled particles to develop their chelation capability toward toxic metals removal from water/wastewater. Triazinyl- β-cyclodextrin was used for the modification of magnetic nanoparticles to enhance its adsorption capacity toward divalent ions such as of cobalt, zinc, copper, and lead. The existence of triazinyl- β-cyclodextrin azinyl nitrogen and hydroxyl groups over magnetic nanoparticles surfaces improved their chelation power toward above mentioned metals. The minimum adsorbed metal was cobalt ions while the maximum adsorbed metal was lead ions due to different chelation ability of surface nitrogen and oxygen [85]. In several studies, biopolymer cellulose was used to functionalize magnetic nanoparticles. Carboxy methyl cellulose-immobilized magnetic nanoparticles were prepared and tested for the removal of divalent lead ions with an enhanced adsorption capacity of 152.1 mg/g. The cellulose chains over magnetic nanoparticles exhibit many functional groups responsible for electrostatic interactions and chelation of lead ions over the nanoadsorbent. It was found that at pH over zero point charge the adsorption capacity was increased while at pH below zero point charge the adsorption capacity was decreased [86]. *Phanerochaete chrysosporium* pellets biomass strain loaded with Ca-alginate and magnetic nanoparticles for the efficient adsorption of divalent lead ions showed an enhanced adsorption capacity of 167.37 mg/g [87]. In another investigation, magnetic nanoparticles bounded in alginate polymer were modified with amino acid glycine in order to improve the removal of lead ions from water due to complexation ability of COOH- and NH_2- groups of glycine toward the toxic metal [28,88].

2.4.6. Polymers Functionalized Magnetic Nanoparticles

In order to get a high adsorption capacity for heavy metals adsorption from water, magnetic nanoparticles were also modified with polymers for many benefits like bio-compatibility, better mechanical strength, and higher chemical stability. Several studies reported the linkage of magnetic nanoparticles and polymers for the improvement of thermal and mechanical properties of nanoadsorbent. Fe_3O_4 nanoparticles were modified using diethylenetriamine and polyacrylic acid for the removal of hexavalent chromium and divalent copper from water with higher adsorption capacity toward copper ions than chromium ions [89]. Additionally, magnetic nanoparticles were coated by the mercaptoethylamino polymer for the removal of divalent cadmium, lead, and mercury ions in addition of monovalent silver ions [90]. Interestingly, divalent zinc and cadmium ions were efficiently removed from water when hydroxyapatite was used to modify magnetic nanoparticles [91]. A green method of synthesis was used for the Fe_3O_4 nanoparticles synthesis via eucalyptus extract. Then, these nanoparticles were modified using chitosan polymer in order to enhance the removal of pentavalent arsenic ions from water. These modified magnetic nanoparticles showed great adsorption capacity of arsenic ions at natural pH and short duration time [32]. Glycidylmethacrylatemaleic anhydride linked iminodiacetic acid was embedded in the fabrication of magnetic nanocomposite for the efficient elimination of divalent cadmium and lead ions from water [92]. In a recent study, the facile hydrothermal method was used for the synthesis of polyethylenimine functionalized magnetic montmorillonite clay (MMT-Fe_3O_4-PEI) for the adsorption of Cr (VI) from aqueous solution [93]. Interestingly, the nanoadsorbent removed chromium ions from water with high adsorption capacity of 62.89 mg/g. The removal mechanism occurred via partial reduction of Cr (VI) to Cr (III) and electrostatic attractions between protonated amino groups with Cr (VI) anions. This nanoadsorbent was reused several times for chromium uptake from water [93].

2.5. Metal Oxides and Metal Based Nanomaterials

Recently, the heavy metals removal from water was successively achieved using metal oxides and metals. However, the bare metals which are unstable, aggregate, and their separation from the processed media is very difficult, so their use is not common in water remediation. These problems are overcome via the functionalization process to ease their separation and enhance their stability to prevent aggregation problems. One of the most important nanoparticles for water treatment is the zero valent nano iron Fe^0 due to its high capacity of adsorption, reducing properties, non-toxicity, high surface area, and high stability. Many studies reported the use of Fe^0 for toxic metals remediation from water. Zero valent nano iron Fe^0 was synthesized for efficient removal of 99% of pentavalent arsenic from water [94]. Similarly, Fe^0 was prepared from ferric chloride using *Syzygium jambos* for the efficient scavenging of hexavalent chromium with an adsorption capacity of 983.3 mg/g [95]. Sometimes, zero valent nano iron Fe^0 was more stabilized by combination with other stabilizing agents. For example, chitosan carboxymethyl β-cyclodextrin complex is a biodegradable non-toxic stabilizer used to envelop Fe^0 for the removal of divalent copper and hexavalent chromium ions from water. The mechanism of heavy metals removal included the reduction of Cu^{+2} to Cu^0 and Cr^{+6} to Cr^{+3} while the Fe^0 oxidized to Fe^{+3} [96]. Additionally, heavy metals elimination has been reported through bimetallic nanoparticles. In this context, kaolinite embedded Fe/Ni nanoparticles were synthesized for the excellent removal of divalent copper ions with efficiency of 99.8% [97]. Furthermore, the metal oxides have been demonstrated for the treatment of water containing heavy metals pollutants. Metal oxides nanoparticles were classified according to their magnetism into magnetic and non-magnetic metal oxides nanoparticles. Heavy metals remediation from water over non-magnetic nanoparticles metal oxides includes the oxides of Mn, Cu, Fe, Al, Ce, Zn, etc. For example, sodium titanate was used for the synthesis of nanofibrous adsorbent useful for the removal of many toxic ions from water including Zn, Pb, Ni, Cd, and Cu. The metals' removal from single and coexisting systems was compared and found higher in the case of single system adsorption owing to the competition among several adsorbates on the adsorption locations [98]. MnO_2 nanofiber was coated by polypyrrole and polyacrylonitrile and aimed at the capture of divalent lead species, with a high adsorption removal capacity of 251.90 mg/g [99]. Likewise, CeO_2 nanofibers were coated with 3-mercaptopropyltrimethoxysilane and vinylpyrrolidone for the removal of divalent ions of lead and copper with higher adsorption to lead ions than copper ions [100]. Additionally, polyvinyl alcohol was used for the coating of ZnO nanofibers for the removal of U (VI), Ni (II), and Cu (II) allowing a minimum adsorption capacity for Ni (II) and maximum adsorption capacity for U (VI) [101]. The deriving factors of this order of adsorption are metallic hydrated ionic radius, electronegativity, and atomic mass. U (VI) has the highest atomic mass, highest hydrated ionic radius and lowest electronegativity that influence the adsorption order over these nanofibers. α-Fe_2O_3 nanofibers were synthesized for the removal of hexavalent chromium from water with an adsorption capacity of 16.18 mg/g [102]. In another study, α-Fe_2O_3 nanoparticles were coated with volcanic rock for the elimination of divalent cadmium toxics from H_2O at 30 °C showing a capacity of adsorption equal to 146.42 mg/g [103]. In separate studies, magnetron sputtering was used for the synthesis of CuO nanoparticles for the elimination of Pb (II) and Cr (VI) from H_2O with adsorption capacity of 37.02 and 15.62 mg/g, respectively [28,87,104]. Additionally, nanoparticles of alkaline metal oxides were reported for the heavy metals' removal. Compared to previously mentioned nanoparticles of metal oxides, these nanoparticles of alkaline metal oxides are eco-friendly and less toxic. The nanoparticles of magnesium oxide MgO are important members of this group, that were studied for heavy metals removal by many researchers. MgO was synthesized for toxic metals removal in addition to bacterial disinfection. The synthesized nanoadsorbent inactivated *Escherichia coli* in addition to the removal of divalent lead and cadmium ions [105]. In another work, flowerlike MgO nanoparticles were synthesized with high surface area for the removal of divalent ions of cadmium and lead with the adsorption capacity of 1500 and 1980 mg/g, respectively. The high adsorption occurred via a dissimilar cation exchange mechanism among metal and magnesium ions [106].

Alumina (Al_2O_3) is another important metal oxide nanoadsorbent for heavy metals removal. Alumina has many structural forms χ, θ, γ, β, and α that are naturally exist in soils [107]. Natural α-Al_2O_3 adsorbent possesses high stability in conventional methods [108]. Al_2O_3 has interesting characteristics such as high electrical insulation, corrosion and water resistance, compressive strength, and thermal conductivity, as well as its strong interatomic bonding which allowed its use as excellent adsorbent [109]. Dehghani et al. [110] benefited from nano γ-alumina properties for the removal of heavy metals from aqueous solution. The synthesized γ-alumina caused the uptake of Co (II) ions from wastewater by maximum adsorption capacity of 75.78 mg/g. This synthesized nanoadsorbent can effectively remove cobalt ions from aqueous solutions within a shorter time and with lower adsorbent dosage than others reported in the literatures. New works indicated the aptitude of alumina nanoparticles to emulate nano-membranes for pollutants adsorption. In this study, the pores of the spent alumina catalyst were filled with 2-amine-1-methyl benzimidazole and dithizone aimed to the efficient elimination of divalent zinc, cadmium, and nickel toxics from aqueous wastes. The adsorption removal capacity over 2-amine-1-methyl benzimidazole/alumina was found 1.01, 0.55, 0.38 mmol/g for Cd (II), Zn (II), and Ni (II), respectively. The adsorption of these mentioned toxic metals occurred via chemical reaction and physisorption over 2-amine-1-methyl benzimidazole/alumina nanosorbents [109]. Polyethersulfone membranes were filled with alumina nanoparticles by several dosages to improve its efficiency toward Cu (II) ions removal. It was concluded that the increased alumina nanoparticles, increased the metal ions elimination from water [111]. The alumina adsorption capacity can also be improved via combination with metallic nanoparticles. In a recent study, Kumari et al. [112] used green synthesized nanoparticles of silver to be stabilized over alumina nanoparticles for heavy metals removal from pharmaceutical effluents. They used bio-based reductant *Bacillus cereus* cell extracts for the green synthesis process environmentally-safe. This nanocomposite caused the removal of Pb and Cr from pharmaceutical effluent with efficiency of 99.5% and 98.44%, respectively [112]. Metal oxides adsorption capacities can be improved through modification with surfactants. Thus, the modification of alumina nanoparticles with surfactants was improved by inhibiting nanoparticles agglomeration and therefore increasing their surface area for better adsorption capacity. γ-Al_2O_3 NPs was modified using sodium dodecyl sulfate (SDS) and sodium tetra decyl sulphate for the elimination of divalent cadmium ions allowing high effect of surfactant on the adsorption process [113]. The adsorption efficiency was found improved by modification from 67% to 95%. Additionally, SDS was used to modify alumina surfaces for better adsorption of Mn (II) ions from manganese-bearing real industrial wastewater and Mn(II)-spiked wastewater. The removal occurred via adsolubilization of manganese ions in the admicelle formed by SDS on the alumina surface. The adsorption of Mn (II) over this modified nanoadsorbent occurred with efficiency of 92% [114].

2.6. Silica Based Nanomaterials

The exceptional properties of silica-based nanomaterials such as definite pore size, manageable surface characters, and high surface area has made these nanomaterials an important metal oxides class for the detoxification of water from heavy metals. Additionally, these nanoadsorbents possess advanced selectivity and adsorption capacity to metals when modified with thiol and amino groups. Additionally, they are eco-friendly and non-toxic that increased their applications in water remediation. The adsorption capacity of functionalized and non-functionalized silica nanoadsorbents toward the chelation of divalent ions of lead, nickel, and cadmium are studied and compared in the literature [115]. The comparison occurred between NH_2-silica gel, non-functionalized nanosilica spheres, and NH_2-silica nano hollow spheres showed higher adsorption capacity of functionalized materials equal to 96.80, 31.40, 40.74 mg/g for the removal of lead, nickel, and cadmium ions, respectively. Additionally, silica nanospheres were modified with phenyl groups and 3-amino propyl in order to enhance the adsorption removal capacity toward divalent copper ions. Interestingly, as the number of amino groups increased, the adsorption efficiency toward copper ions was increased [116]. In a similar research, nitrilotriacetic acid was used to modify silica gel that applied for treatment of wastewater containing divalent ions of lead, cadmium, and copper. The modified nanoadsorbent removed the Pb (II), Cd (II), and Cu (II) with adsorption capacity of 76.23, 53.15, 63.6 mg/g, respectively within only 2–20 min [117]. Recently, Yang et al. used kerf loss silicon waste for the synthesis of 3-aminopropylethoxysilane (APTES)-functionalized nanoporous silicon (NPSi) hybrid materials via nanosilver-assisted chemical etching. The functionalized nanoadsorbent examined for the adsorption of hexavalent chromium Cr (VI) from aqueous solution with a maximum adsorption of 103.75 mg/g within one hour. The adsorption mechanism occurred via the reduction of Cr (VI) to Cr (III) over protonated amino groups. This functionalized nanoadsorbent maintains its efficiency after 5 cycles of adsorption [118]. The several nanomaterials applied for the elimination of toxic metallic ions from wastewater were summarized in Table 1.

Table 1. Nanomaterials for heavy metals removal from water.

Metal Ions	Adsorbent	Contact Time (min)	Initial Metal Conc. (mg/L)	Dosage of Adsorbent (mg/L)	pH	Temp. (K)	Ref.
Cd (II)	Graphene Oxide Nanosheets	-	20	1000	10	303	[35]
	(PVA)/NaX nanozeolite	60	50	500	5	318	[40]
	SPH	-	100	2	7	298	[48]
	HDI-IC-DETA	30	58.9, 112.4	1000	6	333	[50]
	Nitrogen-doped carbon materials	120	40	400	5	298	[56]
	Fe_3O_4/MnO_2	30	10	1000	3	298	[66]
	SNHS, NH_2–SNHS, (NH_2–SG)	1440	140	15	14	298	[115]
	MnO_2/oMWCNTs	150	15	500	5	323	[105]
	NTA-silica gel	720	20	1000	2.9	298	[117]
	DTZ-Al_2O_3 and MAB-Al_2O_3)	720	44.9–865.4	3330	11	298	[109]
	SMA (SDS and STS @ Al_2O_3)	120	-	-	6	298	[113]
	Polystyrene-poly(N-isopropylmethacrylamide-acrylic acid	100	50	0.0011 g	9	298	[119]
	Fe_3O_4/HA	15	0.1	10	6	298	[120]
Co (II)	Graphene Oxide Nanosheets	-	20	1000	10	303	[35]
	Al_2O_3 NPs in zeolite	240	50	-	7	318	[41]
	SPH	-	100	2	7	298	[48]
	HDI-IC-DETA	30	58.9, 112.4	1000	6	333	[50]
Pb (II)	CNTs	360	10	500	6	308	[37]
	MMT/C	50	100	400	5	308	[39]
	PAN nanofibers	30	100	1000	6	298	[49]
	Nitrogen-doped carbon materials	120	40	400	5	298	[56]
	Fe_3O_4	30	100	10,000	5.5	298	[62]

Table 1. *Cont.*

Metal Ions	Adsorbent	Contact Time (min)	Initial Metal Conc. (mg/L)	Dosage of Adsorbent (mg/L)	pH	Temp. (K)	Ref.
	(CMC-Fe_3O_4)	120	200	1000	6	298	[86]
	$CuFe_2O_4$	220	10	20	4.5	298	[29]
	Fe_3O_4–SiO_2-GSH MNPs	120	100	100	5.5	308	[76]
	MNPs–Ca-alginate	60	500	1800	5	308	[87]
	Glycine–MNPs (GF MNPs)	100	10	10,000	2	298	[88]
	PAN/PPy/MnO_2) nanofiber	120	400	60	6	298	[99]
	SNHS, NH_2–SNHS, (NH_2–SG)	1440	140	15	14	298	[115]
	NTA-silica gel	720	20	1000	2.9	298	[117]
	TEMPO	-	-	-	-	298	[109]
	Polystyrene-poly(N-isopropylmethacrylamide-acrylic acid	100	50	0.0011 g	9	298	[119]
	NH_2 MNPs	60	10	100	5	298	[121]
	Fe_3O_4@C	120	50	1000	5.5	293	[122]
	Fe_3O_4/HA	15	0.1	10	6	298	[120]
Ni (II)	(PVA)/NaX nanozeolite	60	50	500	5	318	[40]
	SPH	-	100	2	7	298	[48]
	HDI-IC-PEHA	30	58.6, 63.8, 51.9	1000	6	333	[50]
	Alg-CS	30	70	3000	3	298	[58]
	(PVA)/zinc oxide (ZnO) nanofiber	300	50	1000	5	318	[101]
	DTZ-Al_2O_3 and MAB-Al_2O_3)	720	17.5–368.5	3330	7	298	[109]
	SNHS, NH_2–SNHS, (NH_2–SG)	1440	140	15	14	298	[115]
Cr (III)	Al_2O_3 NPs in zeolite	240	50	-	7	318	[41]
	HDI-IC-PEHA	30	58.6, 63.8, 51.9	1000	6	333	[50]
	Polystyrene-poly(N-isopropylmethacrylamide-acrylic acid	100	50	0.0011 g	9	298	[119]
Cr (VI)	TEPA-NMPs	180	50	5000	3	318	[83]
	EVOH nanofiber membranes	100	150	100	2	318	[52]
	SJA-Fe	90	50	500	5.5	298	[95]
	PAA-coated amino-functionalized Fe_3O_4	-	600	20,640	5	298	[89]
	α-Fe_2O_3 nanofibers	5	200	2000	-	298	[102]
	CuO	10	20	1600	3	298	[104]
	BiOBr/Ti_3C_2	80	20	40	-	298	[123]
Cu (II)	SPH	-	100	2	7	298	[48]
	PAN nanofibers	30	100	1000	6	298	[49]
	HDI-IC-PEHA	30	58.6, 63.8, 51.9	1000	6	333	[50]
	Fe_3O_4	30	2	10	7	298	[82]
	PAA-coated amino-functionalized Fe_3O_4	-	600	20,640	5	298	[89]
	Kaolinite (K-Fe/Ni)	30	200	-	4.7	298	[97]
	(PVA)/zinc oxide (ZnO) nanofiber	300	50	1000	5	318	[101]
	NH_2-silica	150	-	1000	5	298	[116]
	NTA-silica gel	720	20	1000	2.9	298	[117]
	(TPC-CNFs)	-	-	-	-	298	[109]
	Cobalt ferrite, Titanate NTs, alginate	120	10	0.15	6	298	[124]
	Polystyrene-poly(N-isopropylmethacrylamide-acrylic acid	100	50	0.0011 g	9	298	[119]
	GA–APTES	15	15.88	1250	4	293	[125]
Zn (II)	PAN nanofibers	30	100	1000	6	298	[49]
	DTZ-Al_2O_3 and MAB-Al_2O_3)	720	58.8–117.6	3330	11	298	[109]
Mo (II)	$ZnFe_2O_4$	-	110	2000	3	298	[67]
Hg (II)	Fe_3O_4/ SiO_2/APTES	-	50×10^3	-	-	298	[75]
	Fe_3O_4/HA	15	0.1	10	6	298	[120]
As (V)	NZVI	720	1	100	6.5	298	[94]
As (III)	Cobalt ferrite, Titanate NTs, alginate	120	10	0.15	6	298	[124]
Fe (III)	Cobalt ferrite, Titanate NTs, alginate	120	10	0.15	6	298	[124]
U (VI)	(PVA)/zinc oxide (ZnO) nanofiber	300	50	1000	5	318	[101]

3. Factors Affecting Adsorption Process

The adsorption process of heavy metals ions over different surfaces is controlled by various factors including the initial ion concentration, the temperature, the contact time, the adsorbent dosage, and the pH of reaction medium. Thus, in the following sections the effect of the mentioned factors on the heavy metals removal from water/wastewater will be discussed.

3.1. Ionic Strength Effect

The adsorption process over the adsorbate surface is greatly affected by the existence of extra ions in the solution. So, the ionic strength of solution plays a vital role in the adsorption efficiency.

The extra added ions affected significantly the adsorption efficiency through their competition with the main adsorbed ions on the adsorbent chelation sites. Commonly, the influence of experimental ionic strength on chelation performance was studied by the addition of Cl^- and Na^+ ions to the solution. Additionally, the adsorbent affinity toward added ions and their concentrations affected the adsorption efficiency. Sometimes, the effect of ionic strength on the adsorption process is neglected due to the low attraction of adsorbent to the extra ions compared to the target metal ions as reported in several studies. For example, in the study of the adsorption of lead divalent ions over the surface of $Fe_3O_4/SiO_2/GSH$ nanocomposite [76], the adsorption was increased when sodium chloride concentration in the solution was 0.025 mM that enhances the dispersion of functional groups on the adsorbent surfaces. However, when the sodium chloride concentration reached 0.2 mM there was a decrease in the adsorption of lead ions due to the competition on the chelation sites. In another study [92], during the adsorption of divalent cadmium and lead ions over magnetic nanocomposite, the addition of sodium chloride ions up to 3 mol/L has no effect on the adsorption of toxic metals This indicates the great attraction of metal ions over the nanocomposite surface more than the added ions. Consequently, the influence of ionic strength on the adsorption reaction may be significant or neglected depending on the attraction between the target adsorbate and the adsorbent in addition to sodium chloride concentration.

3.2. Initial Ion Concentration Effect

Generally, it is recognized that the increasing of toxic species concentration increases the adsorption process to a certain point after which the adsorption decreases due to the metal ions optimum concentration that must be exist in the reaction medium for the applied adsorbent. Low numbers of toxic metals are available at low concentration allowing the decrease of the scavenging efficiency, but at a higher metallic concentration the available ions for adsorption process are increased leading to the enhancement of the removal efficiency. Nevertheless, above a certain initial concentration, ions with the equal amount of adsorption locations are available, which therefore decreases their elimination adsorption capacity. For instance, nanofibrous adsorbent PVA/ZnO adsorption capacity was improved for the chelation of Ni (II), Cu (II), and U (VI) when the metallic initial concentration increased from 90 to 500 mg/L [101]. Similarly, the adsorption of divalent mercury ions over chitosan-alginate nanoparticles (CANPs) surface enhanced when initial metallic concentration elevated from 4 to 12 mg/L then reduces successively [59].

3.3. Temperature Effect

The adsorption process is greatly influenced by the temperature of the solution. Generally, the increasing of the temperature reduced the solution viscosity and, consequently, improved the ions diffusion to reach the adsorption allowing the enhancement of the adsorption competence. Therefore, the adsorption process of metal ions from aqueous solution can be affected by solution temperature in two ways depending on heat emitted or absorbed. The increasing temperature causes the increase of the adsorption rate for endothermic adsorption while causing the decrease of the adsorption rate for exothermic adsorption. Moreover, the adsorption study under temperature change introduces many thermodynamic parameters about the adsorption process including entropy changes (ΔS), enthalpy change (ΔH), and Gibbs free energy change (ΔG). In the literature [59], for the CANPs adsorption of divalent mercury ions, the adsorption process was exothermic. So, when the temperature is raised from 10 to 20 °C there is an enhancement of adsorption process, but over this temperature the adsorption process decreased significantly due to the increased kinetic energy that caused the desorption of metallic ions from adsorbent sites. Hence, the increased temperature decreased the chelated ions over CANPs surface. Another reason for adsorption decrease is the weak electrostatic interactions between mercury and adsorbent at increased temperature as it is exothermic adsorption. The process is known to be endothermic and spontaneous when ΔH = positive value and ΔG = negative value [126–131]. In another work, Fe_3O_4 nanoparticles chelated divalent lead ions through endothermic adsorption process. So, the increasing of the temperature increased the mobility of ions to reach a large number of binding sites that enhanced the adsorption capacity to lead ions [62].

3.4. Contact Time Effect

The water/wastewater economical remediation from toxic heavy metals significantly depends on the contact time between nanosorbent and metal ions. The adsorption efficiency greatly increases by increasing the time of contact between pollutants and adsorbent due to the increase of interaction time between active sites of chelation and metals. Normally, at the beginning of the adsorption the removal efficiency occurs quickly and then increases gradually. This occurs because of the availability of the free active sites at initial adsorption that are gradually occupied with time by chelated metals. This phenomenon was reported for the adsorption of divalent mercury ions on surfaces of CANPs. When the contact time increased from 0 to 90 min, the adsorption greatly increased. At the first half-hour the adsorption increased rapidly then delayed with reaching equilibrium at 90 min [59]. Similarly, the study reported the chelation of divalent cadmium and lead ions over surfaces of MNCPs [92]. The adsorption efficiency within 20 min reached 91% and 100% of maximum adsorption (48.54 and 53.35 mg/g) for Cd (II) and Pb (II) ions respectively. Subsequently, the adsorption becomes independent on the contact time when the equilibrium reached between desorption and adsorption.

3.5. Adsorbent Dosage Effect

The dosage of the added adsorbent plays a vital role in the removal of pollutants via adsorption. Increasing the adsorbent dosage caused an increase in the number of existing active sites to chelate toxic metals, and greatly enhances the adsorption capacity. So, the amount of nanoparticle dosage efficiently improves the adsorption capacity. Nonetheless, the active sites decrease as a result of surface area decrease by additional increase in the nanoparticles amount due to the agglomeration of the adsorbent that greatly reduces the adsorption capacity. The adsorbent dosage effect on the heavy metals removal from the aqueous solution has been studied in several studies. For instance, nanofibrous TiO_2 coated chitosan was used for the elimination of divalent copper and lead ions from water with an optimal dosage of 2000 mg/L. After this optimal dosage of nanosorbent, there were nanoparticles agglomerations and the decrease of surface area caused a reduction of adsorption capacity toward toxic metals [57]. In another study, Fe_3O_4@C nanoadsorbent used for the removal of lead ions from water improved the efficiency from 41% to 92% when the adsorbent dosage increased from 0.5 to 2 g/L, due to the increase of the active site number and to the simple contact of lead ions with these sites. However, the increase of the adsorbent dosage reduced the removal efficiency from 41% to 22% due to the adsorbent agglomeration that reduces the existed active sites to chelate lead ions [123].

3.6. pH Effect

The adsorption reaction and adsorption capacity greatly depend on the reaction medium pH. So, the effect of pH on metal ions scavenging must be determined. In a study of Hg (II) divalent ions removal via CANPs nanoparticles, there was an increase in the adsorption capacity when the pH changed from 2 to 5. However, at lower pH, the functional groups over nanoparticles surfaces became protonated, which led to electrostatic repulsion that prevents the adsorption of the target heavy metal. Additionally, there was a competition between the heavy metal and H_3O^+ for the active sites of the adsorbent. Sometimes, the adsorption enhanced at moderate values of pH due to the deprotonation of functional groups and to the decrease of repulsion forces that enhances the adsorption capacity [59].

Similarly, chromium reached its highest adsorption on functionalized magnetic polymer at pH 2. This adsorption exhibited a dependence on the electrostatic interactions related to the presence of functional groups. However, it is very important to take into consideration the speciation chromium oxidation state in order to determine the adsorption mechanism [82–84,132]. The heavy metals adsorption, as determined in several studies, is favored at moderate pH values than at lower pH values. For instance, nanofiber chitosan/TiO_2 chelated divalent ions of copper and lead with minimum removal at 2–4 pH values and maximum removal at 6 pH values [57]. Similarly, the cadmium ions removal increased by increasing the pH from 4 to 6, then remained constant at pH 9. However, the increase of pH to 11 decreased the removal of cadmium to 80%, indicating that cadmium removal is favored at moderate pH values due to the deprotonation of functional groups that reduces repulsion forces and increases the attraction between metal ions and functional active sites [88]. Though, at low pH the protonated active sites numbers increase and caused a great repulsion with positive charged toxic pollutants that greatly reduces the adsorption capacity of the nanoadsorbent [87]. At very high pH values, several complexes between metal species and OH groups formed that blocked the large numbers of adsorbent active sites and reduced their adsorption capacity.

4. Desorption and Recyclability of the Nanoadsorbents

The viability of adsorbents in practical applications is depending on their evaluation for the reusability due to the reduced cost of adsorption by the successful successive use of the adsorbent [133–135]. The reusability of nanoadsorbents occurred by the successful separation from the experimental medium using several methods, as reported in the literatures. For example, ion-selective polyacrylonitrile nanofibers were used for the removal of divalent ions of zinc, lead, and copper in a reusability study with up to four cycles of excellent efficiency of more than 90%, by using 0.5 M H_3PO_4 and1.0 M HNO_3 [49]. Stable adsorption results were found for the removal of zinc and lead using these nanoadsorbents indicating the ease of reusability and recyclability, while the adsorption results decreased for copper ions with the reusability. Additionally, CMC-Fe_3O_4 used for the removal of divalent lead ions was tested for reusability through cleaning with 0.1 M Na_2HPO_4 and H_2O [84]. The adsorption capacity of this nanosorbent was slightly decreased during the reusability cycles. After 5 cycles the adsorption capacity was still more than 85%. This nanoadsorbent has a great tendency for reusability, as shown in a magnetism study in which the magnetization before and after five cycles was still the same that made these nanomaterials valued for water treatment. In another study, diluted HNO_3 solution as a desorbent was used to study the reusability of HNC-3 for lead ions removal. The results indicated excellent reusability even after 5 cycles for lead removal. This complete recovery with nitric acid indicated that the lead ions were completely desorbed from the nanomaterial active sites, due to the competition between lead ions and protons for the active sites on the nanoadsorbent surface [56]. Additionally, core/shell gel particles of polystyrene-poly (N-isopropylmethacrylamide-acrylic acid) used for the adsorption of Cr (III), Cd (II), Cu (II), and Pb (II) ions were reused via washing with 0.5 M HCl. The reusability of these core/shell gel particles was successful up to 3 repetitive cycles, but with a small noted decrease in the Cr (III), Cd (II), and Cu (II) removal in the 3rd cycle [119]. MnO_2 nanotubes@reduced graphene oxide hydrogel (MNGH) nanoadsorbent was applied for the removal of divalent lead ions from water in a reusability study. The desorption of metal ions from adsorbent was conducted using 0.5 M HNO_3. Pb (II) adsorption capabilities decrease somewhat with reuse cycles. After eight cycles of reusability, the adsorbent retained 87% of its adsorption capacity. Thus, this nanoadsorbent is suitable for toxic metals removal with ease reusability [34].

5. Conclusions

Recently, great developments have been observed in the field of nanotechnology and nanoscience that produce environmentally-safe, economical, and efficient materials for environmental engineering. These engineered nanomaterials are promising for water treatment due to their unique physicochemical properties that enable the heavy metals scavenging with high adsorption capacity and selectivity even at very low concentrations. Efficient heavy metals removal from water was reached via the simple adsorption technique. The current review discussed the heavy metals removal using different nanomaterials like zeolite, polymers, chitosan, metal oxides, and metals under different conditions. The new studies focused on nanomaterials functionalization in order to enhance properties of separation, stability, and adsorption capacity. The functionalization process was reached using different molecules such as biomolecules, polymers, inorganic materials etc. By providing the magnetic properties to the adsorbent, its separation becomes easier via magnetic separation techniques. The effect of various experimental conditions like ionic strength, initial concentration of metal, contact time, temperature, adsorbent dosage, and pH of the solution on the adsorption was discussed. For initial metal concentration, the increased metal ion concentration initially increases the adsorption process then decreases due to the occupation of adsorbent active sites. For the effect of time, the adsorption occurs rapidly at the beginning of adsorption then slows down at equilibrium. Additionally, the temperature greatly affects the metal ions capture from water. For endothermic adsorption, the increase of the temperature caused an increase of the adsorption capacity while for exothermic adsorption the increase of the temperature caused a decrease of adsorption capacity. The metal/binding sites ratio that showed the adsorbent dosage effect on adsorption was found to greatly affect the adsorption capacity. The metals removal was found greatly depending on the pH solution. Maximum metals removal was reached at moderate pH. However, at low or high pH values, the removal was lower. This review provided precious information regarding the use of engineered nanomaterials for the removal of toxic heavy metals that will guide the researchers intending to fabricate new nanomaterials for water/wastewater remediation. It should be noted, that all experiments are conducted at lab-scale for toxic metals in aqueous solutions and studies are needed to evaluate the process efficiency at pilot and large scale using real wastewater.

Author Contributions: Conceptualization, M.A.T., N.S.A., F.B.R., W.M., and S.M.S.; investigation, N.S.A.; data curation, M.A.T., F.B.R., and S.M.S.; writing—original draft preparation, F.B.R. and W.M.; writing—review and editing, W.M., and N.S.A.; supervision, W.M.; project administration, S.M.S. All authors have read and agreed to the published version of the manuscript.

Funding: This research was funded by Deanship of Scientific Research at King Khalid University.

Acknowledgments: The authors extended their appreciation to the Deanship of Scientific Research at King Khalid University for funding this work through general research project under grant number GRP-106-41. This research was funded by the Deanship of Scientific Research at Princess Nourah bint Abdulrahman University through the Fast-track Research Funding Program.

Conflicts of Interest: The authors declare no conflict of interest.

References

1. Siddeeg, S.M.; Tahoon, M.A.; Rebah, F.B. Agro-industrial waste materials and wastewater as growth media for microbial bioflocculants production: A review. *Mater. Res. Express* **2019**, *7*, 012001. [CrossRef]

2. Ben Rebah, F.; Mnif, W.; M Siddeeg, S. Microbial flocculants as an alternative to synthetic polymers for wastewater treatment: A review. *Symmetry* **2018**, *10*, 556. [CrossRef]

3. Siddeeg, S.M. A novel synthesis of TiO$_2$/GO nanocomposite for the uptake of Pb^{2+} and Cd^{2+} from wastewater. *Mater. Res. Express* **2020**, *7*, 025038. [CrossRef]

4. Ndimele, P.; Kumolu-Johnson, C.; Chukwuka, K.; Ndimele, C.; Ayorinde, O.; Adaramoye, O. Phytoremediation of iron (Fe) and copper (Cu) by water hyacinth (Eichhornia crassipes (Mart.) Solms). *Trends Appl. Sci. Res.* **2014**, *9*, 485.

5. Peng, K.; Li, X.; Luo, C.; Shen, Z. Vegetation composition and heavy metal uptake by wild plants at three contaminated sites in Xiangxi area, China. *J. Environ. Sci. Health Part A* **2006**, *41*, 65–76. [CrossRef] [PubMed]

6. Gabal, E.; Chatterjee, S.; Ahmed, F.K.; Abd-Elsalam, K.A. Carbon nanomaterial applications in air pollution remediation. In *Carbon Nanomaterials for Agri-Food and Environmental Applications*; Elsevier: Cambridge, MA, USA, 2020; pp. 133–153.

7. Wu, H.; Wei, W.; Xu, C.; Meng, Y.; Bai, W.; Yang, W.; Lin, A. Polyethylene glycol-stabilized nano zero-valent iron supported by biochar for highly efficient removal of Cr (VI). *Ecotoxicol. Environ. Saf.* **2020**, *188*, 109902. [CrossRef]

8. Mallikarjunaiah, S.; Pattabhiramaiah, M.; Metikurki, B. Application of Nanotechnology in the Bioremediation of Heavy Metals and Wastewater Management. In *Nanotechnology for Food, Agriculture, and Environment*; Springer: Cham, Switzerland, 2020; pp. 297–321.

9. Sutton, D.J.; Tchounwou, P.B.; Ninashvili, N.; Shen, E. Mercury induces cytotoxicity and transcriptionally activates stress genes in human liver carcinoma (HepG2) cells. *Int. J. Mol. Sci.* **2002**, *3*, 965–984. [CrossRef]

10. Tchounwou, P.B.; Yedjou, C.G.; Foxx, D.N.; Ishaque, A.B.; Shen, E. Lead-induced cytotoxicity and transcriptional activation of stress genes in human liver carcinoma (HepG 2) cells. *Mol. Cell. Biochem.* **2004**, *255*, 161–170.

11. Patlolla, A.K.; Barnes, C.; Hackett, D.; Tchounwou, P.B. Potassium dichromate induced cytotoxicity, genotoxicity and oxidative stress in human liver carcinoma (HepG2) cells. *Int. J. Environ. Res. Public Health* **2009**, *6*, 643–653.

12. Tchounwou, P.B.; Ishaque, A.B.; Schneider, J. Cytotoxicity and transcriptional activation of stress genes in human liver carcinoma cells (HepG2) exposed to cadmium chloride. *Mol. Cell. Biochem.* **2001**, *222*, 21–28.

13. Yedjou, C.G.; Tchounwou, P.B. Oxidative stress in human leukemia (HL-60), human liver carcinoma (HepG2), and human (Jurkat-T) cells exposed to arsenic trioxide. *Met. Ions boil. Med.* **2006**, *9*, 298–303.

14. Agudosi, E.S.; Abdullah, E.C.; Mubarak, N.; Khalid, M.; Pudza, M.Y.; Agudosi, N.P.; Abutu, E.D. Pilot study of in-line continuous flocculation water treatment plant. *J. Environ. Chem. Eng.* **2018**, *6*, 7185–7191. [CrossRef]

15. Rahimi, S.; Ahmadian, M.; Barati, R.; Yousefi, N.; Moussavi, S.P.; Rahimi, K.; Reshadat, S.; Ghasemi, S.R.; Gilan, N.R.; Fatehizadeh, A. Photocatalytic removal of cadmium (II) and lead (II) from simulated wastewater at continuous and batch system. *Int. J. Environ. Health Eng.* **2014**, *3*, 31.

16. Motsi, T.; Rowson, N.; Simmons, M. Adsorption of heavy metals from acid mine drainage by natural zeolite. *Int. J. Miner. Process.* **2009**, *92*, 42–48. [CrossRef]

17. Srivastava, V.; Weng, C.; Singh, V.; Sharma, Y. Adsorption of nickel ions from aqueous solutions by nano alumina: Kinetic, mass transfer, and equilibrium studies. *J. Chem. Eng. Data* **2011**, *56*, 1414–1422. [CrossRef]

18. Naghdali, Z.; Sahebi, S.; Ghanbari, R.; Mousazadeh, M.; Jamali, H.A. Chromium removal and water recycling from electroplating wastewater through direct osmosis: Modeling and optimization by response surface methodology. *Environ. Health Eng. Manag. J.* **2019**, *6*, 113–120. [CrossRef]

19. Wang, L.K.; Vaccari, D.A.; Li, Y.; Shammas, N.K. Chemical precipitation. In *Physicochemical Treatment Processes*; Springer: Hoboken, NJ, USA, 2005; pp. 141–197.

20. Song, Y.; Lei, S.; Zhou, J.; Tian, Y. Removal of heavy metals and cyanide from gold mine waste-water by adsorption and electric adsorption. *J. Chem. Technol. Biotechnol.* **2016**, *91*, 2539–2544. [CrossRef]

21. Huang, Q.; Liu, M.; Chen, J.; Wan, Q.; Tian, J.; Huang, L.; Jiang, R.; Wen, Y.; Zhang, X.; Wei, Y. Facile preparation of MoS_2 based polymer composites via mussel inspired chemistry and their high efficiency for removal of organic dyes. *Appl. Surf. Sci.* **2017**, *419*, 35–44. [CrossRef]

22. Oatley-Radcliffe, D.L.; Walters, M.; Ainscough, T.J.; Williams, P.M.; Mohammad, A.W.; Hilal, N. Nanofiltration membranes and processes: A review of research trends over the past decade. *J. Water Process Eng.* **2017**, *19*, 164–171. [CrossRef]

23. Cao, C.-Y.; Qu, J.; Yan, W.-S.; Zhu, J.-F.; Wu, Z.-Y.; Song, W.-G. Low-cost synthesis of flowerlike α-Fe_2O_3 nanostructures for heavy metal ion removal: Adsorption property and mechanism. *Langmuir* **2012**, *28*, 4573–4579. [CrossRef]

24. Sun, S.; Wang, L.; Wang, A. Adsorption properties of crosslinked carboxymethyl-chitosan resin with Pb (II) as template ions. *J. Hazard. Mater.* **2006**, *136*, 930–937. [CrossRef] [PubMed]

25. Park, H.G.; Kim, T.W.; Chae, M.Y.; Yoo, I.-K. Activated carbon-containing alginate adsorbent for the simultaneous removal of heavy metals and toxic organics. *Process Biochem.* **2007**, *42*, 1371–1377. [CrossRef]

26. Saad, A.H.A.; Azzam, A.M.; El-Wakeel, S.T.; Mostafa, B.B.; El-latif, M.B.A. Removal of toxic metal ions from wastewater using ZnO@ Chitosan core-shell nanocomposite. *Environ. Nanotechnol. Monit. Manag.* **2018**, *9*, 67–75. [CrossRef]

27. Siddeeg, S.M.; Alsaiari, N.S.; Tahoon, M.A.; Rebah, F.B. The Application of Nanomaterials as Electrode Modifiers for the Electrochemical Detection of Ascorbic Acid. *Int. J. Electrochem. Sci.* **2020**, *15*, 3327–3346. [CrossRef]

28. Verma, R.; Asthana, A.; Singh, A.K.; Prasad, S.; Susan, M.A.B.H. Novel glycine-functionalized magnetic nanoparticles entrapped calcium alginate beads for effective removal of lead. *Microchem. J.* **2017**, *130*, 168–178. [CrossRef]

29. Tu, Y.-J.; You, C.-F.; Chen, M.-H.; Duan, Y.-P. Efficient removal/recovery of Pb onto environmentally friendly fabricated copper ferrite nanoparticles. *J. Taiwan Inst. Chem. Eng.* **2017**, *71*, 197–205. [CrossRef]

30. Liaw, B.-S.; Chang, T.-T.; Chang, H.-K.; Liu, W.-K.; Chen, P.-Y. Fish scale-extracted hydroxyapatite/chitosan composite scaffolds fabricated by freeze casting—An innovative strategy for water treatment. *J. Hazard. Mater.* **2020**, *382*, 121082. [CrossRef]

31. Danesh, N.; Hosseini, M.; Ghorbani, M.; Marjani, A. Fabrication, characterization and physical properties of a novel magnetite graphene oxide/Lauric acid nanoparticles modified by ethylenediaminetetraacetic acid and its applications as an adsorbent for the removal of Pb (II) ions. *Synth. Met.* **2016**, *220*, 508–523. [CrossRef]

32. Martínez-Cabanas, M.; López-García, M.; Barriada, J.L.; Herrero, R.; de Vicente, M.E.S. Green synthesis of iron oxide nanoparticles. Development of magnetic hybrid materials for efficient As (V) removal. *Chem. Eng. J.* **2016**, *301*, 83–91. [CrossRef]

33. Islam, M.S.; San Choi, W.; Nam, B.; Yoon, C.; Lee, H.-J. Needle-like iron oxide@ $CaCO_3$ adsorbents for ultrafast removal of anionic and cationic heavy metal ions. *Chem. Eng. J.* **2017**, *307*, 208–219. [CrossRef]

34. Zeng, T.; Yu, Y.; Li, Z.; Zuo, J.; Kuai, Z.; Jin, Y.; Wang, Y.; Wu, A.; Peng, C. 3D MnO_2 nanotubes@ reduced graphene oxide hydrogel as reusable adsorbent for the removal of heavy metal ions. *Mater. Chem. Phys.* **2019**, *231*, 105–108.

35. Zhao, G.; Li, J.; Ren, X.; Chen, C.; Wang, X. Few-layered graphene oxide nanosheets as superior sorbents for heavy metal ion pollution management. *Environ. Sci. Technol.* **2011**, *45*, 10454–10462. [CrossRef] [PubMed]

36. Stafiej, A.; Pyrzynska, K. Adsorption of heavy metal ions with carbon nanotubes. *Sep. Purif. Technol.* **2007**, *58*, 49–52. [CrossRef]

37. Rahbari, M.; Goharrizi, A.S. Adsorption of lead (II) from water by carbon nanotubes: Equilibrium, kinetics, and thermodynamics. *Water Environ. Res.* **2009**, *81*, 598–607. [CrossRef] [PubMed]

38. Liu, D.; Zhu, Y.; Li, Z.; Tian, D.; Chen, L.; Chen, P. Chitin nanofibrils for rapid and efficient removal of metal ions from water system. *Carbohydr. Polym.* **2013**, *98*, 483–489. [CrossRef] [PubMed]

39. Zhu, K.; Jia, H.; Wang, F.; Zhu, Y.; Wang, C.; Ma, C. Efficient removal of Pb (II) from aqueous solution by modified montmorillonite/carbon composite: Equilibrium, kinetics, and thermodynamics. *J. Chem. Eng. Data* **2017**, *62*, 333–340. [CrossRef]

40. Rad, L.R.; Momeni, A.; Ghazani, B.F.; Irani, M.; Mahmoudi, M.; Noghreh, B. Removal of Ni2+ and Cd2+ ions from aqueous solutions using electrospun PVA/zeolite nanofibrous adsorbent. *Chem. Eng. J.* **2014**, *256*, 119–127. [CrossRef]

41. Deravanesiyan, M.; Beheshti, M.; Malekpour, A. The removal of Cr (III) and Co (II) ions from aqueous solution by two mechanisms using a new sorbent (alumina nanoparticles immobilized zeolite)—Equilibrium, kinetic and thermodynamic studies. *J. Mol. Liq.* **2015**, *209*, 246–257. [CrossRef]

42. Yurekli, Y. Removal of heavy metals in wastewater by using zeolite nano-particles impregnated polysulfone membranes. *J. Hazard. Mater.* **2016**, *309*, 53–64.

43. Al Sadat Shafiof, M.; Nezamzadeh-Ejhieh, A. A comprehensive study on the removal of Cd (II) from aqueous solution on a novel pentetic acid-clinoptilolite nanoparticles adsorbent: Experimental design, kinetic and thermodynamic aspects. *Solid State Sci.* **2020**, *99*, 106071. [CrossRef]

44. Abou-Zeid, R.E.; Dacrory, S.; Ali, K.A.; Kamel, S. Novel method of preparation of tricarboxylic cellulose nanofiber for efficient removal of heavy metal ions from aqueous solution. *Int. J. Biol. Macromol.* **2018**, *119*, 207–214. [PubMed]

45. Mautner, A.; Maples, H.; Kobkeatthawin, T.; Kokol, V.; Karim, Z.; Li, K.; Bismarck, A. Phosphorylated nanocellulose papers for copper adsorption from aqueous solutions. *Int. J. Environ. Sci. Technol.* **2016**, *13*, 1861–1872. [CrossRef]

46. Choi, H.Y.; Bae, J.H.; Hasegawa, Y.; An, S.; Kim, I.S.; Lee, H.; Kim, M. Thiol-functionalized cellulose nanofiber membranes for the effective adsorption of heavy metal ions in water. *Carbohydr. Polym.* **2020**, *234*, 115881. [PubMed]

47. Zhang, D.; Xu, W.; Cai, J.; Cheng, S.-Y.; Ding, W.-P. Citric acid-incorporated cellulose nanofibrous mats as food materials-based biosorbent for removal of hexavalent chromium from aqueous solutions. *Int. J. Biol. Macromol.* **2020**, *149*, 459–466. [CrossRef] [PubMed]

48. Shah, L.A.; Khan, M.; Javed, R.; Sayed, M.; Khan, M.S.; Khan, A.; Ullah, M. Superabsorbent polymer hydrogels with good thermal and mechanical properties for removal of selected heavy metal ions. *J. Clean. Prod.* **2018**, *201*, 78–87.

49. Martín, D.M.; Faccini, M.; García, M.; Amantia, D. Highly efficient removal of heavy metal ions from polluted water using ion-selective polyacrylonitrile nanofibers. *J. Environ. Chem. Eng.* **2018**, *6*, 236–245. [CrossRef]

50. Cegłowski, M.; Gierczyk, B.; Frankowski, M.; Popenda, Ł. A new low-cost polymeric adsorbents with polyamine chelating groups for efficient removal of heavy metal ions from water solutions. *React. Funct. Polym.* **2018**, *131*, 64–74. [CrossRef]

51. Efome, J.E.; Rana, D.; Matsuura, T.; Lan, C.Q. Effects of operating parameters and coexisting ions on the efficiency of heavy metal ions removal by nano-fibrous metal-organic framework membrane filtration process. *Sci. Total Environ.* **2019**, *674*, 355–362. [CrossRef]

52. Xu, D.; Zhu, K.; Zheng, X.; Xiao, R. Poly (ethylene-co-vinyl alcohol) functional nanofiber membranes for the removal of Cr (VI) from water. *Ind. Eng. Chem. Res.* **2015**, *54*, 6836–6844.

53. Saeed, K.; Haider, S.; Oh, T.-J.; Park, S.-Y. Preparation of amidoxime-modified polyacrylonitrile (PAN-oxime) nanofibers and their applications to metal ions adsorption. *J. Membr. Sci.* **2008**, *322*, 400–405. [CrossRef]

54. Hasanzadeh, R.; Najafi Moghadam, P.; Samadi, N. Synthesis and application of modified poly (styrene-alt-maleic anhydride) networks as a nano chelating resin for uptake of heavy metal ions. *Polym. Adv. Technol.* **2013**, *24*, 34–41. [CrossRef]

55. Sohail, I.; Bhatti, I.A.; Ashar, A.; Sarim, F.M.; Mohsin, M.; Naveed, R.; Yasir, M.; Iqbal, M.; Nazir, A. Polyamidoamine (PAMAM) dendrimers synthesis, characterization and adsorptive removal of nickel ions from aqueous solution. *J. Mater. Res. Technol.* **2020**, *9*, 498–506. [CrossRef]

56. Yuan, X.; An, N.; Zhu, Z.; Sun, H.; Zheng, J.; Jia, M.; Lu, C.; Zhang, W.; Liu, N. Hierarchically porous nitrogen-doped carbon materials as efficient adsorbents for removal of heavy metal ions. *Process Saf. Environ. Prot.* **2018**, *119*, 320–329.

57. Razzaz, A.; Ghorban, S.; Hosayni, L.; Irani, M.; Aliabadi, M. Chitosan nanofibers functionalized by TiO_2 nanoparticles for the removal of heavy metal ions. *J. Taiwan Inst. Chem. Eng.* **2016**, *58*, 333–343.

58. Esmaeili, A.; Khoshnevisan, N. Optimization of process parameters for removal of heavy metals by biomass of Cu and Co-doped alginate-coated chitosan nanoparticles. *Bioresour. Technol.* **2016**, *218*, 650–658. [CrossRef]

59. Dubey, R.; Bajpai, J.; Bajpai, A. Chitosan-alginate nanoparticles (CANPs) as potential nanosorbent for removal of Hg (II) ions. *Environ. Nanotechnol. Monit. Manag.* **2016**, *6*, 32–44. [CrossRef]

60. Hussain, M.S.; Musharraf, S.G.; Bhanger, M.I.; Malik, M.I. Salicylaldehyde derivative of nano-chitosan as an efficient adsorbent for lead (II), copper (II), and cadmium (II) ions. *Int. J. Biol. Macromol.* **2020**, *147*, 643–652. [CrossRef]

61. Khalil, T.E.; Elhusseiny, A.F.; El-dissouky, A.; Ibrahim, N.M. Functionalized chitosan nanocomposites for removal of toxic Cr (VI) from aqueous solution. *React. Funct. Polym.* **2020**, *146*, 104407.

62. Nassar, N.N. Rapid removal and recovery of Pb (II) from wastewater by magnetic nanoadsorbents. *J. Hazard. Mater.* **2010**, *184*, 538–546. [CrossRef]

63. Giraldo, L.; Erto, A.; Moreno-Piraján, J.C. Magnetite nanoparticles for removal of heavy metals from aqueous solutions: Synthesis and characterization. *Adsorption* **2013**, *19*, 465–474. [CrossRef]

64. Iconaru, S.L.; Guégan, R.; Popa, C.L.; Motelica-Heino, M.; Ciobanu, C.S.; Predoi, D. Magnetite (Fe_3O_4) nanoparticles as adsorbents for As and Cu removal. *Appl. Clay Sci.* **2016**, *134*, 128–135. [CrossRef]

65. Karami, H. Heavy metal removal from water by magnetite nanorods. *Chem. Eng. J.* **2013**, *219*, 209–216. [CrossRef]

66. Kim, E.-J.; Lee, C.-S.; Chang, Y.-Y.; Chang, Y.-S. Hierarchically structured manganese oxide-coated magnetic nanocomposites for the efficient removal of heavy metal ions from aqueous systems. *ACS Appl. Mater. Interfaces* **2013**, *5*, 9628–9634. [CrossRef] [PubMed]

67. Tu, Y.-J.; Chan, T.-S.; Tu, H.-W.; Wang, S.-L.; You, C.-F.; Chang, C.-K. Rapid and efficient removal/recovery of molybdenum onto ZnFe2O4 nanoparticles. *Chemosphere* **2016**, *148*, 452–458. [PubMed]

68. Asadi, R.; Abdollahi, H.; Gharabaghi, M.; Boroumand, Z. Effective removal of Zn (II) ions from aqueous solution by the magnetic $MnFe_2O_4$ and $CoFe_2O_4$ spinel ferrite nanoparticles with focuses on synthesis, characterization, adsorption, and desorption. *Adv. Powder Technol.* **2020**. [CrossRef]

69. Vamvakidis, K.; Kostitsi, T.-M.; Makridis, A.; Dendrinou-Samara, C. Diverse Surface Chemistry of Cobalt Ferrite Nanoparticles to Optimize Copper (II) Removal from Aqueous Media. *Materials* **2020**, *13*, 1537. [CrossRef]

70. Ma, Z.; Zhao, D.; Chang, Y.; Xing, S.; Wu, Y.; Gao, Y. Synthesis of $MnFe_2O_4$ @Mn–Co oxide core–shell nanoparticles and their excellent performance for heavy metal removal. *Dalton Trans.* **2013**, *42*, 14261–14267.

71. Guo, L.M.; Li, J.T.; Zhang, L.X.; Li, J.B.; Li, Y.S.; Yu, C.C.; Shi, J.L.; Ruan, M.L.; Feng, J.W. A facile route to synthesize magnetic particles within hollow mesoporous spheres and their performance as separable Hg2+ adsorbents. *J. Mater. Chem.* **2008**, *18*, 2733–2738.

72. Mokadem, Z.; Mekki, S.; Saïdi-Besbes, S.; Agusti, G.; Elaissari, A.; Derdour, A. Triazole containing magnetic core-silica shell nanoparticles for Pb2+, Cu2+ and Zn2+ removal. *Arab. J. Chem.* **2017**, *10*, 1039–1051.

73. Wang, P.; Shen, T.; Li, X.; Tang, Y.; Li, Y. Magnetic Mesoporous Calcium Carbonate-Based Nanocomposites for the Removal of Toxic Pb (II) and Cd (II) Ions from Water. *ACS Appl. Nano Mater.* **2020**, *3*, 1272–1281. [CrossRef]

74. Li, N.; Li, Z.; Zhang, L.; Shi, H.; Li, J.; Zhang, J.; Zhang, Z.; Dang, F. One-step fabrication of bifunctional self-assembled oligopeptides anchored magnetic carbon nanoparticles and their application in copper (II) ions removal from aqueous solutions. *J. Hazard. Mater.* **2020**, *382*, 121113. [CrossRef] [PubMed]

75. Girginova, P.I.; Daniel-da-Silva, A.L.; Lopes, C.B.; Figueira, P.; Otero, M.; Amaral, V.S.; Pereira, E.; Trindade, T. Silica coated magnetite particles for magnetic removal of Hg2+ from water. *J. Colloid Interface Sci.* **2010**, *345*, 234–240. [CrossRef] [PubMed]

76. Xu, P.; Zeng, G.M.; Huang, D.L.; Yan, M.; Chen, M.; Lai, C.; Jiang, H.; Wu, H.P.; Chen, G.M.; Wan, J. Fabrication of reduced glutathione functionalized iron oxide nanoparticles for magnetic removal of Pb (II) from wastewater. *J. Taiwan Inst. Chem. Eng.* **2017**, *71*, 165–173. [CrossRef]

77. Ghaemi, N.; Madaeni, S.S.; Daraei, P.; Rajabi, H.; Zinadini, S.; Alizadeh, A.; Heydari, R.; Beygzadeh, M.; Ghouzivand, S. Polyethersulfone membrane enhanced with iron oxide nanoparticles for copper removal from water: Application of new functionalized Fe3O4 nanoparticles. *Chem. Eng. J.* **2015**, *263*, 101–112. [CrossRef]

78. Ge, F.; Li, M.-M.; Ye, H.; Zhao, B.-X. Effective removal of heavy metal ions Cd2+, Zn2+, Pb2+, Cu2+ from aqueous solution by polymer-modified magnetic nanoparticles. *J. Hazard. Mater.* **2012**, *211*, 366–372. [CrossRef]

79. Zhang, T.; Wei, S.; Waterhouse, G.I.; Fu, L.; Liu, L.; Shi, W.; Sun, J.; Ai, S. Chromium (VI) adsorption and reduction by humic acid coated nitrogen-doped magnetic porous carbon. *Powder Technol.* **2020**, *360*, 55–64. [CrossRef]

80. Li, Z.; Pan, Z.; Wang, Y. Preparation of ternary amino-functionalized magnetic nano-sized illite-smectite clay for adsorption of Pb (II) ions in aqueous solution. *Environ. Sci. Pollut. Res.* **2020**, *27*, 11683–11696. [CrossRef]

81. Wang, X.; Liu, X.; Xiao, C.; Zhao, H.; Zhang, M.; Zheng, N.; Kong, W.; Zhang, L.; Yuan, H.; Zhang, L. Triethylenetetramine-modified hollow Fe$_3$O$_4$/SiO$_2$/chitosan magnetic nanocomposites for removal of Cr (VI) ions with high adsorption capacity and rapid rate. *Microporous Mesoporous Mater.* **2020**, *297*, 110041. [CrossRef]

82. Gao, J.; He, Y.; Zhao, X.; Ran, X.; Wu, Y.; Su, Y.; Dai, J. Single step synthesis of amine-functionalized mesoporous magnetite nanoparticles and their application for copper ions removal from aqueous solution. *J. Colloid Interface Sci.* **2016**, *481*, 220–228. [CrossRef]

83. Shen, H.; Chen, J.; Dai, H.; Wang, L.; Hu, M.; Xia, Q. New insights into the sorption and detoxification of chromium (VI) by tetraethylenepentamine functionalized nanosized magnetic polymer adsorbents: Mechanism and pH effect. *Ind. Eng. Chem. Res.* **2013**, *52*, 12723–12732. [CrossRef]

84. Shen, H.; Pan, S.; Zhang, Y.; Huang, X.; Gong, H. A new insight on the adsorption mechanism of amino-functionalized nano-Fe$_3$O$_4$ magnetic polymers in Cu (II), Cr (VI) co-existing water system. *Chem. Eng. J.* **2012**, *183*, 180–191. [CrossRef]

85. Abdolmaleki, A.; Mallakpour, S.; Borandeh, S. Efficient heavy metal ion removal by triazinyl-β-cyclodextrin functionalized iron nanoparticles. *RSC Adv.* **2015**, *5*, 90602–90608.

86. Fan, H.; Ma, X.; Zhou, S.; Huang, J.; Liu, Y.; Liu, Y. Highly efficient removal of heavy metal ions by carboxymethyl cellulose-immobilized Fe$_3$O$_4$ nanoparticles prepared via high-gravity technology. *Carbohydr. Polym.* **2019**, *213*, 39–49. [PubMed]

87. Xu, P.; Zeng, G.M.; Huang, D.L.; Lai, C.; Zhao, M.H.; Wei, Z.; Li, N.J.; Huang, C.; Xie, G.X. Adsorption of Pb (II) by iron oxide nanoparticles immobilized Phanerochaete chrysosporium: Equilibrium, kinetic, thermodynamic and mechanisms analysis. *Chem. Eng. J.* **2012**, *203*, 423–431. [CrossRef]

88. Verma, M.; Tyagi, I.; Chandra, R.; Gupta, V.K. Adsorptive removal of Pb (II) ions from aqueous solution using CuO nanoparticles synthesized by sputtering method. *J. Mol. Liq.* **2017**, *225*, 936–944.

89. Huang, S.-H.; Chen, D.-H. Rapid removal of heavy metal cations and anions from aqueous solutions by an amino-functionalized magnetic nano-adsorbent. *J. Hazard. Mater.* **2009**, *163*, 174–179.

90. Madrakian, T.; Afkhami, A.; Zadpour, B.; Ahmadi, M. New synthetic mercaptoethylamino homopolymer-modified maghemite nanoparticles for effective removal of some heavy metal ions from aqueous solution. *J. Ind. Eng. Chem.* **2015**, *21*, 1160–1166.

91. Feng, Y.; Gong, J.-L.; Zeng, G.-M.; Niu, Q.-Y.; Zhang, H.-Y.; Niu, C.-G.; Deng, J.-H.; Yan, M. Adsorption of Cd (II) and Zn (II) from aqueous solutions using magnetic hydroxyapatite nanoparticles as adsorbents. *Chem. Eng. J.* **2010**, *162*, 487–494.

92. Hasanzadeh, R.; Moghadam, P.N.; Bahri-Laleh, N.; Sillanpää, M. Effective removal of toxic metal ions from aqueous solutions: 2-Bifunctional magnetic nanocomposite base on novel reactive PGMA-MAn copolymer@ Fe_3O_4 nanoparticles. *J. Colloid Interface Sci.* **2017**, *490*, 727–746. [CrossRef]

93. Fayazi, M.; Ghanbarian, M. One-Pot Hydrothermal Synthesis of Polyethylenimine Functionalized Magnetic Clay for Efficient Removal of Noxious Cr (VI) from Aqueous Solutions. *Silicon* **2020**, *12*, 125–134.

94. Kanel, S.R.; Greneche, J.-M.; Choi, H. Arsenic (V) removal from groundwater using nano scale zero-valent iron as a colloidal reactive barrier material. *Environ. Sci. Technol.* **2006**, *40*, 2045–2050. [CrossRef] [PubMed]

95. Xiao, Z.; Zhang, H.; Xu, Y.; Yuan, M.; Jing, X.; Huang, J.; Li, Q.; Sun, D. Ultra-efficient removal of chromium from aqueous medium by biogenic iron based nanoparticles. *Sep. Purif. Technol.* **2017**, *174*, 466–473. [CrossRef]

96. Sikder, M.T.; Mihara, Y.; Islam, M.S.; Saito, T.; Tanaka, S.; Kurasaki, M. Preparation and characterization of chitosan–caboxymethyl-β-cyclodextrin entrapped nanozero-valent iron composite for Cu (II) and Cr (IV) removal from wastewater. *Chem. Eng. J.* **2014**, *236*, 378–387. [CrossRef]

97. Cai, X.; Gao, Y.; Sun, Q.; Chen, Z.; Megharaj, M.; Naidu, R. Removal of co-contaminants Cu (II) and nitrate from aqueous solution using kaolin-Fe/Ni nanoparticles. *Chem. Eng. J.* **2014**, *244*, 19–26.

98. Sounthararajah, D.; Loganathan, P.; Kandasamy, J.; Vigneswaran, S. Adsorptive removal of heavy metals from water using sodium titanate nanofibres loaded onto GAC in fixed-bed columns. *J. Hazard. Mater.* **2015**, *287*, 306–316. [CrossRef]

99. Luo, C.; Wang, J.; Jia, P.; Liu, Y.; An, J.; Cao, B.; Pan, K. Hierarchically structured polyacrylonitrile nanofiber mat as highly efficient lead adsorbent for water treatment. *Chem. Eng. J.* **2015**, *262*, 775–784. [CrossRef]

100. Yari, S.; Abbasizadeh, S.; Mousavi, S.E.; Moghaddam, M.S.; Moghaddam, A.Z. Adsorption of Pb (II) and Cu (II) ions from aqueous solution by an electrospun CeO_2 nanofiber adsorbent functionalized with mercapto groups. *Process Saf. Environ. Prot.* **2015**, *94*, 159–171.

101. Hallaji, H.; Keshtkar, A.R.; Moosavian, M.A. A novel electrospun PVA/ZnO nanofiber adsorbent for U (VI), Cu (II) and Ni (II) removal from aqueous solution. *J. Taiwan Inst. Chem. Eng.* **2015**, *46*, 109–118. [CrossRef]

102. Ren, T.; He, P.; Niu, W.; Wu, Y.; Ai, L.; Gou, X. Synthesis of α-Fe_2O_3 nanofibers for applications in removal and recovery of Cr (VI) from wastewater. *Environ. Sci. Pollut. Res.* **2013**, *20*, 155–162.

103. Zhu, X.; Song, T.; Lv, Z.; Ji, G. High-efficiency and low-cost α-Fe_2O_3 nanoparticles-coated volcanic rock for Cd (II) removal from wastewater. *Process Saf. Environ. Prot.* **2016**, *104*, 373–381. [CrossRef]

104. Gupta, V.K.; Chandra, R.; Tyagi, I.; Verma, M. Removal of hexavalent chromium ions using CuO nanoparticles for water purification applications. *J. Colloid Interface Sci.* **2016**, *478*, 54–62. [CrossRef] [PubMed]

105. Cai, Y.; Li, C.; Wu, D.; Wang, W.; Tan, F.; Wang, X.; Wong, P.K.; Qiao, X. Highly active MgO nanoparticles for simultaneous bacterial inactivation and heavy metal removal from aqueous solution. *Chem. Eng. J.* **2017**, *312*, 158–166. [CrossRef]

106. Cao, C.-Y.; Qu, J.; Wei, F.; Liu, H.; Song, W.-G. Superb adsorption capacity and mechanism of flowerlike magnesium oxide nanostructures for lead and cadmium ions. *ACS Appl. Mater. Interfaces* **2012**, *4*, 4283–4287. [PubMed]

107. Chu, T.P.M.; Nguyen, N.T.; Vu, T.L.; Dao, T.H.; Dinh, L.C.; Nguyen, H.L.; Hoang, T.H.; Le, T.S.; Pham, T.D. Synthesis, characterization, and modification of alumina nanoparticles for cationic dye removal. *Materials* **2019**, *12*, 450. [CrossRef]

108. Pham, T.D.; Tran, T.T.; Pham, T.T.; Dao, T.H.; Le, T.S. Adsorption characteristics of molecular oxytetracycline onto alumina particles: The role of surface modification with an anionic surfactant. *J. Mol. Liq.* **2019**, *287*, 110900. [CrossRef]

109. Hojamberdiev, M.; Daminova, S.S.; Kadirova, Z.C.; Sharipov, K.T.; Mtalo, F.; Hasegawa, M. Ligand-immobilized spent alumina catalyst for effective removal of heavy metal ions from model contaminated water. *J. Environ. Chem. Eng.* **2018**, *6*, 4623–4633. [CrossRef]

110. Dehghani, M.H.; Yetilmezsoy, K.; Salari, M.; Heidarinejad, Z.; Yousefi, M.; Sillanpää, M. Adsorptive removal of cobalt (II) from aqueous solutions using multi-walled carbon nanotubes and γ-alumina as novel adsorbents: Modelling and optimization based on response surface methodology and artificial neural network. *J. Mol. Liq.* **2020**, *299*, 112154. [CrossRef]

111. Ghaemi, N. A new approach to copper ion removal from water by polymeric nanocomposite membrane embedded with γ-alumina nanoparticles. *Appl. Surf. Sci.* **2016**, *364*, 221–228. [CrossRef]

112. Kumari, V.; Tripathi, A. Remediation of heavy metals in pharmaceutical effluent with the help of Bacillus cereus-based green-synthesized silver nanoparticles supported on alumina. *Appl. Nanosci.* **2020**. [CrossRef]

113. Nguyen, T.M.T.; Do, T.P.T.; Hoang, T.S.; Nguyen, N.V.; Pham, H.D.; Nguyen, T.D.; Pham, T.N.M.; Le, T.S.; Pham, T.D. Adsorption of anionic surfactants onto alumina: Characteristics, mechanisms, and application for heavy metal removal. *Int. J. Polym. Sci.* **2018**, *2018*, 2830286. [CrossRef]

114. Khobragade, M.; Pal, A. Adsorptive removal of Mn (II) from water and wastewater by surfactant-modified alumina. *Desalin. Water Treat.* **2016**, *57*, 2775–2786. [CrossRef]

115. Najafi, M.; Yousefi, Y.; Rafati, A. Synthesis, characterization and adsorption studies of several heavy metal ions on amino-functionalized silica nano hollow sphere and silica gel. *Sep. Purif. Technol.* **2012**, *85*, 193–205. [CrossRef]

116. Kotsyuda, S.S.; Tomina, V.V.; Zub, Y.L.; Furtat, I.M.; Lebed, A.P.; Vaclavikova, M.; Melnyk, I.V. Bifunctional silica nanospheres with 3-aminopropyl and phenyl groups. Synthesis approach and prospects of their applications. *Appl. Surf. Sci.* **2017**, *420*, 782–791. [CrossRef]

117. Li, Y.; He, J.; Zhang, K.; Liu, T.; Hu, Y.; Chen, X.; Wang, C.; Huang, X.; Kong, L.; Liu, J. Super rapid removal of copper, cadmium and lead ions from water by NTA-silica gel. *RSC Adv.* **2019**, *9*, 397–407. [CrossRef]

118. Yang, Z.; Chen, X.; Li, S.; Ma, W.; Li, Y.; He, Z.; Hu, H. Effective removal of Cr (VI) from aqueous solution based on APTES modified nanoporous silicon prepared from kerf loss silicon waste. *Environ. Sci. Pollut. Res.* **2020**, *27*, 10899–10909. [CrossRef]

119. Naseem, K.; Begum, R.; Wu, W.; Usman, M.; Irfan, A.; Al-Sehemi, A.G.; Farooqi, Z.H. Adsorptive removal of heavy metal ions using polystyrene-poly (N-isopropylmethacrylamide-acrylic acid) core/shell gel particles: Adsorption isotherms and kinetic study. *J. Mol. Liq.* **2019**, *277*, 522–531. [CrossRef]

120. Liu, J.-F.; Zhao, Z.-S.; Jiang, G.-B. Coating Fe$_3$O$_4$ magnetic nanoparticles with humic acid for high efficient removal of heavy metals in water. *Environ. Sci. Technol.* **2008**, *42*, 6949–6954. [CrossRef]

121. Tan, Y.; Chen, M.; Hao, Y. High efficient removal of Pb (II) by amino-functionalized Fe$_3$O$_4$ magnetic nano-particles. *Chem. Eng. J.* **2012**, *191*, 104–111. [CrossRef]

122. Kakavandi, B.; Kalantary, R.R.; Jafari, A.J.; Nasseri, S.; Ameri, A.; Esrafili, A.; Azari, A. Pb (II) adsorption onto a magnetic composite of activated carbon and superparamagnetic Fe3O4 nanoparticles: Experimental and modeling study. *Clean Soil Air Water* **2015**, *43*, 1157–1166. [CrossRef]

123. Huang, Q.; Liu, Y.; Cai, T.; Xia, X. Simultaneous removal of heavy metal ions and organic pollutant by BiOBr/Ti$_3$C$_2$ nanocomposite. *J. Photochem. Photobiol. A Chem.* **2019**, *375*, 201–208. [CrossRef]

124. Esmat, M.; Farghali, A.A.; Khedr, M.H.; El-Sherbiny, I.M. Alginate-based nanocomposites for efficient removal of heavy metal ions. *Int. J. Biol. Macromol.* **2017**, *102*, 272–283. [CrossRef] [PubMed]

125. Ozmen, M.; Can, K.; Arslan, G.; Tor, A.; Cengeloglu, Y.; Ersoz, M. Adsorption of Cu (II) from aqueous solution by using modified Fe3O4 magnetic nanoparticles. *Desalination* **2010**, *254*, 162–169. [CrossRef]

126. Gomaa, E.A.; Tahoon, M.A. Ion association and solvation behavior of copper sulfate in binary aqueous–methanol mixtures at different temperatures. *J. Mol. Liq.* **2016**, *214*, 19–23. [CrossRef]

127. Gomaa, E.A.; Tahoon, M.A.; Negm, A. Aqueous micro-solvation of Li+ ions: Thermodynamics and energetic studies of Li+-(H$_2$O) n (n = 1–6) structures. *J. Mol. Liq.* **2017**, *241*, 595–602. [CrossRef]

128. Gomaa, E.A.; Tahoon, M.A.; Shokr, A. Ionic association and solvation study of CoSO$_4$ in aqueous-organic solvents at different temperatures. *Chem. Data Collect.* **2016**, *3*, 58–67. [CrossRef]

129. Gomaa, E.A.; Negm, A.; Tahoon, M.A. Conductometric and volumetric study of copper sulphate in aqueous ethanol solutions at different temperatures. *J. Taibah Univ. Sci.* **2017**, *11*, 741–748. [CrossRef]

130. Tahoon, M.; Gomaa, E.; Suleiman, M. Aqueous Micro-hydration of Na+ (H$_2$O) n = 1–7 Clusters: DFT Study. *Open Chem.* **2019**, *17*, 260–269.

131. Ben Rebah, F.; Siddeeg, S.M.; Tahoon, M.A. Thermodynamic Parameters and Solvation Behavior of 1-Ethyle-3-methylimidazolium Tetrafluoroborate and 1-Butyl-3-methylimidazolium Tetrafluoroborate in N, N-Dimethylformamide and Acetonitrile at Different Temperature. *Egypt. J. Chem.* **2019**, *62*, 393–404.

132. Oliveira, E.A.; Montanher, S.F.; Andrade, A.D.; Nóbrega, J.A.; Rollemberg, M.C. Equilibrium studies for the sorption of chromium and nickel from aqueous solutions using raw rice bran. *Process Biochem.* **2005**, *40*, 3485–3490. [CrossRef]

133. Siddeeg, M.S.; Tahoon, A.M.; Ben Rebah, F. Simultaneous Removal of Calconcarboxylic Acid, NH4+ and $PO_4{}^{3-}$ from Pharmaceutical Effluent Using Iron Oxide-Biochar Nanocomposite Loaded with Pseudomonas putida. *Processes* **2019**, *7*, 800. [CrossRef]

134. Siddeeg, S.M.; Tahoon, M.A.; Mnif, W.; Ben Rebah, F. Iron Oxide/Chitosan Magnetic Nanocomposite Immobilized Manganese Peroxidase for Decolorization of Textile Wastewater. *Processes* **2020**, *8*, 5. [CrossRef]

135. Siddeeg, S.M.; Amari, A.; Tahoon, M.A.; Alsaiari, N.S.; Rebah, F.B. Removal of meloxicam, piroxicam and Cd+ 2 by Fe_3O_4/SiO_2/glycidyl methacrylate-S-SH nanocomposite loaded with laccase. *Alex. Eng. J.* **2020**, *59*, 905–914. [CrossRef]

Review

Current Advances in Biofouling Mitigation in Membranes for Water Treatment: An Overview

Daniela Pichardo-Romero, Zahirid Patricia Garcia-Arce, Alejandra Zavala-Ramírez and Roberto Castro-Muñoz *

Tecnologico de Monterrey, Campus Toluca, Avenida Eduardo Monroy Cárdenas 2000 San Antonio Buenavista, Toluca de Lerdo 50110, Mexico; dany.pichardo99@gmail.com (D.P.-R.); A01363407@itesm.mx (Z.P.G.-A.); A01421769@itesm.mx (A.Z.-R.)
* Correspondence: food.biotechnology88@gmail.com or castromr@tec.mx

Received: 9 January 2020; Accepted: 27 January 2020; Published: 5 February 2020

Abstract: Membranes, as the primary tool in membrane separation techniques, tend to suffer external deposition of pollutants and microorganisms depending on the nature of the treating solutions. Such issues are well recognized as biofouling and is identified as the major drawback of pressure-driven membrane processes due to the influence of the separation performance of such membrane-based technologies. Herein, the aim of this review paper is to elucidate and discuss new insights on the ongoing development works at facing the biofouling phenomenon in membranes. This paper also provides an overview of the main strategies proposed by "membranologists" to improve the fouling resistance in membranes. Special attention has been paid to the fundamentals on membrane fouling as well as the relevant results in the framework of mitigating the issue. By analyzing the literature data and state-of-the-art, the concluding remarks and future trends in the field are given as well.

Keywords: membrane technologies; biofouling; composite membranes; polymer blending

---

1. Introduction

To date, different new ways of water processing have been proposed to optimize the production and yield, considering also the reduction of production costs and time [1,2]. Nowadays, one of these ways has been membrane technology, which was, for the first time, introduced by Bechold in 1907, who used ultrafiltration processes [3]. Since that time, membrane-based technologies began to gain popularity as separation processes. They are among the most significant advances in chemical and biological process engineering. Membrane processes are well defined due to the membrane being a primary tool for separating different types of molecules [4]. In principle, membranes can be classified into two different categories according to the membrane material, organic (based on polymers) and inorganic (e.g., glass, ceramic, silica, graphene, metal oxide, among others) [5,6]. Moreover, membrane-based technologies are classified based on their driving force, especially for the pressure-driven membrane processes, where the pore size in the membrane is the key characteristic that distinguishes between Microfiltration (MF) (pore size between 100 and 10,000 nm), Ultrafiltration (UF) (pore size between 2 and 100 nm) and Nanofiltration (NF) (pore size 0.5–2 nm)[7]. For instance, Table 1 enlists the main pressure requirements needed for these processes to carry out the separation, and the separation mechanism that may take place.

Table 1. Classification and main parameters of Microfiltration (MF), Ultrafiltration (UF) and Nanofiltration (NF).

Membrane Process	Required Pressure (bar)		Typical Separation Mechanism
	Min.	Max.	
Microfiltration	0.1	2	Sieving
Ultrafiltration	0.1	7	Sieving
Nanofiltration	3	25	Sieving and charge effect

The membrane, as the only separation barrier, is always in contact with the treating solutions, and consequently, they are prone to present chemical or biological deposition of matter [8,9]. Such deposition is the so-called fouling phenomenon. Biofouling implies the adhesion of micro- or macro-organisms as membrane foulant, and it represents the "vulnerable" part of the membrane process since microorganisms can proliferate over time. It has been established that biofouling contributes to over 45% of membrane fouling [10]. Microfouling originated by unicellular or pluricellular microorganisms, such as bacteria, yeast, or fungi, that may form a complex biofilm by mono-species or multi-species, while macrofouling was associated with bigger or visible organisms [11].

Typically, there are four types of fouling mechanisms, which were established by Hermia [12], that can take place in membrane processes: complete pore blocking, partial pore blocking, internal pore blocking, and cake formation. In order to inhibit or, at least control such mechanisms, the chemical nature of the foulants (e.g., organic, inorganic, or biological) must be known due to more than one type of fouling taking place simultaneously in a process, and beyond the classification, the pollutant's nature can provide an overview on the type of fouling and its impact on membrane properties [13]. For instance, the mechanisms can involve adsorption, accumulation, or precipitation either on the surface or inside the membrane's pores. This can have an influence on the separation performance of the membrane, e.g., decrease permeate flux and membrane selectivity, and membrane lifetime [14,15]. Based on such negative effects of biofouling on membranes, several authors have hardly worked on developing new strategies in preparing new concepts of membranes, which may offer enhanced fouling resistance. Therefore, the goal of this review paper is to give an outlook of the ongoing development works at mitigating the biofouling in pressure-driven membrane processes. To better understand the biofouling in membranes, a brief background of such a phenomenon is given, providing the main factors that play a key role in biofouling. Special emphasis has been paid to relevant results in the framework of reducing the issue. By reviewing the literature data and current state-of-the-art, the concluding remarks and future trends in the field are also given.

2. The Main Factors Playing an Important Role in the Biofouling Phenomenon

As mentioned previously, four mechanisms have been used to denote the fouling in pressure-driven membrane processes, such as (a) complete pore blocking, (b) partial pore blocking, (c) internal pore blocking, and (d) cake formation [12]. Figure 1 depicts a representation of such fouling mechanisms. In theory, these models follow three fundamentals: (i) constant pressure filtration, (ii) membrane pores are parallel to each other, cylindrical and equal in diameter, and (iii) foulant particles are spherical and non-deformable [12]. Complete pore blocking considers that the fouling occurs on the membrane surface area, and the particles "close" the pores, but no particle is placed on the top of another one [16]. In the case of partial pore blocking, it is also a surface phenomenon, but in this case, particles can be particularly placed on the top of the other [12]. For the internal pore blocking, the foulant particles are deposited inside the pores attaching themselves into the pore walls, and consequently, they change the pore diameter [12,16]. Finally, the cake formation is due to the complete coverage of the membrane surface, where the foulants can compose layers leading to the increment of hydraulic resistance in proportion to the cake layer thickness [12].

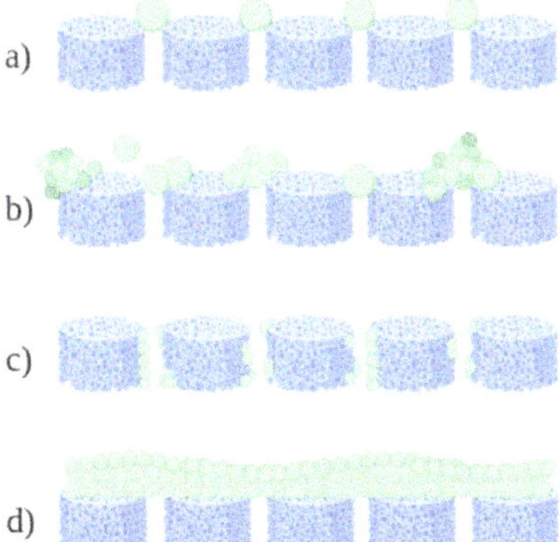

Figure 1. Graphical representation of the fouling mechanisms: (**a**) complete pore blocking, (**b**) partial pore blocking, (**c**) internal pore blocking and (**d**) cake formation [12].

When dealing with biofouling, the most important elements are microorganisms. Herein, the first step is the adhesion of the microbial cells to the membrane surface inducing the formation of a biofilm layer. The population of microorganisms, that may form the biofilm, are different types of bacteria, protozoa, fungi, and algae [10]. The adhesion to the membrane surface can be promoted by the membrane material since it can act as a substrate (e.g., cellulose acetate membranes). However, some other intrinsic properties of the membrane material and membrane itself, such as hydrophobicity, roughness of the membrane and membrane surface charges, are also crucial for biofilm formation. It is important to mention that such physico-chemical characteristics may change due to the extracellular polymeric substances (EPS) that are secreted by the microorganisms. In particular, this phenomenon contributes to a decrease in the permeation rates [10]. The steps of biofilm formation are depicted in Figure 2 [11]. Initially, proteins are fundamentally important factors in developing biofouling due to the fact that a protein layer provides an optimum environment for microbial colonization. As it is well-known, proteins are composed by amino acids, which possess several functional groups, such as carboxyl, amino and methyl groups. Such functional groups provide a specific hydrophilic or hydrophobic nature to the proteins, and thus, the retention of microorganisms on the membrane occurs through diverse intramolecular forces, such as van der Waals forces, hydrophobic interactions, hydrogen bonding, among others [11].

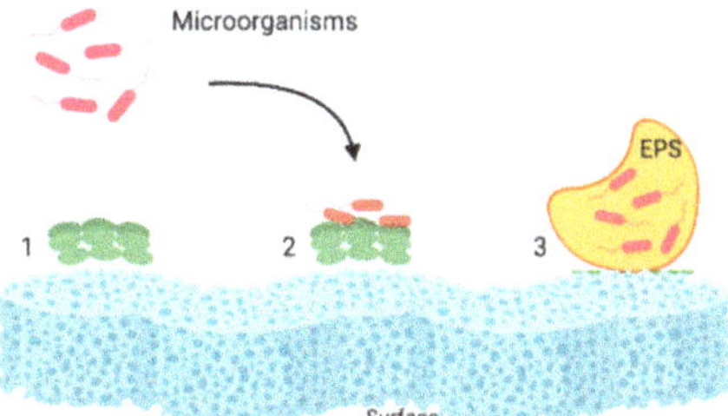

Figure 2. Steps of the biofilm formation: (1) Conditioning film (protein layer). (2) Transport to the surface and immobilization. (3) Attachment to the substrate [11].

Furthermore, different types of fouling may occur during a filtration process depending on multiple factors, including the physico-chemical composition of the feed solution (including pH, nature, etc.), operating parameters (such as feed flow rate, transmembrane pressure, temperature, among others) and the properties of the membrane (such as membrane material, membrane cut-off, charge).

The simultaneous appearance of the different fouling mechanisms can also take place. For instance, during the ultrafiltration of palm oil mill effluent (POME), Said et al. [17] stated that internal pore blocking, cake formation, and partial pore blocking were taking place, and importantly, they were dependent on the operating conditions. Therefore, this section addresses the main factors which play an important role in fouling formation.

2.1. Physico-Chemical Composition of the Feed Solution

According to various studies, the physico-chemical composition of the feed bilk is identified as the most critical issue on membrane biofouling [18]. As the primary element of any composition of feed solutions, the foulants can be categorized into four categories: organic, inorganic, colloidal, and biological, which can also be found among the composition of the feed depending on its origin [19]. These foulants can raise the following severe issues: (i) decrease in membrane flux and thus decrease of permeate production, (ii) degradation of the membrane due to the biofilm forming on its surface, and (iii) increase in pressure differences [20].

Among the different types of foulants, it is likely that the organic ones are the major constituents of wastewater streams. Usually, the organic matter contained in effluents provokes fouling coming from three primary sources: (i) synthetic organic compounds discharged by consumers, or disinfection by-products derived from disinfection processes, (ii) natural organic matter (NOM) generally presented in drinking water, and (iii) soluble microbial products generated during wastewater treatment [19]. In general, some of the main organic compounds found in wastewater are listed in Table 2.

Table 2. Size ranges of typical organic components in biological treated wastewater.

Classification	Compound	MW (Da)	Size (μm)
Dissolved Organic Matter (DOM)	Nutrients	10–10^2	$<10^{-4}$
	Amino acids	>10–$10^{2.5}$	$<10^{-4}$–10^{-4}
	Recalcitrant Matter	>10–10^3	$<10^{-4}$–$10^{-3.8}$
	Carbohydrate Fatty acid	10^2–10^3	$<10^4$–$10^{-3.6}$
	Chlorophyl	10^3–$10^{3.5}$	10^{-4}–$10^{-3.3}$
	Vitamin	10^3–10^4	$10^{-3.8}$–$10^{-3.3}$
	Humic acid	$10^{3.3}$–$10^{6.5}$	$10^{-3.5}$–$10^{-1.8}$
	Proteins	$10^{3.6}$–$10^{7.5}$	$10^{-3.5}$–$10^{-1.2}$
	RNA	10^4–$10^{6.5}$	$10^{-3.1}$–$10^{-2.4}$
	Extracellular enzyme	10^4–10^5	$10^{-4.8}$–$10^{-3.5}$
	Polysaccharide	10^4–10^7	10^{-3}–$<10^{-1}$
	Virus	10^6–10^9	10^{-2}–<10
	Cell fragment	$>10^7$–$<10^9$	$>10^{-2}$–<10
	DNA	$>10^7$–10^9	10^{-1}–<10
Particular/Colloidal Organic Matter (POM/COM)	Organic debris	$>10^8$	$>10^{-1}$–10^3
	Bacteria	$>10^9$	<10–$<10^2$
	Algae and protozoa	$>10^9$	10–10^3

As it can be seen most organic compounds can be classified into two major groups: dissolved organic matter (DOM) and particular/colloidal organic matter (POM) [19]. Particularly, some papers establish that DOM comprises a mixture of several ill-defined aliphatic and phenolic compounds [20], which possess a molecular weight in the range 5000 to 50,000 Da. Such compounds tend to generate an undesirable yellow-black colored water [21].

DOM are typically humic-based molecules and substances which constitute around 90% of the total of foulants in the feed. These substances represent meaningful and variable proportions of organic matter contained in soils and fresh seawaters, and they are known for affecting the aesthetic quality of water by changing the color and acting as a complexing agent for inorganic compounds. The fact that various authors have supported that humic substances are the most common reaction precursors to trihalomethane (chloroform) formation is extremely worrying and has brought many scientists to investigate ways to eliminate these and other foulants from the feed solutions and water [20].

To date, relevant research has been done regarding different pretreatments to prevent biofouling, including coagulation/flocculation (in-line coagulation, pre-coagulation/sedimentation and pre-flocculation), oxidation, ion exchange, adsorption, prefiltration, biofiltration, as well as their possible coupling [19].

2.2. Effect of Transmembrane Pressure

As a pressure-driven membrane process, a membrane filtration system should be operated at specific transmembrane pressures, which may provide higher productivity in terms of permeate flux and rejection capacity [22]. However, transmembrane pressure has an important influence on membrane fouling. For instance, when the pressure increases, the permeate flux follows a linear tendency, in other words, the permeate flux is pressure-dependent but there is a critical point where this can no longer happen, and this is well-known as limiting transmembrane pressure. In order to provide a good performance, it is recommended that the membrane systems operate below such critical points in which the fouling is promoted [23]. In the work presented by Jepsen et al. [22], the constant flux operation was shown less fouling with a fixed permeate flux during a desalination process; while less fouling appeared to occur at a constant transmembrane pressure for surface water treatment.

When the pressure is above the critical zone, the flux declines as the pressure increases due to the deposition of particles on the membrane surface; furthermore, the pressure raises the concentration polarization and the collision of particles promoting their attachment to the membrane pores, and consequently, a cake formation occurs [17]. During POME ultrafiltration, authors have compared the type of fouling that can occur varying the transmembrane pressure from 2 to 5 bar. Specifically, at 2 to 4 bar, the fouling was identified as partial or internal pore blocking, while at a pressure of 5 bar, the fouling occurred was related to cake formation [17].

As described above, increasing the driving force results in the growth of a solute layer, and this is well known as the mass transfer-controlled region; this detains the increase of the permeation rate of the components [24]. According to He et al. [24], the operation at high pressure does not necessarily guarantee high permeate flux. This statement was proven in their study on the separation of reactive dye solution, finding the optimum operating pressure for UF membrane below 1.5 MPa.

2.3. Effect of pH

The membranes could be positively or negatively charged depending on the type and nature of feed solution. At this point, the pH remarkably produces changes on the charges of the membrane due to the disassociation of functional groups [24]. Therefore, depending on the pH used, the solution can possess charges that are similar or different compared to the membrane, which may cause the repulsion or affinity of the particles towards the membrane. The functional groups that generate negatively charged membranes are carboxylic and sulfonic acid groups, while positively charged membranes can originate from amino groups that accept hydrogen ions in solutions of acidic pH [25].

It has been observed that pH has negligible effects on the rejection of dye and salt [24], however, when there is an attractive force between the membrane and the particles, an internal pore-blocking occurs if the size of the particles are smaller than the pore size, then, they can generally go through the pores and then attach to the wall [17]. During the filtration of the calcium sulfate solution, the rejection of calcium sulfate was affected by the pH of the feed solution, leading to the comparison between a high pH of the feed solution and a lower pH. Here, an increase in pH showed higher retention; and a lower pH caused a lower repulsive force at the membrane surface when the dissociation of functional groups was withheld [25].

Importantly, the pH increase in the feed impacts the physico-chemical characteristics of the water, as well as the membrane characteristics. In addition to this, the pH certainly influences the charge or solubility of some components in the feed solution; however, these changes could be used to improve rejection expressed as recovery increase and reduce fouling. It is known that at a high pH, organic compounds are more soluble, specifically at a pH above 10, where the filtration systems that operate at this value or near will help to decrease organic fouling. [26]. On the other hand, higher ionic strength and acidic pH promote bacterial adhesion by modifying the membrane surface in terms of pore size or shape [27].

2.4. Effect of the Feed Flow Rate

The influence of the feed flow velocity on the fouling rate is fundamental to guaranteeing that the system operates at the optimal conditions [28]. In particular, feed flow and hydrodynamic conditions promote the adhesion of particles to the membrane in most of the desalination processes used at an industrial level [27].

When dealing with the filtration of organic matter, the water permeation through the membrane causes the fast formation of a fouling layer, Based on Choi's analysis [29], this comes from the compulsive transport of foulants into the membrane surface by the drag force of permeate flow, which leads to the sealing of pores provoking the fouling layer. Likewise, the moist and low shear environment facilitate the proliferation of bacteria and biofilm formation on the membrane's surface [27].

Another key factor is the concentration polarization since the concentration of microorganisms and nutrients available in the layer directly affects the adhesion and formation of biofilm. Basically, when these elements increase their concentration, the rate of biofouling will increase as well [27]. When the carbon concentration increases in the bulk solution, this can contribute to obtaining higher rates of biofouling, which increases the mass of organic matter (such as microorganisms) present in the biomass [27]. At this point, the element related to the attachment of microorganisms, as well as other organic compounds in the concentration polarization layer, is the lack of convective flow near the membrane surface [27]. Additionally, it is known that both lower crossflow rates and high-water flux increase the rate of pollute accumulation in the boundary layer [27].

Finally, aimed at the reduction of the concentration polarization, it is relevant to understand its dependence on the Reynolds numbers. According to Rezaei et al. [30], the concentration polarization decreases considerably as the Reynolds number increases. This is because high shear stress enhances the mixing phenomenon, allowing to decrease the thickness of the concentration boundary layer and therefore improving the membrane's performance.

2.5. Effect of Feed Temperature

Typically, an increase in temperature results in a higher permeate flux, which does not imply the controlled region of the transmembrane pressure. This could be under the critical point or above the critical point. Generally, high temperatures decrease the viscosity of the feed solution, which reduces the resistance to flow and provokes turbulence. Moreover, the increase in temperature also raises the diffusivity and therefore the rate of transport of solutes carried away from the membrane surface and back into the bulk stream [23]. Herein, the phenomenon of temperature reducing feed viscosity also displays a linear increment between the temperature and the permeation flux. The diffusion coefficient increases while the temperature does, causing the mass transfer resistance to decrease [24].

Experimentally, a higher temperature distributes dyes in a uniform way between both phases of the solution and the membrane, which results in a lower rejection. Nonetheless, salt rejection tends to show a different pattern. He et al. [24] attributed such a phenomenon to the minimal temperature effect on charge repulsion and sieving effect. Jin et al. [31] found out that as the temperature increased, the mass transfer was enhanced, and concentration polarization reduced. Additionally, there was a relation between pressure and temperature, e.g., at a higher feed temperature, lower pressure was needed to deliver the desired flux. In Jin's study, the filtration experiments were tested at 15, 25 and 35 °C, and they used a humic acid solution to evaluate the fouling. At those temperatures, the colloids were on average a size of about 200, 80 and 70 nm, respectively; the larger size of humic acid (at 15 °C) resulted in a vast cake layer, and as a consequence, the flux decline was bigger [31].

2.6. Effect of the Intrinsic Membrane Properties

At this point, we have seen that the physical and chemical interactions between foulants and membrane make the understanding of biofouling more complicated. As the fundamental element used for separation, the properties of the membrane are also crucial in understanding the process and thus finding the possibility of mitigating the biofouling phenomenon. Generally, membrane properties can be conceptualized into membrane structure parameters, which include roughness, porosity and pore size, shape and distribution, and membrane/effluent coupling parameters, including membrane material, surface charge, and hydrophobicity/hydrophilicity. [19]. Regarding the roughness, which refers to the topology of the membrane, it is known that it increases the surface area and thus affects macroscopic properties influencing fouling [32].

On the other hand, when dealing with the pores' features, such as shape, size, and distribution, the membrane features must ideally display the right balance aiming to diminish fouling. Usually, membranes with larger pores are more prone to irreversible fouling due to allowing colloids to penetrate more easily [33]. Membranes, which possess low porosity, demonstrate severe fouling, and hence, suitable pore size and narrow distribution are recommended to control the fouling [19].

Concerning the surface charge, membrane surfaces can display hydrophilic or hydrophobic properties depending on the interfacial tension between water and membrane material. In several works, hydrophobic membranes have been regarded as more prone to fouling compared with hydrophilic ones. This happens because the particles that come from the feed water will accumulate on the hydrophobic surface, minimizing the interfacial tension between water and membrane [34]. Another factor that has a meaningful impact on fouling is electrostatic charge due to membranes that are commonly negatively charged or modified to generate repulsive forces against organic fouling [19]. By considering all these main factors together with operating parameters that influence the biofouling phenomenon, researchers have made a lot of effort to develop new proposals and concepts to face the mitigation of such issues in membrane applications towards water treatment. The next section describes the evolution of the beginnings and current advances in such developments works.

3. Beginnings of the Development Works Aimed at the Mitigation of Biofouling in Membranes

Figure 3 shows the first application of membrane processes for several applications and its evolution until the 1990s. The usage of membrane technologies was sparked off during the discovery of the osmosis phenomenon by J. Abbe Nollet in 1784, and after the following years the discovery became of great interest [3]. A few years later, they found out that the biggest bottleneck of this technology was biological fouling or "biofouling", which is, in fact, the current issue. At the early stages of the development of this technology, the uses ranged from water desalination to disinfection, decarbonization, membrane bioreactors, among others [35].

In 1949, when the concept of seawater desalination became of great interest, several scientists began to investigate the application of membrane technologies to this process. However, since seawater is full of contaminants, such as sand, mud and, most importantly, biological matter, the fouling became a huge problem for this application [36]. In such a period, several techniques, including chlorination, coagulation, acid addition, multi-media filtration and dichlorination, were involved in the pretreatment process to prevent/eliminate undesired contaminants [37,38]. Particularly, to prevent the growth of microorganisms, sodium chlorite was also added [39].

In a more advanced way of solving biofouling, the modification of membranes based on a thin polyamide (PA) layer was performed. This was aimed at mitigating the interaction between the foulants and the barrier layer. By adding macromolecules (e.g., poly (ethylene glycol)) to the surface, the membrane was synthesized to be more hydrophilic. Regarding the membrane preparation protocol, the phase inversion protocol was used for asymmetric membranes, in which thin-film-composite (TFC) membranes were prepared by interfacial polymerization. [40]. These modification protocols to face biofouling have been studied for different types of water sources, but their implementation and efficiency should be dependent on the type of water source [41].

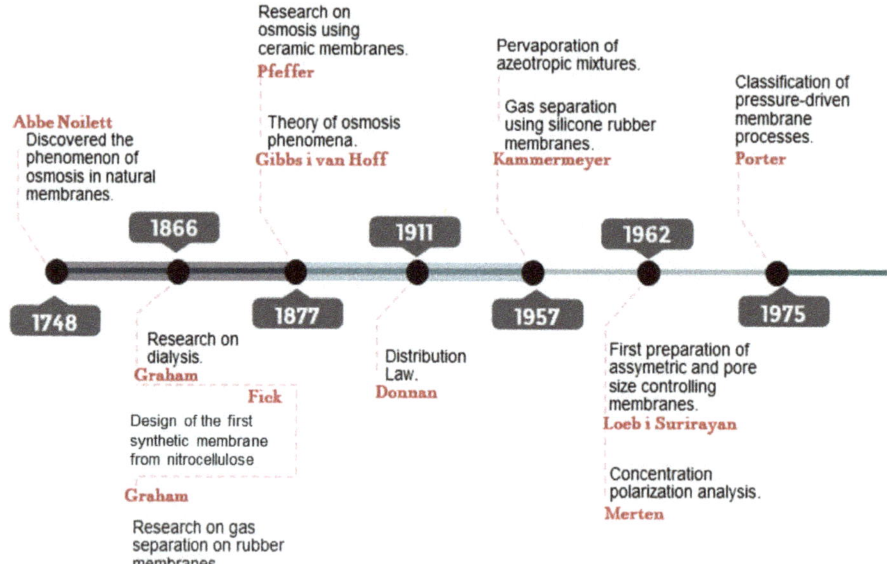

Figure 3. Milestones on the development of membranes from its early beginnings until the 1990s [3].

For a long time, pretreatment with free chlorine was used to prevent and mitigate biofilm and microbial growth but later was avoided since TFC membranes have a low resistance to oxidants. It was then substituted with monochloramine, which resulted in the same oxidizing effect with lower disinfection and it required more frequent cleaning. This was studied by Vikesland and Valentine [41], who suggested the promising use of monochloramine as an oxidant for Fe (II) removal in the production of drinking water.

Efforts have also been made on coating the membranes' surface with antimicrobial products, which could modify its physical–chemical structure to diminish the biofouling effect. Here, an analysis of the biological part that triggers biofouling has been widely studied. For instance, it has been reported that by inhibiting ATP synthesis using chemical uncoupling, the granular sludge biofilm cannot be formed. Besides this, recommendations have been made to study the intercommunication between cells, where the analysis of preventing such signals could avoid biofilm formation [35].

Another way of mitigating biofouling deals with physico-chemical methods, such as sonication, backwashing, and chemical washing, in which acids are the chemicals most commonly used for attempting the removal of multivalent cationic species, metal chelating agents [42], enzymes, and surfactants [35,43]. Membrane bioreactor technology (MBT), which is used in various wastewater treatments, has been used for evaluating the transmembrane pressure (TMP), charge variation, different pH values and salt concentrations, crossflow, and membrane hydrophilicity. The effect of such parameters has also been studied in order to decrease fouling [44]. As a summary, Table 3 reports the most remarkable studies in which the first attempts were directly focused on membrane fouling.

Table 3. Remarkable studies regarding membrane fouling from early 2010.

Year	Authors	Remark of the Study	Reference
1784	J. Abbe Nollet	Discovery of the osmosis phenomenon in natural membranes	[4]
1999	Durham and Walton	Description of the early stages of pretreatment in desalination processes	[36]
1997	Amjad	Starting solutions for fouling	[37]
2001	Isaias	Pretreatment for fouling in desalination processes	[38]
2002	Vikesland and Valentine	Studies in monochloramine as an oxidant for Fe (II) removal in drinking water treatment.	[41]
2006	Le-Clench, Chen and Fane	Early stages of studies in membrane fouling for bioreactors used in wastewater treatment	[44]
2008	Khawaji, Kutubkhanah and Wie	Basic aspects and advances in seawater desalination, and fouling.	[39]
2008	Abu, Tarboush, Rana, Matsuura, Arafat and Narbaitz.	Research in polyamide membranes via surface modification for desalination	[40]
2010	Porcelli and Judd	Cleaning of drinking water using membranes	[42]
2011	Xu and Liu	Membrane fouling and cleaning.	[35]

4. Current Advances in Biofouling Mitigation in Membranes

4.1. Polymer Blending

Figure 4 provides an overview of the evolution of the current advances in biofouling mitigation in membranes. To avoid superfluous fouling, modified membranes have been fabricated in which typical modifications include zwitterions, composite nanomaterials and polymer blending [15]. To date, there is unanimity in research for the first definition and stage of fouling, which refers to the adhesion of foulants to the surface. Such adhesion is mainly attributed to van der Waals attractions, hydrogen bonding, and hydrophobic interactions [14]. Therefore, one of the most forthright and effective strategies comprise surface modification and the design of novel membranes by rendering antifouling properties [13]. Firstly, it must be considered that a copolymer is the merging of a matrix polymer and hydrophilic blocks enable the intermolecular interactions accordingly facilitating the blending. These membranes are sometimes weak, identified as one drawback, and if the polymers are not well interconnected, their permeability will not be suitable [45].

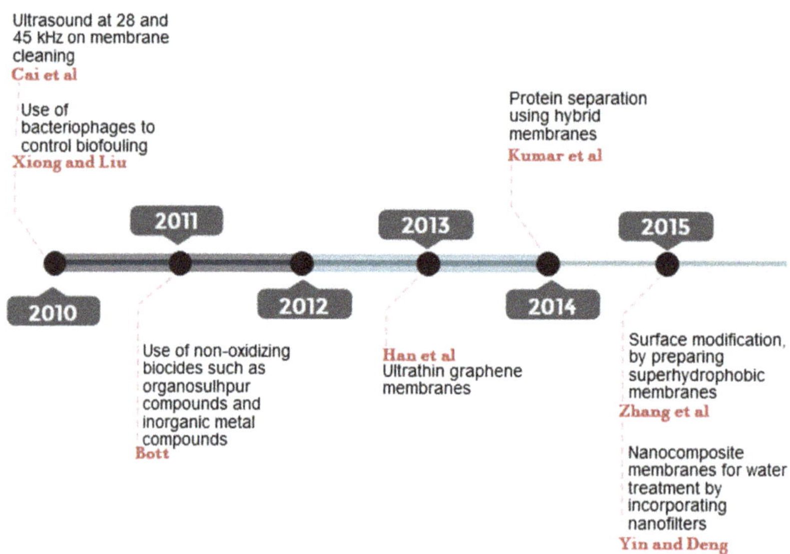

Figure 4. Milestones on the most recent studies on membrane fouling until 2015 [46–50].

An example of polymer blending is the modification of the conjugation of a polyvinyl alcohol membrane with gum arabic for water desalination. Such blending improved the membrane's performance by providing excellent permeation, salt rejection and biofouling resistance, also having more mechanical strength. The improvement was associated with the enhanced hydrophilicity due to the hydrophilic nature of gum arabic and its OH groups. When the surface is more hydrophilic, it displays more water affinity and prevents the adsorption of biofoulants. In this way, changes in hydrophilicity properties by polymer blending is a good approach to prepare a membrane surface with better biofouling resistant properties [51].

Carretier et al. [45] explored the synthesis of a triblock polymer. The used polymers were one block of styrene and two of ethylene glycol. The resulting polymer was achieved to improve the hydrophilicity of the membrane, and consequently, inhibit biofouling in dynamic conditions for three water filtration cycles. However, one of its disadvantages relies on the difficulty in controlling the membrane formation mechanisms when another material is added. Likewise, the lack is related to the stability of the membrane when it has surface modifiers. Despite this, authors still believed these membranes could be promising methods for water treatment and even blood filtration. To date, there are several types of commercial polymers that have been modified aimed at the improvement of their physico-chemical properties. For instance, polyethersulfone (PES)/cellulose acetate phthalate blends were used for the manufacture of ultrafiltration membranes [52]. Such membranes were fabricated by phase inversion-induced, and using polyvinylpyrrolidone as a pore former. The aim was to improve the hydrophilicity of the PES membrane by using cellulose acetate phthalate. By analyzing the water contact angle measurements, the results concluded that the hydrophilicities of blend membranes were enhanced, and such hydrophilic properties were increased by increasing the cellulose acetate phthalate concentration in the casting solution.

Finally, the authors suggested that cellulose acetate phthalate could play a role as an antifouling agent. More recently, Ma et al. [53] proposed another commercial polymer, like poly(vinylidene fluoride) (PVDF), to evaluate the antifouling properties by blending with synthesized amphiphilic poly(poly(ethylene glycol) methyl ether methacrylate-methyl methacrylate) [P(PEGMA-MMA)] copolymers with different initial PEGMA/MMA monomer ratios and PEG side chain lengths. After analysis, it was found that the higher O/C ratio in PEGMA (900) comprised more hydrophilic groups on the surface of the blend membranes and thus enhanced interaction with water molecules. In general, the water molecules released stronger hydration on the membrane surface, and therefore reduced the propensity of foulants to interact with the membrane surface [53]. At the end, the authors stated that the antifouling properties were dependent on the membrane surface hydrophilicity of the copolymer.

Improved antifouling properties on PES membranes by blending the amphiphilic surface modifier with crosslinked hydrophobic segments was reported by Liu et al. [54]. The amphiphilic modifier (MF-g-PEGn) was carried out by etherification of melamine formaldehyde prepolymer with PEG, and later blended in PES polymer to fabricate membranes by means of a nonsolvent induced phase separation method. These blend membranes (MF-g-PEGn) demonstrated a superior antifouling property due to the surface segregation of the hydrophilic polyethylene oxide (PEO). Interestingly, during the filtration testing, the flux recovery ratios after bovine serum albumin (BSA) separation of the PES control membrane and PES/MF-g-PEG6000 (0.36 wt. %) were about 70.8% and 91.6%, respectively. All membranes displayed 100% rejection efficiency. The pure water fluxes were enhanced from 60.7 L m^{-2} h^{-1} for the pristine PES membrane to 164.7 L m^{-2} h^{-1} for PES/MF-g-PEG6000 (0.36 wt. %) [54]. Besides the study reported by Liu et al. [54], there have been other researchers interested in developing the amphiphilic surfaces on UF membranes with different antifouling mechanisms [55]. Ruan et al. [55], reported the fabrication of an amphiphilic NF membrane by a two-step surface modification of a polyamide NF membrane, that implements two mechanisms, one of them was the fouling resistance defense while the other was the fouling release defense. The experiments were performed with BSA solution, HA solution and SA solution; they showed that this modified membrane displayed a better antifouling property than the pristine one, i.e., polyamide NF membrane. Gao et al. [56], also carried out a study of modified PES membranes by incorporating the amphiphilic comb copolymer, where the modified membranes had a very low flux decline rate (15.6%) and the flux recovery rate was up to 96.6%; in addition, Gao concluded that the modified membrane had a stable and durable antifouling property after three cycles of BSA filtration [56]. As a preliminary conclusion of this section, the polymer blending generally aims to shift of the nature of the polymers to a more hydrophilic one which does not favor the interaction of foulants and the membrane surface.

4.2. Nanocomposite Materials

In addition to the polymer blending, the preparation of nanocomposite membranes is also a new trend on preparing membranes with better antifouling properties. A nanocomposite membrane is defined as the next generation of advanced membranes, in which nanomaterials are embedded and ideally well dispersed into a polymer matrix [57]. This concept of membranes has received great attention over the recent years. Typically, a composite membrane, as well as a mixed matrix membrane, combines the strengths of a polymer and inorganic materials to ideally reach a synergistic effect [58]. In principle, the embedding of nanomaterials into polymers can modify the structure, as well as physico-chemical properties of membranes, such as hydrophilicity, porosity, roughness, pore morphology, charge density, thermal, chemical, and mechanical stability. However, some important properties can also be enhanced, including flux permeation, foulant rejection, and antifouling properties [14,59,60]. For instance, Table 4 reports some examples of nanomaterials that have been incorporated into polymers, and their effect on the resulting membranes. Nowadays, considering the advancement of nanotechnology, nanomaterials represent a novel opportunity to mitigate biofouling. One of their greatest advantages is that they provide durability under high operating pressure conditions. In recent years, various porous nanomaterials were categorized as a new class of additives in the membranes. To date, several types of inorganic nanomaterials have been embedded in polymer membranes, such as zeolites [61], metal-organic frameworks (MOFs) [62], carbon nanotubes (CNTs) [63], porous organic cages (POCs) [64], silicas (mesoporous MCM-41) [65], graphene oxide (GO) [66], MOF-silica hybrid particles [67], titanium dioxide (TiO$_2$) [68], magnesium oxide [69] and Ag-based particles [70]. All these materials possess specific intrinsic features that enable them to provide enhanced properties to composite membranes. e.g., zeolites present competitive adsorption and diffusion properties, and cation exchange behavior in desalination processes. TiO$_2$ can alter the membrane structure; for example, Ong et al. [60] mentioned that the use of titanium dioxide nanoparticles exhibited great results by improving the substrate properties instead of modifying the layer properties. Both TiO_2 and silver particles offer the possibility of tuning the hydrophilicity of the membranes, as well as increase their porosity. In 2015, Homayoonfal et al. [71] showed that when using polysulfide/alumina nanocomposite membranes for bioreactors, the water flux could be increased while the membrane fouling was reduced by 83%. Alumina nanoparticles were specifically synthesized by the co-precipitation method and their homogenous dispersion was carried by an ultrasonic bath. The authors concluded that the nanoparticles had a strong influence on the surface properties [70].

Over the last decade, graphene-based materials have attracted special attention for different types of applications. Such two-dimensional materials can elevate the surface area and weight ratio, giving a great mechanical and thermal stability to the membranes. Graphene oxide (GO) is especially preferred due to its hydrophilic nature, owing to the presence of polar hydroxyl and carboxyl groups [72]. GO has certainly been used in the fabrication of nanocomposite membranes for water treatment, including water desalination, removal of toxic ions and organic molecules in polluted water. There is evidence that carbon nanotubes, GO and other carbon allotropes can provide better antifouling and antioxidative properties than normal polyamide membranes, suggesting that when incorporating functionalized multi-walled carbon nanotubes, membrane performance is also improved [72,73].

Table 4. Nanomaterials embedded in composite membranes for water treatment.

Nanomaterial	Polymer	Remark of the Study	Reference
TiO_2	Polyamide (PA)	Good flux recovery by incorporating TiO_2. Enhanced foulant removal than pristine membrane.	[60]
Al_2O_3	Polysulfone (PS)	Water flux increase. Membrane fouling was reduced by 83%.	[71]
GO-Ag	Thin-film composite (TFC)	Static antimicrobial assays showed a significant inhibition to the attachment of *Pseudomonas aeruginosa* cells.	[74]
Cu	Thin-film composite (TFC)	The nanomaterial was deposited via spray- and spin-assisted layer-by-layer. The method was efficient and improved the distribution compared to conventional dip coating techniques. Cu nanoparticles improved the anti-biofouling properties. Cu nanoparticles effectively inhibited the permeate flux reduction caused by bacterial deposition.	[75]
NH_2-TNTs	Polyamide (PA)	The water flux of the membrane was significantly increased. The nanomaterial significantly mitigated the BSA fouling and achieved a promising water flux recovery rate after rinsing.	[76]
Fe_3O_4	Polyethersulfone (PES)	Iron oxide nanoparticles resulted in an increase in hydrophilicity and growth in the membrane sub-layer porosity. The pore radius was affected.	[77]
Silver-based MOF	Thin-film composite (TFC)	The MOF improved both the biocidal activity and the hydrophilicity of the membrane active layer. No effect was observed on the membrane selectivity.	[78]
Silica/QA/POM	Thin-film composite (TFC)	Membrane with 0.2 wt. % nanoparticle incorporation showed superior water flux in forward osmosis processes and minimal increase in reverse salt flux. Moreover, enhanced antifouling propensity toward BSA and sodium alginate foulant was noted.	[79]
QAC/Carbon	Polyvinylidene Fluoride (PVDF)	The introduction of Quatery Ammonium Compound assembled on Carbon into polymeric membranes was an effective way to prepare anti-biofouling membranes for water and wastewater treatment.	[80]
ZnO	Polyaniline (PANI)	The resulting membranes showed a good mechanical strength with moderate elasticity. The membranes showed good antifouling properties toward marine bacteria *V. harveyi* and *B. licheniformis*.	[81]

Abbreviations: TiO_2: titanium oxide, Al_2O_3: aluminium oxide, GO-Ag: Graphene Oxide with silver, Cu: Copper, NH_2-TNTs: Amino functionalized titanate nanotubes, Fe_3O_4: magnetite, GG/AO: Guar Gum with Aluminum oxide, Silver-based MOF: silver-based metal organic frameworks, Polyoxometalates: POM, QAC/carbon: Quaternary Ammonium compound with carbon, ZnO: Zinc Oxide.

GO is also recognized as one of the most promising nano-sized materials that has been applied for the elimination of pharmaceuticals from water and wastewater [82]. Chang et al. [83] evaluated the synergistic influence of GO and polyvinylpyrrolidone (PVP) on the separation performance of PVDF membranes. It was demonstrated that the membrane hydrophilicity and the antifouling properties were enhanced by embedding GO into PVP. The authors reported that the enhancement was attributed to the formation of hydrogen bonds between PVP and GO. Besides improving the hydrophilicity of membranes; GO has been also identified as a potential candidate to enhance the water transport of membranes according to its unimpeded water permeation properties [83–85]. Finally, some other nanomaterials are also providing good insights to mitigate the fouling in membranes. Zinc oxide (ZnO), as a multifunctional inorganic material, is particular due to its relevant physical and chemical properties, e.g., catalytic, antibacterial and bactericide activities. Moreover, this nanomaterial can absorb polar hydroxyl groups (−OH) and its surface area is relatively higher than other inorganic materials [86]. When dealing with its use for preparing nanocomposite membranes, ZnO can enhance specific properties in polymers, including the hydrophilicity, mechanical and chemical properties [87]. The embedding of ZnO also results in the enhancement of the hydrophilicity of PES NF membranes, which gives higher permeabilities in ZnO-filled nanocomposite membranes. In addition, fouling resistance during the filtration of humic acid solutions has been documented [88]. As a final remark from this section, Figure 5 shows in particular the milestones of nanocomposite membranes over the last years.

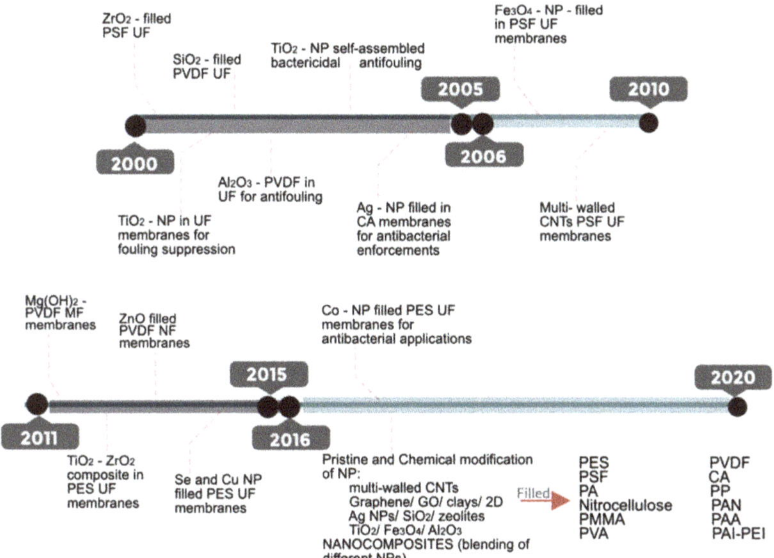

Figure 5. Milestones showing the progress of the nanocomposite membranes over the last 20 years.

4.3. Chemical Modification

Depending on the membrane material, the membrane surface of reverse osmosis, nanofiltration and ultrafiltration membranes is usually negatively charged according to the presence of sulfonic or carboxylic groups. These groups are relevant when using zwitterions, which are defined as molecules with two or more functional groups, in which at least one possesses a positive and one possesses a negative electrical charge, and consequently the net change of the entire molecule becomes zero [89]. In principle, when the foulant and the membrane display the same charge, specific forces, such as electrostatic repulsion, tend to appear, and the presence of such molecules reduces such a phenomenon. Few studies recognize some bacterial solutions carry negative charges, then if the membrane has a negative charge, there is less probability of showing bio-adhesion rather than if it is positively charged [14]. Based on this statement, the chemical modification of membranes via zwitterions has become a promising alternative. For example, Venault et al. [90] designed alternative copolymers of p(MAO–DMEA) (synthetized via the reaction between poly(maleic anhydride-alt-1-octadecene) and N,N-dimethylenediamine) and p(MAO–DMPA) (synthetized via the reaction between poly-(maleic anhydride-alt-1-octadecene) and 3-(dimethylamino)-1-propylamine) of different carbon space length (CSL) using a ring-opening zwitterionization [90]. Such copolymers were later coated on poly (vinylidene fluoride) (PVDF) membranes by means of a self-assembled procedure. The authors studied the antifouling properties of the modified membranes treating several protein, cell, and bacterial assays, concluding that both zwitterionic modified membranes with different coating densities exhibited enhanced membrane hydrophilicity, and better resistance to blood cells, bacteria, platelet and protein adsorption. Over the course of this paper, we have mentioned that hydrophilicity is needed for the non-fouling behavior of an interface. Interestingly, the membranes prepared by Venault et al. [90] indicated that zwitterionic molecules enabled saturated surface hydration, which was in agreement with the higher contact angle measurements and hydration capacity of the membranes. This is, in fact, an impressive approach to preparing smart antifouling membranes. The use of antimicrobial membranes is able to inactivate bacterial cells at the contact by using a biocide, decreasing the rate of biofilm formation. Even though it is an efficient method, their long-term functionality was limited by the accumulation of dead cells; thereby, the best way to increase their efficiency was to design a membrane with both antimicrobial and antifouling properties [15].

It is important to point out that the cost of using chemical additives will definitely increase the membrane production cost but may also increase toxicity. Here, authors should start to evaluate the environmental implications in terms of possible release to the environment and discharge to the water. e.g., chlorine, as a typical biocide, is recognized as the most feasible due to its low cost and efficiency at low concentrations. However, it is under observation since organo-chloro compounds are a result of its use [91]. Table 5 enlists some examples of chemicals used for chemical modification, as well as their effect when used in membrane modifications.

Table 5. Chemicals used for chemical modification of membrane's surface.

Material	Remarks	References
Divalent Cations	Calcium enhanced the fouling properties due to its bridging effects between carboxylic active groups contained in NOM and the negatively charged functional groups in the membrane surface.	[19]
Metal ions (Al^{3+} and Fe^{3+})	They are being used to form large precipitating complexes with the Humic acid and fulvic acid, and thus to facilitate their elimination.	[20]
Sulfonic groups	The attaching of sulfonic groups to the aromatic backbone of polysulfone and polyethersulfone membranes generated an electrophilic aromatic substitution reaction, in which hydrogen is replaced by sulfonic acid.	[92]
Carboxylation	The presence of carboxylic groups increased the membrane hydrophilicity.	[92]
Plasma treatment	The bombarded surface of the membrane with ionized plasma components generated radical sites. Active components generated by such plasma contributed to increasing the hydrophilicity without affecting the bulk of the polymer.	[92]
CO_2-plasma	The addition of oxygen into the membrane' surface, in the form of carbonyl, acid and ester groups, increased in hydrophilicity.	[92]
D-Tyrosine	D-amino acids inhibited the microbial attachment. D-tyrosine enhanced the membrane hydrophilicity and provided a smoother surface to the membrane without modifying its transport properties, and also reduced the propensity for biofouling.	[93]

Table 5. *Cont.*

Material	Remarks	References
GO-pDA	The attaching of graphene oxide nanosheets to the membrane surface, by chemical modification with polydopamine through an oxidative polymerization, reduced the loss of the draw solution and increased both membrane water flux and biofouling resistance.	[94]
Charged hydrogel	Anti-biofouling properties of neutral (polyHEMA-co-PEG10MA), cationic (polyDMAEMA) and anionic (polySPMA) hydrogels in feed spacers were tested with *E. coli*. The membranes showed reduced attachment and biofouling in the spacer-filled channels, resulting in delayed biofilm growth.	[95]
Antimicrobial peptides	A polycyclic antimicrobial peptide, like nisin, decreased the viability of *Bacillus sp.*, and the dislodging of *P. aeruginosa P60*. Nisin served as a biological agent for the mitigation of membrane biofouling.	[96]

Of course, according to the mentioned risks, there is today's interest in alternative and new methods that do not imply the use of chemical additives. The next section addresses some other novel strategies to mitigate the fouling in membranes.

4.4. Alternative Novel Strategies in Biofouling Mitigation

In addition to membrane surface modification, polymer bending or composite membranes, the research community has directed the approaches of mitigating the fouling by applying appropriate preliminary treatments of the bulk feed. Conventionally, the pretreatments comprise different methods, including disinfection, coagulation flocculation, and a filtration process, where all of them aim to remove foulants or their precursors [14]. As an example, Katalo et al. [97] used *Moringa oleifera* seeds during pretreatment prior to microfiltration of river water. *Moringa oleifera* represents a non-toxic natural coagulant, which can be used in a conventional coagulation process. According to the authors' findings, the results of using this biomaterial can be comparable when using aluminum-based coagulants. Therefore, this approach not only represents a promising alternative in water treatment, but also its potentiality regards to the non-use of high-cost chemical coagulants, such as $Al_2(SO_4)_3$ and $FeCl_3$. Moreover, the seeds contain dimeric cationic proteins that substantially reduced membrane fouling by removing suspended solids and dissolving organic matter [97].

During the review and current state-of-the-art provided by Gule et al. [47], several approaches to reducing biofouling have been addressed. Interestingly, one of them is proposing the limitation of the nutrients. Thereby, when the quantity of biodegradable dissolved organic carbon and assimilable organic carbon is reduced, even though the correlation between them does not exist, the biofouling is decreased. Additionally, the reduction of availability of phosphates in the bulk feed may limit the biofouling. Such a reduction could be done by ion exchange, precipitation, and electrochemical coagulation, which do not require the addition of chemicals, representing an environmentally friendly strategy [47]. The application of ultrasound is another approach which can also favor the cleaning of the membrane. There are many studies on the usage of this technique, which are dedicated to studying the cavitation waves and their effect. In theory, ultrasound produces physical phenomena, such as shock waves, acoustic streaming and microstreaming, which may help to release the particles from a fouled membrane. It is proven that at higher values of frequency and power intensity, the flow velocity increases and thus are better at detaching the foulants [98]. On the other hand, there are other physical cleaning methods that comprise the application of forces, such as mechanic and hydraulic, for the detachment of the foulants. Table 6 summarizes some of the novel and current strategies implemented by scientists aimed at the mitigation of biofouling in membranes.

Table 6. Alternative novel approaches for biofouling mitigation.

Approach	Description	References
Addition of bacteriophages as biocidal agents	T4 bacteriophage-facilitated biofouling control in the membrane ultrafiltration to inhibit the propagation of *E. coli* in situ.	[99]
Bio-electrochemistry	Silver was bioelectrochemically recovered from wastewater. It is an eco-friendly method showing the potential in anti-biofouling applications with recovered nano-flakes, particularly in membrane bioreactors.	[100]
Quorum quenching	The quorum quenching caused to prevent biofouling since quorum sensing interrupts the biological communication mechanism between microorganisms. This was achieved with rotational membrane filtration modules	[101]
UV light	Ultraviolet (UV) light penetrates the cell wall and damages the DNA and RNA, thus stopping the microorganism from reproducing. Furthermore, the main advantage is that it does not produce chemical by-products that can affect health.	[102]
Metazoans	The presence of an oligochaete (*Aelosoma hemprichi*), and a nematode (*Plectus aquatilis*) strongly affected the formation of biofilm.	[103]

It is well known that static mixers can be an effective and efficient way to reduce membrane fouling since they divert the fluid, which provokes the increase of the shear rate at the membrane surface. This enhances the back-transport of retained matter. Nevertheless, this implementation of mixers within the flow channel of a membrane implies an extra pressure drop. To diminish this effect, the group of Professor Wessling very recently developed the shortened and twisted tape mixers (see Figure 6) and analyzed the way shortening was translated into the reduction of fouling mitigation.

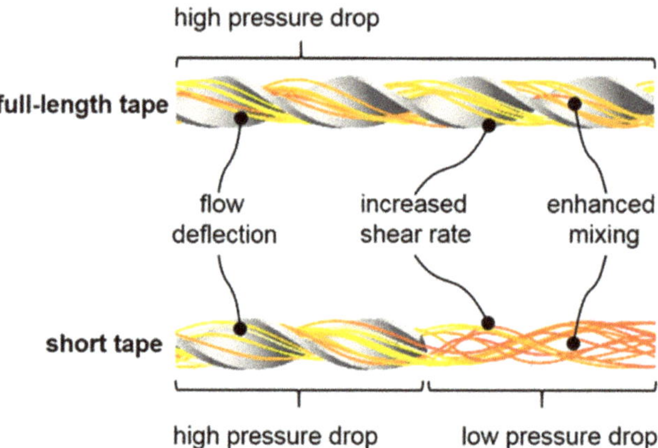

Figure 6. Shortening and spaced twisted tapes in tubular membranes aimed at biofouling mitigation [104].

In such a proposal, they followed the following stages: (i) short total length of the twisted tape, subsequently (ii) the use of spaced short twisted tape elements, which were maintained at their location by smooth rods placed between the twisted elements. As interesting findings, the selection of modified tape mixers presented with lower pressure loss, but enough flow properties toward fouling mitigation [104]. In addition, the influence of foulant concentration in this approach was studied by the authors, who found out that for low silica concentrations (in the range of 0.03 g/L), the short and space twisted tapes mitigated fouling were as similar as the full-length twisted tape. While at high silica concentrations and fluxes, the full-length mixer decreased the fouling and was even stronger than the short and spaced twisted tapes.

5. Concluding Remarks and Future Trends in the Field

Over the course of this paper, we have reviewed the milestones of the research community focused on the strategies and approaches to face the fouling phenomenon on membranes, which in fact, is the primary drawback of pressure-driven membrane processes. Several approaches to preventing the adhesion of matter, translated to biofouling on membranes deals with the shift of their physico-chemical properties. In general, the preparation of highly hydrophilic membranes is sought using different approaches, including the preparation of nanocomposite membranes, membrane modification, and polymer blending. However, diverse options have also come out to be more effective in the removal of organic matter, e.g., combination of different techniques, resulting in efficient strategies for fouling mitigation. Nowadays, it is likely that the concept of nanocomposite membranes is the most explored approach, which comprises the embedding of nanomaterials (including clays, zeolites, metal oxides, graphene-based materials, carbon nanotubes, metal-organic frameworks, to mention just a few) into the polymer matrix. To obtain high performing membranes (in terms of permeation and rejection) with better antifouling resistance, herein, it is crucial the right selection of the nanomaterials according to their intrinsic properties, such as type of material, surface charge, composition, surface area, size, material loading, hydrophilic/hydrophobic nature, among others. However, the type of polymer and its compatibility will also play an important factor not only in the performance but also in the fabrication of the membranes. Even though there are already great advances in the field, there is still a strong need to work on the enhancement of the intrinsic properties of the membrane surface, including hydrophilicity and electrical surface charge, to improve the antifouling/biofouling and antimicrobial properties of membranes. Moreover, it is recommended to new researchers in the field the analysis of the separation performance and biofouling properties of the novel membranes using real complex solutions (such as industrial by-products and wastewaters). In this sense, the developed membranes can provide more realistic insights, which may give a clear overview of the potentiality of those membranes in water treatment applications.

Author Contributions: D.P.-R., Z.P.G.-A. and A.Z.-R. wrote the original draft paper. R.C.-M. conceived, designed, reviewed and edited the manuscript. All authors have read and agreed to the published version of the manuscript.

Funding: This research received no external funding.

Acknowledgments: R. Castro-Muñoz acknowledges the School of Engineering and Science and the FEMSA-Biotechnology Center at Tecnológico de Monterrey for their support through the Bioprocess (0020209I13) Focus Group.

References

1. Castro-Muñoz, R. Pressure-driven membrane processes involved in waste management in agro-food industries: A viewpoint. *AIMS Energy* **2018**, *6*, 1025–1031. [CrossRef]
2. Galanakis, C.M.; Cvejic, J.; Verardo, V.; Segura-Carretero, A. Food Use for Social Innovation by Optimizing Food Waste Recovery Strategies. In *Innovation Strategies in the Food Industry: Tools for Implementation*; Academic Press: Cambridge, MA, USA, 2016. [CrossRef]
3. Liu, B.; Wang, D.; Yu, G.; Meng, X.; David Giraldo, J.; Thakur, V.K.; Gutiérrez, E. The History and State of Art in Membrane Technologies Tarragona, Erasmus 2005. *J. Membr. Sci.* **2013**, *16*, 1–28. [CrossRef]
4. Kabsch-Korbutowicz, M.; Kutylowska, M. The Possibilities of Modelling the Membrane Separation Processes. *Environ. Prot. Eng.* **2008**, *34*, 15.
5. Kayvani Fard, A.; McKay, G.; Buekenhoudt, A.; Al Sulaiti, H.; Motmans, F.; Khraisheh, M.; Atieh, M. Inorganic membranes: Preparation and application for water treatment and desalination. *Materials* **2018**, *11*, 74. [CrossRef]

6.   Goh, P.S.; Ismail, A.F. A review on inorganic membranes for desalination and wastewater treatment. *Desalination* **2018**, *434*, 60–80. [CrossRef]

7.   Castro-Muñoz, R.; Barragán-Huerta, B.E.; Fíla, V.; Denis, P.C.; Ruby-Figueroa, R. Current role of membrane technology: From the treatment of agro-industrial by-products up to the valorization of valuable compounds. *Waste Biomass Valorization* **2018**, *9*, 513–529. [CrossRef]

8.   Madaeni, S.S.; Zinadini, S.; Vatanpour, V. A new approach to improve antifouling property of PVDF membrane using in situ polymerization of PAA functionalized TiO2 nanoparticles. *J. Membr. Sci.* **2011**, *380*, 155–162. [CrossRef]

9.   Ursino, C.; Castro-Muñoz, R.; Drioli, E.; Gzara, L.; Albeirutty, M.; Figoli, A. Progress of Nanocomposite Membranes for Water Treatment. *Membranes* **2018**, *8*, 18. [CrossRef]

10.   Nguyen, T.; Roddick, F.A.; Fan, L. Biofouling of water treatment membranes: A review of the underlying causes, monitoring techniques and control measures. *Membranes* **2012**, *2*, 804–840. [CrossRef] [PubMed]

11.   Ozzello, E.; Mollea, C.; Bosco, F.; Bongiovanni, R. Factors Influencing Biofouling and Use of Polymeric Materials to Mitigate It. *Adhes. Pharm. Biomed. Dent. Fields* **2017**, 185–206. [CrossRef]

12.   Kirschner, A.Y.; Cheng, Y.H.; Paul, D.R.; Field, R.W.; Freeman, B.D. Fouling mechanisms in constant flux crossflow ultrafiltration. *J. Membr. Sci.* **2019**, *574*, 65–75. [CrossRef]

13.   Goh, P.S.; Lau, W.J.; Othman, M.H.D.; Ismail, A.F. Membrane fouling in desalination and its mitigation strategies. *Desalination* **2018**, *425*, 130–155. [CrossRef]

14.   Kochkodan, V.; Hilal, N. A comprehensive review on surface modified polymer membranes for biofouling mitigation. *Desalination* **2015**, *356*, 187–207. [CrossRef]

15.   Perreault, F.; Jaramillo, H.; Xie, M.; Ude, M.; Nghiem, L.D.; Elimelech, M. Biofouling Mitigation in Forward Osmosis Using Graphene Oxide Functionalized Thin-Film Composite Membranes. *Environ. Sci. Technol.* **2016**, *50*, 5840–5848. [CrossRef]

16.   Zheng, Y.; Zhang, W.; Tang, B.; Ding, J.; Zheng, Y.; Zhang, Z. Membrane fouling mechanism of biofilm-membrane bioreactor (BF-MBR): Pore blocking model and membrane cleaning. *Bioresour. Technol.* **2018**, *250*, 398–405. [CrossRef]

17.   Said, M.; Ahmad, A.; Mohammad, A.W.; Nor, M.T.M.; Sheikh Abdullah, S.R. Blocking mechanism of PES membrane during ultrafiltration of POME. *J. Ind. Eng. Chem.* **2015**, *21*, 182–188. [CrossRef]

18.   Castro-Muñoz, R.; Conidi, C.; Cassano, A. Membrane-based technologies for meeting the recovery of biologically active compounds from foods and their by-products. *Crit. Rev. Food Sci. Nutr.* **2019**, *59*, 2927–2948. [CrossRef]

19.   Zheng, X. Major Organic Foulants in Ultrafiltration of Treated Domestic Wastewater and their Removal by Bio-Filtration as Pre-Treatment. Doctoral Thesis, Technische Universität Berlin, Berlin, Germany, 2010.

20.   Abdul Ghani, D.; Radwan, A.-R.; Mohammad, J. Studies on organic foulants in the seawater feed of reverse osmosis plants of SWCC. *Desalination* **2000**, *132*, 217–232.

21.   Maartens, A.; Swart, P.; Jacobs, E.P. Feed-water pretreatment: Methods to reduce membrane fouling by natural organic matter. *J. Membr. Sci.* **1999**, *163*, 51–62. [CrossRef]

22.   Jepsen, K.L.; Bram, M.V.; Pedersen, S.; Yang, Z. Membrane fouling for produced water treatment: A review study from a process control perspective. *Water (Switzerland)* **2018**, *10*, 847. [CrossRef]

23.   Cassano, A.; Conidi, C.; Ruby-Figueroa, R.; Castro-Muñoz, R. Nanofiltration and tight ultrafiltration membranes for the recovery of polyphenols from agro-food by-products. *Int. J. Mol. Sci.* **2018**, *19*, 351. [CrossRef] [PubMed]

24.   He, Y.; Li, G.; Wang, H.; Zhao, J.; Su, H.; Huang, Q. Effect of operating conditions on separation performance of reactive dye solution with membrane process. *J. Membr. Sci.* **2008**, *321*, 183–189. [CrossRef]

25.   Nanda, D.; Tung, K.L.; Li, Y.L.; Lin, N.J.; Chuang, C.J. Effect of pH on membrane morphology, fouling potential, and filtration performance of nanofiltration membrane for water softening. *J. Membr. Sci.* **2010**, *349*, 411–420. [CrossRef]

26.   Franks, R.; Bartels, C.; Nagghappan, L.N.S.P. Performance of a reverse osmosis system when reclaiming high Ph—High temperature wastewater. In Proceedings of the 2009 AWWA Membrane Technology Conference and Exposition, Memphis, TN, USA, 15–18 March 2009; pp. 1–16.

27.   Kucera, J. Biofouling of polyamide membranes: Fouling mechanisms, current mitigation and cleaning strategies, and future prospects. *Membranes* **2019**, *9*, 111. [CrossRef]

28. Koo, C.H.; Mohammad, A.W.; Suja', F. Effect of cross-flow velocity on membrane filtration performance in relation to membrane properties. *Desalin. Water Treat.* **2015**, *55*, 678–692. [CrossRef]
29. Choi, H.; Zhang, K.; Dionysiou, D.D.; Oerther, D.B.; Sorial, G.A. Effect of permeate flux and tangential flow on membrane fouling for wastewater treatment. *Sep. Purif. Technol.* **2005**, *45*, 68–78. [CrossRef]
30. Rezaei, H.; Ashtiani, F.Z.; Fouladitajar, A. Fouling behavior and performance of microfiltration membranes for whey treatment in steady and unsteady-state conditions. *Braz. J. Chem. Eng.* **2014**, *31*, 503–518. [CrossRef]
31. Jin, X.; Jawor, A.; Kim, S.; Hoek, E.M.V. Effects of feed water temperature on separation performance and organic fouling of brackish water RO membranes. *Desalination* **2009**, *239*, 346–359. [CrossRef]
32. Wong, P.C.Y.; Kwon, Y.N.; Criddle, C.S. Use of atomic force microscopy and fractal geometry to characterize the roughness of nano-, micro-, and ultrafiltration membranes. *J. Membr. Sci.* **2009**, *340*, 117–132. [CrossRef]
33. Costa, A.R.; de Pinho, M.N.; Elimelech, M. Mechanisms of colloidal natural organic matter fouling in ultrafiltration. *J. Membr. Sci.* **2006**, *281*, 716–725. [CrossRef]
34. Li, Q.; Xu, Z.; Pinnau, I. Fouling of reverse osmosis membranes by biopolymers in wastewater secondary effluent: Role of membrane surface properties and initial permeate flux. *J. Membr. Sci.* **2007**, *290*, 173–181. [CrossRef]
35. Xu, H.; Liu, Y. Control and cleaning of membrane biofouling by energy uncoupling and cellular communication. *Environ. Sci. Technol.* **2011**, *45*, 595–601. [CrossRef] [PubMed]
36. Durham, B.; Walton, A. Membrane pretreatment of reverse osmosis: Long-term experience on difficult waters. *Desalination* **1999**, *122*, 157–170. [CrossRef]
37. Amjad, Z. RO systems current fouling problems & solutions. *Desalin. Water Reuse* **1997**, *6*, 55–60.
38. Isaias, N.P. Experience in reverse osmosis pretreatment. *Desalination* **2001**, *139*, 57–64. [CrossRef]
39. Khawaji, A.D.; Kutubkhanah, I.K.; Wie, J.M. Advances in seawater desalination technologies. *Desalination* **2008**, *221*, 47–69. [CrossRef]
40. Abu Tarboush, B.J.; Rana, D.; Matsuura, T.; Arafat, H.A.; Narbaitz, R.M. Preparation of thin-film-composite polyamide membranes for desalination using novel hydrophilic surface modifying macromolecules. *J. Membr. Sci.* **2008**, *325*, 166–175. [CrossRef]
41. Vikesland, P.J.; Valentine, R.L. Modeling the kinetics of ferrous iron oxidation by monochloramine. *Environ. Sci. Technol.* **2002**, *36*, 662–668. [CrossRef]
42. Porcelli, N.; Judd, S. Chemical cleaning of potable water membranes: A review. *Sep. Purif. Technol.* **2010**, *71*, 137–143. [CrossRef]
43. Surface, E.; Treatment, W. Committee Report: Recent advances and research needs in membrane fouling. *J. Am. Water Work. Assoc.* **2005**, *97*, 79–89. [CrossRef]
44. Le-Clech, P.; Chen, V.; Fane, T.A.G. Fouling in membrane bioreactors used in wastewater treatment. *J. Membr. Sci.* **2006**, *284*, 17–53. [CrossRef]
45. Carretier, S.; Chen, L.A.; Venault, A.; Yang, Z.R.; Aimar, P.; Chang, Y. Design of PVDF/PEGMA-b-PS-b-PEGMA membranes by VIPS for improved biofouling mitigation. *J. Membr. Sci.* **2016**, *510*, 355–369. [CrossRef]
46. Cai, M.; Zhao, S.; Liang, H. Mechanisms for the enhancement of ultrafiltration and membrane cleaning by different ultrasonic frequencies. *Desalination* **2010**, *263*, 133–138. [CrossRef]
47. Gule, N.P.; Begum, N.M.; Klumperman, B. Advances in biofouling mitigation: A review. *Crit. Rev. Environ. Sci. Technol.* **2016**, *46*, 535–555. [CrossRef]
48. Homaeigohar, S.; Elbahri, M. Graphene membranes for water desalination. *NPG Asia Mater.* **2017**, *9*, e427. [CrossRef]
49. Solís Carvajal, C.A.; Vélez Pasos, C.A.; Ramírez-Navas, J.S. Tecnología de membranas: Ultrafiltración. *Entre Cienc. E Ing.* **2017**, *11*, 26–36. [CrossRef]
50. Yin, J.; Deng, B. Polymer-matrix nanocomposite membranes for water treatment. *J. Membr. Sci.* **2015**, *479*, 256–275. [CrossRef]
51. Falath, W.; Sabir, A.; Jacob, K.I. Novel reverse osmosis membranes composed of modified PVA/Gum Arabic conjugates: Biofouling mitigation and chlorine resistance enhancement. *Carbohydr. Polym.* **2017**, *155*, 28–39. [CrossRef]
52. Rahimpour, A.; Madaeni, S.S. Polyethersulfone (PES)/cellulose acetate phthalate (CAP) blend ultrafiltration membranes: Preparation, morphology, performance and antifouling properties. *J. Membr. Sci.* **2007**, *305*, 299–312. [CrossRef]

53. Ma Wenzhong Rajabzadeh, S.; Shaikh, A.R.; Kakihana, Y.; Sun, Y.; Matsuyama, H. Effect of type of poly(ethylene glycol) (PEG) based amphiphilic copolymer on antifouling properties of copolymer/poly (vinylidene fluoride) (PVDF) blend membranes. *J. Membr. Sci.* **2016**, *514*, 429–439. [CrossRef]

54. Liu, Y.; Su, Y.; Zhao, X.; Li, Y.; Zhang, R.; Jiang, Z. Improved antifouling properties of polyethersulfone membrane by blending the amphiphilic surface modifier with crosslinked hydrophobic segments. *J. Membr. Sci.* **2015**, *486*, 195–206. [CrossRef]

55. Ruan, H.; Li, B.; Ji, J.; Sotto, A.; Van Der Bruggen, B.; Shen, J.; Gao, C. Preparation and characterization of an amphiphilic polyamide nanofiltration membrane with improved antifouling properties by two-step surface modification method. *RSC Adv.* **2018**, *8*, 13353–13363. [CrossRef]

56. Gao, F.; Zhang, G.; Zhang, Q.; Zhan, X.; Chen, F. Improved Antifouling Properties of Poly(Ether Sulfone) Membrane by Incorporating the Amphiphilic Comb Copolymer with Mixed Poly(Ethylene Glycol) and Poly(Dimethylsiloxane) Brushes. *Ind. Eng. Chem. Res.* **2015**, *54*, 8789–8800. [CrossRef]

57. Castro-Muñoz, R.; Fíla, V. Progress on incorporating zeolites in matrimid®5218 mixed matrix membranes towards gas separation. *Membranes* **2018**, *8*, 30. [CrossRef] [PubMed]

58. Castro-Muñoz, R.; De La Iglesia, Ó.; Fila, V.; Téllez, C.; Coronas, J. Pervaporation-assisted esterification reactions by means of mixed matrix membranes. *Ind. Eng. Chem. Res.* **2018**, *57*. [CrossRef]

59. Esfahani, M.R.; Aktij, S.A.; Dabaghian, Z.; Firouzjaei, M.D.; Rahimpour, A.; Eke, J.; Escobar, I.C.; Abolhassani, M.; Greenlee, L.F.; Esfahani, A.R.; et al. Nanocomposite membranes for water separation and purification: Fabrication, modification, and applications. *Sep. Purif. Technol.* **2019**, *213*, 465–499. [CrossRef]

60. Ong, C.S.; Goh, P.S.; Lau, W.J.; Misdan, N.; Ismail, A.F. Nanomaterials for biofouling and scaling mitigation of thin film composite membrane: A review. *Desalination* **2016**, *393*, 2–15. [CrossRef]

61. Zhu, B.; Myat, D.T.; Shin, J.W.; Na, Y.H.; Moon, I.S.; Connor, G.; Maeda, S.; Morris, G.; Gray, S.; Duke, M. Application of robust MFI-type zeolite membrane for desalination of saline wastewater. *J. Membr. Sci.* **2015**, *475*, 167–174. [CrossRef]

62. Gao, R.; Zhang, Q.; Lv, R.; Soyekwo, F.; Zhu, A.; Liu, Q. Highly efficient polymer–MOF nanocomposite membrane for pervaporation separation of water/methanol/MTBE ternary mixture. *Chem. Eng. Res. Des.* **2017**, *117*, 688–697. [CrossRef]

63. Khalid, A.; Abdel-Karim, A.; Ali Atieh, M.; Javed, S.; McKay, G. PEG-CNTs nanocomposite PSU membranes for wastewater treatment by membrane bioreactor. *Sep. Purif. Technol.* **2018**, *190*, 165–176. [CrossRef]

64. María Arsuaga, J.; Sotto, A.; del Rosario, G.; Martínez, A.; Molina, S.; Teli, S.B.; de Abajo, J. Influence of the type, size, and distribution of metal oxide particles on the properties of nanocomposite ultrafiltration membranes. *J. Membr. Sci.* **2013**, *428*, 131–141. [CrossRef]

65. Wang, L.; Han, X.; Li, J.; Zheng, D.; Qin, L. Modified MCM-41 silica spheres filled polydimethylsiloxane membrane for dimethylcarbonate/methanol separation via pervaporation. *J. Appl. Polym. Sci.* **2013**, *127*, 4662–4671. [CrossRef]

66. Sun, X.; Qin, J.; Xia, P.; Guo, B.; Yang, C.; Song, C.; Wang, S. Graphene oxide—Silver nanoparticle membrane for biofouling control and water purification. *Chem. Eng. J.* **2015**, *281*, 53–59. [CrossRef]

67. Huang, J.; Zhang, K.; Wang, K.; Xie, Z.; Ladewig, B.; Wang, H. Fabrication of polyethersulfone-mesoporous silica nanocomposite ultrafiltration membranes with antifouling properties. *J. Membr. Sci.*. [CrossRef]

68. Sotto, A.; Boromand, A.; Balta, S.; Darvishmanash, S.; Kim, J.; Van der Bruggen, B. Nanofiltration membranes enhanced with TiO 2 nanoparticles: A comprehensive study. *Desalin. Water Treat.* **2011**, *34*, 179–183. [CrossRef]

69. Matteucci, S.; Kusuma, V.A.; Kelman, S.D.; Freeman, B.D. Gas transport properties of MgO filled poly(1-trimethylsilyl-1-propyne) nanocomposites. *Polymer* **2008**, *49*, 1659–1675. [CrossRef]

70. Yang, Z.; Wu, Y.; Wang, J.; Cao, B.; Tang, C.Y. In situ reduction of silver by polydopamine: A novel antimicrobial modification of a thin-film composite polyamide membrane. *Environ. Sci. Technol.* **2016**, *50*, 9543–9550. [CrossRef]

71. Homayoonfal, M.; Mehrnia, M.R.; Rahmani, S.; Mohades Mojtahedi, Y. Fabrication of alumina/polysulfone nanocomposite membranes with biofouling mitigation approach in membrane bioreactors. *J. Ind. Eng. Chem.* **2015**, *22*, 357–367. [CrossRef]

72. Jhaveri, J.H.; Murthy, Z.V.P. A comprehensive review on anti-fouling nanocomposite membranes for pressure driven membrane separation processes. *Desalination* **2016**, *379*, 137–154. [CrossRef]

73. Zhao, H.; Qiu, S.; Wu, L.; Zhang, L.; Chen, H.; Gao, C. Improving the performance of polyamide reverse osmosis membrane by incorporation of modified multi-walled carbon nanotubes. *J. Membr. Sci.* **2014**, *450*, 249–256. [CrossRef]

74. Faria, A.F.; Liu, C.; Xie, M.; Perreault, F.; Nghiem, L.D.; Ma, J.; Elimelech, M. Thin-film composite forward osmosis membranes functionalized with graphene oxide–silver nanocomposites for biofouling control. *J. Membr. Sci.* **2017**, *525*, 146–156. [CrossRef]

75. Ma Wen Soroush, A.; Van Anh Luong, T.; Brennan, G.; Rahaman, M.S.; Asadishad, B.; Tufenkji, N. Spray- and spin-assisted layer-by-layer assembly of copper nanoparticles on thin-film composite reverse osmosis membrane for biofouling mitigation. *Water Res.* **2016**, *99*, 188–199. [CrossRef]

76. Emadzadeh, D.; Lau, W.J.; Rahbari-Sisakht, M.; Daneshfar, A.; Ghanbari, M.; Mayahi, A.; Matsuura, T.; Ismail, A.F. A novel thin film nanocomposite reverse osmosis membrane with superior anti-organic fouling affinity for water desalination. *Desalination* **2015**, *368*, 106–113. [CrossRef]

77. Ghaemi, N.; Madaeni, S.S.; Daraei, P.; Rajabi, H.; Zinadini, S.; Alizadeh, A.; Heydari, R.; Beygzadeh, M.; Ghouzivand, S. Polyethersulfone membrane enhanced with iron oxide nanoparticles for copper removal from water: Application of new functionalized Fe3O4 nanoparticles. *Chem. Eng. J.* **2015**, *263*. [CrossRef]

78. Zirehpour, A.; Rahimpour, A.; Arabi Shamsabadi, A.; Sharifian, M.G.; Soroush, M. Mitigation of Thin-Film Composite Membrane Biofouling via Immobilizing Nano-Sized Biocidal Reservoirs in the Membrane Active Layer. *Environ. Sci. Technol.* **2017**, *51*, 5511–5522. [CrossRef] [PubMed]

79. Shakeri, A.; Salehi, H.; Ghorbani, F.; Amini, M.; Naslhajian, H. Polyoxometalate based thin film nanocomposite forward osmosis membrane: Superhydrophilic, anti-fouling, and high water permeable. *J. Colloid Interface Sci.* **2019**, *536*, 328–338. [CrossRef] [PubMed]

80. Zhang, X.; Wang, Z.; Chen, M.; Ma, J.; Chen, S.; Wu, Z. Membrane biofouling control using polyvinylidene fluoride membrane blended with quaternary ammonium compound assembled on carbon material. *J. Membr. Sci.* **2017**, *539*, 229–237. [CrossRef]

81. Mooss, V.A.; Hamza, F.; Zinjarde, S.S.; Athawale, A.A. Polyurethane films modified with polyaniline-zinc oxide nanocomposites for biofouling mitigation. *Chem. Eng. J.* **2019**, *359*, 1400–1410. [CrossRef]

82. Carmalin Sophia, A.; Lima, E.C.; Allaudeen, N.; Rajan, S. Application of graphene based materials for adsorption of pharmaceutical traces from water and wastewater—A review. *Desalin. Water Treat.* **2016**, *3994*, 1–14. [CrossRef]

83. Chang, X.; Wang, Z.; Quan, S.; Xu, Y.; Jiang, Z.; Shao, L. Exploring the synergetic effects of graphene oxide (GO) and polyvinylpyrrodione (PVP) on poly (vinylylidenefluoride) (PVDF) ultrafiltration membrane performance. *Appl. Surf. Sci.* **2014**, *316*, 537–548. [CrossRef]

84. Castro-Muñoz, R.; Buera-Gonzalez, J.; de la Iglesia, O.; Galiano, F.; Fíla, V.; Malankowska, M.; Rubio, C.; Figoli, A.; Téllez, C.; Coronas, J. Towards the dehydration of ethanol using pervaporation cross-linked poly(vinyl alcohol)/graphene oxide membranes. *J. Membr. Sci.* **2019**, *582*, 423–434. [CrossRef]

85. Nair, R.R.; Wu, H.A.; Jayaram, P.N.; Grigorieva, I.V.; Geim, A.K. Unimpeded Permeation of Water Through Helium-Leak–Tight Graphene-Based Membranes. *Science* **2012**, *335*, 442–445. [CrossRef] [PubMed]

86. Shen, L.; Bian, X.; Lu, X.; Shi, L.; Liu, Z.; Chen, L.; Hou, Z.; Fan, K. Preparation and characterization of ZnO / polyethersulfone (PES) hybrid membranes. *DES* **2012**, *293*, 21–29. [CrossRef]

87. Zhao, S.; Yan, W.; Shi, M.; Wang, Z.; Wang, J. Improving permeability and antifouling performance of polyethersulfone ultra fi ltration membrane by incorporation of ZnO-DMF dispersion containing nano-ZnO and polyvinylpyrrolidone. *J. Membr. Sci.* **2015**, *478*, 105–116. [CrossRef]

88. Balta, S.; Sotto, A.; Luis, P.; Benea, L.; Van der Bruggen, B.; Kim, J. A new outlook on membrane enhancement with nanoparticles: The alternative of ZnO. *J. Membr. Sci.* **2012**, *389*, 155–161. [CrossRef]

89. Ahn, S.; Cong, X.; Lebrilla, C.B.; Gronert, S. Zwitterion formation in gas-phase cyclodextrin complexes. *J. Am. Soc. Mass Spectrom.* **2005**, *16*, 166–175. [CrossRef]

90. Venault, A.; Huang, W.Y.; Hsiao, S.W.; Chinnathambi, A.; Alharbi, S.A.; Chen, H.; Zheng, J.; Chang, Y. Zwitterionic Modifications for Enhancing the Antifouling Properties of Poly(vinylidene fluoride) Membranes. *Langmuir* **2016**, *32*, 4113–4124. [CrossRef]

91. Bott, T.R. Potential physical methods for the control of biofouling in water systems. *Chem. Eng. Res. Des.* **2001**, *79*, 484–490. [CrossRef]

92. Van der Bruggen, B. Chemical modification of polyethersulfone nanofiltration membranes: A review. *J. Appl. Polym. Sci.* **2009**, *114*, 630–642. [CrossRef]

93. Jiang, B.B.; Sun, X.F.; Wang, L.; Wang, S.Y.; Liu, R.D.; Wang, S.G. Polyethersulfone membranes modified with D-tyrosine for biofouling mitigation: Synergistic effect of surface hydrophility and anti-microbial properties. *Chem. Eng. J.* **2017**, *311*, 135–142. [CrossRef]

94. Hegab, H.M.; ElMekawy, A.; Barclay, T.G.; Michelmore, A.; Zou, L.; Saint, C.P.; Ginic-Markovic, M. Effective in-situ chemical surface modification of forward osmosis membranes with polydopamine-induced graphene oxide for biofouling mitigation. *Desalination* **2016**, *385*, 126–137. [CrossRef]

95. Wibisono, Y.; Yandi, W.; Golabi, M.; Nugraha, R.; Cornelissen, E.R.; Kemperman, A.J.B.; Ederth, T.; Nijmeijer, K. Hydrogel-coated feed spacers in two-phase flow cleaning in spiral wound membrane elements: Anovel platform for eco-friendly biofouling mitigation. *Water Res.* **2015**, *71*, 171–186. [CrossRef] [PubMed]

96. Jung, Y.; Alayande, A.B.; Chae, S.; Kim, I.S. Applications of nisin for biofouling mitigation of reverse osmosis membranes. *Desalination* **2018**, *429*, 52–59. [CrossRef]

97. Katalo, R.; Okuda, T.; Nghiem, L.D.; Fujioka, T. Moringa oleifera coagulation as pretreatment prior to microfiltration for membrane fouling mitigation. *Environ. Sci. Water Res. Technol.* **2018**, *4*, 1604–1611. [CrossRef]

98. Qasim, M.; Darwish, N.N.; Mhiyo, S.; Darwish, N.A.; Hilal, N. The use of ultrasound to mitigate membrane fouling in desalination and water treatment. *Desalination* **2018**, *443*, 143–164. [CrossRef]

99. Ma Wen Panecka, M.; Tufenkji, N.; Rahaman, M.S. Bacteriophage-based strategies for biofouling control in ultrafiltration: In situ biofouling mitigation, biocidal additives and biofilm cleanser. *J. Colloid Interface Sci.* **2018**, *523*, 254–265. [CrossRef]

100. Ali, J.; Wang, L.; Waseem, H.; Sharif, H.M.A.; Djellabi, R.; Zhang, C.; Pan, G. Bioelectrochemical recovery of silver from wastewater with sustainable power generation and its reuse for biofouling mitigation. *J. Clean. Prod.* **2019**, *235*, 1425–1437. [CrossRef]

101. Ergön-Can, T.; Köse-Mutlu, B.; Koyuncu, İ.; Lee, C.H. Biofouling control based on bacterial quorum quenching with a new application: Rotary microbial carrier frame. *J. Membr. Sci.* **2017**, *525*, 116–124. [CrossRef]

102. Al-Abri, M.; Al-Ghafri, B.; Bora, T.; Dobretsov, S.; Dutta, J.; Castelletto, S.; Rosa, L.; Boretti, A. Chlorination disadvantages and alternative routes for biofouling control in reverse osmosis desalination. *Npj Clean Water* **2019**, *2*. [CrossRef]

103. Klein, T.; Zihlmann, D.; Derlon, N.; Isaacson, C.; Szivak, I.; Weissbrodt, D.G.; Pronk, W. Biological control of biofilms on membranes by metazoans. *Water Res.* **2016**, *88*, 20–29. [CrossRef]

104. Armbruster, S.; Stockmeier, F.; Junker, M.; Schiller-Becerra, M.; Yüce, S.; Wessling, M. Short and spaced twisted tapes to mitigate fouling in tubular membranes. *J. Membr. Sci.* **2020**, *595*, 117426. [CrossRef]

MDPI
St. Alban-Anlage 66
4052 Basel
Switzerland
Tel. +41 61 683 77 34
Fax +41 61 302 89 18
www.mdpi.com

Processes Editorial Office
E-mail: processes@mdpi.com
www.mdpi.com/journal/processes